Colyer's Variations and Diseases of the Teeth of Animals

Colyer's Variations and diseases of the teeth of animals

REVISED EDITION BY

A. E. W. MILES
Honorary Curator
Odontological Museum of the Royal College of
Surgeons of England

AND

CAROLINE GRIGSON
Osman Hill Curator
Odontological Museum of the Royal College of
Surgeons of England

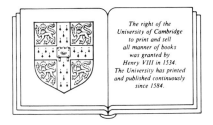

CAMBRIDGE UNIVERSITY PRESS
CAMBRIDGE
NEW YORK   PORT CHESTER
MELBOURNE   SYDNEY

Published by the Press Syndicate of the University of Cambridge
The Pitt Building, Trumpington Street, Cambridge CB2 1RP
40 West 20th Street, New York, NY 10011, USA
10 Stamford Road, Oakleigh, Melbourne 3166, Australia

First edition copyright N. Colyer 1936
Revised edition © Cambridge University Press 1990

First edition published by John Bale, Sons & Danielsson Ltd 1936
This revised edition first published by Cambridge University Press 1990

Printed in Great Britain at The Bath Press, Avon

British Library Cataloguing in Publication Data

Colyer, *Sir* James Frank
Colyer's variations and diseases of the teeth
of animals. – Rev. ed.
1. Veterinary dentistry
I. Title II. Miles, A. E. W. III. Grigson,
Caroline IV. Variations and diseases of
the teeth of animals.
591.2′132   SF867

Library of Congress Cataloging in Publication Data

Colyer, James Frank, 1866–1954.
[Variations and diseases of the teeth of animals]
Colyer's Variations and diseases of the teeth of animals. – Rev.
ed./by A. E. W. Miles and Caroline Grigson.
  p. cm.
Bibliography: p.
1. Veterinary dentistry. I. Miles, A. E. W. II. Grigson,
Caroline. III. Title.
SF867.C65 1990
636.089′76–dc 19   87–3532 CIP

ISBN 0 521 25273 3 hardback

SIR FRANK COLYER, K.B.E., LL.D., F.R.C.S., F.D.S.R.C.S. 1866 – 1954

This portrait, which hangs in the Odontological Museum of the Royal College of Surgeons of England, was presented to Sir Frank by his friends on the occasion of his 80th birthday, 25 September 1946. Painted by Clarence White.

Contents

	Preface to the revised edition	ix
	Preface to the original edition	xv
1	General Introduction	1
	Section 1: Variations in Number, Size and Shape	**17**
2	Introduction	17
3	Order Primates	19
4	Order Carnivora	62
5	Orders Pinnipedia and Cetacea	95
6	The Ungulates	106
7	Orders Rodentia, Lagomorpha, Edentata, Insectivora and Chiroptera	130
8	Order Marsupialia	140
	Section 2: Variations in Position	**153**
9	Introduction	153
10	Order Primates	161
11	Order Carnivora	238
12	Order Pinnipedia	280
13	The Ungulates	282
14	Orders Rodentia, Lagomorpha and Insectivora	316
15	Order Marsupialia	320
	Section 3: Abnormalities of Eruption	**331**
16	Variations and disturbances of eruption	331
17	Overgrowth of teeth	355
	Section 4: Other Disorders of Teeth and Jaws	**371**
18	Injuries of the jaws	371

19 Injuries of the teeth	394	24 Periodontal disease ... 521
20 Enamel hypoplasia	437	25 Odontomes ... 574
21 Caries of the teeth	455	
22 Tooth destruction from causes other than caries	486	*References* ... 607
		Index ... 645
23 Dento-alveolar abscess	506	

Preface to the revised edition

Sir Frank Colyer (1866–1954) was a man who, during an exceptionally long working life, accomplished a great deal. Following his appointment to the consultant staff at the Royal Dental Hospital and School in London, he engaged in the scientific activities of the time and in the associated controversies. He quickly gained a reputation as an educator and, in 1893 when he was 27 years of age, he collaborated with Morton Smale, his senior at the Royal Dental Hospital, in producing a book, *Diseases and Injuries of the Teeth*. This book became the standard undergraduate textbook in British dental schools until the 1950s, being carried through a total of nine editions, at first by Colyer single-handed, who for the third edition (1910) changed its title to *Dental Surgery and Pathology*, and then, in 1931 (sixth edition), began a collaboration with Evelyn Sprawson to produce the subsequent editions.

Colyer served as Dean of the Royal Dental Hospital from 1904 to 1909. During the first World War, he was one of the pioneers who, mainly through their experience in the treatment of jaw fractures, collaborated with the growing nucleus of plastic surgeons in the repair of wounds of the head and face. It was for this work that he was knighted in 1920.

The Museum of the Odontological Society of Great Britain began its life in 1859. In 1900, when Colyer was appointed Honorary Curator in succession to W. C. Storer Bennett, who had died suddenly, the Odontological Society and the Royal Dental Hospital, whose premises it shared and with which it had been closely associated for many years, were seeking new premises to suit their expanding activities; a few months later the Museum was moved temporarily to the Royal College of Surgeons. Then, about 18 months after that, in 1902, it was rehoused with the Society and Hospital in Hanover Square. Negotiations were under way during the whole of this period to amalgamate various medical societies, including the Odontological Society, and this was finally achieved in 1907 with the formation of the Royal Society of Medicine. However, in 1909, when the new building at Wimpole Street was ready for this Society to move into, the Museum was transferred formally to the Royal College of Surgeons, on extended loan, to be housed as an integral part of its museum system (Colyer, 1943a; Miles and White, 1959; Miles, 1972). The collection was thus united with many specimens already possessed by the College, including for many years (until 1963), the odontological specimens of the Hunterian Collection.

Sir Frank's curatorship of the Odontological Museum to which he devoted so much of his time lasted for 54 years; the time he spent at the Museum increased when he retired from his other appointments and from his private practice. Over this long period, the collection grew at a pace much greater than hitherto largely through Sir Frank's personal efforts. He was indefatigable in seeking out material, establishing contacts and acquiring specimens by persuasion or purchase, using the small annual grant, of we believe £75 (Colyer, 1943a), derived ultimately from the Royal Society of Medicine and quite often his own money. The early part of his curatorship overlapped with the period when hunters were collecting, and it now seems to us ravaging, the wild species in Africa and in other parts of the world. Sir Frank had close contact with the grass roots of that situation and later with the game wardens who came into existence when policies of conservation of the wild began to take over. Many of the specimens referred to in this book came into the Museum by these routes. One of the special interests of Sir Frank was injury of elephant tusks from which he realized there was much to be learned about the response of developing dental tissues to injury. In the days of the British Empire, the discovery of elephants with double or misshapen tusks was commonly the subject of notes or letters in The Times or the Illustrated London News. Such a note would be sure to generate a letter from Sir Frank, often leading to the acquisition of another specimen.

In 1947 the size of the Odontological Museum collection was much increased when Sir Frank donated to it his own private collection of 2600 animal skulls. The Odontological Collection as it exists today is thus a memorial to him (Miles, 1972).

In 1936, Sir Frank Colyer produced the first edition of this book which turned out to be yet another memorial to his efforts. The book was based mainly on material in the College supplemented by specimens in museums all over the world that Colyer saw on many journeys made predominantly for that purpose. He gave further interesting information about how the book came into existence in the Preface, which is reproduced here unchanged apart from the omission of a short paragraph explaining certain abbreviations for teeth that he used. As we have used a slightly different system, the retention of this paragraph could have led to confusion.

Judging from the many times it has been and still is cited today, the book came to be widely known and used and it would seem that it is available to most centres over the world. However, it is clear that, almost since it first appeared, it became quite rare because in 1940 the publisher's stocks were destroyed during an air raid. Thus, it seemed to us that consideration should be given either to a facsimile edition or to one supplemented by a record of literature since 1936. However, with the support of Cambridge University Press, we came to the conclusion that it would be better to produce a new edition in which, where new knowledge had made total revision necessary, that should be undertaken. Nevertheless it has been our intention to preserve the main framework of Sir Frank's original and, of course, nearly all the original figures, though with the addition of many others associated with the advances in knowledge over the past 50 years or so.

Sir Frank was noted for his integrity and forthright attitudes and was universally respected. His youthful exuberance, which remained with him into advanced age, made him well-liked, though on occasion, for example, if his beloved museum was in any way threatened, he could reveal unsuspected fierceness. One of us (AEWM) was privileged to be associated with him in his care of the Collection over the last eight years of his Curatorship and thus to acquire a deep affection and respect for him; and furthermore to learn from him the value of observation with the naked eye, mainly of specimens from which the soft parts had been lost, leaving the teeth and bone for better viewing.

We both have a high regard for the value of the specimens that are preserved in museums, which are to a large extent a heritage of past years, and which it is hoped will be preserved for ever. Collections of this sort will never again be gathered, for one thing many of the species so represented are either lost for ever or are endangered, so that deliberate collection from the wild must be strictly forbidden. We are sure that, even if these collections are preserved, no-one will ever attempt *de novo* such encyclopaedic descriptive work as Sir Frank produced in 1936. We certainly could not have contemplated such an undertaking without the foundation stone of all the illustrations and data Sir Frank had accumulated.

We are glad to say again that almost all the Odontological Museum specimens that Sir Frank referred to are still in existence, partly because of his foresight in packing them up in the first year of the last war, and partly because at that time the Collection was housed in the basement and survived the severe damage that the Royal College of Surgeons suffered in 1941. Unfortunately that is not true of all the specimens in the Hunterian Museum of the College and some we have had to record in the text as no longer in existence.

It is appropriate here to pay tribute to the Wellcome Trust which, in 1973, came to the aid of the Odontological Museum with a generous financial grant at a very critical period.

We have done our best to discover whether specimens in other collections that are referred to are still housed where they were, and as far as possible this is indicated in the text; on the whole we were relieved to find that so much material still survives.

The main substance of this new edition consists of Colyer's text rearranged taxonomically. The reference numbers and locations of all the specimens referred to in museums all over the world have been checked as far as possible. Thus the basis of each of the chapters consists of records of specimens that Colyer saw or, in some instances, gathered from the literature. Most of the Tables represent analyses based on Colyer's own observations. We have added a great deal of material from the post-1936 literature but it has been impossible to incorporate all of these additions into the Tables because in many cases subsequent workers have examined at least some of the same specimens as Colyer. The individual instances of variation in the first two sections that we have added from the literature can be distinguished from those described by Colyer by the reference added to each.

We have added a General Introduction (Chapter 1) which contains a great deal of conceptual material which was not available in Colyer's time but needs to be taken account of in thinking of the significance and underlying causes of much of the more factual material that follows. The overall structure of the original has been retained but division of the book into several sections that was present in the original has now been formalized into four Sections. As a result of some re-ordering within the first two Sections, there are now 25 Chapters instead of 23.

Sound arguments occurred to us in favour of some larger reorganization of the material stemming from consideration of advances of knowledge and differences of view that now pertain from those of 1936. However, the only one we have made is to place the chapter, Overgrowth of Teeth, which it is evident is a disorder of eruption, into the Section, Abnormalities of Tooth Eruption. Odontomes, now regarded as predominantly disturbances of tooth development, might have been integrated into the first part of the book. However, we decided not to do so in order to preserve as much as possible of the structure and substance of the original.

Three of the Sections, Variations in Number, Size and Shape, Variations in Position, and Abnormalities of Eruption, are also now introduced by material that we considered more appropriate to place there than in the General Introduction.

We are sure that few would expect us to have attempted to cover systematically the literature of palaeozoology for examples of disorders of dentition. It is only because Sir Frank included a few examples and we happened upon a few others that we mention the matter.

However it is necessary to say that, although the title of the book might lead readers to expect non-mammalian animals to be dealt with, Sir Frank not surprisingly restricted his attention to mammals. We have not attempted systematically to remedy this and indeed a superficial search of the literature revealed a scarcity of references to disorders of dentition in fishes, amphibia and reptiles. This does not necessarily indicate that such disorders are rare, because the difficulties are very great in detecting variations of numbers, shape and position and disorders of eruption in the multiple-generation dentitions composed of large numbers of teeth of simple form, sometimes spread over areas rather than in the rows that are characteristic of mammalian vertebrates; especially as, in non-mammalian vertebrates, even when the teeth are arranged in rows, the number of teeth in each row tends to increase with age by additions to the posterior ends of the rows. Non-mammals are of too much interest and importance to ignore and the observations or the examples we found have been added in places that seemed appropriate.

Because we know, from the innumerable references to it that occur in a diverse range of literature, that Colyer (1936) was, and is, consulted by workers in a wide range of disciplines, probably much wider than Sir Frank himself envisaged, in writing the introductory material, we have tried to predict the needs of a readership of similar width that will include zoologists and veterinarians as well as those trained in human dentistry whose general or research interests lead them beyond the study of man. We have had a mind to the needs of researchers into human dental disease who may be seeking an animal model for experiments designed to throw light on dental or oral disease. Some readers therefore will be already familiar with what the introductory material attempts to explain and may find it over-simplified but others we hope will find it adequately helpful.

In general, the text has had to assume some familiarity with tooth structure and the processes of tooth development, and with comparative dental anatomy. For those needing refreshment in those areas, we recommend *Dental Anatomy and Embryology* edited by J. W. Osborn (1981a). *Recent Mammals of the World* by S. Anderson and J. K. Jones (1967) contains good descriptions of dentitions and dental formulae of the genera. We have also found it necessary to consult some of the older works such as the most recent edition of C. S. Tomes *A Manual of Dental Anatomy, Human and Comparative* prepared by H. W. M. Tims and C. B. Henry (1923) and *The Microscopic and General Anatomy of the Teeth* by J. H. Mummery (1924).

The taxonomic nomenclature used in the First Edition has been updated by reference to *A World List of Mammalian Species* by G. B. Corbet and J. E. Hill (1986), *The Mammals* by D. Morris (1965) and *The Field Guide to the Large Mammals of Africa* by J. Dorst and P. Dandelot (1972). *The Catalogues of the Primates in the British Museum (Natural History)* by P. H. Napier (1976, 1981, 1985) were relied upon for the Primates.

We have tried to ascertain the present whereabouts of all the specimens mentioned in the First Edition, as well as others found in the literature. Almost every one has a unique number beginning with the abbreviated name of the Museum or Institution where it is housed. A list of abbreviations used in the text and the names and addresses of the museums and institutions is given below. The abbreviations are those used by the institutions concerned.

A.E.L.	Anatomisch-Embryologisch Laboratorium, Academisch Medisch Centrum, Meibergdreef 15, Amsterdam, Netherlands.
AGL	Ian Clunies Ross Animal Research Laboratory, Blacktown, New South Wales, Australia.
AHS	A. H. Schultz Collection, Anthropologisches Institut und Museum der Universität Zürich-Irchel, Winterthurerstrasse, Zürich, Switzerland.
AIMUZ	Collection of the Anthropologisches Institut und Museum der Universität Zürich-Irchel, address as above.
AMNH	American Museum of Natural History, Central Park West at 79th Street, New York, U.S.A.
B.I.	Biological Institute, Siberian Centre of the U.S.S.R. Academy of Sciences.
BMNH	Mammal Section, British Museum of Natural History, Cromwell Road, London.
Cambridge	University Museum of Zoology, Downing Street, Cambridge.
FMNH	Field Museum of Natural History, Roosevelt Road at Lake Shore Drive, Chicago, Illinois, U.S.A.
Göteborg Naturhistoriska Museet	Natural History Museum of Gothenburg, Slottskogen, Gothenburg, Sweden.
HZM	Harrison Zoological Museum, Sevenoaks, Kent.
J	Modern mammal collection, Queensland Museum, Brisbane, Queensland, Australia.
KMMA	Koninklijk Museum voor Midden-Afrika, Tervuren, Belgium.
KU	Museum of Natural History of the University of Kansas, U.S.A.
MCZ	Museum of Comparative Zoology, Agassiz Museum, Harvard University, Cambridge, Mass., U.S.A.

MHNG	Muséum d'Histoire naturelle, Casa postale 284, Route de Malagnou, Geneva, Switzerland.	TM	Transvaal Museum, Pretoria, South Africa.
ML	Musée Guimet d'Histoire naturelle, 28 Bd. des Belges, Lyon, France.	USNM	National Museum of Natural History, Smithsonian Institution, Washington, D.C., U.S.A.
MNHN	Muséum National d'Histoire naturelle, 55 Rue de Buffon, Paris, France.	WAM	Modern mammal collection, Western Australian Museum, Perth, Western Australia.
MPEG	Museu Paranese Emilio Goeldi, Av. Magalhaes 376, Belem, Brazil.	ZI	Institute of Zoology, U.S.S.R. Academy of Sciences.
MRI	Mammal Research Institute, Bialowieza, Poland.	ZM	Zoological Museum, Lomonosov University, Moscow, U.S.S.R.
MVZUC	University of California Museum of Vertebrate Zoology, Berkeley, California, U.S.A.	ZMA	Zoölogisch Museum, Universiteit van Amsterdam, Plantage Middenlaan 53, Amsterdam, Netherlands.
NMBE	Naturhistorisches Museum, Bernastrasse 15, CH-3005 Bern, Switzerland.	ZMB	Zoologisches Museum, Museum für Naturkunde der Humboldt-Universität zu Berlin, Invalidenstr. 43, 1040 Berlin, D.D.R.
NMW	Naturhistorisches Museum Wien, Burgring 7, Vienna, Austria.	ZMUC	Zoologisk Museum, Universitetsparken 15, Copenhagen, Denmark.
NRM	Section for Vertebrate Zoology, Naturhistoriska Riksmuseet, Stockholm, Sweden.	Zool. Coll., Oxford Univ. Mus.	The Zoological Collections, University Museum, Parks Road, Oxford.
P-C. Mus.	The Powell-Cotton Museum, Quex Park, Birchington, Kent.	Zoo. Soc. Coll.	Disbanded collection of the Zoological Society of London, Regents Park, London NW1.
RCS Odonto. Mus.	Odontological Museum, Royal College of Surgeons of England, 35–43 Lincoln's Inn Fields, London WC2.		
RCS Osteo. Series	Osteological Series,		
RCSP	Pathological Series, Hunterian Museum, Royal College of Surgeons of England, 35–43 Lincoln's Inn Fields, London WC2.		
RMNH	Rijksmuseum van Natuurlijke Historie, Leiden, Netherlands.		
Rothschild Museum	Lord Rothschild's Museum, Tring. Collection incorporated in British Museum (Natural History).		
Schultz Collection	See AHS above.		

The specimens that have been destroyed or lost since Colyer examined them are listed in the text as 'formerly' in the relevant collection.

Acknowledgements

We are very grateful to members of staff of many museums and institutions who have answered tedious questions on the age, sex, status and catalogue numbers of specimens in their care. They are too numerous to mention individually, but special thanks go to Miss Elizabeth Allen (Hunterian Museum, Royal College of Surgeons of England); Dr Renate Angerman (Museum für Naturkunde, Humboldt-Universität, Berlin); Professor J. Anthony (Muséum d'Histoire naturelle, Paris); Drs A.J. de Haas (Anatomisch-Embryologisch Laboratorium); Dr K.A. Joysey (University Museum of Zoology, Cambridge); members of the Mammal Section of the Natural History Museum. Dr Juliet Clutton-Brock, Daphne Hill, Martin Sheldrick and Paula Jenkins; Rita Larje (Naturhistoriska Riksmuseet, Stockholm); and Ms Suely A. Marques (Museu Paranese Emilio Goeldi, Belem, Brazil). Those who have helped us with general problems of dental anatomy and disease, as well as taxonomy, include J.E. Hill (Natural History Museum), Dr P.J.H. van Bree (Zoologisch Museum, Amsterdam), Dr R.G. Every (Christchurch Hospital, New Zealand), Dr J.A. Sofaer, Dr A.G.S. Lumsden and Dr Grace Suckling (New Zealand Medical Research Council).

As our main task has been to gather information and then to distil it into a small compass, it will be appreciated that we have had to lean on the resources of many libraries. The staff of the following libraries have been particularly helpful: Royal College of Surgeons (Michelle Lelliott and Matthew Derrick), British Library – Science Reference Library (Aldwych Reading Room), British Dental Association, Royal College of Veterinary Surgeons, the Zoology and General Libraries of the British Museum (Natural History), Institute of Archaeology, Zoological Society of London (Susan Bevis and Paul Jeorrett) and the Royal Society of Medicine.

We pay tribute to the largely unsung compilers of the *Zoological Record*, *Index Veterinarius* and *Veterinary Bulletin*, without whose diligent labours it would have been virtually impossible to tap systematically the advances of knowledge published over the past 50 years.

We thank Mrs Lilian Rubin and Caroline Osbourne who bore the brunt of typing the script, Annette Serrant for much general assistance, Dr W. Tschernesky for his translation of some Russian texts and John E. Linder for help with many German ones.

Above all, we gratefully appreciate the generosity of the Anatomical Society of Great Britain in making a grant towards the costs of production which helped and encouraged us to undertake the task.

We have enjoyed continuous help and support from Cambridge University Press during the period of gestation of the book.

June 1990

A.E.W.M.
C.G.

Preface to the Original Edition

During the autumn of 1930 I delivered for the Dental Board of the United Kingdom a course of four lectures on 'Abnormal Conditions of the Teeth of Animals in their Relationship to Similar Conditions in Man'. These lectures were published in book form [Colyer, 1931]. In this volume I have given a fuller account of the subjects dealt with in those lectures, more especially the one on 'Positional Variations of the Teeth' and have added chapters on 'Variations of the Teeth in Number and Shape', 'Abnormal Eruption and Growth of Teeth' and on those interesting abnormalities usually classed as 'Odontomes'. I trust that the facts recorded may be of some value to those interested in Comparative Dental Pathology.

I wish to record my thanks to the authorities of the Natural History Section of the British Museum for the facilities afforded me to examine the extensive collection of mammals' skulls and for permission to photograph several of the specimens. I am also indebted to Major Powell-Cotton for placing at my disposal, for the purposes of study, the valuable collection of Apes and Monkeys in his museum at Quex Park and to Lord Rothschild for allowing me to examine his collection at Tring Park.

During my visits to museums in Europe and America I was invariably received and assisted in the most friendly spirit by those in charge of the collections, and I wish to place on record my gratitude to Professor Dr Einar Lönnberg and Count Nils Gyldenstople of the Natural History Museum of Sweden; Professor Dr Nils Holmgren of the Högskola, Stockholm; Professor Hellman of Lund University; Dr Nic. Peters of the Zoological Museum of Hamburg; Dr H. Schoutenden of the Congo Museum, Belgium; Dr

F. Voss of the Natural History Museum, Berlin; Professor Dr H. Boschma of the Natural History Museum, Leiden; Professor R. Anthony of the Natural History Museum, Paris; Dr A. Hrdlička and Mr G. S. Miller of the United States National Museum, Washington; Mr T. D. Carter of the American Museum of Natural History, New York; Mr Stephen C. Simms of the Field Museum, Chicago; Dr Magnus Degerbøl of the Natural History Museum, Copenhagen; Professor Dr A. M. Ribero of the National Museum, Rio Janeiro.

I desire to express my most sincere thanks to Mr Martin A. C. Hinton, F.R.S., of the British Museum, for the ready help at all times accorded to me; my thanks are also due to Mr R. I. Pocock, F.R.S., for many valuable suggestions, more especially in connexion with the difficult question of classification; to Dr W. L. H. Duckworth of Cambridge University for his unfailing kindness. I also wish to record the help I have received from Dr J. W. Woerderman, Dr Th. E. de Jonge-Cohen, Captain Guy Dollman, Dr G. C. A. Junge, Dr G. M. Vevers, Dr Karl Jordan, F.R.S., Dr E. Schwarz, Mr C. Forster Cooper, F.R.S. and Professor D. Axel Palmgren.

My thanks are also due to Mr R. W. Hayman for the ready and useful assistance he has always given to me during my visits to the British Museum; Mr S. A. Sewell for his valuable help in connexion with the illustrations, and to Mr E. J. Manly for the trouble he has taken in photographing many of the specimens figured in the text.

I am deeply indebted to Mrs L. Lindsay for many valuable suggestions and for her great kindness in reading through the manuscript and the proof sheets.

I am indebted for the loan of blocks to the Council of the Royal Society of Medicine; the Publishing Committee of the British Dental Association; the Dental Board of the United Kingdom; the Dental Manufacturing Co.; the Amalgamated Dental Manufacturing Co.; the S.S. White Dental Manufacturing Co. for Figs. 846 to 848 [21.2 and 21.3] which are taken from the *Dental Cosmos*; Messrs Macmillan and Co. for permission to reproduce Figs. 67 and 68 [3.59 and 3.63] from Bateson's Material for the Study of Variation; *The Field Newspaper* for Figs. 738 and 739 [6.45 and 6.46]; Messrs Cassell and Co. for Figs. 1006 and 1007 [25.35 and 25.34]; and to the Bombay Natural History Society.

Lastly I must express my gratitude to Mr A. E. Bale for the great trouble he has taken with the production of the book.

J. F. COLYER.

CHAPTER 1

General Introduction

Variation

Teeth may vary in number, size, shape and position within the same species. Variation in number, size and shape tends to be of largely genetic origin or at least to involve genetic mechanisms; it used to be thought that this variation, which involves quantity of tooth material, was beyond the influence of environmental factors such as nutrition, but new evidence suggests that this is not so. Variation in position on the other hand is much influenced by environmental factors and, because tooth position depends on bone development and bone growth, is frequently secondary to conditions that affect growth.

Before considering the evidence that bears on the mechanisms that bring about these types of variation, which is one of the main purposes of this introductory chapter, it is necessary to discuss the term variation itself.

Variation in the biological sense comprises differences of every kind, morphological and other, that exist between individuals of the same species. Variation in the observable characteristics of an organism (phenotype) is the product of individual differences in genetic constitution (genotype) and of various environmental influences. Heritable variation is 'the material on which natural selection acts to bring about the evolution of species' (Bateson, 1894).

Variation implies deviation from a mean or norm; when the observed difference is slight, the term normal variation is sometimes applied. When the deviations are more gross and uncommon and amount to abnormalities, they may be said to constitute abnormal variation. The distinction between normal variation and abnormality is quite arbitrary, especially in continuous variation, that is variation in characters, like stature and tooth size, that can be measured against a continuous scale, even an ordinal one. Variation of tooth morphology is broadly continuous in the sense that it shows a great range in degree; not only is it hard to draw a distinction between variation and abnormality but, when frank abnormalities are grouped together, they form a more or less continuous spectrum of increasing disorganization of tooth form with tumour-like masses, odontomes, at the extreme end of the spectrum (see Chapter 25.)

Discrete or discontinuous variation is where there are no intermediate types; a character is either present or absent; for instance, blood types and primary sex characters where either an ovary or a testis is present. The terminology of variation is used with most rigour in the science of genetics in which another term, quasi-continuous variation, is employed for characters which are discrete, that is present or absent, but when present vary continuously in degree. Some dental variants, for example supernumerary cusps and absence of third molars (Grüneberg, 1951) in laboratory-bred animals, have been shown to exhibit quasi-continuous variation (Sofaer, 1975, 1976).

It is only possible to give here a condensed summary of the wealth of observation, terminology and concept which is relevant to systematic thinking about the processes that could have brought about the variation described in the various following sections. More extensive coverage of the early works is available in Gaunt and Miles (1967) and, for more recent concepts, we recommend J.W. Osborn (1978, 1981*b*, 1984) and Lumsden (1979). Sir Frank Colyer, in writing the first edition of this book, was influenced by the encyclopaedic authority of Bateson's classic work *Materials for the Study of Variation* (1894), which still deserves to be consulted, not only for the descriptions given of abnormalities of dentitions in various species, much of which Colyer incorporated into the present work, but also for Bateson's conceptual analyses. Bateson pointed out (pp. 195–198) that the tooth arches can be regarded, like the vertebral column, as a linear meristic series, that is a series of repeating parts; variation in the number, shape or size of the parts can be referred to as meristic variation. Bateson coined the term *homoeosis* (Gk. being alike) for where one part in a series assumes the shape characteristic of another part; for example, when a cervical vertebra possesses a rib element and thus resembles a vertebra in the thoracic region, or the sacralization of a lumbar vertebra. Homoeotic variation is a term that could be applied to a tooth in the canine region of the jaw which may, by virtue of cusps arising from its cingulum, partially resemble a premolar (see also Butler, 1967).

Morphogenetic fields

Our understanding of variation in size, number and shape and of position is greatly helped by the embryological concept of morphogenetic fields, whereby areas in the embryo develop the capacity to become organized in a particular way, provided that an evocator of appropriate strength acts upon them. Gradients of capacity have been postulated within such fields; capacity to react may be maximal at some point and fall off on either side of it, though not necessarily equally, so that a field can have eccentrically distributed gradients of capacity. Butler (1939, 1963, 1967, 1978*b*, 1982) was the first to adapt these field theory concepts to the mammalian dentition and to postulate that the dentition, or each quadrant of it, develops as a series of units within a continuous morphogenetic field. Each unit develops in a particular way according to its position in the field. The field has an antero-posterior axis with a morphogenetic capacity that differs qualitatively at each end. The field is also differentiated into regions corresponding to the incisor, canine and molar regions, and these may have some degree of independence, as when the canine is reduced or enhanced without a corresponding change in the rest of the dentition (Fig. 1.1).

In the dog, for example, the molarization influence can be said to be greatest in the region of the carnassial lower first molar and upper fourth premolar and becomes progressively weaker anteriorly through the premolars and posteriorly through the other molars.

The dentition is thus composed of a series of units which, although morphologically separate, arise from a morphogenetic field as a continuous system which may be altered by various processes as a whole, or at least as groups of teeth, rather than as single units. This concept has been widely used to describe the evolution of dentitions and the relation of one to another (Henderson and Greene, 1975) and also to explain in systematic terms variation in size and shape of groups of teeth. The concept of a morphogenetic field of odontogenic capacity also provides a basis for explanation of supernumerary teeth, and indeed for absence of teeth. Where there are supernumerary teeth, the capacity for tooth formation was abnormally strong and, where teeth are absent, the capacity, perhaps in a localized area and for reasons that may not be understood, did not develop.

Variability, including differences between corresponding structures on the right and left sides of the body that constitute asymmetry, tends to be greatest in the later-developing teeth in each morphological class. Furthermore, heritability, the proportion of the observed variation due to additive genetic effects, tends to be lower in the later-developing teeth. These observations can be explained in terms of the field concept, or the clone concept to be described later, as due to the progressive exhaustion of the field substance responsible for initiating tooth development (Sofaer, 1977). There are some mouse mutants (Grüneberg, 1966; Sofaer, 1969) in which a supernumerary molar is sometimes associated with smallness of the true molars or absence of the third molar. Furthermore, if in a given morphological segment, e.g. the molar segment, the teeth which develop early are large, those that develop late tend to be small or absent or vice versa (Sofaer, MacLean and Bailit, 1972). There is some kind of competitive interaction between the tooth units within the segment. These phenomena can be explained in terms of a fixed potential for the total quantity of tooth material. Supernumerary teeth, to be dealt with in the next section, in general can be explained in terms of this concept or slight elaborations of it. For example, supernumerary teeth which are morphological duplications, or twins, can be thought of as resulting from a dichotomy of the tooth primordium. Even a simple conical (haplodont) supernumerary could be explained as the product of an unequal dichotomy.

The morphogenetic field concept has been elaborated by J. W. Osborn (1978, 1981b) to explain how individual teeth, each bearing some basic morphological resemblance to one another, may arise within the fields. The field influence, exerted from a generation point, can be assumed to be a field substance which diffuses through the tissues of the growing region. For purely physical reasons, there would be a gradient of concentration of the substance at increasing distances from the generation point but, in order to explain an eccentricity or peripheral limitation of the gradient, it is possible to postulate the destruction of the substance by the cells among which it is diffusing. To account for the production of tooth primordia at separate points in the field, it is necessary to postulate a further secondary field effect (Fig. 1.2), or a number of oscillations in field strength (Fig. 1.3), and inhibitory influences around the newly developing primordium. There are three primordial field areas in each jaw quadrant, the most anterior forming the incisors, the middle primordial area forming the canine, and the most posterior the cheek teeth. Each primordial field area is thought to give rise to a tooth, termed the stem

Fig. 1.1. A hypothesis concerning the nature of the differentiation of the dentition. (a) The dental lamina with undifferentiated rudiments. (b) The morphogenetic field which, acting upon the rudiments, determines them to develop in different ways. (c) The differentiated dentition that results. From Gaunt and Miles (1967), after Butler (1939).

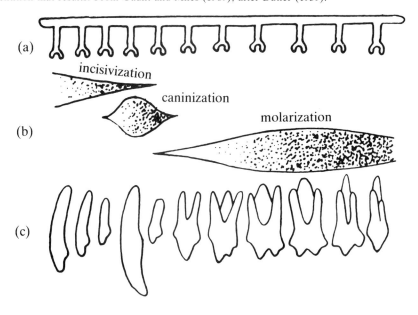

progenitor (Schwartz, 1984), which shows the characteristics of its tooth class; for the molar field this, in general, is the first permanent molar.

Osborn postulated a further idea, known as the odontogenic clone concept, which has been elaborated upon by Lumsden (1979). The concept postulates that each primordium in a tooth field develops from a single cell mass (clone) which initially consists perhaps of only two cells, one for the mesodermal parts of the tooth and one for the ectodermal cap (multiclone), which after a certain number of cell divisions become competent to initiate a primordium. Then, as the clone grows and space becomes available, it gives rise to further primordia (Fig. 1.4). Obviously this idea has to incorporate concepts of local morphogenetic thresholds and of inhibitory influences which develop in the surrounding mesoderm or even within the primordia themselves, otherwise a single enlarging tooth mass would develop.

The differences between various concepts are subtle but important. Some can be tested by experiment; for example, the validity of the odontogenic clone concept received strong support from the work of Lumsden (1979) who showed that early primordia of the lower first molar tooth from mouse embryos cultured in the anterior eye chambers of adult mice generated the perfectly shaped and mineralized crowns of the entire three-molar tooth row.

The concepts expressed here in a simple form are undergoing rapid development and change; for instance, Osborn (1984) has renamed the cell units involved in tooth initiation, clades, in place of the previous clones.

There is a great deal of other evidence that tooth

Fig. 1.2. An explanation, in stages I–V, in terms of morphogenetic fields for the development of five structures of differing size and shape graded A–E. A field generator (heavy stipple) is generated in I and produces a gradient of diffusing field substance (light stipple in II–V). Primordia develop at points within the field by a secondary influence. All primordia (indicated by crossed open boxes) initially have identical potential. Each primordium develops a size and shape determined by the concentration of field substance. The differing sizes and shapes are indicated by the height of each structure (black rectangles) which corresponds with the level of concentration of field substance. Inhibitory influences around each primordium are postulated.
From: Osborn, J. W. (1978).

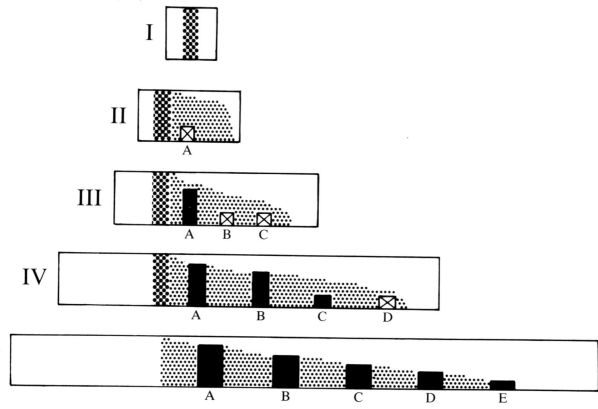

primordia acquire at an early stage the inherent capacity to form teeth of predetermined potential size and shape. For example, each half of a bisected rabbit molar tooth primordium explanted *in vitro* develops the full cusp pattern of a molar (Glasstone, 1952) though it remains much smaller than *in vivo*, presumably because *in vitro* it is not sufficiently nourished to develop its full size; in general, tooth primordia grow larger when cultured *in vivo*, as in the anterior chamber of the eye where a blood supply develops, than they do *in vitro* (Lumsden, 1978).

Of course, during the conversion of a primordium into a tooth, elaborate processes of differentiation of cell types and differential growth rates occur in various parts of the primordium or tooth germ, all under the control of between-cell influences (Slavkin, 1974). See also Reif (1980) on reversal of polarity in tooth germs referred to in the next section.

Although the matter cannot be fully discussed here, it is appropriate to mention that tooth germs, explanted at stages before they are likely to be contaminated with osteogenic cells of the jaw and cultured either *in vivo* or *in vitro*, often develop at least traces of an ensheathing periodontal ligament and alveolar bone. This finding seems to indicate that the periodontium, including alveolar bone, develops from, or under the influence of, the same ectomesenchymal primordium that gives rise to the teeth themselves (Ten Cate, 1972; Ten Cate and Mills, 1972; Lumsden, 1984). See also Chapter 9, p. 155.

Fig. 1.3. Gradient-field theory explanation for the formation of a 7-unit (4 deciduous molars and 3 molars) mandibular post-canine dentition in a hypothetical carnivore. Where oscillations in the field strength of a special substance produced by the field generator (Fg) rise above a certain threshold, indicated by the interrupted line, a tooth primordium (solid circle) is initiated. The field strength is greatest towards the centre where the primordium of M_1, the largest and most carnassial tooth, develops and it diminishes in strength both distally and mesially where teeth of diminishing size develop.

The open circles indicate hypothetical potential primordia that could arise if the field strength were sufficient or if some suppressive influence were lifted. Other (supernumerary) primordia could arise between the solid circles if for some reason an additional local area or oscillation of concentration of generator substance above the threshold occurred. If such a concentration arose at the expense of an adjacent oscillation, but was not sufficient to reduce it below the threshold, a supernumerary tooth associated with smallness or cusp deficiency in a tooth of the normal series would occur.
Based on Lumsden, A.G.S. (1979).

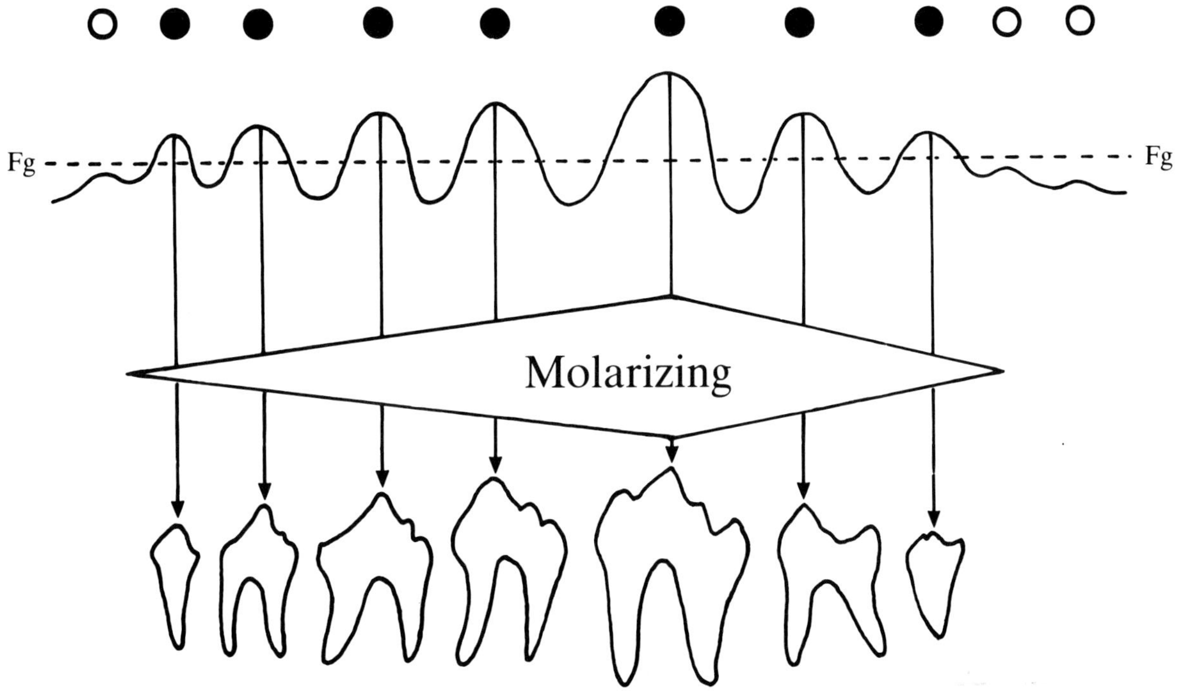

Supernumerary teeth

Supernumerary teeth can be loosely categorized morphologically into three kinds: (1) Supplemental teeth which resemble teeth of the normal series in both crown and root morphology though not always in size. They may sometimes resemble ones already in the arch so closely that it may be impossible to be sure which is the supernumerary element. (2) Haplodont supernumerary teeth have simple, usually conical, crowns and single roots. (3) Tuberculate supernumerary teeth have more complex crowns, and have what can be called an occlusal surface bearing several tubercles, often with deep indentations. Like haplodont supernumerary teeth, these tuberculate supernumeraries usually have single roots.

Supernumerary teeth usually develop more or less synchronously with the teeth adjacent to them but occasionally their development is out of step (Miles, 1954). For example, a supernumerary tooth in the incisor region may have a fully-formed root whereas the roots of the incisors are still forming; conversely the supernumerary may be in an earlier stage of development than the adjacent teeth. Obviously sometimes this asynchrony is explicable on the basis that the supernumerary tooth is associated with a different dentition, deciduous or permanent.

Reference has already been made (p. 2) to the explanation of the origin of supernumerary teeth in terms of the morphogenetic field theory; namely, that they are a manifestation of a localized excess of odontogenic capacity. There is some evidence that, in general, the prevalence of numerical variation is greatest at the peripheries of the tooth-class fields. The evidence is strongest in respect of the distal end of the molar field; for example, disto-molars, either larger or smaller than normal, and absence of third molars are particularly common in primates (Schwartz, 1984).

A further idea is that supernumerary teeth arise from the splitting or dichotomy of a tooth primordium (Bateson, 1894, p. 268; Sprawson, 1937; Wolsan, 1984c). Supplemental teeth would be explained on the basis of the primordium dividing into equal parts each with the capacity to form a tooth of normal morphology. Haplodont and tuberculate supernumerary teeth would be explained by division of the primordium into parts which, although having the capacity to form a complete tooth, did not have the full regulative capacity to form a tooth of normal morphology and size. Certain observations, such as the existence of tooth forms which strongly suggest partial dichotomy, and some *in-vitro* experiments lend plausibility to the idea of dichotomy. This evidence is elaborated on in the next section, Connate teeth.

Some observations made in sharks are of particular importance to the dichotomy hypothesis of origin of supernumerary teeth as well as to the mechanisms of control of tooth morphogenesis. Compagno (1967) described three abnormal dentitions in three different taxonomic families of sharks. The teeth are normally blade-like with a central cusp which is distinctly distally inclined. Other morphological details distinguish the mesial and distal surfaces of the teeth; for instance, in *Galeorhinus zyopterus* the distal margins of the central cusps are serrated. In the lower jaw of two and the upper jaw of the other of the three abnormal specimens, one tooth as well as all its successors were of

Fig. 1.4. Explanation in terms of clone theory for the formation of the mouse molar dentition. The stippled region represents the growing margin of the clone; the large circles represent zones of inhibition. The gradient of final form is related to the times at which the primordia (black dots) are initiated. Small circles represent tissues which have reached the critical mass to form a primordium (Lumsden, 1979).
From: Lumsden, A. G. S. (1979).

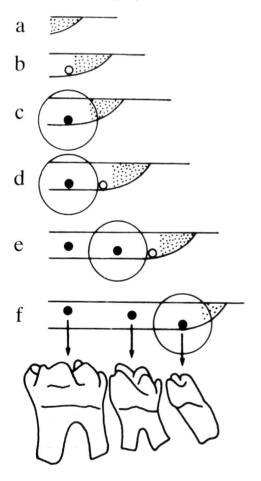

reversed morphology; that is, the central cusp was inclined mesially and the serrated edge faced mesially. In other words, at least superficially, the affected tooth families had the morphology of the teeth of the opposite side.

Reif (1980) described additional examples in other species of shark and noted that many were associated with evidence of trauma and of initial splitting of a tooth germ at a very early stage (protogerm), producing two teeth instead of one. He suggested that sometimes the mesio-distal polarity of one part of the divided germ undergoes reversal. Although the first generation of teeth derived from the divided germ would be smaller than normal, successive generations are larger until eventually only the tooth with reversed morphology would be recognizable. The dental lamina and newest generations of teeth in the elasmobranchs are susceptible to trauma because they are situated beneath the mucosa in the inner aspects of the cartilaginous jaws. An instance was illustrated by Andre (1784; see also Gudger, 1937) in a tiger shark (*Galeocerdo tigrinus*) in which a spine of a sting-ray had stuck in the lower jaw and produced a family of several generations of malformed teeth.

In the older literature on supernumerary teeth, suggestions are common that they may be examples of atavism, that is they represent the recurrence of a character possessed by an ancestral form which had apparently been lost in the process of evolution. Atavism, with the original authority of for example Bateson (1894), is a concept that now appears to be out of favour with geneticists in general, and is rarely discussed in the general literature on abnormal development. Atavism cannot for example be offered as an explanation for polydactyly because no ancestor of the mammals with more than five digits on each extremity has ever been postulated.

Nevertheless, geneticists do recognize in breeding experiments a process called reverse mutation in which the mutation appears to go back to what is called the wild-type. Furthermore, it is well established that the genome can carry genetic information in the form of recessive genes for many generations without the information being expressed; then, if the carrier of the recessive gene happens to be crossed with an individual that carries the same gene, the information gains expression in the offspring.

Horns occasionally appear in hornless breeds of domestic cattle and sheep whose ancestors were horned (Darwin, 1859, p. 454 and 1868, p. 315). An example in the dentition is the occasional occurrence of lower second molars, absent from the felid phylogeny since the Miocene (Kurtén, 1963). In such cases, it would seem to be legitimate to consider atavism, especially if there was evidence that the condition was heritable (Wolsan, 1984c). Similarly, it is probably wise only to give serious consideration to atavism as a cause of supernumerary teeth where there is at least a probability that the condition is heritable or, in exceptional circumstances, such as the occurrence of a supernumerary tooth in the diastema in the Rodentia.

Although supernumerary teeth often appear to arise as spontaneous accidents of development, in the few instances where the familial relationships of the animals are known or where numerical variation within particular populations has been quantified, there is evidence that the supernumerary teeth tend to occur more frequently in isolated groups and particularly in some domesticates. The controlling mechanism then is almost certainly genetic drift and many instances will be mentioned in the relevant chapters.

Although true duplication of continuously-growing rodent incisors does occur (pp. 131, 133 and p. 134), that is the development of more than one incisor where there is normally only one, some conditions have been described in which apparently-extra incisors may erupt beside the normal ones during adult life; for example, following the administration to rats of drugs that interfere with cell proliferation (Vahlsing, Kim and Feringa, 1977), or carcinogenic drugs to hamsters (Edwards, 1980) or of ^{224}Ra to mice (Humphreys, Robins and Stones, 1985), and in aged mice (p. 405) (Robins and Rowlatt, 1971). In the rats described by Vahlsing *et al.*, the drug seems to have stopped the proliferation of the formative tissues at the base of the incisor and its eruption had slowed or ceased. However, in due course as the cytotoxic effect wore off, proliferation restarted and tooth formation recommenced. Because of the period of non-formation, the new portion of tooth was not in continuity with the old and erupting with normal vigour grew, like a new tooth, alongside the old tooth and eventually erupted beside it. In the instance described by Edwards, the carcinogen appeared to have caused a proliferative overgrowth of some of the formative tissues at the base of the incisor which gave rise to a complex mass of dysplastic dentine, cementum and enamel. It seems that part of this proliferative tissue budded off and, exhibiting a more properly regulated growth of dental formative tissue, formed as an incisiform structure alongside the parent tooth.

Connate teeth

There is a variety of tooth anomaly, bearing evidence of being composed of two or more tooth elements, for which terms such as gemination, double teeth (Miles, 1954, 1966) connation, incomplete dichotomy, fused teeth, synodonty and odontopagy have been employed. We favour the term connation (developed or born together), which was employed in this context at least as early as the middle of last century (J. Tomes, 1859), because we agree with Hitchin and Morris (1966) that the term exactly describes the anomaly. Many of the alternative terms express one or other of two divergent views of the origin of the anomaly. According to one view, it arises from a process of fusion or joining up of tooth germs; according to the other, it arises from a partial splitting or dichotomy of a tooth primordium. Insufficient evidence is available to determine this question but there are a few observations which may be significant.

It may be that some anomalies arise from fusion and others from a process more like dichotomy. An occurrence that suggests that a process of dichotomy of a tooth germ can produce multiple teeth, and that likewise a process of partial dichotomy could produce connate teeth, is that of the multiple small tusks that sometimes arise in connection with physical injuries to elephant tusks (p. 414 *et seq.*). The work of Glasstone (1952), who found that each half of a rabbit-molar tooth germ cut in two formed a rudimentary whole tooth when grown in tissue culture, also seems relevant to the mechanism of production of both connate and supernumerary teeth.

Most connate teeth show various degrees of separation of the crowns of the tooth elements whereas the roots are single as in Figure 1.5A,B. Examples with joined crowns and separate roots are rare (Fig. 1.5C). As it is the tips of the crowns that are formed first, this means that in the initial stages of development of the majority of connate teeth the elements must have had separate identities and union could only have occurred as development proceeded. This observation would appear to be more consistent with the theory of fusion rather than that of dichotomy. In many connate teeth, the separateness of the crown elements consists only of a notch and groove. There is no difficulty in envisaging such teeth as developing within a single tooth germ that showed only some

Fig. 1.5. Three examples of connate teeth.
A and B: *Canis familiaris* (dog). Left and right sides of jaw (from Bateson 1894, p. 211). The left upper canine is bifid (A). In the right upper canine (B), the disturbance of development is similar; however not only is the crown bifid but an axial groove extends on to the root. These teeth are remarkably similar to those of a fox shown in Figure 4.49. Both these specimens support, or are consistent with, the hypothesis of dichotomy of tooth germs.
C: *Cercopithecus aethiops tantalus* (tantalus monkey). ♂. Captive. RCS Odonto. Mus., G 13.4. Mandible. A much more uncommon type of connation in which, although the incisive edges are bifid, each crown element has a separate root; a specimen which favours the idea of fusion for a time of two elements which were separate at an extremely early stage of development.

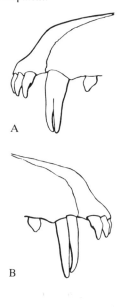
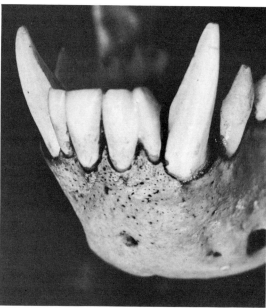

degree of division into elements in its internal structure. However, connate teeth occasionally occur which consist of more than one crown springing from a single root element and the distance is so great between the tips of the crowns, where tooth formation is initiated, that it is inconceivable that the whole connate tooth was formed from a single germ. It must have developed from two separate germs which later united. A close study of some connate teeth shows a ridge of tissue at the point of junction of the two elements producing an appearance suggestive of distortion by pressure.

It seems reasonable to conclude that not all connate teeth arise in the same way; some may arise by dichotomy and others by a process of fusion of tooth germs or primordia. It must be borne in mind that a quite different kind of tooth fusion can also occur, caused by the overgrowth of cementum after tooth formation is complete. This can happen when the roots of adjacent teeth are in contact and the cementum of the roots undergoes hyperplasia as a result of some chronic inflammatory process.

Connation of teeth occurs sporadically in many species; it has been observed in the median egg-teeth in the viper, *Vipera berus* (Smith, Bellairs and Miles, 1953).

Many examples of connation are described in later sections of the book. Whether any of these are heritable can only be determined by studies of families or controlled breeding experiments. Certainly some are; for example, Hitchin and Morris (1966) noted that connated upper incisors turned up in some Lakeland Terrier dogs being bred for show. Connation was then deliberately bred for over several generations, mostly by sire-daughter matings. Connate incisors, mostly in the upper jaw and often bilaterally, occurred commonly in these terriers and eventually occasionally in the mandible also. Connation in the deciduous dentition was often repeated in the permanent successors. The heritability in this instance was amply demonstrated but the exact mode of inheritance was unclear.

These experiments produced suitable material for the study of the embryonic stages in the development of the anomaly. In normal development, tooth germs arise along the course of a continuous two-layered epithelial tooth band; the parts of the band between the germs normally degenerate and disappear, but it seems that, where connate teeth are to be formed, the band persists. Then, as the formation of the originally-separate adjacent teeth proceeds and they become contiguous, stellate reticulum spreads from the enamel organs of both tooth germs between the two layers of the persistent band. Gradually not only does the stellate reticulum become continuous but so does the ameloblast layer. Although amelogenesis may cease at the point of contact of the two tooth germs, differentiation of odontoblasts continues, the two odontoblast layers join and then proceed to form the dentine of what is now a single connate tooth.

The morphology of the connate tooth, whether for instance it is simply a large incisor with a deep notch at the incisive edge, or has two separate crowns joined only at the cervices to a single root, depends on how much of the originally separate teeth has been formed before the stellate reticulum grows across within the persistent tooth band.

Connation, both of an upper incisor with a supernumerary incisor and of a first permanent molar with a mesially-situated supernumerary molar, is common in the mutant tabby-mouse (Sofaer, 1969). The histology of the early stages of tooth development in these mice tended to be similar to that just described. Crowding together of the tooth germs, as when supernumerary teeth appear, seemed conducive to what must be called a process of fusion. Appearances suggestive of dichotomy were not met with.

Sofaer and Shaw (1971) noted that connate molars and supernumerary molars occurred sporadically in a colony of rice rats (*Oryzomys palustris*). Selection of animals for the presence of these abnormalities produced a strain with a high incidence of fusion of the first and second molars, and even of the third molar as well, to produce a molar of giant size. Although rudimentary supernumerary molars occurred sometimes, in this instance crowding of the teeth did not appear to be a factor in the connation. The autogenetic mechanism appeared to be as described above. The mode of inheritance appeared to be by a single autosomal recessive gene with variable penetrance in homozygotes.

Knudsen (1965, 1966a,b) reported many varieties of tooth fusion associated with exancephaly in mice induced by teratogenic agencies, mainly maternal hypervitaminosis A. There were even bizarre examples of fusion of upper molar germs with lower molar germs on the same side. Ritter (1963) induced fusion of the lower incisors in mice with X-radiation. In these cases, it was argued that the underlying cause is a deficiency of the growth of the connective tissue which normally separates individual tooth germs during development.

Genetic and environmental causes of variation

It used to be thought that tooth size is genetically determined and is not at all influenced by environment. However, evidence (reviewed by M. L. Moss,

1978) now suggests that the size of the crowns of teeth may be affected by the maternal diet; for instance Holloway, Shaw and Sweeney (1961) found that, in the offspring of rats fed on a diet deficient in protein, the crowns of the molar teeth were smaller than those of offspring of rats on a normal diet. By appropriate measurements, they established that the smaller size was due to the dentine cores of the crowns being smaller, rather than due to the enamel covering being thinner. This is extremely important because the size of the crown of a tooth is determined primarily by the size of the dentine part of it which, once formed, cannot be added to externally and therefore cannot increase in size; the ultimate size of the crown is then determined secondarily by the thickness of enamel laid down upon the dentine core which depends upon the rate and duration of amelogenesis, both of which could be affected by environmental factors such as nutrition and hormone levels. M. L. Moss (1978) adduced evidence that male-female differences in the duration of amelogenesis may be responsible at least in part for the larger tooth crowns so often found in the male. Similarly, minor sex differences in tooth crown morphology could be produced by sexually different rates or durations of amelogenesis affecting localized areas of the crowns.

The discovery in 1944 that maternal infection with the rubella virus in the early months of pregnancy leads to various human birth defects (Kraus, Ames and Clark, 1969; Smith, Soskolne and Ornoy, 1978) gave impetus to experimental work on laboratory animals which has now provided ample evidence that a wide variety of conditions, such as vitamin deficiencies or the presence of noxious substances, can disturb embryonic development (Robens, 1970). One of the commonest malformations is cleft palate in which the development of the teeth on either side of the cleft is also disturbed (Shah, 1979). However, malformed teeth and absence of teeth (anodontia), in the absence of cleft palate, have been described as products of experiments that disturb pregnancy (Miles, 1954).

There is now a vast literature on experimental teratogenesis, or abnormal development, stimulated by the discovery that maternal administration of many new drugs during the early stages of pregnancy have a teratogenic side-effect. In such experiments, developmental anomalies show a marked tendency to be multiple, as indeed do those that occur spontaneously; for instance anencephaly and lesser abnormalities of development of the head region such as cleft palate are commonly associated with polydactyly and failure of the caudal end of the neural tube to close (spina bifida).

The genetic potential and influence of the general environment would be expected to be the same on both sides of the body; hence, when an abnormally high level of asymmetry occurs between paired structures, it is regarded as an indicator of developmental instability resulting either from poor genetic control or from unusually large local environmental differences between sides. Thus inbreeding, which is known to increase sensitivity to environmental variation, is associated with a tendency for abnormally high levels of asymmetry of tooth-crown size (Bader, 1965; Sofaer, 1978); a similar increase in asymmetry of teeth and limb bones has been observed in rats and mice following disturbance of their prenatal and perinatal environment by noise, heat, cold or protein deprivation (Sciulli, *et al*, 1979). When populations are available for study, data on dental symmetry can throw light on the relative contributions of genetic and environmental sources of instability of the developmental processes (Sofaer, 1975, 1976).

Many examples of the high incidence of dental anomalies in isolated populations and on islands are given in the text. Such characters can be produced by the processes known as genetic drift and founder effect. Genetic drift is the fluctuation of gene frequencies resulting from random combinations of alleles from generation to generation. In a large population, its effect is small compared with that of selection, but in an isolated, finite population its effect may be more marked, leading to the appearance of characters that may have no selective advantage. In founder effect, a population is initiated from a small number of individuals as when a few animals are introduced on to an island. Genetic drift and the absence of some alleles present in the parent population may produce genetic change in the island population (Hedrick, 1983).

Large mammals on islands tend to be smaller than in the parent population on the mainland (Foster, 1964; Sondaar, 1977; M. Williamson, 1981). The extent to which this is genetically or environmentally determined is uncertain, but it may be related to founder effect and to the absence of predators whose activities would result in the selection of larger individuals, as well as to the exhaustion of preferred food plants. Teeth seem to diminish in size more slowly than the jaws; thus such isolation may encourage tooth crowding. The process of domestication (which entails the isolation of a small segment of the wild population) may have a similar effect, resulting in small size in many domestic animals; in canids, small size and tooth crowding are characters that can be used to distinguish the sub-fossil dog (*Canis familiaris*) from the ancestral wolf (*C. lupus*) (Degerbøl, 1961; Clutton-Brock, 1981).

Disturbances of tooth eruption and variation in position

Not all of the foregoing is directly relevant to disturbances of tooth eruption and variations in the position of teeth though some of it is; for example variations in the size and shape of the jaws and of their relation to one another, and even of eruption, occur as deviations from the normal that result from genetic mechanisms. However, many of them are not variations or abnormalities of the dentition as such but are secondary to variations or defectiveness of bone or jaw development or growth. It is more appropriate to consider the causes of these in the introductions to the chapter on eruption (Chapter 16) and to the chapters in Section 2 on variation in tooth position.

Variations in man

Because of our dominant interest in our own species, there is a considerable literature on variation and diseases of the teeth in man, in which much of the terminology has been established, and from which many of the ideas about aetiology arise. It will be appropriate to indicate some signposts to this literature at various points in this book.

In man, there is a continuous spectrum from what can be regarded as normal variation in the morphology of individual teeth to grosser variants which must be regarded as accidents of development sufficient to produce something that has a crown and root, and can therefore be called a tooth. These grosser variants themselves form a continuous spectrum at the extreme end of which are odontomes (Chapter 25); that is, tumour-like masses with little morphological organization, only related to teeth because they are composed of dental tissues of recognizable structure, enamel, dentine and cementum. Although complete records of such a spectrum do not appear to exist for any non-human species, it seems reasonable to believe that similar degrees of disordered tooth development occur in them also.

Tooth nomenclature

Incisors, both permanent and deciduous, are the teeth which are anterior to the canines. In the upper jaw, they are usually further defined as the teeth in the premaxilla; in the embryo, the upper incisors are usually regarded as developing in the fronto-nasal process and the canine as the most anterior tooth to develop in the maxillary process. However, it cannot be taken for granted that the premaxilla-maxillary suture corresponds exactly with the junction between the embryological fronto-nasal and maxillary processes. For instance, when union between the two proces-

Fig. 1.6. A hypothetical eutherian mandibular dentition (Row A) containing three incisors, a canine and seven post-canine (molar) teeth, some of which are permanent, others are deciduous and are replaced by permanent teeth (Row B). Replacements for deciduous molars are called premolars. In each region of the jaw, teeth are graded in size.

The first post-canine tooth is not replaced; it is uncertain whether it is a deciduous tooth (m1) whose successor has been suppressed or a permanent tooth (P1) with no deciduous predecessor.
From: Osborn, J.W. (1978).

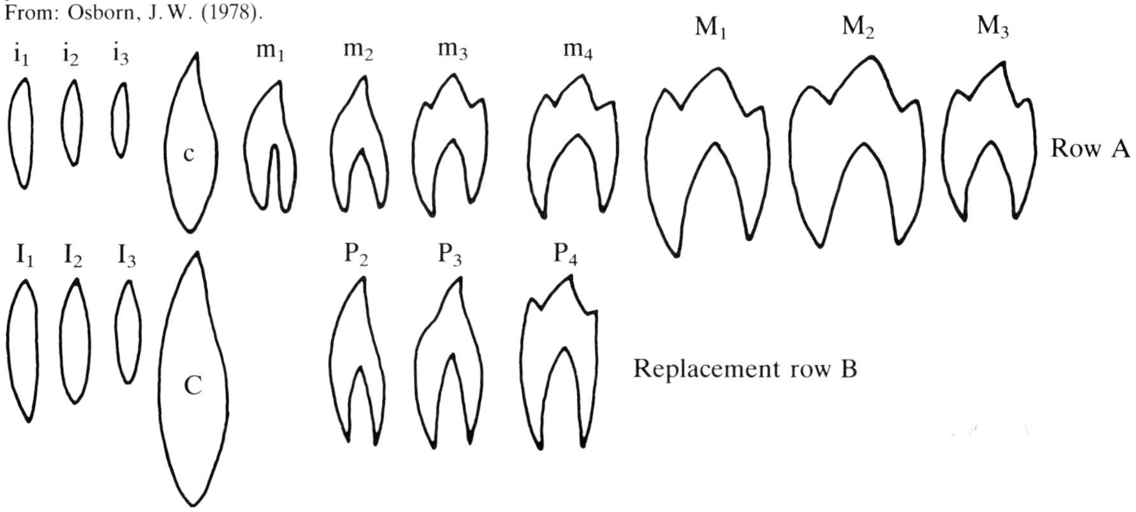

ses is defective, as in the cleft alveolar process associated with cleft lip, at least in man a supernumerary tooth is commonly present on the lateral side of the cleft medial to the canine; the supernumerary tooth sometimes has the morphology characteristic of an incisor (Miles, 1954).

Although the deciduous and permanent canines are typically prominent, pointed teeth, this is by no means always so, and the canine is therefore better defined as the first tooth in the upper arch behind the suture between the premaxilla and the maxilla. The definition of the mandibular canine is less satisfactory, nothing better having been devised than the first tooth in the arch to articulate immediately in front of the upper canine. When the homologies and phylogenies of individual dentitions are considered, we have to be prepared for proposals based on four incisors, incisiform canines and so on. A stimulating discussion of the problems that can only be touched on here has been provided by Schwarz (1982, 1983).

Simple definitions of the cheek teeth are: deciduous molars are post-canine teeth that are normally shed, they are replaced by premolars; molars are post-canine teeth that lack precursors (see Fig. 1.6).

Although it is customary to follow the nomenclature of Figure 1.6 which implies that the replacement teeth form a single series with the molars that are not replaced, the whole being called the permanent dentition, there is good evidence for regarding all those teeth in row A of the Figure as belonging to a single series and the replacing ones in row B as a second series.

Tooth symbols

There is no uniformity in the literature in the use of symbols to indicate particular teeth; for example Pm is commonly used for premolars and Di and Dm (or Dpm) to indicate deciduous incisors and molars respectively. We have chosen the convention of single lower case letters for deciduous teeth, e.g. i for each deciduous incisor, and single capital letters for permanent teeth, e.g. P for each premolar, where necessary combined with a numeral to indicate the particular tooth in the series. This simple convention seems sufficient to avoid ambiguity. Furthermore, we have used the numeral superscript, as M^1, for upper teeth, and subscript, as M_1, for lower teeth. When the numeral is neither above nor below the symbol, M1, it indicates the tooth in both upper and lower jaws.

None of the ways of indicating symbolically which side a tooth belongs to is completely satisfactory and we shall spell out in most instances left or right but, in Tables and where information is presented in semitabular form, left and right will be reduced to L and R.

A complication arises when one or more of a tooth series is normally absent. For example, Old World monkeys and apes have only the last two of the normal four premolars of the ancestral eutherian. In this case, the premolars are usually designated P3 and P4 (and the corresponding deciduous teeth as m3 and m4) but, because the apes are so often compared with man, in which these teeth are known as first and second premolars, these names are used in apes and Old World monkeys also. We will follow this convention by referring to first premolars or P3 and second premolars or P4 in apes and Old World monkeys. Using the same convention, the three premolars of New World monkeys are first (P2), second (P3) and third (P4), but in all other mammals the same verbal and symbolical designations are utilized, dependent upon the phylogenetic homology of the tooth.

The dental formula

The dental formulae of some species are still the subject of disagreement and debate, resting as they do upon fossil evidence which has been greatly supplemented during the past 100 years or so since the first definitive descriptions of the dentitions of most living forms were made. It is beyond the scope of this book to recount these differences of view, except occasionally.

The basic dental formula of adult eutherian mammals is regarded as comprising a row of 11 tooth positions in each quadrant: 3 incisor, 1 canine and 7 post-canine (or premolar and molar) positions; it is depicted in the following form:

$$I\tfrac{3}{3} : C\tfrac{1}{1} : P\tfrac{4}{4} : M\tfrac{3}{3} = 44$$

i.e. 3 incisors, 1 canine, 4 premolars and 3 molars in each quadrant. Many groups of mammals have in the course of evolution lost some of these teeth; only in a few taxa, such as the bat-eared fox (*Otocyon*) and the cetaceans, regarded as a specially evolved group, is this number exceeded.

Most eutherian mammals are diphyodont with eight deciduous teeth (Row A in Fig. 1.6):

$$i\tfrac{3}{3} : c\tfrac{1}{1} : m\tfrac{4}{4} = 32$$

However, some teeth anterior to the fourth deciduous molar, usually the first post-canine tooth, may not be replaced (Row B in Fig. 1.6). There is some doubt whether these are retained deciduous molars or premolars whose deciduous predecessors have been suppressed. The former interpretation, which is used in this book, can be described by referring to these post-canine teeth as m1, m2, m3, and m4; the latter interpretation by using P1, m1, m2, and m3. Many groups of mammals have lost some or all of their deciduous teeth; in seals, for example, the entire deciduous dentition is shed before birth; in a few taxa, the number is exceeded.

In the Metatheria (marsupials), the basic dental formula appears to be:

$$I\frac{5}{4} : C\frac{1}{1} : p\frac{3}{3} : M\frac{4}{4} = 50$$

There are, however, difficulties with nomenclature and homology which are discussed at length by C.S. Tomes (1904, pp. 600–609). Although rudimentary precursors of other teeth occur in some species, such as the brush-tailed phalanger (*Trichosurus vulpecula*) and the short-tailed wallaby (*Setonyx biachyurus*) (Berkovitz, 1968a, c), usually it is only one of the apparently permanent teeth, the third premolar, that replaces a deciduous precursor. The functional metatherian dentition can either be thought of as a permanent one, with a deciduous dentition that has been almost entirely suppressed (evidence for this view has been presented by Berkovitz, 1972), or Lydekker (1899) can be followed and the whole dentition, apart from the third premolar, be regarded as homologous with the eutherian deciduous dentition plus the permanent molars, in which case the correct adult formula would be:

$$I\frac{5}{4} : C\frac{1}{1} : \frac{mmPm}{mmPm} : M\frac{3}{3}$$

This is the same formula as that implied when the teeth in Row A (Fig. 1.6) are regarded as a single series, except for the greater number of incisors and the replacement of m3, and sometimes of m2 as well, by P3. Although it is probably best, like C.S. Tomes (1904) and Lumsden (1981) to follow the late-nineteenth century pioneer palaeontologist, F. Ameghino, and to consider the post-canine teeth (P.C.) as a single series with the basic formula:

$$I\frac{5}{4} : C\frac{1}{1} : P.C.\frac{7}{7}$$

it is more convenient for descriptive purposes to use the basic formula given above (Chapter 8, p. 140).

The numbers of teeth vary greatly in the different metatherian species; the opossums (Didelphidae) have the formula given above but, in the other families, the numbers of incisors, canines and post-canine teeth are reduced in various degrees. The reduction in numbers is most extreme in the wombats (Vombatidae) with only one large, rodent-like incisor in each quadrant, which, as there is no canine, is separated from the first of the five post-canine teeth by a long diastema.

Little is known about the dental development in the primitive, but highly specialized, Prototheria (monotremes). In the Tachyglossidae (echnida and spiny anteater), dental functions are taken over by a sticky tongue and by cornified spines on the tongue and palate. In the duck-billed platypus (*Ornithorhynchus*), the dental formula is:

$$\frac{0}{5} : \frac{1}{1} : \frac{2}{2} : \frac{3}{3}$$

at least one upper and one lower premolar on each side having replacement buds beneath; in the adult, all the teeth are replaced by horny plates on the gum.

Special terms

The incisors and canines as a group can be referred to as anterior teeth and cheek teeth as posterior and, in long-snouted mammals such as the toothed whales where the teeth are in long more-or-less straight rows, contiguous tooth surfaces can be referred to as anterior and posterior. However, in the majority of mammals, the teeth are arranged as an arch or arcade in which the incisors are medial or lateral to one another rather than anterior and posterior. To avoid this difficulty the practice has grown up of naming the contiguous surfaces of teeth according to their relationship to the midline of the tooth arch. Thus, surfaces that face that midline are mesial and those that face away from the midline are distal. Similarly, a tooth can be referred to as situated mesially or distally to another according to whether it is further from or nearer to the midline of the arch. We shall in general adopt this terminology; where there are transgressions from its strict application, we hope that it will be evident that there is reason for doing so.

Surfaces that face the inner aspect of the tooth arch, that is, face the tongue, will be lingual. Surfaces which face the outer aspect of the tooth arch will be called buccal, although often the corresponding surfaces of the incisors face the lips rather than cheeks and will therefore be referred to as labial.

Tooth shape

It is unnecessary to list here every other term used in comparative dental anatomy; many are standard terms and some are explained when used. Only three need to be explained here. Teeth with prominent bluntly-conical cusps are bunodont (Gk. *bunos* = hill), teeth

14 *Colyer's Variations and Diseases of the Teeth of Animals*

with low crowns are brachydont (Gk. *brachy* = short or low) and teeth with high crowns, a feature that evolved as an adaptation to wear, are hypsodont (Gk. *hypso* = high).

There are considerable differences in tooth shape among the various taxa; these differences play an important role in determining taxonomy. Crown morphology, because it is so easily seen, naturally plays a

Fig. 1.7. The occlusal surfaces of mammalian molars.
A: The three basic cusps of the tribosphenic upper molar are the protocone, paracone and metacone. A hypocone develops from the cingulum in many groups, and so do minor cusps, some of which are shown here (after Butler, 1981). Particular cusps may be lost in the course of evolution of some groups. In bilophodont teeth, ridges connect the protocone to the paracone and the hypocone to the metacone.
B: The basic cusps in the lower molar are the protoconid, paraconid and metaconid on the trigonid; the talonid is lower, but is usually well developed and bears two main cusps, the hypoconid and the entoconid (from Butler, 1981). Cusps may be added or subtracted in the course of evolution.

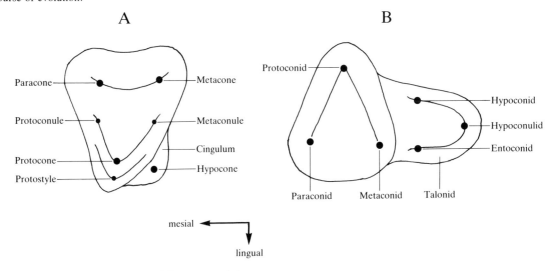

Fig. 1.8. Occlusal surfaces of upper (A) and lower (B) cheek teeth of a horse. In these very hypsodont teeth, cementum covers the enamel of the crown and both layers are invaginated together with the dentine. As the tooth wears away, a complicated pattern of cementum, enamel and dentine is exposed. The worn areas of the teeth still correspond to the cusps on the unworn teeth which are homologous with the cusps of other mammals (Fig. 1.7). The paraconid has been lost from the lower molars, but the talonid bearing the hypoconid and the entoconid is well developed and stands at the same level as the trigonid.

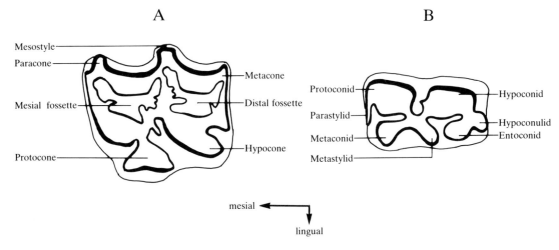

larger part than root morphology, which is more variable and also less studied and less well-recorded.

It is beyond the scope of this book to describe systematically the normal morphology of dentitions in various taxa, although naturally many aspects of this will be touched on. As far as possible, an indication will be given where good descriptions of the normal are to be found.

In general, incisors and canines are of more simple form than cheek teeth and tend to be single-rooted. Cheek teeth usually have more than one root and their occlusal surfaces are often of considerable complexity, presenting systems of cusps and crests (that sometimes carry minor cusps) separated by valleys or fissures. The cusp patterns present homologies among various taxa and, for this reason and for purposes of description, are named systematically. The most widely used nomenclature is given in Figure 1.7, based upon that introduced by H. F. Osborn in 1907 and elaborated by Butler (1978a, 1981, 1982). Ancestral mammals developed cheek teeth in the Cretaceous period with crowns that have a tribosphenic pattern consisting of three basic cusps. The term tribosphenic seems to derive from *tribo* – Gk. rubbing together – and *sphenic* – Gk. wedge-shaped – and only by chance carried the connotation of the number three. The three basic cusps in the upper teeth are named protocone, paracone and metacone, and in the lower teeth are protoconid, paraconid and metaconid. This pattern persists in modern mammals with varying degrees of modification in the different taxonomic groups. The commonest modification is the development of additional cusps, the corresponding nomenclature of which is indicated in Figure 1.7.

Some tribosphenic molars have a ridge of enamel partially or completely surrounding the crown just above the cervix known as the cingulum (Lat. girdle); it is thought that the talonid (Fig. 1.7) and many of the cusps peripheral to the three basic cusps arise from the cingulum, for example the hypocone and in primates the so-called cusps of Carabelli (Chapter 3, p. 19–20).

In many groups of mammals, the grinding efficiency of the teeth has been improved by the development of enamel ridges or lophs (Gk. *lophos* = yoke) connecting the cusps. Where this is a predominant feature of the occlusal surface, the teeth are said to be of lophodont type. In such teeth, and particularly in the herbivorous ungulates where the complexity of the occlusal pattern is increased by deep invaginations of the enamel into the dentine, the cusps and their homologies are difficult to recognize. Figure 1.8 provides, as an example, the nomenclature applied to the cheek teeth of equids.

SECTION 1
Variations in Number, Size and Shape

CHAPTER 2

Introduction

The conceptual aspects of the significance and possible mechanisms of production of variations in size, number and shape of teeth have been elaborated in the General introduction. In general, smallness of teeth, reduction in their complexity and total absence are different degrees of the same thing, explicable in terms of the morphogenetic field concept.

In this subdivision of the book, composed of six chapters, each of which is devoted to a different mammalian order or group of orders, examples will be systematically described. Much of the account has to be anecdotal, that is it consists of a catalogue of examples, but wherever possible the information is collated in order to make generalizations.

For some supernumerary teeth, it is not possible to record more than that they were present and where they were situated. Their shape, whether they were simple, conical or tuberculate in their morphology, or whether they resembled teeth of the normal series (supplemental teeth, Chapter 1, p. 6) is often not recorded. Where, for example, the primary account mentions that an extra premolar was present, it is not always clear whether it was so designated because it was in the premolar region of the arch, or whether it had the morphology of a premolar, in which case, of course, there would be difficulty in deciding which of the premolars was supernumerary. It is unusual for such teeth to resemble those of the normal series exactly, so they are often recognizable by some degree of abnormal morphology.

In the incisor and canine regions, it is sometimes necessary to distinguish truly supernumerary teeth from deciduous teeth that have been retained beyond their proper time, especially as occasionally the permanent tooth may erupt beside its deciduous predecessor.

It goes almost without saying that when a supernumerary tooth is present there is not always space available for it in the arch, and it, or adjacent teeth, may erupt to one or other side of the arch. It does not seem appropriate therefore to describe the exact position of supernumeraries, or whether they are rotated, or incompletely erupted.

Where teeth are absent and no contrary circumstantial evidence is present, it can be assumed that the teeth failed to develop. Loss of teeth from trauma or disease often leaves some evidence of a healed tooth socket; furthermore, it is not appropriate to consider loss from periodontal disease or dental caries as a cause if the other teeth are free of disease and if dental disease is rare in that species.

Variation in size of teeth is of course difficult to assess without detailed research into the normal degree of variation of each and every type of tooth of each and every mammalian species; by and large, this research has not been done, so only extreme and obvious examples are described in the text. However, in the case of primates, tooth size has been quantified (for example Swindler, 1976; Gingerich and Schoeninger, 1979) and is often referred to in discussions of the relationship of human species to each other and to other primates (Ashton and Zuckerman, 1950*a*, *b*; 1951*c*; 1952).

Because the roots, unlike the crowns, of teeth are not, while *in situ*, exposed to direct scrutiny, much less is known about their variability. The roots of multi-rooted teeth, such as those of the cheek teeth of primates and carnivores, appear to be particularly variable in that extra roots and degrees of fusion between roots seem to be common.

Fusion of roots, although the commonest term employed, is not a good one because it implies that the root elements were originally separate, whereas it is more a matter of a failure of the roots to separate or be formed as separate elements during development. To use connation (p. 7) for joined roots would lead to ambiguities so we shall continue to refer to fused roots hoping that it will be evident that the term carries with it an invisible 'so-called'.

It is useful to distinguish two types of fused roots; in one, the roots are close to each other and may resemble a single root partially divided by longitudinal grooves and often with a bifid apex; in the other type, webbed roots, they are further apart and are joined by a thin web of root tissue rather like the webbed feet of a water-fowl. Webbed roots appear to be particularly common in primates.

CHAPTER 3

Order **PRIMATES**

Illustrated systematic descriptions of normal primate dentitions are given by James (1960) and Swindler (1976; 1978).

For obvious reasons, much less is known about variations in the roots of teeth than in the crowns. However, it is well-established that in man the roots of the teeth, particularly the molars, are no less variable than the crowns. This variability is least in M1 but is markedly greater in M2 and greater still in M3. The M2 quite frequently have extra rootlets and some degree of fusion between roots is also common. M3 are much more affected in this way and, in addition, the overall length of their roots is very variable. Descriptions of the roots of the teeth of the non-human primates, for example Bennejeant (1936), suggest that root morphology is probably as variable as in man and that the types are similar, variability being greatest in M3. Bennejeant (1936, pp. 115, 182) drew attention to longitudinal grooves on the labial surfaces of the roots of some primate upper incisor teeth and less commonly on upper canines. An example in a baboon is shown in Figure 20.5. These grooves cannot be regarded as a tendency for the formation of two roots because the upper incisors and canines never have two roots. The grooves could indicate minor degrees of connation or dichotomy.

In man, M^1 often has a small cusp, the cusp or tubercle of Carabelli, on the lingual aspect of the protocone in addition to the normal four. If the cusp is well-developed, it is usually present on m^4, M^2 and M^3 as well. The prevalence of the cusp on M^1 varies between different racial groups (Sofaer, 1981, p. 153). It is naturally of interest to discover whether the cusp or a trace of it, such as a groove in the same position, occurs in apes or in fossil hominids and other primates. Examples, mostly controversial, both in hominids and apes have been described (see Frisch, 1965; Lavelle, Shellis and Poole, 1977) as well as in ancestral early tertiary primates, but few workers, except Sakai (1982), have referred to them as cusps of Carabelli. However, it is easy to see that, when additional cusps, even as evident anomalies, are found on upper molars, in what can be called the Carabelli position, they excite special interest.

The lower molars of most of the early Hominoidea, i.e. the ancestors of modern pongids and man, have five cusps (three buccal and two lingual) and this is invariably the case in the Miocene ape, *Dryopithecus*, which is agreed to be either an ancestor from which apes and man derive or at least close to it. Dryopithecine lower molars have five cusps separated by fissures (grooves) which form a Y pattern, which is a feature by which closeness to the evolutionary hominoid line is recognized. The dryopithecine Y-5 occlusal pattern is found in varying degrees in the lower molars of modern pongids and in the majority of human M_1 (Frisch, 1965, p. 102). Modern Old World monkeys have five cusps on M_3 but they lack the Y-5 arrangement; their other lower molars and all those of New World monkeys are typically four-cusped. As in the case of additional cusps in the Carabelli position, five-cusped lower molars, particularly if arranged in the Y-5 pattern, arouse interest when they occur in apes or monkeys.

We have already mentioned (Chapter 1, p. 10) that developmental anomalies tend to be multiple. Bearing in mind how specimens tend to be collected, for example often it is only the head that is collected in the field and supernumerary and absent teeth are rarely noticed until the skull is prepared, it is not surprising that it is rarely mentioned whether dental anomalies were associated with abnormalities elsewhere. A notable exception is that Schultz (1956) described a gibbon with a supernumerary premolar which also had a sutural bone in the occipito-parietal suture; also a gibbon, with all M3 and an incisor absent, which exhibited polydactyly.

Sub-order **ANTHROPOIDEA**

Family **Pongidae**. Great apes and gibbons

$$\frac{2}{2}\cdot\frac{1}{1}\cdot\frac{2}{2}\cdot\frac{3}{3} = 32$$

The incisors of the great apes are large and spatulate. Both the upper and lower canines project beyond the general occlusal plane and are markedly larger in males than in females. The cheek teeth are brachydont and bunodont. The upper molars have four cusps and the lower molars four or five (see above).

In gibbons and siamangs, the incisors are smaller than in the other apes and I^2 may be pointed. The canines are long and slender with little sex dimorphism.

The numbers of specimens of each species examined and the numerical variations recorded in the pongids by Sir Frank Colyer in the original edition of this book are summarized in Table 3.1; a summary of variations for the family as a whole is given in Table 3.2. Details of individual specimens are given below.

Among the many studies that have been made of pongid dentitions, the work of A. H. Schultz is outstanding and will be frequently referred to. He built up a large collection of primate material, which is now housed in the Anthropologisches Institut in Berne. In one particularly valuable paper (1944), he described two large series of gibbons. His Chiengmai Series consists of 233 *Hylobates lar* skulls which, all being from the same locality, provide a unique opportunity to assess variation within a single population; his General Series II is composed of 579 specimens of *H. lar* (including the Chiengmai Series) as well as other specimens of *Hylobates moloch* and *Hylobates gabrielli*.

Classic works to which Colyer sometimes referred include those by Selenka (1898, 1899). Selenka had a large collection of skulls of orang-utans which is now in the Zoologische Staatsammlung of the Schloss Nymphenburg in Munich.

In general, supernumerary incisors, canines and premolars are rare in the pongids but fourth molars are quite common. In Bolk's (1916) terminology, extra molars are termed disto-molars when situated distally to M3 and para-molars when they are buccal to the line of the arch. In the sample studied by Colyer, disto-molars occurred in 3.7% of gorillas, 2.5% of chimpanzees, 6.1% of orang-utans but in only 0.8% of gibbons; similar conclusions were reached by Lavelle and Moore (1973). The most comprehensive figures, however, are those of Krapp and Lampel (1973) as they brought together all the previously published work on disto-molars; their figures are: gorilla 3.95% ($n = 1409$), chimpanzee 2.71% ($n = 1040$), orang-utan 11.24% ($n = 1808$) and gibbon 0.66% ($n = 1276$). The only discrepancy between their findings and Colyer's is in the percentages of disto-molars in orang-utans which Krapp and Lampel found to be so high. There is no doubt that theirs is the more reliable finding. All authors agree about the rarity of disto-molars in gibbons which was first noted by Bateson in 1894. Some

Table 3.1 Numerical variations in the PONGIDAE

Genus		No. of specimens	Extra Teeth					Specimens affected		Absence of teeth				Specimens affected		Totals of specimens showing extra teeth or absence of teeth	
			I	C	P	M		No.	%	I	P	M		No.	%	No.	%
Gorilla (gorilla)	Skulls with mandibles	546	6	—	—	22		27 ⎫	4.4	2	1	1		4	0.6	34	5.0
	Skulls only	129	—	—	—	3		3 ⎭		—	—	—		—			
Pan (chimpanzee)	Skulls with mandibles	467	3	—	1	13		15 ⎫	2.9	—	2	8		9	1.6	24	4.3
	Skulls only	93	—	—	—	1		1 ⎭		—	—	—		—			
Pongo (orang-utan)	Skulls with mandibles	229	—	1	—	16		17 ⎫	6.8	1	1	5		7	2.4	25	8.5
	Skulls only	38	—	—	—	—		— ⎬		—	—	—		—			
	Mandibles only	28	—	1	—	2		3 ⎭		—	—	—		—			
Hylobates (gibbon)	Skulls with mandibles	306	1	—	—	1		2	0.7	3	1	7		11	3.6	13	4.2
Symphalangus (siamang)	Skulls with mandibles	85	1	—	1	2		4	4.7	1	—	—		1	1.2	5	5.9

In skulls in which two or more classes of teeth varied in number, each class is recorded individually; for example, if a skull has an extra incisor as well as an extra molar, both columns would be credited with one and the column 'number of specimens affected' with one only.

early workers were much excited by the occasional presence of disto-molars in man and as well as in the other hominoids, believing this to be an atavistic phenomenon indicating descent from an ancestor with more than three molars; as no evidence of such an ancestor has ever been found however, the fourth molar came to be regarded as an occasional variation, as originally suggested by Bateson (1894). Bolk (1916) considered that the disto-molars are homologous with M3 of the Eocene progenitors of the Anthropoidea, the hominoid M1 in his opinion being the homologue of the ancestral last deciduous molar.

Mijsburg (1931) had a similar interpretation for additional maxillary teeth in two skulls of siamang, but each of these teeth was a deciduous molar situated between the deciduous canine and the first of the two normal deciduous molars. He considered them to be homologous with the first of the four deciduous molars that would be expected to occur in a primitive ancestral primate. Similarly, he interpreted an additional lower premolar in a siamang as an atavistic P_4.

Absence of molars is much less common in pongids than the presence of extra molars (Table 3.1). Lavelle and Moore (1973) found that only one (0.5%) of 190 gorilla skulls lacked any teeth (a lower premolar) which agrees closely with the figure of 0.6% for the present sample (Table 3.1). Schultz (1944) found one or more M3 absent in 3.8% of his General Series gibbons and in 4% of his Chiengmai Series. In two of the General Series, there were no molars at all; thus there is a clear tendency to suppression of M3 in gibbons, which seems to be related to its frequent reduction in size and complexity (Frisch, 1965), and to the rarity of the disto-molars mentioned above.

Teeth of normal shape but larger or smaller than usual are found occasionally in the pongids. Their occurrence was not quantified by Colyer but Schultz (1944) noted that a size reduction in the M3 occurred in about 5% of his gibbon skulls (see below).

Various types of abnormality of shape occur in pongid teeth, most commonly in molars, but only rarely in premolars. Incisors and canines may be reduced to small pegs or may be connate. In the molars, these variations take the form of supernumerary or reduced cusps and of abnormalities in the positions of the cusps. The number of roots may vary, and some of the roots may be fused.

When molars have large extra cuspal elements, the additional tooth mass is sometimes associated with an extra root as in a gorilla (MNHN 1898–316) described by Bennejeant (1936, p. 131).

The prevalence of variations in number, size and shape may be increased in isolated populations of pongids. For example, amongst 60 adult chimpanzee skulls from Batouri District in Cameroun which are preserved in the Powell-Cotton Museum, two had an upper disto-molar, one a lower disto-molar, one lacked both P_4, two lacked the left M_3, in one the left M_3 was abnormal in shape and in another the left M^3 was about half the normal size. Incisor underjet was particularly common in the same group (Chapter 10, p. 168) and so was caries (Chapter 21, p. 465).

Genus *Gorilla*, gorilla

EXTRA TEETH

(a) Incisors

(i) ♂. A.E.L. Amsterdam, 1912/116. (Fig. 3.1). Three teeth in the right premaxilla; the outer one is conical; the remaining two are incisor-like. The left I^2 is larger than the right and has a longitudinal groove on the lingual aspect of the crown suggestive of partial dichotomy.

(ii) ♂. A.E.L. Amsterdam, 1917/2. An extra tooth between the left I^2 and upper canine.

(iii) BMNH 1939.938. ♂. An extra tooth in the right premaxilla situated to the mesial side of I^1 (Wegner, 1910).

(iv) Cambridge University Museum. An extra tooth between the right I_1 and I_2.

Table 3.2 The classes of teeth showing numerical variations in the PONGIDAE as a whole

| Class | Extra teeth | | | | Absence of teeth | | | |
| | Maxilla | | Mandible | | Maxilla | | Mandible | |
	Right	Left	Right	Left	Right	Left	Right	Left
Incisors	3	3	3	2	1	2	4	3
Canine	—	—	—	2	—	—	—	—
Premolars	1	—	—	1	1	3	2	2
Molars	31	30	27	23	10	10	7	8

(v) A Private Collection. An extra incisor between the right I_1 and the mid-line, and bilateral distomolars in the upper jaw.

(vi) ♂. Formerly RCS Osteo. Series, 21a. (Fig. 3.2). Five incisors in the mandible. The extra tooth is probably the rotated one. This specimen was described by Bateson (1894, p. 203) and by Wegner (1910).

(vii) Institut de Paléontologie humaine, Paris. Two supernumerary incisors in the premaxilla situated lingually to the first incisors. All the lower incisors are absent (Bennejeant, 1936, p. 186).

Schultz (1950) found extra incisors in two out of 267 gorilla skulls.

(b) Canines

Supernumerary canines are rare. Selenka (1899, p. 141) described a conical supernumerary tooth situated lingually to the upper canine in an adult male gorilla in his collection and interpreted it as a supernumerary canine, an interpretation with which Remane (1921, p. 13) agreed. However, Bennejeant (1936, p. 195) argued, for reasons that are difficult to follow, that the conical tooth is a supernumerary incisor. We can only say that in Selenka's original illustration the conical tooth seems to be on the maxillary side of the premaxillary-maxillary suture and is therefore a canine.

(c) Premolars

Extra premolars are rare in gorillas. One specimen in the Museum für Naturkunde in Berlin has a supernumerary tooth between the canine and the premolars in the right mandible. The crown is conical but there are two roots and Remane (1921, p. 13) regarded it as a premolar. In a male skull in the Naturhistorischen Museum in Vienna (No. 3114), Weninger (1948) observed a pair of supernumerary premolars in the maxilla.

Two gorilla skulls in the Schultz Collection in the Anthropologisches Institut in Zurich have additional premolars; a male, AIMUZ 7081, has one in each mandible and a female, AIMUZ 7118, has one in the left mandible. Schultz (1964), because of the rarity of extra premolars in gorillas and because the two specimens were from practically the same locality and of nearly the same date, suggested that the condition might be hereditary. Bennejeant (1936, p. 201) showed a tiny conical supernumerary between the upper left canine and P^3 in a gorilla (MNHN 1869–142).

(d) Molars

Disto-molars are common in gorillas, ranging in number from one to one in each quadrant. In shape, they vary from conical forms to duplicates of normal M3. Occasionally they are larger than M3. Schultz (1950) found disto-molars in 15 out of 267 gorilla skulls; five were impacted or retarded in eruption. Remane (1921) described 11 gorilla skulls with disto-molars in German museums and reviewed 15 more previously published instances, one of which had five molars in the right maxilla. Disto-molars, especially bilateral examples, were more common in the upper jaw.

Bennejeant (1936, p. 217) described bilateral maxillary disto-molars in the upper jaw and another in the left mandible in a sub-adult male gorilla (MNHN 1962–1493). The upper disto-molars were tricuspid and slightly smaller than M^3. The lower tooth could only be studied in a radiograph because it was buried in the jaw but it appeared to be about the same size as M_3 and to have five cusps.

A rare condition in a male gorilla skull, originally described by Bennett (1885, 1886) and commented on by Bateson (1894, p. 204) and Colyer (1915, 1931), is shown in Figure 3.3. Two conical supernumerary teeth are embedded in the right ramus of the mandible, with their crowns pointing backwards. The tips of the

Fig. 3.1. *Gorilla gorilla* (gorilla). ♂. A.E.L. Amsterdam, 1912/116. In the right premaxilla, there is an extra incisor; the left I^2 is unusually large and is grooved longitudinally, suggestive of dichotomy. × c. 0.5.

Fig. 3.2. *Gorilla gorilla* (gorilla). ♂. Formerly RCS Osteo. Series, 21a. Mandible with an additional incisor.

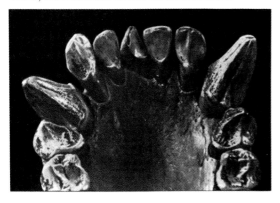

crowns have emerged slightly through the surface of the bone. A further point of interest is that, whereas the lower tooth, like most conical supernumerary teeth, has a single root, the upper tooth has three roots. Because these teeth are not within the normal tooth-bearing area of the jaw, they are categorized as ectopic; the most obvious explanation is that they arose from an extension backwards of the ectodermal dental lamina that gives rise to the tooth germs.

Bennejeant (1936, p. 211) referred to a small paramolar situated in the buccal embrasure between M^1 and M^2 in a skull which was formerly in the Zoologisches Museum, University of Hamburg, no. 11. A similar example was noted by Remane (1921).

ABSENCE OF TEETH

(i) ♂. Formerly RCS Osteo. Series, 23.21. Both I^2 absent.
(ii) *Gorilla gorilla berengei*. NRM 45. Right I_2 absent.
(iii) P-C. Mus., Z vi 32. An adult. Both P_4 absent. It is noteworthy that there was still space in the dental arch for these teeth which suggests that the m_4 may not have been shed until shortly before death. On the other hand, the opposing upper teeth would then have shown signs of wear; instead they were unworn.
(iv) ♀. A.E.L., Amsterdam, 1913/263. Left M_3 absent.

SIZE

Variations in the size of individual teeth are uncommon in the gorilla. Normally, the molars increase in size from front to back but M3 may be smaller than M2; more rarely, M1 may be the largest of the series; for example in one specimen the lengths of the crowns of the molars were: M^1 15 mm, M^2 11.3 mm and M^3 12.8 mm. A male gorilla skull (AIMUZ 7081) described by Schultz (1964) had exceptionally large teeth, as well as supernumerary premolars. Contralateral teeth may also occasionally differ in size.

SHAPE

The following types of variations in shape were noted:
(i) I^2 sometimes peg-shaped and small.
(ii) (Fig. 3.4). A ridge on the lingual aspect of the I_2 extends from the cingulum to the incisive edge.
(iii) A small cusp on the lingual aspect of the upper molars is common, particularly in the mountain gorilla (*G. g. beringei*); it was present in 5 out of 578 specimens. Extra cusps in this subspecies were mentioned by Gyldenstolpe (1928).
(iv) Bennejeant (1936, p. 141) mentioned the skull of an immature gorilla (ML, M50) that had two tubercles on the lingual aspect of the protocone of both M^1 resembling the cusp of Carabelli, which is a regular feature of many human molars (p. 19).
(v) *Gorilla gorilla beringei* (mountain gorilla). KMMA 9424. An extra cusp on the disto-buccal portion of each M^3.
(vi) ♀. RCS Odonto. Mus., A 64.3. (Fig. 3.5). On the left M^3, the paracone and the protocone are not, as is normal, united by a ridge but, between them, there is a small extra cusp mesially. On the lingual aspect, there are three cusps in a row and buccally to them there is a well-developed conical cusp. The metacone and the hypocone are normal in shape. The paracone and the small extra cusp are carried on the mesio-buccal root; the protocone and the row of extra cusps are on the lingual root; the metacone, the

Fig. 3.3. *Gorilla gorilla* (gorilla). ♂. RCS Odonto. Mus., G 10.1. The bone has been removed to display two conical teeth projecting from the inner surface of the ascending ramus of the mandible.

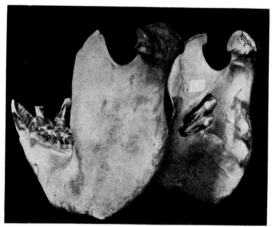

Fig. 3.4. *Gorilla gorilla* (gorilla). MNHN 1900/52. Left I^2 with a ridge on the lingual surface.

hypocone and the one extra cusp are on the disto-buccal root. The right M^3 is normal in shape.

Fig. 3.5. *Gorilla gorilla* (gorilla). ♀. RCS Odonto. Mus., A 64.3. Maxillary third molars. (i) Left M^3 abnormal in shape; (ii) right M^3, normal. (a) paracone; (b) metacone; (c) protocone; (d) hypocone; (e), (f), (g), (h) and (i) extra cusps.

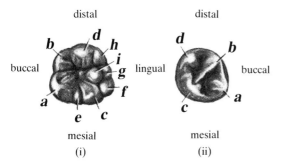

(vii) An extra cusp or cusps on the occlusal surface of M_3 is common and is usually situated between the two distal cusps. In the mandible depicted in Figure 3.6, the distal part of the right M_3 is composed of seven cusps in addition to the two mesial cusps. On the left side, there is a disto-molar. There is thus a degree of symmetry of presence of what might be called basic tooth material at the distal end of the molar morphogenetic field, on the right only forming an addition to M_3 but on the left forming a larger amount of separately-existing tooth material.

(viii) ♀. BMNH 1923.11.29.8 (Fig. 3.7). The left M_2 is very large and has three roots instead of two. The mesial cusps, the protoconid and the paraconid are in normal position. The cusp which is displaced buccally is probably the metaconid. The protoconid and the paraconid are carried on one root, the metaconid and the hypoconid on another and the entoconid on another.

Fig. 3.6. *Gorilla g. beringei* (mountain gorilla). FMNH 27551. A disto-molar in the left mandible; the right M_3 has nine cusps and is larger than the contralateral. Reduced.

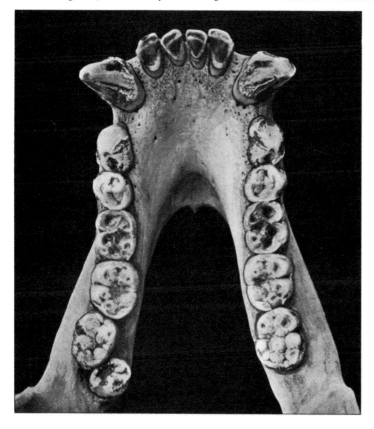

26 *Section 1: Variations in Number, Size and Shape*

Under-development of cusps is seen in the gorilla, the cusp most usually affected being the entoconid on M_3.

Systematic observations of variation in the number and shape of roots has never been undertaken but a common variation is fusion of the disto-buccal and lingual roots of M^3 and, less frequently, fusion of the three roots.

Amongst the specimens with the deciduous dentition present, there is one of connation of the right i_1 and i_2 (BMNH 1861.5.14.2).

Genus *Pan*, chimpanzee

EXTRA TEETH

Extra teeth are uncommon in the chimpanzee (*Pan troglodytes*).

(a) Incisors
 (i) Figure 3.8. A small extra tooth lingual to the left I^1. There are also supernumerary molars (p. 28).
 (ii) Figure 3.9. In the position of the left I_2, there is a connate tooth composed of two incisors completely separated only at the incisive edges. This separation continues as a groove for the whole length of the labial surface of the root and as a similar deeper groove on the lingual surface.
 (iii) ♂. Figure 3.10. Five mandibular incisors and an extra premolar on the left lingual to the normal premolars which does not quite duplicate either in shape. Both M^3 absent.
 (iv) ♀. NMBE, Congo 1934. An extra incisor in the right upper jaw shows all the characteristics of an I^2 but is much smaller than either of the normal I^2 (Bennejeant, 1936, p. 189).

Fig. 3.7. *Gorilla gorilla* (gorilla). ♀. BMNH 1923.11.29.8. The left M_2 is abnormal. (a) protoconid; (b) metaconid; (c) probably the hypoconid in an abnormal position; (d) entoconid; (e) hypoconid.

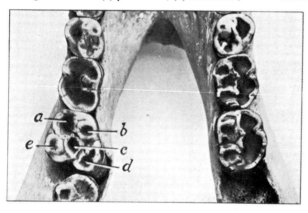

Fig. 3.8. *Pan troglodytes schweinfurthii* (Cotton's chimpanzee). NRM 160. Three supernumeraries on the left side: a small tooth lingual to I^1, another buccal to M^2 and a distomolar in the mandible. ×c. 0.6.

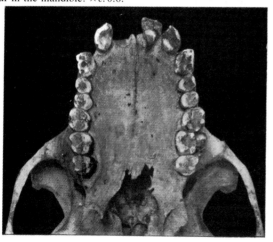
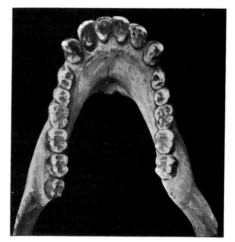

(b) Molars

Extra teeth in the molar region are often diminutive, sometimes conical and are most often disto-molars. A typical example is seen in Figure 3.11, where there are diminutive bilateral disto-molars; the one on the right has three cusps and that on the left only two.

Fig. 3.9. *Pan troglodytes* (chimpanzee). Museo Zoologico Florence, 19. In the position of the left I_2, there is a connate tooth composed of two similar incisors on a single root. Slightly reduced.

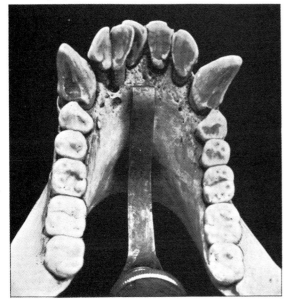

Fig. 3.10. *Pan troglodytes* (chimpanzee). ♂. Horniman Museum, 9.28. Five mandibular incisors and an extra left premolar. Slightly reduced.

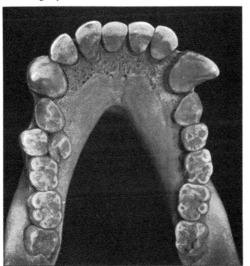

Extra teeth situated buccally to the molars (paramolars) also occur, as in the skull illustrated by Duckworth (1907) in which the supernumerary tooth was conical and situated at the mesio-buccal border of the right M^2.

Of the fourteen skulls of chimpanzees with extra molars, two show unusual conditions. In the first (Fig. 3.12), there are three additional teeth in the left maxilla and a disto-molar immediately distal to M^3 which is tilted under M^2; two others are situated

Fig. 3.11. *Pan troglodytes schweinfurthii* (Cotton's chimpanzee). ♀. NRM 8484. Small mandibular disto-molars. $\times c.\,0.5$.

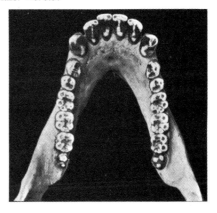

Fig. 3.12. *Pan troglodytes* (chimpanzee). Formerly in the Stockholm Högskola, 3358. Three extra teeth in the left maxilla, and a right maxillary disto-molar. $\times c.\,0.8$.

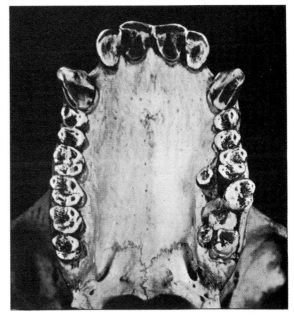

lingually to the molars, the mesial one being opposite the interval between M^1 and M^2 and the distal one opposite the interval between M^2 and M^3. The disto-molar is molariform but small; the other extra molars are small with cylindrical crowns. On the right, there is a disto-molar of normal size. The mandible has bilateral disto-molars.

The second skull (Fig. 3.8) has a small tooth on the buccal aspect of the left M^2; distal to this supernumerary is a tiny fragment of tooth tissue. There is a left mandibular disto-molar as well as an extra tooth in the left premaxilla (p. 26).

Remane (1921) described three chimpanzee skulls with disto-molars in museums in Germany and collated eight additional previously published reports. About half were bilateral in either the upper or lower jaw.

Bennejeant (1936, p. 218) recorded a disto-molar with two cusps and a single root in the mandible of a female chimpanzee (MNHN 1961–1477).

ABSENCE OF TEETH

The following examples of absence of teeth in chimpanzees were noted:

(a) *Incisors and premolars*
   (i) ♀. AHS 5717. Right I^2, two lower premolars and all upper premolars absent (Schultz, 1956, p. 1001).

(b) *Premolars*
   (i) P-C. Mus., M 440. Both P_4 absent.
   (ii) KMMA 181. Left P^3 and right M^3 absent.

(c) *Molars*
   (i) KMMA 3465. Left M^3 absent.
   (ii) ZMB 29472. Both M^3 absent.
   (iii) Horniman Museum, 9.28. Both M^3 absent, associated with an extra premolar and an extra incisor in the left mandible.
   (iv) Formerly Rothschild Coll., BMNH. Both M^3 absent, associated with a diminutive left M_3.
   (v) ♂. BMNH 1939.3362. Both M_3 absent.
   (vi) P-C. Mus., M 184. Left M_3 absent.
   (vii) P-C. Mus., M 450. Left M_3 absent.

SIZE

Differences in the size of contralateral pairs of teeth were seen in several specimens, particularly between M3 (Figs. 3.13, 3.14).

SHAPE

Extra cusps on molars were recorded in four chimpanzees:

(i) One on the disto-buccal aspect and another on the mesial portion of the buccal aspect of left M^2.
(ii) At the disto-lingual borders of both M^2 and M^3.
(iii) On the distal portion of the buccal aspect of the left M^3 and a minute one in the same position on the contralateral tooth.
(iv) On the buccal aspect of the right M^3.

The mandibular premolars normally have one mesial and one distal root. An example of three roots in a chimpanzee was shown by Bennejeant (1936, p. 152).

The maxillary premolars in the pongids have three roots, two buccal and one lingual; in the chimpanzee, the two buccal roots may be fused. The maxillary molars have three roots. Various degrees of fusion of the disto-buccal and the lingual roots occur (Fig. 3.13).

Extra roots are common on M^3. The contralateral pair depicted in Figure 3.14 are of special interest because the overall size of one, including the extra root, is much larger than the other. The pair depicted in Figure 3.15 also differ in overall size and the roots of one are more webbed than the other. As pointed out (p. 22), extra roots are often associated with additional crown elements.

Differences of the cusp morphology are also common between contralateral M^3 (Figs. 3.16, 3.17, 3.18).

Fig. 3.13. *Pan troglodytes* (chimpanzee). ♀. RCS Odonto. Mus., A 63.5. Left maxillary molars showing progressive stages in the fusion of the disto-buccal and the lingual roots.

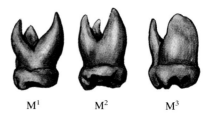

M^1 M^2 M^3

Fig. 3.14. *Pan troglodytes* (chimpanzee). ♀. BMNH 1923.3.1.1. Contralateral M^3 differ in size and each has an extra root. (A) left tooth; (B) right tooth; (a) mesio-buccal root; (b) disto-buccal root; (c) lingual root; (d) extra root.

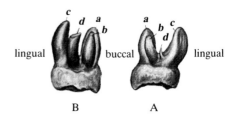

Two chimpanzee skulls (formerly A.E.L., Amsterdam 1914/52 and Stuttgart 511) showed connation between i_1 and i_2. In the Stuttgart skull, there was a slight notch between the incisive edges which continued as a longitudinal groove for the length of the crown and on to the single root (Hilzheimer, 1908). Schultz (1956, p. 1000) referred to a deciduous lower canine in a chimpanzee which consisted of a duplicated crown on a single root.

Genus *Pongo*, orang-utan

EXTRA TEETH

(a) Incisors

Table 3.1 shows that extra incisors seem to be rare in orang-utans (*Pongo pygmaeus*), but Schultz (1941) found one female (USNM 142171) with an extra deciduous incisor. Hübner (1930) described the skull of an immature orang-utan with supplemental i^2 in both upper and lower jaws on the left side situated lingually to the other incisors.

(b) Canines

Throughout the mammals, supernumerary canines are rare. Hrdlicka (1907) described the mandible of a male orang-utan in which there is an extra tooth in the region of the left canine (Fig. 3.19). The crown of the extra tooth is broken and cannot be compared with the true canines because they have been lost. Its root shows a degree of flattening more like that of an incisor but, because it is much larger than an incisor and because of its position, it can perhaps be regarded as a canine. The right M_3 is absent.

Figure 3.20 shows an immature orang with two deciduous canines in the left mandible; the teeth are not exactly similar in shape, the one adjacent to i_2 showing to a slight degree some characteristics of an incisor. There are two developing permanent canines, slightly unequal in size.

(c) Premolars

Selenka (1898) found additional premolars in three old male orang-utans. In two specimens, the extra teeth were between the canine and P^4 and both had maxillary disto-molars. In the third, there were three upper premolars on each side. According to Schultz (1956), a male orang-utan (AIMUZ 1986) had an additional P^3 on both sides and the left M^3 was diminutive.

Berkovitz and Musgrave (1971) described a male orang skull in the Anatomy Department of Bristol University that had three premolars in each maxilla.

Fig. 3.15. *Pan troglodytes* (chimpanzee). ♂. Formerly RCS Osteo. Series, 10. Contralateral M^3 differ in size; the roots of the right tooth are more webbed. (A) right tooth, disto-buccal view; (B) right tooth, mesio-lingual view; (C) left tooth, distal view; (a) mesio-buccal root; disto-buccal root; (c) lingual root; (d) extra root; (e) tongue of enamel; (f) sickle-shaped projection.

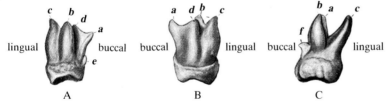

Fig. 3.16. *Pan troglodytes* (chimpanzee). Formerly RCS Osteo. Series, 12.7. Contralateral M^3; on the right tooth, there is an extra cusp on the buccal aspect. (a) paracone; (b) metacone; (c) protocone; (d) hypocone; (e) and (f) extra cusps.

Fig. 3.17. *Pan troglodytes* (chimpanzee). ♂. Formerly RCS Osteo. Series, 11.219. Right M^3 showing a variation in crown shape. (a) paracone; (b) metacone; (c) protocone; (d) hypocone; (e) extra cusp; (f) groove between the lingual cusps.

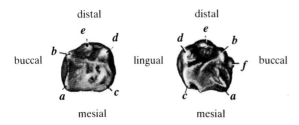
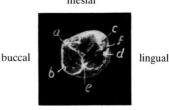

Section 1: Variations in Number, Size and Shape

(d) Molars

Supernumerary molars are common in orang-utans, varying from one to one in each quadrant. Hooijer (1948) found disto-molars in five out of 45 adult skulls (three males and two females) in museums in the Netherlands. They are more common in the mandible than in the maxilla. Extra molars may be connate with M^3 (Fig. 3.21). A specimen with two disto-molars in the right mandible (♂ USNM 142199) was mentioned by Schultz (1941, 1956) and by Schwartz (1984); Selenka (1898, p. 90) referred to a specimen with two on one side of the upper jaw. Disto-molars on the left side of both jaws of a captive male orang were described by Hübner (1930).

ABSENCE OF TEETH

The following were seen:
(i) USNM 143596. Left I_1 absent.
(ii) ♂. RCS Odonto. Mus., G 11.11. Left P_3 absent; bilateral mandibular disto-molars present.

Fig. 3.18. *Pan troglodytes* (chimpanzee). ♀. Formerly RCS Osteo. Series, 11.222. Contralateral M^3 showing differences in crown shape. (a) paracone; (b) metacone; (c) protocone; (d) hypocone; (e) extra cusp.

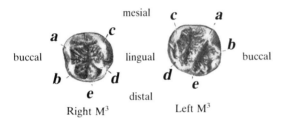

Fig. 3.19. *Pongo pygmaeus* (orang-utan). USNM 142181. An extra tooth (broken) to the lingual side of the empty socket of the left mandibular canine.

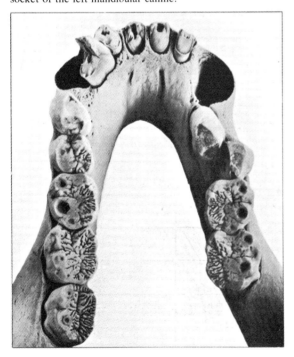

Fig. 3.20. *Pongo pygmaeus* (orang-utan). RCS Odonto. Mus., G 11.1. Two deciduous canines in the left mandible. Reduced.

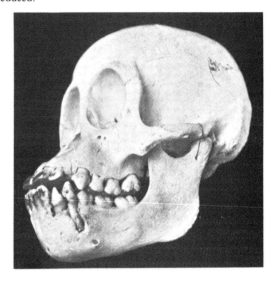

Fig. 3.21. *Pongo pygmaeus* (orang-utan). Connation between the right M_3 and a disto-molar (Duckworth, 1907).

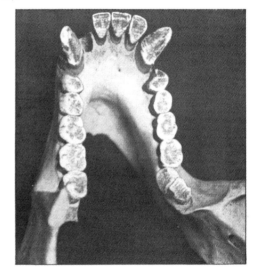

(iii) ♂. Formerly RCS Osteo. Series, 37. Both M^3 and right M_3 absent.
(iv) ♀. Formerly RCS Osteo. Series, 44. Right M^3 absent (Bateson, 1894, p. 200).
(v) USNM 142181. Right M_3 absent and extra tooth in the region of the mandibular left canine.
(vi) ♀. Formerly RCS Osteo. Series, 40b. Both M_3 absent.
(vii) Nat. Hist. Mus., Madrid. Right M_3 absent.

SIZE AND SHAPE

Schultz (1941) mentioned two orang-utan skulls in which both I^2 were vestigial. In one specimen (formerly USNM 49854), the M^2 was larger on the right than the left, and in another (USNM 143588) the left I^2 was a peg-shaped tooth. The left M_3 was small in the skull AIMUZ 1986 mentioned above because of its supernumerary premolars (Schultz, 1956).

No examples were seen of teeth with extra cusps. An example of variation in the roots of an m_2 is shown in Figure 3.22, in which there is an extra root between the mesial and the distal roots.

Genus *Hylobates*, gibbon

EXTRA TEETH

There were two examples of extra teeth in the gibbons:
(i) *Hylobates agilis* (dark-handed gibbon). ♂. BMNH 1955.1484. There are six upper incisors; the extra ones seem to be the outermost. Both M_1 have six cusps, the extra one being situated between the hypoconid and hypoconulid.
(ii) *Hylobates moloch albibarbis* (grey gibbon). USNM 145328. Bilateral maxillary disto-molars about one-third the size of M^3 and shaped like P^3; also a disto-molar in the left mandible.

Lönnberg (1930) recorded two instances of extra incisors in gibbons:
(i) *Hylobates lar* (white-handed gibbon). NRM 2324. Six upper incisors in an orderly row.
(ii) *Hylobates moloch*. NRM 4. An extra right I^2.

Schultz (1944) mentioned two instances of additional incisors in his Chiengmai Series of gibbons and three in his General Series. There was one additional premolar in the Chiengmai Series and four instances of additional molars in the General Series; in one skull of *Hylobates lar*, the left maxilla had five molars; the right maxilla and each mandible had four.

Remane (1921) described a *Hylobates lar* with disto-molars in both maxillae, and another with an additional conical denticle fused to the lingual surface of the right M_3.

Supplemental deciduous canines on one side of the upper jaw are shown in Figure 3.23. This is probably the specimen referred to by Schultz (1956, p. 1,000). A skull of a black and white gibbon (*Hylobates concolor haihanus*), juvenile, (ZMB A59.09) has an additional conical tooth in the right mandible, as well as an alveolus for a missing, but probably similar, tooth distal to the erupted left M_1. The M_2 and M_3 are normal but unerupted and partly formed (Inke and Ernst, 1965–66).

ABSENCE OF TEETH

Of the anthropoid apes, the gibbon shows the greatest tendency to absence of teeth (Table 3.1). The following were noted:

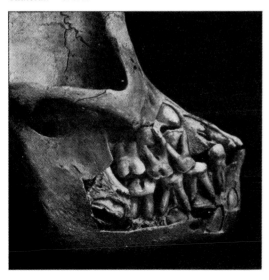

Fig. 3.22. *Pongo pygmaeus* (orang-utan). RCS Odonto. Mus., G 11.2. The right m_2 has an extra root situated between the mesial and distal roots. The condition is bilateral. × c. 0.7.

Fig. 3.23. *Hylobates hoolock* (hoolock gibbon). AMNH 19400. Two identical teeth in the position of the right deciduous canine. Reduced.

(a) Incisors
  (i) *Hylobates moloch*. NRM 998. (Fig. 3.24). Left I^2 absent; the crown of the left canine is much larger than the right one and has deep axial grooves on its labial and lingual aspects which divide it into a smaller mesial and a larger distal part. It is possible that this is an example of connation between I^2 and C.
  (ii) *Hylobates agilis*. USNM 123151. Both I_1 absent.
  (iii) *Hylobates* sp. MNHN A 558. Both I_2 absent.

Schultz (1944) noted five examples of absent incisors in his Chiengmai Series and eight in his General Series.

(b) Premolars
  (i) *Hylobates* sp. Formerly ZMB 7852. One premolar absent in each maxilla.

(c) Molars
  (i) *Hylobates hoolock* (Hoolock gibbon). AMNH 70398. Left M^3 absent.
  (ii) *Hylobates lar lar*. ♂. BMNH 1924.9.2.1. Both M^3 absent.
  (iii) *Hylobates agilis*. USNM 144090. Both M_3 absent.
  (iv) *Hylobates lar*. USNM 124292. Left M_3 absent.
  (v) *Hylobates lar*. Formerly ZMB 7801. All M3 absent. In the left mandible, the tooth in the position of P_4 appears to be a deciduous molar associated with absence of its successor. In the left maxilla, at least one premolar is absent because there is only one empty socket between the deciduous canine and M^1. On the right between P^3 and M^1, there are two small empty sockets which possibly contained remnants of a deciduous molar.
  (vi) *Hylobates agilis*. USNM 114499. Both M_3 absent and right M^3 diminutive.

SIZE

In *Hylobates*, the M_3 is often smaller than normal. Schultz (1944) in his General Series found that 22 skulls (4.4%) had small M_3, some being small pegs; in his Chiengmai Series, there were 12 (5.4%).

SHAPE

Variations in shape of the premolars are uncommon in the Pongidae. In a mandible of *Hylobates lar* (USNM 143567), the left P_3 differs in shape from the contralateral one and has three cusps.

Additional cusps are rare in gibbons but a few have been described by Frisch (1963) who commented that they wear faster than the normal cusps and are therefore easily identifiable only in immature animals. In 32 immature dentitions of *Hylobates lar* from a single population (housed in the Schultz collection in the Anthropologisches Institut in Zurich), he found that 16 had teeth with one or several extra cusps.

The lower first premolar (P_3) of pongids normally has two roots and is bucco-lingually flattened, but occasionally it may be triangular in shape and have three roots; Frisch (1963) found this variation in 38 out of 96 specimens of *Hylobates lar* from the same locality. The variation was much less common in *Hylobates lar* from other localities and in other species of *Hylobates*, suggesting that the frequency Frisch described was the genetic effect of isolation of the population. This suggestion is supported by the presence of triangular three-rooted P_3 in two *Hylobates lar pileatus* from another isolated population (MCZ 41448 and MCZ 41447).

Cusps may be suppressed; in the specimen shown in Figure 3.25, the right M_3 has only four cusps, the hypoconid being missing; the contralateral M_3 has a poorly developed entoconid. The distal root of this tooth is divided into two, a buccal and a lingual root,

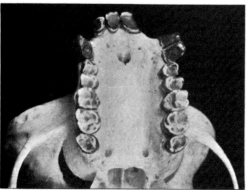

Fig. 3.24. *Hylobates moloch muelleri* (Bornean gibbon). NRM 998. Left I^2 absent; the left maxillary canine is much larger than the right and is partially divided by deep longitudinal grooves.

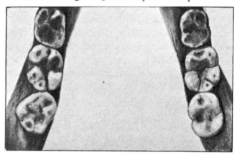

Fig. 3.25. *Hylobates hoolock* (hoolock gibbon). BMNH 1848.1.27.46. The right M_3 has only four cusps.

but their relationship to the cusps is not clear; the buccal root seems to carry the hypoconid and the lingual one the entoconid and the hypoconulid.

In the mandible shown in Figure 3.26, the left M_3 is normal but on the right M_3 there is a large cusp on the buccal side which appears to be the hypoconulid which has displaced the hypoconid lingually. There are three roots; the mesial cusps are carried on one, the entoconid and the hypoconid are carried on another which is quite small and the hypoconulid is on the third root which is large. Both M_1 have an extra cusp mesial to the entoconid.

Schultz (1944, p. 48) noted four instances of connation in his samples of gibbons. In two, the right I^1 is double and both root and crown are fused; in one, the crowns of both lower canines are doubled and partially connate; in the fourth, the left M_2 is incompletely doubled with partial connation at the root.

Amongst the specimens with the deciduous dentition present, there was one in which on the right i^1 and i^2 were connate (A.E.L. Amsterdam, 1908/43).

Genus *Symphalangus*, siamang

EXTRA TEETH

There were four specimens of siamang (*Symphalangus syndactylus*) showing supernumerary teeth:
(i) USNM 141163. An extra incisor between the left I_2 and canine.
(ii) ♀. BMNH 1919.11.12.3. (Fig. 3.27). A small

Fig. 3.26. *Hylobates hoolock* (hoolock gibbon). BMNH 1937.3.24.7. The right M_3 has an additional buccal element.

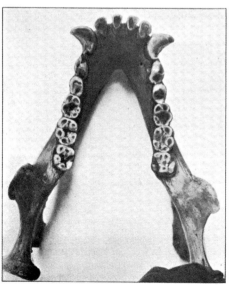

tooth mesial to the right P_3. Both M^2 have two extra cusps between the metacone and the hypocone; the one abutting the hypocone is the larger.
(iii) ♀. BMNH 1919.11.12.2. An additional molar high up in the bone on the buccal aspect of each M^3.
(iv) MNHN A 10936. A left mandibular disto-molar.

Remane (1921) described bilateral mandibular disto-molars in the siamang, and another specimen with a single-cusped tooth in the left maxilla between the canine and the first premolar (P^3).

ABSENCE OF TEETH
(i) A.E.L. Amsterdam, 1910/141. Right I_2 absent.

SHAPE

In the siamang, M_3 normally has five cusps. In the mandible in Figure 3.28, the right M_3 has a poorly-

Fig. 3.27. *Symphalangus syndactylus* (siamang). ♀. BMNH 1919.11.12.3. There is a small extra tooth between P^3 and the canine.

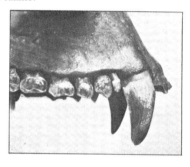

Fig. 3.28. *Symphalangus syndactylus* (siamang). ♂. BMNH 1881.3.15.1. The right M_3 is abnormal in shape; (a) metaconid; (b) protoconid; (c) small extra cusp. In the left M_3, the entoconid is suppressed.

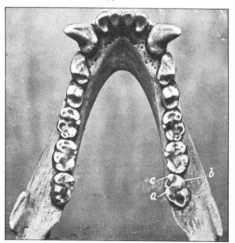

developed hypoconid, hypoconulid and entoconid. The protoconid and metaconid have a small extra cusp on their mesial aspects and are carried on a single mesial root. In the left M_3, the entoconid is suppressed; the metaconids of the M_1 and M_2 on both sides are slightly notched as if partially divided.

Family **Cercopithecidae**. Old World monkeys

$$\frac{2}{2} \cdot \frac{1}{1} \cdot \frac{2}{2} \cdot \frac{3}{3} = 32$$

The incisors of cercopithecids have proportionately longer crowns than in pongids and the canines are larger in males than in females. The molars are bilophodont, that is the four cusps are arranged to form a mesial and a distal transverse ridge; M_3 usually has a third ridge, or talonid, distally formed from the hypoconulid.

The numbers of specimens of each genus examined and the numerical variations recorded in the Old World monkeys by Colyer in the original edition of this book are given in Table 3.3; a summary of variations for the family as a whole is given in Table 3.4. Both Tables have been modified in the light of Colyer's subsequent work on *Colobus* (1943b) and the modern taxonomic nomenclature. Details of individual specimens are given later.

Table 3.3 Numerical variations in the CERCOPITHECIDAE

Genus	No. of specimens	Extra teeth			% of specimens affected	Absence of teeth			% of specimens affected	Totals of specimens showing extra teeth or absence of teeth	
		I	P	M		I	P	M		No.	%
COLOBINAE											
Colobus (Colyer, 1943)	1485	7	3	6	1.1	1	25	1	2.4	51	3.4
Presbytis											
entellus & *vetulus* groups	289	1	—	1	0.7	—	—	—	—	2	0.7
cristata group	385	1	1	—	0.5	—	—	—	—	2	0.5
melalophus group	311	1	1	—	0.3	2	1	4	1.9	7	2.3
Pygathrix	16	—	—	—	—	—	—	—	—	—	0
Rhinopithecus	17	—	—	—	—	—	—	—	—	—	0
Simias	10	—	—	—	—	—	—	—	—	—	0
Nasalis	83	—	—	—	—	—	—	—	—	—	0
CERCOPITHECINAE											
Cercopithecus	1823	3	1	10	0.8	1	2	7	0.6	24	1.3
Erythrocebus	95	—	—	3	3.2	—	1	—	1.1	4	4.2
Cercocebus	311	—	—	—	—	—	—	—	—	—	0
Papio	410	2	—	2	1.0	1	1	1	0.7	7	1.7
Mandrillus	56	—	—	—	—	—	—	1	1.8	1	1.8
Macaca	901	2	—	2	0.4	—	—	1	0.1	5	0.6
Theropithecus	7	—	—	—	—	—	—	—	—	—	0

In skulls in which two or more classes of teeth varied in number, each class is recorded individually; for example, if a skull has an extra incisor as well as an extra molar, both columns would be credited with one and the column 'totals of specimens showing extra teeth or absence of teeth' with one only.

Table 3.4 The classes of teeth showing numerical variations in the CERCOPITHECIDAE

	Extra teeth				Absence of teeth			
	Maxilla		Mandible		Maxilla		Mandible	
Class	Right	Left	Right	Left	Right	Left	Right	Left
Incisors	6	9	1	1	—	1	3	2
Premolars	3	1	1	2	10	17	1	2
Molars	11	9	9	8	10	10	6	6

In one case, the position of the extra incisor in the mandible could not be determined.

It is often justifiable, especially perhaps when absence of certain classes of teeth and the presence of supernumerary teeth of the same tooth class are about equal, to consider them as a combined statistic, numerical variation, because this may be regarded as evidence of instability of the tooth fields which may have evolutionary significance. Table 3.3 shows that numerical variations occur in only a small percentage of Old World monkeys; this has been confirmed by subsequent workers. Schultz (1958) studied a sample of 309 western black-and-white colobus (*Colobus polykomos polykomos*), red colobus (*Colobus badius badius*) and olive colobus (*Procolobus verus*) from one small area of Liberia and found only one supernumerary tooth and eight congenitally absent teeth (2.6%). In another sample from a small area of Liberia consisting of 270 specimens of greater white-nosed monkey (*Cercopithecus nictitans buëttikoferi*), Campbell's monkey (*Cercopithecus campbelli campbelli*) and Diana monkey (*Cercopithecus diana diana*) (now in the Zoologisches Institut of the University of Freiburg, Switzerland), Lampel (1963) found one supernumerary tooth and seven absent teeth (2.6%). Hooijer (1952) looked at more than 700 skulls of leaf monkeys (*Presbytis*), langurs (*Nasalis*) and macaques (*Macaca*) in various museums in the Netherlands and the United States and found no supernumeraries at all and only one or two absent teeth (all premolars).

However, it does seem as though numerical variations tend to be more prevalent in some geographically isolated populations. In 1940 Colyer published his study of a group of 67 skulls of the rare Preuss's red colobus, *Colobus badius preussi*, collected in the same locality of Cameroun. There were three skulls with congenitally absent teeth (4.5%) compared with 2.3% for the remaining *Colobus* specimens which he studied later (Colyer, 1943*b*).

Variation in number is more marked in the green monkeys (*Cercopithecus sabaeus*) from the island of St Kitts. Ninety two skulls of this geographically isolated population (which is described in more detail on pp. 195–197) are in the Odontological Museum of the Royal College of Surgeons. In analysing the data of Colyer (1948), Ashton and Zuckerman (1950*a*) showed that four out of 76 adult skulls had numerical variations (5.3%) compared with only 1.9% of the 104 skulls of the same species originating in West Africa; in Colyer's sample of *Cercopithecus* as a whole (1823 specimens), only 1.3% varied.

One cercopithecid, *Rhinopithecus*, seems to be more variable than Colyer realized. He found no specimens with numerical variations, but Hylander and Kay (1975) found that 35% of 17 skulls of snub-nosed monkey (*Rhinopithecus roxellanae*) in various museums in the United States lacked one or both P^3 and Hooijer (1952) found the same in two out of nine adult females of the same species; there were too few males for valid comparison.

Table 3.4 shows that (i) in the region of the upper incisors, extra teeth are more common than absence of teeth, (ii) absence of upper premolars is more common than extra premolars, (iii) the molars are more variable in number than the other teeth.

Both Schultz (1958) and Colyer (1943*b*) noted that variations in shape of the molars (particularly in the numbers of cusps on the third molars) were common in *Colobus* monkeys. Table 3.5 shows that 36.5% of the 515 *Colobus* skulls studied by Colyer had variations in shape of the third molar. The prevalence of this variation in the other cercopithecid species was not quantified.

Variations in shape also seem to be more frequent in isolated populations. Thirty six out of 67 skulls of Preuss's red colobus (*Colobus badius preussi*) from Yabassi in Cameroun had supernumerary cusps on at least one M3 (54%) (Colyer, 1940).

Three out of four albino eastern black-and-white colobus (*Colobus guereza kikuyuensis*) collected on Mount Kenya had numerous extra cusps; so did all five of the same sub-species from Uplands in Kenya. Seven out of 10 western black-and-white colobus (*Colobus polykomos palliatus*) from E. Usambara, Tanganyika (Tanzania) had additional M3 cusps (Colyer, 1943*b*), but it should be noted that in the general sample of *Colobus* monkeys (Table 3.5) these high figures also occur in mixed, but small, samples of *Colobus badius waldroni* and *Colobus polykomos palliatus*.

Another type of variation in the shape of the M3 is fusion of the roots; again it seems to be more common in isolated populations. In the St Kitts population of green monkeys (*Cercopithecus sabaeus*), Ashton and Zuckerman (1950*a*) noted it in 15 out of 36 adult skulls (26.3%) compared with three out of 33 (9.1%) in a sample of the same species derived from West Africa; the difference was statistically significant.

Although Swindler *et al.* (1963) described normal variability in tooth size in African monkeys, there are few quantified studies of abnormal variation. Ashton and Zuckerman (1950*a*) noted that the deciduous and permanent teeth of both sexes of green monkey on St Kitts were significantly larger than those in the sample derived from W. Africa; when adult males and females were compared, the difference was significant in all tooth types in females, and in the cheek teeth, but not the incisors and canines of males. The larger teeth in the St Kitts sample as a whole was clearly related to the larger size of their skulls (Ashton and Zuckerman, 1951*b*). At the same time, the degree of variability in

Sub-family **Colobinae**

Genus *Colobus*, colobus monkeys

EXTRA TEETH

(a) Incisors
  (i) *Colobus guereza* (eastern black-and-white colobus). MNHN A 3841. An extra tooth in the left premaxilla.
  (ii) *C. polykomos satanas* (black colobus). ♂. P-C. Mus., Cam II 69. (Fig. 3.29). An extra incisor in the left premaxilla; on the right side, the deciduous incisor has been retained labially to I^2.
  (iii) *C. kirkii* (Kirk's colobus). Formerly ZMB 6908. A tiny conical tooth in the midline beween the I^1 (a so-called mesio-dens), but its root appears to be in the left premaxilla.
  (iv) *C. kirkii*. ZMB 6907. A tiny conical tooth lingual to the left I^1 and I^2.
  (v) *C. polykomos cottoni* (Powell-Cotton's Angolan colobus). ♂. BMNH 1930.11.11.9. A supernumerary incisor, a little smaller than I_1, between the left I_1 and I_2 (Colyer, 1943b).
  (vi) *C. guereza caudatus*. Six incisors in the premaxilla; Wegner (1910, p. 355, Fig. 3) considered the extra teeth to be those between the normal I^1 and I^2 on each side.
  (vii) *C. badius preussi* (Preuss's red colobus). ♂. P-C. Mus., M 178. An extra incisor in each premaxilla (Colyer, 1943b, p. 110).

(b) Premolars
  (i) *C. polykomos angolensis* (Angolan black-and-white colobus). ♂. BMNH 1927.3.1.2. (Fig. 3.30). An extra tooth slightly lingual to the space between P_3 and the canine. The supernumerary resembles a premolar but is much smaller than the

size of some of their teeth has been significantly lessened (Chapter 10, pp. 195–197).

Table 3.5 Numbers of *COLOBUS* monkeys of various sub-species showing variations of shape of the molars (after Colyer, 1943b)

Sub-species	No. of specimens	Specimens affected upper molars no.	lower molars no.	totals no.	%
Colobus badius					
waldroni	9	0	7	7	77.8
preussi	63	16	35	40	63.5
powelli	16	2	1	3	18.8
rufomitratus	8	0	0	0	0.0
tephrosceles	11	0	1	1	9.1
ellioti	49	2	2	4	8.2
temminckii	14	0	1	1	7.1
Colobus kirkii	8	2	2	2	25.0
Colobus polykomos					
palliatus	14	6	8	10	71.4
angolensis	8	5	0	5	62.5
adolfi-friederici	12	6	0	6	50.0
satanas	68	22	18	33	48.5
Colobus guereza					
occidentalis	43	15	15	25	58.1
kikuyuensis	51	15	12	21	41.2
matschiei	15	3	2	5	33.3
caudatus	31	8	5	9	29.0
dodingæ	18	0	5	5	27.8
guereza	23	1	3	4	17.4
uellensis	54	7	0	7	13.0
Totals	**515**	**110**	**117**	**188**	**36.5**

premolar adjacent to it. Both M^3 bear an extra distal cusp.

(ii) *C. guereza uellensis*. ♂. ZMB 38736. On the right, an extra tooth between the upper premolars and another buccal to the lower premolars.

(iii) *C. badius*. ♀. AIMUZ 6039. An extra premolar is in place of an upper canine presumed to be missing congenitally (Schultz, 1958). It is possible that it could be derived from the canine tooth germ and that the morphogenetic field influence (Chapter 1, p. 2) was insufficient for it to develop the full characteristics of a canine.

(iv) *C. polykomos satanas*. A supplemental left P_4 (Bennejeant, 1936, pp. 200–202).

(c) Molars

(i) *C. polykomos angolensis*. KMMA 5747. A disto-molar in the right mandible.

(ii) *C. guereza occidentalis*. ♀. KMMA 2798. A disto-molar in each maxilla.

(iii) *C. guereza kikuyuensis* (Kenyan eastern black-and-white colobus). ♂. Albino. RCS Odonto. Mus., G 68.8217. A minute additional tooth on the buccal side of left M^3 (Colyer, 1943b).

(iv) *C. polykomos adolfifriederici*. ♂. RCS Odonto. Mus., G 11.121. Two extra teeth in the left maxilla, one buccal and the other distal to the left M^3 (Colyer, 1943b).

(v) *Rhinopithecus roxellanae* (snub-nosed monkey). ♀. USNM 268886. A disto-molar on the left maxilla and both P^3 absent (Hooijer, 1952).

(vi) Colobinae (*Colobus*?) Laboratoire d'Anthropologie, Geneva. A disto-molar in the right mandible (Bennejeant, 1936, p. 216).

Fig. 3.29. *Colobus polykomos satanas* (black colobus). P-C. Mus., Cam II 69. An extra incisor in the left premaxilla and, on the right, the i^2 has been retained labially to I^2.

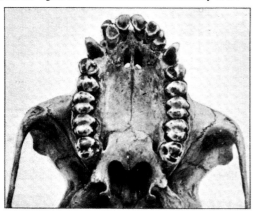
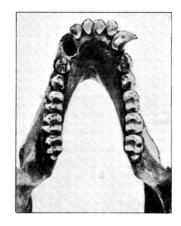

Fig. 3.30. *Colobus polykomos angolensis* (Angolan black and white colobus). ♂. BMNH 1927.3. 1.2. An extra tooth lingual to P_3.

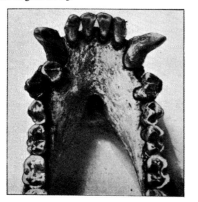

Section 1: Variations in Number, Size and Shape

ABSENCE OF TEETH

(a) Incisors
 (i) *Colobus badius preussi*. P-C.M., M246. Both I^2 absent in a poorly-developed premaxilla (Colyer, 1940, p. 765). All the M3 lack some cusps (see below, p. 40).
 (ii) *Procolobus verus*. ♂. AIMUZ 6271. One I^2 absent (Schultz, 1958, p. 102).
 (iii) *C. polykomos*. ♂. AIMUZ 6005. One I_2 absent (Schultz, 1958, p. 102).
 (iv) *C. badius*. ♂. AIMUZ 6042. One retained lower deciduous incisor. Its permanent successor was presumed to be absent (Schultz, 1958, p. 102).
 (v) *C. polykomos adolfifriederici*. ♂. RCS Odonto. Mus., G 11.12. Right I_1 absent (Colyer, 1943b).

(b) Canines
 (i) *C. badius*. ♀. Lower left canine missing; there is no space for it in the arch (Schultz, 1958).

(c) Premolars
 (i) *C. badius rufomitratus*. ♀. BMNH 1923.10.15.2. Left P^4 absent.
 (ii) *C. badius rufomitratus*. ♂. BMNH 1923.10.15.1. Right P^4 absent.
 (iii) *C. badius rufomitratus*. ♂. BMNH 1923.10.15.5. Both P^4 absent.
 (iv) *C. badius rufomitratus*. ♂. BMNH 1936.12.30.1. Left P^4 absent (Colyer, 1943b, p. 113).
 (v) *C. badius rufomitratus*. ♀. BMNH 1936.12.30.2. Left P^4 absent (Colyer, 1943b, p. 113).
 (vi) *C. badius rufomitratus*. ♀. BMNH 1936.12.30.5. Right P^4 absent (Colyer, 1943b, p. 113).
 (vii) *C. badius preussi*. ♀. RCS Odonto. Mus., G 11.7. Left P^3 absent (Colyer, 1940, 1943b).
 (viii) *C. badius preussi*. ♀. RCS Odonto. Mus., G 169.21. An adult in which both lower and upper right m^4 are retained. All four P4 absent (Colyer, 1940, 1943b).
 (ix) *C. badius tholloni* (South Congo red colobus). KMMA 5701. Right P^4 absent (Colyer, 1943b, p. 112).
 (x) *C. polykomos palliatus* (mantled colobus). ♀. BMNH 1902.1.2.20. Right P^3 absent.
 (xi) *C. guereza kikuyuensis*. ♀. BMNH 1900.2.1.3. Right P^3 absent.
 (xii) *C. guereza kikuyuensis*. USNM 164584. The left m^3 is present and its successor is absent. The right m^4 is still present though its successor is erupting beneath it. The right m_3 is still present and its successor has erupted beside it buccally (Colyer, 1943b, p. 115).
 (xiii) *C. guereza occidentalis*. ♀. RCS Odonto. Mus., G 68.76. Left P^4 absent (Colyer, 1943b, p. 113).
 (xiv) *C. guereza occidentalis*. ♂. RCS Odonto. Mus., G 11.6. Both P^4 and left P_3 absent (Colyer, 1943b, p. 113).
 (xv) *C. guereza uellensis*. ♀. P-C. Mus., M446. Left P_3 absent (Colyer, 1943b, p. 113).
 (xvi) *C. kirkii*. ♀. BMNH 1937.5.10.1. Left P^4 absent (Colyer, 1943b, p. 113).
 (xvii) *Procolobus verus*. ♀. AIMUZ 6335. One P^4 missing; no space for it in the arch (Schultz, 1958).
 (xviii) *Rhinopithecus roxellanae*. ♀. AMNH 119648. Both P^3 absent (Hylander and Kay, 1975).
 (xix) *R. roxellanae*. ♀. USNM 268886. Both P^3 absent associated with a left upper disto-molar (Hooijer, 1952).
 (xx) *R. roxellanae*. ♀. USNM 268894. Both P^3 absent (Hooijer, 1952).

(b) Molars
 (i) *Colobus badius ellioti*. ♂. BMNH 1930.11.11.27. Both M_3 absent.
 (ii) *C. polykomos satanas*. ♂. BMNH 1932.8.1.11. Right M^3 absent.
 (iii) *C. polykomos satanas*. ♀. P-C. Mus., Cam II 476. Left M^3 absent.
 (iv) *C. polykomos satanas*. ♀. P-C. Mus., Cam II 485. Left M_3 absent.
 (v) *C. polykomos satanas*. ♀. P-C. Mus., Cam II 39. Both M_3 absent.
 (vi) *C. polykomos satanas*. ♀. P-C. Mus., Cam II 483. (Fig. 3.31). Right M_3 and right M^3 absent, associated with abnormality of shape of lower molars (p. 40).
 (vii) *C. polykomos*. ♂. AIMUZ 5991. Left M_3 absent (Schultz, 1958).
 (viii) *C. polykomos*. ♂. AIMUZ 6038. Both M^3 absent (Schultz, 1958).

SIZE AND SHAPE

(a) General

The *Colobus guereza caudatus* in Figure 3.32 was described by C.S. Tomes (1898). Several teeth are congenitally absent and most of the teeth present are unusually small. Both I^2 are absent, but shallow sockets show that both I^1 were lost post-mortem. The upper canines are well-formed; the right P^3 is absent and the left is a conical tooth situated close to the canine. The upper molars are small and both M^1 have only three cusps instead of four. In the mandible, the left canine is small and is probably deciduous; the right canine is absent. Both P_3 are well-formed, the right P_2 is missing and the root of the left shows some resorption. The molars are dwarfed and cylindrical with imprecise cusp patterns. M_1 are the largest of the three whereas normally M_3 are the largest. Microscopic examination showed that the enamel is unusually thin but well-mineralized; the dentine is normal.

The colobus skull in Figure 3.33, described by Remane (1926), shows a similarly abnormal dentition. In the upper jaw, both I^1 are of unusual shape and both I^2 are very small. The upper canines are normal but, as in the previous specimen, the right P^3 is absent and the left P^3 has a small cylindrical crown. The remaining premolars and molars are dwarfed. In the mandible, all the incisors are missing but shallow sockets suggest that some may have been lost post-mortem. Both lower canines are present but they are deciduous. Shallow sockets show that both P_3 were present in life. Both m_4 are still present so their successors may be absent. The permanent molars, both upper and lower, are dwarfed and have only three cusps.

The enamel hypoplasia of the dentitions of these

Fig. 3.31. *Colobus polykomos satanas* (black colobus). P-C. Mus., Cam II 483. On the right, M^3 and M_3 are absent; the left M_3 has a spine-like projection at the centre of the occlusal surface surrounded by five peripheral cusps.

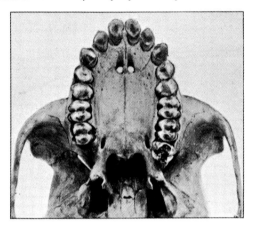
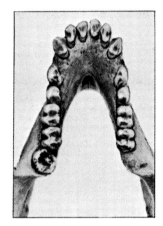

Fig. 3.32. *Colobus guereza caudatus* (Kilimanjaro guereza). ♂. BMNH 1897.3.14.2. Many teeth are absent; those present are small and malformed, especially the left P^2 which has a simple conical crown; the enamel is thin and defective.

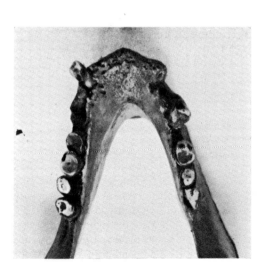
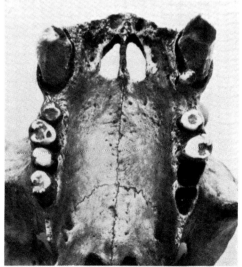

40 Section 1: Variations in Number, Size and Shape

two specimens is of a type known in man to be hereditary and sometimes associated with hypoplasia of ectodermal structures such as hair and sweat glands. However, Tomes (1898) mentioned that the hair and claws of the animal he described were normal.

(b) Incisors
  (i) *Colobus kirkii*. Left I^1 larger than the contralateral and with a conical projection from the centre of the cingulum (Colyer, 1943b).
  (ii) *C. guereza caudatus*. Left I^2 much larger than the contralateral tooth with features suggesting connation (Colyer, 1943b).

Bennejeant (1936, p.182) referred to a juvenile *Rhinopithecus roxellanae* (MNHN 1962–1380) in which one deciduous upper canine is larger than the other and the tip of the crown is bifid with a groove extending towards the root, an example of so-called dichotomy.

(c) Molars
The occlusal surfaces of normal unworn M3 of *Colobus* are shown in Figure 3.34. This pattern varies frequently; examples are given below:
  (i) *Colobus badius preussi*. P-C. Mus., M246. Both M_3 lack one cusp and the hypocones of both M^3 are poorly developed. This is associated with a reduction in the number of incisors (p.38) (Colyer, 1940).
  (ii) *C. badius preussi*. ♀. RCS Odonto. Mus., G 68.782. Each M^3 bears an additional distobuccal cusp (Colyer, 1940) which radiographs show is carried on an additional root; the teeth can be regarded as connation between distomolars and M^3.
  (iii) *C. kirkii*. ♀. BMNH 1924.12.15.2. (Fig. 3.35). Left M^3 has an extra cusp on the lingual aspect of the hypocone. Each M_3 has a well-defined additional cusp.
  (iv) *C. polykomos palliatus*. RCS Odonto. Mus., G 68.8262. M^2 and M^3 on both sides have grooves on the buccal aspect of the metacone (Colyer, 1943b).
  (v) *C. polykomos palliatus*. The bucco-lingual diameter of the left M_3 is less than that of the contralateral (Colyer, 1943b).
  (vi) *C. polykomos sharpei*. BMNH 1897.7.3.1. (Fig. 3.36). Each M^3 has an elevation in the centre of the distal margin. Each M_3 has an extra cusp between the hypoconulid and the entoconid.
  (vii) *C. polykomos satanas*. P-C. Mus., Cam II 483. (Fig. 3.31). Some teeth absent (p.38). The

Fig. 3.33. *Colobus* sp. ZMB 38737. As in the specimen of Figure 3.32, some teeth are absent and those present are small; an adult animal but the mandibular canines are deciduous. Many of the teeth are unusually small and the enamel is thin and defective (Remane, 1926).

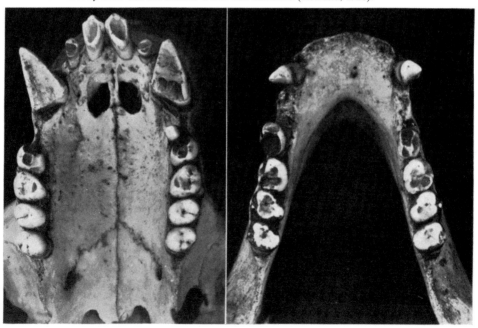

occlusal surface of the left M_3 has five peripheral cusps with ridges which join a spine-like projection at the centre.

(viii) *C. guereza guereza*. Right M^3 smaller than left (Colyer, 1943b).

(ix) *C. guereza caudatus*. ♂. BMNH 1900.2.1.2. (Fig. 3.37). Clusters of minute cusps on the distal margin of M_3, five on the left, two on the right.

(x) *C. guereza uellensis*. ♀. BMNH 1930.8.1.10. (Fig. 3.38). The distal margin of the right M^3 has not united with the metacone and so forms an extra distal cusp. This is a common abnormality in *Colobus*.

(xi) *C. guereza kikuyuensis*. RCS Odonto. Mus., G51.91. Each M^3 has large additional disto-buccal cusps, each of which is carried on an additional root fused to the normal disto-buccal root, so probably they represent small disto-molars connate with the normal M^3.

(xii) *Procolobus verus* (olive colobus). Several examples of accessory cusps on the lingual side of M^3 were noted by Schultz (1958).

Genus *Presbytis*, leaf monkeys

Presbytis entellus and *Presbytis vetulus* groups

EXTRA TEETH

(i) *Presbytis entellus lania* (Hanuman langur). ♀. BMNH 1909.7.16.1. Five mandibular incisors, the extra tooth being probably the middle one which resembles I_1.

(ii) *P. entellus*. NRM Z2815. Bilateral disto-molars in both upper and lower jaws.

SHAPE

(i) *P. entellus ajax* (Ajax langur). ♀. BMNH 1933.12.1.1. (Fig. 3.39). Connation between the right I^2 and an additional smaller incisiform tooth on its distal side.

(ii) *P. vetulus monticola* (purple-faced langur). ♀. BMNH 1852.5.9.16. Connation of the left i^1 and i^2; the corresponding teeth on the right are lost but

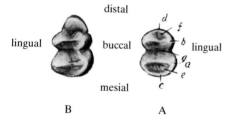

Fig. 3.34. *Colobus*. Diagrams of the occlusal surfaces of normal third molars.
A: the M^3: this tooth is bilophodont, that is the cusps on the buccal side (the paracone and metacone) are joined to the cusps on the lingual side (the protocone and hypocone) by transverse ridges (a and b); the ridges are separated by a deep trough (g). The mesial and distal margins (c and d) are raised slightly so that there are shallow depressions (e and f) between the transverse ridges and these margins.
B: the M_3: this has a mesial ridge connecting the protoconid to the metaconid, and a distal ridge connecting the hypoconid to the entoconid. There is also a fifth distal cusp, the hypoconulid.

Fig. 3.35. *Colobus kirkii* (Kirk's colobus). ♀. BMNH 1924.12.15.2. In both M^3, the distal depression (see Fig. 3.34f) extends on to the buccal surface; the left tooth has an additional disto-lingual cusp. Each lower third molar has a well-defined extra distal cusp.

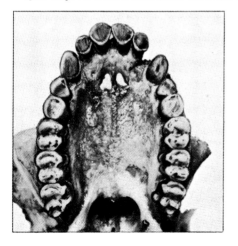
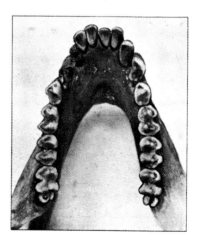

42 Section 1: Variations in Number, Size and Shape

the shape of the sockets suggests that they too were joined.

Presbytis cristata group

EXTRA TEETH

(i) *Presbytis pileata durga* (capped langur). ♀. BMNH 1921.7.16.3. Three incisors in the right premaxilla; the extra tooth appears to be the distal one, which is more pointed than the others.

(ii) *P. cristata* (silvered leaf monkey). NRM 3714. An extra tooth wedged between the lower left canine and P_3.

SHAPE

(i) *P. phayrei crepuscula* (Phayre's leaf monkey). ♀. BMNH 1917.8.4.1 (Fig. 3.40). On the buccal surface of the right M^3, there is an additional conical element which has a separate root.

(ii) *P. pileata durga*. ♂. BMNH 1875.1.8.1. An extra cusp on the mesial portion of the buccal surface of the left M^3.

(iii) *P. cristata pyrrhus*. ♂. BMNH 1910.7.11.1. An extra cusp on the distal aspect of the left M^3 and two extra cusps on the right M^3.

(iv) *P. pileata shortridgei*. ♂. BMNH 1915.5.5.10. (Fig. 3.41). An extra cusp on the buccal side of the protoconid of the left M_3.

Presbytis melalophos group

EXTRA TEETH

(i) *Presbytis melalophos rhionis*. ♀. BMNH 1909.4.1.19. An extra tooth between the two I_1.

Fig. 3.36. *Colobus polykomos sharpei* (Sharpe's colobus). BMNH 1897.7.3.1. An extra cusp on the distal aspect of each M^3. Each M_3 has six cusps (right).

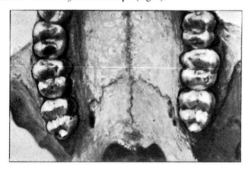

Fig. 3.37. *Colobus guereza caudatus* (Kilamanjaro guereza). ♂. BMNH 1900.2.1.2. (A) There are two small cusps on the distal margin of the right M_3, and (B) five on that of the left M_3. $\times c. 2.0$.

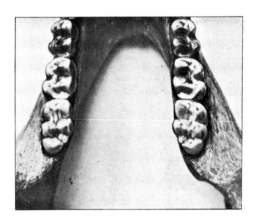

Fig. 3.38. *Colobus guereza uellensis* (eastern black and white colobus). ♀. BMNH 1930.8.1.10. The occlusal surface of the right M^3 is abnormal in shape. $\times c. 1.3$.

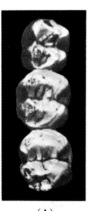

(A)   (B)

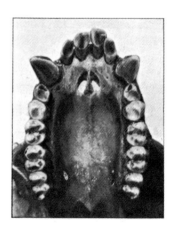

Fig. 3.39. *Presbytis entellus ajax* (Ajax langur). ♀. BMNH 1933.12.1.1. In the position of the right I², there is a connate tooth composed of two incisor crowns on a single root.

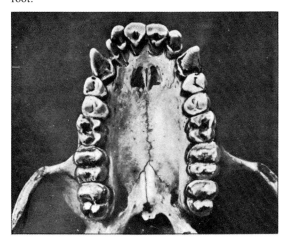

ABSENCE OF TEETH
  (i) *P. aygula hosei* (grey leaf monkey). ♀. BMNH 1893.6.2.1. Left i₁ retained. Both I₁ absent, with no sign of injury.
  (ii) *P. aygula hosei*. Private collection. Right I₁ and left P⁴ absent. Right I₂ has moved mesially so that it occludes with the left I¹.
  (iii) *P. rubicunda rubicunda* (maroon leaf monkey). USNM 145336. Left M₃ absent.
  (iv) *P. rubicunda ignita*. USNM 105908. Both M³ absent.
  (v) *P. melalophos femoralis* (banded leaf monkey). Formerly NRM 38735. Both M³ absent.
  (vi) *P. rubicunda*. Formerly NRM 7757. Left M₃ absent.

SHAPE
  (i) *P. aygula canicrus*. USNM 197647. An extra cusp at the mesio-buccal border of the left P⁴.
  (ii) *P. melalophos femoralis*. Formerly USNM

Fig. 3.40. *Presbytis phayrei* (Phayre's leaf monkey). ♀. BMNH 1917.8.4.1. (1) Part of maxilla showing an extra cusp on the buccal surface of the right M³. (2) Drawings of the upper third molar. (I) left M³; (II) right M³; (A) mesial surface; (P) distal surface; (a) mesio-buccal root; (b) disto-buccal root; (c) root supporting the extra cusp.

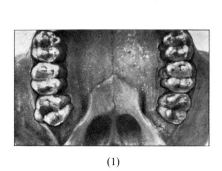
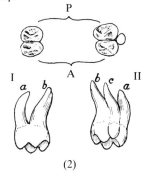

Fig. 3.41. *Presbytis pileata shortridgei* (Shortridge's capped langur). ♂. BMNH 1915.5.5.10. The left M₃ has an extra cusp on the buccal side of the protoconid. Slightly reduced.

Fig. 3.42. *Presbytis melalophos rhionis* (Rhio Archipelago banded leaf monkey). ♀. BMNH 1909.4.1.19. An extra cusp on the buccal surface of each upper molar. ×c. 2.0.

112709. Extra cusp on the buccal surface of each M^1.
- (iii) *P. melalophos rhionis*. ♀. BMNH 1909.4.1.19. (Fig. 3.42). Extra cusp on the buccal surfaces of all the maxillary molars.
- (iv) *P. aygula sabana*. ♂. BMNH 1893.3.4.2. (Fig. 3.43). The left M^3 has an extra cusp on the buccal surface of the metacone; the cusp has a separate root which is fused to the disto-buccal root which is itself united with the lingual root. In both M^3, the metacone and protocone are united by a ridge.
- (v) *P. melalophos cana*. ♂. BMNH 1909.4.1.14. A small extra cusp on the distal surface of both M^3, smaller on the left than on the right.
- (vi) *P. melalophos chrysomelas*. ♂. BMNH 1891.8.28.9. (Fig. 3.44). Maxillary third molars show curious variations. The disto-buccal roots of both M^3 are hood-shaped in that they are unusually wide and bent round to form a deep hollow on their distal surfaces.
- (vii) *P. melalophos chrysomelas*. UISNM 142203. Left M^3 small and conical.
- (viii) *P. melalophos*. MNHN A 1410. An additional cusp buccal to the metacone of right P^4 (Bennejeant, 1936, p. 211).

Sub-family **Cercopithecinae**

Genus ***Cercopithecus***, guenon monkeys

EXTRA TEETH

(a) Incisors
- (i) *Cercopithecus aethiops johnstoni* (vervet monkey). ZMB A90.06/100. Conical extra tooth lingual to the left I^1.
- (ii) *C. wolfi pyrogaster* (fire-bellied guenon). KMMA 5744. (Fig. 3.45). There is disordered dental development in the premaxilla on both sides. On the left I^1, arising from the lingual cingulum region, there is a large cusp which forms a buttress extending towards the incisive edge. This additional element is associated with a lingual root so that the tooth can be regarded as an unusual example of connation in which the elements, 'born together' (p. 8), are in labio-lingual relationship whereas in the common connation of incisors they are side by side.

   In the right premaxilla, both incisors are smaller than on the left and there is evidence of two supernumerary teeth; one is a small conical tooth situated on the lingual side of I^1, that is in the same position as the connated element of the left I^1. A small socket medial to I^1 shows that there had been an extra tooth.
- (iii) *C. pogonias grayi* (crowned guenon). ♂. RCS Odonto. Mus., G 11.323. In place of the right I^2, two incisors similar in size and shape to I^2 of the opposite side.
- (iv) *C. sabaeus* (green monkey). ♂. RCS Odonto.

Fig. 3.43. *Presbytis aygula sabana* (leaf monkey). ♂. BMNH 1893.3.4.2. Semi-diagrammatic views of the upper third molars. (A) Right M^3 (B) left M^3. (a) paracone. In both teeth, the metacone (b) and the protocone (c) are joined by a ridge which is separated from the hypocone (d) by a groove. The left M^3 has an extra cusp (e). × c. 2.0.

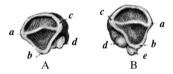

Fig. 3.45. *Cercopithecus wolfi pyrogaster* (fire-bellied guenon). KMMA 5744. The left I^1 has a ridge extending from the cingulum like a buttress to the lingual surface. On the right, a small supernumerary tooth lingual to I^1. An empty socket near the midline shows that another tooth had been present.

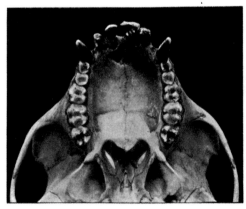

Fig. 3.44. *Presbytis melalophos chrysomelas* (crested lutong). ♂. BMNH 1891.8.28.9. Right and left M^3; the disto-buccal roots are expanded to hollow hood shapes. × c. 2.0.

Mus., A 72.63. A supplemental left I_2 (Colyer, 1948).
(v) *C. sabaeus*. ♀. RCS Odonto. Mus., A 72.6712. A supplemental left I_2. This is a St Kitts specimen (Colyer, 1948) that has a misplaced M_3 (pp. 196, 197).

(b) Premolars
  (i) *Cercopithecus ascanius whitesidei* (Whiteside's guenon). KMMA 8106. (Fig. 3.46). There are four premolars in the right maxilla, two erupting under the m^4, and one between M^1 and M^2. In the left maxilla, there is an empty socket between M^1 and M^2 so that there had been a supernumerary tooth there also.

(c) Molars
  (i) *C. aethiops tantalus* (tantalus monkey). ♀. BMNH 1923.1.22.4. A small incompletely erupted disto-molar in the left maxilla.
  (ii) *C. aethiops aethiops* (grivet monkey). BMNH 1913.10.18.6. Bilateral disto-molars in the mandible.
  (iii) *C. aethiops aethiops*. Formerly in the Zoologisches Museum of the University of Hamburg, 40645. Bilateral disto-molars in the mandible and a conical disto-molar in the right maxilla.
  (iv) *C. aethiops*. Formerly Zoo. Soc. Coll. A right mandibular disto-molar.
  (v) *C. sabaeus*. ♂. Captive. RCS Odonto. Mus., G 11.3. (Fig. 3.47). Embedded in the right maxilla in place of M^2 and M^3 are two teeth; the morphology of one suggests connation between M^2 and M^3 and the more distal one is a tuberculate supernumerary. On the left, a disto-molar is embedded in the tuberosity.
  (vi) *C. sabaeus*. ♂. RCS Odonto. Mus., A 72.615. An unerupted disto-molar in the right mandible. A St Kitts specimen (Colyer, 1948).

Fig. 3.46. *Cercopithecus ascanius whitesidei* (Whiteside's guenon). KMMA 8106. Four premolars in the right maxilla; in the left maxilla, there had been an extra tooth between M^1 and M^2.

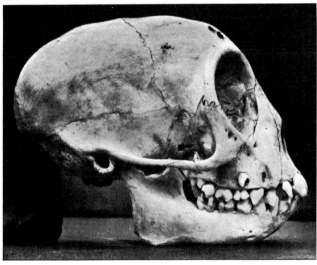
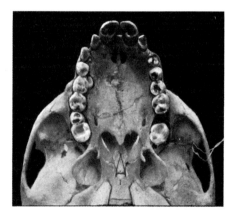

Fig. 3.47. *Cercopithecus sabaeus* (green monkey). ♂. Captive. RCS Odonto. Mus., G 11.3. Semi-diagrammatic drawing showing abnormally-shaped teeth in the position of the right M^2 and M^3. (A) The M^2 removed from jaw.

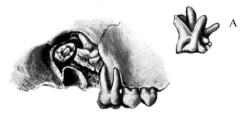

46 Section 1: Variations in Number, Size and Shape

(vii) *C. aethiops centralis*. ♂. BMNH 1930.3.6.1. A disto-molar in the left mandible.

(viii) *C. erythrotis erythrotis* (red-eared monkey). ♂. BMNH 1904.7.1.1. Bilateral disto-molars in the mandible; the left M^3 has an extra buccal cusp.

(ix) *C. cephus cephus* (moustached monkey). ♂. RCS Odonto. Mus., G 11.313. A small disto-molar in the right maxilla.

(x) *C. campbelli* (=*mona*) *campbelli* (Campbell's monkey). ♂. RCS Odonto. Mus., G 68.134. A small unerupted disto-molar in the right maxilla.

(xi) *C. campbelli* (=*mona*) *campbelli*. ♀. A disto-molar in the left maxilla (Lampel, 1963).

(xii) *C. mitis albogularis* (diadem monkey). Formerly in the Zoologisches Museum of the University of Hamburg, 38098. Bilateral mandibular disto-molars.

(xiii) *C. aethiops callidus* (Naivasha vervet). ♀. RCS Odonto. Mus., G 11.321. Right maxillary disto-molar (Colyer 1945b).

(xiv) *C. aethiops callidus*. ♂. RCS Odonto. Mus., G 11.322. One left and two right mandibular disto-molars (Colyer 1945b, Fig. 1).

(xv) *C. mona mona* (mona monkey). ♂. RCS Odonto. Mus., G 11.23. An extra tooth on each side lingual to M^2 and M^3. (Colyer 1945, Fig. 2).

ABSENCE OF TEETH

(a) Incisors

(i) *Cercopithecus aethiops*. NRM 28. Left I^2 absent.

(ii) *C. campbelli* (=*mona*) *campbelli*. ♂. Both I_2 absent. (Lampel, 1963).

(b) Canines

(i) *C. diana diana* (Diana monkey). Two females with congenitally absent lower right canines were described by Lampel (1963).

(ii) *C. campbelli* (=*mona*) *campbelli*. ♂. Right mandibular canine absent (Lampel, 1963).

(c) Premolars

(i) *C. aethiops marjoriae*. ♂. BMNH 1923.5.9.1. Left P^3 absent.

(ii) *C. ascanius whitesidei*. KMMA 7700. Fully adult but the left m^4 present with no sign of P^4.

(iii) *C. diana diana*. ♀. Both P_4 absent (Lampel, 1963).

(d) Molars

(i) *C. aethiops rufoviridis*. ♀. BMNH 1901.6.26.1. (Fig. 3.48). Left M_3 absent and right M_3 smaller than normal and malformed (p. 47).

(ii) *C. aethiops aethiops*. Formerly Zoo. Soc. Coll. Both M^3 absent.

(iii) *C. petaurista petaurista* (lesser white-nosed monkey). Private collection. Left M_3 absent.

(iv) *C. ascanius ascanius* (black-cheeked white-nosed monkey). AMNH 52550. Both M^3 absent.

(v) *C. nictitans nictitans* (greater white-nosed monkey). P-C. Mus., Cam 9. Both M^3 absent.

(vi) *C. aethiops hilgerti*. FMNH 27179. Left M^3 and right M_3 absent.

(vii) *C. diana diana*. ♀. Left M^3 absent (Lampel, 1963).

(viii) *C. diana diana*. In two males, right M^3 absent and left M^3 with three cusps instead of the normal four (Lampel, 1963). These are two of many examples where abnormalities tend to be bilaterally similar. Here, absence on one side with reduced simple morphology on the other.

SIZE

(i) *C. nictitans*. NRM 100. The right maxillary canine larger than the left.

(ii) *C. nictitans nictitans*. P-C. Mus., Cam No. 9. Left I^1 larger than the right.

Fig. 3.48. *Cercopithecus aethiops rufoviridis* (Mozambique monkey). ♀. BMNH 1901.6.26.1. The left M_3 is absent; the right M_3 is smaller than normal.

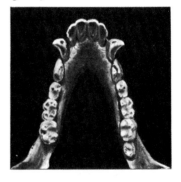

Fig. 3.49. *Cercopithecus erythrotis erythrotis* (red-eared monkey). ♂. BMNH 1904.7.1.1. The right M^3 has an extra cusp on the buccal surface.

SHAPE

Additional cusps on premolars are rare in cercopithecids. In one specimen, there was an extra cusp on the disto-buccal border of the left P^4 (*Cercopithecus mitis kandti*, formerly ZMB 33574).

Extra cusps were seen on molars in various positions, the lingual surface of M^1, the middle of the buccal surface of M_2 and the buccal surface of the right M^3 (Fig. 3.49). Reduction of the metacone of M^3 is common.

Lampel (1963, pp. 57, 88) gave a detailed account of variations in the number of cusps in his sample of *Cercopithecus* skulls: *C. nictitans buettikoferi* (no. of adults = 34), *C. mona* (= *campbelli campbelli*) (no. of adults = 146), and *C. diana diana* (no. of adults = 76). An additional cusp on the M_3 was noted in one skull of each species. In one *C.m.(c.) campbelli*, the left M^3 was exceptionally broad and both the metacone and the paracone were divided into two. In one *C.n. buettikoferi*, both M^3 were unusually broad, the metacone was much reduced and there was a small extra cusp mesially to the paracone.

Lampel (1963) also noted reductions in the number of M3 cusps and that, though they were rare in the mandible, they were common in the maxilla, particularly in *C.m.(c) campbelli*, only 33% of which had the complete complement of cusps (p. 192).

Abnormalities of shape were as follows:
(i) *Cercopithecus pogonias grayi* (crowned guenon). ♀. BMNH 1900.2.5.6. Right I^2 is a conical tooth; left I^2 lost.
(ii) *C. pogonias grayi*. A private collection. Left P^3 conical.
(iii) *C. aethiops rufoviridis*. ♀. BMNH 1901.6.26.1. (Fig. 3.48). Right M_3 malformed with one cusp; left M_3 absent (p. 46).
(iv) *C. nictitans buettikoferi*. ♂. Left I^2 reduced to a small peg (Lampel, 1963, p. 80).

Connation of incisors was present in the following:
(i) *Cercopithecus aethiops tantalus* (tantalus monkey). RCS Odonto. Mus., G 13.4. (Fig. 1.5). Between the lower parts of the the crowns of I_1 and I_2, the greater parts of the roots are separate. Degrees of separation of the crowns of connate teeth are common and sometimes may be complete, together with parts of the roots. However, separation of the roots with union of the crowns is rare.
(ii) *C. mitis kibonotensis*. BMNH 1917.8.21.5. Connation between right i_2 and deciduous canine.
(iii) *C. sabaeus*. RCS Odonto. Mus., A 72.636. The right i_2 has two crowns of similar size and shape side by side, but the root is single with a shallow groove labially and lingually extending from the division between the crowns (Colyer, 1948).
(iv) *C. aethiops cynosuros*. MNHN Paris, A 1346. (Fig. 3.50). A connate P^4 in which the mesio-buccal root and the associated part of the crown are partially separated from the rest of the crown (Bennejeant, 1936, p. 180).

Colyer (1948) described many examples of fusion of the roots of third molars in the St Kitts green monkey *C. sabaeus*, 20.8% compared with 8.3% in West African examples of this species. In one St Kitts example (♀, RCS Odonto. Mus., A 72.6712), the buccal margins of the mesial and distal roots of both M_3 are joined by a web and the three roots of both M^3 are fused. In another (RCS Odonto. Mus., A 72.6713), the roots of both M_1 are completely fused to form a cylindrical root tapering to a blunt apex.

Genus ***Erythrocebus***, patas monkey

EXTRA TEETH
(i) *Erythrocebus patas* (patas monkey). ♂. BMNH 1912.8.1.1. Bilateral maxillary disto-molars (Colyer, 1928).
(ii) *E. patas*. ♂. BMNH 1934.4.7.1. Connation between the distal aspect of the right M^3 and a disto-molar.
(iii) *E. patas*. ♀. Formerly RCS Odonto. Mus., G 68.1481. Connation between the unerupted right M^3 and a very small disto-molar (Colyer, 1928).

ABSENCE OF TEETH
(i) *E. patas*. ♀. RCS Odonto. Mus., G 11.36. Right P^4 absent (Colyer, 1928).

SHAPE
The maxillary premolars normally have three roots: mesio-buccal, disto-buccal and lingual. In a juvenile patas monkey (Fig. 10.73), the roots of the unerupted

Fig. 3.50. *Cercopithecus aethiops cynosuros* (Malbrook monkey). MNHN A 1346. Right P^2 to which a denticle is fused.

48 Section 1: Variations in Number, Size and Shape

premolars are joined on the lingual side by a web (Colyer, 1928). In another specimen of the same species, ♂, RCS Odonto. Mus., G 68.1486, (Figs. 3.51, 10.75) there are several variations in shape, mainly in the form of webbed roots (Colyer, 1928).

Genus *Cercocebus*, mangabeys

There were no numerical variations in the 311 specimens of *Cercocebus* examined. The one instance of variation in shape was an additional cusp on both M_2 (Fig. 3.52), on one side large and on the other a trace.

Genus *Papio*, baboons

EXTRA TEETH

(a) Incisors
  (i) *Papio ursinus* (chacma baboon). Formerly in the Zoology Museum, University of Cape Town. (Fig. 3.53). An extra small incisiform tooth lingual to the right I^1. The left I^1 is exceptionally large and has a longitudinal groove suggestive of connation with a supernumerary.
  (ii) *P. anubis* (olive baboon). AMNH 54288. A supplemental left I^2; the two teeth are so similar it is impossible to decide which is the extra tooth.

Fig. 3.51. *Erythrocebus patas* (patas monkey). ♂. RCS Odonto. Mus., G 68.1486. Maxillary teeth showing variations in shape. (A) Right P^3, disto-buccal and lingual roots fused. (B) Right P^4, all roots fused; extra cusp on distal side of crown. (C) and (D) Right M^2 lingual and buccal views, disto-buccal and lingual roots fused; additional cusps at mesio-lingual edge of crown. (E) Right M^3, roots incompletely formed but joined to form a web containing a single horseshoe-shaped cavity. (F) Left P^3, disto-buccal and lingual roots fused; apex of mesio-buccal root bent distally. (G) Left P^4, all three roots fused and several areas of enamel hypoplasia. (H) and (I) Left M^2, disto-buccal and lingual roots fused; a small projection on the crown in the same position as the extra cusp in the right M^2. (J) Left M^3, all three roots fused; (see also Colyer, 1928).

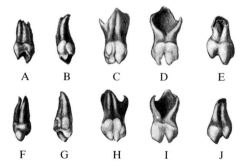

(b) Molars
  (i) *Papio cynocephalus* (yellow baboon). BMNH 1897.10.1.9. A right mandibular disto-molar similar to M_3.
  (ii) *P. hamadryas* (hamadryas baboon). Formerly ZMB 11613. A left mandibular disto-molar.
  (iii) *P. hamadryas*. ♂. BMNH 1932.7.6.1. A left upper disto-molar (Hill, 1970).
  (iv) *P. hamadryas*. ♂. ZMB 7744. Immature. Small conical deciduous disto-molars on both sides of the mandible (Inke and Ernst, 1965/6).

Fig. 3.52. *Cercocebus torquatus atys* (sooty mangabey). ♀. BMNH 1911.6.10.4. An extra cusp on the buccal surface of the left M_2 and a smaller one in the same situation on the contralateral tooth (Colyer, 1928).

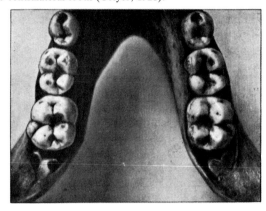

Fig. 3.53. *Papio ursinus* (Chacma baboon). Formerly in Zoology Museum, University of Cape Town. A supernumerary incisiform tooth lingual to the displaced I^1. Slightly enlarged.

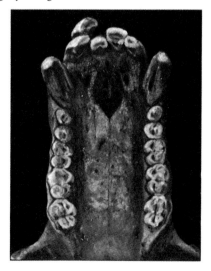

(v) *P. ursinus*. Zool. Coll. Oxford Univ. Mus., 2011b. Bilateral maxillary disto-molars (Bateson, 1894, p. 204).
(vi) *P. ursinus*. TM 11712. Bilateral upper disto-molars (Freedman, 1957).
(vii) *P. ursinus*. TM 757. Bilateral disto-molars in both jaws (Freedman, 1957).
(viii) *P. ursinus*. TM 11709. A left upper disto-molar (Freedman, 1957).
(ix) *P. ursinus*. TM 11713. Small bilateral molars in both jaws (Freedman, 1957).
(x) *P. anubis* (olive baboon). ♂. Anthropological Research Laboratory, Texas A & M University, 691–24. Bilateral upper disto-molars resembling, but smaller than, M^3. In the right mandible, a disto-molar much smaller than M_3 and with only three cusps (Rice and Oyen, 1979).

J.C.M. Shaw (1927) illustrated four instances of disto-molars among 112 South African baboon skulls in the Museums of Albany and Pretoria and in the Anatomical Museum of Witwatersrand University. Three were single examples but in the others they were bilateral in both jaws. There were in addition shallow fossae behind the M^3 but not behind the M_3, in ten of the remaining skulls; the fossae could be crypts similar to those described by Zuckerkandl (1891) in man and interpreted by him as having contained developing supernumerary teeth.

Bennejeant (1936, p. 212–213) described a skull of a female *Papio ursinus* (Naturhistorisches Museum, Bern) from a menagerie that was affected by an osteodystrophy (Chapter 16, pp. 333–334, 350) of the jaws. Several deciduous teeth were retained and their successors had erupted beside them. Small conical supernumerary teeth had erupted distal to both M_2 and radiographs showed that the true M_3 were buried in the jaw beneath the supernumerary teeth.

ABSENCE OF TEETH
(i) *Papio* sp. ♂. RCS Odonto. Mus., G 11.5. Left I_2 absent.
(ii) *P. anubis*. ♂. BMNII 1901.2.13.1. Right P^4 absent and contralateral tooth particularly small.
(iii) *P. cynocephalus*. ♂. BMNH 1924.1.1.7. Right M_2 and M_3 absent; left M_2 peg-shaped.
(iv) *P. ursinus*. Immature. MNHN 1894–697. In the upper left incisor region, there is a single deciduous tooth which resembles the contralateral i^1, but is much wider and is probably an example of connation between i^1 and i^2 (Bennejeant, 1936, p. 177).

SHAPE
(i) *P. cynocephalus*. Root and crown of upper left canine with longitudinal grooves suggesting a triple origin (Bennejeant, 1936, pp. 119, 123).

Genus *Mandrillus*, mandrills

The only specimen which showed numerical variation of the teeth was a male mandrill (*Mandrillus sphinx*, BMNH 1905.11.27.13) in which M^3 was absent on the left and dwarfed on the right. Figure 3.54 shows a rare variation of the mandibular right canine; attached to the labial surface of the crown, there is a cylindrical denticle, the root of which is fused with the root of the canine; most of the crown of the denticle is free. The other canines were normal.

Hill (1953a) found among 24 skulls of drill (*M. leucophaeus*) from the same locality many in which the I_2 had wide or sometimes bifid roots independently of sex.
(i) *Mandrillus sphinx* (mandrill). ♀. NMBE 873. Left P^4 two-rooted and conical (Bennejeant, 1936, pp. 162–163).
(ii) *M. leucophaeus* (drill). ♀. Lower canines much smaller than usual (Vogel, 1966).

Genus *Macaca*, macaques
EXTRA TEETH
(i) *Macaca mulatta* (rhesus monkey). Captive. BMNH 1855.12.26.48. A socket for an extra tooth between the right I^1 and I^2.
(ii) *M. sylvanus* (Barbary ape). MNHN 1872/104.

Fig. 3.54. *Mandrillus sphinx* (mandrill). RCS Odonto. Mus., G 11.41. The distal surfaces of the two mandibular canines. The left canine is normal but the right canine shows an unusual type of connation. Most connations of incisors or canines are joined in a labio-lingual plane. Here a supernumerary denticle is attached to the labial surface of the canine; the join is in the mesio-distal plane.

(Fig. 3.55). In the premaxilla on the left, there are two incisiform teeth in the position of I^2 but neither resemble the right I^2. The mesial one has a prominent ridge on its lingual surface.

(iii) *Macaca* sp. An extra tooth in the midline of the mandible (mesio-dens) between the two I_1 (Wegner, 1910, p. 355, figs. 4, 6).

(iv) *Macaca* sp. ZMB 14445. A conical tooth between I_2 and the second deciduous molar (m_4) on each side of the mandible. No typical m_3 is present. Mainly because the conical teeth are single-rooted, whereas both m_4 and any premolar would be expected to be two-rooted, Inke and Ernst (1965) took the view that the conical teeth are duplicated permanent canines rather than either m_3 or P_3.

(v) *M. mulatta*. Zoologisches Institut Halle, K 181. Bilateral conical deciduous mandibular disto-molars (Inke and Ernst, 1965/6).

(vi) *M. mulatta*. Musée d'Histoire naturelle de Bordeaux. Bilateral disto-molars in both jaws (Kunstler and Chaine, 1906).

(vii) *M. fascicularis fascicularis* (crab-eating macaque). BMNH 1910.4.5.19. Bilateral maxillary and left mandibular disto-molars.

(viii) *M. mulatta*. ♂. BMNH 1848.2.1.26. Upper and lower disto-molars on the right (Bateson, 1894, p. 204).

ABSENCE OF TEETH

(i) *M. sinica* (toque monkey). ♀. RCS Odonto. Mus., G 8.2. Right M^3 absent.

SHAPE

One extra cusp was observed on the M_1 in a Barbary ape (*Macaca sylvanus*, formerly Zoo. Soc. Coll.) and in a female rhesus monkey (*M. mulatta*, BMNH 1855.12.26.48) on the distal part of the buccal surface and, quite commonly, on the distal part of various mandibular molars in the pig-tailed macaque (*M. nemestrina*). Some examples in macaques are shown in Figures 3.56, 3.57 and 3.58. Bennejeant (1936, p. 210) showed a connate M_2 from a toque macaque (*M. sinica*) with a denticle fused to the lingual surface. A skull of a crab-eating macaque (*M. fascicularis*, NMBE 54A) has an additional cusp buccally on an M_3 (Bennejeant, 1936, p. 212).

Extra cusps were present on the lingual surfaces of all the maxillary molars in a female *Macaca andamensis* (BMNH 1920.7.3.5); an extra cusp on the inner

Fig. 3.55. *Macaca sylvanus* (barbary ape). MNHN 1872/104. Two incisiform teeth in the position of the left I^2. (A) labial surface; (B) lingual surface.

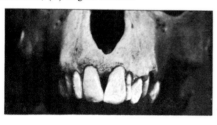
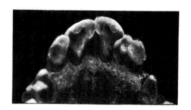

A　　　　　　　　　　B

Fig. 3.56. *Macaca nigra* (black ape). RCS Odonto. Mus., G 13.1. An extra cusp on the buccal surface of (A) the left M_3 (B) the normal right M_3.

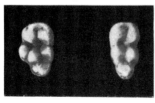

A　　　　B

Fig. 3.57. *Macaca mulatta mulatta* (rhesus monkey). ♂. BMNH 1845.1.8.223. Both M_3 have supernumerary distal cusps.

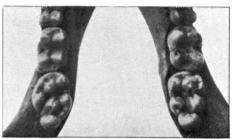

surface of M³ is so common in *M. nemestrina* that its presence is the rule rather than the exception.

Rather more is recorded about the roots of the cheek teeth of macaques than for other species of cercopithecines. The upper premolars normally have three roots like the upper molars and the lower premolars two roots. Webbed and more solidly fused roots, like the examples shown in Figures 3.58 and 3.59, are common.

Among the 310 specimens of *Macaca* with the deciduous dentition present, there were three instances of connation between upper incisors, including RCS Odonto. Mus., G 13.31 (♀) and G 13.41, described by Colyer (1921).

Fig. 3.58. Mandibular third molars in *Macaca*. (i) normal; (ii) abnormal right molar of *M. mulatta mulatta* (BMNH 1845.1.8.223) showing an upgrowing cingulum on the buccal side of the protoconid (a) and a web uniting the buccal aspects of the roots; (iii) abnormal left molar of same specimen showing two additional cusps (f and g) and an upgrowth of the cingulum around the protoconid (a).

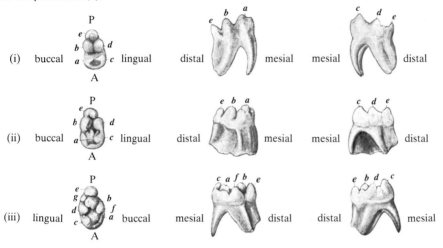

Fig. 3.59. A: *Macaca mulatta mulatta* (rhesus monkey). The right P³ (a) has three roots; the left P³ (b), the right P⁴ (c) and the left P⁴ (d) have the disto-buccal and disto-lingual roots fused.

B: *M. mulatta mulatta*. Both P³ (e and f) have three roots; the right P⁴ (g) has the buccal roots joined for about three quarters of their length; in the left P⁴ (h) the disto-buccal root is united with both the lingual and mesio-buccal roots. Enlarged.

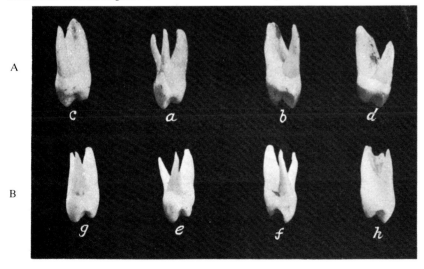

Family **Cebidae**. New World monkeys

$$\frac{2}{2}\cdot\frac{1}{1}\cdot\frac{3}{3}\cdot\frac{3}{3} = 36$$

The incisors of New World monkeys are similar to those of cercopithecids. The canines are large and project beyond the occlusal plane but sex dimorphism is marked only in the genus *Cebus*. There are three molars in each quadrant each with four cusps. In woolly monkeys (*Lagothrix*), spider monkeys (*Ateles*) and the woolly spider monkey (*Brachyteles*), M_3 has a fifth cusp, the hypoconulid.

The numbers of specimens of each genus examined and the numerical variations recorded in the original edition are summarized in Table 3.6, and a summary of variations for the Family as a whole is given in Table 3.7. Details of individual specimens are given below.

Schultz (1960) and Smith, Genoways and Jones (1977) examined skulls of the same three cebids from fairly limited areas in Central America, between the Panama Canal Zone and Southern Mexico, and in Nicaragua. In his sample of 203 adult black-handed spider monkeys (*Ateles geoffroyi*), Schultz found only one skull (0.5%) with a supernumerary, a molar; among 118 white-throated capuchins (*Cebus capucinus*), there were two instances of additional premolars and two of additional molars (1.1%). In 285 adults of the three same cebids, Smith and his colleagues found 3 (1%) with supernumerary teeth.

These figures are similar to Colyer's (Table 3.6); for *Ateles*, his figure is higher, but it must be remembered that he was dealing with several species of this genus from a variety of areas.

In the collection studied by Schultz (1960), incisors were absent in only one of 203 black-handed spider monkeys (*Ateles geoffroyi*) (0.5%) and in 18 of 378 mantled howlers (*Alouatta palliata* (4.8%); none were absent in 118 capuchins (*Cebus capucinus*). One *Cebus* lacked a canine; one *Ateles*, one *Cebus* and one

Table 3.6 Numerical variations in the CEBIDAE

| Genus | No. of specimens | Extra teeth | | | % of specimens affected | Absence of teeth | | | % of specimens affected | Totals of specimens showing extra teeth or absence of teeth | |
		I	P	M		I	P	M		No.	%
Lagothrix	94	—	—	—	0	—	—	1	1.1	2	2.1
Brachyteles	25	—	1	—	4.0	1	—	1	8.0	3	12.0
Ateles	232	2	2	7	4.7	—	—	8	3.4	19	8.2
Saimiri	110	—	—	1	1.0	—	—	—	0	1	1.0
Cebus	651	—	—	4	0.6	1	3	16	3.1	23	3.5
Alouatta	787	4	—	3	0.9	—	—	3	0.4	10	1.3
Cacajao	23	—	—	—	0	—	—	—	0	—	0
Pithecia	155	—	—	—	0	1	—	2	1.9	3	1.9
Callicebus	122	—	1	—	0.8	—	—	—	0	1	0.8
Aotus	10	—	—	—	0	—	—	—	0	—	0

In skulls in which two or more classes of teeth varied in number, each class is recorded individually; for example, if a skull has an extra incisor as well as an extra molar, both columns would be credited with one and the column 'number of specimens affected' with one only.

Table 3.7 The classes of teeth showing numerical variations in the CEBIDAE as a whole

| | Extra teeth | | | | Absence of teeth | | | |
| | Maxilla | | Mandible | | Maxilla | | Mandible | |
Class	Right	Left	Right	Left	Right	Left	Right	Left
Incisors	2	3	1	—	1	1	1	1
Premolars	3	1	1	—	1	1	—	2
Molars	8	6	4	7	17	16	5	6

Alouatta lacked one or more premolars. One or more M3 were congenitally absent in 9 *Ateles* (4.4%), three *Cebus* (2.5%) and two *Alouatta* (0.5%).

Absence of one or more teeth occurred in 11 *Ateles* (0.5%), 5 *Cebus* (4.2%) and 21 *Alouatta* (5.6%); in the sample examined by Smith, *et al.* (1977), comparable figures were 3.1% in *Ateles*, none in *Cebus* and 3.6% in *Alouatta*. Clearly, frequency of absence of teeth must be very variable in cebids (compare Colyer's results in Table 3.6). This is particularly true for M3 in different populations from limited areas. For example, Schultz (1956) found that one or more M3 were absent in 3.5% among *Ateles geoffroyi* from one locality in Panama but, in 40 specimens from Nicaragua, one or more M^3 were absent in 15% (Schultz, 1925). Colyer's figure for 651 skulls of various species of *Ateles* was 2.5%.

Schultz (1960) considered that congenital absence of teeth was quite common in the cebids that he examined and, though the rates are not high, Table 3.7 suggests that absence of one or more upper teeth occurs more commonly than other numerical variations.

Genus *Lagothrix*, woolly monkeys

(i) *Lagothrix lagothrica cana* (common woolly monkey). ♀. BMNH 1900.11.5.17. Bilateral maxillary disto-molars.
(ii) *L. lagothrica lugens*. BMNH 1890.2.22.1. ♂. Both M^3 absent.

(iii) *L. lagotricha*. ♀. NRM 1166. Left M^3 absent and right M^3 very small (Lönnberg, 1940).

Genus *Brachyteles*, woolly spider monkey

(i) *Brachyteles arachnoides* (woolly spider monkey). ♀. BMNH 1845.4.2.7. (Fig. 3.60). A supplemental right P^3 (Colyer, 1919; Bateson, 1894, p. 205).
(ii) *B. arachnoides*. Formerly in the Zoologisches Museum of the University of Hamburg. Both I_1 absent.
(iii) *B. arachnoides*. NRM 1477. Left M^3 and left M_3 absent. The right M^3 is only about one-quarter the size of M^2.

Genus *Ateles*, spider monkeys

EXTRA TEETH
(a) Incisors
(i) *Ateles paniscus* (black spider monkey). ♀. BMNH 1872.4.11.5. (Fig. 10.123). An empty socket between the right I^1 and I^2 (Bateson, 1894, p. 207).
(ii) *A. geoffroyi vellerosus* (Mexican spider monkey). FMNH 13900 (Fig. 3.61). An extra tooth labial to the left I^2 appears to consist of two connate denticles.
(iii) *A. paniscus chamek* (black-faced spider monkey). ♂. BMNH 1897.10.3.1. Left I^2 much larger than right and has a lingual groove suggestive of connation.

Fig. 3.60. *Brachyteles arachnoides* (woolly spider monkey). ♀. BMNH 1845.4.2.7. Four premolars in the right maxilla.

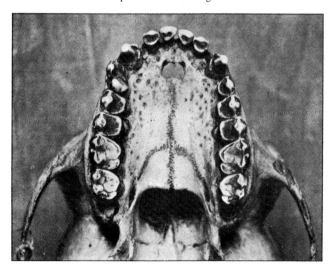

54   *Section 1: Variations in Number, Size and Shape*

(b) Premolars
  (i) *A. belzebuth marginatus* (white-whiskered spider monkey). ♀. BMNH 1853.3.19.73. (Fig. 3.62). Four premolars in each maxilla; they resemble each other too closely for it to be possible to identify which is supernumerary (Bateson, 1894, p. 206).
  (ii) *A. fusciceps* (brown-headed spider monkey). ♂. BMNH 1872.4.30.2. An empty socket between the right P^3 and P^4 shows that a single-rooted supernumerary had been present.
  (iii) *A. paniscus*. NRM 928. Four well-formed premolars on each side of the mandible (Lönnberg, 1940).
  (iv) *A. paniscus*. NRM 929. Four premolars in the left upper jaw (Lönnberg, 1940).

In Dr G. Hagmann's private collection at Taperinha, River Amazon (now in the Museu Paranese Emilio Goldi, Belem), there were 30 specimens of white-whiskered spider monkeys (*Ateles belzebuth marginatus*); extra premolars were present in three of them:
  (i) ♂. MPEG 5.144. Four premolars in the right maxilla all crowded together, with the canine a

Fig. 3.61. *Ateles geoffroyi vellerosus* (Mexican spider monkey). FMNH 13900. An extra tooth labial to the left I^2 consists of two connate denticles.

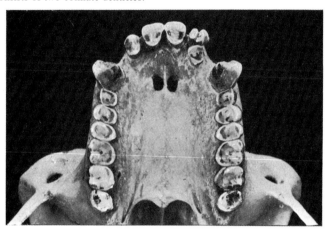

Fig. 3.62. *Ateles belzebuth marginatus* (white-whiskered spider monkey). ♀. BMNH 1853.3.19.73. A supplemental premolar in each maxilla; it is impossible to distinguish which is supernumerary.

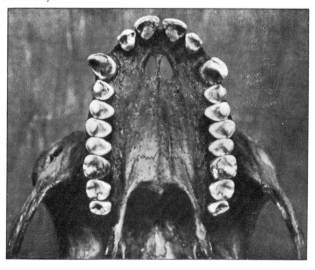

little displaced buccally. Dr Hagmann thought that the second of the four premolars was the extra tooth because it was a little smaller than the others.
  (ii) ♀. MPEG 5.148. Four premolars in each maxilla.
 (iii) ♀. MPEG 5.143. Four premolars in each maxilla.

(c) Molars
   (i) *Ateles paniscus chamek*. FMNH 21386. A disto-molar in right maxilla.
  (ii) *A. paniscus*. A.E.L. Amsterdam, 1908/189. A disto-molar in left maxilla. On the right, a conical tooth lingual to M^2 and M^3.
 (iii) *A. fusciceps*. NRM 6. Bilateral disto-molars in both jaws.
  (iv) *A. geoffroyi vellerosus*. ♂. BMNH 1889.12.7.1. A disto-molar in left mandible (Bateson, 1894, p. 205).
   (v) *A. paniscus paniscus* (red-faced spider monkey). ♂. BMNH 1910.9.29.1. A disto-molar in left maxilla.
  (vi) *A. geoffroyi vellerosus*. AMNH 17218. Bilateral mandibular disto-molars.
 (vii) *A. paniscus*. ♀. BMNH 1927.8.11.27. A small disto-molar in left maxilla.
(viii) *A. geoffroyi*. ♀. KU 104,439. Bilateral disto-molars in both jaws; those in the mandible and the right maxilla are miniature replicas of the M^3 but, in the left maxilla, the disto-molar is a small spicule to the buccal side of M^3 (Smith *et al.*, 1977).
  (ix) *A. paniscus*. ♀. NRM 1596. A small disto-molar in the left maxilla (Lönnberg, 1940).
   (x) *A. paniscus*. A disto-molar in both upper and lower jaws on the right (Bennejeant, 1936, p. 215).

ABSENCE OF TEETH
   (i) *Ateles geoffroyi geoffroyi* (Nicaraguan spider monkey). BMNH 1848.10.26.3. Both M^3 absent (Bateson, 1894, p. 207).
  (ii) *A. paniscus chamek*. FMNH 21387. Right M^3 absent.
 (iii) *A. paniscus*. Formerly ZMB 38734. Right M_3 absent.
  (iv) *A. fusciceps robustus*. AMNH 38106. Left M_3 absent.
   (v) *A. belzebuth*. RMNH 31771. Right M^3 absent.
  (vi) *A. belzebuth*. RMNH 31772. Both M^3 absent.
 (vii) *Ateles* sp. Formerly Stockholm Högskola, 2235. Right M^3 absent.
(viii) *A. belzebuth hybridus* (brown spider monkey). USNM 216663. Right M_3 absent.
  (ix) *A. geoffroyi*. ♂. AHS 750. Both M^3 and right P_3 absent (Schultz, 1960).
   (x) *A. geoffroyi.*, ♂. USNM 337702. Right M_3 absent (Smith *et al.*, 1977).
  (xi) *A. geoffroyi*. ♂. KU 104,447. Right M^3 absent (Smith *et al.*, 1977).
 (xii) *A. geoffroyi*. ♀. AHS 754. Right M^3 absent (Schultz, 1925).
(xiii) *A. geoffroyi*. ♂. AHS 733. Both M^3 absent (Schultz, 1925).
(xiv) *A. paniscus*. ♀. NRM 1222. Both M_3 absent (Lönnberg, 1940).

SIZE
   (i) *Ateles geoffroyi*. ♀. AHS 709. Both M^3 unusually small (Schultz, 1925).
  (ii) *A. paniscus*. Formerly ZMB 38734. Left M^3 unusually small; right M^3 absent (see above).
 (iii) *A. fusciceps*. MNHN 1812–12. M_1 and M_2 on both sides have five cusps (Bennejeant, 1936, p. 133).
  (iv) *A. belzebuth*. MNHN 1962–1372. Both M_1 and the right M_2 have five cusps; the other molars are lost from the specimen (Bennejeant, 1936, p. 133).

SHAPE
   (i) *Ateles geoffroyi*. ♂. AHS 1009. The root of the abscessed, partially-erupted, upper left canine is considerably distorted and, according to Schultz (1960), is divided so that the apex of one branch of the root projects into the nasal cavity.

Genus ***Saimiri***, squirrel monkeys

Among 100 specimens, there was only one with an additional tooth, a disto-molar in the left mandible (RCS Odonto. Mus., G 12.3); Blüntschli (1913) mentioned disto-molars in *Saimiri sciureus*.

Genus ***Cebus***, capuchin monkeys

EXTRA TEETH
   (i) *Cebus capucinus* (white-throated capuchin). ♂. MHNG 611.60. There are six upper incisors; the outermost pairs are like normal I^2 in this species, slightly caniniform (Bennejeant, 1936, p. 189).
  (ii) *C. capucinus*. ♀. KU 104,435. An additional incisor between left I^1 and I^2 (Smith *et al.*, 1977).
 (iii) *C. apella* (brown capuchin). RMNH 31770. A minute disto-molar in the left mandible; left I^2 absent.

(iv) *C. apella*. A disto-molar in right upper jaw (Bennejeant, 1936, p. 215).
(v) *C. apella pallidus*. AMNH 40839. A small disto-molar in the left mandible.
(vi) *C. apella pallidus*. ♂. NRM 70. A disto-molar in the right mandible (Lönnberg, 1939).
(vii) *C. apella pallidus*. ♀. NRM 118. A disto-molar in the right mandible (Lönnberg, 1939).
(viii) *C. capucinus*. FMNH 18241. (Fig. 3.63). Two tuberculate supernumerary teeth have erupted lingually to the right M^2 and M^3.
(ix) *Cebus* sp. MNHN A2696. Bilateral maxillary disto-molars (Geoffroy St Hilaire, 1838; Bennejeant, 1936, p. 214).

ABSENCE OF TEETH
(i) *Cebus apella*. Formerly in the Zoologisches Museum, University of Hamburg, 22335. A premolar absent from each maxilla.
(ii) *Cebus* sp. Muséum National d'Histoire naturelle, Paris, A 12917. Left P_2 absent.
(iii) *Cebus* sp. ♀. RCS Odonto. Mus., G 12.1. Left P_2 absent.
(iv) *C. albifrons* (brown pale-fronted capuchin). AHS 761. Left M^3 and right M_3 absent (Schultz, 1925).
(v) *C. apella*. RMNH 31770. Left ^{2} absent. Disto-molar in left mandible (see above).

SIZE
(i) *Cebus albifrons*. ♂. AHS 7804. Both M^3 small. (Schultz, 1925).

Fig. 3.63. *Cebus capucinus* (white-throated capuchin). FMNH 18241. Two tuberculate supernumerary teeth lingual to the right M^2 and M^3.

(ii) *C. capucinus*. ♂. AMNH 176,637. Left P^4 is a single-rooted peg (Smith *et al.*, 1977).

SHAPE
The M^3 is often small and may be reduced to a small cone as in *Cebus capucinus* (AMNH 176,637) above. The tooth may be flattened mesiodistally as in Figure 3.64 and the right M^3 in Figure 3.65.

In Figure 3.64, the metacones of both M^2 are displaced a little lingually; in both M^3, this displacement is more marked, the metacones being almost distal to the protocones. There is a small extra cusp on the mesial part of the lingual surface of the right M^1, in the Carabelli position.

Figure 3.65 shows a skull with extra cusps on the lingual surfaces of the M^1 and M^2 that appear to arise from the cingulum, the one on the right M^1 being best developed. Extra cusps were seen in two other specimens; in one (*C. apella*, ♀, BMNH 1846.10.13.5), on the lingual surfaces of both M^2; in the other (in Lund University), on the lingual surface of a right M^2. In one specimen (Fig. 3.66), the metacones of both M^2 were much smaller than normal.

Genus *Alouatta*, howler monkeys

EXTRA TEETH
(a) Incisors
(i) *Alouatta caraya* (black howler). ZMUC CN 179 (Fig. 3.67). The left I^2 has been lost but, between its socket and the canine, there is an empty socket for a tooth which must have been twice the size of

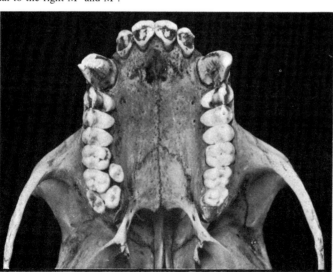

I^2. As this socket is in the premaxilla, it must be regarded as that of an incisor rather than canine.
(ii) *A. palliata* (mantled howler). NRM 6. An extra tooth in the right premaxilla.
(iii) *A. seniculus arctoidea* (red howler). AMNH 36172. An extra incisor in the right mandible, larger than the I^2 beside it.

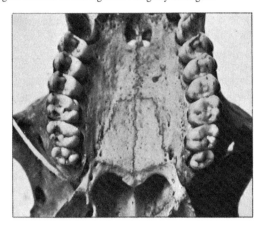

Fig. 3.64. *Cebus capucinus* (capuchin). ♀. BMNH 1924.12.6.3. Both M^3 are oval; a small extra cusp on the lingual surface of the right M^1. Slightly enlarged.

Fig. 3.65. *Cebus apella* (brown capuchin monkey). ♂. BMNH 1913.6.13.2. Extra cusps on the lingual surfaces of M^1 and M^2.

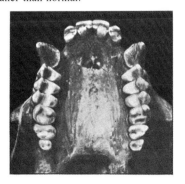

Fig. 3.66. *Cebus apella* (brown capuchin monkey). ♀. BMNH 1897.10.3.7. The metacones of both M^2 are smaller than normal.

(iv) *A. seniculus straminea*. AMNH 40034. Three incisors in the left premaxilla; the supernumerary appears to be the one between the I^1 and the mid-line.

(b) *Premolars*
  (i) *A. palliata*. ♂. An additional conical tooth in the right upper premolar region (Schultz, 1960).
  (ii) *A. palliata*. ♀. An additional premolar lingual to the lower left premolars (Schultz, 1960).

(c) *Molars*
  (i) *A. caraya*. ♀. BMNH 1846.10.16.1. An extra tooth buccal to the right M^3 (Bateson, 1894, p. 207).
  (ii) *A. seniculus*. A.E.L. Amsterdam, 1910/123. A left mandibular disto-molar.
  (iii) *A. seniculus straminea*. AMNH 30638. Bilateral mandibular disto-molars.

ABSENCE OF TEETH
(a) *Incisors*
  (i) *Alouatta palliata*. i^1 and i^2 retained on both sides. Both I^1 and all lower incisors absent (Schultz, 1960) (p. 58).
  (ii) *A. palliata*. ♀. One upper incisor absent (Schultz, 1960).
  (iii) *A. palliata*. ♀. AHS 657. One of five specimens lacking one lower incisor (Schultz, 1925).
  (iv) *A. palliata*. USNM 337856. Both I_1 absent (Smith et al., 1977).
  (v) *A. palliata*. ♀. KU 108,139. Both I_1 absent (Smith et al., 1977).

Fig. 3.67. *Alouatta caraya* (black howler). ZMUC CN 179. There is an extra socket mesial to the left canine; the socket is in the premaxilla, so it must have accommodated an unusually large incisor and not a canine.

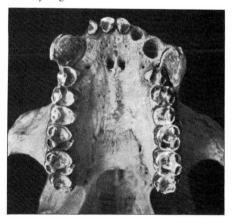

Section 1: Variations in Number, Size and Shape

(b) *Premolars*
 (i) *A. palliata*. ♀. AHS 585. Right P^4 absent (Schultz, 1925).

(c) *Molars*
 (i) *A. seniculus*. NRM 1. Both M^3 absent.
 (ii) *A. seniculus straminea*. AMNH 48120. Left M^3 absent.
 (iii) *Alouatta* sp. RCS Odonto. Mus., G 12.2. Left M^3 absent.

SIZE
 (i) *Alouatta palliata*. ♀. AHS 777. Left M_3 conical (Schultz, 1960).

SHAPE
 (i) *Alouatta palliata*. ♀. On the right, I^1 is exceptionally wide; although there is a single crown, it is divided by longitudinal grooves into a larger mesial part and a smaller distal part. The left I^1 is absent (p. 57) (Schultz, 1960).
 (ii) *A. palliata*. ♀. The crown of the lower right canine appears to be divided into three conical parts, the largest, lingual part having a bifid tip. There is a single root with longitudinal grooves corresponding with the crown divisions (Schultz, 1960).
 (iii) *A. palliata*. ♂. KU 97,863. Right M_3 malformed; fused to the postero-lingual aspect of its crown is the crown of a conical denticle with a separate root (Smith *et al.*, 1977).

Genus *Pithecia*, saki monkeys

ABSENCE OF TEETH
 (i) *Pithecia monachus* (monk saki). NRM, no number. Right I^2 absent.
 (ii) *P. pithecia* (white-faced saki). ♀. Formerly BMNH 1920.7.14.7. Left M_3 and right M^3 absent.
 (iii) *P. pithecia*. ♀. BMNH 1925.12.11.5. Right M^3 absent.

Genus *Callicebus*, titi monkeys

 (i) *Callicebus personatus* (masked titi). NRM 1. Four premolars in the right mandible.
 (ii) *C. moloch* (dusky titi). ♂. BMNH 1894.3.6.1. A small cusp on the buccal surface of each M^1.
 (iii) In Dr G. Hagmann's collection (now in the Museu Paranese Emilio Goldi, Belem) of 42 specimens of dusky titi monkeys (*C. moloch*), three showed variation in number and shape of teeth: in one, right I_2 was missing, in another right M^3 missing; in a third, right I_1 and I_2 were connate with a deep groove on the labial aspect.

Family **Callitrichidae**. Marmosets and tamarins

$$\frac{2}{2}:\frac{1}{1}:\frac{3}{3}:\frac{2}{2}=32 \text{ or } \frac{2}{2}:\frac{1}{1}:\frac{3}{3}:\frac{3}{3}=36$$

The upper and lower incisors are long and spatulate. The canines are of the same height as the incisors in the so-called short-tusked species but, in the long-tusked species, they extend beyond the occlusal surface. There are two molars in each quadrant except in Goeldi's marmoset (*Callimico goeldi*) which has a small M3 as well. All molars have four cusps but the hypocone in the upper molars is small.

The Callitrichidae are unusually free of numerical variations; Colyer found only one in the 247 specimens he examined. Herschkovitz (1970) found this to be so also: in 904 wild-caught marmosets (*Callithrix*) and tamarins (*Saguinus*), supernumerary teeth occurred in 0.8% and there were no congenitally absent teeth.

EXTRA TEETH
 (i) *Callithrix jacchus penicillata* (black-plumed marmoset). ♀. AHS 1835. Five well-developed mandibular incisors (Schultz, 1925).
 (ii) *Saguinus midas* (negro tamarin). ♂. FMNH 86956. Three incisors in the right premaxilla (Herschkovitz, 1970).
 (iii) *S. oedipus* (cotton top tamarin). ♂. FMNH 69964. Four premolars in the left maxilla (Herschkovitz, 1970).
 (iv) *S. midas*. ♂. FMNH 96500. Additional bilateral M^2. (Herschkovitz, 1970).
 (v) *S. mystax* (moustached tamarin). ♂. FMNH 86952. An additional right M^2 (Herschkovitz, 1970).
 (vi) *Callithrix jacchus* (common marmoset). ♂. BMNH 1903.9.4.26. An upper left disto-molar (Herschkovitz, 1970).
 (vii) *Saguinus oedipus*. ♂. FMNH 69937. A conical supernumerary on the lingual side of the left M_2 (Herschkovitz, 1970).
 (viii) *Saguinus oedipus*. ♀. USNM 302337. A conical supernumerary, connate with the crown of the left P_4 but with a separate root (Herschkovitz, 1970).
 (ix) *Callimico goeldi* (Goeldi's marmoset). ♂. USNM 303322. A left maxillary disto-molar (Herschkovitz, 1970).

ABSENCE OF TEETH
 (i) *Saguinus oedipus geoffroyi* (Geoffroy's tamarin). AHS 1498. One premolar absent from each maxilla (Schultz, 1956, p. 1001).
 (ii) *Saguinus midas*. ♀. BMNH 1906.1.1.3. Right M_2 absent.

Sub-order **PROSIMII**

The number of premolars varies in the prosimians; when three premolars are present, they will be named P2, P3 and P4; when two are present P2 and P3.

Table 3.8 summarizes Colyer's records of the number of numerical variations in prosimians. So few Tarsidae and Daubentonidae were examined that the absence of variation cannot be regarded as carrying much weight. The same may apply to the Indriidae. Nevertheless, the frequency of variations in the prosimians as a whole is very low.

Family **Lemuridae**. Lemurs

$$\frac{0-2}{2}:\frac{1}{1}:\frac{3}{3}:\frac{3}{3} = 32 \text{ or } 36$$

The lower incisors and the incisiform lower canine are long, slender and procumbent; they are close-set forming a dental comb, often used in grooming. When present, the upper incisors are small and peg-like. The upper canines are large and sexually dimorphic; P^2 may be caniniform.

Genus *Lepilemur*, weasel lemurs and sportive lemurs

This genus normally has no incisors in the premaxilla. The exceptions found are:
 (i) *Lepilemur mustelinus microdon* (weasel lemur). ♀. BMNH 1897.9.1.23. One incisor in each premaxilla.
 (ii) *L. mustelinus microdon*. BMNH 1897.9.1.18. One incisor in each premaxilla.
 (iii) *L. mustelinus*. BMNH 1897.9.1.24a. One incisor in each premaxilla.

Genus *Lemur*, lemurs

 (i) *Lemur mongoz* (mongoose-lemur). MNHN 1903–757. The left I^1 and I^2 were connate (Bennejeant, 1936, p. 178).
 (ii) *L. variegatus* (ruffed lemur). BMNH 1867.4.12.79 (Fig. 3.68). The crown of each P_4 is larger than

Table 3.8 Numerical variations in the PROSIMII

Family or sub-family	No. of specimens	Specimens with extra teeth No.	%	Specimens with absence of teeth No.	%	Totals of specimens showing extra teeth or absence of teeth No.	%
Lemurinae	289	4	1.4	2	0.7	6	2.1
Cheirogaleinae & Galaginae*	244	4	1.6	1	0.4	5	2.0
Indriidae	91	0	0	0	0	0	0
Daubentoniidae	9	0	0	0	0	0	0
Lorisinae	133	3	2.3	0	0	3	2.3
Tarsidae	9	0	0	0	0	0	0
Total	**775**	**11**	**1.4**	**3**	**0.4**	**14**	**1.8**

* In the first edition, Colyer included *Cheirogaleus* and *Microcebus* under the Galaginae, whereas they are now usually classified as Cheirogalidae; he included the Galaginae in the family Lemuridae, whereas they are now believed to belong to the Lorisidae.

Fig. 3.68. *Lemur variegatus* (ruffed lemur). BMNH 1867.4.12.79. The crown of P^4 is bifid and unusually large, suggesting connation.

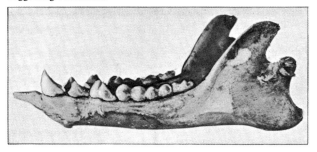

normal and the principal cusps, the buccal ones, are bifid. This must be regarded as partial dichotomy or connation.
(iii) *L. macaco albifrons* (black lemur). BMNH 1846.3.23.15. Four premolars in the right maxilla. The extra tooth is probably the one of about the same size as P^2 which is situated lingually to it.
(iv) *L. macaco rufus*. ♀. BMNH 1875.7.20.6. Left M^3 absent.
(v) *L. macaco rufus*. BMNH 1855.12.24.54. Left M_3 absent.
(vi) *L. catta* (ring-tailed lemur). Muséum d'Histoire naturelle, Bâle, 169. Five cusps instead of the usual four on both M_1 (Bennejeant, 1936, p. 132).

Family **Cheirogaleidae**. Mouse lemurs and dwarf lemurs

$$\frac{2}{2}:\frac{1}{1}:\frac{3}{3}:\frac{3}{3}=36$$

The dentition is similar to that of the Lemuridae and indeed some authorities include mouse lemurs and dwarf lemurs in the Lemuridae.
(i) *Chirogaleus major* (greater dwarf lemur). ♂. Captive. BMNH 1885.10.8.1. Only one incisor in each premaxilla.
(ii) *Chirogaleus trichotis* (hairy-eared dwarf lemur). BMNH 1875.1.29.2. An extra tooth between the left P^2 and P^3. There is a gap between the right P^2 and P^3 and, in the mandible, a space between the P_3 and P_4 on each side, but nothing otherwise to suggest that supernumerary teeth had been present in these spaces.
(iii) *Microcebus murinus* (lesser mouse lemur). ♂. BMNH 1925.12.9.10. Four premolars in each maxilla.
(iv) *Chirogaleus major*. RCS Odonto. Mus., G 14.1. Four left mandibular premolars.

Family **Indriidae**. Indrisoid lemurs

$$\frac{2}{2}:\frac{1}{0}:\frac{2}{2}:\frac{3}{3}=30$$

The lower incisors form a dental comb. There are no lower canines but P_3 is caniniform and occludes with the short upper canine.

Although Colyer did not find any numerical variations in the Indriidae, small supernumerary teeth sometimes occur: Friant (1935) noted one in *Indri indri* and Bennejeant (1936, p. 199) one in *Propithecus verreauxi*, but Bennejeant showed that, in *P. verreauxi* at least, these were actually persistent deciduous molars.

(i) *Indri indri* (indri). In the embrasure on the lingual side between the left P^3 and the upper canine, there was a tiny conical denticle (Friant, 1935).

Family **Daubentonidae**. Aye-aye

$$\frac{1}{1}:\frac{0-1}{0}:\frac{1}{0}:\frac{3}{3}=18 \text{ or } 20$$

No numerical variations.

Family **Lorisidae**. Lorises, pottos and galagos

$$\frac{1-2}{2}:\frac{1}{1}:\frac{3}{3}:\frac{3}{3}=34 \text{ or } 36$$

The lower incisors form a dental comb which is shorter and less procumbent than in lemurs. The upper incisors are small and peg-like and the upper canine is long and short.

Sub-family **Lorisinae**

(i) *Loris tardigradus* (slender loris). ♂. RCS Odonto. Mus., G 14.3. In the right premaxilla, there are three incisors; the extra tooth seems to be the one nearest the middle line.
(ii) *Perodicticus potto* (potto). ♂. BMNH 1902.7.12.1. Four premolars in the left maxilla.
(iii) *Nycticebus coucang bengalensis* (slow loris). BMNH 1843.3.25.13. Four premolars on each side of the mandible. One is situated to the lingual side of each P_3 and may be the supernumerary.

In the majority of the sub-species of *Nycticebus coucang*, there are two incisors in each premaxilla but

Fig. 3.69. *Galago crassicaudatus* (greater bushbaby). ♂. BMNH 1908.1.1.132. Both upper incisors are unusually large and are partially divided by longitudinal grooves, suggesting connation with a supernumerary element. × c. 2.0.

sometimes there is only one. Schultz (1956) noted that this lack was sometimes unilateral. According to Hill (1953b), I^2 is always present in *N. coucang insularis* and in *N. coucang naturae*, but *N. coucang hilleri* has four upper incisors in adolescence and only two in older individuals, whereas *N. coucang javanicus* may have two, three or four upper incisors.

Absence of P^3 in *N. coucang borneanus* was noted in the first edition of this book and by Hill (1953b), but Schultz (1956) maintained that it is occasionally present in this sub-species though much reduced in size.

Variation in the size of teeth in *Periodicticus potto* was described by Allen (1922–25).

Sub-family **Galaginae**

(i) *Galago crassicaudatus* (greater bushbaby). ♂. BMNH 1908.1.1.132. (Fig. 3.69). The incisors in each premaxilla are connate.
(ii) *G. senegalensis moholi* (lesser bushbaby). BMNH 1845.6.17.1. Four premolars in the right maxilla.

Family **Tarsidae**. Tarsiers

$$\frac{2}{1}:\frac{1}{1}:\frac{3}{3}:\frac{3}{3}=34$$

No numerical variations.

CHAPTER 4

Order CARNIVORA

A guide to and description of most carnivore dentitions is available in Ewer (1973).

The problem of deciding in carnivores whether teeth were congenitally absent, or lost during life leaving completely healed sockets, is particularly acute. Colyer felt that the problem was insoluble and therefore only systematically recorded extra teeth and variations in shape. Table 4.1 shows the numbers of extra teeth he found in the various carnivore families.

Family **Felidae**

$$\frac{3}{3}:\frac{1}{1}:\frac{3}{2}:\frac{1}{1} = 34$$

In felids, the only tooth distal to the carnassial M_1 and P^4 is the very small M^1, the rest of the molars having been lost in the course of evolution.

Colyer found extra teeth in 2.7% of his sample of 1308 felids (excluding domestic cats). The extra tooth was most frequently a premolar, although even the canine can be duplicated (Fig. 4.1).

Among the skulls of 465 *Felis lynx canadensis* (North American lynx) in the United States National Museum, Manville (1963) found four (0.9%) with extra teeth, two with incisors and two with an extra P. He noted extra teeth in 21 (1.1%) of 1983 *Felis rufa* (bobcats) (8 skulls with extra incisors, 10 with extra premolars and 3 with extra molars). Lüps (1977) found an extra tooth between the canine and the first premolar in six of his sample of 251 skulls of feral cats in Berne, three (1.2%) in the maxilla and three in the mandible. A much higher incidence was found by Pocock (1916a) amongst 11 skulls of the wild cat (*F.*

silvestris) from Inverness-shire and Ross-shire in Scotland; four skulls had an extra tooth between the lower canine and P_3 on one or both sides; one of the four had an extra tooth between P^2 and the upper canine as well; this is perhaps related to their geographical isolation.

Kurtén (1963) pointed out that the fossil evidence is strong that the ancestral felids in the Miocene and the lynx-like *Felis issiodorensis* in the Villafranchian had no M_2. However, it was present in about 10% of his sample of 60 modern European lynxes (*F. lynx lynx*) and Kvam (1985) found an M_2 in about 10% of 550 Norwegian lynxes. Kurtén regarded its occasional presence as the return of a lost structure, related to the reactivation of the molarization field that had never been genotypically lost. Werdelin (1987) has pointed out that M_2 is absent from all other living and fossil species of lynx and claims that its occurrence in *F. lynx lynx* indicates a new character evolving in response to evolutionary pressure.

According to Hall (1940), absence of teeth may be less common in felids than in other carnivores because their dentition is already so much reduced. He found no teeth absent in 123 N. American lynxes and 372 bobcats and in only two (1.7%) of 115 pumas (*F. concolor*). In the sample mentioned above, Manville (1963) found only one lynx (0.7%) and 14 bobcats (0.7%) with any teeth absent. Colyer on the other hand found that P^2 was often absent in the cheetah (*Acinonyx jubatus*), the North American lynx, the caracal (*F. caracal*) and the rusty-spotted cat (*Felis rubiginosus*). However, the alveolar process was in some instances rough and he believed that many of these small premolars may have been lost during life.

The presence or absence of P^2 varies in different regions. Lüps (1980) compared large samples of feral and domestic cats from several parts of Europe, Singapore, Mexico, Venezuela and the Antarctic Kerguelen islands, and found that absence of P^2 varied from 3.4% in Britain to 28.4% in the Kerguelen islands, a difference which he explained as a geographical cline from north to south. Glass and Todd (1977) found significant intraspecific differences in the presence and absence of P^2 in the leopard cat (*Felis bengalensis*). It was absent in 26 (40%) of adults to the north of the Himalayas and in 7 (11%) of those to the south. They interpreted this as an example of quasi-continuous variation, but were uncertain whether it was related to latitude or genetic isolation.

Kratochvil (1965) found one absent M^1 among 10 *Felis lynx* from the Carpathians; the M^1 was also variable in size and shape.

Elzay and Hughes (1969) described total absence

Fig. 4.1. *Panthera tigris* (tiger). RCS Odonto. Mus., G15.1. Two canines in the right maxilla. ×*c*. 0.33.

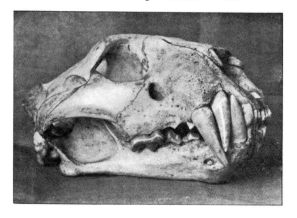

Table 4.1 Extra teeth in CARNIVORA

Family	No. of specimens	Extra teeth				Totals of specimens showing extra teeth	
		I	C	P	M	No.	%
Felidae	1308	4	1	26	4	35	2.7
Hyaenidae	142	0	0	1	0	1	0.7
Viverridae	1585	6	0	14	6	26	1.6
Mustelidae	1844	10	1	7	2	20	1.1
Procyonidae	413	1	0	5	1	7	1.7
Ursidae	379	4	0	5	0	9	2.4
Canidae	1280	3	1	16	15	35	2.7
Total	**6951**	28	3	74	28	133	1.9

Domestic cats and dogs are not included; neither are badgers (*Meles meles*) or otters (*Lutra lutra*), which show considerable variation in the first premolars.

of the dentition, confirmed by histological examination of the jaws, in a domestic cat. In this case, there was no evidence of generalized ectodermal dysplasia, such as deficiency of hair and claws, which is a known cause of absence of teeth in man and in cats (Burroughs, Miller and Harvey, 1985).

Variation in size of P^2 is marked in felids and sometimes the contralaterals differ in size. Graf et al. (1976) gave coefficients of variation of 14.1 for males and 17.7 for females in the mesio-distal length of P^2 obtained by measuring the tooth sockets in 75 feral cats from Berne. They considered this high variability to be related to the incipient evolutionary loss of this tooth in cats.

In felids, an extra root may be present on P_4 lingually between the mesial and the distal roots. The maxillary first molars may have either one or two roots.

In primitive felids, the lower carnassial (M_1) consists of a mesial portion with two cusps, the paraconid and the protoconid, and a distal portion, the talonid bearing the metaconid. In the course of felid evolution, the metaconid has been lost. According to Kurtén (1963), the Villafranchian *Felis issiodorensis* had no metaconid and the talonid was reduced to a small distal bulge. He showed that the metaconid is sometimes present in various modern felids and was invariably so in his sample of 60 European *F. lynx*. Like the occasional presence of the lost M_2 mentioned above, Kurtén described this as a surprising exception to the dictum that structures lost in the course of evolution never reappear in their original form.

Pocock (1940) described a male puma (*Felis concolor*, BMNH 1926.1.12.2) which had an extra cusp that he believed to be the homologue of the primitive metaconid.

Sub-family **Pantherinae**

Genus *Panthera*

Panthera leo (lion). 150 specimens examined.

(i) ♀. RCS Odonto. Mus., G15.2. A small premolariform extra tooth on each side of the mandible between P_3 and the canine.
(ii) BMNH 1918.5.23.2. An extra premolar mesial to the left P_3.
(iii) RCS Odonto. Mus., G15.11. The left M^1 is duplicated; one is in the normal position; the other is lingual to it.
(iv) In the right maxilla, a tooth similar to the P^2, but smaller, was erupting medially to it (Herrick, 1951).
(v) An extra P^3 between the normal P^3 and P^4 on the right (Keep, 1985).

Panthera tigris (tiger). 97 specimens examined.

(i) RCS Odonto. Mus., G15.1 (Fig. 4.1). Two canines in the right maxilla, slightly unequal in size and occupying the same socket (mentioned by M'Intosh, 1929).
(ii) ♀. BMNH 1932.4.16.3. (Fig. 4.2). In the position of the left P^4, there are three teeth. The lingually-situated one is similar to the contralateral P^4 but smaller. Buccal to it are two much smaller teeth. The mesial one is single-rooted with a tricuspid crown, the buccal cusp being the largest; the mesial cusp is very small. The distal tooth is smaller and single-rooted with a bluntly conical crown.
(iii) ♂. Formerly RCS Osteo. Series, 308. An empty socket between the right P^2 and P^3 from which a supernumerary tooth has been lost.

Panthera pardus (leopard). 136 specimens examined.

(i) BMNH 1880.2.16.1. Four incisors in the left premaxilla. The tooth distal to I^2 and resembling it, though it is smaller, is the probable supernumerary.
(ii) Captive. RCS Odonto. Mus., G15.31. The right I^3 is duplicated; the distal one is incompletely erupted.
(iii) BMNH 1887.4.25.1. An additional left P^2; the mesial one is slightly larger and is similar in size to the contralateral P^2.
(iv) ♀. Formerly RCS Osteo. Series, 365. An additional socket, from which the supernumerary tooth has been lost, mesial to the left P^2.
(v) BMNH 1867.1.8.2. A small left mandibular disto-molar.
(vi) ♀. BMNH 1935.1.6.71. In the position of the left M^1, there are three teeth. The mesial one is similar to the contralateral M^1 though smaller; distal to it, there are two small haplodont teeth.

Sub-family **Acinonychinae**

Genus *Acinonyx*

Acinonyx jubatus (cheetah).

Both P^2 were missing from several specimens, but there were indications that the teeth had been present. In only one instance (RCS Odonto. Mus., G15.4) was it probable that a p^2 was congenitally absent.

Sub-family **Felinae**

Genus *Felis*

Felis catus (domestic cat). 102 specimens examined.

The number of extra teeth recorded does not indicate

the frequency of this type of variation because some of the specimens were presented to museums expressly as abnormalities.

  (i) RCS Odonto. Mus., G 16.2. In place of the left P^2, there are two similar teeth, the mesial being the smaller.
  (ii) RCS Odonto. Mus., G 16.21. The left P^2 is duplicated.
  (iii) ♂. Feral. BMNH 1923.10.4.1. Supplemental P^2 on both sides, the mesial being the smaller.
  (iv) RCS Odonto. Mus., G 16.23. A supplemental P^2 on both sides and a supplemental left I^3.
  (v) Formerly RCS Osteo. Series, 414. (Fig. 4.3). In addition to the normal left P^3, identified by being similar in shape and size to the contralateral P^3, there is on its lingual side a supplemental smaller tooth.
  (vi) ♀. BMNH 1885.8.1.20. Two minute extra conical teeth between P^2 and C on both sides.
  (vii) BMNH 1888.3.10.1. (Fig. 4.4). Lingual to each P^4, there is a tooth of similar shape but much smaller.
  (viii) ♀. ZMUC 105. An extra tooth distal to the right M_1.

Fig. 4.2. *Panthera tigris* (tiger). ♀. BMNH 1932.4.16.3. In place of the left P^4, there are three teeth. ×c.0.5.

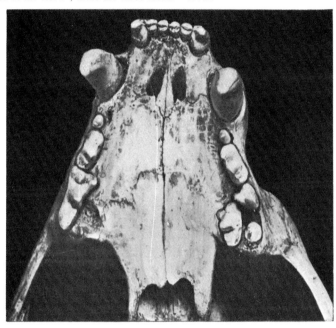

Fig. 4.3. *Felis catus* (domestic cat). Formerly RCS Osteo. Series, 414. There are two similar teeth in the position of the left P^3.

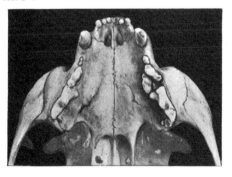

Fig. 4.4. *Felis catus* (domestic cat). BMNH 1888.3.10.1. A supplemental tooth lingual to each P^4.

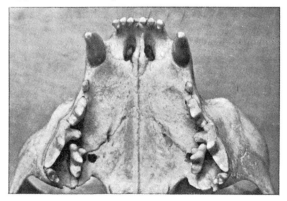

(ix) Department of Zoology, Oklahoma Agricultural and Mechanical College. An additional conical tooth in each jaw in the premolar region distal to the canines (Van Eaton, 1947).

(x) A connate left P^3 (Verstraete, 1985).

Felis silvestris (European wild cat). 67 specimens examined.

(i) BMNH 1902.6.3.4. A supplemental right P^2.

(ii) BMNH 1919.7.7.2901. The left P^2 is duplicated.

(iii) ♂. RCS Odonto. Mus., G 16.24. A tiny conical supernumerary between P_3 and C on both sides.

(iv) ♂. BMNH 1859.9.6.55. An extra tooth between the left P_3 and C.

(v) ♂. BMNH 1847.7.22.2. Bilateral conical supernumerary between P^2 and C.

(vi) BMNH 1916.11.27.2. An extra tooth between P_3 and C on both sides, larger on the right than on the left.

Felis libyca (African and Indian wild cat). 111 specimens examined.

(i) BMNH 1881.1.6.1. Buccal to the right P^2 is a smaller, supplemental tooth.

(ii) ♂. BMNH 1925.5.12.6. Two similar teeth in the position of the right P^2; the distal one is larger.

Felis chaus (jungle cat). 144 specimens examined.

(i) ♂. BMNH 1915.11.1.67. A small tooth distal to the left I_3.

(ii) ♂. BMNH 1932.2.1.6. A supplemental right P^2.

(iii) ♂. BMNH 1932.2.1.2. An extra, conical single-rooted tooth mesial to the right P^2. Both P^2 are larger than usual and have two roots, whereas P^2 in felids is usually single-rooted.

(iv) P-C. Mus., K.9. A supplemental right P^2. On the left there is an empty socket for a supernumerary tooth mesial to P^2.

(v) ♂. BMNH 1913.19.18.36. P^2 duplicated on both sides; the distal ones are the larger.

Felis lynx (lynx). 44 specimens examined.

(i) Captive. BMNH 1855.12.26.176. A supernumerary haplodont tooth between I^3 and C on each side.

(ii) Four incisors in each premaxilla (Taylor, 1965).

(iii) USNM 135631. Four incisors in the right mandible (Manville, 1963).

(iv) ♂. ZMB. Four incisors in each premaxilla (Herold, 1956).

Felis rufa (bobcat). No specimens examined by Colyer.

(i) USNM 232474. Eight upper incisors (Manville 1963).

(ii) USNM 44606. Extra tooth (P^1) on both sides (Manville, 1963).

(iii) USNM 203900. Supplemental P^2 on both sides (Manville, 1963).

(iv) USNM 214965. A supplemental left M^2 (Manville, 1963).

(v) USNM 221919. A supernumerary cusp on the buccal aspect of the right P^4 (Manville, 1963).

Felis marmorata (marbled cat).

(i) Captive. BMNH 1855.12.24.254. (Fig. 4.5). The distal part of the left P^4 is normal but the mesial part is malformed, principally because there are two cusps in place of the paracone.

Felis bengalensis (leopard cat). 41 specimens examined.

(i) Captive. RCS Odonto. Mus., G 16.321. Supplemental P^2 on both sides.

Prionailurus chinensis (Chinese leopard cat). 4 specimens examined.

(i) ♀. BMNH 1908.2.26.20. A premolariform supernumerary between left P_3 and C.

Felis viverrina (fishing cat). 19 specimens examined.

(i) ♂. RCS Odonto. Mus., G 16.3. Supplemental right P^2.

Fig. 4.5. *Felis marmorata* (marbled cat). Captive. BMNH 1855.12.24.254. The left P^4 has two mesial cusps instead of one. Part of the mesial part of the tooth has been destroyed by caries.

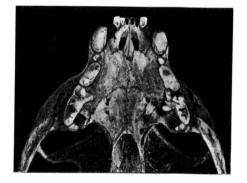

Felis pardalis (ocelot). 65 specimens examined.

(i) ♂. BMNH 1904.12.4.5. A tiny two-rooted premolariform extra tooth between P_3 and C on both sides.

(ii) BMNH 1954.6.25.1. Left I^1 absent; I^2 and I^3 larger on the left than on the right. The left I_1 is also absent, but there are signs that it was lost during life so that it is possible that the left I^1 and I_1 were lost from a common injury.

Felis wiedii (Margay cat). 18 specimens examined.

(i) ♂. BMNH 1908.5.9.16. A small single-rooted haplodont extra tooth between the right P_3 and C.

(ii) ♀. BMNH 1926.1.12.5. A minute haplodont extra tooth distal to the left M^1.

Felis yagouaroundi (jaguarundi). 34 specimens examined.

(i) BMNH 1875.2.25.1. Two premolariform teeth in place of the right P^2, the distal being the larger; neither is as large as the contralateral P^2.

(ii) NRM 343. A left mandibular disto-molar (Werdelin, 1987).

Felis geoffroyi (Geoffroy's cat). 15 specimens examined.

(i) ♂. BMNH 1909.12.1.9. There are two single-rooted teeth side by side in place of the right P^2. The contralateral P^2 is larger than either and has two roots.

Felis geoffroyi salinarum. 6 specimens examined.

(i) ♂. Captive. RCS Odonto. Mus., G 16.3. Two teeth in the region of the right P^2; the distal is the larger and neither tooth is exactly opposite the left P^2.

Felis concolor (puma). No specimens examined by Colyer.

(i) MVZUC 4919. An extra tooth in the region of the left P^3 (Hall, 1940).

Family **Hyaenidae**

$$\frac{3}{3}:\frac{1}{1}:\frac{4}{3}:\frac{1}{1} = 34$$

In hyaenas, all the cheek teeth including the carnassial are large and well developed for crushing bones. An extra tooth between P_2 and the canine was present in one of the 142 specimens of the spotted hyaena (*Crocuta crocuta*) (RCS Odonto. Mus., G 16.5). M^1 in *Hyaena* is larger in proportion to the rest of the dentition than in the felids and is rarely absent. In *Crocuta*, the M^1 is a smaller tooth and is often absent; for instance, in 51 specimens M^1 was absent on both sides in 27, and absent on one side in four.

Family **Viverridae**

$$\frac{3}{3}:\frac{1}{1}:\frac{3-4}{3-4}:\frac{1-2}{1-2} = 36-40$$

In civets, genets and mongooses the premolars are small, the upper carnassial (P^4) lacks the mesial lobe, but the lower carnassial (M_1) has a well-developed talonid and the other molars are large. Eighteen specimens were examined in the sub-families Hemigalinae, Galidiinae and Cryptoproctinae; no extra teeth were present.

Sub-family **Viverrinae**

$$\frac{3}{3}:\frac{1}{1}:\frac{4}{4}:\frac{2}{2} = 40$$

Genus *Viverra*. 87 specimens examined.

(i) *Viverra civetta* (African civet). ♀. Captive. RCS Odonto. Mus., G 16.42. A supplemental I_3.

(ii) *V. civetta*. Captive. RCS Odonto. Mus., G 16.41. There are two teeth in place of the right P_1; the distal one matches the left P_1; the mesial one is a small, single-rooted supernumerary between the right P_1 and C; there is also a minute empty socket in the same position on the left indicating that the abnormality had been bilateral.

(iii) *V. civetta*. BMNH 1910.6.19.1. Between the right P_4 and M_1, there is a large tooth with a globular, approximately molariform crown. On the left in the same situation, there is a supernumerary tooth which resembles a premolar.

(iv) As shown by the examples in Figures 4.6 and 4.7, the size, shape and number of roots may vary between right and left sides. This was so in four specimens of African civet. However, it is not established that this variability is greater than in carnivores generally.

Genus *Viverricula*. 101 specimens examined

(i) *Viverricula indica* (small Indian civet). BMNH 1884.6.3.12. An extra tooth distal to the right M_2.

68 Section 1: Variations in Number, Size and Shape

(ii) *Viverricula indica*. BMNH 140e. In place of the left P_2, there are two teeth similar in morphology to the right P_2.

Genus ***Genetta***. 196 specimens examined

(i) *Genetta genetta* (common genet). RCS Odonto. Mus., G 21.1. In place of the left P^2, there are two teeth of which the distal resembles the contralateral P^2; the mesial one is smaller.
(ii) *Genetta tigrina stuhlmanni* (large spotted genet). ♂. BMNH 1921.3.2.5. In place of the left P^1, there are two teeth, the mesial of which resembles the contralateral P^1. The distal one is larger.

Sub-family **Paradoxurinae**

$$\frac{3}{3}:\frac{1}{1}:\frac{4}{4}:\frac{2}{2}=40$$

Genus ***Paradoxurus***. 180 specimens examined

(i) *Paradoxurus hermaphroditus* (common palm civet). BMNH 1845.4.21.11. An empty socket shows that there had been an extra tooth distal to the left M^2.
(ii) *Paradoxurus* sp. BMNH 1036e. The right M^2 is absent.
(iii) *P. hermaphroditus minor* (small palm civet). ♀. BMNH 1908.7.20.16. The crown of the right M_3 shows evidence of connation with a supernumerary element.
(iv) *P. hermaphroditus*. ♀. BMNH 1879.11.21.279. The left M_2 is small and conical.

Sub-family **Herpestinae**

Genus ***Suricata***. 39 specimens examined

$$\frac{3}{3}:\frac{1}{1}:\frac{3}{3}:\frac{2}{2}=36$$

(i) *Suricata suricatta lophurus* (grey meerkat). ♂. BMNH 1897.11.5.11. Buried in the bone distal to the right I^3 is an extra tooth of similar morphology.
(ii) *S. suricatta hahni* (Hahn's grey meerkat). ♂. BMNH 1926.12.7.57. Bilateral upper disto-molars.
(iii) *Suricata* sp. (meerkat). Kongelingen Naturalienkabinetts Stuttgart, 5938. Instead of the normal four cusps, the right M^1 has seven and has a strongly-developed mesio-buccal root (Hilzheimer, 1908).

Genus ***Herpestes***. 567 specimens examined

$$\frac{3}{3}:\frac{1}{1}:\frac{3\text{--}4}{3\text{--}4}:\frac{2}{2}=36\text{--}40$$

(i) *Herpestes ichneumon* (Egyptian mongoose).

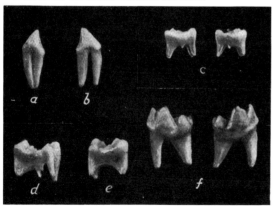

Fig. 4.6. *Viverra civetta* (African civet). Captive. RCS Odonto. Mus., G 16.41. Differences between contralateral pairs of teeth. Buccal surfaces of: (a) the right P^2 with fused roots; (b) normal left tooth; (c) mesial surfaces of M^2; each has a short additional root between the lingual and mesio-buccal roots. (d) and (e) mesial surfaces of M^1. The right tooth (d) has an extra root; the left tooth (e) is normal. (f) Buccal surfaces of M_1; each has an enamel nodule in the bifurcation between the roots.

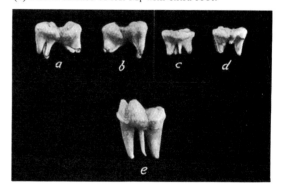

Fig. 4.7. *Viverra civetta* (African civet). Captive. RCS Odonto. Mus., G 16.43. Differences between contralateral pairs of teeth. Distal surfaces of (a) right M_1 with one extra distal root; there is another (not shown) between the buccal roots; (b) normal left tooth.
(c) and (d) distal surfaces of M^2. The right tooth (c) has an extra root; the left tooth (d) has a slight prominence in the same position; the lingual root is bifurcated (not shown).
(e) buccal surface of left M_1 with extra root.

♂. RMNH 34478. An extra tooth between P^3 and P^4 on both sides.
(ii) *H. microcephalus*. BMNH 34479. The right P^1 is duplicated.
(iii) *H. ichneumon*. Captive. RCS Odonto. Mus., G 16.47. In place of the left P^1, there are two teeth both smaller than the right P^1.
(iv) *H. pulverulentus pulverulentus* (Cape grey mongoose). ♀. BMNH 1895.9.3.8. A supplemental left I_1.
(v) *H. pulverulentus ruddi* (Cape grey mongoose). ♂. BMNH 1904.2.3.29. A small right upper disto-molar.
(vi) *H. sanguineus* (slender mongoose). ♀. BMNH 1923.5.9.42. Two extra mandibular incisors, one lingual to the left I_2 and I_3, the other lingual to the right I_1 and I_2.
(vii) *H. sanguineus*. BMNH 1848.11.10.7. Seven mandibular incisors, the extra one probably being in the midline.
(viii) *H. sanguineus erongensis*. BMNH 1863.7.7.18. A supplemental right P_3.
(ix) *H. sanguineus erongensis*. BMNH 1898.3.9.3. Both P_3 absent.
(x) *H. sanguineus ornatus*. BMNH 1869.2.2.6. A small disto-molar in the left mandible.

Genus *Helogale*. 87 specimens examined

$$\frac{3}{3}:\frac{1}{1}:\frac{3}{3}:\frac{2}{2}=36$$

(i) *Helogale parvula* (dwarf mongoose). BMNH 1916.1.15.8. A supplemental right I^2.

Fig. 4.8. *Atilax paludinosus* (marsh mongoose). BMNH 1903.11.7.10. Bilateral supplemental P_4 but the right one is now lost.

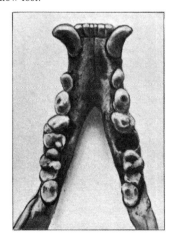

Genus *Atilax*. 71 specimens examined

$$\frac{3}{3}:\frac{1}{1}:\frac{3}{3}:\frac{2}{2}=36$$

(i) *Atilax* shows a considerable degree of variation in the number of upper and lower premolars.
(ii) *Atilax paludinosus* (marsh mongoose). BMNH 1903.11.7.10. (Fig. 4.8). There was a supplemental P_4 on each side although one is now lost.

Genus *Mungos*. 31 specimens examined

$$\frac{3}{3}:\frac{1}{1}:\frac{3}{3}:\frac{2}{2}=36$$

(i) Variations of the maxillary second molars. In *Mungos*, M^2 has a triangular crown with one lingual and two buccal cusps; in eight skulls, it had an extra cusp (parastyle) at the mesio-bucal border and in one skull both M^1 and M^2 had additional cusps (Fig. 4.9). Two other skulls also had abnormally-shaped M^2.
(ii) *Mungos mungo* (banded mongoose). BMNH 1873.2.24.18. Bilateral maxillary disto-molars.

Genus *Crossarchus*. Number of specimens examined by Colyer not recorded

In *Crossarchus*, M^2 has a triangular crown with one lingual and two buccal cusps; in one skull, it had an extra cusp (parastyle) at the mesio-buccal border (Fig. 4.10). Figure 4.11 shows a *Crossarchus* skull in which the occlusal surface of the right M^2 is crossed by a fissure which, towards the lingual aspect, is well marked and runs down the root almost dividing it into

Fig. 4.9. *Mungos mungo* (banded mongoose). BMNH 1892.10.28.9. Each M^1 and M^2 has an extra cusp on the buccal aspect. $\times c.\ 2.0$.

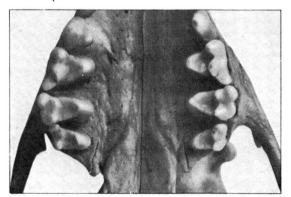

two. This tooth appears to be an example of connation (see Chapter 25, pp. 586–587).

Genus ***Ichneumia***. 62 specimens examined

$$\frac{3}{3}:\frac{1}{1}:\frac{4}{4}:\frac{2}{2}=40$$

(i) *Ichneumia albicauda* (white-tailed mongoose). ♂. BMNH 1927.8.14.5. (Fig. 4.12). Two conical teeth in place of the left P^1 (Fig. 4.12A). The right P^1 has a bifid crown (Fig. 4.12B); this is an example of formation of two separate teeth on one side with connation on the other.

Genus ***Rhynchogale***. 8 specimens examined

$$\frac{3}{3}:\frac{1}{1}:\frac{4}{4}:\frac{2}{2}=40$$

(i) *Rhynchogale melleri* (Meller's mongoose). BMNH 1437a. Two teeth in place of each P^1, the distal ones being the smaller.

Fig. 4.10. *Crossarchus obscurus* (long-nosed mongoose). BMNH 1871.7.8.4. Each M^1 and M^2 has an extra cusp on the buccal aspect. ×c. 2.1.

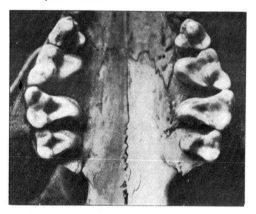

Fig. 4.11. *Crossarchus obscurus* (long-nosed mongoose). ♂. BMNH 1911.11.3.10. The right M^2 is divided by an abnormal fissure into mesial and distal parts. ×c. 1.25.

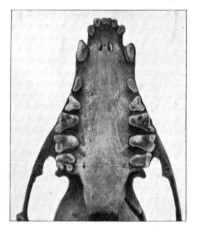

Fig. 4.12. *Ichneumia albicauda* (white-tailed mongoose). ♂. BMNH 1927.8.14.5
A: left side showing two conical teeth in place of the P^1.

B: right side showing bifid P^1 and indications of connation or dichotomy.

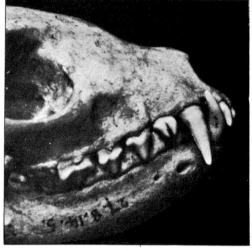

A

B

Genus **Cynictis**. 53 specimens examined

$$\frac{3}{3}:\frac{1}{1}:\frac{3}{3}:\frac{2}{2} = 36$$

(i) *Cynictis penicillata* (yellow mongoose). ♂. BMNH 1897.11.5.6. An extra tooth lingual to the right P^2 and P^3, resembling P^3 but smaller.

Family **Mustelidae**

In badgers, otters, skunks and weasels most of the premolars are small but the carnassials P^4 and M_1 are well developed.

Marshall (1952) studied two series of skulls of *Martes americana* (American marten), 107 from central Idaho and 97 from Western Montana. No molars were absent but first premolars and lower incisors were frequently absent (Table 4.2); in addition, the P^1 or P_1, or both, of 18 of the Idaho skulls were smaller than normal. In these populations at least, it seems that there is a tendency for disappearance of P1, particularly in the lower jaw. However, in 220 skulls of pine marten (*M. martes*) from Poland, Wolsan (1984b) found no absent premolars. Two skulls (0.9%) lacked a lower incisor, one an upper molar and one a lower canine; in one (♂, MRI 66633), the right I^1 and I^2 were connate. Amongst 43 skulls of beech marten (*M. foina*), Wolsan, Ruprecht and Buchalczyk (1985) found one P^1 absent in one or both P_1 absent in eight. There were no extra teeth in these beech martens nor in pine martens mentioned above.

Hall (1940) found no extra teeth in North American mustelids, but teeth were absent in as many as 5.1% of 257 *Mephitis mephitis* (striped skunks).

Bateman (1970) examined 936 mustelids in various museums in Britain; 270 were of *Mustela* spp.

In the sample as a whole, there were 38 skulls with supernumerary incisors, of which 36 were *Mustela putorius*. This taxon consists of the wild polecat, the domestic albino ferret (*M. putorius furo*) and hybrids between them. Bateman found that all 36 skulls with supernumerary incisors were of albino ferrets or hybrids. Although these data cannot be quantified on a percentage basis, they do point to the frequent occurrence of extra incisors in ferrets and their hybrids, in contrast with the incidence of extra incisors in only 1.9% of 801 skulls of polecats in the wild in Poland found by Ruprecht (1978c). As Berkowitz (1968b) has shown, albino ferrets have the potential for eight deciduous incisors, sometimes succeeded by only seven permanent incisors; extra incisors in polecats may thus indicate hybridization with the domestic form. Berkovitz and Thomson (1973) observed 13 permanent supernumerary incisors in the I2 position in albino ferrets, all of which were associated with corresponding deciduous supernumeraries; seven of them showed various degrees of fusion with the adjacent I1, suggesting that the deciduous supernumerary arose as the result of the dichotomy of i1 and that the permanent supernumerary may be lost as a result of fusion with I1 during development.

Dental variations are common in other species of *Mustela*; among 93 weasels (*Mustela nivalis*) from W. Germany, Neuenschwander and Lüps (1975) found that 6.4% had supernumerary premolars or molars, 4.3% showed variations in the shape of P_2 and M_2; P_2 or M_3 were absent in 6.4%.

Fabian (1933, p. 38) illustrated connation in a ferret (*Mustela putorius furo*) between two supernumerary teeth. They were situated between the right and left I^1, so the connation was across the midline.

Van Gelder and McLaughlin (1961) described a

Table 4.2 Number and location of missing incisors and premolars in 204 AMERICAN MARTEN *(Martes americana)* skulls (Marshall, 1952)

Tooth type	Incisors			Premolars		
	1 R L	2 R L	3 R L	1 R L	2 R L	3 R L
Upper jaws						
Idaho	0 0	0 0	0 0	4 4	1 0	0 0
Montana	0 0	0 0	0 0	4 4	0 0	0 0
Lower jaws						
Idaho	2 1	1 1	0 0	7 9	0 2	0 0
Montana	1 5	2 1	1 0	11 14	0 0	0 0

connate tooth in a skunk (*Mephitis mephitis*, Los Angeles County Museum 8428) in which the roots of the right I^1 and I^2 were fused, whereas the crowns were partially separated.

Sub-family **Mustelinae**

Genus *Martes*. 209 specimens examined

$$\frac{3}{3}:\frac{1}{1}:\frac{4}{4}:\frac{1}{2}=38$$

(i) *Martes foina* (beech marten). ♀. BMNH 1919.7.7.2551. Four lower left incisors.
(ii) *M. martes* (pine marten). Formerly RCS Osteo. Series, 681. All four P1 absent.
(iii) *M. gwatkinsi* (south Indian yellow-throated marten). BMNH 1879.11.21.621. All four P1 absent.

Genus *Mustela*. 593 specimens examined

$$\frac{3}{3}:\frac{1}{1}:\frac{3}{3}:\frac{1}{2}=34$$

(i) *Mustela erminea cicognani* (N. American stoat). ♂. BMNH 1902.4.2.7. Four lower right incisors.
(ii) *M. nivalis* (weasel). Zool. Mus. Univ., Oslo, M-1443. Both M_2 have three well-developed cusps giving them a carnassial appearance (Mazak, 1964).
(iii) *M. nivalis*. RCS Odonto. Mus., G 24.3. A small tooth distal to the right M^1.
(iv) *M. nivalis*. ♂. BMNH 1911.1.3.205. Left M_2 absent.
(v) *M. erminea* (stoat). ♂. MRI 30731. In place of I_2, a tooth with a connate crown; the distal part had the size and shape of I_2; the mesial part was smaller but incisiform (Wolsan, 1984a).
(vi) *M. putorius* (polecat). Formerly RCS Osteo. Series 4230b. All the teeth have been lost, but there is a socket for an extra tooth adjacent to the left I^1.
(vii) *M. putorius furo* (ferret). Formerly RCS Osteo. Series, 702. An extra tooth resembling I^1 (though smaller) distal to I^2.
(viii) *M. putorius* (polecat). RCS Odonto. Mus., G 24.4. An extra incisiform tooth mesial to the right I^1.
(ix) *M. putorius furo* (ferret). BMNH 1856.12.10.704. (Fig. 4.13). A supplemental right I^1. The left I^1 is unusually large and its crown consists of two elements which are quite separate for the incisive half; the division continues as longitudinal labial and lingual grooves on to the root. This is an example of connation (see also Bateman, 1970).
(x) *M. putorius* (polecat) – possibly *M.p. furo*, the ferret. BMNH 1839.17.15.2. The left I^1 is unusually large with longitudinal grooves labially and lingually and a bifid incisive edge; an example of connation.
(xi) *M. putorius furo* (ferret). Formerly RCS Osteo. Series, 707. The crown of the right I^1 has a longitudinal groove labially, suggestive of dichotomy or connation.
(xii) *M. putorius furo* (ferret). ♂. BMNH 1935.2.18.1. Supplemental right I^1.
(xiii) *M. putorius* (polecat). BMNH 192m. Left I^1 is larger than the contralateral and has longitudinal grooves suggestive of dichotomy or connation.
(xiv) *M. putorius furo* (ferret). ♂. BMNH 1932.5.9.1. A supplemental right I^1.
(xv) *M. putorius* (polecat/ferret hybrid). Nat. Mus. Wales, 63.443. Four upper left incisors (Bateman, 1970).
(xvi) *M. putorius furo* (albino ferret). K.C. Walton, Aberystwyth, Private Collection, FM 3. Crown of left I^1 bifurcated (Bateman, 1970).
(xvii) *M. putorius* (polecat/ferret hybrid). K.C. Walton, Aberystwyth, Private Collection FM 4. Crown of right I^1 bifurcated (Bateman, 1970).
(xviii) *M. putorius furo* (ferret). Seven upper incisors (Reinwaldt, 1958).
(xix) *M. putorius furo* (ferret). Eight upper incisors (Reinwaldt, 1958).
(xx) *M. putorius* (polecat). A supplemental left P_3 (Ruprecht, 1965b).
(xxi) *M. steppe eversmanni* (polecat). BMNH 1867.4.12.458. A supplemental left I^2.
(xxii) *M. vison* (domestic mink). ♀. Only four mandibular incisors (Reinwaldt, 1958).
(xxiii) *M. vison* (mink). On both sides in place of the normal peg-shaped M_2, there is a tooth of carnassial morphology (Parmalee and Bogan, 1977).

Genus *Vormela*. 17 specimens examined

$$\frac{3}{3}:\frac{1}{1}:\frac{3}{3}:\frac{1}{2}=34$$

(i) *Vormela peregusna* (marbled polecat). BMNH 1908.4.10.1. In place of the right I_1, there are two minute conical teeth.

Genus *Eira*. 61 specimens examined

$$\frac{3}{3}:\frac{1}{1}:\frac{3}{3}:\frac{1}{2}=34$$

(i) *Eira barbara* (tayra). ♂. BMNH 1901.2.1.2. Left P^2 absent.
(ii) *E. barbara*. An extra tooth adjacent to P^2 on both sides.

Genus **Galictis**. 60 specimens examined

$$\frac{3}{3}:\frac{1}{1}:\frac{3}{3}:\frac{1}{2}=34$$

(i) *Galictis vittata* (greater grison). ♂. BMNH 1904.12.4.8. (Fig. 4.14). Lingual to the right maxillary canine, there is another tooth which is caniniform, but much smaller than the true canines; it is rotated so that its lingual surface faces the lingual surface of the canine.
(ii) *G. vittata*. BMNH 1911.6.7.23. Tiny bilateral maxillary disto-molars.
(iii) *G. cuja* (little grison). ♂. BMNH 1905.2.4.4. Both P_2 absent and no space for them in the arch.
(iv) *G. cuja*. ♂. BMNH 1912.7.12.2. P^2 and P_2 absent on both sides.

Sub-family **Mephitinae**

Genus **Conepatus**. 167 specimens examined

$$\frac{3}{3}:\frac{1}{1}:\frac{2}{3}:\frac{1}{2}=32$$

(i) *Conepatus semistriatus* (striped hog-nosed skunk). BMNH 1847.4.20.14. Four left upper incisors. The extra tooth is probably a supplemental I^1.
(ii) *C. semistriatus*. ♀. BMNH 1903.3.3.19. Left P_2 absent.
(iii) *Conepatus* sp. In four specimens an extra tooth was present mesial to P^3 on both sides.

Fig. 4.13. *Mustela putorius furo* (ferret). BMNH 1856.12.10.704. Supplemental right I^1 and connate left I^1. ×*c*. 2.0.

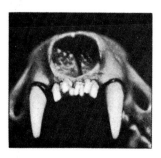

Sub-family **Melinae**

Genus **Meles**. 95 specimens examined

$$\frac{3}{3}:\frac{1}{1}:\frac{4}{4}:\frac{1}{2}=38$$

Meles meles (badger). The upper first premolar is so often absent that Colyer considered the normal premolar formula to be $\frac{3}{4}$. Among 95 badger skulls, 70% lacked one or both P^1 and P_1 was absent in 6%. Corresponding figures given by Heran (1971) for European badgers were 40% lacking P^1 and 30% P_1, but the incidence of P1 seems to vary from place to place, for example (Ratcliffe, 1970) found that in 21 badgers from Argyll only 24% lacked P^1 and 12% P_1. Fullagar, Rogers and Mansfield (1960) described a skull in the Chelsea College of Science and Technology in which both M^1 were unusually small and both M_2 were exceptionally large.

Among 1050 badger skulls from various parts of Europe and Asia, Hancox (1988) found important regional variations but the absence of P1 was consistently common. There was also instability of P2, for instance in having one, two or three roots. In two skulls, bilateral supernumerary maxillary molars, or disto-molars were present.

Sub-family **Mellivorinae**

Genus **Mellivora**. 52 specimens examined.

$$\frac{3}{3}:\frac{1}{1}:\frac{3}{3}:\frac{1}{1}=32$$

(i) *Mellivora capensis* (honey badger). ♂. Odonto. Mus., G 24.5. An additional lower left premolar similar in shape to P_2 but less well formed (Colyer 1945b, Fig. 6).

Fig. 4.14. *Galictis vittata* (greater grison). ♂. BMNH 1904.12.4.8. Palatal and right buccal views. Supernumerary caniniform tooth linguo-distal to the right maxillary canine. ×*c*. 2.0.

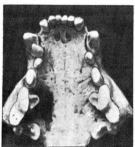 

Colyer (1945b) figured an extra lower premolar in the skull of a honey badger, *Mellivora capensis*, from East Africa.

Genus *Melogale*. 23 specimens examined

$$\frac{3}{3}\cdot\frac{1}{1}\cdot\frac{4}{4}-\cdot\frac{1}{2}=38$$

(i) *Melogale orientalis* (Burmese ferret-badger). BMNH 824a. An extra tooth, similar in size to P^2 and P^3.

Genus *Taxidea*. 18 specimens examined

$$\frac{3}{3}\cdot\frac{1}{1}\cdot\frac{3}{3}\cdot\frac{1}{2}=34$$

Colyer's sample of American badger (*Taxidea taxus*) showed no dental variations but they do occur occasionally. Among 900 skulls (mostly in the USNM), Long and Long (1965) found extra premolars in four (0.4%), absent premolars in 11 (1.2%) and an absent molar in one (0.1%). Hall (1940) also noted a bilateral extra premolar in a mandible of this species (MVZUC 41500).

No variations were seen in the following genera: *Mephitis* (24), *Spilogale* (11), *Poecilogale* (21), *Ictonyx* (74), *Poecilictis* (22), *Arctonyx* (19) and *Mydaus* (29).

Sub-family **Lutrinae**

Genus *Lutra*. 183 specimens examined

$$\frac{3}{3}\cdot\frac{1}{1}\cdot\frac{4}{3}\cdot\frac{1}{2}=36$$

The maxillary first premolar is a small tooth but more regularly present than in *Meles*. There were two examples of absence of P^1 in 63 *Lutra lutra* (European otter): in *Lutra perspicillata* (smooth-coated otter), no P^1 were absent in 40 specimens but, in *Lutra felina* (South American otter), one or both P^1 were absent in 7 out of 168 specimens.

In 202 skulls of the Canadian otter (*L. canadiensis*) from the Chesapeake Bay region of Maryland and Virginia, U.S.A., Beaver, Feldhamer and Chapman (1981) found five lacking a P^1 and one lacking one P^2 and lacking both P^2.

Both lower first premolars were absent in one *Lutra canadensis* from Chesapeake Bay (Beaver, Feldhamer and Chapman, 1981) and in another (University of Massachusetts, Museum of Vertebrate Zoology, 2321), the second premolars were small mesio-distally (Dearden, 1954).

Two skulls of *Lutra lutra* from the Peleponnese in Greece had small additional teeth (Douma-Petridou, 1984).

Genus *Amblonyx*. 24 specimens examined

$$\frac{3}{3}\cdot\frac{1}{1}\cdot\frac{3}{3}\cdot\frac{1}{2}=34$$

Although P^1 is normally absent in the oriental small-clawed otter (*Amblonyx*), in four skulls it was present on both sides and in one on the left. One skull (BMNH 1909.1.5.664) had a supernumerary premolar intermediate in size between the right P^2 and P^3 and situated between them (Fig. 4.15).

Genus *Aonyx*. 11 specimens examined

$$\frac{3}{3}\cdot\frac{1}{1}\cdot\frac{4}{3}\cdot\frac{1}{2}=36$$

In one skull of the African clawless otter (*Aonyx capensis*) (BMNH 836b), both P^1 were absent.

There were no variations in the 25 *Enhydra* or 3 *Pteronura* examined.

Family **Procyonidae**

$$\frac{3}{3}\cdot\frac{1}{1}\cdot\frac{3-4}{3-4}\cdot\frac{2}{2}=36-40$$

In the racoons and their allies, the premolars are small and pointed, the carnassials are weakly developed and the molars are broad and somewhat bunodont.

Genus *Bassaricus*. 10 specimens examined

Colyer found no dental variations in *Bassaricus*, but Hall (1940) described a skull of *Bassaricus astutus*

Fig. 4.15. *Amblyonyx cinerea* (oriental small-clawed otter). BMNH 1909.1.5.664. An extra tooth between the right P^2 and P^3.

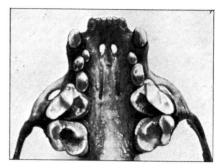

raptor (ring-tailed cat, MVZUC 21561) in which there was a supplemental left P_1.

Genus ***Procyon***. 77 specimens examined

(i) *Procyon lotor* (common racoon). Formerly Høgskola, Stockholm, 4972. Eight mandibular incisors.
(ii) *P. cancrivorous* (crab-eating racoon). Captive. RCS Odonto. Mus., G 24.2 (Fig. 4.16). A supplemental left P^3.

Genus ***Nasua***. 233 specimens examined.

(i) *Nasua* sp. (coati). ♂. BMNH 1903.7.1.19. In place of the left P^1, there are two teeth. The distal one resembles the right P^1 in size and shape, in particular in being two-rooted. The mesial tooth is smaller and single-rooted and must therefore be assumed to be the supernumerary.
(ii) *Nasua* sp. ♀. BMNH 1926.1.2.4. (Fig. 4.17). On the right, there is a supernumerary tooth lingual to P_4 and resembling it, although it is smaller and has three roots, whereas P_4 has two (Fig. 4.17 B and C); furthermore, the buccal surface of the supernumerary resembles the lingual surface of P_4 to which it is adjacent; the lingual surface resembles the buccal surface of P_4. In other words, the two teeth are to some extent mirror images morphologically. This is of special interest to those who might regard these two teeth as the product of dichotomy of a single tooth primordium.

On the left, P_4 is malformed; on the lingual aspect of the crown, there is an additional tooth-like element partially separated from the main tooth by longitudinal grooves suggestive of connation (Fig. 4.17D). The specimen shows a feature common in developmental anomalies, a tendency for both sides of the body to be affected, though not always equally.
(iii) *Nasua nasua* (southern coati). ♂. MPEG 4321. In place of the left P_1, there are two teeth; the mesial tooth is much smaller than the distal. Unfortunately the contralateral P_1 has been lost.

Fig. 4.16. *Procyon cancrivorous* (crab-eating raccoon). RCS Odonto. Mus., G 24.2. Extra tooth between the left P^3 and P^4. ×c. 0.6.

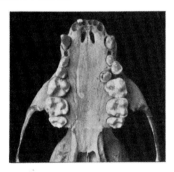

Fig. 4.17. *Nasua* sp. ♀. BMNH 1926.1.2.4.
A: bilateral anomalous development in P_4 region. On the right, there is a supplemental P_4; on the left, P_4 has additional elements.
B: the supplemental lingually-situated right P_4, lingual aspects and buccal aspects. ×c. 2.0.
C: corresponding aspects of the buccally-situated right P_4. ×c. 2.0.
D: left P_4, buccal and occlusal aspects. ×c. 2.0.

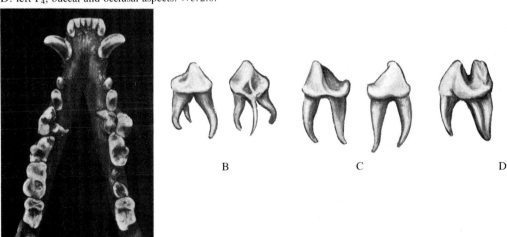

(iv) *Nasua* sp. ♂. BMNH 1882.9.30.11. (Fig. 4.18). Between the right P_4 and M_1, there is an extra tooth with a crown of cuboidal shape which does not resemble the deciduous molars in this species.

(v) *Nasua nasua*. NRM. An extra tooth distal to right M_1.

(vi) *N. nelsoni* (Cozumel coati). BMNH 1886.10.8.1. (Fig. 4.19). The left P^2 shows features of connation; its distal part resembles the right P^2 in size and shape and is separated from the smaller mesial part by deep longitudinal grooves which become a complete cleft occlusally. The mesial part has a separate root.

(vii) *N. nasua* (southern coati). ♀. BMNH 1903.7.7.54. Both P_1 absent.

(viii) *N. nasua*. ♂. BMNH 1903.3.3.27. Both M^2 absent.

(ix) *N. nasua*. Captive. Formerly MPEG 33. The left P_3 is not fully erupted and is wedged between the adjacent teeth; the crown is longitudinally grooved, suggesting connation or dichotomy. The right M^2 has a wart-like projection on the distal aspect of the crown and the left M^2 is abnormal by virtue of the occlusal surface having no well-defined cusps.

Dr G. Hagmann informed Colyer that in his private collection, now in the Museu Paranese Emilio Goeldi, Belem, Brazil, there were the following variations in 74 specimens of *Nasua nasua*:

(i) The left P_1 was a double tooth (connation).
(ii) The left P^3 was double with two cusps and a small accessory root on the lingual side. The right P^3 was normal.
(iii) The left M_1 had an accessory buccal root.

Genus *Potos*. 55 specimens examined

(i) *Potos flavus* (kinkajou). ♂. BMNH 1905.7.5.6. Right M^2 absent.

Genus *Ailurus*. 21 specimens examined

(i) *Ailurus fulgens* (red panda). In two out of 21 skulls, both P_1 were absent and there was no space for them in the arch; in three others, both P_1 were probably absent but there was space for them and the evidence of absence was not conclusive.

There were no dental variations among 17 specimens of *Nasuella*.

Fig. 4.18. *Nasua* sp. ♂. BMNH 1882.9.30.11. Extra tooth (arrow) between the right P_4 and M_1.

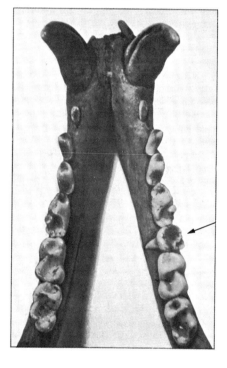

Fig. 4.19. *Nasua nelsoni* (Cozumel coati). BMNH 1886.10.8.1. A: buccal view of the left upper premolars; B: lingual view. P^2 is larger than normal and is bifid. The smaller mesial part has a separate root. ×c. 2.0.

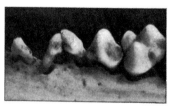

A   B

Family **Ursidae**

$$\frac{3}{3} \cdot \frac{1}{1} \cdot \frac{4}{4} \cdot \frac{2}{3} = 42$$

Most bears are omnivorous and their dentition is specialized for crushing food. The carnassials are reduced in size, P^4 having rounder cusps and smaller cusps than in canids and M^1 and M^2 are expanded and bunodont.

In his sample of 159 skulls of American black bear (Ursus americanus), Rausch (1961) found two specimens with additional upper incisors, but no extra canines, premolars or molars. There does seem to be a definite tendency for the elimination of P^2 and P_2 in *Ursus americanus*. Rausch (1961) found that only three (2%) had the full complement of premolars and analysis of his figures shows that in 149 cases (93.7%) one or more second premolars were absent. He found no difference in the loss of premolars in animals of different ages and so concluded that the absence was congenital. There was one instance of a missing incisor and one of an absent right M_3.

Genus **Ursus**. 144 specimens examined

(i) *Ursus arctos* (brown bear). Formerly Høgskola, Stockholm, 5908. (Fig. 4.20). Supplemental right I^2.
(ii) *Ursus arctos*. Göteborg Naturhistoriska Museet, Coll. an. 4583. An extra tooth between the left I^1 and I^2, more similar to I^1 than to I^2 in shape.
(iii) *Ursus arctos isabellinus* (Isabelline bear). RCS Odonto. Mus., G 25.1. Four incisors on each side of the mandible. On the left, there are two teeth of I_3 morphology, the mesial one being slightly the smaller. On the right, there is an I_3 next to the canine, but it is larger than either of the corresponding left teeth. Mesially to it is a much smaller tooth, slightly more similar to I_3 in morphology than I_2.

This is an example of developmental anomalies tending to be bilateral although not always similar in degree. However, here there is an excessive formation of teeth in number at the distal end of the incisor tooth field and it can be said to be similar in quantity because the mass of tooth tissue formed on the left, that is the combined mass of two similar-sized I_3, is about the same as the mass on the right where one tooth is much smaller than the other.

(iv) *Ursus arctos sheldoni* (grizzly bear). MVZUC 970. Bilateral extra premolars immediately mesial to P^4 (Hall, 1928).
(v) *Ursus americanus emmonsi* (American black bear). MVZUC 4721. An additional tooth in right mandible mesial to M_1. P_2 absent in both jaws (Hall, 1928).
(vi) *Ursus arctos gyas* (grizzly bear). MVZUC 4386. Disto-molars in both maxillae (Hall, 1928).

Genus **Selenarctos**. 84 specimens examined

(i) *Selenarctos thibetanus* (Asiatic black bear). BMNH 1891.11.26.1. A premolariform tooth between the right P^2 and P^3.
(ii) *Selenarctos thibetanus*. ♀. BMNH 1926.10.8.40. An extra tooth between the left P^3 and P^4. It has a crown with a single cusp more blade-like than normal, and two divergent roots instead of the usual one; it is larger than the adjacent P^3 but smaller than P^4.
(iii) *Selenarcos thibetanus*. ♀. BMNH 1897.5.6.1. Between P_1 and P_4 on both sides, are three instead of the usual two premolars; the central one is premolariform, the other two are conical. There is an empty socket for a supernumerary between the right P^1 and P^2. Both P^3 are two-rooted and have more blade-like crowns than normal, resembling the extra premolar in the previous specimen (BMNH 1926.10.8.40).
(iv) *Selenarctos thibetanus*. Captive. RCS Odonto. Mus., G 25.11. (Fig. 4.21). Both P^4 absent. The right P_4 has three roots instead of two and shows connation involving an additional element mesially which consists of an additional sectorial cusp associated with an additional mesial root (Fig. 4.21, below). The additional mesial crown element is defined from the rest of the crown by

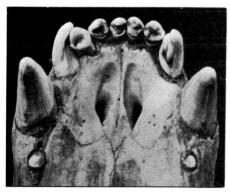

Fig. 4.20. *Ursus arctos* (brown bear). Formerly Högskola, Stockholm, 5908. A supplemental right I^2.

78 Section 1: Variations in Number, Size and Shape

prominent grooves. In place of the right P_2, there are two tiny conical teeth.

Genus *Tremarctos*. 20 specimens examined

(i) *Tremarctos ornatus* (spectacled bear). Formerly Rothschild Museum, 1259. An extra tooth between the left I^3 and the upper canine.
(ii) *Tremarctos ornatus*. Formerly RCS Osteo. Series, 815. An additional premolar in the left maxilla.

Genus *Melursus*. 35 specimens examined

$$\frac{2}{3} \cdot \frac{1}{1} \cdot \frac{4}{4} \cdot \frac{2}{3} = 40$$

In *Melursus*, the first two premolars are single-rooted and so small as to have little function.

(i) *Melursus ursinus* (sloth bear). BMNH 220a. There are three similar teeth in place of P^1 and P^2 on both sides.
(ii) *Melursus ursinus inornatus*. ♂. BMNH 1931.9.24.1. The right P^3 is duplicated. One is situated lingual to the arch and is much worn. The crown of the mandibular right first premolar is much larger than the contralateral tooth and lingual and buccal grooves extending nearly to the cervix divide it into mesial and distal halves. Connation.

VARIATIONS IN ROOT MORPHOLOGY. 20 bear skulls examined

The M^2 in bears normally has four roots, two of which are regarded as buccal and two lingual (Fig. 4.22). The mesio-buccal root is situated close to the mesial margin; the disto-buccal and the mesio-lingual roots are about opposite one another at the middle of the tooth; the disto-lingual root is flatter and extends mesially toward the space between the mesio-lingual and disto-buccal roots but inclining a little towards the disto-buccal. In the triangular space between the

Fig. 4.21. *Selenarctos thibetanus* (Asiatic black bear). RCS Odonto. Mus., G 25.11.
Above left: both P^4 are absent.
Above right: the crown of the right P_4 has an extra mesial sectorial cusp (arrow) associated with an extra root.
Below: buccal (A) and lingual (B) surfaces of right P_4 showing the extra mesial root. $\times c.\,1.5$.

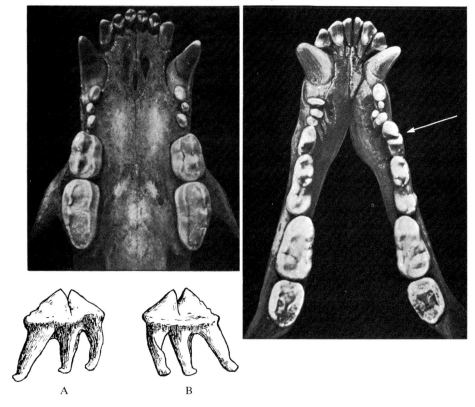

mesio-lingual and two buccal roots, there is sometimes a slight elevation.

(i) *Ursus arctos*. RCS Odonto. Mus., A 171.23. (Fig. 4.23). The disto-lingual root of the left M² extends mesially to fuse with the lingual aspect of the disto-buccal root at the origin of the roots at the base of the crown (Fig. 4.23 C). There is an elevation mesial to the mesio-lingual and mesio-buccal roots. There are two nodules of enamel close to the origin of the roots which sections through the tooth show contain minute cores of dentine projecting from the tooth surface (Fig. 4.23 A). The right M² (Fig. 4.23 B) shows similar enamel nodules and more complete fusion between the disto-buccal and mesio-buccal roots.

(ii) *Ursus americanus*. RCS Odonto. Mus., A 171.33. (Fig. 4.24). Both M² have the mesio-buccal and the mesio-lingual roots united by a web at their bases. The left tooth (Fig. 4.24 B) has two tiny extra roots springing from the disto-lingual aspects of the base of the disto-buccal root. The buccal surfaces of the disto-buccal roots of both teeth are deeply longitudinally grooved (Fig. 4.24 C).

(iii) *Ursus arctos*. RCS Odonto. Mus., G 139.9. (Fig. 4.25). The right M² has a small extra root between the mesio-buccal and mesio-lingual roots; the left M² has two extra roots in the same position.

(iv) *Ursus arctos syriacus*. RCS Odonto. Mus., A 171.28. (Fig. 4.26). The left M² has an extra root between the mesio-buccal and the mesio-

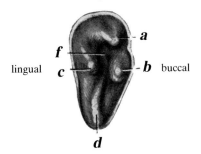

Fig. 4.22. *Ursus* sp. (bear). The normal arrangement of the roots in a right M². (a) mesio-buccal root; (b) disto-buccal root; (c) mesio-lingual root; (d) disto-lingual root; (f) elevation, sometimes present.

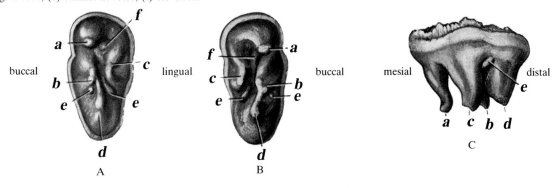

Fig. 4.23. *Ursus arctos* (brown bear). RCS Odonto. Mus., A 171.23. Left and right M² with anomalous roots. (A) root surface of left M², (B) root surface of right M², (C) lingual surface of left M². (a) mesio-buccal root; (b) disto-buccal root; (c) mesio-lingual root; (d) disto-lingual root; (e) enamel nodules; (f) elevation.

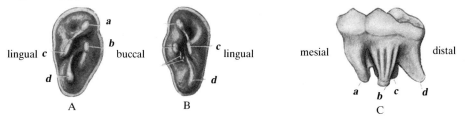

Fig. 4.24. *Ursus americanus* (American black bear). RCS Odonto. Mus., G 171.33. Both M² with anomalous roots. (A) root surface of left M²; (B) root surface of right M²; (C) buccal surface of left M². (a) mesio-buccal root; (b) disto-buccal root; (c) mesio-lingual root; (c) disto-lingual root; (e) extra root.

80 *Section 1: Variations in Number, Size and Shape*

lingual roots and the disto-buccal, mesio-lingual and disto-lingual roots are webbed at their bases. The apex of the mesio-lingual root is bifid.

(v) *Ursus americanus.* RCS Odonto. Mus., A 171.32. (Fig. 4.27). The right M^2 has an extra root between the mesio-buccal and mesio-lingual roots. There is partial fusion of the mesio-lingual and the disto-buccal roots.

(vi) *Selenarctos thibetanus.* RCS Odonto. Mus., A 171.43. (Fig. 4.28). The right M^2 has mesio-buccal, mesio-lingual and disto-buccal roots that are grooved deeply on their lingual aspects. The buccal surface of the disto-buccal root has several longitudinal grooves, one of which is deep enough to separate off to form a short extra root (Fig. 4.28).

(vii) *Ursus americanus.* RCS Odonto. Mus., G 143.32. (Fig. 4.29). Both M^2 have an extra root between the disto-buccal and the disto-

Fig. 4.25. *Ursus arctos* (brown bear). RCS Odonto. Mus., G 139.9. Both M^2 with anomalous roots. (A) lingual surface of right M^2; (B) lingual surface of left M^2. (a) mesio-buccal root; (b) disto-buccal root; (c) mesio-lingual root; (d) disto-lingual root; (e) extra roots.

Fig. 4.27. *Ursus americanus* (American black bear). RCS Odonto. Mus., A 171.32. Right M^2 with anomalous roots, lingual surface. (a) mesio-buccal root; (b) disto-buccal root; (c) mesio-lingual root; (d) disto-lingual root; (e) extra root.

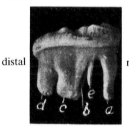

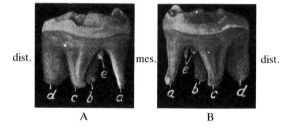

Fig. 4.26. *Ursus arctos syriacus* (Syrian brown bear). RCS Odonto. Mus., A 171.28. Left M^2 with anomalous root, lingual surface. (a) mesio-buccal root; (b) disto-buccal root; (c) mesio-lingual root; (d) disto-lingual root; (e) extra root.

Fig. 4.28. *Selenarctos thibetanus* (Asiatic black bear). RCS Odonto. Mus., A 171.43. Right M^2 with anomalous roots. (A) root surface; (B) buccal surface. The buccal aspect of the disto-buccal root is longitudinally grooved in three places. (a) mesio-buccal root; (b) disto-buccal root; (c) mesio-lingual root; (d) disto-lingual root; (e) extra root.

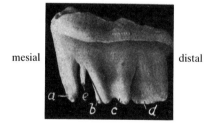

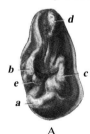
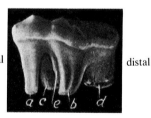

Fig. 4.29. *Ursus americanus* (American black bear). RCS Odonto. Mus., G 132.32. Both M^2 with anomalous roots. (A) buccal surface of left M^2; (B) buccal surface of right M^2; (C) root surface of right M^2. (a) mesio-buccal root; (b) disto-buccal root; (c) mesio-lingual root; (d) disto-lingual root; (e) extra root.

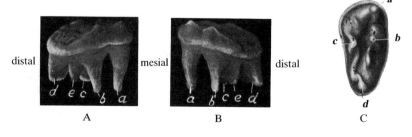

lingual roots fused by a web to the base of the disto-buccal root.

(viii) *Ursus arctos isabellinus*. (Fig. 4.30). Both M² have an extra root in the area between the mesio-lingual and the two buccal roots where a slight elevation referred to above (p. 79) is commonly present. The two lingual roots of the left tooth are fused and the apex of the disto-buccal one is bent towards the mesio-lingual root. On the right, the two lingual and the disto-buccal roots are fused.

(ix) *Ursus arctos*. (Fig. 4.31). The left M² has fused lingual roots; the mesio-lingual and disto-buccal roots are separated only by a groove. The right tooth has the mesio-lingual root fused with the disto-buccal.

(x) *Selenarctos thibetanus*. (Fig. 4.32). Both M² have the disto-buccal root fused with the disto-lingual; in the right tooth, these roots are also fused with the mesio-lingual.

(xi) *Selenarctos thibetanus*. (Fig. 4.33). In both M², the disto-lingual roots are very flattened mesio-distally and present a buccal convexity corresponding to that of the buccal surface of the crown. A web of root runs distally for the disto-buccal roots.

(xii) *Helarctos malayanus*. RCS Odonto. Mus., G 75.3. (Fig. 4.34). In this genus, the lingual roots of M² are usually fused but, in this specimen, they are also united with the mesio-buccal root.

(xiii) *Ursus arctos*. RCS Odonto. Mus., A 171.22.

Fig. 4.30. *Ursus arctos isabellinus* (isabelline bear). Root surfaces of (A) right M²; (B) left M². (a) mesio-buccal root; (b) disto-buccal root; (c) mesio-lingual root; (d) disto-lingual root; (e) extra root between the two buccal roots where an elevation is commonly present (p. 79 and Fig. 4.23A).

Fig. 4.32. *Selenarctos thibetanus* (Asiatic black bear). RCS Odonto. Mus., A 171.42. Root surfaces of M² with partially fused roots. (A) right M²; (B) left M². (a) mesio-buccal root; (b) disto-buccal root; (c) mesio-lingual root; (d) disto-lingual root.

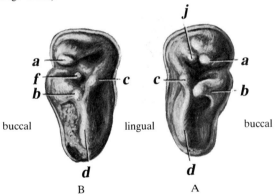

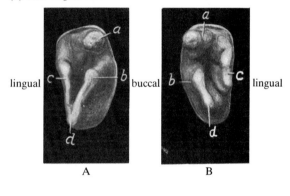

Fig. 4.31. *Ursus arctos* (brown bear). Root surfaces of (A) right M²; (B) left M². (a) mesio-buccal root; (b) disto-buccal root which on the right M² is fused with (c) mesio-lingual root, on the left M² these roots are separated by a narrow groove; (d) disto-lingual root.

Fig. 4.33. *Selenarctos thibetanus* (Asiatic black bear). Root surfaces of (A) right M², (B) left M². (a) mesio-buccal root; (b) disto-buccal root from which (f) a web of root extends distally; (c) mesio-lingual root; (d) disto-lingual root.

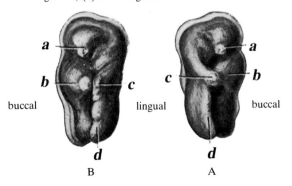

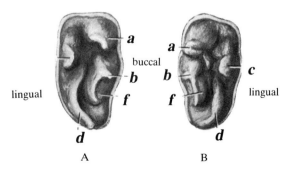

(Fig. 4.35). Normally M² have two roots. Here a small extra root is present on the lingual aspect between the mesial and the distal roots.

(xiv) *Ursus arctos*. RCS Odonto. Mus., G 139.9. (Fig. 4.36). Normally the roots of M₃ are fused. Here both M₃ show a partially-separated extra lingual root.

(xv) *Ursus arctos*. RCS Odonto. Mus., A 171.24.

(Fig. 4.37). Both M₃ have two distinct roots, the mesial ones being deeply grooved on their distal aspects; the right tooth has a small root springing from the lingual surface of the distal root.

(xvi) *Selenarctos thibetanus*. RCS Odonto. Mus., A 171.44. (Fig. 4.38). Both M₃ show a partial separation into two roots.

(xvii) *Ursus arctos* (brown bear). Captive. Formerly RCS Osteo. Series, 838. (Fig. 4.39). The crowns of both M₃ are more or less symmetrically distorted as if by pressure, the distal part being folded over the occlusal surface. The teeth were only partially erupted from under the ascending ramus of the mandible. It seems possible that the crypts failed to grow sufficiently to accommodate the growing M₃.

Fig. 4.34. *Helarctos malayanus* (sun bear). RCS Odonto. Mus., G 75.3. (A) right M² (printed in reverse) and (B) left M². Both show the fusion of the disto-lingual with the mesio-lingual root, which is normal in this species, but in the left tooth they are united with (a) mesio-buccal root also; (b) disto-buccal root; (c) mesio-lingual root; (d) disto-lingual root.

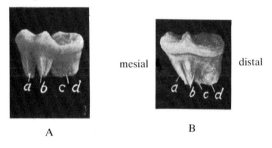

Fig. 4.35. *Ursus arctos* (brown bear). RCS Odonto. Mus., A 171.22. Lingual surfaces of (A) left M₂ and (B) right M₂. There are small extra roots (e) between (a) mesial root, and (p) distal root.

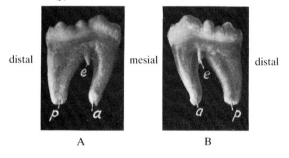

Fig. 4.36. *Ursus arctos* (brown bear). RCS Odonto. Mus., G 139.9. Lingual surfaces of (A) right M₃ and (B) left M₃. Both teeth have a partially separated extra lingual root. (a) mesial root; (b) distal root; (e) extra root.

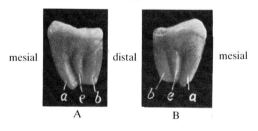

Fig. 4.37. *Ursus arctos* (brown bear). RCS Odonto. Mus., A 171.24. Lingual surfaces of (A) right M₃ and (B) left M₃. Both mesial roots (a) are deeply grooved on their distal surfaces; (p) distal root; the right M₃ (B) has a small extra root (e).

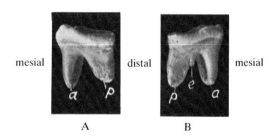

Fig. 4.38. *Selenarctos thibetanus* (Asiatic black bear). RCS Odonto. Mus., A 171.44. Lingual surfaces of both M₃ showing a partial fusion of the roots.

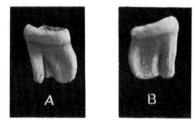

Fig. 4.39. *Ursus arctos* (brown bear). Captive. Formerly RCS Osteo. Series, 838. Both M₃ show gross distortion of the crowns as if compressed.

Family **Canidae**. Dogs, wolves, foxes

$$\frac{3}{3}:\frac{1}{1}:\frac{4}{4}:\frac{2}{3}=42$$

The dentition in the dog family is more primitive than in other carnivores in the sense that it almost reaches the full eutherian formula, lacking only M^3. As in other carnivores, P^4 and M_1 are carnassials, but M^1 is a large, more crushing tooth and the remaining molars are small. When extra premolars occur, they are usually in the area of P1 and it is not always possible to decide which is the extra tooth. Extra teeth in the molar region are almost invariably disto-molars.

Table 4.3 shows the incidence of extra teeth in various genera of canids studied by Colyer; extra teeth were present in 5 out of 1280 skulls of wild canids (2.7%) and in 69 out of 799 domestic dogs (8.6%). The high figures for the S. American genera, *Speothos*, and perhaps *Nyctereutes*, *Dusicyon*, *Cerdocyon* and *Otocyon*, are probably exaggerated by the small sample sizes, although Bateson (1894) did suggest that extra teeth are more frequent in these genera.

As far as the other canid genera are concerned, the figure of 3.5% of animals with extra teeth for wild *Canis* (wolves, coyotes and various jackals) is higher than the 1.3% for *C. latrans* (coyote) in the combined samples of Hall (1940) and Paradiso (1966) and the 1.8% found in *C. lupus* (wolf) by Dolgov and Rossolimo (1964). It can be explained by the high proportion (18.2%) of specimens with extra teeth in one of the *Canis* species studied by Colyer, the side-striped jackal (*C. adustus*), and again this may be an effect of the small sample size (22). The high figure for domestic *Canis* is discussed below.

In 3734 skulls of the coyote *Canis latrans* in the United States National Museum examined by Paradiso (1966), and in 939 in the University of California's Museum of Vertebrate Zoology (Hall, 1940), there were no supernumerary incisors or canines, but Nellis (1972) found one lower incisor in 489 skulls from Central Alberta. Paradiso found four skulls with additional premolars (0.1%), Hall found seven (0.7%), and Nellis found eight (1.6%). Paradiso found 0.5% with additional lower molars in various positions and there were 0.3% in Hall's and 0.6% in Nellis's. Maxillary disto-molars were present in 0.8% of Paradiso's sample, in 0.2% of Nellis's and there were none in Hall's. Paradiso regarded their occasional presence in *C. latrans* as an atavistic phenomenon that linked this species to more primitive Pliocene canids. Although extra teeth are more frequent in Nellis's sample than in Hall's, Nellis found that the difference was not statistically significant.

Red foxes (*Vulpes vulpes*) have been extensively studied. Van Bree and Sinkeldam (1969) examined 293 skulls from mainland Western Europe (Table 4.4), Reinwaldt (1962) 291 from Scandinavia and Rantanen and Pulliainen (1970) 75 from Finnish Lapland. The only additional teeth that Reinwaldt found were incisors in 0.7% of his skulls; Van Bree and Sinkeldam found only additional cheek teeth, premolars in 2% and molars in 1%; Rantanen and Pulliainen found one extra premolar (1.3%). Overall, extra teeth occurred in 1.9% of these foxes which is quite close to Colyer's figure of 1.6% for the genus *Vulpes* as a whole.

Table 4.3 The incidence of extra teeth in various genera of canids (including domestic CANIS) noted by Colyer

Genus	No. of specimens	Extra teeth				Totals of specimens showing extra teeth	
		I	C	P	M	No.	%
Domestic *Canis*	799	19	0	39	13	69	8.6
Wild Canidae							
Canis	315	1	0	9	1	11	3.5
Vulpes	493	2	1	3	2	8	1.6
Alopex	130	0	0	1	0	1	0.8
Cerdocyon	79	0	0	1	3	4	5.1
Dusicyon	139	0	0	1	3	3	2.9
Nyctereutes	29	0	0	1	1	2	6.8
Otocyon	31	0	0	0	2	2	6.4
Cuon	50	0	0	1	0	1	2.0
Speothos	14	0	0	0	3	3	21.0
Total wild	**1280**	**3**	**1**	**16**	**15**	**35**	**2.7**

Although *Otocyon megalotis* (the bat-eared fox) possesses three upper and four lower molars, there is some doubt as to the nature of the additional teeth. Guilday (1962) noted that the M^1 and M^2 resemble one another in size and shape and so do the M_2 and M_3; he contended that the presence of an M^3 and an M_4 is the result of duplication of the M^1 and M_2 and that the M^3 and M_4 are therefore homologues of the normal M^2 and M_3. So what otherwise would be regarded as the anomalous symmetrical development of supernumerary teeth prospered and became fixed under the pressure of natural selection. On the other hand, Van Valen (1964) suggested that the extra molars in *Otocyon* are due to the expansion of the field of molarization and are not homologous to any normal teeth in recent canids.

Congenitally absent teeth are rare in wild canids. In the 3734 *Canis latrans* (coyote) skulls studied by Paradiso, there was only one possible congenital absence of an incisor. None lacked canines. In Hall's 939 skulls, there were no absent incisors or canines. Paradiso found 0.08% of missing premolars and 1.1% of missing molars. In contrast, 4% in Hall's sample lacked one or more premolars or molars; to explain this higher prevalence, Paradiso suggested that Hall in some cases may have mistaken absence due to injury or disease for congenital absence.

Bateson (1894) mentioned missing incisors in two red foxes (*Vulpes vulpes*), but in only one was it probable that the tooth was congenitally absent. In studying red foxes from the Netherlands, western France and southern France for taxonomic purposes, Van Bree and Sinkeldam (1969) were struck by the high percentage of skulls with an abnormal tooth formula and concluded that there is an evolutionary trend in this species towards a reduction in the dentition, particularly of P_1 and M_3 (Table 4.4). Their figure of 10.9% of absent M_3 is close to that of Reinwaldt (1962) for Scandinavian foxes (10%). A similar prevalence was observed in racoon dogs *Nyctereutes procyonides* in western Japan (Asahi and Mori, 1980).

However, among 1192 red foxes from the wild in the Museum für Naturkunde in Berlin, Döcke (1959) found only 3.3% lacking P_1 and 4% lacking M_3. His main purpose was to compare wild and domestic foxes and he examined 3103 skulls of farm silver foxes (a variety of *Vulpes vulpes* bred for its fur) from a single population collected over a period of 60 years (representing about 100 generations). Here the incidence of lack of P_1 was 0.47%, but the lack of M_3 was 19.65%. His method of presentation of his figures makes statistical comparisons difficult, but the high percentage of missing M_3 seems likely to be significant; Döcke suggested that it was related to the high degree of inbreeding amongst the farm foxes rather than to domestication as such.

Variations in the size of pairs of teeth occur. A lobe is at times seen on the lingual aspect of P^2 and P^3; in the Shri Lankan jackal (*Canis aureus lanka*) it is normally present. Examples of this variation were seen in other sub-species of jackal, in the Indian wolf (*Canis pallipes*) and in the Cape fox (*Vulpes chama*).

An extra root to the premolars is rare; an example was seen in a jackal where the left P_2 had an extra root on the lingual aspect. Examples of extra roots in coyotes are given by Nellis (1972) and in red foxes by Döcke (1959) and Lüps (1974). Lüps found four skulls with an additional root on at least one P_2 in a sample of 488 red foxes, in the Naturhistorischen Museum in Bern. All the animals came from the same small area so Lüps considered the abnormality to be genetically controlled. Nellis (1972) described one connate M^1 among 489 coyotes from Central Alberta. An example of connation between I^1 and I^2 in a farm silver fox is illustrated by Fabian (1933, p. 38).

Pavlinov (1975) studied fox skulls in several collections in USSR and described anomalies in eight skulls of *Vulpes vulpes*, in 11 of the Arctic fox (*Alopex lagopus*) and in one of the grey fox (*Vulpes cinereoargenteus*). Most of the connate teeth, supernumeraries and unusual malformations are described here in the appropriate sections.

Genus *Canis*

Dental variations in large numbers of canids (mostly *Canis* sp.) were studied by Hensel (1879), Bateson (1894), Nehring (1882), Hilzheimer (1906) and Bradley (1902). The figures that Hensel arrived at are rather high and there is some suggestion that his

Table 4.4 Numbers of teeth missing in 293 skulls of RED FOXES from continental western Europe (Van Bree and Sinkeldam, 1969)

I^1	6	= 2.0%	I_1	3	= 1.0%
I^2	—		I_2	1	= 0.3%
I^3	—		I_3	—	
C	1	= 0.3%	C	—	
P^1	4	= 1.4%	P_1	19	= 6.5%
P^2	3	= 1.0%	P_2	9	= 3.1%
P^3	—		P_3	1	= 0.3%
P^4	—		P_4	—	
M^1	1	= 0.3%	M_1	—	
M^2	—		M_2	1	= 0.3%
			M_3	32	= 10.9%

collection was not random and that he confused ante-mortem loss with congenital absence. We suspect that when Colyer wrote, 'In the large majority of specimens where teeth appear to be absent, scrutiny of the bone with a magnifying glass shows that in most cases it is rough and that the teeth have been lost during life. I am inclined therefore to regard absence of teeth as far less common than is usually stated', he had Hensel's finding in mind. It is likely that the samples studied by Bateson and Colyer included many of the same canid skulls. These authors used different combinations of species to constitute their wild canid group, so comparison between their results is impossible.

The effect of domestication on dental variation can be assessed in canids by comparing the domestic dog with its wild ancestor, the wolf (*Canis lupus*) (Clutton-Brock, 1981). When as is now usual dingos are included with domestic dogs, re-analysis of Colyer's figures shows that extra teeth occur in 8.6%. This is much higher than in wild *Canis* and other canids studied by Colyer (Table 4.3). Colyer's sample of wolves was rather small, but it may be significant that one of the two skulls in which he found extra teeth was from a captive animal. Dolgov and Rossolimo (1964) studied dental variations in 324 wild wolves and found only 1.8% with extra teeth (Table 4.5). However, extra teeth occurred in 5.5% of 234 wolves from the Bialowieza Forest in Poland and the USSR, as well as two instances of connate I^2 and one connate P_1 (Buchalczyk, Dynowski and Szteyn, 1981).

Canis familiaris (domestic dog). 799 specimens examined

An excess in number of the teeth is common in the domestic dog and this is especially the case in certain breeds; absence of teeth is far less frequent.

The breeds were grouped according to the classification adopted by Pocock (1929); in about a third of the specimens, the breed could not be ascertained.

EXTRA TEETH

Summaries of the occurrence of extra teeth in dogs are given in Tables 4.6 and 4.7.

Table 4.5 The incidence of extra teeth in DOMESTIC DOGS studied by Colyer compared with their wild progenitors, WOLVES (Dolgov and Rossolimo, 1964)

	No.	I	C	P	M	Total	%
Domestic dog, *Canis familiaris*	799	19	0	39	13	69	8.6
Wolf, *Canis lupus*	324	2	0	4	0	6	1.8

Table 4.6 Numerical variations in DOMESTIC DOGS

Group	No. of specimens	Extra teeth			Totals of specimens showing extra teeth	
		I	P	M	No.	%
Eskimo	58	—	1	—	1	1.7
Sheep dogs	42	—	3	2	5	11.9
Greyhounds	48	—	6	2	8	16.7
Mastiffs	143	10	4	2	16	11.2
Spaniels	72	3	9	2	14	19.4
Hounds	48	4	4	1	8	16.7
Terriers	76	—	4	—	4	5.3
Poodles	17	—	1	—	1	5.9
Dingos	42	—	—	1	1	2.4
Not classified	253	2	7	3	11	4.3
Total	**799**	**19**	**39**	**13**	**69**	**8.6**

In one hound skull, extra teeth were present in both the incisor and the molar regions, and in one unclassified skull in the premolar and molar regions.

86 Section 1: Variations in Number, Size and Shape

(a) Incisors

Extra incisors were seen in the mastiffs, spaniels and hounds. Dogs in the mastiff group are characterized by the more or less shortened muzzle. A striking feature was the frequency of extra incisors in this group particularly in bulldogs; in 73 specimens, there were nine instances; in five of these an extra tooth was present in each premaxilla. The tendency of the bulldog to vary in the incisor region was also noticeable in two further specimens; in one (Royal Dick Veterinary College, Edinburgh, 53), the left I^1 and I^2 were fused (M'Intosh, 1929; Bradley, 1902) and in the other the crown of the right I^2 showed connation (Bateson 1894). The shortening of the muzzle of the bulldog has been accompanied by an increase in the width of the premaxillae and it may be that this increase is in some way related to the development of extra teeth.

Bodingbauer and Hager (1959) and Hauck (1942) showed that extra incisors were particularly common in dogs with shortened muzzles and Aitchison (1964) confirmed this: 39.4% of the 73 bulldogs examined in 1962/3 had one or more extra upper incisors, and so did 26.4% of 140 boxers, compared with only 1.3% of 74 Pekingese and 3.8% of 105 pugs. Aitchison based his results on a questionnaire sent to breeders listed in Cruft's catalogues; this evoked an emotional response from some of the breeders who were clearly sensitive to any suggestion that over-breeding could be leading to congenital defects of any sort. In an earlier paper (1963), Aitchison showed that shortening of the muzzle was less extreme in 19th century bulldogs and none of these had extra incisors.

In addition to the example of connation in the bulldog, this type of variation was present in two other specimens; in one, a bull-terrier, the crown of the left I^1 was divided into two parts of which the mesial was the smaller (Fig. 4.40); in the other (Fig. 4.41), of which the breed was not known, the crown of the left I^2 was divided into two equal parts.

(b) Premolars

The additional premolars were all present in the region of the first premolar. Greyhounds were the only breed showing extra premolars in both the maxilla and the mandible; in two skulls, there were extra teeth in both maxillae and in both sides of the mandible, and in one skull in the right maxilla and the right side of the mandible.

Skrentny (1964) recorded extra teeth in 10.8% of 102 Doberman dogs in the United States; the additional teeth were all first premolars (Table 4.8).

Additional lower premolars or third molars occurred in 4% of 250 dogs of the prehistoric and early historic periods reviewed by Bodingbauer (1963).

When an extra tooth is present in each maxilla, the teeth may differ in size as shown in Figure 4.42, so that there is an asymmetrical arrangement of the right and left teeth. In the mandible of this specimen, there was an extra premolar in each side, the left being larger than the right.

Table 4.8 The positions of absent teeth in DOBERMAN DOGS (Skrentny, 1964)

Incisors		Premolars		Molars		
lower	upper	lower	upper	lower	upper	No.
14	0	17	4	0	0	103

Fig. 4.40. *Canis familiaris* (domestic dog, bull terrier). ♂. BMNH Lataste Coll., 3535. The crown of the left I^1 is connate. Of its two parts, the mesial is the smaller.

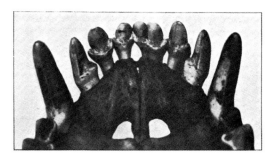

Fig. 4.41. *Canis familiaris* (domestic dog). RCS Odonto. Mus., G 17.12. The crown of the left I^2 is connate.

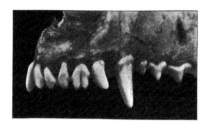

Table 4.7 The distribution of the 94 extra teeth seen in DOGS

	Maxilla		Mandible	
	Right	Left	Right	Left
Incisors	11	15	—	—
Premolars	18	13	9	7
Molars	2	5	7	7

(c) Molars

Extra molars are more common in the mandible than in the maxilla. In the 12 specimens with additional molars, there was an extra molar on both sides in seven, on one side in four; in one, there was an extra tooth in the left maxilla and in the left mandible. Allo (1971) found alveoli in the position of disto-molars in about 7% of 389 dog mandibles from prehistoric sites in New Zealand; this may be attributed to founder effect.

ABSENCE OF TEETH

Absence of P^1 and P_1 is common in the Eskimo dog but other breeds seldom lack these teeth or the last molars. With a view to confirming his doubts as to the absence of teeth in breeds other than Eskimo, Colyer submitted 60 specimens, in which the first premolar or the last molar was not present, to a critical examination; in two cases only did the condition of the bone warrant the assumption that the premolar had not been present and in two cases the absence of the last molar was doubtful.

In two adult mandibles, an m_2 had been retained and radiographic examination showed that P_2 was absent. In Skrentny's sample of Doberman dogs, 17 animals lacked at least one tooth, most commonly the last premolar (Table 4.8). He suggested that in this sample of Dobermans a single autosomal recessive inheritance pattern was responsible for the absence of teeth.

Bodingbauer and Hager (1959) examined a large number of skulls of dogs of various breeds and found the absence of one or more premolars was more prevalent in modern breeds with long snouts than in prehistoric dogs. This was particularly true of P_4 in greyhounds and their conclusion was that the cause was genetic. Later, Bodingbauer (1963) reviewed the literature on supernumerary and absent teeth in early dogs many of which, though early, were not actually prehistoric. It was often not possible to show whether absence was actually congenital but, of about 250 mandibles, 30 (12%) lacked one or more teeth, and of two Upper Pleistocene specimens one lacked a P_3 and another a P_4. This finding seems to have received editorial approval by the *Veterinary Record* (Anon, 1963), in which it was published, as it showed that tooth loss is not necessarily the product of modern overbreeding. However, loss of teeth is particularly common as a result of overcrowding in modern dog breeds with shortened snouts such as bulldogs (St Clair

Fig. 4.42. *Canis familiaris* (domestic dog, greyhound). RCS Odonto. Mus., G 18.35. Bilateral duplication of P^1. ×c.0.75

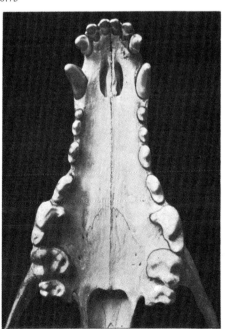

Fig. 4.43. *Canis familiaris* (domestic dog). A skull that belonged to A. T. Hopwood. Left P^2 replaced by a conical tooth.

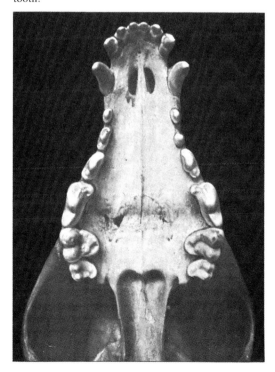

and Jones, 1957). Absence of teeth related to congenital ectodermal defects, such as a deficiency of hair, was described in dogs by Burrows, Miller and Harvey (1985).

SHAPE

P^2 in dogs, as in other canids, may have an extra cusp on its lingual aspect carried on an extra root. In one case (Fig. 4.43), the left P^2 was replaced by a conical tooth; in another, the position of the P_4 was also occupied by a conical tooth (Fig. 4.44). Fusion of the roots of P^2 premolar and occasionally of P^3 occurs. Verstraete (1985) figured a connate left P^3 in a female Afghan hound. More examples of dental variations in dogs were described by Bradley (1907).

Fleischer (1967) showed that the shape of the premolars can be very variable in the genus *Canis*. Stockhaus (1962) found that it can vary even within one race of domestic dog and tooth shape is more variable in domestic dogs than in their wild progenitors *Canis lupus*.

Canis lupus (wolf). 75 specimens examined

(i) Captive. RCS Odonto. Mus., G 19.6. A supernumerary tooth between the right P_1 and C which closely resembles P_1, although its root is larger; unlike any of the other teeth present, it has a circumferential band of enamel hypoplasia near the cervix.
(ii) *C. lupus arabs*. BMNH 1924.8.13.1. Two teeth in the position of the right P_1.
(iii) *C. lupus lycaon*. MVZUC 77344. A small distomolar in each maxilla (Hall, 1940).

Canis aureus (jackal). 141 specimens examined

(i) A supplemental left I_2.
(ii) MNIIN A13255. Two teeth in the position of the right P^1; the mesial one is the smaller and both are smaller than the contralateral.

Canis latrans (coyote). 278 specimens examined

(i) Captive. RCS Odonto. Mus., G 19.7. Between the left P^1 and the upper canine, there is a smaller supplemental P^1 that is incompletely erupted.
(ii) University of Michigan Museum of Paleontology 45222. Supplementary P_2 (Fine, 1964).
(iii) USNM 289064. Bilateral maxillary disto-molars smaller than M^2 (Paradiso, 1966).

The supernumerary teeth noted by Hall (1940) among *Canis latrans* in the Museum of Vertebrate Zoology, University of California are:

(iv) *C. latrans clepticus*. 3247. There were two small sockets mesio-buccally to the left P^2, presumed to have been occupied by a two-rooted tooth, now lost. Judging from the size of the sockets, the tooth had been about half the size of P^2. It was possibly its retained deciduous predecessor.
(v) *C.l. ochropus*. 61394. Bilateral supplemental P^2.
(vi) *C.l. ochropus*. 8970. Bilateral supplemental P_2.
(vii) *C.l. thamnos*. 77340. A supplemental left P_1.
(viii) *C.l. ochropus*. 11732. A supplemental left P_1.
(ix) *C.l. lestes*. 23796. A supplemental left P_3.
(x) *C.l. mearnsi*. 25900. A supplemental left P_2.
(xi) *C.l. mearnsi*. 51651. Bilateral mandibular disto-molars.

Fig. 4.44. *Canis familiaris* (domestic dog, bloodhound). ♀, juvenile. BMNH D 159. A small conical right P_4.

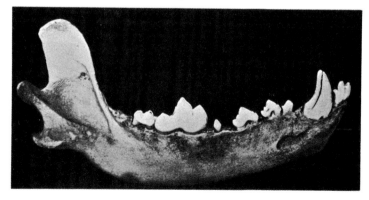

(xii) *C.l. lestes*. 24574. A small conical disto-molar in the right mandible.
(xiii) *C.l. lestes*. 23790. A disto-molar in the right mandible.

Canis adustus (side-striped jackal). 22 specimens examined.

(i) BMNH 1900.10.3.3. (Fig. 4.45). A small supplemental P^1 mesial to the left P^1.
(ii) BMNH 1891.5.27.8. (Fig. 4.46). Two teeth in place of the right P^1; the distal one is a little smaller than either P^1.
(iii) ♀. BMNH 1926.6.11.6. (Fig. 4.47). Supplemental left P_3 and P_4. The distal ones in each case are slightly smaller than the presumed P_3 and P_4. The length measurements of the tooth crowns are shown in Table 4.9.
(iv) ♂. P-C. Mus., C400. An extra tooth distal to the left M^2. The left M^2 is smaller than the contralateral. Both M_3 are unusually large.

Canis mesomelas (black-backed jackal). 50 specimens examined.

(i) ♂. BMNH 1899.7.5.3. A supplemental left P_1.
(ii) Formerly RCS Osteo. Series 634. (Fig. 4.48). Two teeth in place of the right P^3; the larger is in the normal position in the arch and is the same size as the contralateral P^3. The other is smaller and situated lingually. (Bateson, 1894, p. 212).

Fig. 4.45. *Canis adustus* (side-striped jackal). BMNH 1900.10.3.3. A supplemental left P^1. × c. 0.75.

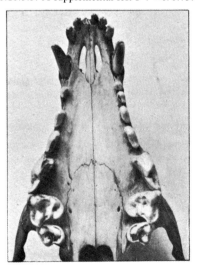

(iii) A small bicuspid disto-molar in the left maxilla (Dönitz, 1872).

Genus *Vulpes*. 492 specimens examined

Vulpes vulpes (red fox)

(i) ZMUC 307. A supplemental left I^3.
(ii) RCS Odonto. Mus., G 19.8. (Fig. 4.49). The tips of both maxillary canines are bifid. The division involves about half the height of the crown and is continued as a groove which, on the right canine, is particularly prominent and extends on to the root. These are examples of connation or so-called dichotomy. Bateson (1894, p. 211) figured the skull of a dog, which has proved impossible to trace, in which there

Table 4.9 The lengths (in mm) of the crowns of supernumerary and normal lower premolars in a SIDE-STRIPED JACKAL *(Canis adustus)* (BMNH 1926.6.11.6.)

	Left normal	Left supernumerary	Right normal
P_2	7.1	—	7.45
P_3	8.0	6.2	8.3
P_4	9.5	8.8	9.8

Fig. 4.46. *Canis adustus* (side-striped jackal). BMNH 1871.5.27.8. Two similar teeth in place of right P^1; the distal one is the smaller.

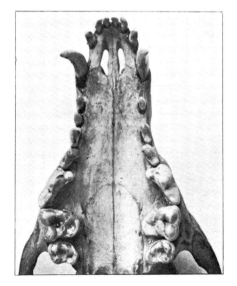

was a remarkably similar anomaly of both upper canines (see Fig. 1.5 A and B).

(iii) Two teeth, each in separate sockets, in the position of each maxillary canine. This is a very rare example of supernumerary canines (Tegetmeier, 1901).

(iv) BMNH 1919.7.7.2942. A pair of premolariform teeth in place of the left P^1; both are smaller than the right P^1. The distal one of the pair resembles the right P^1 more closely because it has two roots.

(v) ZMA 8797. A supplemental right P^1 (Van Bree and Sinkeldam, 1969).

(vi) ZMA 8765. A supplemental left P^1 (Van Bree and Sinkeldam, 1969).

(vii) ZMA 9020. Supplemental P^1 on both sides (Van Bree and Sinkeldam, 1969).

(viii) RMNH 11.229. A supplemental right P^1 (Van Bree and Sinkeldam, 1969).

(ix) *V. vulpes montana* (Himalayan red fox). P-C. Mus., K71. The crown of the left P^1 is larger than the right and is smaller than the left tooth.

(x) BMNH 1859.9.6.86. The right P^1 is larger than the left; the tip of the crown is bifid, the mesial portion being smaller than the distal (illustrated by Bateson, 1894, Fig. 41).

(xi) ♀. ZI 9144. (Fig. 4.50). Both P^1 show evidence of a double origin or dichotomy. The tip of the cusp of the left P^1 is made bifid by a deep transverse groove (Fig. 4.50A). The crown of the right P^1 is composed of two almost completely separate mesial and distal elements, joined only at the cervix. Each element is borne on a separate root (Fig. 4.50B) (Pavlinov, 1975).

(xii) ♀. ZM S-94276. In place of the right P^1, there are two teeth. The mesial one is single-rooted and resembles a normal P^1. The distal one is similar but much larger mesio-distally and the root is grooved longitudinally and ends in two blunt apices (Pavlinov, 1975).

(xiii) ZMA 8798. Supplemental P^2 on each side, both with extra roots (Van Bree and Sinkeldam, 1969).

(xiv) RMNH 11,430. Supplemental P_2 on both sides and a mandibular right disto-molar (Van Bree and Sinkeldam, 1969).

(xv) A supplemental left P^3 (Gasson, 1980).

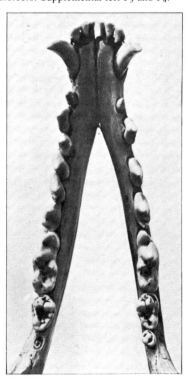

Fig. 4.47. *Canis adustus* (side-striped jackal). ♀. BMNH 1926.6.11.6. Supplemental left P_3 and P_4.

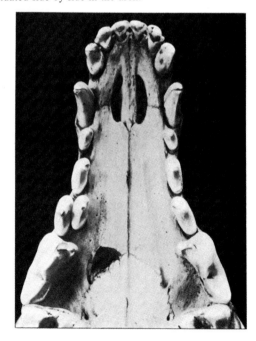

Fig. 4.48. *Canis mesomelas* (black-backed jackal). Formerly RCS Osteo. Series, 634. Supplemental P^3 situated side by side in the arch.

(xvi) Formerly in the Museum of the College of Veterinary Medicine, Uppsala, 19605. Supplemental left P^3; the two teeth are side by side.

(xvii) BI 7132. (Fig. 4.51). In place of the right P^3, there is a three-rooted connate tooth. The mesial element resembles a normal two-rooted P^3 but is joined distally at the cervical part of the crown to the crown of an extra element which resembles P^3 but has only a distal root. The two elements, as it were, share the middle root. Mesial to this tooth, there is an extra premolar which resembles, but is not identical with, P^2 and P^3 (Pavlinov, 1975).

(xviii) Formerly Schultz Coll., 186. (Fig. 4.52). Supplemental bilateral P^4 lingual to and smaller than the normal P^4.

(xix) RMNH 11,141. A supplemental left M_1 (Van Bree and Sinkeldam, 1969).

(xx) ZM S–1984. The crown of the right M^2 is unusually large and complex but its roots are normal. There are three buccal cusps; from the tip of the middle one, a slender pointed projection arises. The lingual part of the occlusal surface is divided by a mesio-distal groove. A process like an extra root projects from the distal surface of the crown (Pavlinov, 1975).

(xxi) Formerly Naturhistorisch Museum, Vienna. From the island of Mytilene, Greece. A distomolar in the left maxilla (Breuer, 1943).

(xxii) ZMA 10,628. Right mandibular disto-molar (Van Bree and Sinkeldam, 1969).

(xxiii) *V. vulpes arabica* (Arabian fox). ♂. BMNH 1914.2.26.4. Four left upper incisors; the most mesial one is smaller and more conical than the others and is the probable supernumerary; labial to it is a probable retained fragment of its deciduous predecessor.

(xxiv) *V. vulpes bengalensis* (Bengal fox). ♂. BMNH

Fig. 4.49. *Vulpes vulpes* (red fox). RCS Odonto. Mus., G 19.8. Lateral views of the upper jaws. Both canines are bifid. On the right the division extends as a groove on to the root. Connation.

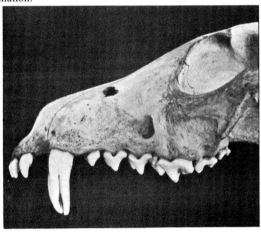

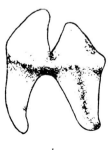

Fig. 4.50. *Vulpes vulpes* (red fox). ZI9144. (a) left P_1 with a bifid crown; (b) connate right P_1 composed of two almost completely separate mesial and distal elements. (After Pavlinov, 1975.).

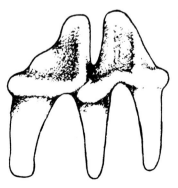

Fig. 4.51. *Vulpes vulpes* (red fox). BI 7132. Three-rooted, connate right P^3. (After Pavlinov, 1975.)

1915.4.3.67. Bilateral small mandibular disto-molars.

(xxv) *V. macrotis arsipus* (kit fox). MVZUC 61938. Lingual to each carnassial P^4 is an extra tooth of similar morphology but smaller and rotated with the presumptive buccal surface facing distally (Hall, 1940).

(xxvi) *V. vulpes atlantica.* ♀. BMNH 1898.7.4.8. Bilateral mandibular disto-molars; the left one is connated, the crown being divided by a deep cleft into two parts.

(xxvii) *V. cinereoargenteus scottii* (grey fox). MVZUC 55221. A two-rooted extra premolar between the right P^1 and P^2 and intermediate in size between them (Hall, 1940).

(xxviii) *V. cinereoargenteus californicus.* MVZUC 4539. A right mandibular disto-molar smaller than M_2 (Hall, 1940).

(xxix) *V. vulpes.* BMNH 1903.12.4.26x. (Fig. 4.53). Variations in the size of contralateral teeth. The length measurements of some of the upper premolars are shown in Table 4.10.

Gingerich and Winkler (1979) measured the teeth of 51 red foxes in North America and found that in size the last molars (i.e. M_3 and M^2) were much more variable than the rest of the dentition.

Pavlinov (1975) described two skulls of red foxes in which M^2, in one bilaterally, was reduced in size, and four skulls in which an extra very small molar was situated distal to M^2 either on one side or both.

Genus *Alopex*. 130 specimens examined

(i) *Alopex lagopus* (arctic fox). BMNH 1934.10.12.1. (Fig. 4.54). A supplemental right P^3.

(ii and iii) ZM S–5090. (Fig. 4.55a). A small extra root springs from the lingual aspect of both P^3 close

Table 4.10 The lengths (in mm) of some of the upper premolars of a FOX *(Vulpes vulpes)* (BMNH 1903.12.4.26x) showing differences in size between the contralaterals

	Left	Right
P^1	5.5	5.1
P^2	8.4	7.5
P^3	8.6	9.1

Fig. 4.52. *Vulpes vulpes* (red fox). Formerly Schultz Coll., 186. Bilateral smaller supplemental P^4 lingual to the normal P^4.

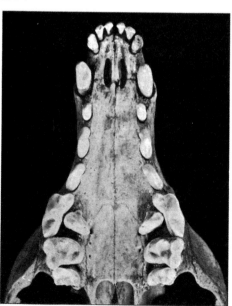

Fig. 4.53. *Vulpes vulpes* (red fox). BMNH 1903.12.4.26x. The three left P^2, P^3 and P^4 are larger than the right.

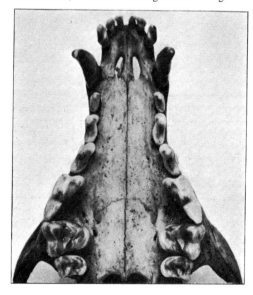

to the furcation between the two normal roots. ZM S–30373 has a similar right P^3 (Pavlinov, 1975).

(iv) ZI 9490. (Fig. 4.55b). An extra tooth between the left P_2 and P_3. It has a single cusp and two main mesial and distal roots with a third smaller root arising from close to the cervix lingually. The tooth does not resemble exactly any permanent or deciduous teeth but, according to Pavlinov (1975), could be an anomalous deciduous tooth.

(v and vi) ZI 2171 and ZI 2321. In both specimens, lingual to the left P^4 there is an extra tooth which to some extent resembles the P^4 but is only about one-third as large. In ZI 2321, empty sockets in the right maxilla suggest that the condition had been bilateral (Pavlinov, 1975).

(vii) ZM S–35694. On both sides, there is an extra premolar lingual to P^4 which, as in the previous examples, resembles it. However, what morphologically appear to be the distal cusps are situated mesially so that both extra teeth appear

Fig. 4.54. *Alopex lagopus* (arctic fox). BMNH 1934.10.12.1. A supplemental right P^3.

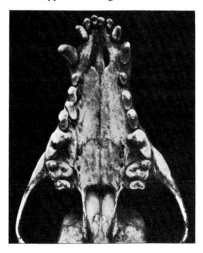

Fig. 4.56. *Cerdocyon thous* (crab-eating fox). BMNH 1926.1.12.9. Bilateral supplemental P_3.

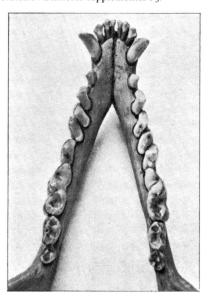

Fig. 4.55. (a): *Alopex lagopus* (arctic fox). ZM S-5090. Lingual view of a three-rooted P^3. (b): *Alopex lagopus* (arctic fox). ZI 9490. An extra three-rooted tooth that was between the P_2 and P_3 in the left mandible. Possibly an anomalous deciduous tooth. (After Pavlinov, 1975).

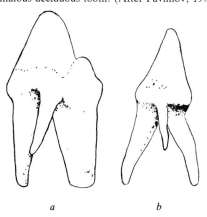

Fig. 4.57. *Cerdocyon thous* (crab-eating fox). BMNH 1846.6.15.3. In place of the right M_3, there is a cluster of three teeth, one being similar to the left M_3.

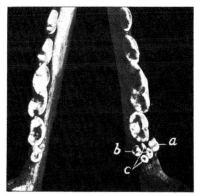

to have undergone 180° rotation (Pavlinov, 1975).

Genus ***Cerdocyon***. 79 specimens examined

(i) *Cerdocyon thous* (crab-eating fox). BMNH 1926.1.12.9. (Fig. 4.56). Bilateral supplemental P_3. The teeth vary in size, the mesio-distal lengths being: supernumerary, R; 7.9, L: 7.4. P_3, R: 8.0, L: 8.6 mm.
(ii) BMNH 1846.6.15.2. Small bilateral mandibular disto-molars.
(iii) BMNH 1856.4.12.5. Minute bilateral mandibular disto-molars.
(iv) BMNH 1846.6.15.3. (Fig. 4.57). In place of the right M_3, there is a cluster of three teeth, one of which is similar to the left M_3; another is smaller with a single root; the third resembles three joined denticles on a single root. This specimen could be regarded as the result of a disturbance of development extreme enough to be regarded as an odontome (Chapter 25, p. 586). It was mentioned by Bateson (1894, Fig. 44 III) as well as by Huxley (1880) and Mivart (1890).
(v) BMNH 1849.1.12.44. The right M_2 is abnormally large; possibly an example of connation with an extra tooth.

Genus ***Dusicyon***. 139 specimens examined

(i) *Dusicyon vetulus* (hoary fox). BMNH 1884.2.21.1. Small right mandibular disto-molar (Mivart, 1890).
(ii) *D. vetulus*. ♂. BMNH 1903.7.7.40. Bilateral maxillary disto-molars. The left one, and the left M^1 and M^2, are larger than the contralaterals.

(iii) *D. gymnocercus* (*Pseudalopex azarae*) (Azara's fox). ♂. RMNH 17694. Bilateral maxillary disto-molars.

Genus ***Nyctereutes***. 29 specimens examined

(i) *Nyctereutes procyonoides* (racoon dog). RMNH 34480. An extra tooth, cone-shaped, between the right P^1 and P^2.
(ii) *N. procyonoides*. ♀. BMNH 1923.4.1.20. A right maxillary disto-molar.

Genus ***Cuon***. 50 specimens examined

(i) *Cuon alpinus* (Asiatic wild dog). ♂. BMNH 1934.9.18.3. An extra premolar between right P_2 and P_3, smaller than either.

Genus ***Otocyon***. 31 specimens examined

$$\frac{3}{3}:\frac{1}{1}:\frac{4}{4}:\frac{3}{4}=46$$

(i) *Otocyon megalotis* (bat-eared fox). Formerly RCS Osteo. Series, 675. Bilateral maxillary disto-molars.
(ii) *O. megalotis*. RCS Odonto. Mus., G120.2. Right maxillary disto-molar.

Genus ***Speothos***. 14 specimens examined

$$\frac{3}{3}:\frac{1}{1}:\frac{4}{4}:\frac{1}{2}=38$$

Three skulls of *Speothos venaticus* (bush dog) had bilateral maxillary disto-molars: ♀. BMNH 1923.11.6.2; ♂. BMNH 1920.1.1.8; ♀. BMNH 1902.1.1.9.

CHAPTER 5

Orders **PINNIPEDIA** and **CETACEA**

Although the Pinnipedia (seals) and Cetacea (whales) are not related to each other, they are grouped for convenience in the same chapter because they are both aquatic.

Order **PINNIPEDIA**

The pinnipedes (sealions, walruses and seals) bear many features indicative of a relationship to the Carnivora; they first appeared in the Miocene as an offshoot of the Oligocene carnivores, fully adapted to an aquatic life. They used to be placed in a sub-order of the Carnivora but the trend today is to give them the status of a separate order.

In general, their dentitions consist of single-rooted teeth with crowns of simple morphology adapted to a sea-food diet prehended and consumed within water. There is little distinction between molars and premolars and the dental formulae quoted here are based partly on the fossil evidence.

The deciduous dentition is non-functional and tends to be shed soon after birth; in some seals it does not erupt at all and is resorbed *in utero*.

The first two upper incisors of the Otaridae (sealions and fur seals) differ from those of all the other pinnipedes in having a mesio-distal groove across the occlusal surface (Harrison and King, 1965). The grooved incisors can be seen in Fig. 5.1.

Pinnipedes show considerable individual variation in the number of teeth and, to quote Bateson (1894), 'illustrate nearly all the principles observed in the numerical variation of the teeth. In both premolars and molars there are examples of the replacement of

Section 1: Variations in Number, Size and Shape

one tooth by two, and in some of these the resulting teeth stand in series while in others they do not. Besides these there are numerous instances of extra premolars and molars belonging to various categories'. Colyer's findings were similar to Bateson's, but there was considerable overlap between their samples because both included the pinnipedes from the same collections, particularly those in the British Museum of Natural History.

The numbers of extra and absent teeth in the Otaridae and the Phocidae are given in Table 5.1. The most common variation is the presence of extra cheek teeth in the Phocidae, a result which agrees with the results of Sahlertz's (1877–8) study of a large number of seal dentitions in Danish collections.

Kubota and Togawa (1964), on the basis of the number of extra and absent teeth they found in skulls of sealions (*Eumetopias jubatus*), Northern fur seals (*Callorhinus ursinus*) and seals of the genus *Phoca*, concluded that the dentition of the pinnipedes in general is unstable, by which they meant particularly liable to variation, mainly in number. Their study is an impressive one because it included many fetal specimens and many specimens were radiographed.

Family **Otaridae**. Sealions, earred seals

Genus ***Otaria***

$$\frac{3}{2} : \frac{1}{1} : \frac{4}{4} : \frac{2}{1} = 36$$

Table 5.1 Numerical variations in the PINNIPEDIA (excluding Odobaenidae)

Family	No. of specimens	Extra teeth			Specimens affected		Absence of teeth			Specimens affected		Totals of specimens showing extra teeth or absence of teeth	
		I	C	Cheek teeth	No.	%	I	C	Cheek teeth	No.	%	No.	%
Otaridae	298	2	—	25	15	5.0	—	—	28	17	5.7	32	10.7
Phocidae	417	7	—	45	33	7.9	6	—	8	4	1.0	36	8.6
Total	**715**	**9**	**—**	**70**	**48**	**6.7**	**6**	**—**	**36**	**21**	**2.9**	**68**	**9.5**

In skulls in which two or more classes of teeth varied in number, each class is recorded individually; for example, if a skull has an extra incisor as well as an extra cheek tooth, both columns would be credited with one and the column 'number of specimens affected' with one only.

Fig. 5.1. *Otaria byronia* (southern sealion). RCS Odonto. Mus., G26.11. Five extra cheek teeth. Note the transverse grooves on the occlusal surfaces of the incisors.

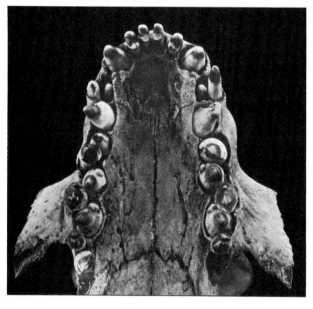

Otaria byronia (southern sealion). 104 specimens examined (22 without mandibles)

(i) BMNH Discovery Committee Coll., 1021. An extra incisor on the right between I^2 and I^3; it resembles I^2 but is larger.
(ii) ♂. Juvenile. RCS Odonto. Mus., G 26.11. (Fig. 5.1). The right maxilla has seven cheek teeth and empty sockets for two others; the left has six cheek teeth and an empty socket for another.

As there is little difference in shape between premolars and molars in sealions and, in this skull, some teeth are only partially erupted, it is particularly difficult to determine which are the extra teeth.

In the right maxilla, the supernumeraries are probably the tooth buccal to the arch between the presumptive P^2 and P^3, a similarly placed socket between P^3 and P^4, and a small tooth with a stumpy bifid cusp between the presumptive P^4 and M^1.

The supernumeraries in the left maxilla seem to be the tooth buccal to the arch between P^3 and P^4, and the tooth distal to P^4 and similar in size and shape to the contralateral tooth with a bifid cusp. This interpretation assumes that the empty socket distal to that tooth contained both M^1 and M^2.

Bilateral mandibular disto-molars are also present.
(iii) RCS Odonto. Mus., G 26.12. Bilateral mandibular disto-molars.
(iv) Formerly RCS Osteo. Series, 975. The crown of the left P^3 is partially divided longitudinally into two connated elements.
(v) In five specimens, both M^2 were absent and, in three, one M^2.

Genus *Eumetopias*

$$\frac{3}{2}:\frac{1}{1}:\frac{4}{4}:\frac{1}{1}=34$$

Eumetopias jubatus (northern sealion). 25 specimens examined

(i) RMNH 13439. An extra tooth, probably an incisor, on the lingual side of the lower left canine.
(ii) Formerly BMNH 1859.11.5.2. Most of the teeth have been lost, but the left P^2 was connate.
(iii) In 68 skulls examined by Kubota and Togawa (1964), one had an additional upper molar, one lacked an upper and one a lower molar.

Genus *Zalophus*

$$\frac{3}{2}:\frac{1}{1}:\frac{4}{4}:\frac{2}{1}=36$$

Zalophus californianus (Californian sealion). 15 specimens examined.
(i) In three of the skulls, there were five instead of six upper cheek teeth.

Genus *Arctocephalus*

$$\frac{3}{2}:\frac{1}{1}:\frac{4}{4}:\frac{1}{1}=34$$

Arctocephalus pusillus (Afro-Australian fur seal). 51 specimens examined.

(i) ♀. Juvenile. RMNH 13445. Mesial to the left P_1, there is an extra tooth about two-thirds the size of P_1. Opposite the space between the right P^3 and P^4 but buccal to them, is an extra similar tooth (Bateson, 1894, no. 323).
(ii) ♂. BMNH 1889.2.20.1. Six cheek teeth in each maxilla.
(iii) ♂. BMNH 1927.7.2.3. An extra premolar between the right P^1 and P^2 resembling P^2 but smaller.
(iv) ♂. BMNH 1927.7.2.4. Both M^2 absent.
(v) ♂. University Museum of Zoology, Cambridge K 7426. A supplemental tooth lingual to each M^1, the right being a little larger than the left. This specimen is referred to by Bateson (1894, no. 328) under the name *Otaria cinerea*.

Arctocephalus australis (South American fur seal). 17 specimens examined (3 without mandibles)

(i) ♂. Juvenile. Formerly RCS Osteo. Series, 984. Both M^2 absent.
(ii) RMNH 13455. Left M^2 absent.

Arctocephalus fosteri (New Zealand fur seal). 5 specimens examined (2 without mandibles)

(i) MNHN 1877–32. On the buccal side of each P^1, an extra smaller tooth.

Genus *Callorhinus*

$$\frac{3}{2}:\frac{1}{1}:\frac{4}{4}:\frac{2}{1}=36$$

Callorhinus ursinus (northern fur seal). 81 specimens examined

98 Section 1: Variations in Number, Size and Shape

(i) ♂. BMNH 1928.4.21.29. In each maxilla buccal to the space between the P^2 and P^3, an extra tooth smaller than either premolar.
(ii) ♀. BMNH 1928.4.21.51. An extra premolar partially erupted between the right P^2 and P^3 and smaller than either.
(iii) ♂. BMNH 1928.4.21.36. An extra premolar between the left P^2 and P^3. The right P^2 is unusually large and has a bifid central cusp with longitudinal grooves extending from the division on to the root; another example of two separate teeth being formed on one side and a connate tooth on the other.
(iv) ♂. BMNH. 1928.4.21.40. An extra premolar between the right P^2 and P^3 but smaller than either.
(v) Possibly *Arctocephalus gazella* (Kerguelen fur seal). ♂. University Museum of Zoology, Cambridge, K7223.2. A supplemental left P^2 (Bateson, 1894, no. 325).
(vi) ♂. Formerly RCS Osteo. Series, 990. Several teeth have been lost post mortem, but there is an extra tooth between P^2 and P^3 on both sides and an empty socket for an extra tooth between P^4 and M^1 on both sides.
(vii) BMNH. 1928.4.21.42. Both M^2 absent.
(viii) ♂. BMNH.1928.4.21.30. Right M^2 absent.
(ix) BMNH.1891.12.18.7. Left M^2 absent.
(x) In 38 female fur seals, Chiasson (1955) found five with one cheek tooth missing from one or both maxillae. In 120 skulls, Kubota and Togawa (1964) found one with an extra incisor, six with extra teeth and 11 lacking both M^2. In one of their skulls, the right P^2 was connate with a supernumerary premolar and there was another supernumerary between P^2 and P^3 on the left; this can be interpreted as a bilateral disturbance of development.

Family **Odobaenidae**. Walruses

$$\frac{1-2}{0}:\frac{1}{1}:\frac{3-4}{3-4}:\frac{0}{0} = 18-24$$

Extra canines in the walrus were recorded by Degerbøl (1929/30). In one (Fig. 5.2), two canines on the right share the same socket; the mesial one is small. In another (Fig. 5.3), there are two teeth of about equal size on the right side, both being smaller than the single canine. In the other specimen (Fig. 5.4), there are two teeth, unequal in size on the left side, both being

Fig. 5.2. *Odobaenus rosmarus* (walrus). ♀. ZMUC 477. Two canines in the right maxilla (Degerbøl, 1929/30).

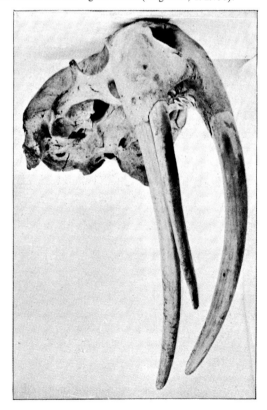

Fig. 5.3. *Odobaenus rosmarus* (walrus). ♀. ZMUC 276. Two canines in the right maxilla (Degerbøl, 1929/30).

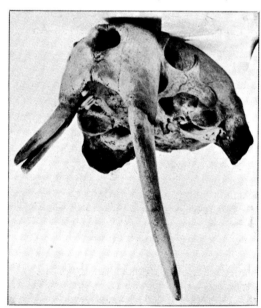

smaller than the single right canine. Caldwell (1964) illustrated a Pacific walrus head in which both upper canines were duplicated.

Family **Phocidae**. Earless seals

Sub-family **Phocinae**

$$\frac{3}{2}:\frac{1}{1}:\frac{4}{4}:\frac{1}{1}=34$$

Halichoerus grypus (grey seal). 50 specimens examined

(i) RCS Odonto. Mus., G 26.1. An extra tooth mesial to the left I^1.
(ii) In seven skulls examined by Colyer, there were six cheek teeth in each maxilla. Bateson (1894, p. 242) found additional upper molars in 12 out of 47 specimens.

Phoca vitulina (common seal). 71 specimens examined (7 without mandibles)

(i) BMNH 1891.12.18.5. An extra tooth betweeen the left P^4 and M^1, similar in shape to both but smaller.
(ii) University Museum of Zoology, Cambridge, K 8082. There is an extra tooth lingual to the space between the right P^2 and P^3, and another between the right P_3 and P_4 (Bateson, 1894, no. 329).
(iii) University Museum of Zoology, Cambridge, K 8086.2. An extra tooth lingual to the right P_3 and P_4 (Bateson, 1894, no. 334).

Phoca groenlandica (harp seal). 66 specimens examined

(i) MNHN A 2897. In place of the right P^2, there are two similar teeth, two-rooted like normal premolars. In place of the left P^2, there is a connate tooth; the crown is separated into two elements by oblique longitudinal grooves. The distal element has two roots; the mesial is smaller and has a single root (Bateson, 1892, and 1894, no. 326).
(ii) RMNH 34481. A small extra tooth between the right P^3 and P^4 (Bateson, 1892 and 1894, no. 324).
(iii) BMNH 1843.6.23.4. An extra tooth beween the left P^4 and M^1, similar in shape to both but smaller.
(iv) RMNH 37782. A small disto-molar in the left maxilla.
(v) In two cases, there was a disto-molar in each maxilla.

Fig. 5.4. *Odobaenus rosmarus* (walrus). ♂. ZMUC 478. Two canines in the left maxilla (Degerbøl, 1929/30).

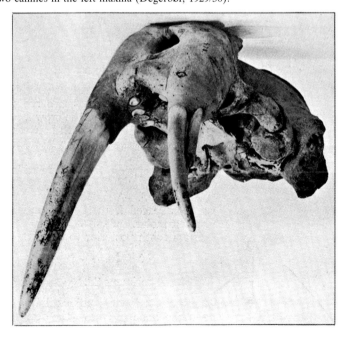

Phoca hispida (ringed seal). 103 specimens examined

(i) NRM A 612117. An extra tooth buccal to the right P_2.
(ii) NRM A 613510. An extra tooth buccal to the left P_2.
(iii) Formerly Högskola, Stockholm, 2124. Only four cheek teeth in the right maxilla, but which tooth is missing it is impossible to say. An extra tooth between the right P_3 and P_4 resembles M_1 but is a little smaller.
(iv) BMNH 1890.8.1.6. Only two upper incisors on each side and one lower incisor on each side.
(v) Formerly Högskola, Stockholm. Two incisors only in each premaxilla.

Histriophoca fasciata (ribbon seal).

Colyer did not examine any skulls of ribbon seals, but Kubota and Togawa (1964) found a connate P^3 and P^4 in the right maxilla of a skull of this species.

Erignathus barbatus (bearded seal). 28 specimens examined

(i) BMNH 1896.9.23.6. Bilateral maxillary disto-molars; they are almost completely buried in the bone and have the occlusal surfaces facing almost directly mesially.
(ii) BMNH 1890.8.1.6. Only one lower and two upper incisors present on each side (Bateson, 1894, no. 318).

Sub-family **Monachinae**

$$\frac{2}{2}:\frac{1}{1}:\frac{4}{4}:\frac{1}{1}=32$$

Monachus monachus (monk seal). 11 specimens examined

(i) ♂. BMNH 1894.7.27.1. Two extra upper incisors on the right; one is mesial to I^1 and smaller, the other is distal to I^2 and smaller.
(ii) BMNH 1934.8.5.4. An extra incisor mesio-lingual to the left I^1 and smaller.

Leptonychotes weddelli (Weddell's seal). 44 specimens examined

(i) ♀. BMNH 1908.2.20.62. A small left maxillary disto-molar.
(ii) ♂. BMNH 1908.2.20.5. Bilateral maxillary disto-molars, each larger than M^1.
(iii) ♂. BMNH 1908.2.20.15. Empty socket for a left maxillary disto-molar.
(iv) BMNH 1908.2.20.25. Bilateral maxillary disto-molars; the right M^1 is much smaller than the contralateral.

Ommatophoca rossi (Ross seal). 9 specimens examined

(i) BMNH 1843.11.16.7. Both P_1 and the right P^1 have the buccal and lingual surfaces of the crowns deeply grooved. The roots are also grooved throughout their entire length, but to a lesser degree. In the position of the left P^1, there are two teeth similar in shape but the distal one is slightly larger; their sockets are separate. This specimen is described and illustrated by Bateson (1892 and 1894, no. 320).
(ii) ♂. BMNH 1908.2.20.48. An empty socket for a small supernumerary tooth lingual to the left I^2. In the mandible, a small conical supernumerary lingual to the left I_1 and a left disto-molar.
(iii) ♂. BMNH 1908.2.20.49. An extra premolar between each P_1 and P_2 and smaller than either.
(iv) MNHN 1926–76. A right maxillary disto-molar.

Sub-family **Cystopherinae**

$$\frac{2}{1}:\frac{1}{1}:\frac{4}{4}:\frac{1}{1}=30$$

Cystaphora cristata (hooded seal). 35 specimens examined (3 without mandibles)

(i) RMNH 34483. An empty socket distal to the right upper canine. In the mandible, a small tooth on the lingual aspect of the left P_1 (Bateson, 1894, no. 322).
(ii) University Museum of Zoology, Cambridge, K7744. An extra tooth in the region of the mandibular right I_1 (Bateson, 1894, no. 321).
(iii) ♂. Formerly RCS Osteo. Series, 1101. An upper right cheek tooth absent.

Order **CETACEA**. Whales

The order Cetacea consists of two sub-orders, the Mysticeti, whalebone or baleen whales, which have only rudimentary unerupted teeth, and the Odontoceti, the toothed whales.

Rudimentary unerupted teeth are common, not only in the jaws of the whalebone whales (Van Dissel-

Scherft and Vervoort, 1954), but also in the regions of the jaws without functional teeth in the many Odontoceti in which the dentition is much reduced; for example in the normally toothless upper jaw of the sperm whales (*Physeter catodon*) (Boschma, 1938a; Slijper, 1979). Indeed Ritchie and Edwards (1913) found rudimentary upper teeth which were worn, that is functional, in two out of seven sperm whales caught in the vicinity of Rockall. Rudimentary teeth are also common in the jaws of the beaked whales *Mesoplodon* spp. (Robson, 1975) in which the functional dentition consists only of a single tooth in each mandible. Fraser (1953, p.36) demonstrated by radiography the presence of a considerable number of unerupted vestigial teeth in both upper and lower jaws of Sowerby's beaked whales (*Mesoplodon bidens*) and the bottle-nosed whale (*Hyperoodon ampullatus*). Although these teeth are usually small, as Boschma (1950a, 1951) showed, they often project from the gum a little and may be fairly large; they may therefore function in grasping the slippery squids which are the main items of diet. It is of interest that Arvy (1977) mistakenly referred to rudimentary 'bifid teeth' in a specimen of *Mesoplodon grayi* described by Boschma (1950a). In fact the teeth had split lengthwise which, as Boschma pointed out, is common in cetacean teeth preserved as museum specimens.

Whether these rudimentary teeth in whales properly belong to the category of supernumerary teeth is not easy to decide and depends partly upon how regularly they occur. However, they are of interest because, at any rate in the Mysticeti, they are commonly compound or connate (Fig. 5.5).

The numbers of teeth in most genera of toothed whales are variable, as was first noted by John Hunter (1787). They often exceed the basic eutherian number of 44 but, in some groups, there may be only one pair of teeth either in the maxilla or in the mandible. The teeth

Fig. 5.5. Teeth from embryos of whalebone whales.
Upper row: Developing double teeth from the upper jaw of *Balaenoptera musculus* (blue whale) (left) and *B. rostrata* (right). Only the tips of the crowns have been mineralized. (Drawn from Kükenthal, 1893.)
Lower row: Teeth from the lower jaw of *B. physalus* (common rorqual). Left is a simple conical tooth showing some resorption. The crowns of both the other teeth are composed of three conical elements. (Drawn from Van Dissel-Scherft and Vervoort, 1954, fig. 12.)

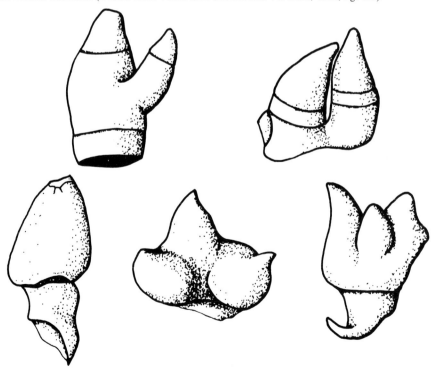

are monophyodont, homodont and single-rooted. The crowns of the teeth are either conical or spade-shaped (squalodont). According to Fordyce (1982), supernumerary teeth are common in fossil odontocetes.

In many toothed whales, for example sperm whales and porpoises, the crowns of the teeth have no enamel but are covered with cementum. In other taxa, for example *Steno bredanensis* (rough-toothed dolphin) and *Inia* (Amazon porpoise), the crowns are covered with enamel. In *Hyperoodon* (bottle-nosed whale), the teeth are said to be tipped with enamel.

The bottle-nosed dolphin (*Tursiops truncatus*) has 21–25 teeth on each side of both jaws as well as two rudimentary teeth at the rostral end of each lower jaw. Among 15 stranded on British coasts between 1927 and 1931, Fraser (1934) found that in five the rudimentary teeth were completely absent.

Although the average number of teeth in each mandible of the sperm whale is 22, there are sometimes as many as 27 (Matthews, 1938) and Fraser (1946) mentioned a stranded whale with 30.

Variations in the shape of cetacean teeth have not been described systematically, perhaps because they occur only rarely, but connate teeth seem to be common. Fraser (1946) described a porpoise (*Phocaena phocaena*) (Fig. 5.6) with a connate tooth in the right mandible. The upper part of the crown is bifid consisting of two spade-shaped parts that merge into a common basal part with a single root.

The mandible of a common dolphin (*Delphinus delphis*) (Fig. 5.7) has three extra teeth, each beside a normal tooth; another of its teeth is slightly larger than the others and is connate with a bifid tip similar to the tooth in Fig. 5.6. It has a single root.

Arvy (1977) gathered 5–6 lesser known examples of double teeth in sperm whales from the literature. She gained the impression that connate teeth are particularly common in the sperm whale; however, as she points out, the large teeth of the sperm whale have been collected in large numbers throughout the world over many years, so abnormal specimens may have attracted special attention.

Pouchet and Beauregard (1889, p. 75) described two unusual teeth in the sperm whale. One was composed of two teeth united only at the tips of the crown (Fig. 5.8). Neuville (1932) described the same specimen in greater detail and pointed out that wear of the tips of the teeth had exposed two separate islands of dentine with cementum between. He sectioned the specimen through the joined area and confirmed that it consisted of two teeth joined by coronal cementum. It may be inferred that this secondary fusion by cementum-formation occurred before the teeth erupted and that the teeth were then in unusually close contiguity. Boschma (1938b) described a similar pair of teeth joined by cementum. A deep hole extended into the worn surface of the cementum of fusion. Either cementum had not been formed in this area or some destructive process had affected the worn surface exposed to the mouth. Boschma also described a single sperm whale tooth that on one side of the crown had a rough facet of cementum where it had been broken away from another tooth to which it had been attached.

The other specimen described by Pouchet and Beauregard was a sperm whale tooth with two divergent roots (Fig. 5.9); all the other teeth were said to be similarly two-rooted. The way in which such roots were formed can be inferred from what is known of the mode of formation of multi-rooted teeth in general. Tongues of dentine are laid down under the influence of the epithelial root sheath across the tooth axis. The sheath, thus divided into more than a single sheath, then forms the appropriate number of separate roots. A similar process must have applied in the formation of this two-rooted whale tooth; it is not to be regarded as an example of fusion or dichotomy of tooth germs or of connation as discussed by Neuville (1928, 1929, 1932), Boschma (1938b) and Arvy (1977).

Neuville (1928) described a sperm whale tooth in

Fig. 5.6. *Phocaena phocaena* (common porpoise). BMNH 1933.18. A connate tooth in the left mandible. (Drawn from Fraser, 1946.)

Ch.5. Orders Pinnipedia and Cetacea   103

which on the buccal side there was a cleft at the growing edge of the root. A small rootlet had been formed at the margin of this cleft (Fig. 5.10). There was a longitudinal groove on the opposite, lingual side of the root. This specimen can be interpreted in terms of behaviour of the epithelial root sheath as above. Three examples described by Boschma of whale teeth with longitudinal grooves on the roots, in one case associated with ragged clefts at the growing margin, can be interpreted as lesser examples of a tendency for the formation of two roots. However, one of these specimens could be regarded as a true connate tooth

Fig. 5.7. *Delphinus delphis* (common dolphin). BMNH 1931.8. Four extra teeth in the mandible, one of which is connate. (Drawn from Fraser, 1946.)

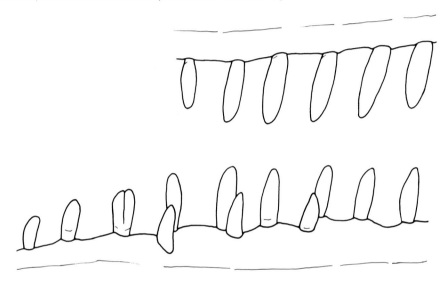

Fig. 5.8. *Physeter catodon* (sperm whale). MNHN A 5719.
Left: two adjacent teeth joined at their tips by cementum. The dentine has been exposed by wear.
Right: a section through the join. (Drawn from Neuville, 1932, plates 3 and 4.)

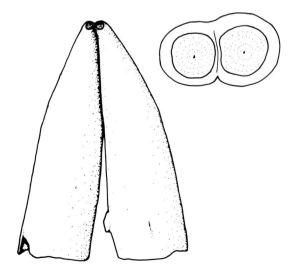

because transverse sections showed that the pulp in the tip of the tooth was bifid. Similar grooves are met with on the teeth of the killer whale (*Orcinus orca*, formerly *Orca gladiator*) (Neuville, 1932). Among about 60 sperm whales, Matthews (1938) noted three, all males, in which teeth bore secondary cusps which gave the appearance of double teeth.

Neuville (1932) showed a connate last tooth in the right mandible of a fresh-water La Plata dolphin (*Pontoporia blainvillei*) about twice the size of the normal teeth and composed of a mesial and a distal crown of equal size separated by a deep groove which did not extend on to the root. In another *Pontoporia* mandible, both the last and penultimate teeth on the right were connate teeth like the one just described.

In the mandible of a female La Plata dolphin, described by Pilleri and Gihr (1976), the 47th tooth in the row of 53 on the right and the 48th in the row of 54 on the left were connate teeth composed of three elements (Fig. 5.11). On the right, the crown was divided almost completely into three elements arranged in the line of the arch; on the left, the three elements were less distinct and separate and were arranged as a cluster. On both sides, there was a single large root.

Trebbau and van Bree (1974) figured without comment a mandible of an Amazon porpoise (*Inia*

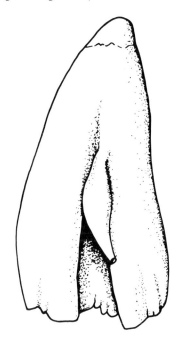

Fig. 5.10. *Physeter catodon* (sperm whale). Labial surface of a tooth with a cleft to the edge of which a toothlet is attached. Both the tooth and toothlet have open growing ends. The opposite lingual surface of the tooth has a longitudinal groove. (Drawn from Neuville, 1929, fig. 1.)

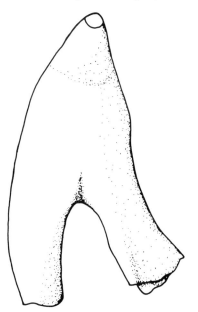

Fig. 5.9. *Physeter catodon* (sperm whale). Nantucket Museum. A tooth with two divergent roots. (Drawn from Pouchet and Beauregard, 1889, fig. 5.)

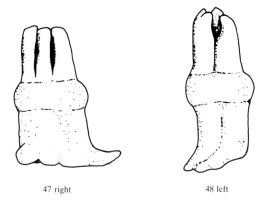

Fig. 5.11. *Pontoporia blainvillei* (La Plata dolphin). Lingual surfaces of connate teeth in left and right mandibles. Each is composed of three elements, which in the right tooth A (47th in the tooth row) are arranged in the line of the arch; in the left tooth B (48th in the tooth row) the three elements are clustered together. After Pilleri, G. and Gihr, M. (1976).

47 right   48 left

geoffrensis) in which the 15th tooth in the left row appears to be double or connate.

The teeth of the rough-toothed dolphin (*Steno bredanensis*) are of distinctive character and need to be mentioned because some features could be mistaken for abnormalities. The enamel is wrinkled with irregular ridges disposed mostly longitudinally and there is a particularly prominent ridge at the centre of both mesial and distal surfaces of the crowns. The enamel in *Inia* is similarly wrinkled. The roots in *Steno* are disproportionately long and, near the centre of the mesial and distal surfaces, there are often prominent longitudinal ridges which appear to be thickenings of cementum but which, according to Neuville (1928), are associated with prolongations of the pulp. He also described a connate tooth in the lower jaw of this species (Fig. 5.12).

Fig. 5.12. *Steno bredanensis* (rough-toothed dolphin). The bone has been cut away to show the roots of the teeth. Only the connate tooth in the mandible is depicted in any detail. The wrinkled enamel is indicated. (Drawn from Neuville, 1928, plate 3.)

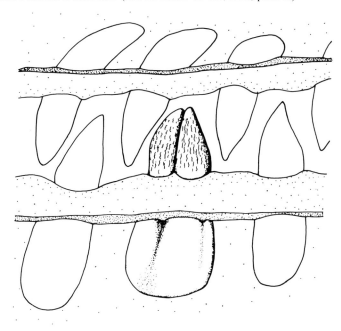

CHAPTER 6

The Ungulates: Orders ARTIODACTYLA, PERISSODACTYLA, HYRACOIDEA AND PROBOSCIDEA

Although it is convenient to follow the custom of grouping the ungulates, that is hoofed mammals, together, they are a disparate group of Orders with little in common beyond the fact that the claws of the primitive mammals evolved into hooves as part of a remarkably effective swift-running mechanism. It seems highly likely that hooves evolved by parallel evolution in a variety of forms although it is also possible that there was some ultimate common Tertiary ancestor. In other respects, however, the ungulate Orders show large fundamental differences, particularly in respect of their dentitions and gut morphology. The consensus view was expressed succinctly by Romer (1962) as 'A cow is not improbably as closely related to a lion as to a horse'.

Even the most typically ungulate Orders, the Perissodactyla (odd-toed) and Artiodactyla (even-toed), comprise notably disparate forms, a principal disparity being between their dentitions. In most genera in the artiodactyl Sub-order Ruminantia, the upper incisors are replaced by a firm mucosal pad against which the lower incisors bite; this incisor mechanism is associated with the complex stomach typical of ruminants. The incisor mechanism in the sheep is described in Chapter 24 (p. 552).

The equids (Perissodactyla) and the ruminants have very hypsodont cheek teeth of basically similar strucure adapted for triturating a strictly herbivorous diet; however, the equids have not lost their upper incisors.

The pigs and peccaries (Sub-order Suina), although also artiodactyls, have dentitions that are totally different from those of the ruminant artiodactyls, and suited to their entirely different, omnivorous habit. There is an incisor-to-incisor mechanism, the canines form tusks and the cheek teeth are brachydont and bunodont. In dentition and in many other features the Suina have remained relatively unspecialized and closer to the primitive hypothetical ancestor of the ungulates.

The Hyracoidea and Proboscidea show some morphological affinities with each other. They are categorized as sub-ungulates with the indistinct implication that they sprang from a proto-ungulate ancestor before the true ungulates arose.

The basic arrangement of cusps is remarkably similar in the different ungulate groups. The tribosphenic molar pattern has given way to a square one due to the presence of a hypocone on the disto-lingual side of the upper molars, and the development on the talonid of the lower molars of two cusps, the hypoconid and entoconid, which form a pair distal to the metaconid and protoconid, the paraconid having been lost.

Order **ARTIODACTYLA**

Sub-order **SUINA**

Family **Suidae**. Old World pigs

Pigs of the genus *Sus* have the full eutherian complement of 44 teeth, but in the other genera the numbers have been reduced. The molars are brachydont and bunodont; the canines are of continuous growth, usually forming tusks outside the mouth and, except in the wart hog (*Phacochoerus aethiopicus*), are much larger in males than females.

In *Sus*, the tooth in the position of the first premolar is not replaced and it is uncertain whether it is a permanent tooth without a deciduous precursor or a persistent deciduous tooth. It is quite frequently absent in domestic pigs. For example, Habermehl (1957) found P_1 absent in six out of 14 skulls and P^1 absent in two out of 18. Otto and Schumacher (1978) found one or all P1 absent in 37% of 56 Vietnamese pot-bellied pigs. The only supernumerary tooth was one duplicated lower canine.

Colyer found no dental variations in his sample of 79 wart hogs (*Phacochoerus aethiopicus*); however, Child, Sowls and Mitchell (1965) found much variability in both the deciduous and permanent tooth formulae. The deciduous formula appeared to be

$$\frac{1}{2-3}:\frac{1}{1}:\frac{2-4}{1-3}.$$

In 108 skulls with fully erupted deciduous incisors, 35% had three incisors in each mandible, 59% had two; the remaining 6% had three on one side and two on the other. In 129 with fully erupted deciduous molars, 81% had three in each maxilla and two in each mandible. The permanent tooth formula was particularly difficult to determine because there was a tendency for incisors to be lost in life and, in particular, for the first and second molars to be lost with the coming into wear of the third molar which is a disproportionately large tooth; its mesio-distal length is often larger than that of all the other cheek teeth together and it moves mesially as it becomes worn, rather like the molars in the elephant. Child *et al.* (1965) summarized the permanent formula as

$$\frac{1}{2-3}:\frac{1}{1}:\frac{1-3}{1}:\frac{3}{3}.$$

Accessory roots are common in the cheek teeth of pigs. In addition to the principal roots that correspond to the cusps they support, there tend to be other smaller roots variable in number and position (Habermehl, 1957).

Genus ***Potamochoerus***. 65 specimens examined

$$\frac{3}{3}:\frac{1}{1}:\frac{3}{3}:\frac{3}{3}=40$$

Potamochoerus porcus (bush pig)

(i) ♂. BMNH 1934.4.1.235. A small extra tooth between the left I^1 and I^2, resembling I^2 but smaller.
(ii) BMNH 1893.5.6.4. P_1 present on both sides.
(iii) BMNH 1838.4.16.81. P_1 present on both sides.
(iv) ♀. NRM A595832. (Fig. 6.1). In place of the right M_2, there are two teeth side by side in the line of the arch. Neither tooth is worn as the upper tooth occludes between them.
(v) ♂. BMNH 1913.3.19.1. Both I_3 absent.

Genus ***Sus***. Specimens examined: wild 191, captive 64

$$\frac{3}{3}:\frac{1}{1}:\frac{4}{4}:\frac{3}{3}=44$$

(i) *Sus scrofa* (domestic pig). RCS Odonto. Mus., A 221.31. Partially erupted bilateral mandibular

Section 1: Variations in Number, Size and Shape

Fig 6.1. *Potamachoerus porcus* (bush pig). ♀. NRM A595832. Two teeth in the position of the right M_2. Neither resembles the left M_2 exactly ×c. 0.43.

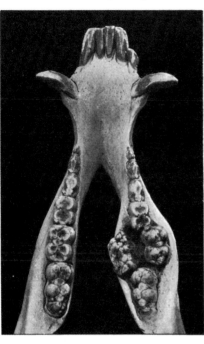

disto-molars, similar in shape to M_3. A radiograph of this specimen is reproduced in Henry (1935).

(ii) *Sus scrofa* (domestic pig). Museum of the Anatomical Institution of the Veterinary College, Stockholm. (Fig. 6.2). In the mandible, there are two canines on each side. On the right, they emerge from a common socket and their roots curve distally one above the other. On the left, the mesial canine curves buccally more or less normally; the distal one arises from a socket in the position of P_1 and P_2, which are absent, and is directed almost horizontally mesially and buccally to the other canine.

(iii) *Sus scrofa cristatus* (Indian wild boar). Formerly RCS Osteo. Series, 1750. In the position of the left P^2, there were two teeth, both similar in shape to the right P^2, though the mesial one was a little larger.

(iv) *Sus verrucosus* (Java pig). ♂. BMNH 1909.1.5.832. An extra tooth mesial to the right P^1 and similar to it but smaller.

(v) *Sus verrucosus celebensis* (Celebes pig). Juvenile. BMNH 1859.4.6.4. The right M^3 appears to be absent but the left M^3 is erupting.

In *Sus scrofa papuensis*, P1 shows a tendency to be missing; there was satisfactory evidence in five out of 17 specimens that one or more of the premolars was absent.

Fig. 6.2. *Sus scrofa* (domestic pig). Formerly Museum of the Anatomical Institution of the Veterinary College, Stockholm. An additional canine in each mandible. Reduced.

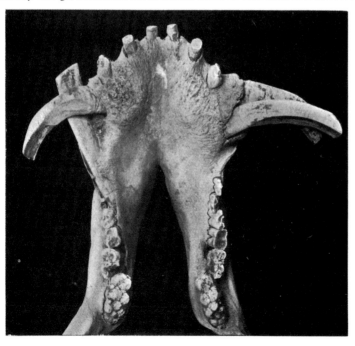

Absence of the left P_1 was noted in two specimens of the bearded pig (*S. barbatus*) (Formerly RCS Osteo. Series, 1760.3 and 1760.31).

Genus *Hylochoerus*. 18 specimens examined.

$$\frac{1}{2}:\frac{1}{1}:\frac{3}{1}:\frac{3}{3}=30$$

Hylochoerus meinertzhageni (giant forest hog)

There is variation in the number of mandibular incisors. In 18 skulls, five had three lower incisors on each side and one had four on the right and two on the left (Fig. 6.3).

In the two skulls with deciduous teeth, there were three incisors in each side of the mandible.

Genus *Babyrousa*. 58 specimens examined

$$\frac{2}{3}:\frac{1}{1}:\frac{2}{2}:\frac{3}{3}=34$$

Babyrousa babyrossa (babirusa)

In two skulls, P^2 was present and in one both P_2 and P^2. Among 11 skulls from the island Taliabu, Indonesia (BMNH 1919.11.23.1–11), six had a small tooth mesial to one or more of the P3. The tooth was double-rooted, much worn and probably deciduous; unfortunately there were no specimens from Taliabu with the rest of the deciduous dentition present.

Family **Tayassuidae**. New World pigs or peccaries

$$\frac{2}{3}:\frac{1}{1}:\frac{3}{3}:\frac{3}{3}=38$$

Upper canines of peccaries are small and the molars are brachydont and bunodont.

Genus *Tayassu*

Tayassu tajacu (collared peccary). 42 specimens examined

(i) Captive. RCS Odonto. Mus., G 26.31. (Fig. 6.4). A smaller than normal supplemental right P^2.
(ii) ♂. BMNH 1903.7.7.125. Both I_3 absent.
(iii) ♂. BMNH 1918.3.28.2. Both P_2 absent.
(iv) Captive. RCS Odonto. Mus., G 26.3. Right P^2 absent.
(v) ♂. BMNH 1921.1.1.78. (Fig. 6.5). Left P^2 absent. To the lingual side of the upper right canine, there is a small tooth with mesial and distal roots. Its crown does not resemble a canine but has the shape of a cone flattened bucco-lingually.
(vi) ♀. The left mandibular canine curves sharply downwards and outwards, instead of upwards and outwards, and has an additional smaller similarly-curved tusk-like element on its buccal surface. This double tooth has a single continuously-growing root. It must be categorized as an unusual example of connation as it must be inferred that the two elements arose from two contiguous, but separate tooth germs; then, after the formation of a centimetre or two of tusk, their formative ends must have fused into one, which then went on to form a single root. Mesial to this abnormal tusk, there is a separate supernumerary conical tooth (Neal and Kirkpatrick, 1957).

There were no variations in 20 specimens of white-lipped peccary (*Tayassu albirostris*).

Fig. 6.3. *Hylochoerus meinertzhageni* (great forest hog). ♂. BMNH 1904.11.5.14. Part of mandible. Two incisors on the left and four on the right. Several teeth are worn and chipped.

Section 1: Variations in Number, Size and Shape

Family **Hippopotamidae**

Genus *Hippopotamus*. 38 specimens examined

$$\frac{2}{2}:\frac{1}{1}:\frac{4}{4}:\frac{3}{3}=40$$

In hippopotamuses, the incisors as well as the canines are tusk-like and are of continuous growth. The molars are brachydont and bunodont.

In some of the specimens, one or both I_2 were missing but in every instance the condition of the bone suggested that the teeth had been present. P1 is sometimes absent but there is usually evidence that it was lost during life.

A skull was formerly in the Royal College of Surgeons Museum in which the left M_1 was absent (Fig. 6.6). On this side, M_2 and M_3 had moved mesially. There was a space of about 22mm between the P_2 and P_3 on both sides; there was some variation in the size of contralateral teeth in this specimen. It was evident from the sockets of P_3 that the right tooth was smaller than the left; the right P_4 was slightly larger than the left tooth and there was a considerable difference in the size of the M_3; the distal cusp of the left M_3 had not developed; its length was 56.5mm compared with the 65.1mm of the right tooth.

Another example of variation in the size of contralateral teeth is shown in Figure 6.7, where the left P^3 is larger than the right and the right P^4 is larger than the left. The left P^2 has an extra root.

The mandible in Figure 6.8 has three teeth in the position of each P_4, two being lingual and one buccal. The distal ones on both sides, in juxtaposition to M_1, have distal cusps and resemble normal P_4; the others are premolariform but lack the distal cusp.

Sub-order **RUMINANTIA**

Family **Tragulidae**. Chevrotains

$$\frac{0}{3}:\frac{1}{1}:\frac{3}{3}:\frac{3}{3}=34$$

Genus *Tragulus*. 106 specimens examined

Chevrotains have no upper incisors, well-developed upper canines and incisiform lower canines; their molars are selenodont and brachydont.

(i) *Tragulus javanicus fulviventer* (lesser Malay chevrotain). BMNH 53a. (Fig. 6.9). Both M^3 conical.

Family **Cervidae**. Deer

$$\frac{0}{3}:\frac{0-1}{1}:\frac{3}{3}:\frac{3}{3}=32-34$$

In deer, the lower canine is incisiform and researchers on dental variations usually refer to it as a fourth incisor. The upper canines are normally lacking, but are present in the deer taxa that lack antlers, musk deer (*Moschus*) and Chinese water deer (*Hydropotes*). In taxa in which the males have small antlers, muntjacs (*Muntiacus*) and tufted deer (*Elaphodus*), upper

Fig. 6.4. *Tayassu tajacu* (collared peccary). RCS Odonto. Mus., G 26.31. An extra tooth mesial to the right P^2 and similar to it in shape. ×c. 0.5.

Fig. 6.5. *Tayassu tajacu* (collared peccary). ♂. BMNH 1921.1.1.78. Left P_2 absent; an extra tooth lingual to the right canine. ×c. 0.67.

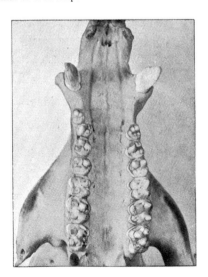

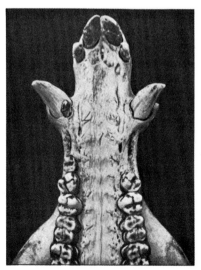

canines are usually present in both sexes, and they are often present in brocket deer (*Mazama*) (Helminen, 1958). The molars of all deer are selenodont and brachydont.

Variations in the number of teeth are uncommon in deer, for example, in 191 sika deer (*Cervus nippon*) in Maryland, USA, Feldhamer (1982) found that none had either supernumerary or absent teeth.

The absence of one or more incisiform teeth occurs in varying degrees in different populations. Meyer (1975) found only one example in 6000 north German roe deer (*Capreolus capreolus*); Miller and Tessier (1971) found that an incisor was absent in 2% of 1176 caribou (*Rangifer tarandus*); Jackson (1976) noted absence in only 1.8% of 399 New Forest fallow deer (*Cervus dama*) but, in confined populations, one or more incisors can be absent in as many as 18.7%, as Chapman and Chapman (1969) found in the fallow deer of Richmond Park. This is highly suggestive of genetic drift (p. 10).

Although supernumerary incisiform teeth are usually less common, Peterson (1955) found one or more in 7.5% of 320 adult moose (*Alces alces*) in Ontario. Single instances in white-tailed deer (*Odocoileus virginianus*) were noted by Fowle and Passmore (1948) and Mech *et al.* (1970), in mule deer (*O. hemionus*) by Robinette (1958) and by De Calesta, Zemlicka and Cooper (1980), in moose (*Alces alces*) by Steele and Parama (1979), in caribou (*Rangifer tarandus*) by Miller and Tessier (1971) and in roe deer by Meyer (1975).

Although most deer normally lack the upper canine, it does occur quite often in both sexes, though it is usually shed early in life. Heath (1976) found it in 10 out of 34 fallow deer (*Cervus dama*) in the English midlands, and Chapman and Chapman (1973) in 25% of new-born fallow deer fawns in Richmond Park, but in only 0.5% of older animals, a figure similar to the 0.8% occurrence that they noted for 389 fallow deer in other parts of southern England. In wapiti (American elk, *Cervus canadensis*), small rudimentary upper canines with deciduous predecessors occur in both males and females (O.J. Murie, 1951); they do not always erupt fully, are incompletely mineralized (Seliger, Erickson and Denney, 1969) and are smaller in females, according to Greer and Yeager (1967). In reviewing the literature of the occurrence of upper canines in white-tailed deer, Krausman (1978) showed that there was considerable geographic variation (from 1.7% to 17.9%) in the proportion of deer with additional canines in different localities (Southwick, 1954); it was particularly high in one isolated population in Texas. The findings of Brokx (1972) were similar,

Fig. 6.6. *Hippopotamus amphibius* (hippopotamus). RCS Osteo. Series, 1863. Left M_1 absent. The specimen was prepared from an animal shot in Central Africa by E.D. Young, Commander of the Livingstone Search Expedition in 1867. $\times c. 0.29$.

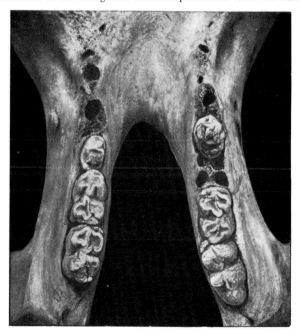

112 Section 1: Variations in Number, Size and Shape

suggesting genetic control of this character. Upper canines also occur quite frequently in roe deer (Kratochvil, 1984; Meyer, 1975; Borg, 1967).

Upper canines are occasionally not only present but duplicated; single examples are known in *Cervus* (Kinghorn and Yeager, 1951; Greer and Yeager, 1967) and in roe deer (De Beaufort, 1964).

Most cervids lack P1, but it is sometimes present in the mandible. Moran and Fairbanks (1966) found three instances in the Michigan herd of wapiti (*Cervus canadensis*) which was imported into the area in the early 1900s. It is common in roe deer (Prior, 1968; Virchow, 1940; Jackson, 1975; Kratochvil, 1984; Meyer, 1985) and in *Odocoileus* (Mech et al., 1970); Wing (1965) found several examples in sub-fossil white-tailed deer in Florida.

The incidence of absence of P_2 is variable. None of the 33 337 white-tailed deer (*Odocoileus virginianus*) examined by Ryel (1963) in Michigan lacked P_2; one or both were absent in 1.2% of 422 deer in New York State (Free *et al.*, 1972) and in 2.0% of 401 deer in northern Minnesota (Mech *et al.*, 1970) but, among 24 skulls from a small area of Maryland, three lacked P_2 on both sides and one on one side, an incidence of 16.7% suggesting genetic drift in an isolated population (Feldhammer and Chapman, 1980).

In his study of 580 *Cervus elaphus* (red deer) and 130 *C. canadensis* (wapiti) introduced into New Zealand, Pekelharing (1968) suggested that the few dental anomalies found might be useful for tracing the spread of rare genes in *Cervus* after introduction. He found only two anomalies, one a supernumerary M_3 in *C. elaphus*, which resembled the normal M_3 in having three pillars but they were conical and separate above

Fig. 6.7. *Hippopotamus amphibius* (hippopotamus). Formerly RCS Osteo. Series, 1860.
Left: the right P^3 is smaller than the left and the left P^4 is smaller than the right. ×c.0.31.
Right: the two P^2 are of equal size, but the left one has an extra root on its lingual aspect. Reduced.

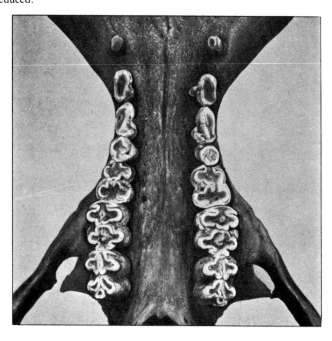

the gum line, the other a supernumerary M^3 in *C. canadensis*. Miller and Tessier (1971) found one upper and one lower disto-molar in 1176 caribou in N. America. Meyer (1979a) described a fallow deer with a large, undifferentiated tooth distal to the left M_3 as well as enamel pearls on the lateral surface of the M_3. Single instances of mandibular disto-molars in white-tailed deer are given by Abler and Scanlon (1975), Verme (1968) and Free *et al.* (1972).

Tooth shape can be variable in deer; for example, in 271 sub-fossil white-tailed deer from Florida, Wing (1965) found that the ectostylid was present on P_4 in 23%. In the same species, the distal column of M_3 is sometimes lacking (Guilday, 1961), as it is in roe deer (Kratochvil, 1984). Kierdorf and Kierdorf (1987) described a roe deer skull in which the right P_4 had been replaced with a large haplodont tooth.

Fig. 6.8. *Hippopotamus amphibius* (hippopotamus). RCS Odonto. Mus., G 26.2. Three premolariform teeth in the position of each P_4. One has been lost from the right side. ×c. 0.26.

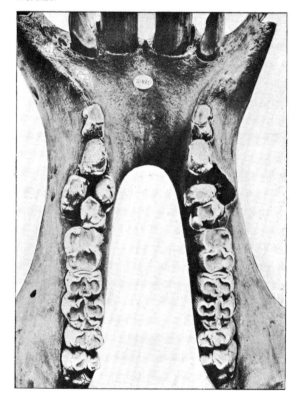

Fig. 6.9. *Tragulus javanicus fulviventer* (lesser Malay chevrotain). BMNH 53a. Both M^3 conical.

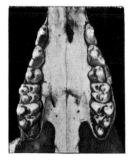

Genus *Muntiacus*. 98 specimens examined

(i) *Muntiacus muntjak vaginalis* (Indian muntjac). ♂. BMNH 1922.5.16.72. (Fig. 6.10). Bilateral supplemental P^2. At about the centre of the buccal surface of the right M^2 is a prominent flange-like ridge extending from the occlusal surface to the cervix.
(ii) *M. muntjak grandicornis*. ♂. BMNH 1914.12.8.239. Both M_3 absent.
(iii) *M. reevesi reevesi* (Chinese muntjac). ♂. BMNH 1910.5.26.2. The left M_3 is absent, the right M_3 is smaller than normal, less broad than the right M_2 and has no distal column.
(iv) *M. crinifrons* (black muntjac). BMNH 1991a. (Fig. 6.11). In some bovids, an accessory column is present on the buccal surface of the mandibular molars; in this specimen, the column is unusually well-developed on the left M_3 and is supported on a separate root. The right M_3 also shows in a lesser degree a similar column supported on a slender root.

Genus *Cervus*

A skull of *Cervus axis* (spotted deer) described by Dönitz (1872) had several additional teeth – a disto-molar in the left maxilla and, in the mandible, a molar on each side situated lingually to M_2, as well as a small disto-molar consisting of two columns on the right.

Genus *Hydropotes*. 11 specimens examined

(i) *Hydropotes inermis* (Chinese water deer). ♂. Formerly RCS Osteo. Series, 1651. (Fig. 6.12). Bilateral supplemental P^3.

Genus *Capreolus*. 59 specimens examined

Capreolus capreolus (roe deer)

(i) ♂. Anatomischen Institut der Tierärztlichen Hochschule, Hannover. Four supernumerary

114 Section 1: Variations in Number, Size and Shape

incisiform teeth of which three are probably incisors and one a canine (Meyer, 1979c).
(ii) A second row of incisors, consisting of two teeth, immediately lingual to the normal incisors (Geiger, 1980).
(iii) ♂. Right I_1 absent. (Meyer, 1975).
(iv) ♀. BMNH 1911.2.1.264. Right P_2 absent.
(v) P_2, P_4, P^2 and P^3 absent on both sides (Meyer, 1979d).
(vi) ♂. A malformed extra molar between the right M_2 and M_3 (Meyer, 1975).

Fig. 6.10. *Muntiacus muntjak vaginalis* (Indian muntjac). ♂. BMNH 1922.5.16.72. Bilateral supplemental P^2. The left M^2 has a prominent additional buccal ridge (pointer).

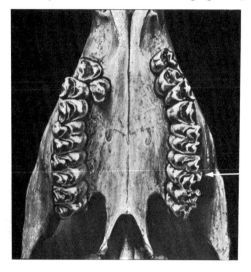

Fig. 6.11. *Muntiacus crinifrons* (black muntjac). BMNH 1991a. An extra element on the buccal surface of the left M_3 associated with an accessory root. The right M_3 has the same condition to much less degree, but cannot be seen.

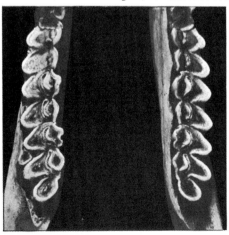

(vii) ♂. BMNH 1911.12.5.1. Both M_3 absent; the right M^3 is also absent but may have been lost from disease.
(viii) ♂. Anatomischen Institut der Tierärztlichen Hochschule, Hannover. Both M_3 and both M^3 absent (Meyer, 1979b).

Family **Bovidae**. Cattle, antelope, sheep, goats

$$\frac{0}{3} \cdot \frac{0}{1} \cdot \frac{3}{3} \cdot \frac{3}{3} = 32$$

Bovids lack upper incisors and upper canines, the lower canines are incisiform and the molars are selenodont and hypsodont.

As in the Cervidae, upper canines occur occasionally in bovids. Benson (1943) found them in two out of 53 skulls of bighorn sheep (*Ovis canadensis*) and Bunch, Hoefs and Glaze (1984) found them in two out of 130 males of dall sheep (*Ovis dalli*). In the royal antelope (*Nesotragus pygmaeus*), oribi (*Ourebia orebi*), dik-dik (*Madoqua* spp.) and gazelles (*Procapra* sp. and *Gazella* spp.), they seem to be present only in young animals (Major, 1904; Ritchie, 1940; Dekeyser and Derivot, 1956; Lönnberg, 1937), and so presumably are soon lost.

Henrichsen (1981) found that 24% of 131 musk oxen (*Ovibos moschatus*) from eight localities in N.E. Greenland lacked one or both lower permanent canines; he considered this to be genetically controlled and related to the isolation of the various herds (p. 450).

Although P_1 and its deciduous precursor are normally lacking, teeth thought to be m_1 were found in both mandibles of one skull of American bison (*Bison*

Fig. 6.12. *Hydropotes inermis* (Chinese water deer). RCS Osteo. Series, 1651. Bilateral supplemental P^3; one of each pair is displaced from the arch and is therefore less worn. × c. 0.66.

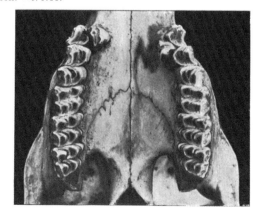

bison) and a right P_1 was found in another among about 626 animals in northern Alberta (Fuller, 1954).

There seems to be a tendency in bovids for P_2 to be lacking as well as P_1. This has been noted occasionally in domestic cattle (Garlick, 1954a), in American bison (*Bison bison*) (Van Vuren, 1984), in feral goats (*Capra hircus*) (Rudge, 1970), in impala (*Aepyceros melampus*) (Child, 1965, 1969) and in wildebeest (*Connochaetes taurinus*) (Talbot and Talbot, 1963). Its absence is particularly common in isolated populations, as Ratti and Habermehl (1977) found in 19.4% of males and 17.8% of females in a herd of ibex (*Capra ibex*) in a Swiss population. The phenomenon is well known in domestic cattle from British prehistoric sites (Andrews and Noddle, 1975); in a Roman site in London, 7% of 188 mandibles had this condition (Meek and Gray, 1911). That the control is genetic is emphasized by numerous occurrences in the small and totally isolated feral herd of cattle at Chillingham Park (Grigson, forthcoming). Founder effect is probably implicated in the high incidence of absence of one or more cheek teeth found by G. Niethammer (1971) among chamois (*Rupicapra rupicapra*) living on Mount Cook, New Zealand, as they are descended from only one male and six females, introduced from Europe in 1907.

There are few records of supernumerary teeth in bovids, but Van Vuren (1984) found three in 27 skulls from a single herd of American bison in Utah; one had a supplemental P_3 and m_3 and each of the two others had an unerupted, deformed tooth; in one it was beneath the right M_1 and in the other beneath the right P^3.

Genus **Bos**. Number of specimens not noted

Bos taurus (domestic cattle)

There were no variations in the specimens examined but missing premolars and deficient molars are discussed above.

Morot (1887) referred to a calf with nine deciduous incisiform teeth and Cadéac (1910) and Kitt (1921) figured mandibles of cattle with 12 and 10 incisors respectively. Morot (1886b, p. 321) recorded a supplemental I_2. A mandible with bilateral supplemental I_2 was figured by Joest (1926, fig. 84); on each side, one of the two I_2 was rotated so that the lingual surfaces faced each other. An extraordinary example of extra teeth in a cow was brought to Colyer's notice by J. C. M. Shaw in which there were 24 teeth in the incisor part in the mandible (Fig. 6.13). Baldi (1948) described a similar mandible in which there were 12 incisors in a double row.

Joest (1926, fig. 111) figured a mandible of an ox in which both I_3 and both canines were absent and their deciduous precursors retained. The right I_1, both I_2 and a supernumerary tooth in the position of the left I_3 were enormous. The incisive edge of each consisted of a row of globular protuberances. The complexity of the distortion of shape and size is such that these teeth could be categorized as hamartomas or odontomes (Chapter 25, p. 575). However, as we have pointed out, no hard and fast distinction can be drawn between a disturbance of development that gives rise to a supernumerary tooth of abnormal shape and an odontome.

According to Morot (1886c), who described a calf with a caniniform tooth in the upper jaw just behind the incisive pad, rudiments of upper canines are found in embryos of domestic cattle.

Rieck (1985) described a calf in which absence of much of the dentition was associated with generalized ectodermal dyplasia, as in man and in cats (Burroughs, Miller and Harvey, 1985).

Joest (1926, fig. 105) illustrated a mandible in which both P_3 and both canines were absent and their deciduous precursors had been retained. Morot (1898c) noted absence of an incisor in seven cattle skulls, in five of which there was a persistent deciduous incisor in place of the absent permanent tooth. He described a cow in which the canines and I_3 were absent on both sides, the deciduous canines had been retained, but on one side there was an additional conical tooth. In another cow, one of the right incisors had a simple conical (haplodont) crown; a similar instance in a bull, in which both I_1 were haplodont, was figured by Cadéac (1910). A haplodont lower canine was figured by Kitt (1892).

One example of connation in a cow was described by Morot (1898a). All the incisors were normal except for the right I_2, the crown of which was partially divided by a groove on the labial surface.

Genus **Oryx**

(i) *Oryx dammah* (scimitar-horned oryx). Bilateral supplemental P_4 (Oboussier and Storkmann, 1975).

Genus **Alcelaphus**. 67 specimens examined.

(i) *Alcelaphus buselaphus cokei* (Coke's hartebeeste). RCS Odonto. Mus., G29.1. (Fig. 6.14). In the mandible, there are two similar teeth in the position of each P_4. Situated side by side, they are similar in pattern, those on the lingual side being slightly smaller than those on the buccal side.

Section 1: Variations in Number, Size and Shape

Genus ***Ourebia***. 58 specimens examined.

Orebia ourebi (oribi)

(i) ♀. BMNH 1846.19.26.18. (Fig. 6.15). A supplemental right I_2.

(ii) ♂. BMNH 1915.3.6.40. Both P_2 smaller and more conical than normal.

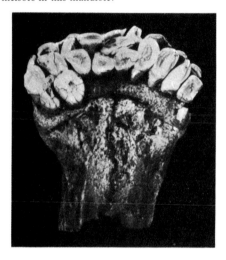

Fig. 6.13. *Bos taurus* (domestic ox). Ouderstepoort Veterinary College, South Africa. There were said to be 24 incisors in this mandible.

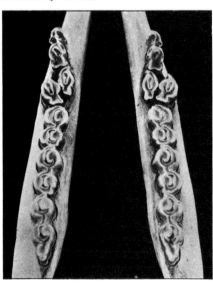

Fig. 6.14. *Alcelaphus buselaphus cokei* (Coke's hartebeeste). RCS Odonto. Mus., G20.1. Bilateral supplemental P_4. ×c. 0.66.

Genus ***Nesotragus***. 38 specimens examined.

(i) *Nesotragus moschatus* (suni). ♂. BMNH 1893.2.1.2. (Fig. 6.16). To the buccal side of each mandibular canine, there is a stunted tooth with three small cusps.

Genus ***Madoqua***. 130 specimens examined.

(i) *Madoqua piacentinii* (dik-dik). ♂. BMNH 1912.12.28.38. Both P_2 absent.

(ii) *M. kirkii* (Damaraland dik-dik). ♀. BMNH 1916.6.23.1. (Fig. 6.17). An accessory conical cusp on the lingual aspect of each P^4.

Genus ***Ammodorcas***. 5 specimens examined

In the mandible of a male dibatag *Ammodorcas clarkei* (BMNH 1891.12.19.4), both presumptive P_2 are small conical teeth in the diastemata about 15 mm mesial to P_3.

Fig. 6.15. *Ourebia ourebi* (oribi). BMNH 1846.10.26.18. A supplemental right I_2. ×c. 2.0.

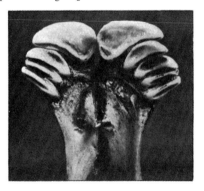

Fig. 6.16. *Nesotragus moschatus* (suni). ♂. BMNH 1893.2.1.2. An extra tooth with three small cusps distal to the left mandibular canine. The condition is bilateral. ×c. 3.0.

Genera *Gazella* and *Procapra*. 170 specimens examined.

(i) *Procapra picticaudata* (Tibetan gazelle). ♂. BMNH 1922.1.21.1. Right P_2 absent.
(ii) *Gazella rufifrons kanuri* (red-fronted gazelle). Supplemental left P_4 (Oboussier and Storkman, 1975).
(iii) *Gazella rufifrons* (red-fronted gazelle). BMNH 1900.8.6.10. (Fig. 6.18). The distal column in both M_3 is exceptionally large.

Genus *Ovis*. Specimens examined: 44 wild, 70 captive (mandibles were missing in 46).

Extra incisors occur occasionally (Magitot, 1877, pl. VI, fig. 3; Morot, 1887), but there were none in the

Fig. 6.17. *Madoqua kirkii* (Damaraland dik-dik). ♀. BMNH 1916.6.23.1. Each P^4 has a cusp on its lingual aspect and is symmetrically rotated with buccal surface turned lingually.

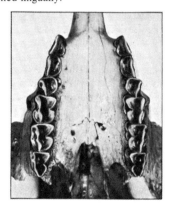

Fig. 6.18. *Gazella rufifrons* (red-fronted gazelle). BMNH 1900.8.6.10. The lingual aspect of an M_3 with an exceptionally large distal column which is also seen in the occlusal view below.

specimens examined. Hoefs (1974) found an extra P^3 in each maxilla of a dall sheep (*Ovis dalli*) that lacked both P_2. Lower disto-molars were seen in two domestic sheep (*Ovis aries*), one of which is RCS Odonto. Mus. G 28.1, and a bilateral example was recorded by Magitot (1877, pl. V.10). R. M. S. Taylor (1986) mentioned upper disto-molars in two sheep in New Zealand.

Four nodules of enamel were present in the mesial fosette on the left M^3 of a merino sheep (Fig. 6.19). Morot (1886a, p. 319) mentioned instances in sheep in which the canines were more conical than incisiform; later (1888c) he referred to a triple tooth in the lower incisor region, that is a tooth composed of a cluster of three crown-elements arising from a single root.

Genus *Capra*. No specimens examined by Colyer.

(i) *Capra hircus* (domestic goat). Two kids lacking both deciduous canines (Morot, 1887).
(ii) *C. hircus*. Two kids lacking one deciduous canine (Morot, 1887).
(iii) *C. hircus*. Incisors and canines absent; associated with a shortened mandible (Emele-Nwanbani and Ihemelandu, 1984).
(iv) *C. hircus*. An extra deciduous incisiform tooth (Morot, 1887).
(v) *C. hircus*. An extra incisor in the left mandible (Morot, 1886b, p. 321).
(vi) *Capra ibex* (ibex). An extra canine in the right mandible (Eschler and Lüps, 1984).

Family **Antilocapridae**. Pronghorns

$$\frac{0}{3} \cdot \frac{0}{1} \cdot \frac{3}{3} \cdot \frac{3}{3} = 32$$

The teeth are similar to those of bovids. Eskelsen and Fichter (1985) described a skull of *Antilocapra americana* with an additional M_2 and M_3 in the left mandible.

Fig. 6.19. *Ovis aries* (merino sheep). RCS Odonto. Mus., A 180.8. Occlusal surfaces of left (A) and right (B) M^3. Four nodules of enamel are present in the mesial fossette of (A). ×c. 2.0.

A B

Section 1: *Variations in Number, Size and Shape*

Order **PERISSODACTYLA**

Family **Equidae**. Horses, zebras, asses

$$\frac{3}{3}:\frac{0-1}{0-1}:\frac{3}{3}:\frac{3}{3} = 36\text{--}40$$

Equids have broad hypsodont incisors; canines are usually present only in males and are situated in the diastema; the premolars are molariform and, like the molars, are markedly hypsodont.

Genus *Equus*.

Equus caballus (horse). 484 specimens examined

EXTRA TEETH AND VARIATIONS IN SHAPE
Incisors
According to Bateson (1894), supernumerary incisors are common in the horse but he may have been misled by the prevalence of such specimens in veterinary museums where specimens tend to be selected for abnormalities. Colyer (1906) found only three instances (0.6%) among the 484 specimens he examined systematically.

Supernumerary incisors are more common in the maxilla than in the mandible. The extra teeth may be in regular alignment with the other incisors, as in Figure 6.20. However, the majority of extra teeth cause a distorted arrangement of the other incisors, as in Figures 6.21, 6.22 and 6.23.

In most of the examples shown, the teeth are supplemental and consequently it is impossible to be

Fig. 6.21. *Equus caballus* (horse). RCS Odonto. Mus., G 76.11. There are seven upper incisors in a crowded arch so that one is rotated and another displaced labially. × *c*. 0.66.

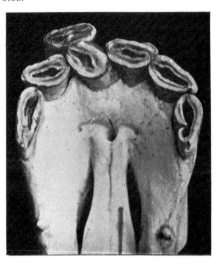

Fig. 6.22. *Equus caballus* (horse). RCS Odonto. Mus., G 27.23. A supplemental I^1 on each side, lingual to the normal I^1. × *c*. 0.66.

Fig. 6.20. *Equus caballus* (horse). RCS Odonto. Mus., G 27.1. Seven upper incisors, instead of six, in regular alignment. × *c*. 0.66.

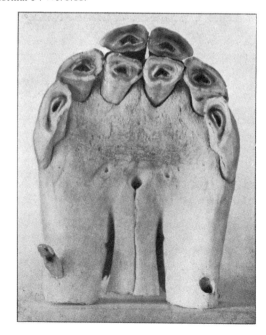

sure which are supernumerary. In other instances, one or more of the teeth are of abnormal morphology Morot (1888b) and can therefore be assumed to be the supernumeraries (Figs. 6.24 and 6.25).

Unilateral or asymmetrical abnormalities of incisor development are met with; for example, in the upper jaw depicted in Figure 6.26, there is an extra incisor on the right side; on the left, the I^3 is abnormally large. The upper jaw in another specimen (Fig. 6.27) has four incisors on each side. Those in the position of the right I^2 and left I^1 are both of similar

abnormal shape; their labial surfaces are irregular and concave instead of smoothly convex. On the left, the distal tooth is abnormally large and bears evidence of double origin, that is of connation. The mark is divided

Fig. 6.25. *Equus caballus* (horse). RCS Odonto. Mus., G 27.22. Five upper incisors on the right side. The left I^1 is broken. *a* is probably I^1 and *b* is similar in morphology to I^2 on the opposite side. *e* is similar to I^3; *c*, *d* are similar, slightly smaller than any of the other incisors and are of slightly distorted morphology. They are probably the supernumerary teeth. ×c. 0.66.

Fig. 6.23. *Equus caballus* (horse). RCS Odonto. Mus., G 27.2. The four upper incisors on each side are so similar in shape that it is not possible to ascertain whether the two displaced labially are supernumerary. ×c. 0.66.

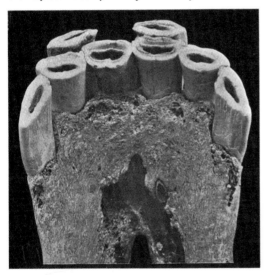

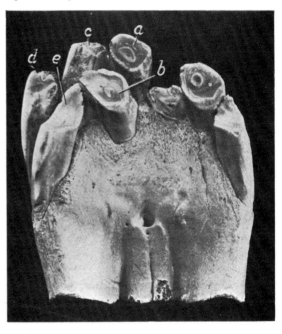

Fig. 6.26. *Equus caballus* (horse). A supernumerary upper incisor on the right; left I^3 abnormally large. Palatal mucosa *in situ*. ×c. 0.66.

Fig. 6.24. *Equus caballus* (horse). Four upper incisors on each side. The two displaced lingually are of slightly abnormal morphology and are therefore probably the supernumerary teeth. Palatal mucosa *in situ*. ×c. 0.66.

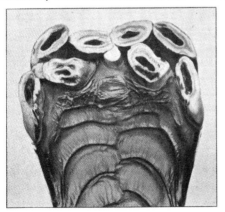

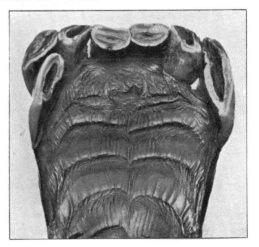

by a bucco-lingual septum opposite which, on the labial surface, is a longitudinal groove.

Some other examples of connation of incisors are shown in Figures 6.28, 6.29, 6.30 and 6.31. That depicted in Figure 6.29 is of special interest because the crown has the morphology of two teeth separated mainly by longitudinal grooves, but it is borne on two separate roots. It has already been pointed out that, in connation, the degree of separation tends to be greater towards the incisive edge or occlusal surface (p. 8). The specimen in Figure 6.30 is similar in that there are two crowns joined together and yet the roots are separate. It is furthermore an example of connation in the deciduous dentition. In the specimen depicted in Figure 6.31, the two incisor elements are situated

Fig. 6.27. *Equus caballus* (horse). RCS Odonto. Mus., G27.21. Four incisors on each side. *a* and *b* are abnormally shaped, especially their labial surfaces which are irregularly concave. *a* is abnormally small. The most distal incisor on the left is unusually large and the mark is divided by a septum. ×*c*. 0.66.

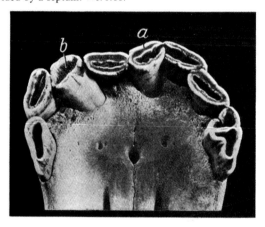

Fig. 6.28. *Equus caballus* (horse). RCS Odonto. Mus., G27.5. Both I^3 show connation. ×*c*. 0.66.

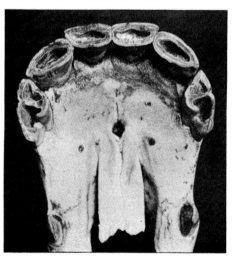

Fig. 6.29. *Equus caballus* (horse). RCS Odonto. Mus., G27.41. Right I^2 enormous. The crown is composed of two elements. Each of these has a separate root which is unusual in connation. ×*c*. 0.66.

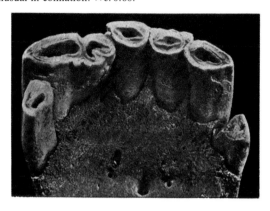

Fig. 6.30. *Equus caballus* (horse). RCS Odonto. Mus., G27.4. Upper deciduous incisors showing connation between the crowns of the left i^1 and i^2. However, the roots are separate. Bone removed from palate has been replaced. Reduced slightly.

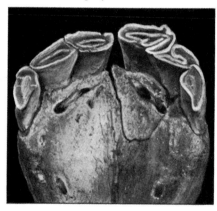

Fig. 6.31. *Equus caballus* (horse). Connation between the right I_2 and a supernumerary element A on its labial surface (Dupas, 1903).

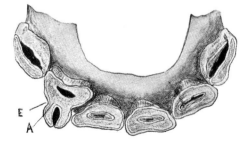

bucco-lingually to each other whereas nearly always in connation the composite elements lie in the line of the arch.

There are some notable examples of large numbers of extra incisors in equids. Lafosse (1772) referred to horses with a double row of incisor teeth and one is figured by Joest (1926); Goubaux (1854) claimed to have seen a horse with such a double row in both the upper and lower jaws. The lower jaw of a horse in which there are 12 well-formed incisors arranged as a double row was figured by Cadéac (1910).

Instances of extra incisors are listed in Kitt (1921, p. 459), M'Intosh (1929, p. 361) and Bradley (1907); five examples recorded by Magitot (1877) are worthy of description here because they are so clearly depicted in beautifully reproduced drawings.

(i) An extra incisor labial to the left I^1 but lying across the midline obliquely so that the crown is labial to the right I^1 (Magitot, 1877, pl. IV, figs. 14 and 15).

(ii) ♀. An extra incisor on the lingual aspect of left I_2 (pl. VI, fig. 1).

(iii) Three extra upper incisors. In the figure (pl. VI, fig. 9), all the incisors are similar but, according to Magitot, there was a supernumerary tooth lingual to the right I^1, another in the position of the left I^1 and a third lingual to it.

(iv) ♀. Two extra upper incisors which, from the figure (pl. VI, fig. 10), were more or less in the positions of I^2, the left one is much larger than normal.

(v) ♀. Two extra upper incisors. One appears to be a supplementary right I^2; the one on the left is lying transversely across the palate (pl. VI, fig. 11).

Morot (1888c) described a double lower incisor in a horse. It was composed of a large cylindrical incisiform tooth to which was attached a smaller one with about 3 mm separating the two crowns. The two were fused for about half their length and demarcated only by the lingual and labial grooves; there was a single conical root-end. Two other examples of connate permanent incisors were recorded by Cadéac (1910), one in the upper jaw and one in the lower; another example was figured by Kitt (1892, p. 173).

(b) Canines

Canines are a characteristic of male equids but rudimentary canines are commonly present in females. Colyer found one or more in 27.8% of 173 mares (Table 6.1).

There are few records of extra canines. Morot (1889) described two instances; in one, there were two canines in the right maxilla of an ass (*Equus asinus*), the two teeth being in the same socket; in the second, the extra canine was in the left maxilla. Figure 6.32 shows two caniniform teeth in each maxilla distal to the incisors.

(c) Premolars

The presence of P_1 was noted in only three skulls, but P^1 occurred more commonly; it was present on one or both sides in 16% of the horse skulls (Table 6.2). These so-called wolf teeth were rudimentary in form and occasionally associated with supernumerary incisors or molars.

In the specimens in which the sex was noted, P1 was present in 21.4% of 173 females, and in 14.9% of 208 males.

P1 varies in size from a diminutive to a fairly well-developed tooth. The morphology of wolf teeth was described in some detail by Petit (1938). Many seem to be retained deciduous teeth. Bradley (1907) found them in 14 foals aged less than 78 months, in the upper or lower jaws or in both. In one mare, the right P^1 was

Table 6.1 Canines in FEMALE HORSES. The numbers of specimens with rudimentary canines amongst 173 skulls

Jaw	Side	With canine No.	%
maxilla and mandible	both	10	5.8
maxilla	both	1	0.6
mandible	both	22	12.7
	right	10	5.8
	left	5	2.9
Total with canine		**48**	**27.8**

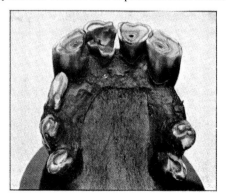

Fig. 6.32. *Equus caballus* (horse). Upper jaw with bilateral supplemental canines. Part of palatal mucosa *in situ*.

Section 1: Variations in Number, Size and Shape

well-developed and situated on the lingual aspect of P².

Other supernumerary premolars are rare but some instances were described by Kitt (1892). The skull in Figure 6.33 has bilateral supplemental P², and an extra incisor on the right side.

Figure 6.34 shows a P³ of abnormal shape. The mesial surface has a deep complex infolding which during development has displaced the mesial buccal ridge and made it more prominent.

(d) Molars
Disto-molars were present in 12 out of 484 skulls, mostly bilaterally, five in the mandible, three in the maxilla, and four in both jaws. Of the instances where the sex was known, four were female and three male.

Menveux (1898), Morot (1896) and Kitt (1892) recorded instances of unerupted bilateral upper disto-molars in horses. A mule with bilateral upper disto-molars was described by Cadéac (1910); this may be the same animal referred to by Morot (1898b, p. 49).

Connation involving cheek teeth sometimes occurs. Section of the massive M¹ in Figure 6.35 showed it to have the morphology of a molar connate with a supernumerary molar.

ABSENCE OF TEETH

Absence of teeth is rare in horses but a few instances have been recorded Morot (1888a). Morot (1885) referred to the absence of teeth in two half-sisters with the same mother. In one, there were only four deciduous upper incisors; at four years of age, they had been replaced by four permanent teeth. In the mandible, m₂ and m₃ had been replaced by permanent teeth, but i₃ had not. The other half-sister had only four upper deciduous milk incisors. At five years of age, there were six mandibular permanent incisors but the left I³ was absent; on the other hand, a well-formed supernumerary was present lingual to the right I³.

Bourdelle (1904) described a mare in which there was total absence of the lower left cheek teeth which appears to have been congenital because the slight wear of the teeth in the upper jaw could be accounted for by contact with the edentulous gum. The absence of I₂ in a stallion was described by Dupas (1904) and of the lower left canine in a stallion in Joest (1926).

Family **Tapiridae**. Tapirs

$$\frac{3}{3} \cdot \frac{1}{1} \cdot \frac{4}{4} \cdot \frac{3}{3} = 44$$

The incisors of tapirs are chisel-shaped, the canines are

Fig. 6.33. *Equus caballus* (horse). ♀. Bilateral supplemental P². On the right both P² are in the line of the dental arch and P¹ is absent; on the left P¹ is present and the two P² are side by side; also an extra right incisor (Cadéac, 1910, fig. 53).

Table 6.2 First premolars (wolf-teeth) in HORSES: the number of specimens with first premolars amongs 484 skulls of both sexes

Jaw	Side	With P1 No.	%
maxilla and mandible	left	1	0.2
maxilla	both	37	7.6
	right	23	4.8
	left	16	3.3
mandible	side unknown	2	0.4
Total with P1		**79**	**16.3**

well-developed and conical and, except for P1, the premolars are molariform. The cheek teeth are brachydont, the uppers have simple lophs and the lowers are bilophodont.

Hooijer (1961) described a rare anomaly in shape in the South American tapir (*Tapiris terrestris*) (RMNH 11632). The left P^1 was normal but, in the right P^1, the usually prominent ectoloph was entirely absent and the paracone stood as a large, isolated cone in the centre of the crown.

Family **Rhinocerotidae**. Rhinoceroses

The Asian rhinoceroses (*Rhinoceros unicornis*, *R. sondaicus* and *Dicerorhinus sumatrensis*) have one large upper incisor in each premaxilla and *R. unicornis* may have a small mesial incisor as well. They have no upper canines. They sometimes have a small incisor in each mandible, mesial to the large tooth in the position of the lower canine; although it is uncertain whether this large tooth is a canine or an incisor, it is usually referred to as a canine.

In the African species (*Diceros bicornis* and *D. simum*), the anterior part of the mandible and the premaxillae are much reduced, the incisors are small and often lost in life and there are no canines. In all rhinoceroses, the premolars are molariform and the cheek teeth are brachydont. P1 is often lost early in life.

Genus ***Rhinoceros***. 45 specimens examined

$$\frac{1-2}{0-1}.\frac{0}{1}.\frac{3-4}{3-4}.\frac{3}{3} = 28-34$$

(i) *Rhinoceros unicornis* (Indian rhinoceros). Formerly RCS Osteo. Series, 2124. Three small incisors in the mandible, instead of the usual pair between the canines.

(ii) *R. unicornis*. Formerly RCS Osteo. Series, 2125 (Fig. 6.36). No mesial upper incisor on the left but the right premaxilla had a large malformed mesial incisor. The mesial lower left incisor had a bilobed crown.

(iii) *Rhinoceros sondaicus* (Javan rhinoceros). Formerly RCS Osteo. Series, 2134. Three deciduous incisors in each mandible.

Fig. 6.34. *Equus caballus* (horse). RCS Odonto. Mus., G 27.7. An abnormal left P^4 with a deep hollow on the mesial surface associated with a prominent mesio-buccal ridge. × c. 0.66.

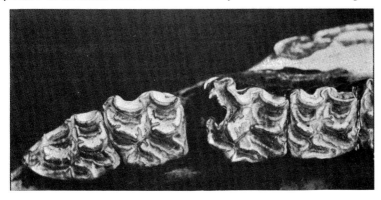

Fig. 6.35. *Equus caballus* (horse). RCS Odonto. Mus., G 27.8. An enormous molar in the position of the left M^1. The external morphology and a section through the tooth suggest connation between M^1 and a supernumerary smaller molar element × c. 0.4.

Section 1: Variations in Number, Size and Shape

Genus *Diceros*

$$\frac{1}{1-2}:\frac{0}{0}:\frac{3-4}{3-4}:\frac{3}{3} = 28-34$$

(i) *Diceros bicornis* (black rhinoceros). Left P^2 absent (Schaurte, 1965).

Order **HYRACOIDEA**

Family **Procavidae**. Hyraxes. 354 specimens examined (178 without mandibles)

$$\frac{1}{2}:\frac{0}{0}:\frac{4}{4}:\frac{3}{3} = 34$$

Hyraxes have long, continuously-growing, pointed upper incisors, premolars tending to be molariform, molars lophodont and either hypsodont or brachydont.

VARIATIONS IN NUMBER

(i) *Procavia johnstoni matschiei* (Tanzanian rock hyrax). BMNH 1924.2.25.15. (Fig. 6.37). An extra tooth of roughly premolariform morphology lingual to the right P^3.

(ii) *P. capensis* (large-toothed rock hyrax). BMNH 1869.12.23.1. Bilateral conical maxillary distomolars (Flower, 1869).

(iii) *Dendrohyrax dorsalis* (Beecroft's hyrax). ♂. BMNH 1904.7.1.119. A small conical tooth to the distal side of the left I_2.

(iv) *D. validus neumanni* (Zanzibar tree hyrax). BMNH 1906.6.5.23. (Fig. 6.38). Left M_2 and both

Fig. 6.37. *Procavia johnstoni matschiei* (Tanzanian rock hyrax). BMNH 1924.2.25.15. An extra nondescript tooth between the right P^2 and P^3.

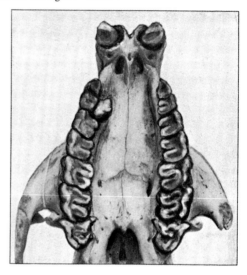

Fig. 6.36. *Rhinoceros unicornis* (Indian rhinoceros). Formerly RCS Osteo. Series, 2125. A supernumerary incisor of abnormal shape erupting mesially to the right I^1.

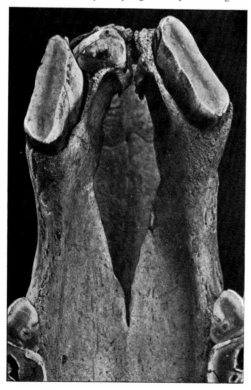

Fig. 6.38. *Dendrohyrax validus neumanni* (Zanzibar tree hyrax). BMNH 1906.6.5.23. Left M_2 and both M_3 absent. Left M_1 malformed.

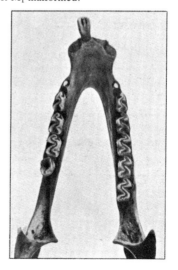

M_3 absent, almost certainly congenitally so because the opposing upper molars are unworn.

VARIATIONS IN SHAPE

The mandibular incisors in hyrax have a peculiar shape. The cutting edges are comb-like with three cusps; the two notches between these cusps are continued as grooves down the lingual and labial aspects for about 3mm. In a few specimens, the grooves amount to actual divisions as shown in Figure 6.39 (right). I_2, at times, shows a curious formation; the upper portion of the tooth is cleft, the central cusp is absent and the mesial and distal cusps diverge as seen in Figure 6.39 (left). In one male hyrax (*Procavia capensis welwitschi*, BMNH 1928.9.11.407), this occurred only in the right I_2. In the following specimens, abnormally-shaped teeth were present:
(i) *Dendrohyrax arboreus crawshayi* (tree hyrax). BMNH 1911.12.8.111. (Fig. 6.40). The right M^3 has been replaced by a conical tooth.
(ii) *D. dorsalis* (Beecroft's hyrax). ♂. BMNH 1903.1.6.3. (Fig. 6.41). The right P_2 has been replaced by a conical tooth.

Order **PROBOSCIDEA**

Family **Elephantidae**. Elephants

$$I\frac{1}{0} : C\frac{0}{0} : m\frac{3}{3} : M\frac{3}{3} = 24$$

The only incisors of elephants are the continuously-growing upper tusks (I^3). There are no canines but three milk molars and no premolars. Only one or two cheek teeth are functional at any one time, beginning with m2. As they are worn away, they are replaced horizontally by a developing tooth from behind. The molars are hypsodont, lengthened mesio-distally, lophodont, and consist of transverse plates of enamel-covered dentine held together by cementum.

Fig. 6.39. *Procavia ruficeps marrensis* (western rock hyrax). Mandibles.
Left: BMNH 1923.1.1.482. ♂. I_2 has only two abnormally divergent cusps.
Right: BMNH 1923.1.1.480. ♀. I_1 and I_2 have divisions beneath the cusps rather than the normal grooves.

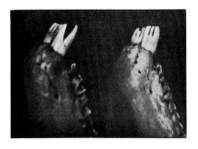

Fig. 6.40. *Dendrohyrax arboreus crawshayi* (tree hyrax). BMNH 1911.12.8.111. A small conical tooth in place of the right M^3.

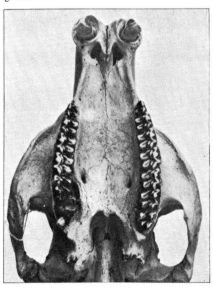

Fig. 6.41. *Dendrohyrax dorsalis* (Beecroft's hyrax). ♂. BMNH 1903.1.6.3. A conical tooth (arrow) in place of the right P_2.

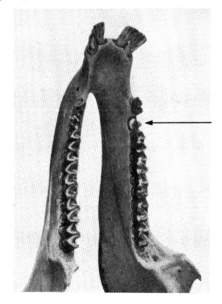

EXTRA TEETH

From time to time, reports appear of elephants with multiple tusks. There are some well-recorded examples of male African elephants with bilateral paired tusks (Anthony and Prouteaux, 1929; Anthony, 1933; Lippens, 1948; Stanton, 1949), each of near-normal size. The pairs grew either from a common socket or from two closely-adjacent sockets which were only partially separated by a bony septum. Colyer (1952) gave an account of a four-tusked African elephant and mentioned that such elephants are said often to become leaders of the troop.

Unilateral supplementary tusks are more common. One of the most remarkable was described by Junge (1942) in an Indian elephant. There was a pair of tusks on the right side, one was almost straight and the other spiralled $2\frac{1}{2}$ times around it. The morphology of the growing ends was normal as was the rest of both tusks, except that the spiral one was a little flattened against the straight one. Hence there was nothing to suggest injury as the cause.

Resembling this example to some extent was a double tusk from a male African elephant (Fig. 6.42) described by Friedlowsky (1869). The outer tusk was twisted in front of the inner one so that at the free ends, which were broken, it came to lie medially to it. A section through the two tusks showed that they were separate, though so moulded to each other that they appear to be blended to each other. The general distortion of the tusks, including their growing ends, suggests the possibility of dichotomy of the tooth germ by trauma.

Figure 6.43 shows a double tusk in an African elephant described by Anthony (1933). There is a pair of tusks of normal size and, on the right side, growing from a socket mesio-lingual to that of the definitive tusk, there is a smaller additional one that curves distally and buccally.

Examples occur of the normal tusk being replaced by a cluster of small tusks. The six small tusks shown in Figure 6.44 were removed from one side of an Indian elephant which had been thought to be tuskless. There was also a seventh tusk which was too firmly embedded in the bone to be recovered. The letter of gift says that 'the tusks were jammed together and between them a substance like marrow was found. The tusks protruded 2 or 3 inches from the socket'.

The dimensions of the tusks in cm are given in Table 6.3. Sections through four of them showed that they were composed of well-formed ivory surrounded by a layer of cementum.

The specimen in Figure 6.45 is made up of eight small tusks which vary in length from 23 cm to 37 cm. The tusks were to some extent functional as the tissues show signs of wear.

All the tusks in the two skulls just described have

Fig. 6.42. *Loxodonta africana* (African elephant). ♂. A double tusk the two spirally twisted parts of which can be separated but fit together congruously as one. Their distorted form, including that of the growing ends to the left, suggests dichotomy of the tooth germ by trauma (Friedlowsky, 1869).

Fig. 6.43. *Loxodonta africana* (African elephant). A supernumerary smaller tusk is growing from a separate socket. (From Anthony, 1933).

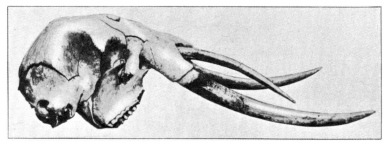

the funnel-ended roots characteristic of continuous growth.

The portion of the premaxilla shown in Figure 6.46 has six tusks in position. Five of the tusks are exposed, with flat smooth ends which suggest that they have been subjected to wear. The tusks are rounded in form and slightly curved, the largest being about 20 mm in diameter. Enough of their structure can be seen to show that they have the normal morphology of dentine surrounding a central pulp cavity.

A multiple-tusked African elephant was reported by Murray (1868). There were four tusks on one side and five on the other. From the sketches made in the field, these tusks were much larger than those in the specimens just described and projected well beyond the mouth, some curving forwards and others downwards. However, the field description leaves open the possibility that one or more of the tusks on each side grew from the mandible. Mandibular tusks in living elephants seem never to have been described before; they would be of special interest because some of the fossil ancestors or relatives of the elephants had mandibular tusks as well as premaxillary ones. Anthony (1933) described a specimen of fossil elephant (*Tetrabelodon turicensis*) with an extra tusk beside the normal ones on the premaxilla.

In interpreting examples of multiple tusks, it must be borne in mind that the elephant tusk has a transitory deciduous predecessor which is normally shed by the second year. The deciduous predecessor is not of continuous growth and has a blunt, cone-shaped end to the root. Its retention beside the definitive tusk would

Table 6.3 The dimensions in cm of six small tusks removed from one side of the jaw of an INDIAN ELEPHANT *(Elephas maximus)* (Fig. 6.44)

Tusk no.	Length	Circumference
1	61	19
2	56	15.3
3	42	12.7
4	42	7
5	28	7.3
6	26	7.6

Fig. 6.44. *Elephas maximus* (Indian elephant). RCS Odonto. Mus., G132.21. Left: six tusks from the same side of a rogue elephant from Assam. ×c.0.14. Right: their probable relationship to one another.

simulate a double tusk. It is noteworthy that two at least of the tusks in the cluster of six depicted in Figure 6.46 had conical roots and therefore must be regarded as deciduous.

In considering the causative mechanism of multiple tusks, Anthony (1933) discussed the possibility of supernumerary tooth germs, which could be of atavistic significance, and dichotomy of tooth germs as a cause of duplication of tusks. Figure 6.47 shows an early stage of formation of a tusk which has a deep longitudinal groove partially dividing it. Had formation of the tusk continued, presumably the morphological growth would have produced a tusk like that depicted in Figure 6.48 which is much longer and is deeply grooved on both sides for the whole of its length. It could, of course, be speculated that such specimens arise from the fusion of two primordial tooth germs, one being a supernumerary germ. Either of these specimens could, had the grooving deepened as formation proceeded, have come to resemble the tusk in Figure 6.49 which consists of two tusks which are irregularly flattened on their contiguous sides; these have longitudinal grooves and ridges which are closely congruous. Fitted together, the two tusks form a cylinder about 12 cm in greatest diameter and nearly circular in cross-section. The appearance is strongly suggestive of a tusk growing in a confined space from a tooth germ that had divided into two.

Reade (1930) described an Indian elephant with the right tusk straight, compared to the normally curved one on the left, with a deep longitudinal groove on one side, containing a fleshy mass.

The argument in favour of trauma as a cause of multiple tusks and of partially dichotomous tusks depends largely on the formation of tusklets which is a

Fig. 6.46. *Elephas maximus* (Indian elephant). RCS Odonto. Mus., G132.22. Palatal view of a portion of a maxilla with six small tusks. The tusk (5) is quite small, 9.5 cm long, and the root tapers to a pointed extremity, as does the root of tusk (4) which is implanted in the bone about 5 cm. The largest tusk (1) is 33 cm long and tusk (2) is 29 cm long. Both have growing funnel-shaped ends. In the space between (1) and (4) there is a fragment of tusk (6) lying loose; its exact relationship cannot be determined. ×c. 0.43.

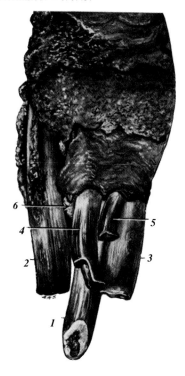

Fig. 6.45. Elephant. RCS Odonto. Mus., G132.2. Eight small tusks from the same side of an elephant.

Fig. 6.47. *Elephas maximus* (Indian elephant). RCS Osteo. Series, 2258. A tusk at an early stage of its growth with a deep longitudinal groove that partially divides it into two.

common feature of healed severe injuries to the pulps of tusks. Many examples of these are described in Chapter 19. Colyer inclined to the view that the specimens depicted in Figures 6.44 to 6.46 arose in this way. Our view is that, in the absence of a history of trauma or of objective evidence of trauma in the specimen itself, specimens of multiple tusks are best described in the category of developmental variations. Figure 6.42 depicts a specimen that shows a little of such evidence. Colyer (1944) described an example in which the history of injury as the cause was much stronger. There was a group of four separate small tusks in the right premaxilla which, together with the maxilla, had received a severe injury at some time in the past. For a full appreciation of Sir Frank Colyer's views, we refer readers to the first edition of this work.

ABSENCE OF TEETH

Absence of teeth in elephants seems to be rare, but Owen-Smith (1966) noted that 10% of 150 adult and immature elephants in a game reserve in Africa were tuskless; even though some were too young to carry tusks, this seems to be a surprisingly high proportion.

FUSION OF MOLARS

Fusion of the molars occurs occasionally in elephants. The skull of a young elephant (formerly RCS Osteo. Series, 2254) had this rare abnormality in both the maxillae and the mandible. The tooth in use, the antepenultimate molar (M1), was joined to the one behind by a thick mass of cementum, the division between them being indicated by a slight constriction on each side (Colyer, 1926a).

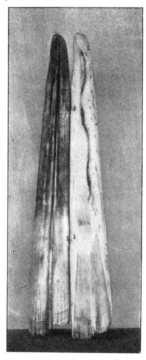

Fig. 6.48. Elephant. RCS Odonto. Mus., G 27.9. A tusk with a bifid tip from which extends a deep groove for the whole length of the tusk to the growing end (Colyer, 1952). $\times c. 0.14$.

Fig. 6.49. Elephant. RCS Odonto. Mus., G 27.91. A double tusk consisting of two separate tusks which have developed in such close contiguity that they have flattened longitudinally grooved surfaces which fit closely together. The growing ends are to the right. $\times c. 0.14$.

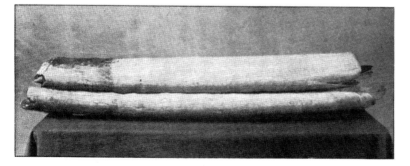

CHAPTER 7

Orders **RODENTIA**, **LAGOMORPHA**, **EDENTATA**, **INSECTIVORA** and **CHIROPTERA**

Apart from the Rodentia and Lagomorpha which have dentitions of similar type, these five Orders have little in common and are only grouped here because what there is to say about variations of the teeth in number and shape among them is too little to justify separate chapters.

Order **RODENTIA**

The Rodentia comprises a large number of diverse forms divided taxonomically into three sub-orders, Sciuromorpha, Myomorpha and Hystricomorpha, which contain about 30 living families. Their rodent or gnawing habit depends upon possessing continuously-growing, chisel-shaped incisors with enamel on the labial surface only. They are placed well forward on the jaws and separated from the cheek teeth by a diastema. The affinities between the different sub-orders are not nearly as close as their dentitions would suggest.

The number of cheek teeth varies from 2 to 5. The arrangement of the cusps is basically bilophodont but, in most families, additional cusps increase the grinding efficiency of the molars. In some families, for example Cavidae and Hydrochoeridae, the cheek teeth are of continuous growth.

It is impossible to give here a systematic account of the dentitions of the Rodentia. These are available in such texts as Anderson and Jones (1967).

The dental formula varies in the different rodent families. Most commonly, it is:

$$\frac{1}{1}:\frac{0}{0}:\frac{1-2}{1}:\frac{3}{3} = 20\text{--}22$$

but, of the histricomorph families, gundis (Ctenodactylidae) have two premolars in each quadrant and African mole-rats (Bathygeridae) have either two or three premolars and molars varying from 0–3. The Muridae, one of the largest myomorph families, which includes the Old World rats and mice, have no premolars; unlike the other murids, the island water-rats (Hydromyinae) have only two molars in each quadrant.

Although Colyer found only three (0.1%) variations in number in the 2554 rodents which he examined, subsequent workers have recorded higher prevalences. Few instances of extra incisors are recorded; Hansen (1956) found one in each mandible of a captive collared lemming (*Dicrostonyx groenlandicus*) and Gupta (1978) described one guinea pig with additional lower incisors associated with part of an additional mandible (Chapter 25, p. 601). According to Danforth (1958), in laboratory mice the presence of duplicate incisors can, at least occasionally, be attributed to genetic mutation.

Pilleri (1983) noted a small extra lower premolar in three out of 14 skulls of American beavers (*Castor fiber canadensis*) in his collection. Similar instances in three out of 142 skulls in the Museum of Zoology, University of Michigan appeared to be retained deciduous teeth. Supernumerary cheek teeth (usually disto-molars) have been recorded in various isolated instances in a surprising number of species: in the porcupine *Erithizon dorsatum* (Lechtleitner, 1958), in the southern pygmy mouse *Baiomys musculus* (Hooper, 1956), in *Peromyscus truei* (Hooper, 1957), in *P. maniculatus* (Sheppe, 1964), in *Saccostomus hildae* (Schwann, 1906), in *Microtus longicaudus* (Harris and Fleharty, 1962) and in *M. agrestis*, *Hystrix leucura* and *Mesembriomys gouldi* (Johnson, 1952), *Apodemus sylvaticus* (Ruprecht, 1978a; Van Laar, 1980; Kowalski, 1987), *Zapus princeps* (Krutsch, 1953), as well as *Myocastor coypus* (Schitoskey, 1971). Sheppe (1966) described four mice (*Mus musculus*), each with a narrow, cylindrical supernumerary molar that possibly originated from an M^1 tooth bud; as they were among 43 mice collected at the same time in the same farm building, a genetic cause was suggested.

Absence of teeth in rodents is recorded only rarely. The absence of a lower incisor in *Rattus exulans* is referred to in Chapter 17 (p. 360). In 2084 wild-caught house mice in the New World, Wallace and Bader (1966) found that only 0.82% lacked any molars; the missing teeth were always M^3 or M_3. Similarly, in a sample of 1573 *Mus musculus* of various wild sub-species in eastern Europe, Herold and Zimmerman (1960) found only one mouse (0.06%) lacking any cheek teeth (an M^3) but, in 2071 specimens of commensal forms of the same species, 63 (3.04%) lacked one or both M^3. They suggested that this was related to the smaller size of the commensal mice, a phenomenon which has also been noted in the mice of North Africa and the Middle East (Tchernov, 1984). Absence of one or both M_3 has been noted in beavers (*Castor fiber*). Zakrzewski (1969) found it in only 1.5% of 133 North American beavers but, in Europe, the figure was 32% of 252 animals (Piechocki, 1977). It is possible that European beavers are more genetically European beavers are more genetically isolated, or smaller, than their North American relatives.

Variation in molar size in a population of wild-caught house mice was described by Wallace (1968), and Searle (1954) showed that in a certain strain of laboratory mice such size differences are related to environmental factors, like diet. However, the influence of both environment and genetic background is well illustrated by Grüneberg (1963) in his classic work on the genetics of the mouse. He showed that, in certain strains of laboratory mice, small size and absence of the M_3 is caused by a complicated interrelation of both factors; he wrote (1963, p. 845), 'Absence of teeth tends to happen if tooth size falls below a certain critical level. The size of the teeth is partly determined by that of the mother; it is, however, also influenced by maternal environment (such as diet); the influence of the mother on the molar size of her offspring is partly prenatal, but mainly postnatal (lactation)'. His data are illustrated in Figure 7.1. He described these traits as an example of quasi-continuous variation (see Chapter 1, p. 2).

Most species of the genus *Rattus* are omnivorous but, according to Braithwaite (1979), some sub-species of the Australian swamp rat (*Rattus lutreolus*) are distinctly herbivorous in certain environments and this is reflected in dental adaptations; namely, long molar rows and the presence of a distal cingulum on M_1 which adds to the total molar area (Table 7.1).

Some species of microtine voles have grooves running the full length of the labial surfaces of the upper incisors but G. S. Jones (1978) found that such grooves occurred occasionally in other microtines, for example, among 631 skulls of *Microtus longicaudus* (from most of the species' range in the western USA), grooved upper incisors occurred in 11 individuals

(1.7%); Whitaker (1971) found one example among 43 pine voles (*M. pinetorum*) in Illinois and Indiana.

Variations in molar cusp pattern are common in wild rodents (Farbiszewska and Makarzec, 1960; Kelly and Fairley, 1982; Grummt, 1961; Petter and Tostain, 1981; Zejda, 1960; Guilday, 1982; and Haft, 1963). Sometimes these cusp variations can be used in studies of population genetics; for example, Wolfe and Layne (1968) showed up differences and similarities between four populations of *Peromyscus floridianus* in Florida

Table 7.1 Percentage occurrence of distal cingulum on lower first molars of *RATTUS LUTREOLUS* in Victoria, Australia *(R.1. lutreolus)* and Tasmania *(R.1. velutinus)*. (Braithwaite, 1979)

Subspecies	Locality	Habitat	% occurrence
R. l. lutreolus	Cranbourne	heathland	0 (50)
	Grampians	heathland	1.6 (68)
	Frankston	heathland	3.3 (60)
	Traralgon	riparian	4.2 (48)
	Coranderrk	riparian	4.3 (46)
	Boneo	riparian	6.3 (48)
	Healesville	human commensal-riparian	9.8 (122)
R. l. velutinus	Green's Beach	heathland	93.8 (32)
	Waratah	rainforest-sedgefield	97.2 (72)
	Cradle Mt.	sedgefield	100 (40)

Fig. 7.1. Absence and variation in size of third molars in CBA and C57BL strains of laboratory mice.
From: Grüneberg, H. (1963). By kind permission of the publisher.

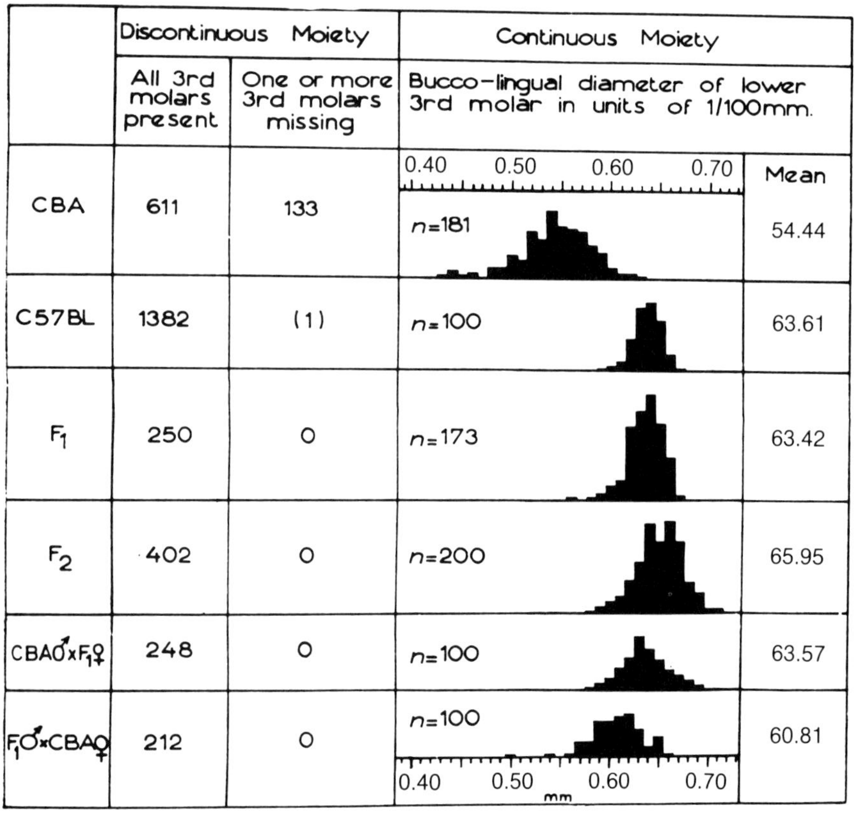

	Discontinuous Moiety		Continuous Moiety	
	All 3rd molars present	One or more 3rd molars missing	Bucco-lingual diameter of lower 3rd molar in units of 1/100mm.	Mean
CBA	611	133	n=181	54.44
C57BL	1382	(1)	n=100	63.61
F$_1$	250	0	n=173	63.42
F$_2$	402	0	n=200	65.95
CBA♂×F$_1$♀	248	0	n=100	63.57
F$_1$♂×CBA♀	212	0	n=100	60.81

on the basis of the morphology of the molars. Corbet (1986) showed that in Orkney voles (*Microtus arvalis*) divergence in the shape of M_1 between those on the islands of Mainland and Westray must have taken place in the last 5000 years and that a similar change in the field vole (*M. agrestis*) took place on the Scottish island of Jura within the last 2000 years but, in an isolated population of bank voles (*Clethrionomys glareolus*) on the Scottish mainland, a similar but temporary change took only 15 years (Corbet, 1975).

Variability in the position and number of roots has received more attention in rodents than in most mammals. Herold (1960) examined 1278 brown and 60 black rats (*Rattus norvegicus* and *R. rattus*) from different parts of the world and found much variation in the number and position of the roots of the upper molars. Hellwing and Ghizela (1963) applied the same criteria to a sample of 67 *Apodemus sylvaticus* in Moldavia; the results of their work and of Herold's are summarized in Table 7.2. Molar root variations in *Apodemus sylvaticus* have also been studied by Herold (1964), Gatineau (1956), Krommenhoek and Slob (1967), Zejda (1965), and by Robel (1971) who found that teeth of field mice on some north European islands were more variable than on the mainland, with a pronounced tendency to fusion of the roots. On some other north European islands, Herold (1955a) found less variability, and Kahmann (1965) found that on the large islands of Sardinia and Corsica variability was the same as on the mainland.

We have found only one example in the literature of connation in rodent teeth from the wild. Herold (1955b) noted bilateral connation between M^1 and M^2 in *Apodemus flavicollis* (yellow-necked mouse); although it was a single example, as it occurred in an isolated population, he suggested that it was a genetic trait. This conclusion received some support from the work of Shaw, Griffiths and Osterholtz (1963), who found connation of M^1 and M^2 in certain family lines in their colony of rice rats (*Oryzomys palustris*); Sofaer and Shaw (1971) found that this character, as well as the presence of disto-molars, was dependent on a single autosomal recessive gene.

2554 specimens examined

(i) *Castor fiber canadensis* (American beaver). BMNH 1855.3.11.4. Mesial to each P_4, a small tooth with a simple columnar crown devoid of the folded structure characteristic of beaver cheek teeth.
(ii) *Proechimys longicaudatus* (long-tailed spiny rat). BMNH 1923.12.12.12. A disto-molar in the right mandible, about half the size of M_3.
(iii) *Proechimys longicaudatus*. BMNH 1910.11.10.6. Tooth sockets show that upper right and lower left disto-molars had been present.
(iv) *Tatera indica* (Indian gerbil). HZM 17.10399. A supplemental left I_2, though possibly a persistent i_2 (Ingles, Bates and Harrison, 1981).
(v) *Gerbillus campestris* (large North African gerbil). ♀. BMNH 79.2177. Bilaterally absent M^3 (Ingles, Bates and Harrison, 1981).
(vi) *Eutamias sibiricus barberi* (Asiatic chipmunk). ♀. USNM 299582. Both P^3 absent (J. K. Jones, 1960).
(vii) *E. s. barberi*. ♀. USNM 299583. Both P^3 absent (J. K. Jones, 1960).

Order **LAGOMORPHA**

In pikas (Ochotonidae), the dental formula is

$$\frac{2}{1}\cdot\frac{0}{0}\cdot\frac{3}{2}\cdot\frac{2}{3}=26$$

In rabbits and hares (Leporidae), it is

$$\frac{2}{1}\cdot\frac{0}{0}\cdot\frac{3}{2}\cdot\frac{3}{3}=28$$

Although their dentitions and life styles resemble those of the Rodentia, the Lagomorpha are a separate offshoot of the early eutherian stock. The most obvious difference between their dentitions is that lagomorphs have a small peg-like I^2 lingual to I^1 and enamel on the lingual as well as the labial surfaces of the incisors. The cheek teeth are of continuous growth and there is no morphological distinction between premolars and molars; they are lophodont with sharp transverse ridges used for cutting rather than grinding.

Table 7.2 Variations in number and position of molar roots in 1879 *RATTUS* (Herold, 1960) and in 67 *APODEMUS SYLVATICUS* (Hellwing and Ghizelea, 1963)

	M^1	M^2	M^3
No. of roots			
R. norvegicus	4–7	4–5	3–5
R. rattus	5–6	4–6	3–5
A. sylvaticus	4–7	3–5	2–5
No. of variations of number and position			
R. norvegicus	7	8	8
R. rattus	7	7	7
A. sylvaticus	9	6	6

In 1265 skulls of lagomorphs, Colyer found no missing teeth and only five with additional teeth. The rarity of dental anomalies in lagomorphs is confirmed by Nachtsheim (1936) who, in 266 skulls of wild rabbits, found only 1.2% with anomalies, one skull lacking both P^2 and two lacking both M^3; but Kratochvil (1987) found that two out of 29 wild rabbits in Czechoslovakia lacked one or both M^3. There were more anomalies amongst his 101 domestic rabbit skulls; 4% lacked one or more teeth and 4% had additional teeth. In one domestic rabbit, both I_1 had longitudinal grooves on the labial surfaces. In his great work, *The Variation of Animals and Plants under Domestication*, Darwin (1868) noted additional upper cheek teeth in a domestic, lop-eared rabbit.

The reduction or absence of one or both M^3 in 7.9% of 140 skulls of hares (*Lepus capensis*) in New Zealand may be attributable to founder effect (Chapter 1, p. 10) as the population is descended from only a few introduced animals (Flux, 1980). Kratochvil (1987) found that only 1.4% of 72 hares in Czechoslovakia lacked an M^3. Among 170 skulls of the Afghan pika (*Ochotona rufescens*), J. Niethammer (1968) found only one dental anomaly, absence of both I^2, in one animal.

1265 specimens examined

EXTRA TEETH
(i) *Lepus timidus hibernicus* (Irish hare). BMNH 1876.7.12.2. Right I^2 duplicated.
(ii) *Lepus* sp. (hare). ♀. BMNH 1909.6.10.9. Right I^2 duplicated.
(iii) *Oryctolagus cuniculus* (rabbit). RCS Odonto. Mus., G 32.1. (Fig. 7.2). An extra tooth in the left maxilla lingual to the P^4 and M^1 and smaller than either; another, about half its size, in the right maxilla lingual to P^4 and M^1.
(iv) *O. cuniculus* (domestic rabbit). RCS Odonto. Mus., G 32.12. An extra tooth between P^4 and M^1 on the left, about half the size of both.
(v) *O. cuniculus*. RCS Odonto. Mus., G 32.11. An extra similar tooth lingual to the left P^3 and P^4.
(vi) *Lepus* sp. KU. 641. An extra tooth in each maxilla, mesial to P^2. The left one resembles a normal cheek tooth but the right is irregular in shape (Dawson, 1956).
(vii) *O. cuniculus*. Both I^2 absent; an extra upper molar on both sides (Geiger, 1976).
(viii) *O. cuniculus* (domestic rabbit). An extra cheek tooth in the right maxilla (Hilzheimer, 1908).

Order **EDENTATA**

Among the recent edentates, only the sloths and armadillos have teeth and these are variable in number both between and within taxa. Colyer did not report any anomalies and we have seen none in the literature, except for two reports in fossil sloths. The first was a Plio-Pleistocene fossil ground sloth *Megalonyx* (Idaho State Museum, 75001/6). It was a duplicate tooth in the left maxilla in the diastema behind the caniniform tooth which it closely resembled (McDonald, 1978). Edmund and Hoffstetter (1970) noted connation of two molariform teeth bilaterally in the mandibles and the maxillae of a ground sloth (*Megatherium americanum*).

Order **CHIROPTERA**

Fruit bats (Megachiroptera) have the dental formula

$$\frac{1-2}{0-2}.\frac{1}{1}.\frac{3}{3}.\frac{1-2}{2-3} = 24-34$$

with small incisors, large, often sexually dimorphic, canines and spaced-out brachydont molars with a few prominent cusps. The dental formulae of insectivorous bats (Microchiroptera) are very varied:

$$\frac{0-2}{2-3}.\frac{1}{1}.\frac{1-3}{2-3}.\frac{2-3}{2-3} = 26-38$$

usually with small incisors, large canines and molars with two V-shaped ridges (dilambdodont).

All bats lack at least one of the three upper incisors of the primitive eutherians and there is some controversy as to which incisor has been lost in the course of evolution. The traditional view is that it is I^1;

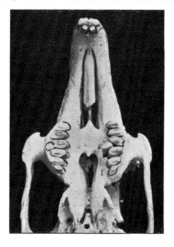

Fig. 7.2. *Oryctolagus cuniculus* (rabbit). RCS Odonto. Mus., G 32.1. An extra tooth in each maxilla.

however, as summarized by Handley (1959), "on the basis of an abnormal supernumerary deciduous incisor in *Myotis lucifer*, Stegeman (1956, p. 60) postulated that the missing tooth may be I^2. G. Allen (1916, p. 335) described a specimen of *Plecotus townsendi* (USNM 150273) with three upper permanent incisors. The extra incisor was the outermost of the three and differed in form from the other two. Allen speculated that that might represent a long lost I^3". But Handley himself felt that such deviations from a deep-seated characteristic in such a large and varied group could not be assumed to represent a reversion to an even more primitive long-lost character, and that the more distal incisors, I^3, had been lost by being crowded against the canine. Very little is known about the dentition of fossil forms, but it is clear that all bats are close to the ancestral eutherian stock that had the full number of four premolars in each quadrant, and all bats normally lack P2 in both maxilla and mandible, though many authors refer to the premolars as first, second and third. Sometimes P^1 or P_1 or both are also absent; hence, the premolar formula can be:

$$\frac{P^1P^3P^4}{P_1P_3P_4} \text{ or } \frac{P^3P^4}{P_1P_3P_4} \text{ or } \frac{P^3P^4}{P_3P_4}.$$

The presence of P3 is variable in some genera. In *Myotis*, this tooth is usually present but, in 75 *M. lucifugus* (little brown myotis) skulls examined by Frum (1946), one or both P^3 were absent in 9.3%. In *M. occultus* (Arizona myotis), Stager (1943) found one or more P^3 absent in 68% of 91 skulls, but only one skull lacked P_3. All 11 *M. fortidens* (cinnamon myotis), from the same locality in Texas, lacked upper and lower P3 on both sides (Hall and Dalquest, 1950). In *M. macrodactylus* in central Japan, Miyao (1973) found three out of 27 skulls lacking one or both P^3. Although it is normally absent in long-eared bats (*Plecotus* spp), Handley (1959) found a P^3 in a few skulls which was so similar in form to P^3 in *Myotis* that he considered its presence in *Plecotus* to be atavistic.

Colyer did not describe variations in number and shape of the teeth of bats and, despite the large number of species, about 950 according to Corbet and Hill (1986), there are few references to dental anomalies in the literature; this probably reflects the small amount of work that has been done on chiropteran dentitions. Where particular groups have been described in detail (Table 7.3), the incidence of variations in number is low.

Ruprecht (1978b) described a serotine bat (*Eptesicus serotinus*) with bilateral supernumerary incisors lingual to I^1, and Miyao (1973) noted an extra incisor in a pipistrelle (*Pipistrellus abramus*). Peterson (1968) found bilateral supplemental I^1 and a supernumerary left upper deciduous incisor in a young greater yellow bat from Vietnam (*Scotophilus heathi*) (Royal Ontario Museum 43202).

Glass (1953) examined 105 skulls of the Mexican freetail bat (*Tadarida mexicana*) of which only one lacked both I_3 and another lacked one I_3.

Stager (1943) mentioned a skull of *Myotis occultus* (no. 846 in his collection) which had two canines in each maxilla and P^3 absent on both sides. One *M. occultus* skull (Fig. 7.3) contained only one premolar in each maxilla and two in each mandible (Mumford, 1963).

In a detailed monograph on the morphology and variability of the teeth of glossophagine bats, Phillips (1971) described the anomalies listed below. In most instances when Phillips noted a tooth in the position of P^2, he considered it to be atavistic because the genera in which the deciduous dentition is known have an equivalent m^2.

Sub-family **Glossophaginae**

Genus *Lonchophylla*. 49 specimens examined

$$\frac{2}{2}:\frac{1}{1}:\frac{2}{3}:\frac{3}{3} = 34$$

(i) *L. robusta* (Panama long-tongued bat). ♀.

Table 7.3 Extra teeth in three families and one sub-family of MICROCHIROPTEAN BATS

	No. of specimens	Extra teeth			% of specimens affected	Absent teeth				% of specimens affected	Connate teeth etc. No.
		I	P	M		I	C	P	M		
Emballonuridae	520	3	6	0	1.7	2	0	6	0	1.5	1
Noctilionidae	50	0	0	0	0	0	0	0	0	0	0
Chilonycteridae	1033	0	1	0	0.1	8	0	6	0	1.3	1
Glossophaginae	2025	9	14	2	1.2	8	2	0	7	0.8	6

Data extracted from Phillips and Jones (1968) and Phillips (1971).

136 *Section 1: Variations in Number, Size and Shape*

FMNH 51732. A small supernumerary premolar in the right maxilla, possibly a P^2.
(ii) *L. mordax*. ♀. USNM 309389. Bilateral minute peg-like supernumeraries mesial to P^3.

Genus *Lionycteris*

$$\frac{2}{2}:\frac{1}{1}:\frac{2}{3}:\frac{3}{3}=34$$

Lionycteris spurrelli. 25 specimens examined.

(i) ♂. USNM 23947. A supplemental incisor between the left I_1 and I_2.
(ii) ♂. AMNH 97267. A supernumerary premolar in the right maxilla, possibly a P^2.
(iii) ♀. USNM 385709. A supernumerary premolar in the left maxilla, possibly a P^2.
(iv) ♂. AMNH 97265. A small peg-like upper right disto-molar.

Genus *Anoura*

$$\frac{2}{0}:\frac{1}{1}:\frac{3}{3}:\frac{3}{3}=32$$

Anoura geoffroyi (Geoffroy's tail-less bat). 364 specimens examined.

(i) USNM 252000. Lower left canine absent.
(ii) ♀. USNM 370119. Lower right canine absent.

No numerical variations were found in 43 specimens of *A. caudifer* and 13 *A. cultrata*.

Genus *Glossophaga*

$$\frac{2}{2}:\frac{1}{1}:\frac{2}{3}:\frac{3}{3}=34$$

Glossophaga sorcina (Pallas' long-tongued bat). 818 specimens examined.

Fig. 7.3. *Myotis occultus* (Arizona myotis). Six premolars absent, two upper and one lower on both sides. From: Mumford, R.E. (1963).

EXTRA TEETH
(i) ♀. KU 60880. An extra upper left incisor labial to I^2, as well as extra premolars (see xiii below).
(ii) ♂. KU 95754. An extra lower incisor near the midline.
(iii) ♂. KU 60935. Bilateral supplemental I_1.
(iv) ♂. KU 64895. A supplemental right I_1.
(v) ♂. KU 102323. An extra incisor in the left mandible.
(vi) ♂. KU 102346. An extra incisor in the left mandible.
(vii) ♂. KU 87312. An extra incisor in the left mandible.
(viii) ♀. KU 60841. A supplemental left P^3.
(ix) ♀. KU 75204. Bilateral supplemental P^3.
(x) ♀. KU 102350. A supplemental P^1 on the right and, on the left, a single-rooted tooth that may be a P^2.
(xi) ♀. KU 91571. A supernumerary mesial to P^1, perhaps a P^2.
(xii) ♂. KU RRP 928. Bilateral supernumeraries mesial to P^1, both perhaps P^2.
(xiii) ♀. KU 60880. Bilateral supplemental P^3.
(xiv) ♀. KU 60847. Bilateral supplemental P^3.
(xv) ♂. KU 60855. A supplemental right P^3.
(xvi) ♂. KU 102312. Bilateral supplemental P^3.

ABSENT TEETH
(i) ♂. KU 60826. Left I^1 and right I_1 absent.
(ii) ♂. KU 102334. All lower incisors and left I^2 absent.
(iii) ♂. KU 70630. All lower right incisors absent.
(iv) ♂. KU 60921. Both I_1 and right M^3 absent.
(v) ♀. KU 82845. Right M^3 absent.
(vi) KU 97604. Left M^3 and M_3 absent.
(vii) ♀. KU 70636. All M3 absent.
(viii) ♀. KU 27989. Right M_3 absent.
(ix) ♀. KU 94093. Both M_3 absent.

VARIATIONS IN SHAPE
(i) ♂. KU 39543. Bilateral connate I_1 and I_2.

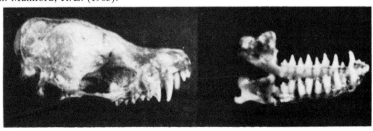

No numerical variations were found in 101 specimens of *G. commissarsi*, nor in 136 of *G. alticola*.

Genus **Monophyllus**. 106 specimens examined

$$\frac{2}{2}:\frac{1}{1}:\frac{2}{3}:\frac{3}{3} = 34$$

(i) *M. plethodon* (Barbados long-tongued bat). ♀. USNM 106098. Both I_2 absent; also and an apparently connate left P_4 consisting of two large cones and three roots along the line of the arch.

Genus **Leptonycteris**

$$\frac{2}{2}:\frac{1}{1}:\frac{2}{3}:\frac{2}{2} = 30$$

L. sanborni (little long-nosed bat). 157 specimens examined

(i) ♂. UNAM 8835. Small bilateral supplemental P^3.
(ii) ♂. KU 23693. There are two connate teeth. The crown of right P_3 consists of mesial and distal parts each of which has the normal two roots. The left P_4 consists of mesial and distal parts each of which resembles a normal P_4, but there are only three roots, the middle one being as it were shared between the two parts.
(iii) ♂. KU 34154. Bilateral connate P^4.
(iv) ♂. KU 34137. A large caniniform tooth in place of the right I^1.

L. nivalis (big long-nosed bat). Number of specimens not stated.

(i) Both I_1 and one upper molar absent.

Genus **Lichonycteris**. 13 specimens examined

$$\frac{2}{0}:\frac{1}{1}:\frac{2}{3}:\frac{2}{2} = 26$$

(i) *L. obscura* (brown long-nosed bat). ♂. USNM 331258. An extra premolar mesial to right P^3, possibly a P^2.
(ii) *L. degener*. ♀. AMNH 95485. An extra premolar mesial to right P^3, possibly a P^2.

Genus **Choeronyctetis**

$$\frac{2}{0}:\frac{1}{1}:\frac{2}{3}:\frac{3}{3} = 30$$

Choeronycteris mexicana (Mexican long-tongued bat). 143 specimens examined

(i) ♂. AMNH 188392. Left I^1 absent. Left M_2 with no metaconid.

Genus **Choeroniscus** (long-tailed bats). 31 specimens examined

$$\frac{2}{0}:\frac{1}{1}:\frac{2}{3}:\frac{3}{3} = 30$$

(i) *C. inca*. ♂. AMNH 67526. Bilateral supernumerary premolars mesial to P^3, possibly P^2.
(ii) *C. godmani*. ♂. AMNH 131765. An extra molariform tooth mesial to M_1, similar to M_1 in morphology but lacking a paraconid.
(iii) *C. minor*. ♀. AMNH 140471. Left P_3 connate.

No dental variations were found by Phillips in 24 skulls of *Hylonycteris underwoodi* (Underwood's long-tongued bat).

Phillips and Jones (1968) studied dental variations in three families of N. American bats; in 45 skulls of the Noctilionidae, they found no variations, but they were present in the Emballonuridae and Chilonycteridae and are listed below:

Family **Emballonuridae**

$$\frac{1}{3}:\frac{1}{1}:\frac{2}{2}:\frac{3}{3} = 32$$

Rhynchonycteris naso (proboscis bat). 72 specimens examined

(i) ♂. KU 19101. A supplemental left P_3.
(ii) ♂. KU 19122. Two small teeth in position of left P^3; they resemble the contralateral but are much smaller.
(iii) ♂. KU 19100. Both I_3 absent.
(iv) ♀. KU 32062. Right P^3 absent.

Saccopteryx bilineata (greater white-lined bat). 97 specimens examined

(i) ♀. KU 91467. Bilateral supplemental incisors in the same alveoli as the normal upper incisors (Phillips and Jones, 1968, fig. 1c).
(ii) ♀. KU 105412. P^3 absent on both sides.

Peropteryx macrotis (lesser sac-winged bat). 48 specimens examined

(i) ♀. KU 23444. Right I_2 and I_3 bilobed rather than trilobed, as normal. Both I_3 smaller than normal.

Balantiopteryx plicata (Peters' sac-winged bat). 170 specimens examined

Section 1: Variations in Number, Size and Shape

(i) ♀. KU 87265. A supplemental right P_1.
(ii) ♀. KU 34070. A small supplemental left P_1.
(iii) ♂. KU 60684. Both upper incisors absent.
(iv) ♀. KU 67302. Left P_1 absent.
(v) ♂. KU 28288. Right P^3 absent.

B. io (Thomas' sac-winged bat). 116 specimens examined

(i) ♂. KU 64634. Two small peg-like teeth in the position of both P^3.
(ii) ♂. KU 64634. Bilateral supplemental P^3.
(iii) ♀. KU 17694. Left P^3 absent.
(iv) ♂. KU 23500. Left P^3 absent.

Family **Chilonycteridae** (= **Mormoopidae**)

$$\frac{2}{2} \cdot \frac{1}{1} \cdot \frac{2}{3} \cdot \frac{3}{3} = 34$$

Pteronotus personatus psilotis (moustached bat). 292 specimens examined

(i) ♂. KU 36405. Bilateral supplemental P^3.
(ii) ♀. KU 91517. Right I^2 absent.
(iii) ♂. KU 91514. Right I^2 absent.
(iv) ♂. KU 95671. Left P_3 unusually large, the enlarged labial cingulum is partially separated from the rest of the tooth by longitudinal grooves, suggestive of connation.

P. davyi (Davy's naked-backed bat). 197 specimens examined

(i) ♂. KU 64747. Right P_3 absent.
(ii) ♀. KU 68636. Left P_3 absent.

None of the 34 specimens of *P. suapurensis* examined showed dental variation.

Mormoops megalophylla (Peters' ghost-faced bat). 338 specimens examined

(i) ♂. KU 64835. Both I^2 absent.
(ii) ♂. KU 64873. Both I^2 absent.
(iii) ♀. KU 64856. Left I_2 absent.
(iv) ♂. KU 85588. Right P^3 absent.

Order **INSECTIVORA**

The dental formula of insectivores is very variable both between families and within some of them but many of the extant species have, or come close to having, the primitive eutherian maximum of 44. The incisors are modified as forceps for picking up small prey. The canines often resemble the incisors or the most mesial premolars. As in most mammalian groups when the premolars are reduced in number, the most mesial are lost first (Kirkdahl, 1967). The arrangement of cusps on the molars is very close to the basic tribosphenic pattern. In many insectivores, the deciduous dentition is erupted and shed before birth.

Colyer did not note any variations in number, shape or size for the insectivores, but they are not unknown and in some groups they are quite common, for example in European hedgehogs as in Ruprecht's (1965c) data in Table 7.4.

In contrast to Ruprecht's figure of 2% of European hedgehogs with absent teeth, 49.3% of 71 hedgehogs in New Zealand lacked one or more teeth (Brockie, 1964). It is said that all New Zealand hedgehogs are descended from about a dozen animals introduced from Britain in the 1890s, so this seems to be a good example of the spread of a genetic character due to founder effect in an island population (Chapter 1 p. 10).

A mandible of the Daurian hedgehog *Hemiechinus dauricus* (Skull Collection of the Zoological Institute of the Academica Sinica in Peking, 39 626/01826) was

Table 7.4 Dental anomalies in a sample of EUROPEAN HEDGEHOGS *(Erinaceus europaeus)*, after Ruprecht (1965c)

Locality ...	Germany	UK	Poland	Switzerland	USSR	Other parts of Europe	Total	%
No. of specimens	31	15	13	3	119	21	202	
extra teeth	2	—	—	1	7	—	10	4.9
connate teeth	—	—	—	—	2	—	2	1.0
absent teeth	1	—	—	1	1	1	4	2.0

figured by Poduschka and Poduschka (1986); the crown of the right I_1 was divided into two sharp points instead of the normal spatulate shape.

Hall (1940) found that dental anomalies were extremely rare in most species of N. American moles (Talpidae). However, 7.9% of 265 broad-footed moles (*Scapanus latimanus*) lacked at least one tooth, this anomaly being much more frequent in the smaller southern races than in the northern, and perhaps correlated with the shorter length of the jaws.

In a detailed study of European moles (*Talpa europaea*) from most of Europe (except Ireland), totalling 8803 skulls, Stein (1963) showed that variations in the numbers of upper premolars are rare. Variations in the lower premolars are more common and seem to be related to sex and body size. Thus females tend to have fewer than normal and so do moles in eastern Europe (the USSR) which tend to be smaller whereas, in moles in Flanders and the Netherlands which are larger, extra lower premolars are more common. Stein suggested that the high incidence (37.1%) of a reduction in numbers of teeth in 35 free-living moles collected within Augsburg Zoo was due to the genetic effects of geographical isolation.

In the Soricidae (shrews), when dental variations occur they are usually among the upper unicuspid teeth, that is I^2, I^3, the canine and all the premolars mesial to P^4, which is molariform. The small conical unicuspid teeth form a continuous row. In most American species of shrews, anomalies are extremely rare (Hall, 1940; Jackson, 1928; Long, 1961; Pruitt, 1957; Hooper, 1946; French, 1984) but Choate (1968) found that, in 712 northern short-tailed shrews (*Blarina brevicauda*), 0.7% had additional teeth, 3.5% lacked at least one tooth and 2.4% had one or more teeth of reduced size. He studied several sub-species of this shrew and again there was a higher incidence of tooth absence in the smaller subspecies.

In Europe, dental anomalies are rare in shrews, the most usual being a reduction in the number or size of P^3, in 0.4% of pygmy shrews (*Sorex minutus*, Hutterer, 1977) and in 1.8% of common shrews (*Sorex araneus*, Reinwaldt, 1961). A further example of a high incidence of reduction in number that might be attributed to founder effect in an island population is the occurrence of 56% of shrews (*Sorex araneus*) lacking one or both P^3 among 16 animals from islands in the Scottish Inner Hebrides (Barrett-Hamilton and Hinton, 1913).

Dippenaar (1978) examined 514 skulls of an African shrew (*Crocidura mariquensis*) and found that 0.6% had extra maxillary premolars and 0.4% had lower disto-molars; in one (0.2%), the right M^3 had a small hypocone and a metastyle, like those normally found only on M^1 and M^2. Setzer (1957) noted an upper left disto-molar in a skull of *Crocidura olivieri* from Egypt. Further examples of dental anomalies in African shrews are given by Meester (1959).

Buchalczyk (1961) described a water shrew (*Neomys fodiens*) in which the left upper canine was connate with the left I^3; a similar abnormality in a white-toothed shrew (*Crocidura lasiura*) was described by J. K. Jones (1957) and, in the northern short-tailed shrew (*Blarina brevicauda*), by Choate (1968). Vesmanis and Vesmanis (1980) found what appears to be a connate I^3 in a house shrew (*Crocidura russula*).

CHAPTER 8

Order **MARSUPIALIA**

The basic metatherian tooth formula seems to be

$$\frac{5}{4}:\frac{1}{1}:\frac{3}{3}:\frac{4}{4}=50$$

but, in most groups, the numbers are reduced, though *Myrmecobius* (the banded ant-eater) has 8 upper and 9 or 10 lower cheek teeth on each side. On the basis of homology with the Eutheria, Colyer and others designated the most distal premolar as a P4 believing that a more mesial premolar has been lost in the course of evolution. However, as almost all species of the Marsupialia (the only order in the sub-class Metatheria) have only three premolars, we follow Van Deusen and Jones (1967) and Lumsden (1981) in numbering the most distal premolar P3 and the two others P1 and P2. The problems of nomenclature and homologies of marsupial teeth are discussed further in Chapter 1 (p. 13). There appear to be no comprehensive accounts of the normal dentition in the various marsupials but much information can be gathered from F. W. Jones (1923), Van Deusen and Jones (1967), Vaughan (1978), and Lawlor (1979).

In the past, marsupials were divided into two sub-orders on the basis of the dentition: the Polyprotodontia with small, unequal incisors and large canines, and the Diprotodontia with large first incisors and the other incisors and canines reduced or absent. The marsupials are usually now divided instead into five super-families, the Phalangeroidea, Perameloidea, Dasyuroidea, Didelphoidea and Caenolestoidea, on the basis of non-dental characters, particularly limb morphology.

Colyer examined about 3500 marsupial skulls and found extra canines in two and connate canines in three, a total of about 0.1%.

Super-family **PHALANGEROIDEA**

Family **Macropodidae**. Kangaroos and wallabies

$$\frac{3}{1}:\frac{0-1}{0}:\frac{2}{2}:\frac{4}{4}=32\text{--}34$$

The single large lower incisor is procumbent and is of continuous growth. In the bettongs (*Bettongia*), the upper first incisor also grows continually. The second premolar (P3) is large and serrate, replacing not only m3 but also the tooth that is immediately mesial to m3 (that is m2, P2 or PC2 depending on the terminology employed). The other cheek teeth are broad and quadritubercular or bilophodont in keeping with the herbivorous diet.

Genera *Macropus* and *Wallabia*. 590 specimens examined (30 without mandibles)

(a) Incisors

 (i) *Macropus dorsalis* (black-striped wallaby). RCS Odonto. Mus., A 342.3. I^2 are absent.
 (ii) *M. irma* (western brush wallaby). BMNH 1906.8.1.365. Left I^2 is absent.
 (iii) *M. giganteus* (eastern grey kangaroo). RCS Odonto. Mus., A 335.41. Although the upper incisors have been lost, the sockets show that only two teeth were present in each premaxilla. Judging from the size of the sockets, I^2 was absent on the right and I^3 on the left.
 (iv) *M. giganteus*. ♀. BMNH 1885.1.3.3. (Fig. 8.1). The slight longitudinal ridge normally present at the centre of the labial surface is grossly exaggerated on the left I^3, forming a prominent flange.
 (v) *M. giganteus*. Queensland Mus., J23087. A small extra incisor on the right between I^2 and I^3. (Archer, 1975, plate 34c, p.256).
 (vi) *M. rufogriseus bennetti* (Bennett's wallaby). Formerly RCS Osteo. Series, 3724. (Fig. 8.2). The right I^3 consists of two parts, a lingual and a buccal, which are separated by a fissure extending

Fig. 8.1. *Macropus giganteus* (eastern grey kangaroo). BMNH 1885.1.3. Left I^3 has an exaggerated longitudinal labial ridge at (c). × c. 1.3.

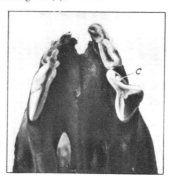

Fig. 8.2. *Macropus rufogriseus bennetti* (Bennett's wallaby). Formerly RCS Osteo. Series, 3724. The crown of the right I^3 consists of lingual and buccal parts separated by a fissure. (A) Right incisors; (B) left incisors; (C) palatal view.

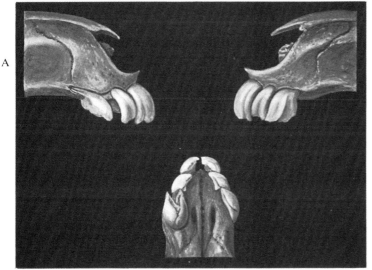

almost to the cervix. The tooth had one root with a single pulp cavity. The buccal part was similar in shape to the mesial portion of the left I^3, the lingual aspect of the lingual part was convex like the buccal aspect of the distal portion of the left incisor. The abnormal tooth gave the impression that it is an over-developed incisor in which the distal part has been bent on the mesial portion.

(b) Premolars
M. eugenii (Dama wallaby). In 51 skulls from various localities, one lacked both P_3, but all four skulls from Middle Island, Western Australia, lacked P_3 on both sides and one lacked P^3 in each maxilla as well, which might be due to founder effect on an island population (Chapter 1, p. 10).
  (i) ♂. BMNH 1905.4.1.1. Middle Island. All P3 absent.
  (ii) ♂. BMNH 1905.4.1.2. Middle Island. Both P_3 absent. Both P^3 present but unerupted.
  (iii) ♀. BMNH 1905.4.1.3. Middle Island. Both P_3 absent and both P^3 small and of simple shape.
  (iv) ♀. BMNH 1905.4.1.4. Middle Island. Both P_3 absent.
  (v) ♂. BMNH 1886.1.26.2. Both P_3 absent.
  (vi) *M. giganteus*. Queensland Mus., M23089. An extra premolar erupting under the right P^3 (Archer, 1975, plate 35G, p. 256).

(c) Molars
  (i) *M. giganteus*. ♂. BMNH 1925.8.1.23. A disto-molar in the left maxilla.
  (ii) *M. giganteus*. ♂. BMNH 254g. An unerupted disto-molar in the right mandible.
  (iii) *M. giganteus*. Queensland Mus., J23083. Bilateral maxillary disto-molars (Archer, 1975, plate 35A, p. 266).
  (iv) *M. giganteus*. Queensland Mus., J23085. Bilateral maxillary disto-molars (Archer, 1975, plate 35F, p. 256).
  (v) *Wallabia bicolor* (swamp wallaby). ♀. BMNH 1908.8.8.65. Bilateral mandibular disto-molars.
  (vi) *M. rufus* (red kangaroo). Formerly in Colyer's private collection, 259.2. Left M_4 absent.
  (vii) *M. rufus*. Queensland Mus., J23091. A three-rooted, tricuspid supernumerary tooth between the right M^3 and M^4 (Archer, 1975, plate 35B, p. 257).
  (viii) *M. rufus*. Queensland Mus., J23088. Right M^2 with only one major loph. A small single alveolus in place of the right M^4 which was probably single-rooted (Archer, 1975, plate 35E, pp. 256–7).
  (ix) *M. rufus*. Queensland Mus., J23084. A disto-molar with a complex metaloph in the left maxilla (Archer, 1975, plate 35D, pp. 256–7).
  (x) *M. rufus*. Queensland Mus., J23086. A conical disto-molar in the left maxilla (Archer, 1975, plate 35E, pp. 256–7).
  (xi) *M. irma*. BMNH 1920.4.29.3. Right M^4 absent.

Genus *Petrogale*. 81 specimens examined (3 without mandibles)
  (i) *Petrogale penicillata* (brush-tailed rock wallaby). RCS Odonto. Mus., G 34.2. Both I^2 absent.

Genus *Lagorchestes*. 46 specimens examined
  (i) *Lagorchestes conspicillatus* (spectacled hare-wallaby). BMNH 1906.11.9.16. Bilateral unerupted maxillary disto-molars.
  (ii) *L. leporides* (eastern hare-wallaby). ♀. BMNH 1846.4.4.23. Both I^2 absent.

Genus *Onychogalea*. 45 specimens examined (2 without mandibles)
  (i) *Onychogalea lunata* (crescent nail-tailed wallaby). ♂. BMNH 1906.8.1.259. Both I^2 absent.
  (ii) *O. unguifera* (northern nail-tailed wallaby). ♀. BMNH 1915.3.5.64. Right P_3 absent and left P_3 conical.
  (iii) *O. unguifera*. Formerly Naturhistoriska Riksmuseet, Stockholm. Left P_3 absent and right P_3 conical.

Genus *Dorcopsulus*. 37 specimens examined
  (i) *Dorcopsulus macleayi* (Papuan forest wallaby). ♀. BMNH 1903.12.1.13. (Fig. 8.3). On the right, the tooth in the position of the canine resembles a small P^2. Just distal to it and separated from P^3 by a wide space, there is a tooth of similar morphology. On the left, there is a P^2 immediately mesial to P^3, as described by Thomas (1888*a*).

Genus *Dendrolagus*. 61 specimens examined
  (i) *Dendrolagus bennettianus* (Bennett's tree kangaroo). ♂. Formerly RCS Osteo. Series, A 3799. A small tooth mesial to the left P^2. The deciduous molar is in position.
  (ii) *D. lumholtzi* (Lumholtz's tree kangaroo). ♂. BMNH 1894.6.10.1. A small extra tooth with

mesial and distal cusps mesial to P_2 on both sides. An opening made in the bone shows P_3 unerupted beneath P_2 and m_3.
(iii) *D. bennettianus*. ZMB 14622. A small cylindrical extra tooth mesial to the left P^3.
(iv) *D. ursinus* (Vogelkop tree kangaroo). ♀. RMNH 13507. The crowns of the maxillary canines have longitudinal grooves suggestive of connation.

Genus *Caloprymnus*. 5 specimens examined

The desert rat-kangaroo was a rare animal and has recently become extinct. One skull, RCS Odonto. Mus., G 34.5, has bilateral maxillary disto-molars.

Genus *Bettongia*. 92 specimens examined

(i) *Bettongia penicillata* (brush-tailed bettong). ♂. BMNH 1907.9.1.71. A small left maxillary and bilateral mandibular disto-molars.
(ii) *B. gaimardi* (eastern bettong). RCS Odonto. Mus., G 34.1. Bilateral unerupted maxillary disto-molars.
(iii) *B. lesueuri* (boodie). BMNH 1841. 1157. Bilateral maxillary disto-molars.
(iv) *B. lesueuri*. ♀. BMNH 1846.4.4.32. Bilateral maxillary disto-molars.
(v) *B. lesueuri*. Captive. BMNH 1855.12.24.393. A left maxillary disto-molar.

Fig. 8.3. *Dorcopsulus macleayi* (Papuan forest wallaby). ♀. BMNH 1903.12.1.13. On the right, the tooth in the position of the upper canine resembles a small P^2, and P^2 is situated more mesially than normal. $\times c.\,1.5$.

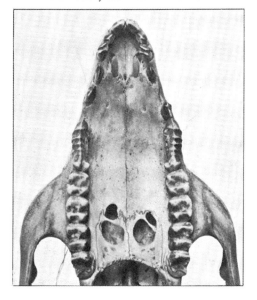

(vi) *B. gaimardi* (eastern bettong). ♀. BMNH 1851.4.24.7. Bilateral disto-molars in each maxilla and mandible.
(vii) *B. gaimardi*. BMNH 1846.8.7.16. Left M_4 absent.
(viii) *B. penicillata*. ♂. BMNH 1846.4.4.41. Bilateral empty alveoli show that M^4 were present in life. A small conical left disto-molar in the tuberosity. These teeth were described by Bateson (1894, p. 258, no. 390).
(ix) *B. penicillata*. ♂. BMNH 1844.4.4.37. Right M_4 and M^4 absent. Left M_4 small.

Genus *Potorous*. 27 specimens examined (five without mandibles)

(i) *Potorous tridactylus* (potoroo). BMNH 1846. 4.4.47. Right M^4 absent.

There were no variations in 74 specimens of *Aepyprymnus* (rufous rat kangaroo) nor in 26 *Lagostrophus* (banded hare-wallabies).

Family **Phalangeridae**. Phalangers

$$\frac{2-3}{1-3}:\frac{1}{0-1}:\frac{1-3}{1-3}:\frac{3-4}{3-4}=24-42$$

In the mandible of phalangerids, between the long, procumbent I_1 and P_3, there is a variable number of small teeth. Here the problem of homology with eutherians and even with other metatherians has defied solution. Thomas (1888a) termed them 'intermediate teeth' because it is by no means certain which, if any of them, are the canines. This term did not acquire wide currency but was used by Bateson (1894) and by Colyer in the original edition of this book. Data on variations in the numbers of intermediate teeth gathered by Colyer are summarized in Table 8.1. Phalangerid molars are quadritubercular, usually with rounded cusps.

Genus *Schoinobates*. 58 specimens examined (9 without mandibles)

(i) *Schoinobates volans* (greater glider). BMNH 1839.6.29.3. A small tooth between the right P^1 and P^2, similar in shape to the P^1.
(ii) *S. volans*. ♀. BMNH 1925.8.1.87. The right upper canine is connate and much larger than the contralateral; its crown is almost completely divided into mesial and distal parts.

Genus *Petaurus*. 92 specimens examined

(i) *Petaurus australis* (fluffy glider). Formerly RCS Osteo. Series, 3694. Both M^4 absent.
(ii) *P. australis*. ♀. BMNH 1914.1.23.7. Right P^1 double-rooted. The left P^1 is represented by a connate tooth with three roots; the mesial part of the tooth simulates the right P^1, the distal part is much smaller and is carried on a separate root.
(iii) *P. australis*. BMNH 1907.1.1.216. The third post-incisor tooth in the right mandible is connate and much larger than the contralateral; its occlusal surface is divided by a groove into mesial and distal parts.

Genus *Dactylopsila*. 31 specimens examined

(i) *Dactylopsila trivirgata* (common striped possum). BMNH 1867.9.17.1. A small conical tooth mesial to the left P^1.
(ii) *D. megalura* (large-tailed possum). Formerly Rothschild Collection. (Fig. 8.4). Right P^2 absent; right P^1 has a deep notch on the upper part of the crown; the groove running from the notch on the buccal aspect extends further down the tooth than that on the lingual aspect. The left P^2 is present, but the left P^1 is smaller than the contralateral and shows no division of the crown. It is possible that the right P^1 represents the normal P^1 connate with either P^2 or P^3.

Genus *Pseudocheirus*. 154 specimens examined (11 without mandibles)

The maxillary premolars vary in size and in their distance from each other. The number of intermediate teeth in the mandible varies.

(i) There were no examples of extra premolars in the maxilla but, in a mongan *Pseudocheirus herbertensis* (formerly RCS Odonto. Mus., A 357.6), the right P^1 was much larger than the contralateral and was grooved on the buccal aspect, suggesting connation.
(ii) *P. forbesi* (moss-forest ringtail). ♂. BMNH 1888.3.16.7. Both I^3 absent. Both P^1 are absent but the surface of the bone is not quite smooth and it is possible that the teeth were lost in life.
(iii) *P. albertisii* (d'Albertis' ringtail). BMNH 1894.2.14.4. Right P^1 absent.
(iv) *P. peregrinus* (common ringtail). ♀. BMNH 1923.5.11.2. Left M^4 absent.

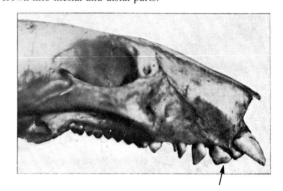

Fig. 8.4. *Dactylopsilia megalura* (large-tailed possum). Formerly Rothschild Collection. Right P^2 absent. The right P^1 (arrow) has a deep groove partly dividing the crown into mesial and distal parts.

Table 8.1 Percentages of specimens of various PHALANGERID taxa with different combinations of 'intermediate teeth' in each side of the mandible

Taxon	No.	L/R 0/2	L/R 0/3	L/R 1/1	L/R 1/2	L/R 2/2	L/R 2/3	L/R 3/3	L/R 3/4	L/R 4/4	L/R 3/1	L/R 4/2
Petaurus spp.	81	0	0	0	0	1.2	4.9	**90.1**	2.5	1.2	0	0
Dactylopsila trivirgata	25	0	0	0	0	4.0	4.0	**80.0**	8.0	4.0	0	0
Pseudochirus peregrinus	58	0	0	0	3.4	**62.1**	10.3	20.7	3.4	0	0	0
Trichosaurus spp.	186	0	0	**86.0**	7.5	4.8	1.6	0	0	0	0	0
Phalanger maculatus	112	0	0	5.4	12.5	**69.6**	6.2	3.6	1.8	0	9	0.5
Phalanger orientalis	205	0.5	0.5	2.9	1.5	5.4	5.8	**78.0**	3.4	1.0	0.5	0.5

The most common combinations are in bold type. Note the wide degree of variation in number in the different taxa and the high degree of asymmetry in numbers.

Genus *Trichosurus*. 186 specimens examined

(i) *Trichosurus vulpecula* (brush-tailed possum). BMNH 1914.11.21.8. (Fig. 8.5). The upper canines are connate, divided into larger mesial and smaller distal parts by grooves which separate the tips of the crown and which, on the left, where a window has been cut in the bone, can be seen to extend as a shallow groove for the whole length of the root.

(ii) *T. caninus* (mountain possum). Formerly RCS Osteo. Series, 3686. (Fig. 8.6). Mesial to the right upper canine, there is an extra tooth which resembles the canine but is much smaller; however, the extra tooth is situated on the premaxilla. P^1 on the right is closer to the canine than on the left.

(iii) Premolars. In *Trichosurus*, there are two premolars in each maxilla, P^1 and P^3. According to Bateson (1894, no. 378), in *Trichosurus vulpecula* var. *fuliginosus*, P^1 is a large tooth about the same size as the canine but is sometimes absent. In six out of 17 specimens, one or both first premolars were missing, but it was impossible to be certain that they were not lost in life.

Fig. 8.5. *Trichosurus vulpecula* (brush-tailed possum). BMNH 1914.11.21.8. Labial views of the upper canines (A, right, B, left) showing partial division of the crowns into mesial and distal parts. Enlarged.

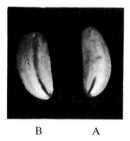

B A

Fig. 8.6. *Trichosurus caninus* (mountain possum). Formerly RCS Osteo. Series, 3686. Mesial to the right canine, an extra tooth which is caniniform, but is on the premaxilla. On the right, P^1 is closer to the canine than on the left, as emphasized by the line.

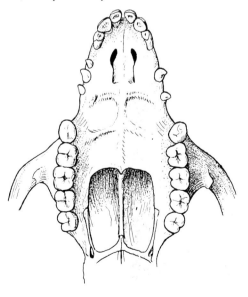

Genus *Phalanger*. 329 specimens examined

P^1 was always present but P^2 was often missing; however, the bone in that situation was often pitted or rough and occasionally partly filled-in sockets could be seen.

Phalanger maculatus (spotted phalanger). 112 specimens examined

(i) BMNH 1879.3.5.8. Both I^3 absent.
(ii) BMNH 1883.3.24.5. Right I^3 absent.
(iii) ♂. RMNH 34484. Both I^3 absent.
(iv) ♂. RMNH 34485. Right I^3 absent.
(v) ♂. RMNH 34486. Left I^3 absent.

P. ursinus (bear phalanger). 6 specimens examined

In four skulls, there were two so-called intermediate teeth in each mandible; in two, there were three on the right and two on the left.

P. orientalis (common phalanger). 205 specimens examined

(i) ♂. BMNH 1888.1.5.42. Left I^3 absent.
(ii) RCS Odonto. Mus., A 360.8. (Fig. 8.7). Two teeth in the region of the left canine; the mesial tooth is similar to the right canine but is a little smaller and is situated more mesially. The distal tooth is caniniform but smaller than the mesial one.

There were five specimens with extra maxillary premolars:

(i) ♀. RMNH 34487. In left maxilla.
(ii) RMNH 34488. In left maxilla.
(iii) ♂. RMNH 34489. In both maxillae.
(iv) ♂. RMNH 34490. In both maxillae.
(v) ♂. BMNH 1883.3.24.7. An extra premolar between the right P^2 and P^3, about two-thirds the size of P^3 and similar in shape.

There were no numerical dental variations in the genera *Acrobates* (7), *Distoechurus* (4) and *Cercatetus* (13).

146 *Section 1: Variations in Number, Size and Shape*

Family **Phascolarctidae**

$$\frac{3}{1}:\frac{1}{0}:\frac{1}{1}:\frac{4}{4}=30$$

There were no numerical dental variations in 27 skulls of koala.

Family **Vombatidae**

$$\frac{1}{1}:\frac{0}{0}:\frac{1}{1}:\frac{4}{4}=24$$

There were no numerical dental variations in 51 skulls of wombats.

Super–family **PERAMELOIDEA**

Family **Peramelidae**. Bandicoots

$$\frac{4-5}{3}:\frac{1}{1}:\frac{3}{3}:\frac{4}{4}=46-48$$

Peramelids are insectivorous; the lower incisors are more or less equal in size and are not procumbent; the canines may be large or as small as the incisors and the molars are tritubercular or quadritubercular.

Genus *Isoodon*. 124 specimens examined (8 without mandibles)

(i) *Isoodon auratus* (golden bandicoot). ♂. BMNH 1904.1.3.89. A small premolariform, two-rooted tooth between the left P^1 and P^2.

There were no variations in *Rhynchomeles* (7), *Macrotis* (26), *Echymipera* (17), *Peroryctes* (7), *Chaeropus* (4), or *Perameles* (29). Pohle (1940) described a skull of a captive long-nosed bandicoot (*P. nasuta*) with no upper molars and fewer than normal lower molars.

Super-family **DASYUROIDEA**

Family **Dasyuridae**. Marsupial cats, marsupial mice, etc.

The dentition of dasyurids is adapted for an insectivorous or carnivorous diet. The incisors are not procumbent, the canines are well-developed and the molars are tritubercular.

Archer (1975) examined many dasyurid and other marsupial dentitions and gave numerous rather sketchy descriptions of abnormal teeth. In summarizing his data on 1184 modern dasyurid skulls, we calculate that the only absent teeth were one or more lower premolars in five skulls (0.42%) and upper premolars in two (0.17%). The only skulls with extra teeth were three (0.25%) with additional upper canines, two (0.17%) with lower extra premolars and four (0.34%) with extra upper premolars. Some of Archer's figured instances are included below.

Genus *Dasyurus*. 124 specimens examined (8 without mandibles)

$$\frac{4}{3}:\frac{1}{1}:\frac{2}{2}:\frac{4}{4}=42$$

In this genus, the two premolars were regarded by Thomas (1888b) as P1 and P3.

(i) *Dasyurus maculatus* (tiger-cat). BMNH 1841.12.2.3 (Fig. 8.8). There is an extra molar in the left maxilla and on both sides of the mandible. This skull excited the attention of both Thomas (1888a, p. 265) and Bateson (1894, pp. 256–7). Thomas took the view that the upper supernumer-

Fig. 8.7. *Phalanger orientalis* (common phalanger). RCS Odonto. Mus., A360.8. A: Left side; two teeth in the position of the canine. B: Right side (normal). ×c. 1.3.

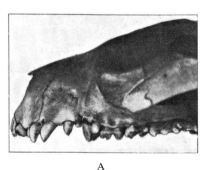

A

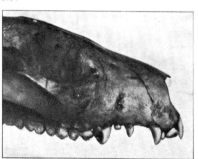

B

ary was between M^3 and M^4 because the last tooth on that side resembles a normal M^4, and the tooth in front of it in size and shape more resembles M^3. Bateson, however, pointed out that the right M^4 in this skull is unusually large and represented, as we would express it today, excessive tooth formation. According to him, there had been a similar excess, though more severe, on the left side leading not only to the formation of an M^4 of excessive size but also to a supernumerary tooth at the end of the arch (a disto-molar) which happened to resemble a normal M^4 in size and shape; in the mandible, similar excessive formative processes had led to extra molars at the ends of the arches similar in size and shape to M_4. Bateson mentioned that Thomas had privately accepted his view. We hope that our paraphrased account of Bateson's view of this specimen will show that he came close to expressing ideas of tooth formative influence that are in accord with present-day field theory.

(ii) *D. geoffroii* (chuditch). ♀. BMNH 1906.8.1.347. A left mandibular disto-molar, about two-thirds the size of M_4.

(iii) *D. maculatus*. Formerly RCS Osteo. Series, 3896 (Fig. 8.9). There is no tooth in the position of the right P^3. The right P^1 is much larger than the contralateral and its crown is divided into two unequal portions, the mesial of which is the smaller; it is probably a P^1 connate with P^3.

(iv) *D. geoffroii*. University Museum of Zoology, Cambridge, A6.12/1. The crown of the right P_3 is partly divided into buccal and lingual parts.

Genera **Phascolosorex**, **Myoictis**, **Antechinus** and **Phascogale**. 114 specimens examined

$$\frac{4}{3}:\frac{1}{1}:\frac{3}{3}:\frac{4}{4} = 46$$

(i) *Phascolosorex dorsalis* (narrow-striped marsupial mouse). ♀. Formerly BMNH 1887.8.17.6. (Fig. 8.10). An extra tooth between the left P^1 and P^2.

Fig. 8.9. *Dasyurus maculatus*. Formerly RCS Osteo. Series, 3896. A: Right; P^3 absent; P^1 connate. B: Left; P^3 present.

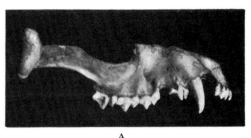

A

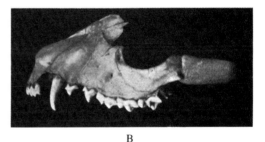

B

Fig. 8.8. *Dasyurus maculatus* (tiger-cat). A: right maxilla of normal specimen.
B: BMNH 1841.12.2.3. Palatal view. The right M^4 is larger and less blade-like than normal; the tooth in the position of the left M^4 is larger still and distal to it is a blade-like disto-molar resembling a normal M^4.
C: BMNH 1841. 12.2.3. Mandible, with bilateral disto-molars nearly as large as M_4. (From Bateson, 1894.).

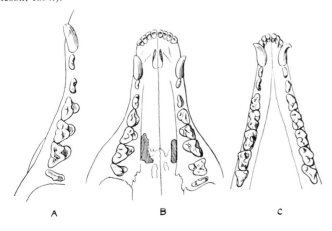

(ii) *P. doriae* (red-bellied marsupial mouse). ♂. BMNH 39.3231. Two premolars in the right maxilla and three in the left. The right P^1 and P^3 are larger than the contralaterals. The left P^2 is the largest of all the premolars.

(iii) *P. dorsalis* (narrow-striped marsupial mouse). ♀. BMNH 1911.11.11.104. Left P^3 distinctly larger than the contralateral.

(iv) *Antechinus flavipes* (yellow-footed antechinus). WAM M6785. An extra premolar mesial to the right P_1. (Archer, 1975, plate 30A and p. 253).

(v) *Phascogale tapoatafa* (tuan). WAM M7453. Left P^2 three-rooted with a partially-divided crown (Archer, 1975, plate 30C and p. 253).

(vi) P_3 is occasionally small and single-rooted, and varies in size within these genera. It was absent as follows: on both sides in 10 out of 14 specimens of the three-striped marsupial mouse (*Myoictis melas*), in 5 out of 40 *Antechinus flavipes*, on one side in 2 out of 17 *Phascogale tapoatafa*.

In the mulgara (*Dasycercus cristicauda*), P_3 was absent in three skulls.

Genus **Dasyuroides**. 4 specimens examined

$$\frac{4}{3}:\frac{1}{1}:\frac{2}{2}:\frac{4}{4}=42$$

In all four specimens of kowari, P_3 was absent on both sides; in one, a P^3 was also absent.

Genus **Sarcophilus**. 22 specimens examined

$$\frac{4}{3}:\frac{1}{1}:\frac{2}{2}:\frac{4}{4}=42$$

Fig. 8.10. *Phascolosorex dorsalis* (narrow-striped marsupial mouse). Formerly BMNH 1887.8.17.6.
Above: Left maxilla from the canine to the first molar, showing an extra premolariform tooth.
Below: the teeth of the right maxilla reversed, to show that the extra tooth is not matched by any of the contralaterals (Bateson, 1894). × *c.* 5.0.

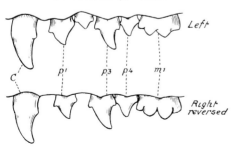

Although Colyer found no numerical dental variation in *Sarcophilus harrisii* (Tasmanian devil), Archer (1975, plate 30B, p. 253) figured a cranium (WAM 71.10.20g) in which the left maxilla had a normal P^1, an abnormally shaped P^2 and no molars.

There were no numerical dental variations in *Thylacinus* (18), *Sminthopsis* (26) and *Antechinomys* (15).

Family **Myrmecobiidae**. Banded anteaters

$$\frac{4}{3}:\frac{1}{1}:\frac{3}{3}:\frac{5-6}{5-6}=50-54$$

The banded anteater has more teeth than any other land mammal. The deciduous molar usually stays in place between P3 and M1. All the teeth are small and delicate; the molars are tritubercular.

Genus **Myrmecobius**. 15 specimens examined (4 without mandibles)

In one skull of *Myrmecobius fasciatus* (banded anteater) (BMNH 314g), there were two connate teeth in the position of the left I^3.

Two out of 11 mandibles had extra teeth in the region of the incisors. In one (BMNH 1876.11.27.3), there were three incisors distal to I_1 on the right; in the other (BMNH 1855.12.24.71), the extra tooth is probably the one mesial to the left canine because it is larger than the other three incisors which match in morphology the contralateral incisors.

In the nine mandibles with the molar regions intact, there were six molars on both sides in six specimens, five on both sides in one, and six on one side and five on the other in two specimens; in the last, M_6 was a small tooth with a single root.

In the right half of another skull, there was a small cylindrical tooth between the right P_3 and M_1.

Eleven of the crania had the maxillary regions intact. There were five molars on each side in nine specimens and four in one (formerly RCS Osteo. Series, 3879); in this skull, the left P_3 was smaller than the contralateral and had only a single cusp. In another (Fig. 8.11), there were in each maxilla six teeth distal to P^3. In the left maxilla, the most mesial of these teeth was small and cylindrical and the most distal tooth was peg-shaped; in the right maxilla, the two mesial teeth were more like premolars than molars, the most distal tooth being similar in shape to the contralateral.

Ch.8. Order Marsupialia   149

Super-family **DIDELPHOIDEA**

Family **Didelphidae**. American opossums.

$$\frac{5}{4} : \frac{0-1}{1} : \frac{3}{3} : \frac{4}{4} = 50$$

The lower incisors are more or less equal in size and are not procumbent; the canines are large and the molars tritubercular.

Genus *Didelphys*. 177 specimens examined

(a) Incisors
 (i) *Didelphys marsupialis* (common opossum). BMNH 1892.11.3.29. Two teeth in the position of the left I_3; they are very worn, are of about the same size and shape, and are both like the right I_3 only a little smaller.
 (ii) *D. marsupialis*. BMNH 1892.11.3.28. The incisors have been lost but there is evidence that six teeth were present in the right premaxilla. The left premaxilla is missing.
 (iii) *D. azarae* (S. American opossum). ♂. BMNH 1898.8.19.1. In the position of right I_4, there are two incisors as little smaller than left I_4.
 (iv) *D. azarae*. ♀. BMNH 1926.1.1.162. Right I^5 absent.
 (v) *D. marsupialis*. ♂. FMNH 41609. Left I^4 and I^5 connate (De la Torre and Dysart, 1966).

(b) Premolars
 (i) *D. marsupialis*. ♀. BMNH 1899.10.3.67. (Fig. 8.12). The tooth in the position of the left P^3 seems to be a P^3 connate at the mesial end with an extra tooth which is situated mesio-lingually. The division between the crowns is complete as far as the cervix where, on the buccal and the lingual aspects, there are ridges on either side of the groove that divides the rest of the tooth.
 (ii) *D. azarae*. ♀. BMNH 1901.6.3.23. (Fig. 8.13). The left P^2 is larger than the contralateral and has a deep fissure on the buccal surface of the crown extending from the distal slope of the cusp.
 (iii) *D. azarae*. ♂. BMNH 1917.1.25.56. The right P^1 is double rooted, as normal; the left P^1 has been lost but evidently was minute and single-rooted.

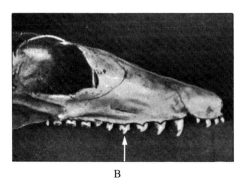

Fig. 8.11. *Myrmecobius fasciatus* (marsupial banded anteater). ♂. BMNH 1906.8.1.357. A: left side. B: right side. Each maxilla has six teeth distally to P^3 (pointers). On the left, the most mesial of these teeth is small and cylindrical and the most distal tooth is peg-shaped; on the right, the two most mesial of these teeth are premolariform and the most distal tooth is peg-shaped. × c. 1.5.

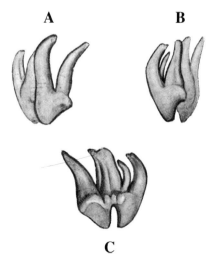

Fig. 8.12. *Didelphys marsupialis* (common opossum). ♀. BMNH 1899.10.3.67. A: Buccal view of a connate left P^3 which appears to be composed of a more or less normal P^3 with a similar smaller tooth fused to its mesio-lingual surface. B: Mesial surface. C: Lingual surface showing that the crowns are separate as far as the cervix where ridges are present on either side of the groove that partially separates the rest of the teeth. The mesial component (left) has a supernumerary root.

150 Section 1: Variations in Number, Size and Shape

(c) *Molars*

(i) *D. marsupialis*. ♀. BMNH 1899.8.1.19. Bilateral maxillary disto-molars.
(ii) *D. marsupialis*. ♂. BMNH 1899.8.1.18. A partially erupted right maxillary disto-molar.
(iii) *D. azarae*. BMNH 1899.2.18.22. A partly erupted right mandibular disto-molar.
(iv) *D. azarae*. RCS Odonto. Mus., G 35.1. Both M^4 absent.
(v) *D. marsupialis*. ♀. BMNH 1903.7.1.107. Right M^4 absent.

Genus *Metachirus*. 123 specimens examined

(i) *Metachirus nudicaudatus* (brown four-eyed opossum). ♂. BMNH 1901.6.7.71. The right P^3 is in position but the left P^3 is absent and the deciduous molar is in its place.

Genera ***Philander*** and ***Caluromys***. 71 specimens examined

(i) *Philander opossum aztecus* (grey four-eyed opossum). ♂. BMNH 1894.12.18.28. Left I^4 and I^5 connate, forming an unusually large tooth with a bifid incisive edge.
(ii) *P. opossum*. ♂. BMNH 1902.7.28.4. Both M^4 absent.
(iii) *P. opossum*. ♂. BMNH 1905.11.2.23. Both M_4 absent.
(iv) *P. opossum*. BMNH 1854.6.3.1. Both M^4 absent.
(v) *P. opossum*. ♀. BMNH 1903.3.1.125. Both M_4 absent.
(vi) *Caluromys philander* (woolly opossum). ♀. BMNH 1910.12.3.10. Both M^4 absent.

Genus *Lutreolina*. 33 specimens examined

(i) *Lutreolina crassicaudata* (little water opossum). BMNH 1879.5.1.3. One incisor absent in each premaxilla.

Genus *Monodelphis*. 76 specimens examined

(i) *Monodelphis maraxina* (short-tailed opossum). ♀. BMNH 1923.8.9.9. Right P^1 absent.
(ii) *M. domestica* (grey short-tailed opossum).

Fig. 8.13. *Didelphys azarae* (S. American opossum). ♀. BMNH 1901.6.3.23. A: the left P^2 larger than the right and the crown has a deep buccal fissure extending from the distal slope of the cusp. B: right side (normal). $\times c.\,1.5$.

   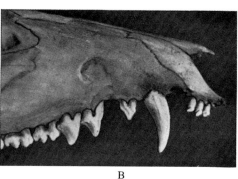

A   B

Fig. 8.14. *Monodelphis domestica* (grey short-tailed opossum). BMNH 1900.9.15.2. On the left side (B), P^3 (pointer) is smaller and more conical than on the right (A). $\times c.\,2.0$.

   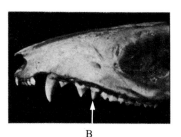

A   B

BMNH 1900.9.15.2. (Fig. 8.14). Left P^3 smaller than contralateral and more conical.

Genus *Marmosa*. 460 specimens examined

(i) *Marmosa robinsoni* (mouse opossum). ♂. BMNH 1899.8.1.51. Left M^4 absent.

There were no numerical variations in *Chironectes* (8), *Glironia* (2) or *Lestodelphys* (5).

Super-family **CAENOLESTOIDEA**

No shrew-opossums were examined by Colyer.

SECTION 2
Variations in Position

CHAPTER 9
Introduction

There is a wide range of types of variations in position and malocclusion. At one end of the range are single teeth which, for various reasons, are outside the regular tooth arch and have persisted in that position, perhaps because they have acquired an abnormal relationship with the cusps of teeth in the opposing arch or because the space they would normally occupy in the arch has closed up due to contiguous tooth movements. At the other end of the range are much more marked abnormalities of occlusal relationship due to defective development or growth of the jaws, especially if one jaw is affected more than the other.

Very rarely, pairs of teeth are found in transposed relationship in the arch, for instance a canine in the position of a first premolar and vice versa. Whether this is the result of movement of tooth germs or is a manifestation of homoeosis, that is, a tooth germ developing the characteristics of another in a meristic series (Chapter 1, p. 2), can only be speculated upon.

Although there are many examples of variation of tooth position, in particular of the relationship of the tooth arches, that are of genetic origin, in general the position of teeth in the jaws is much more influenced by environmental factors than are tooth size or shape because position depends mainly on bone development and bone growth.

Eruption
Apart from the rare exception when teeth develop in ectopic situations away from the normal tooth-bearing area, for instance a molar situated far back in the vertical ramus of the mandible, tooth germs develop in the first instance regularly in rows at or near the

margins of the jaw. The germs tend to maintain this arrangement as an orderly row although, with the increase in size of the germs associated with the actual formation of mineralized teeth, and then the development of a successional series of teeth, the jaws appear to become overcrowded with teeth and their arrangement at first sight may seem to be disorderly (Fig. 9.1). This is not really so, however, and an ordered pattern of pre-eruptive movements is recognizable. The term eruption is used here not simply for the actual emergence of the tooth into the mouth, but for all movements of the tooth in the bone that carry it towards the mouth cavity (see Chapter 16). Once the tooth emerges through the gum, the eruption process continues until the tooth reaches its final position in contact with the corresponding tooth in the opposing jaw. As a tooth cannot move of its own accord, these movements, pre-eruptive and eruptive, depend upon purposive activities in the surrounding bone which are in the nature of bone removal (resorption) and bone deposition (incremental growth). Once the crowns of the teeth have emerged in the mouth, the movement into their normal occlusal relationship with one another involves other co-ordinated mechanisms that are not fully understood; in the first place, the teeth appear to be guided into position by the pressure of the tongue on one side and of the cheek on the other; once the teeth come into contact with those of the opposing jaw, the contours of the opposing tooth surfaces, for instance the inclined planes of molar cusps, contribute guiding forces.

Fig. 9.1. *Pan troglodytes* (chimpanzee). Juvenile. The outer plates of bone have been removed to display the normal positions of the teeth. All the M1 have emerged and are just reaching occlusion. The developing premolars, permanent canines and incisors are in crowded relationships beyond the roots of the deciduous teeth. Note the mesially-tilted crown of the developing M_2.

One feature of overcrowding of the jaws with teeth when a diphyodont dentition is developing is that in many mammals before the permanent molars erupt the maxilla has a posterior projection, the alveolar bulb, which contains the three permanent molars, arranged in a curve, with M^1 below, M^2 above and M^3 high up on the posterior surface of the bone. As growth of the jaws and eruption of the molars proceeds, M^1 tilts down and, from facing disto-buccally, its occlusal surface, as it emerges through the bone, assumes a more horizontal position. With further growth, M^2 and then M^3 undergo similar movements. With the eruption of M^3, all that remains of the bulb is the maxillary tuberosity, a slight extension of the alveolar process behind the functional M^3.

A gross example of what can happen if the maxilla is too small to accommodate the developing cheek teeth is seen in a skull of a new-born lamb affected by cyclopia (Fig. 9.2), a disturbance of development in which the premaxilla is absent and lateral fetal growth of the maxilla is defective, although the number and size of the teeth is unaffected. In this skull, the alveolar bulb almost fills the infratemporal fossa.

In the lower jaw, the growing permanent molars are accommodated in a curve within the ascending ramus of the mandible, with the occlusal surface of M_1 facing slightly mesio-lingually, and M_2 and M_3 are distal to and above it facing in the same direction. The teeth developing in the ramus may form a slight bulge on the medial surface of the bone anterior to the mandibular foramen.

The features just described of the developing molars are seen most characteristically in apes and in man, but similar features can be detected in the young

Fig. 9.2. *Ovies aries* (lamb). Cyclopia. The bones around the central eye are so distorted that there is insufficient space for the developing molar teeth which project into the sub-zygomatic fossa within a thin shell of maxillary bone; an exaggerated alveolar bulb.

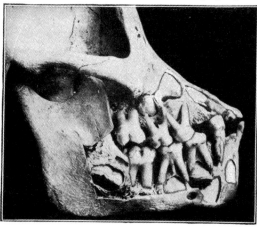

jaws of other mammals; in the pig, the maxillary alveolar bulb is remarkably elongated.

One form of abnormality of position is evidently due to a failure of this progression of the molar teeth in a curve associated with an untilting process; M3 may either erupt with, in the case of the M^3, the occlusal surface facing disto-buccally and, in the case of M_3, facing lingually, or may remain in such a posture but only partially erupted. M_3 may also remain tilted mesially and, because of lack of space, be caught against the distal surface of M_2. It may also be horizontal. In both these cases, the eruption force carries M_3 against the obstructing tooth and it cannot possibly erupt; it is then said to be impacted.

In the Sirenia, for example the manatee, there is a special variation of this alveolar bulb in the mandible (Fig. 9.3), which takes the form of a rounded end of a tube of bone filled with teeth and surrounded by a deep groove which gives the tube the appearance of being separate from the bone of the jaw (Mummery, 1924, p. 460); it is in fact joined to the jaw further forward in the region of M_1. As the molars erupt, they move forward in the bony tube which in the fully adult state is much less prominent.

J.W. Osborn (1984) described a similar appearance suggestive of separateness between the bone of the mandible and the bone enclosing the developing molars in the skull of an elephant.

This separateness of the alveolar tube accords with the evidence for a close relationship between teeth and alveolar bone already referred to (see also Chapter 1, p. 5) and for a separate embryonic origin of the alveolar bone from that of the jaw bone and for its functional separateness.

Growth of the jaws

In a broad sense, most malocclusions are the product of disturbances of jaw growth. Here only some aspects of normal jaw growth can be dealt with.

It is important to realize that the mandible can be resolved into a number of elements (Fig. 9.4), a basic element, two subsidiary elements which are largely for the attachment of muscles and an alveolar process or tooth-bearing element. The existence of the alveolar process appears to depend on the presence of the teeth because loss of the teeth is followed by gradual resorption of that part of the bone.

For the development of the proper functional relationship between the upper and lower tooth arches, there must be accurate co-ordination of development and growth of the two jaws, upper and lower. The processes by which under normal circumstances this is usually achieved are not fully understood and it is beyond the scope of this book to discuss them in detail. It must, however, be mentioned that the upper jaw is an integral part of the cranium in contact with many other bones and its growth depends mainly

Fig. 9.3. *Trichecus* sp. (manatee). Sub-adult. Mandible. On the medial surface of the vertical ramus, the most distal of the row of developing molars can be seen contained within a shell of bone (part of which has been removed to display the tooth). This has the appearance of being separate from the bone proper of the jaw and of projecting through an opening in its surface.

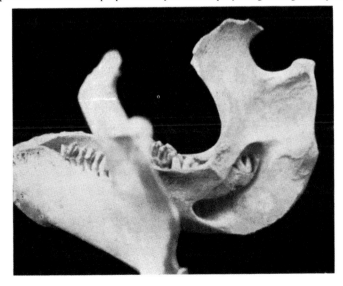

upon sutural growth, whereas the lower jaw is in direct contact only with the squamosal bones, the mandibular joints intervening, and its growth depends largely on endochondral ossification at the mandibular condyles.

Bearing in mind the great complexity of the factors involved, it is not surprising that some of the grosser malocclusions appear to arise from disorders of the process of co-ordination that determines the ultimate sizes of the upper and lower jaws.

If for any reason there is a slight disproportion between the size of the basic elements of maxilla and mandible, adjustments within the alveolar processes of the two jaws can nevertheless lead to a normal occlusal relationship between the teeth. If the disproportion is great, however, full adjustment within the alveolar processes may not be possible and a generalized malocclusion may occur. This is seen for example in certain breeds of dog, which indicates that some variations in tooth position, particularly those associated with jaw disproportions, can be of genetic origin. Lesser degrees of variation in position, such as the eruption of a tooth beside the normal arch as if it were crowded out, seem more likely to be due to some local factor, such as a failure of the intercusp contact mechanism which normally guides a tooth into its ultimate position, or to deficient growth of the alveolar process or of the jaw as a whole. If for some reason the jaws do not grow to a large enough size to accommodate all the teeth, the tooth arches may be crowded and irregular and it would be expected that the teeth most likely to be crowded out would be those that erupt late or last in the series.

Bone growth before birth is not much affected by environmental influences but after birth it is very sensitive to such influences. Bone growth is affected by the body's intrinsic environment; it is for example controlled by the endocrine system. Extrinsic environmental influences that affect bone growth are nutrition, infection and various other disease processes, and local injury.

It is well established that malnutrition adversely affects skeletal growth including growth of the jaws. Although very relevant to malocclusion found particularly in captive animals, it is impossible to review here all the evidence, experimental and otherwise, for this statement. The evidence was evaluated by Brash, McKeag and Scott (1956). Rickets, which in its various forms affects bone growth, is dealt with in connection with eruption in Chapter 16, pp. 333–334 (see also Chapter 25, pp. 602–603).

The hypothesis that bone growth is stimulated by muscle action is of interest; in particular the question whether the use of the jaws in mastication (as would occur if the diet were hard and tough) benefits jaw growth has been a matter of controversy. This is because of the greater prevalence of malocclusion among urbanized people than among primitive people, which matches the high prevalence of irregularities in the position of teeth in captive animals reported in the following chapters. Some animal experiments support the hypothesis.

Moore (1965) showed that, after 17 weeks of feeding rats on water-softened rat cake, the overall measurements of the brain case and facial skeleton and the weights of the muscles of mastication were about 12% less than in animals fed on hard cake. Beecher and Corruccini (1981) found that tooth arch width in particular was larger in young rats kept for four months

Fig. 9.4. Diagrammatic representation of a mammalian mandible (human) illustrating the concept of a basic element, of which the articular condyle is part, two subsidiary muscle-attachment elements, angular and coronoid, and the tooth-bearing alveolar process. From Symons (1951).

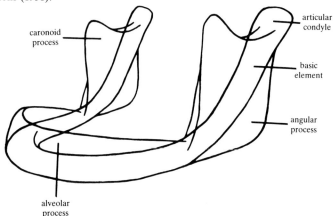

on a hard pelletted diet than on the same diet in semi-liquid form.

In some similar experiments on the squirrel monkey (*Saimiri sciureus*), the facial skeleton was in general smaller in animals maintained on a soft diet than in those on a hard diet during growth and malocclusion was more prevalent (Corruccini and Beecher, 1982; Beecher, Corruccini and Freeman, 1983). Corruccini and Beecher (1984) compared the dentitions and various jaw and cranial measurements in two groups of sub-adult baboons which had been kept in captivity for about two years. One group of 24 animals was provided with a soft non-solid laboratory diet and the other group of 16 with naturally tough food including laboratory pellet. Apart from a much higher prevalence (10 of the 24 animals) of incisor underjet in the soft-diet group, there were no large differences. However, there was in the soft-diet group a statistically greater range of variability in most measurements and a statistically higher prevalence of asymmetry between measurements of the two sides of the skull. Unfortunately, 'to improve manageability', the recently erupted canine teeth were removed from half the soft-diet group. However, this was not done until shortly before the end of the experiment and comparison with the animals that retained their canines suggested that removal of the canines produced no large differences.

Mesial drift
During function, teeth move individually in their sockets under the heavy lateral stresses associated with vigorous mastication; consequently, where adjacent teeth are in contact, the approximal surfaces undergo wear; the teeth drift towards one another, so keeping the approximal surfaces in contact in spite of the wear. In most instances that have been studied, the drift is in a mesial direction.

When there is a diastema in the arcade, and a diastema of some sort is present in most mammalian tooth arches, mesial drift of the cheek teeth would tend to close it. Wood (1938) pointed out that there is no evidence that the diastema does get smaller with age; in fact he was inclined to think that it tended to be longer in old animals.

In the human dentition, which is unusual in having no natural diastemata, mesial drift is well-documented; for example, when a tooth is removed, the space tends to close up by movement forward, often associated with some tilting, of the teeth behind the gap.

In the macaques, it may be the interdigitation of the large canines in the diastema that maintains the diastema. Macaques have been much used experimentally to study tooth drift so there is ample evidence that, as in man, if a cheek tooth is removed, the teeth that are distal to it drift mesially into the space (J. P. Moss and Picton, 1967 and 1982). It could be that mesial drift associated with wear is part of the mesial movement of the whole arch on the body of the jaw bone. There is some evidence that this occurs in man (Yilmaz, Darling and Levers, 1980).

In rodents and some ungulates (e.g. horse), the diastema between incisors and cheek teeth is very long and the cheeks can be made to meet across the midline, so dividing the mouth into two functional compartments. It may be that the forces created by this arrangement of the cheeks prevent the cheek teeth drifting forwards. According to Sicher and Weinmann (1944) and Kronman (1971), the molars of rodents tend to drift in a distal direction.

Schumacher (1929), who made careful observations and measurements on a large number of skulls of chamois and roe deer of a wide age range, confirmed that approximal wear of the cheek teeth was accompanied by mesial drift of the cheek teeth towards the first tooth distal to the diastema. Thus the length of the cheek tooth-row was appreciably shortened; in young roe bucks (*Capreolus capreolus*), it was about 69 mm and, at eight years, about 60 mm (Schumacher, 1929, p. 520). Figure 9.5 shows the mesial movement of the distal cheek teeth with advancing age in the gazelle (*Gazella* sp.). This shortening of the cheek tooth-row with advancing age is also a feature of cattle (Grigson, 1974), sheep (Chapman, 1933) and horses (Gidley, 1901).

Wood (1938) brought together and evaluated much of the evidence in mammals generally, including some from palaeontology, for an association between wear of the approximal surfaces and the drift of the teeth which keeps them together as a functional unit. He concluded that the direction in which teeth move, when space for them to do so is provided by approximal wear, depends upon the forces they are subjected to, and that depends upon the complex details of the morphology of the jaw-tooth mechanism and the way in which it functions; these vary from one species to another. The forces even vary locally so that, in some instances, a tooth may move mesially, or distally, or in some other direction.

Malocclusion
The causes of variation in tooth position and malocclusion are best understood in man because, in industrialized or so-called civilized societies, malocclu-

Section 2: Variations in Position

sions are very common and treatment is sought, mainly for cosmetic reasons. What we know about malocclusion in non-human species is to some extent a spin-off from experimental work on various macaque species, a few other primates, the pig and on various rodents. An important source book which, although orientated to malocclusion in man, contains an appraisal of the literature on animals since the publication of the first edition of the present book in 1936 is by Brash, McKeag and Scott (1956).

Jaw deformities and malocclusion in non-mammalian vertebrates

Bellairs (1981), in his account of anomalous development of the head region in reptiles, which includes cleft palate and the spectacular double-headed snakes which can survive for surprisingly long periods, referred to a condition in crocodilians in which the snout is shortened and pug-like resembling the condition described here in some mammals, mostly domestic (Chapter 11, pp. 266–268 and Chapter 13, p. 294). Hall (1985) described two skulls of crocodiles (*Crocodylus novaeguineae*) which had been in captivity for many years and showed a condition akin to this. The upper jaw was shorter and the lower wider than normal so that the tooth arches were not in proper occlusion. In these skulls and one other, several tooth alveoli were partly filled with masses of mineralized tissue; this needs further investigation.

It is worth mentioning also here that Huchzermeyer (1986) found the jaws of young crocodiles (*Crocodylus niloticus*) fed de-boned chicken were bendable like rubber and the teeth translucent and glass-like. Radiography showed that the whole skeletons were of low density. A change to a mineral-

Fig. 9.5. *Gazella thomsoni* (Thomson's gazelle). Two skulls showing the movement forward of the cheek teeth as age advances. The position of the teeth is emphasized by a line vertical to the anterior margin of the orbit.
Below: A young skull in which the line passes through the centre of M^1 (which is unfortunately split into mesial and distal halves).
Above: An older skull in which the line passes through M^3.

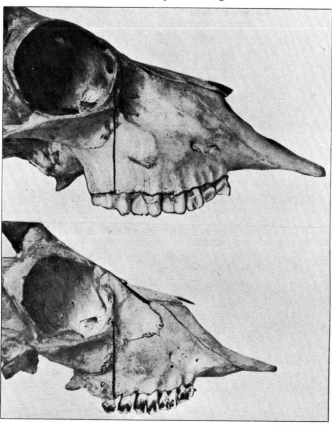

supplemented diet for three months produced recovery, including less translucency of the teeth.

Cooper, Arnold and Henderson (1982) found that about 50% of a captive population of skinks (*Leiolopisma telfairii*) showed severe spinal curvature and jaw disproportion so that the upper jaw projected beyond the lower. They were maintained in Jersey because their only natural habitat, an island in the Indian Ocean, has been almost destroyed. A genetic factor is an obvious possible cause but there are other potential causes related to changes of environment such as a dietary deficiency, an environmental toxin or changes in temperature and humidity.

We have pointed out that such a complex of factors is concerned in producing the co-ordination of development and growth of the two jaws, as well as the change from a funcional deciduous dentition to a permanent one, that it is not surprising that co-ordination can sometimes go wrong. Such appears to be the case with the dentitions of the ayu fish, *Plecoglossus altivelis*, in which the dentition in juveniles is composed of simple conical teeth and in adults is one of complex comb-like teeth. A change in shape of the jaws is associated with the change from juvenile to adult type of dentition. Komada (1980) found that in over one-third of hatchery-reared adults the jaws were malaligned; either they did not close properly or the lower jaw was twisted to one side. The presence of variable numbers of teeth of both types indicated a defect in the process of change from a juvenile to an adult-type dentition.

What is called pugheadedness, usually involving marked shortening of the upper jaw, has been described in a wide variety of fishes; for example perch, *Perca fluviatilis* (Marlborough and Meadows, 1966), landlocked Atlantic salmon, *Salmo salar* (Leggett, 1969), striped bass, *Morone saxatilis* (Grinstead, 1971), comb-fish, *Zaniolepus latipinnis* (Talent, 1975), Japanese char, *Salvelinus leucomaenis* (Honma and Yoshie, 1978) and California roach, *Hesperoleucus symmetricus* (Leidy, 1985) as well as sharks (Hoenig and Walsh, 1983). Bortone (1972) showed that in the pirate perch, *Aphredoderus sayanus*, the shortening is not due to underdevelopment of the various bones that contribute to the upper jaw but to shortening of all the bones of the head, apart from the lower jaw; for instance, the orbits are small so that there is slight exophthalmy. Additional examples can be traced from the series of bibliographies prepared by Dawson (1961–1964; 1965–1969; 1970–1971; Dawson and Heal, 1976) and also from an important study by Riehl and Schmitt (1985) who listed many other species in which the condition has been recorded. They also referred to shortening of both upper and lower jaws which is know as fillister-headedness. Using mainly X-ray projection microscopy, Riehl and Schmitt concluded that in pug-headedness in the mosquito fish, *Heterandria formosa*, some bones of the neurocranium are affected as well as the various bones of the upper jaw.

Other types of jaw deformities also occur in fishes; for example, lateral misplacement of the lower jaw (Merriner and Wilson, 1972; Sinha, Singh and Singh, 1980); Honma and Ikeda (1971) described a black porgy, *Acanthopagrus schlegeli*, in which both jaws were shortened and together formed a bird-like beak.

Osteometrics

It is obvious that, in studies of malocclusion which concern such matters as the relation between jaw size and the size of the dentition, there is a place for measurement; indeed the literature on human malocclusion is full of measurements of various kinds. Even in the recent literature on non-human malocclusion, however, references to actual measurements are all too few. Colyer in the first edition of this book (1936), and later (1943b and 1947a) in respect of primate dentitions, did something to introduce measurement into this area. We have had to modify some of his terms slightly to accord with present practice.

Basal length: basion (the most anterior point on the foramen magnum) to opisthion (the most anterior point of the skull in the midline).

Basicranial axis: basion to nasion.

Palatal length: the distance between the anterior margin of the bone between the I^1 (alveolare, virtually prosthion) and the line joining the distal borders of the last upper molars (Keith, 1931).

Palatal breadth: the distance beween the lingual surfaces of M^1 at the cervix (Colyer, 1943b).

Cheek-tooth length: the sum of the mesio-distal diameters of the post-canine teeth.

Length of mandible: the distance from the mid-point of a line joining the posterior aspects of the condyles to the most anterior point of the bone (gnathion). Where possible, this was measured in the horizontal plane with the mandible resting on a flat surface (Colyer, 1943b).

Glenoid–incisor length: the distance from the mid-point of a line joining the post-glenoid tubercles to the alveolare.

Those who we hope may wish to follow Colyer's pioneering footsteps would have no difficuly in devising additional or alternative measurements and would be helped by the system of zoological osteometrics given by Duerst (1930) and Von den Driesch (1976).

Textbooks of human orthodontics and of physical anthropology are additional sources.

Terminology

It is convenient to employ a few terms borrowed from the literature of human orthodontics. For example, where the whole lower arch is more anterior than normal so that the lower incisors are in front of the upper ones, we shall use the term inferior protrusion; where the lower arch is too posteriorly situated, inferior retrusion will be used, although it must be borne in mind that occasionally the upper arch is too small or more posterior than normal, producing an apparent inferior protrusion. In such cases, a detailed osteometric or cephalometric study, involving measurements of the arch and jaw positions in relation to the base of the skull, is needed to distinguish inferior retrusion from superior protrusion. It is quite common for the incisor relationship to be abnormal while the rest of the arch relationship is unchanged; e.g. so that the upper incisors are situated further forward than normal – or vice versa. It seems necessary to avoid referring to this as superior or inferior protrusion so we shall adopt the terms incisor overjet and incisor underjet.

Colyer chose to include in this section examples of teeth that, because misplaced in the jaws, remained unerupted; this included teeth that were malaligned so that their path of eruption was changed and was obstructed by another tooth, that is impacted. We decided to leave this unchanged although it is obvious that such abnormalities could alternatively be regarded as disturbances of eruption (Chapter 16).

CHAPTER 10

Order **PRIMATES**

Sub-order **ANTHROPOIDEA**

Family **Pongidae**. Great apes and gibbons

The great apes and gibbons are of special interest because they most resemble man in their basic morphology; they have the same dental formula and their teeth are similar in form. One of the main differences is that, in man, the upper incisors overlap the lower incisors labially in occlusion and the inter-incisor relationship only becomes edge-to-edge late in life when the teeth are very worn; in apes, the upper incisors slightly overlap the lowers immediately after eruption but thereafter the relationship is edge-to-edge.

The canines are much larger in apes and, unlike human canines, show marked sex dimorphism. The lower canine is accommodated in occlusion in a diastema between I^2 and the upper canine, and a smaller diastema between the lower canine and the mesial slope of P_3, accommodating the upper canine, is frequently present. Both diastemas are smaller in the female in accord with the smaller canines.

In apes, the incisors form a flat curve, as in man. However, the upper premolars and molars (cheek teeth) are arranged in straight lines which usually converge slightly distally so that the widest part of the upper arch is between the canines and the narrowest between the M^3 (Fig. 10.1). The lower tooth arch is naturally somewhat similar though, as in man, because the upper cheek teeth overlap the lower ones, the lower arch is narrower than the upper.

In man, the sides of the upper arch are slightly rounded with the widest part not between the canines but between M^1, and the narrowest between the M^2

and the M^3, forming what is defined as a parabolic curve.

Divergence, either unilateral or bilateral, from what can be called the perfect arch is so common that, to quantify it, it is necessary to assume criteria on which to decide what is a minimal positional variation. It seems that Colyer regarded moderate rotation of P^4 as the criterion of minimal positional variation in the cheek-tooth part of the arch. He based this on the observation that, in the Pongidae, P^4 is roughly triangular in plan view (in man it is roughly rectangular), the buccal margin being wider than the lingual. Thus, if the tooth is rotated, its mesio-distal diameter becomes less. If there is insufficient space in the arch when P^4 erupts, it is likely to rotate, as it may also do later if there is mesio-distal pressure within the arch produced by the eruption of other teeth, e.g. M^2 and M^3. Hence Colyer regarded moderate rotation of P^4, defined as rotation by one sixteenth to one eighth of a circle, as the minimal criterion of positional variation in the post-canine tooth arch.

The triangularity of P^4 is very evident in hylobatids. In the great apes, it is much less marked and is lost in adult skulls as a result of approximal wear.

Colyer did not explicitly include the lower second premolar in his statement about premolar rotation, but the column head 'minimum rotation of P4' in his Tables do include P_4 as well as P^4. Although P_4 in both pongids and cercopithecids is more regular in plan view than P^4, it may have a similar tendency to rotate when there is insufficient space to accommodate it.

Colyer listed moderate rotation of P4 separately from other positional irregularities of the premolars because he regarded it as indicative of lack of space. Larger degrees of rotation than one eighth of a circle were not necessarily due to lack of space; they could arise as an error of development, and were therefore not separated in his Tables from other positional variations.

Genus *Gorilla*, gorilla

Positional variations in the gorilla are summarized in Tables 10.1 and 10.2. The normal arrangement of the maxillary teeth in an adult gorilla is shown in Figure 10.1. Sometimes there is a slight lingual bend in the

Table 10.1 Positional variations in APES and GIBBONS from the wild

Genus	No. of specimens	Class of tooth					Total specimens affected				Incisor underjet	
					mod. rot.		with P4 column		without P4 column			
		I	C	P	P4	M	No.	%	No.	%	No.	%
Gorilla	689 (129 without mandibles)	80	9	74	45	65	218	31.6	173	25.1	6	1.1
Pan	465 (93 without mandibles)	35	1	46	24	18	103	22.2	79	17.0	24	6.5
Pongo	255 (38 without mandibles)	35	5	15	9	24	64	25.1	55	21.6	3	1.4
Hylobates	277 (5 without mandibles	7	0	34	22	12	65	23.5	43	15.5	9	3.3
Symphalangus	88	4	0	17	15	2	36	40.9	21	23.9	1	1.1
Total	1774 (**265** without mandibles)	161	15	186	115	121	486	27.4	371	20.9	43	2.8

In skulls in which two or more classes of teeth varied in position, each class is recorded individually; for example, if in a skull one or more of the incisors, premolars and molars varied, each appropriate class would be credited with one and the column 'total specimens affected' with one only. Rotation of P4 (by less than one sixteenth of a circle) has been excluded; moderate rotation (one sixteenth to one eighth of a circle) of P4 is listed in a column ('mod.rot. P4') separately from other positional variations of the premolars.

region of P^4 and buccal bend in that of M^3. The pre-canine diastema, in accord with the smaller canines, is smaller in the female.

(a) Incisors
With the wear of the occlusal margins, an edge-to-edge condition is soon acquired, replacing the original incisor overjet. Incisor underjet is common and may be associated with general inferior protrusion as in Figure 10.2.

Positional abnormalities of the incisors are much more common in the lower jaws than the upper. I^1 may show slight rotation with their mesial borders lingually; Krapp and Lampel (1973) noted a female gorilla (Serie Sabater Pi Go go 4, Zoologisch-vergleichend-anatomischen Instituts der Universität Freiburg, Switzerland) in which the right I^1 was rotated through 45°. I^2 may be placed a little lingually to the plane of the I^1, overlap them, or be rotated as in Figure 10.3. In one female skull (RCS Odonto. Mus., G 67.12), the right I^1 projects labially, overlapping the left I^1 (Colyer, 1947a). A plaster cast showing irregularity of the incisors and the premolars is shown in Figure 10.4.

Of the 80 specimens showing positional irregularity of the incisors, 22 were associated with some degree of irregularity of P4, nine with abnormal positions of M^3 and five with abnormal positions of both. These data suggest that insufficient growth of the upper jaw as a whole was the cause of the irregularities.

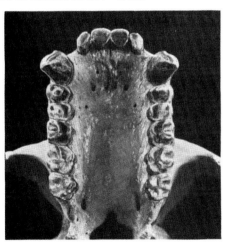

Fig. 10.1. *Gorilla gorilla* (gorilla). ♂. The normal arrangement of the upper dental arch. The arch is widest between the canines, and the cheek teeth form straight rows which converge slightly distally so that the arch is narrowest between the M^3. × c. 0.5.

Fig. 10.2. *Gorilla gorilla* (gorilla). ♂. BMNH 1923. 11.29.6. Lower incisors and canines slightly labial to the upper incisors. The lower cheek teeth also occlude in a pre-normal position on both sides but slightly less on the right. Note that left P^3 and P_3 are opposite one another instead of P^3 occluding between P_3 and P_4. This suggests a generalized inferior protrusion. However, the profile of the premaxilla (compare with Fig. 10.7) suggests that it is underdeveloped so that the abnormal arch relationship is probably due to the whole upper jaw being under-developed or situated further posterior than normal, rather than the mandible or its dental arch being further forward than normal. × c. 0.5.

Table 10.2 Crowding and impaction in wild and captive APES AND GIBBONS (Schultz, 1935)

	No. of specimens	% with incisors crowded		% with premolars crowded		% with impacted teeth
		Upper	Lower	Upper	Lower	
Gorilla	245	0.4	0.4	2.9	0.8	0.8
Pan	126	0.8	0	4.0	2.4	5.6
Pongo	67	1.5	1.5	3.0	3.0	3.0
Hylobates	130	0.8	0	2.3	0.8	0.8

164 *Section 2: Variations in Position*

Incisors are rarely crowded in the gorilla. Schultz (1950) found only 1.1% of marked incisor crowding in 267 gorilla skulls and only 0.8% in his earlier sample (Schultz, 1935) of 245 (Table 10.2).

Fig. 10.3. *Gorilla gorilla* (gorilla). P-C. Mus., Zvi 30. Both P^4 slightly rotated symmetrically, anti-clockwise on the right and clockwise on the left. They are also situated slightly more lingually than normal and the cheek-tooth rows are irregularly curved and crowded. $\times c.\,0.5$.

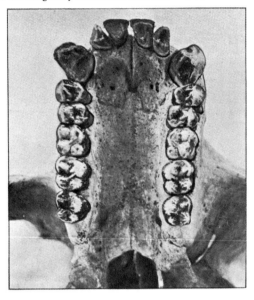

Fig. 10.4. *Gorilla gorilla* (gorilla). ♂. Formerly RCS Osteo. Series, 33. Plaster cast of the maxillae showing abnormal positions of the incisors and the premolars. It was presented to the Royal College of Surgeons in 1848 by the Bristol Philosophical Institution and must have been made from one of the first gorillas brought to Britain. The cast was destroyed when the College was bombed in 1942. $\times c.\,0.5$.

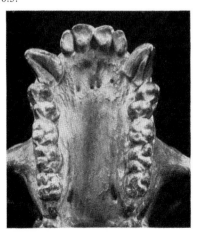

(b) Canines

There were eight examples of various degrees of labial misplacement of the canine in the 560 mandibles (Figs. 10.5 and 10.6). A male gorilla (BMNH 1925.1.4.1)

Fig. 10.5. *Gorilla gorilla* (gorilla). P-C. Mus. Cam I 99. The mandibular left canine has erupted labially to the lower incisors. $\times c.\,0.33$.

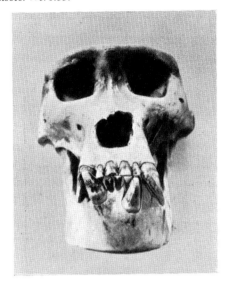

Fig. 10.6. *Gorilla gorilla* (gorilla). ♂. BMNH 1939.928. The right mandibular canine has erupted labially to I_2 which is displaced lingually. The canine impinges on the incisive edge of the right I^2 which is rotated. The left upper canine is not fully erupted and is jammed against P^3. The lower left canine is more lingually situated than normal and in consequence has worn the distal surface of I^2. $\times c.\,0.33$.

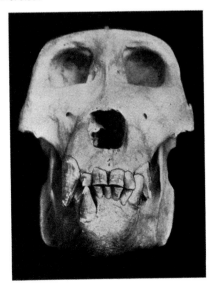

shows a similar condition; both canines have erupted labially to the I_2 with their roots directed towards the midline. Buccal misplacement of the lower canine, making it unusually prominent, is common. In Figure 10.7, a supernumerary incisor has caused the disto-buccal displacement of the lower canine and this in turn has displaced the upper canine.

Neuville (1936) described the skull of a gorilla in which the right upper canine was misplaced and probably deformed. The tip of the crown projected slightly through the facial aspect of the maxilla about where the apex of the root of the canine is normally situated.

(c) Premolars

In contrast to incisors and molars, premolars are commonly crowded in the gorilla. Schultz (1950) recorded crowding in 4.5% of his sample of 267 skulls and suggested that there is a general tendency towards lack of space for P4. The rotation or misplacement of upper premolars is common; some gross examples are shown in Figures 10.8, 10.9 and 10.10. Krapp and Lampel (1973) described a female gorilla in the Zoologisch-vergleichend-anatomischen Instituts of the University of Freiburg, Switzerland. (Series Sabater Phi go go 3) in which both P^4 were misplaced lingually, similar to Figure 10.9.

The mandibular premolars are far less variable in position. Rotation of P_4 occurs and in one skull the left P_4 had turned through 45°, the buccal surface facing mesially. Considerable misplacement of the P_3 may occur (Fig. 10.11). Schultz (1956, p. 1001) figured a mandible (Anthropologisches Institut of the Univer-

Fig. 10.8. *Gorilla g. beringei* (mountain gorilla). KMMA 812. Left P^4 displaced lingually and rotated through nearly 90° so that its buccal surface is in contact with the distal aspect of P^3. × *c*. 0.67.

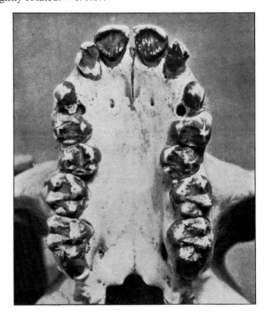

Fig. 10.9. *Gorilla gorilla* (gorilla). USNM 252578. Symmetrical irregularity of the arch; the principal features are that both P^4 are misplaced lingually and both M^1 are slightly rotated. × *c*. 0.67.

Fig. 10.7. *Gorilla gorilla* (gorilla). Formerly RCS Osteo. Series, 21a. An additional lower incisor has caused the left canine to erupt in such a way that it occludes with the distal aspect instead of the mesial aspect of the upper canine. × *c*. 0.5.

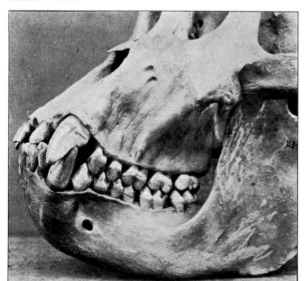

sity of Zurich) in which both P_4 are misplaced lingually and are erupting between the P_3.

(d) Molars

Slight rotation of M^2 or M^3, or both, occur (Fig. 10.12). Krapp and Lampel described a skull of a female gorilla (Amsterdam Anthropobiologisch Laboratorium, 1891 B 8c) in which the right M^3 was rotated through 30° and in a male (Amsterdam Anthropobiologisch Laboratorium, 1919/41) the same tooth was rotated by 45°.

In the adult gorilla, the occluding surfaces of the M^1 and M^2 face directly downwards and that of M^3 a little distally. Sometimes the distal inclination of the M^3 may be increased so that the mesial cusps alone occlude with M_3; this is probably due to insufficient growth of the posterior part of the maxilla. In the well-

Fig. 10.10. *Gorilla gorilla* (gorilla). Formerly Anatomy School, Cambridge University. Right P^4 rotated and misplaced lingually. On the left, the arch curves slightly lingually. Both these features indicate insufficient space for the cheek teeth. $\times c. 0.5$.

Fig. 10.12. *Gorilla gorilla* (gorilla). ♂. BMNH 1923.11.29.2. Both M^2 rotated and both M^3 even more so, producing a buccal curve of the distal ends of the arch. $\times c. 0.67$.

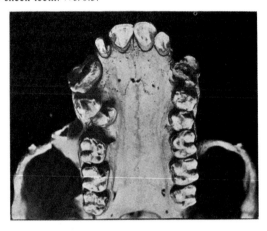

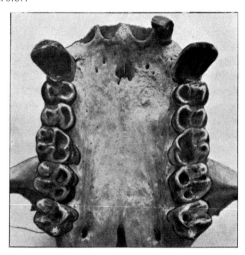

Fig. 10.11. *Gorilla gorilla* (gorilla). ♀. BMNH 1939.925. Right P_3 misplaced buccally so that the lingual slope of its buccal cusp occludes with the buccal cusp of P^3. An unusually large diastema between P_3 and the lower canine. $\times c. 0.37$.

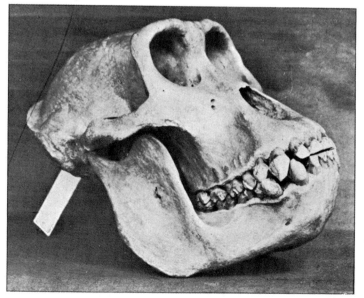

developed jaw, there is a tuberosity beyond the last molar whereas, when the molar is tilted, the tuberosity is lacking.

Extreme irregularity of M^3 is rare, the only example seen being in the skull of a mountain gorilla (Fig. 10.13) in which the marked disto-buccal direction of the tooth is associated with lack of growth of the right maxilla; the left maxilla is normal. Another type of irregularity is a lingual misplacement of M^2; in one specimen, the tooth was out of occlusion to the lingual side of the mandibular molars.

Positional variations of the mandibular molars are limited to M_3 and consist of rotation or a mesial tilt of the crown.

Schultz (1956) found only one example of molar crowding in his sample of 267 gorilla skulls.

(e) Asymmetry of the arch
Asymmetry of the arch is usually associated with a similar condition of the facial portion of the skull. The amount of asymmetry varies from a slight flattening of one side of the arch to marked distortion as seen in Figure 10.14, where the maxillae are bent to the right. The bend has displaced the premolars and molars on the right so that they are distal to their normal relationship with the mandibular teeth. The occlusion of the left premolars and molars is normal. This type of asymmetry is often associated with asymmetry of the base of the skull.

Asymmetry of the facial skeleton is common in the mountain gorilla (*Gorilla g. beringei*). Gyldenstolpe (1928) gave details of 13 adults; in eight of the ten males, there was some asymmetry, the deviation in six of the specimens being to the right. In the majority of these skulls, no injuries could be traced. In a skull of a male mountain gorilla (USNM 239883), the deviation of the bones was to the right. In this specimen, there was evidence of injury to the bones in the regions of the mastoid process and the glenoid fossa, and the condition of the right auditory meatus indicated much damage to the inner ear (Howell, 1925).

Genus *Pan*, chimpanzee

Tables 10.1 and 10.2 include a summary of positional variations in the chimpanzee.

(a) Incisors
Incisor underjet is more common in the chimpanzee than in the gorilla and the orang-utan. In two skulls, there was general inferior protrusion. Incisor underjet was noted in 10 (16.6%) of 60 adult chimpanzees from Batouri District, Cameroun, in the Powell-Cotton Museum, but occurred in only 14 (4.5%) of the 312 skulls from other areas. Many of the chimpanzee skulls from Batouri also showed variations in number and

Fig. 10.13. *Gorilla g. beringei* (mountain gorilla). ♂. BMNH 1920.4.13.4. Lateral view of the right upper jaw showing misplaced M^3 with its occlusal surface facing bucco-distally. Much of the front of the jaw has been destroyed, perhaps at death. Reduced.

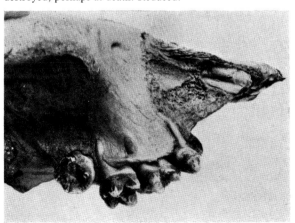

Fig. 10.14. *Gorilla g. beringei* (mountain gorilla). KMMA 2258. Entire left side markedly further forward than the right.

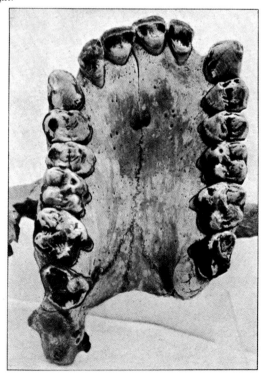

shape (Chapter 3, p. 22). In one of the Batouri skulls, all the teeth had been lost and there was a marked protrusion of the mandible (Fig. 10.15).

In the 19 skulls from Batouri in which the deciduous incisors were in position, five had incisor underjet.

A slightly irregular arrangement of the incisors occurred in about 8% of the chimpanzees and was more frequent in the upper jaw than in the mandible. A marked example is shown in Figure 10.16. The mesial angles of both I^2 are tilted in a disto-buccal direction and the upper incisors as a whole are splayed out labially with their roots arising from a narrow base. The sutures between the premaxillae and the maxillae are not evident, suggesting that growth was not active. The canines are partially erupted and there is insufficient room for them to move into. The cheek teeth are a little irregular.

(b) Canines
The only example of irregularity in position of the canine was in a skull (formerly RCS Osteo. Series, 11.215) in which the canines had erupted to the buccal side of the deciduous canines; on the left, the deciduous canine was still in place but, on the right, it had been lost.

(c) Premolars
Positional variations of the premolars are more common in the maxilla than in the mandible. In the maxilla, the variations are rotation of the teeth or a displacement from the line of the arch, and there is a tendency for P^4 to be placed lingually (Fig. 10.17) whereas P^3 is more commonly misplaced buccally (Fig. 10.18). In the mandible, the most common anomaly is a slight rotation of P_4.

Extreme misplacement of the premolars is rare in wild animals. Figure 10.19 shows an exception in which

Fig. 10.16. *Pan troglodytes* (chimpanzee). Young. Formerly private collection of W. L. H. Duckworth. Incisors splayed out labially. Premaxilla-maxillary sutures are not evident. Canines partially erupted and there is insufficient space for them to move into.

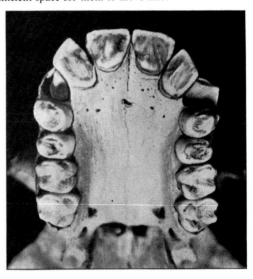

Fig. 10.15. *Pan troglodytes* (chimpanzee). ♂. P-C. Mus., M 165. Extreme protrusive relationship of the mandible. All the teeth have been lost during life and perhaps the jaw relationship is secondary to this edentulousness; atrophy of the upper arch would follow and, in the absence of teeth, the hinge movement of the mandible at the joint would bring the anterior end of the mandible forward. × c. 0.5.

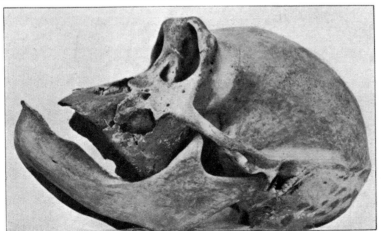

there are irregularities in all parts of the arch but in particular both are rotated with their buccal surfaces facing distally and the P^4 are lingual to the arch. Both M^1 are rotated, as happens when they move mesially.

Krapp and Lampel (1973) described a male chimpanzee in which the right P_4 was tilted distally so that the three mesial cusps were under the mesial surface of M_1.

(d) Molars

The most frequent variation of the molars is a tilting of M_3 so that its occlusal surface faces mesially or lingually. M^3 may be rotated slightly or misplaced a little lingually or buccally.

(e) Effect of captivity

Variations in position occurred in 10 (45.5%) of 29

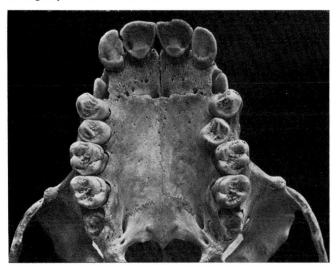

Fig. 10.17. *Pan troglodytes* (chimpanzee). Young. The cheek-tooth rows are irregular, mainly because both P^4 are situated lingually and the left one is rotated. ×c. 0.67.

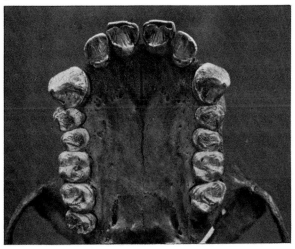

Fig. 10.18. *Pan troglodytes* (chimpanzee). ♂. Formerly RCS Osteo. Series, 7. Cheek teeth crowded and slightly irregular, principally because both P^4 are rotated; right P^3 is placed buccally and projects beyond the mandibular premolars. ×c. 0.75.

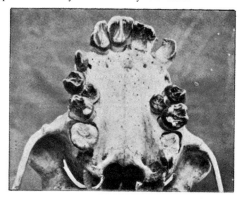

Fig. 10.19. *Pan troglodytes* (chimpanzee). ♀. Young. P-C. Mus., M 805. Gross irregularity in all parts of the arch. In particular, the premolars and M^1 are rotated and in a jumble. Both M^1 are rotated on their lingual roots as happens when they move mesially. Reduced.

170 Section 2: Variations in Position

captive chimpanzees compared with 22.2% of animals from the wild (Table 10.1).

In chimpanzees which have been brought into captivity at an early age, positional abnormalities are usually more severe than in animals from the wild. This is especially noticeable in the case of P^4 which may be completely displaced from the arch (Fig. 10.20).

Among the skulls of captive chimpanzees, there

Fig. 10.20. *Pan troglodytes* (chimpanzee). Captive. Sub-adult. BMNH 1939.908.
A: Right side. P^3 is misplaced so far buccally that it is out of occlusion; P^4 is erupting in the palate lingual to M^1. M^2 is tilted mesially and rotated with its buccal surface facing slightly mesially; both the tilting and rotation suggest that M^1 has drifted mesially into the space that should have been occupied by P^4. $\times c. 0.5$.
B: Left side. Incisor overjet is present. P^3 is erupting over the roots of the deciduous molars with its occlusal surface facing almost directly buccally. On this side, there is incisor overjet which is likely to be due to an upper jaw defect; note that the upper canine and post-canine teeth are in post-normal occlusion. All the permanent teeth show several rows of ring enamel hypoplasia. $\times c. 0.5$.

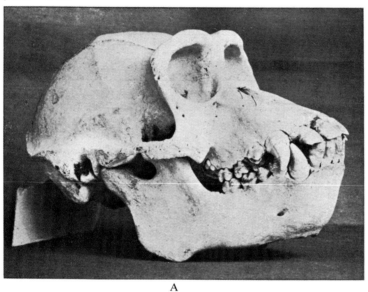

A

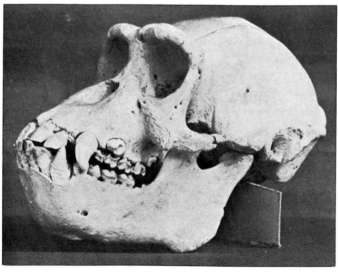

B

were two unusual irregularities of molars. In one (Fig. 10.21), the right deciduous maxillary molars in occlusion are buccal to the mandibular teeth. In the other, M_3 had erupted with its occlusal surface so far lingually that the upper tooth occluded with its buccal surface. This can be regarded as due to the M_3 continuing to be more or less in the orientation it has during an earlier stage in its development before eruption (Fig. 10.22 and Chapter 9, pp. 154–155).

In all the chimpanzee skulls seen from the wild and captive states, there was no example of extreme misplacement of canines. There is, however, in the British Museum of Natural History a plaster cast of an animal that died in the Dresden Zoo. The right maxillary canine has erupted on the buccal aspect of P^3 close to the P^4.

Genus ***Pongo***, orang-utan

Tables 10.1 and 10.2 include a summary of positional variations in the orang-utan.

(a) Incisors

Incisor underjet is common in the deciduous dentition but rare in adults; there were three examples in 217 complete skulls. In one (Fig. 10.23), the lower incisor overlap is associated with evidence of defective growth of the upper jaw. There is some crowding of the cheek teeth, the premolars being inclined mesially and the M^3 distally, so that only their mesial cusps occlude with M_3.

Irregular position of the incisors was present in about 15% of the orang skulls. There may be a slight

Fig. 10.21. *Pan troglodytes* (chimpanzee). Captive. Formerly Museum of the Royal Dental Hospital. The right deciduous lower molars pass within the upper arch.

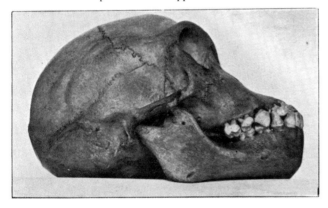

Fig. 10.22. *Pan troglodytes* (chimpanzee). Lateral view of a mandible in which a window has been cut to show the normal position of the developing M_2 and M_3. Their occlusal surfaces are facing lingually, M_3 more than M_2.

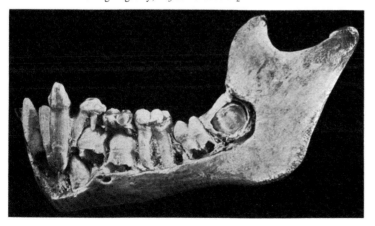

rotation of all the upper incisors but more often the irregularity is limited to I^2 which may be rotated or displaced lingually partially behind I^1. The mandibular incisors vary less than the maxillary teeth; the most common irregularities are slight rotations, but extreme rotation does occur as in Figure 10.24.

Fig. 10.23. *Pongo pygmaeus* (orang-utan). BMNH 1868.4.16.2. The lower incisors pass labially to the upper ones. The lower canine has been lost but the rest of the arch relation appears to be normal; the upper cheek teeth are crowded and irregular, the premolars being inclined mesially and the M^3 distally. $\times c.\ 0.6$.

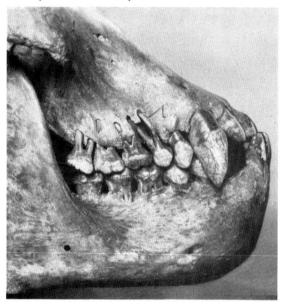

Fig. 10.24. *Pongo pygmaeus* (orang-utan). ♀. Formerly RCS Osteo. Series, 40d. Right I_1 is rotated through 90° so that its lingual surface faces mesially; left I_1 is lost but the shape of the socket shows that it was similarly rotated. $\times c.\ 0.75$.

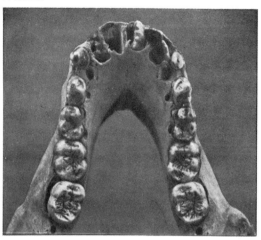

Fig. 10.25. *Pongo pygmaeus* (orang-utan). BMNH 1886.12.20.10.
A: palatal view.
B: side view.
Right maxillary canine and P^3 misplaced mesially so that the canine occupies the diastema normally occupied in occlusion by the lower canine; it is also misplaced buccally. A space exists between P^3 and P^4; the lower canine has been lost.

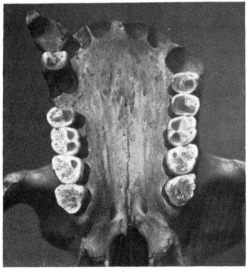

A

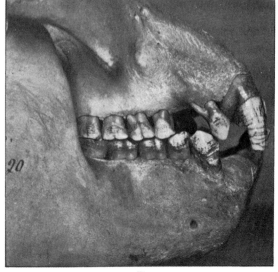

B

(b) Canines

Irregularities in position appear to occur only occasionally.

In the specimen shown in Figure 10.25, the right maxillary canine and P^3 are misplaced mesially, so that the canine is in the diastema normally occupied in occlusion by the lower canine and is in contact with I^2. The lower canine is now missing but a wear facet on the lingual aspect of the upper canine shows where the lower canine occluded with it. As the P^3 erupts before the canines, it is possible that it developed in the jaw and erupted more mesially than normal. The upper canine could then only move mesially likewise on eruption.

Figure 10.26 shows a skull in which the left upper canine is misplaced distally in contact with the P^4 with P^3 misplaced on its lingual side; both P^3 are rotated with their buccal surfaces facing mesially. The symmetrical rotation of P^3, which erupt before the canines, suggests that they developed with that abnormal orientation in the first instance and that the canine assumed its abnormal position secondarily during eruption.

Other examples of misplacement of canines are shown in Figures 10.27, 10.28 and 10.29.

(c) Premolars

Extreme misplacement of P^4 in the arch is common. Examples are shown in Figures 10.30, 10.31 and 10.32.

The upper jaw of an orang in Figure 10.33 is particularly interesting because the teeth that appear to be P^4 are both in the premaxilla. On the right, the tooth is

Fig. 10.27. *Pongo pygmaeus* (orang-utan). Formerly RCS Osteo. Series, 40a. Right maxillary canine is misplaced distally to the lingual aspect of P^3. Right P^3 is rotated so that its buccal surface faces mesially; the rotation of the left P^3 is in the opposite direction and its buccal surface faces bucco-distally. × c. 0.67.

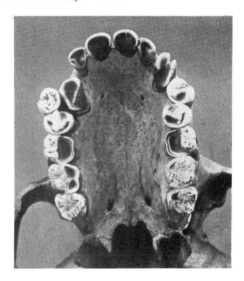

Fig. 10.26. *Pongo pygmaeus* (orang-utan). ♂. RCS Osteo. Series, 41.
A: palatal view.
B: side view.
Left upper canine misplaced distally in contact with P^4, with P^3 misplaced on its lingual side; both premolars are rotated with their buccal surfaces facing mesially.

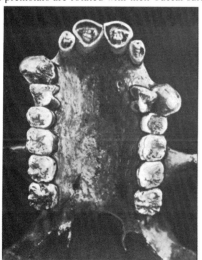

A

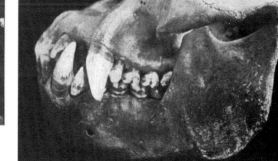

B

174 *Section 2: Variations in Position*

fully erupted lingually to the diastema between I^2 and the upper canine. On the left, the tooth is unerupted lingually to I^1.

Neuville (1936) described the mandible of a young orang in which the crown of P_2 projected slightly through the outer surface of the bone on both sides, about one centimetre below the crown of the predecessor; a similar specimen was figured by Bennejeant (1936, p. 198).

(d) Molars

The main types of positional variation of the maxillary molars are distal inclination of the crown (Fig. 10.23) or rotation. In the mandible, M_3 may be unerupted and

Fig. 10.28. *Pongo pygmaeus* (orang-utan). BMNH 1859.8.16.2. Right canine has erupted to the labial side of I_2, the crown of which is displaced distally in contact with P_3. The lateral view shows that the lower incisors and right canine occlude outside the upper arch and that the mesio-distal relationship of the cheek teeth is normal, but only because the upper cheek teeth are more mesially situated than usual. $\times c. 0.67$.

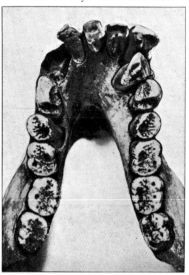
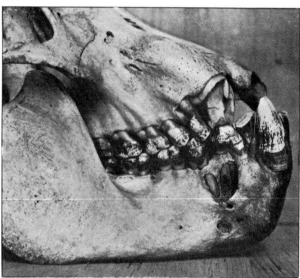

Fig. 10.29. *Pongo pygmaeus* (orang-utan). BMNH 1879.11.21.214. Because the right mandibular canine is misplaced distally, it occludes cusp-to-cusp with the upper canine. $\times c. 0.6$.

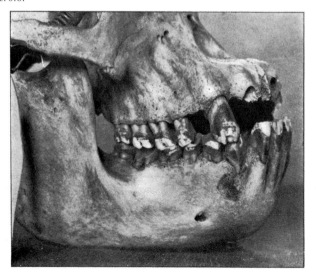

more grossly misplaced, as in Figure 10.34. Amongst young animals, there was a mandible (BMNH 3x) in which the crown of the right M_1 was tilted mesially so that the mesial margin was overlapped by the distal surface of the m_4. On the left, the M_1 had been lost but there was evidence of a similar irregularity.

Krapp and Lampel (1973) described a female orang-utan (Amsterdam Anthropobiologisch Laboratorium, 1911/197) in which both M^1 are misplaced lingually.

(e) Effect of captivity

The only evidence of the effect of captivity, or of the nutritional inadequacy that tends to be associated with it, was provided by Björk (1950) who made a detailed study, including radiography of the skull of Jacob (now in the Zoologisk Museum, Copenhagen), an adult orang-utan reared in the Copenhagen Zoo, and compared it with the skull of a wild orang. The upper face of Jacob was markedly concave rather like that of short-faced dogs. Björk quoted other evidence that rearing in captivity is likely to reduce the length of the jaws in different animals.

Genus *Hylobates*, gibbon

Tables 10.1 and 10.2 include a summary of positional variations in the gibbon.

(a) Incisors

In the gibbon, although the incisors in the adult usually meet edge-to-edge, incisor overjet is quite common. On the other hand, incisor underjet is even more common, occurring in the gibbon about as frequently as in the chimpanzee and more commonly than in the gorilla and the orang-utan. In one skull, there was true inferior protrusion, the whole lower arch being in prenormal relationship with the upper. Irregularity in the position of the incisors is uncommon; I^2 may be rotated and placed a little lingually to the line of I^1 or a space may be present between the I^1.

(b) Canines

There were no instances of irregularity in position of the canines, but Martini (1877) illustrated a mandible (Fig. 10.35) in which the right canine had erupted lingually to the incisors, the deciduous canine being in position.

Fig. 10.30. *Pongo pygmaeus* (orang-utan). ZMA 7. Right P^4 erupted high up above M^2 through the lateral surface of the maxilla. M^1 has moved mesially into contact with P^3. (Duckworth, 1907).

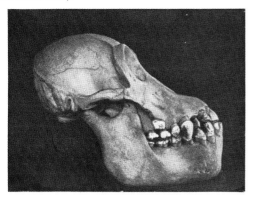

Fig. 10.31. *Pongo pygmaeus* (orang-utan). FMNH 19024. Left P^4 (pointer) erupting high up on the lateral surface of the maxilla above M^1. M^1 is in contact with P^3 which is impacted between it and the canine. On the right, there is less severe crowding, the predominant feature being the lingual misplacement and rotation of P^4. $\times c. 0.5$.

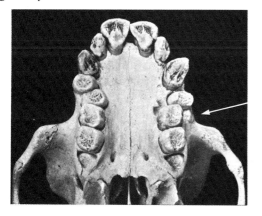

Fig. 10.32. *Pongo pygmaeus* (orang-utan). FMNH 19025. Both P^4 misplaced lingually and rotated so that their buccal surfaces are facing distally. $\times c. 0.5$.

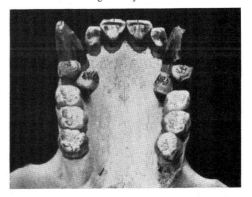

(c) Premolars

Positional variations of the premolars occurred in about 12% of the gibbon skulls examined, rotation being the most common but the teeth may be misplaced lingually (Fig. 10.36).

Misplacements of premolars are more common in the maxilla than in the mandible but they may occur in both jaws as in Figure 10.37, which also shows distal tilting of both M^3.

(d) Molars

It is common for the M^3 to be misplaced lingually so that in occlusion their buccal cusps are edge-to-edge with the buccal cusps of the M_3 or even lingual to them. This is a manifestation of narrowness of the distal part of the upper arch. When distally tilted, M^3 may be rotated or impacted against M^2. Rotation of M_3 occurs, but mesial tilting of the crown of M_3 was not encountered in gibbons from the wild. Irregularity in the position of M_2 is rare in apes and monkeys, but rotation does occur (Fig. 10.38).

Krapp and Lampel (1973) described a gibbon skull (Genf Muséum d'Histoire naturelle, 745/33) in which both M_3 are misplaced lingually. Distal inclination of the crowns of M^3, probably due to insufficient growth of the jaw, is met with.

Fig. 10.33. *Pongo pygmaeus* (orang-utan). Zool. Coll. Oxford Univ. Mus., Mus. Ref. 8787 (O.C. 2043a). The crowns of what morphologically appear to be P^4 are in the premaxilla, on the left behind I^1. $\times c. 0.5$.

Fig. 10.35. *Hylobates lar leuciscus* (white-handed gibbon). Right mandibular canine has erupted lingually to I_1 and I_2. The root of its deciduous predecessor is still present in its normal position in the arch (Martini, 1877).

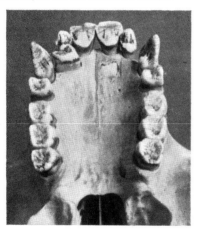

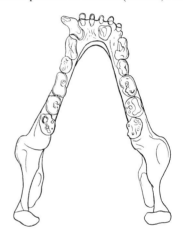

Fig. 10.34. *Pongo pygmaeus* (orang-utan). ♀. RCS Odonto. Mus., G67.3. Lingual aspect of mandible with a window cut to show M_3 buried in the bone and horizontally placed with its occlusal surface impacted against M_2. $\times c. 0.66$.

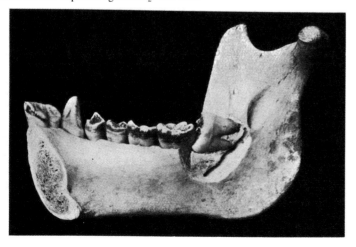

General crowding of groups of permanent teeth is common in wild gibbons. Schultz (1944) noted pronounced degrees of crowding in 6.8% of his Chiengmai Series ($n=233$, all *H. lar*) and in 4.8% of his General Series II ($n=579$, several species of *Hylobates*). Such extreme crowding can result in non-eruption or impaction of one or more teeth; this was noted in 2.2% of the Chiengmai Series, usually affecting second incisors.

(e) Effect of captivity

Variations in position occurred in nine (31.0%) of 29 captive gibbons, compared with 23.5% of animals from the wild (Table 10.1).

Genus *Symphalangus*, siamang

Table 10.1 includes a summary of positional variations in the siamang.

Malocclusion is more common in *Symphalangus* than in *Hylobates*. Abnormalities in position of the premolars was present in about 20% of the specimens. The types of variation are similar to those seen in *Hylobates*. Figure 10.39 shows an example in which the left P^4 had erupted well towards the lingual aspect of the arch and mesially to persistent m^4.

Family **Cercopithecidae**, Old World monkeys

There is as wide a range of positional variations of the teeth in the Old World monkeys as in the great apes and gibbons, but the extreme irregularities seen in the apes are less common. Tables 10.3 and 10.4 show that the Colobinae are more variable than the Cercopithecinae and the incidence of protrusion of the lower incisors is high in the colobines, especially in the leaf-monkeys (*Presbytis*) and in *Colobus badius* and *C. kirkii*; Swindler (1978) suggested that this is related to their leaf-eating habits.

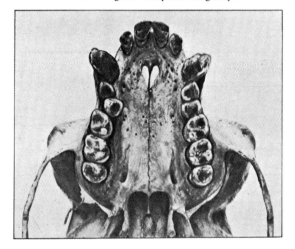

Fig. 10.36. *Hylobates hoolock* (hoolock gibbon). ♂. BMNH 1915.5.5.1. Right P^4 misplaced lingually.

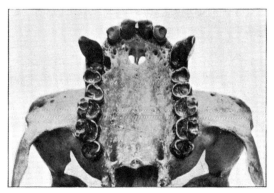
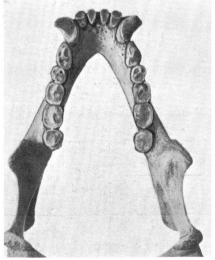

Fig. 10.37. *Hylobates lar lar* (white-handed gibbon). ♂. BMNH 1914.12.8.10. Right P^4 lingually misplaced and rotated. Upper cheek-tooth rows irregular and curved slightly with the arch narrowing between M^3. The lower arch has a similar, more regular curvature. Natural size.

178 Section 2: Variations in Position

Comparison of wild and captive patas monkeys (*Erythrocebus patas*) shows that positional variations are more common and more extreme in the captive state. In particular, M_3 is frequently misplaced, probably as a result of the shortening of the mandible that often occurs in domestic and captive animals (see p. 175, 197–200, 264).

Sub-family **Colobinae**

Genus *Colobus*, colobus monkeys

The normal arrangement of the teeth in colobus monkeys is shown in Figure 10.40. The upper incisors form an even curve; the line of the premolars and the molars curves slightly lingually towards M^3. In the mandible, the incisors and canines form a curve which is narrower than in the upper jaw; the premolars and the molars are in straight parallel lines. The large P_3 is usually set slightly obliquely to the line of the arch.

Table 10.3 shows that the prevalence of positional variations in *Colobus* is high. Schultz's (1958) figure for positional variations (2%) in a population of 309

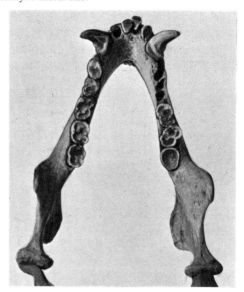

Fig. 10.38. *Hylobates moloch muelleri* (Bornean gibbon). ♂. BMNH 1920.12.4.6. Right M_2 rotated clockwise through about 45° so that its mesial surface faces mesiobuccally. Natural size.

Table 10.3 Positional variations in the CERCOPITHECIDAE from the wild. The figures for *Colobus* are from Colyer (1943b).

Genus	No. of specimens	Class of tooth					Total specimens affected				Incisor underjet	
		I	C	P	mod. rot. P4	M	with P4 column		without P4 column			
							No.	%	No.	%	No.	%
Colobinae												
Colobus	1344	352	6	230	170	22	634	47.2	464	34.5	245	18.2
Presbytis	985	199	3	145	50	16	357	36.2	305	31.0	354	35.9
Pygathrix	16	2	0	0	1	0	3	18.8	2	12.5	0	0
Rhinopithecus	18	4	0	5	3	0	8	44.4	5	27.8	1	5.5
Simias	10	3	0	0	1	4	4	40.0	3	30.0	3	30.0
Nasalis	85	6	1	8	6	0	20	23.5	14	16.5	0	0
Total	2458	566	10	388	231	42	1026	41.7	793	32.3	603	24.5
Cercopithecinae												
Cercopithecus	1473	100	1	162	136	25	390	26.5	254	17.2	5	0.3
Erythrocebus	42	3	0	8	3	2	14	33.3	11	26.2	0	0
Cercocebus	277	21	1	45	83	1	139	50.2	56	20.2	2	0.7
Macaca	588	49	0	87	53	11	182	30.9	129	21.9	1	0.2
Papio	289	10	1	27	52	8	87	30.1	35	12.1	0	0
Mandrillus	35	0	0	3	2	2	8	22.9	6	17.1	0	0
Theropithecus	21	0	1	0	4	0	5	23.8	1	4.8	0	0
Total	2725	183	4	332	333	49	825	30.3	492	18.1	8	0.3
Total	5183	749	14	720	564	91	1851	35.7	1285	24.8	611	11.8

In skulls in which two or more classes of teeth varied in position, each class is recorded individually; for example, if in a skull one or more of the incisors, premolars and molars varied, each appropriate class would be credited with one and the column 'total specimens affected' with only only. Rotation of P4 (by less than one sixteenth of a circle) has been excluded; moderate rotation (one sixteenth to one eighth of a circle) of P4 is listed in a column 'mod. rot. P4' separately from other positional variations of the premolars.

colobus monkeys from Liberia is so low that he may have employed less rigorous criteria of normality than Colyer.

(a) Incisors

The main features of interest in *Colobus* are the frequency of protrusion of the mandibular incisors and of irregularity in the position of the upper incisors.

Specimens have been classed as showing incisor underjet if the mandibular incisors project beyond the upper incisors by at least half the labio-lingual diameter of the tooth, as shown in Figure 10.41. When underjet occurred in monkeys of the genus *Colobus* as a whole, it was usually greater than this and the lower incisors were almost completely clear of the upper teeth; in about one-eighth of colobus monkeys, the lower incisors were completely free. One of the most extreme instances is shown in Figure 10.42; the other was in a female *C.b.preussi* (P-C. Mus., M270) (Colyer, 1943b). The prevalence of incisor underjet varied between the different species and sub-species. Table 10.5 indicates that it was significantly more common in red colobus (*C. badius*), Kirk's colobus (*C. kirkii*) and black colobus (*C. satanas*) than in the various species of black-and-white colobus (*C. angolensis*, *C. guereza* and *C. polykomos*).

Table 10.4 The percentage of positional variations in some genera of CERCOPITHECIDAE from the wild; data from Schultz (1935, p. 547, table 21)

Genus	No. of specimens	Incisors crowded Upper	Lower	Premolars crowded Upper	Lower
Colobinae					
Colobus	59	3.4	1.7	1.7	0
Nasalis	41	2.4	0	0	0
Pygathrix	273	5.5	2.2	0.7	0.4
Total	**373**	**4.8**	**1.9**	**0.8**	**0.3**
Cercopithecinae					
Cercopithecus	134	0.7	0	0	0
Macaca	227	0.5	0	2.3	0.5
Papio	68	1.5	1.5	1.5	0
Total	**429**	**0.5**	**0.2**	**1.4**	**0.2**

Fig. 10.39. *Symphalangus syndactylus* (siamang). A.E.L. 1908/119. Left P^4 misplaced lingually with its deciduous predecessor still in position.

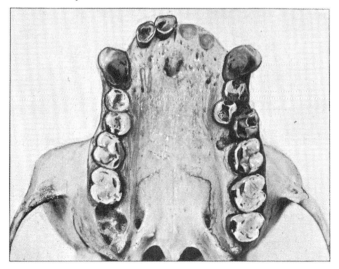

Colyer's (1943b and 1947a) analysis of the dimensions of normal skulls, and skulls with incisor underjet, of *Colobus badius preussi* suggested that underjet is caused by a reduction in the length of the upper jaw rather than lengthening of the lower jaw, that is by superior retrusion (Table 10.6).

Colyer's results for *C. badius* as a whole are similar to those of Schultz (1958, p.101) who found underjet in 32% of 138 skulls of red colobus (*C. badius badius*), but the low figures for black-and-white colobus (*C. angolenisis, C. guereza* and *C. polykomos*) are in direct conflict with Swindler's (1976, p.140) high figure of 78% for *C. polykomos* and Schultz's 60% for *C. polykomos polykomos*. However, the results are not directly comparable because Swindler does not state which sub-species he studied nor whether he included *C. guereza* with *C. polykomos* in his black-and-white sample; Schultz's material was all of *C. polykomos polykomos*, a sub-species not studied by Colyer.

In *Colobus*, irregularity in the position of the

Fig. 10.40. *Colobus guereza kikuyuensis* (Kenya black-and-white Colobus). ♂. BMNH 1904.11.5.1. The normal tooth arches in *Colobus*.

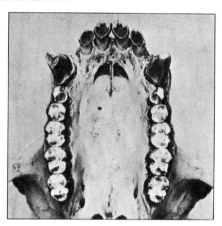

Table 10.5 Incisor underjet in various species and sub-species of *COLOBUS*. The incidence of incisor underjet is greater in *C. badius, C. kirkii* and *C. satanas* than in the other species; this difference is statistically highly significant.

Species and sub-species	No. of specimens	Incisor underjet No.	%
Colobus badius badius	23	2	8.7
C. b. ellioti	119	39	32.8
C. b. oustaleti	96	32	33.3
C. b. foai	23	4	17.4
C. b. tholloni	26	4	15.4
C. b. preussi	65	26	40.0
C. b. temmincki	14	4	28.6
C. b. rufomitratus	12	2	16.7
Total *Colobus badius*	**378**	**113**	**29.9**
Colobus kirkii	33	8	24.2
Colobus satanas	84	17	20.2
Colobus angolensis angolensis	42	3	7.1
C. a. adolfifriederici	26	4	15.4
C. a. palliatus	48	6	12.5
C. a. cottoni	69	7	10.1
Total *Colobus angolensis*	**185**	**20**	**10.8**
Colobus polykomos vellerosus	26	1	3.8
Colobus guereza guereza	115	17	14.8
C. g. occidentalis	217	31	14.3
C. g. matschei	43	5	11.6
C. g. kikuyuensis	90	10	11.1
C. g. caudatus	56	0	0
C. g. dodingae	18	6	33.3
C. g. gallarum	28	0	0
Total *Colobus guereza*	**567**	**69**	**12.2**

The Table does not include taxa for which only a small amount of material was available. From Colyer (1943b) with nomenclature amended according to Napier (1985).

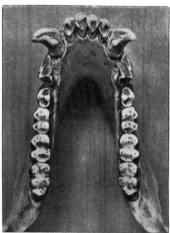

Fig. 10.41. Diagram showing partial underjet of the worn lower incisors in an adult *Colobus*. (Li = lingual, La = labial).

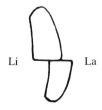

incisors is common and usually consists of a protrusion of both I^1 (Fig. 10.43), or of one I^1 (Fig. 10.44). The irregularity may also take the form of an overlap of the left and right I^1, or tilting of I^2 sufficient to cause overlap with I^1. Table 10.7 shows that the proportion of specimens with positional variations of the incisors

Table 10.6 The relationship between underjet in *COLOBUS BADIUS PREUSSI* to various cranial dimensions (defined on p. 159) and indices, showing that underjet is likely to have been caused by a shortening of the upper jaw rather than the mandible

A	No.	Mean palatal index	Mean mandibular length	Mean basal length			Relation of mandibular to basal length	Relation of palatal to mandibular length
				Basal portion	Palatal portion	Total length		
Without underjet	22	52.3	71.9	36.7	38.7	75.4	95.4%	53.8%
With underjet	26	55.5	70.9	36.4	37.4	73.8	96.1%	52.8%

B	No.	Mean gnathic index	Mean basal length
Without underjet	16	119.4	75.4
With underjet	16	117.9	73.8

A: In both sexes. The palatal index is the palatal breadth × 100/palatal length. (From Colyer, 1943b.)
B: In females alone. The gnathic index is the basal length × 100/the length of the basicranial axis (basion to nasion). (From Colyer, 1947a.)

Fig. 10.42. *Colobus badius tholloni* (South Congo red colobus). ♂. BMNH 1927.3.1.4. Extreme incisor underjet. × c. 0.67.

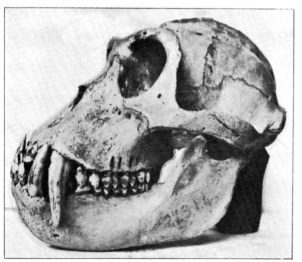

Fig. 10.43. *Colobus angolensis cottoni* (Powell-Cotton's Angolan colobus). BMNH 1906.6.2.1. Labial protrusion of both I^1.

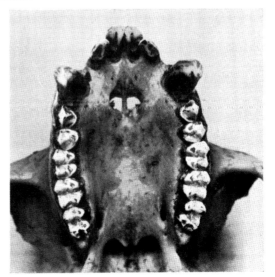

differed considerably between species, and within species, the sub-species *C. badius preussi* and *C. guereza gallarum* having an incidence that is significantly higher than that of the other taxa.

Some measurements that Colyer (1943b, 1947a) made of the facial skeleton of *Colobus badius preussi* and *C. guereza caudatus* suggest that irregularity of the upper incisors was associated with a narrow premaxilla. However, as Colyer pointed out, the number of specimens was too few for reliable inferences to be drawn, so that his table of measurements will not be reproduced here.

Table 10.7 Positional variations of incisors in various species and subspecies of *COLOBUS*. The incidence of positional variations of the incisors is variable, and higher in *C. badius preussi* and *C. guereza gallarum* than in the other taxa. This difference is statistically significant.

Species and sub-species	No. of specimens	Positional variations of incisors No.	%
Colobus badius badius	23	4	17.4
C. b. tephosceles	10	5	50.0
C. b. ellioti	119	27	22.7
C. b. oustaleti	96	32	33.3
C. b. foai	23	7	30.4
C. b. thollani	26	9	34.6
C. b. preussi	61	38	62.3
C. b. temmincki	14	3	21.4
C. b. rufomitratus	10	5	50.0
Total *Colobus badius*	**382**	**130**	**34.0**
Colobus kirkii	33	5	15.2
Colobus satanas	84	13	15.5
Colobus angolensis angolensis	42	8	19.0
C. a. adolffriederici	26	10	38.5
C. a. palliatus	48	15	31.3
C. a. cottoni	69	28	40.6
Total *Colobus angolensis*	**185**	**61**	**33.0**
Colobus polykomos vellerosus	26	6	23.1
C. guereza guereza	115	22	19.1
C. g. occidentalis	217	44	20.3
C. g. matschei	43	7	16.3
C. g. kikuyuensis	90	34	37.8
C. g. caudatus	56	28	50.0
C. g. dodingae	18	2	11.1
C. g. gallarum	28	2	7.1
Total *Colobus guereza*	**567**	**145**	**25.6**

The Table does not include taxa for which only a small amount of material was available. From Colyer (1943b) with nomenclature amended according to Napier (1985).

The prevalence of variation in tooth position in the same or similar taxa in different environments or districts is of interest; unfortunately, suitable material is scarce. The American Museum of Natural History has a good series of the sub-species *Colobus badius oustaleti* collected from two different environmental areas, one heavy rain forest and the other gallery (riverside) forest in the savannah. The *Colobus* from these two areas show certain differences which justify classifying them as different races, *C. b. (oustaleti) powelli* and *C. b. (oustaleti) brunneus* respectively (Napier, 1985).

Of the 41 *C. b. powelli* skulls from the rain forest area, 19 showed variations of the incisors and, of these, 13 were associated with some malposition of the premolars. Amongst the 31 *C. b. brunneus* skulls from the gallery forest, 11 showed variations of the incisors, 5 being associated with malposition of premolars. The percentage of variant specimens from the rain forest was 46.3% and from the gallery forest 35.5%.

According to J. A. Allen (1922–25), *C. b. brunneus* has a larger cranium than *C. b. powelli* (Table 10.8). The females are not appreciably smaller than males. These data show that the length of cheek tooth row in relation to the length of the skull is greater in *C. b. powelli* than in *C. b. brunneus*, although *C. b. powelli* shows a greater degree of variability. So it would seem that the difference in the prevalence of malposition of the teeth may be due to differences between the environments of the two populations.

Fig. 10.44. *Colobus guereza kikuyuensis* (Kenya black-and-white colobus). ♀. BMNH 1900.1.3.11. Labial protrusion of left I^1.

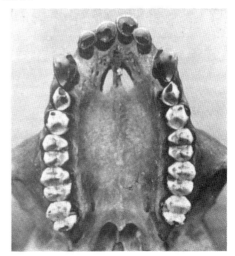

(b) Canines

In six specimens, one or more canines were misplaced. In the skull shown in Figure 10.45, the right P^3 is absent and the canine has moved distally into contact with P^4. In a skull of *C. guereza guereza* (♂, BMNH 1902.9.9.2), the mandibular teeth were erupting buccally to the line of the arch; in another (♀, BMNH 1919.6.1.1), the right mandibular canine projected buccally and, in three others (*C. g. caudatus*, NRM 2400; *C. badius foai*, ZMB A 48.09; and *C. badius foai*, KMMA 9273), the lower canine had moved buccally and so caused displacement of the upper teeth.

Krapp and Lampel (1973) described a skull of a female *Colobus badius* (Zoologisch-vergleichend-anatomischen Instituts of the University of Freiburg, Switzerland, Serie Himmelheber Co ba 13) which had

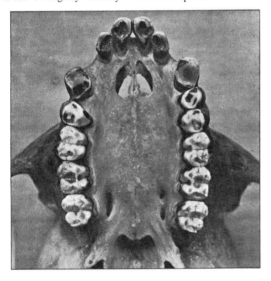

Fig. 10.45. *Colobus guereza kikuyensis* (Kenya black and white colobus). ♀. BMNH 1900.2.1.3. The right maxillary canine is situated more distally than normal associated with the absence of the right P^3; the other cheek teeth have moved slightly mesially to close the space.

a diastema between the canine and P_3 in the right mandible. The lower canine occluded between the right I^1 and I^2 and P_3 was in occlusion with the upper canine.

(c) Premolars

The most common positional variation was rotation, sometimes even to the extent that the lingual surface of P3 faced distally; P^4 may be situated slightly lingually or buccally to the arch; the commonest variation was rotation with the lingual cusp facing mesially. In the mandible, P_3 was commonly slightly oblique to the line of the arch; P_4 was set obliquely in two specimens. It could be that these rotations and other misplacements were due to pressure exerted on them by the eruption of the canines a little later.

In a skull of a female Preuss's red colobus (*C. badius preussi*) (RCS Odonto. Mus., G 68.792), the right P^3 is rotated through about 135° and both P^4 are slightly rotated. The right P_4 is transverse to the line of the arch (Colyer, 1943b, p. 25, fig. 28).

(d) Molars

Misplacements of the molars are seldom marked. M^3 may be misplaced a little lingually (Fig. 10.46) or tilted distally and in abnormal occlusion with the M_3. A slight rotation of M_3 sometimes occurs, as in Figure 10.47; Krapp and Lampel (1973) described a skull of a male *Colobus verus* (Zoologisch-vergleichend-anatomischen Instituts of the University of Freiburg, Switzerland, Serie Himmelheber Co ve 19) in which the right M_3 is rotated through 90° so that its distal surface faces lingually.

(e) Asymmetry of the arch

Among 71 skulls of *C. g. occidentalis* from the lowland primary rain forest of Uganda, Colyer (undated) found three with asymmetry of the arch. In one male (RCS Odonto. Mus., G 68.8161), the upper right cheek teeth and canine are in advance of their normal relationship

Table 10.8 The average length of the upper cheek-tooth row compared with the length of the cranium of two sub-species of *COLOBUS BADIUS* (from Allen, 1922–25)

	No.	A Length upper cheek-tooth row (mm)	B Maximum length of cranium (mm)	A:B (%)
C. badius brunneus	17	41.2	118.7	34.7
C. badius powelli	13	40.3	113.2	35.6

with the mandibular teeth. The arch on the right is straight but on the left it is bowed. The right half of the cranium is rounded and larger than the left, but the area for the attachment of the temporal muscle is larger on the left.

There was one marked example of asymmetry of the arch (Fig. 10.48). Although the right upper jaw and tooth arch are longer antero-posteriorly than the left, the left upper jaw is set further forward than the right.

Fig. 10.46. *Colobus guereza gallarum* (Abyssinian black-and-white colobus). ♀. P-C. Mus., A II 50. Both M^3 are more lingually situated than normal and are rotated slightly.

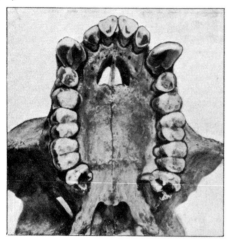

Fig. 10.47. *Colobus guereza occidentalis* (eastern black-and-white colobus). ♂. BMNH 1914.1.24.1. Both M_3 slightly rotated with buccal surfaces turned mesially.

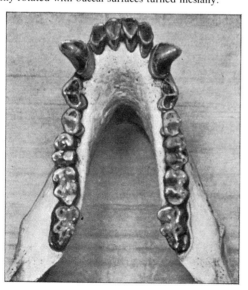

The facial surface of the right jaw is flatter than the left. The asymmetry is likely to be due in part at least to unequal growth at the posterior aspects of the jaws because the spheno-maxillary suture on the right is less evident than on the left, suggesting that growth may have ceased there prematurely. The upper and lower tooth arches occlude normally and the main asymmetry of the mandible is a smaller coronoid process on the right.

Genus *Presbytis*, leaf-monkeys

Tables 10.3 and 10.9 show that the prevalence of positional variations in the genus *Presbytis* is high.

(a) Incisors

The chief features of interest in leaf-monkeys, as in *Colobus*, is the frequency of incisor underjet and of variations in the arrangement of the upper incisors.

Colyer's (1947a) analysis of the dimensions of 25 normal skulls and 15 skulls with incisor underjet of both sexes of *Presbytis entellus thersites* suggested that, as in *Colobus*, underjet was caused by superior retrusion (Table 10.10). Incisor overjet is rare in leaf-monkeys and langurs. Figure 10.49 shows a well-marked example, the whole-arch relationship is abnormal, the lower arch occluding slightly more posteriorly

Fig. 10.48. *Colobus polykomos angolensis* (Angolan black-and-white colobus). ♂. BMNH 1879.11.12.1. Asymmetry of the maxilla and of the dental arch, the left side being more anterior than the right. Note the asymmetry at the distal ends of the tooth rows and that the centre of the arch is a little to the right side.

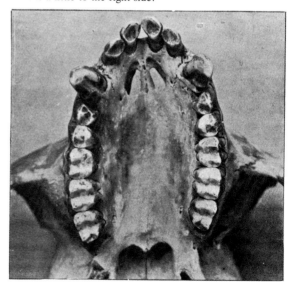

than normal, so that it is a matter of speculation whether the upper arch is pre-normal or the lower arch post-normal.

Irregularity in the arrangement of the incisors is

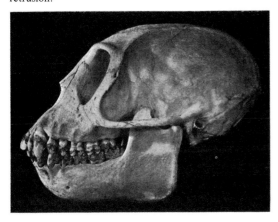

Fig. 10.49. *Presbytis cristata germani* (silvered leaf monkey). AMNH 54661. The upper incisors are prominent; the relationship of the premolars and the molars shows that the whole upper arch is in advance of the lower; alternatively the lower arch is in inferior retrusion.

more common in the maxilla than in the mandible. There may be protrusion of one or both I^1 (Fig. 10.50).

Some measurements that Colyer (1947a) made of the facial skeleton of *Presbytis entellus thersites* suggest that irregularity of the upper incisors was associated with a narrow premaxilla. However, as Colyer pointed out, the number of specimens was too few for reliable inferences to be drawn, so his table of measurements will not be reproduced here. Similar irregularities arising from insufficient space occur in the mandible (Figs. 10.51 and 10.52).

In the four groups of species of *Presbytis*, there were considerable differences in the incidence of mandibular incisor protrusion and to a lesser degree in the irregularity of the incisors (Table 10.9). The melalophus group had a significantly lower proportion of irregularity of the incisors and a higher proportion of incisor underjet. Although the differences among the various species within each group was much less significant (Table 10.9), among sub-species of the same species there were some statistically significant differences (Table 10.11), including a higher incidence of incisor irregularity in *P. entellus thersites* than in *P. e. entellus*; there was also a higher proportion with

Table 10.9 Positional variations in various groups and species of *PRESBYTIS* MONKEYS. The melalophos group differs from the others in having a lower incidence of incisor irregularity and a higher incidence of incisor underjet; these differences are statistically significant. The differences between the species within each group are much less significant.

Species	No. of specimens	Irregularity of incisors no.	%	Incisor underjet no.	%
Entellus group					
P. entellus	233	50	21.5	23	9.9
Vetulus group					
P. vetulus	56	17	30.4	11	19.6
Cristata group					
P. cristata	222	51	23.0	47	21.2
P. pileata	27	7	25.9	2	7.4
P. obscura	88	24	27.3	16	18.2
P. phrayei	31	9	29.0	7	22.6
P. potenziani	12	1	8.3	8	66.7
Total	**380**	**92**	**24.2**	**80**	**21.0**
Melalophos group					
P. melalophos	177	23	13.0	130	73.4
P. rubicunda	55	10	18.2	49	89.1
P. frontata	16	4	25.0	14	87.5
P. comata, thomasi and *aygula*	63	3	4.8	47	74.6
Total	**311**	**40**	**12.9**	**240**	**77.2**

Section 2: Variations in Position

incisor underjet in *P. cristata cristata* than in *P. c. pyrrhus* and in *P. melalophos femoralis* compared with *P. m. melalophos* and *P. m. chrysomelas*. The underlying cause of these differences is unknown.

(b) Canines
Three specimens showed irregularity of the position of the canines:
(i) *Presbytis cristata pyrrhus* (silvered leaf monkey). BMNH 1855.12.17.22 (Fig. 10.50). The left upper canine is misplaced slightly buccally and mesially, probably as a result of the position of the lower canine which is tilted distally and buccally so that there is no diastema distal to it; the left I^1 is protruded.
(ii) *Presbytis pileata shortridgei* (Shortridge's capped langur). ♀. BMNH 1915.5.5.11. The left upper canine is misplaced mesially so that in occlusion it rests over the mandibular canine. The left upper premolars and molars are mesial to their normal positions.
(iii) *Presbytis cristata pyrrhus* (silvered leaf monkey). USNM 155671. The right upper canine is misplaced mesially.

(c) Premolars
The most common positional variation was rotation of P^4 and less commonly of P^3. A more severe rotation is shown in Fig. 10.53.

(d) Molars
The maxillary third molar is often tilted so that its occlusal surface faces a little distally and only its mesial cusps occlude with M_3. This misplacement suggests insufficient growth of the posterior part of the maxilla. Rotation of M_3 is rare.

(e) Asymmetry of the arch
This was seen in three skulls, one being shown in Fig. 10.54. The intermaxillary suture and the left upper arch curve to the left. The curve of the corresponding mandibular arch is even more marked. The right maxilla and the temporal fossa enclosed by the zygomatic arch are smaller than the left. The asymmetry was widespread because the right ascending ramus of

Table 10.10 The relationship between underjet in *PRESBYTIS ENTELLUS THERSITES* to basal length and gnathic index (basal length × 100/length of the basicranial axis) (see p. 159), showing that underjet is likely to have been caused by a shortening of the upper jaw rather than of the mandible. (From Colyer, 1947a.)

Sex	Underjet	No.	Mean gnathic index	Mean basal length (mm)
Males	without	13	115.6	75.1
	with	8	113.9	73.8
Females	without	12	112.5	66.3
	with	7	111.3	65.1

Fig. 10.50. *Presbytis cristata pyrrhus* (silvered leaf monkey). Captive. BMNH 1855.12.17.22. Protrusion of the left I^1. The left canine is also misplaced slightly buccally and mesially, probably secondarily to the position of the lower canine which is tilted distally and buccally so that there is no diastema distal to it; in fact, it overlaps P_3 slightly which is also rotated as occurs when there is insufficient space.

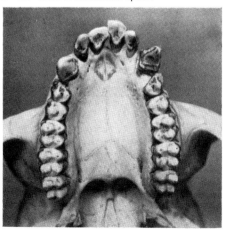
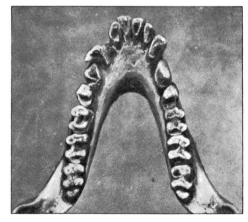

the mandible and its coronoid process are less well-developed than the left. There are no signs of injury.

Marked asymmetry of the arch was present in a skull of a female Mentawai Islands leaf monkey, *Presbytis potenziani* (BMNH 1895.1.9.1). The maxillae were bent to the left, the mesial part of the arch being flattened on that side. The left M^3 was close to the wing of the sphenoid and the spheno-maxillary suture was smaller on the left than on the right. The temporal fossa was a little smaller on the left than on the right, but the left ramus of the mandible was larger than the right. There were no signs of injury.

In the third skull, the maxillae were bent to the left; the spheno-maxillary suture was less evident on the side to which the maxillae deviated, suggesting a deficiency of growth at the suture. There was no appreciable difference in the sizes of the temporal fossae nor asymmetry of the mandible.

Table 10.11 Positional variations in some sub-species of some species of *PRESBYTIS* MONKEYS. There is a higher incidence of incisor irregularity in *P. entellus thersites* than in *P. e. entellus*; also a higher incidence of incisor underjet in *P. cristata cristata* than in *P. c. pyrrhus*, and in *P. melalophos femoralis* compared with *P. m. melalophos* and *P. m. chrysomelas*. These differences are statistically significant.

Species and sub-species	No. of specimens	Irregularity of incisors no.	%	Incisor underjet no.	%
P. entellus entellus	71	8	11.3	2	2.8
P. e. thersites	64	21	32.8	6	9.4
P. cristata cristata	105	32	30.5	33	31.4
P. c. pyrrhus	82	16	19.5	10	12.2
P. melalophos melalophos	49	5	10.2	39	79.6
P. m. femoralis	99	13	13.3	65	65.7
P. m. chrysomelas	29	5	17.2	26	89.7

Fig. 10.51. *Presbytis cristata pyrrhus stresemanni* (Bali silvered leaf monkey). BMNH 1903.3.6.3. The lower incisors are crowded with lingual misplacement of the right I_2.

Fig. 10.52. *Presbytis rubicunda* (maroon leaf monkey). ♀. BMNH 1920.12.4.3. The lower incisors are crowded, both I_2 being misplaced lingually and partially overlapped by I_1.

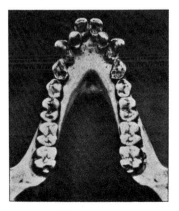

188 Section 2: Variations in Position

Genera *Pygathrix* (Douc langurs), *Rhinopithecus* (snub-nosed monkeys), *Simias* (pig-tailed monkeys) and *Nasalis* (proboscis monkeys)

Table 10.3 shows that the incidence of positional variations is higher in *Rhinopithecus* and *Simias* than in *Pygathrix* and *Nasalis*, but the numbers of specimens of these genera are small.

One of the *Nasalis* skulls showed extreme protrusion of the mandibular incisors and, in three *Nasalis* skulls, there were no canine diastemas in the upper jaw. Hooijer (1952) found one skull of *Rhinopithecus roxellanae* (USNM 268893) in which the lower left canine was rotated. The most marked irregularity was in a *Rhinopithecus* skull in which the left P^3 was rotated through about 60° (Fig. 10.55).

Fig. 10.53. *Presbytis phayrei shanica* (Shan States leaf monkey). ♂. BMNH 1914.7.8.1. Right P_4 (arrow) is rotated 90° with the buccal surface facing distally.

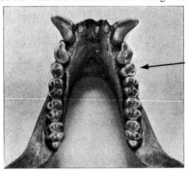

Sub-family **Cercopithecinae**

Genus *Cercopithecus*, guenon monkeys

Apart from spacing of the teeth which occurs quite commonly, the teeth of guenon monkeys do not show a tendency to any particular type of positional variation and the abnormalities are seldom severe. They are summarized in Tables 10.3, 10.12 and 10.13.

The normal arrangement of the maxillary teeth in *Cercopithecus* is shown in Figure 10.56. In both the maxilla and the mandible, the incisors are usually in contact and form a regular curve; the buccal surfaces of the premolars and molars present an even and regular line. There is, however, a wide range of variation in the size and shape of the palate; this is shown in the three skulls in Figure 10.60 which are males from the same district.

(a) Spaces between the teeth

In apes and monkeys, the teeth are in contact along the arch, with the exception of the maxillary canine and P^3 which are separated by a diastema. Variations from this arrangement are fairly common in *Cercopithecus*. The mandibular canine and P_3 or the premolars may be spaced. Spaces are more common between the upper than between the lower incisors; they are usually between I^1 and I^2, but separation of the left and right I^1 occurs (Fig. 10.57). The most marked spacing was in a De Brazza's monkey (*Cercopithecus neglectus*) in

Fig. 10.54. *Presbytis obscura obscura* (dusky leaf monkey). ♀. BMNH 1903.2.6.7. Asymmetry of the arches. Note the position of the intermaxillary suture and the buccal curvature of the upper left arch, which is even more marked in the mandible.

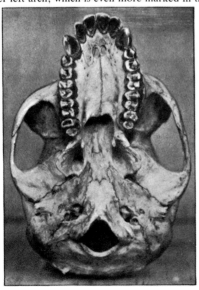
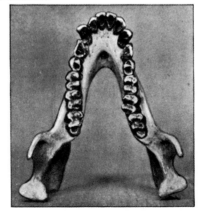

which gaps were present between the incisors, between the maxillary canines and P³ and between the mandibular premolars.

The various species of *Cercopithecus* show a considerable difference in the tendency towards the occurrence of spaces (Table 10.13), which is statistically significant, but that between the various sub-species within the species, *C. mitis*, *C. ascanius*, and *C. aethiops* is not.

(b) Incisors

The occlusion of the incisors in the adult *Cercopithecus* is edge-to-edge. Incisor underjet was present in only 5 skulls, 0.3% (Table 10.12). Incisor overjet was more

Table 10.12 Positional variations in *CERCOPITHECUS* MONKEYS from the wild

Species	No. of specimens	Class of tooth					Total specimens affected				Incisor underjet	
		I	C	P	mod. rot. P4	M	with P4 column		without P4 column			
							No.	%	No.	%	No.	%
C. aethiops	219	18	1	39	32	5	91	36.5	59	23.7	0	0
C. mona	212	5	0	3	17	1	26	12.3	9	4.2	1	0.5
C. lhoesti	15	2	0	4	1	1	7	46.7	6	40.0	0	0
C. neglectus	345	26	0	64	45	6	127	36.9	82	23.8	0	0
C. diana	8	0	0	0	1	0	1	12.5	0	0	1	12.5
C. nictitans	447	33	0	36	20	7	88	19.7	68	15.2	3	0.8
C. cephus	112	11	0	12	11	3	33	29.5	22	19.6	0	0
C. talapoin	15	2	0	0	1	0	3	20.0	2	13.3	0	0
Cercopithecus sp.	70	3	0	4	8	2	14	20.0	6	8.6	0	0
Total	**1473**	**100**	**1**	**162**	**136**	**25**	**390**	**26.5**	**254**	**17.2**	**5**	**0.3**

In skulls in which two or more classes of teeth varied in position, each class is recorded individually; for example, if in a skull one or more of the incisors, premolars and molars varied, each appropriate class would be credited with one and the column 'total specimens affected' with one only. Rotation of P4 (by less than one sixteenth of a circle) has been excluded; moderate rotation (one sixteenth to one eighth of a circle) of P4 is listed in a column 'mod. rot. P4' separately from other positional variations of the premolars.

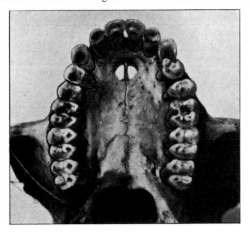

Fig. 10.55. *Rhinopithecus avunculus* (Tonkin snub-nosed monkey). ♀. BMNH 1927.12.1.6. Left P³ is rotated through about 60° with its buccal surface facing distally so that the buccal cusp is in occlusion with the lower premolars instead of being buccal to the mandibular arch.

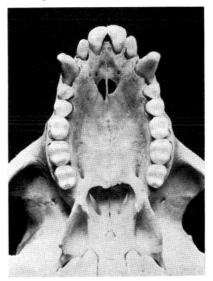

Fig. 10.56. *Cercopithecus* sp. (guenon monkey). The normal arrangement of the upper tooth arch.

common, particularly in *C. sabaeus* and *C. cephus* in a skull of each of which extreme protrusion of the upper incisors was noted. In one (Fig. 10.58), the upper teeth project about 3mm beyond the lower incisors, the occlusion of the premolars and the molars being normal; as the lower incisors are crowded and the upper ones are not unduly spaced, it seems possible that defective growth of the mandible is responsible. In the other (Fig. 10.59), in addition to overjet, there is general inferior retrusion and the lower cheek teeth occlude about 3mm more distally than normal. The upper arch (Fig. 10.60) is narrow, especially between the canines, and both P^4 are displaced lingually. The mandibular incisors are crowded and both P_4 are lingual to the arch. It seems possible that the postnormality of the lower arch, or mandible, is secondary to the narrowness between the upper canines. This would make it difficult to accommodate the lower canines in occlusion and would prevent the mandible, or its arch, from coming forward.

Irregularity of the incisor arch, generally taken to indicate lack of space, was present in about 7% of the *Cercopithecus* skulls and was more common in the upper jaw than the lower (Fig. 10.60). Examples included a slight labial positioning of the I^1 (Fig. 10.61), rotation of I^2 and an overlapping of the left I^1 (Fig. 10.62). Lampel (1963) described two skulls of *C. campbelli* (*mona*) *campbelli* with misplaced lower incisors. Schultz (1935), in a sample of 134 skulls of

Fig. 10.57. *Cercopithecus mitis moloneyi* (diadem monkey). ♀. BMNH 1897.10.1.1. There is a wide space between the I^1. Reduced slightly.

Table 10.13 The prevalence of spaces between teeth in various *CERCOPITHECUS* taxa. *C. cephus*, *C. neglectus* and *C. nictitans* have a higher incidence of spacing than the other species which is statistically significant, but the differences between the sub-species within each species are not.

Species and sub-species	No. of specimens	Totals with spaces No.	%
Cercopithecus cephus cephus	102	15	14.7
Cercopithecus neglectus	80	15	18.8
Cercopithecus mitis stuhlmanni	97	13	13.4
C. m. kolbi	30	2	6.7
C. m. albogularis	30	2	6.7
C. m. kandti	43	1	2.3
Total *Cercopithecus mitis*	**200**	**18**	**9.0**
Cercopithecus nictitans nictitans	159	25	15.7
Cercopithecus ascanius whitesidei	105	3	2.9
C. a. schmidti	78	5	6.1
C. a. katangae	40	5	12.5
Total *Cercopithecus ascanius*	**223**	**13**	**5.8**
Cercopithecus aethiops aethiops	50	0	0
C. ae. centralis	24	0	0
C. ae. sabaeus	35	1	2.9
C. ae. johnstoni	29	1	3.4
C. ae. pygerythrus	34	3	8.8
Total *Cercopithecus aethiops*	**172**	**5**	**2.9**
Cercopithecus denti	37	2	5.4
Cercopithecus wolfi wolfi	66	0	0
Cercopithecus pogonias grayi	83	5	6.0

Cercopithecus, found crowding of the upper incisors in only one and none with crowding of the lower incisors.

(c) Canines

Misplacement of a canine was present in only one skull, a sub-adult vervet monkey (Fig. 10.63). The lower right canine and cheek teeth occlude more distally than normal by the mesio-distal diameter of a premolar; in other words, P_3 occludes between P^3 and P^4 instead of mesially to P^3. There is no pre-canine diastema in either jaw so that arch space must be diminished. The maxillary arch is narrow in this specimen, as in a previous one (Fig. 10.59), and probably the same mechanism applied; the upper arch was too narrow for the lower arch, and for its canine in particular, to move mesially. Mesial movement of the upper canine and cheek teeth which closed the pre-canine diastema would have contributed to the abnormal occlusal relationship. This specimen illustrates how, as the teeth erupt and the arch relationship is established,

Fig. 10.58. *Cercopithecus sabaeus* (green monkey). ♂. BMNH 1920.7.10.6. The upper incisors protrude well beyond the lower ones and overlap them. The rest of arches are in normal relationship. Reduced slightly.

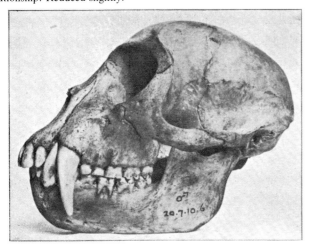

Fig. 10.59. *Cercopithecus cephus cephus* (moustached monkey). P-C. Mus., M 14. Incisor overjet associated with an abnormal arrangement of the cheek teeth, the lower ones occluding about half the mesio-distal diameter of a premolar more distally than normal. Note that the distal surface of P_3 is opposite the distal surface of P^3 instead of opposite the tip of its buccal cusp. A palatal view of the upper arch is shown in Figure 10.60 (right).

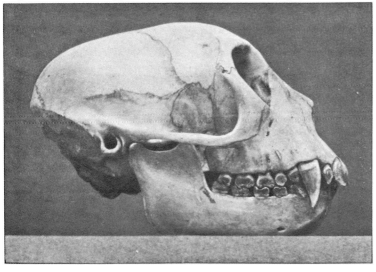

192 Section 2: Variations in Position

if the normal anterior-posterior relationship cannot be established, the tendency is for the discrepancy to be at least by one major cusp or by the width of a canine, that is for a functional inter-cusp relationship to be established even if not a normal one.

(d) Premolars

Some degree of positional abnormality of the premolars was seen in approximately 20% of the *Cercopithecus* skulls. The most common was a rotation of P^4, the lingual cusp usually facing mesially; the rotation was never as much as 45°. Rotation of P^3 was less common, the lingual cusp usually facing distally. Other variants were mesial tilting of P^3 (Fig. 10.64) and lingual misplacement of P^4 (Fig. 10.65).

Lampel (1963) examined 146 adult Campbell's monkeys (*C. mona* (= *campbelli campbelli*), 34 greater white-nosed monkeys (*C. nictitans buettikoferi*) and 76 diana monkeys (*C. diana diana*) and found four with positional variations of the premolars. A female and a male diana monkey had both P^3 rotated mesio-lingually and, in the male, misplaced lingually. In two male Campbell's monkeys, the right P_4 was misplaced with its disto-labial surface to the lingual aspect of M_1. Slight rotation of P_3 is common. P_4 may also be rotated (p. 47).

(e) Molars

Abnormalities in position of the molars were present in 1.7% of the specimens. Examples are rotation of M^2, distal tilt of M^3, impaction of M^3 against M^2, a lingual misplacement and rotation of M^3 (Fig. 10.62), rotation of M_2 and rotation or mesial inclination of M_3. Lampel (1963) also noted buccal misplacement of M^1 and M^2.

Fig. 10.60. *Cercopithecus cephus cephus* (moustached monkey). P-C. Mus. Three skulls of adult males from the same district to show the range of variation in the shape and size of the palate. The skull in the centre has a broad arch with slight spacing of the incisors (compare with Fig. 10.56); that on the left is much narrower, particularly between the canines, and there is no incisor spacing. That on the right (P-C. Mus., M14) is narrower still and this is associated with a general disturbance of the arch relationship (inferior retrusion), as in the view of the same specimen shown in Figure 10.59, and must be categorized as an abnormality.

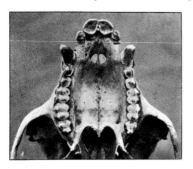  

Fig. 10.61. *Cercopithecus neglectus* (De Brazza's monkey). ♂. BMNH 1912.7.26.2. Both I^1 are slightly labial to the arch; they have an overjet relationship to the lower incisors.

Fig. 10.62. *Cercopithecus erythrotis erythrotis* (red-eared monkey). ♂. BMNH 1904.7.1.4. The I^1 overlap slightly; right M^3 rotated and misplaced slightly lingually.

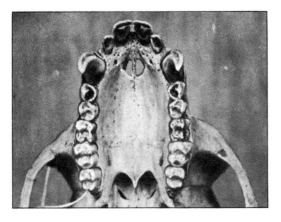 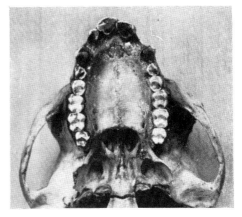

(f) Asymmetry of the arch

There was one well-marked example in a grivet monkey (Fig. 10.66). The intermaxillary suture and the maxilla as a whole are bent to the left so that both I^1 are to the left of the sagittal plane of the skull. The left maxilla, measured from the pterygo-maxillary fissure, is much shorter than the right. Furthermore, whereas on the right there is much space between M^3 and the pterygo-maxillary fissure, on the left there is so little space that M^3 is impacted against M^2. Thus, the asymmetry seems to be due to defective growth of the left maxilla, in particular at the pterygo-maxillary fissure. This view is supported by study of the occlusion on the left (Fig. 10.66 right). There is incisor underjet with the left I^2 in occlusion with the lingual aspect of the lower canine, the upper canine is nearly its whole mesio-distal diameter distal to the lower canine and occludes between P_3 and P_4. The upper cheek teeth occlude the

Fig. 10.63. *Cercopithecus aethiops callidus* (Naivasha vervet monkey). USNM 181890. The jaws are held apart to display them better. The right mandibular canine is distal to the upper one, instead of passing mesial to it. All the lower cheek teeth occlude the width of one premolar more distally than normal.

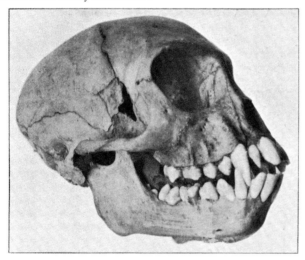

Fig. 10.64. *Cercopithecus aethiops pygerythrus* (savannah monkey). ♂. BMNH 1906.11.8.2. Right P^3 inclined mesially. Defects of the alveolar bone (dehiscence) are present over the roots of the molars.

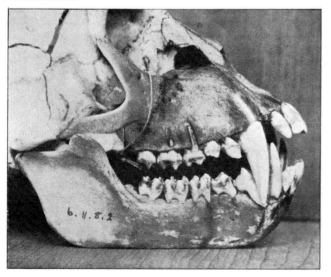

width of a premolar more distally than normal; that is P^3 occludes between P_4 and M_1 instead of between P_3 and P_4.

Slight asymmetry of the arch was present in another specimen, the deviation being to the right. There were no signs of injury to the bones in either specimen.

Fig. 10.65. *Cercopithecus aethiops helvescens* (Cunene grivet monkey). ♀. BMNH 1925.12.4.1. The arch is unduly narrow between the premolars; both P^4 are misplaced lingually and are slightly rotated.

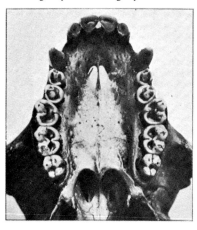

(g) Effects of captivity

In addition to the 1473 skulls of guenon monkeys (*Cercopithecus* spp.) from the wild, 338 skulls of captive animals were examined. Positional abnormalities were different in degree but not in kind. Three skulls (0.9%) of captive animals had misplaced maxillary canines compared with only one (0.7%) in the wild sample. In one (Fig. 10.67), the mandibular right canine had erupted much nearer the mid-line than the left and had pushed the incisors mesially and towards the left so that they projected beyond the upper teeth. The misplaced canine was out of contact with the upper canine which was over-erupted and more prominent than normal. In two other skulls, the maxillary canines occluded as in Figure 10.68.

In captive guenons, irregularity of the premolars was common, misplacement of P^4 being the most frequent, as in Figure 10.69 where the P^4 has erupted above the roots of the deciduous molar with the occlusal surface facing buccally, and P^3 has erupted between its deciduous predecessor and the canine.

Irregularity of the M_3 is significantly more common in the captive than in the wild guenon. Amongst the specimens from the wild state, there was only one mandible (0.7%) in which the tooth was tilted mesially whereas, in the captive specimens, there were four (1.2%). In one (Fig. 10.70), M_3 is impacted under the

Fig. 10.66. *Cercopithecus aethiops rubellus* (grivet monkey). ♂. BMNH 1906.2.1.1.
Left: Palate and base of cranium. There is marked asymmetry with the intermaxillary suture curving to the left, so that both I^1 are to the left of the median plane. Note that the distance between the posterior margin of the pterygoid plate and M^3 on the right is greater than on the left where M^3 is impacted against M^2. × *c*. 0.67.
Right: Lateral view of the left side. There is incisor underjet and the lower canine is nearly the whole width of P_3 mesial to its normal position. The upper cheek teeth are the width of a premolar more distal than normal, P^3 occluding with M_1 instead of with P_3 and P_4.

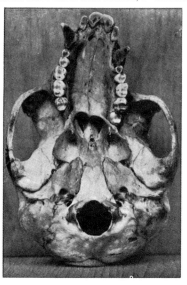
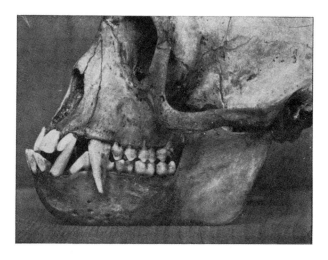

distal surface of M_2 and there is a similar impaction of the left M^3 against the M^2. In another (Fig. 10.71), both M_3 were transverse to the line of the arch.

(h) Effects of isolation

Important evidence of the effects of geographical isolation is provided by a collection of 92 green monkey skulls from St Kitts brought together by Sir Frank Colyer (1948) at the Odontological Museum at the Royal College of Surgeons and studied in depth by Ashton and Zuckerman (1950a, 1951a,b,c; and Ashton, Flinn, Griffiths and Moore, 1979) (Chapter 3, p. 36). A sample of green monkeys (*Cercopithecus sabaeus*) descended from those introduced from West Africa into the island of St Kitts in the West Indies about 300 years ago was compared with a sample of green monkeys derived from the mainland of Africa representing the parent stock. The St Kitts monkeys had highly significantly larger teeth and skulls, more variations in the position of M3 and much asymmetry in the dimensions of the skull and of the incisors and canines. Less statistically significant was the higher incidence of variations in numbers of teeth and in the morphology of the third molar roots in the St Kitts monkeys (Table 10.14). Although some of the differences, particularly those of tooth size, suggest an evolutionary genetic change during the estimated 75–100 generations since the population was installed in the island, environmental effects cannot be discounted, especially with respect to variations in position (Ashton *et al.*, 1979).

The observations on the teeth, apart from the mensuration, were made by Colyer (1948). He found irregularity in the upper-incisor tooth arch, consisting mostly of protrusion of one or both I^1 in 25% of 76 skulls in which the permanent dentition was present. Sixteen (12%) showed irregularities of position of premolars, consisting mostly of rotation or misplacement from the arch of P^4. He implied that these percentages were high in comparison with what he had seen in a collection of specimens derived from Africa but unfortunately gave no precise data. However, he

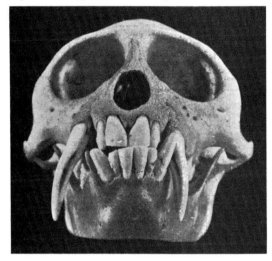

Fig. 10.67. *Cercopithecus sabaeus* (green monkey). ♂. Captive. RCS Odonto. Mus., G 68.611. The mandibular incisors protrude labially to the upper incisors and are crowded, with their centre a little to the left. The right lower canine is inclined disto-labially and appears to have displaced the upper canine buccally.

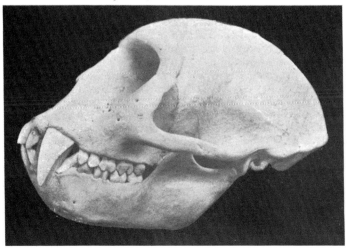

Fig. 10.68. *Cercopithecus aethiops pygerythrus* (savannah monkey). Captive. Formerly RCS Osteo. Series, 113. The upper canine is more mesial than normal. Note that it is not in occlusion with the mesial slope of P_3.

also found among 72 St Kitts skulls four with unerupted, malposed M_3 impacted against M_2, either on one side or both, and one skull with a similarly malposed M^3; among 104 skulls of the green monkey derived from Africa, he found only one with misplaced M_3. The relative prevalences of this disorder were St Kitts 6.9%, African 2.8%, which does suggest a significant difference.

Fig. 10.69. *Cercopithecus petaurista petaurista* (lesser white-nosed monkey). ♀. Juvenile. Captive. RCS Odonto. Mus., G 68.1335. Bone has been removed to reveal the root of the left P^4 which has erupted above the deciduous molar.

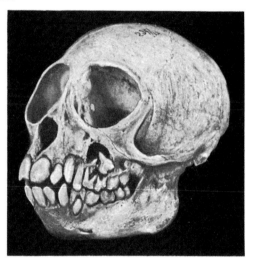

The positional irregularities of the incisors, premolars and M3 can all be regarded as indications of crowding of the arches, in other words, of jaws, or at least the tooth-bearing part of the jaws, being too small in proportion to the size of the teeth. The evidence for a high prevalence of such a condition in the St Kitts skulls is merely tentative, except in respect of M3, but raises a number of questions in view of the substantial evidence that both jaws and teeth in the St Kitts are larger than the African. These questions cannot be answered without a further investigation of the material.

The first question is whether the increase in the size of the jaws that appears to have occurred during the isolation of the St Kitts green monkey population has been less than that of the teeth. The proportional difference between the mean dimensions of the adult skulls of the St Kitts and those of the African controls was indeed smaller than the difference between the size of the crowns of the permanent teeth (Ashton *et al.*, 1979). However, the comparison needs to be made with the size of the jaws and unfortunately Ashton and Zuckerman (1951*a*,*b*,*c*) gave only the dimensions of the mandible and none for the upper jaw. However, from these, something of interest emerges. The means of all nine mandibular dimensions were larger in the adult St Kitts monkeys than in the African but, in the sub-adult females, as well as in the juvenile animals, one dimension, namely length of the horizontal ramus (gnathion-gonion) was smaller in the St Kitts. This

Fig. 10.70. *Cercopithecus mona* (mona monkey). ♂. Captive. RCS Odonto. Mus., G 68.136. A: Right mandible. Bone has been removed to display the teeth. M_3 inclined mesially and impacted under the distal convexity of the crown of M_2. B: Left maxilla. M^3 is similarly impacted against M^2.

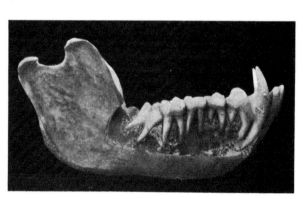

A

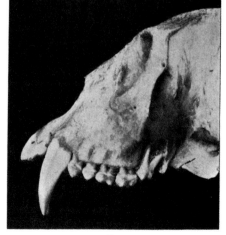

B

dimension is more relevant than most to the amount of space available for the arch of teeth. It is worth mentioning that, of the five St Kitts skulls in which there was malposition and failure of eruption of at least one M3, four were of females. Unfortunately, Ashton and Zuckerman do not give the data on mandibular dimensions for adult males and females separately.

Better measurements, including the overall length of the mandible and the length of the upper jaw could be devised to test the hypothesis that, although the crowns of the teeth have become larger in the St Kitts monkeys, the increase in the size of the tooth-bearing parts of the jaws has not been correspondingly large.

Genus *Erythrocebus*, patas monkeys

Of 42 wild patas monkeys, 14 (33%) had positional variations (Table 10.3); four were well-marked irregularities of premolars and two consisted of tilting of M_3. Among 54 captive animals, however, 26 (48%) showed positional irregularities, mostly pronounced (Fig. 10.72). In some, there was generalized crowding of the dental arches suggesting that the jaws were too small to accommodate the dentition. Several juvenile specimens, including those shown in Figures 10.73 and 10.74, were of this kind and it seems likely that, had the animals survived to become adults, the arch form

Table 10.14 Dental abnormalities in GREEN MONKEYS *(Cercopithecus sabaeus)* derived from Africa and from St Kitts

	African green monkey		St Kitts green monkey		
	Total no.	No. affected	Total no.	No. affected	P
Variations in number	104	2	76	4	0.5–0.3
Variations in roots of M_3	36	3	72	15	0.2–0.1
Positional normalities of M_3	104	1	72	5	0.05–0.1

P = level of probability of finding such differences by chance; the low values show that the differences are significant (after Ashton and Zuckerman, 1950a, b, and Colyer, 1948).

Fig. 10.71. *Cercopithecus aethiops pygerythrus* (vervet monkey). ♀. Captive. RCS Odonto. Mus., G140.43. Both M_3 are rotated through 90° so that their buccal surfaces face mesially. The distal surfaces of both M_2 are carious.

Fig. 10.72. *Erythrocebus patas* (patas monkey). ♂. Captive. RCS Odonto. Mus., G68.146. On this left side, and on the right, P^4 and P_4 have erupted on the lateral surfaces of the jaws with their crowns more or less at the level of the root apices of the other teeth (Colyer, 1928).

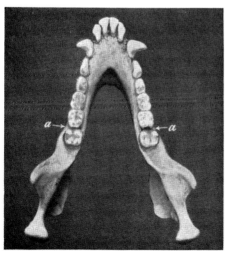
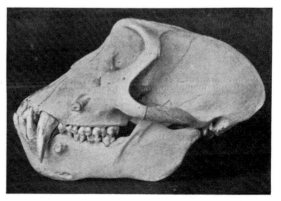

would have been very irregular and some teeth would have remained unerupted. In both these skulls, the arrangement of the incisors and canines is more or less normal, as is the arrangement of the deciduous molars and M1. However, the developing premolars are much more crowded than would be the case if the jaws were growing properly; indeed, several are erupting outside the arch. Furthermore in one (Fig. 10.73),

Fig. 10.73. *Erythrocebus patas* (patas monkey). ♀. Captive. RCS Odonto. Mus., G 68.1485. Juvenile. Bone has been removed to display the teeth.
A: right side. The right P^3 is erupting mesially to its deciduous predecessor. The right P_4 is more buccal than normal and its crown has pierced the outer alveolar plate below the roots of its predecessor. M_2 is unerupted and so tilted mesially that it is impacted against the roots of M_1. The roots of M_3 are beginning to form and it is tilted mesially more than normal.
B: left side. There is a similar crowding of the unerupted teeth; the crown of P_4 is tilted distally, M_2 is impacted against M_1 and seems unlikely to be able to erupt; M_3 is impacted beneath the distal convexity of the crown of M_2. The roots of the maxillary premolars are webbed on their lingual aspects (Colyer, 1928).

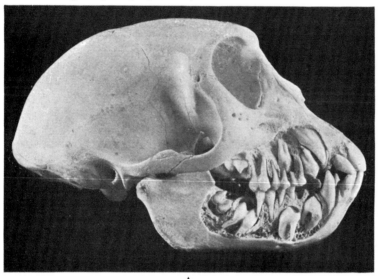

A

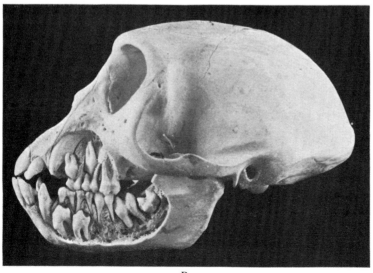

B

the M_2 are impacted against the roots of M_1 and it is doubtful whether the M_3 would have erupted. In the other (Fig. 10.74), at least one M^3 is unlikely ever to have erupted. Figure 10.75 shows an example of misplaced and impacted right M_3 and left M^3; the right M^3 was similarly impacted.

The skull shown in Figure 10.76 has slight incisor underjet, both P_3 are slightly rotated and, on the right, the lower cheek teeth are in slightly pre-normal occlusion. On the left, the premolar occlusion has not yet been established because the deciduous predecessor of P_4 has not been completely shed. Both M_3 are tilted mesially to an unusual extent and are unerupted. However, there does not appear to be any lack of space for them to erupt into.

To establish whether in these instances there is truly a disproportion between the size of the jaws and the size of the dentition, and whether such disproportion is associated with captivity, perhaps to some dietary deficiency, a careful osteometric comparison

Fig. 10.74. *Erythrocebus patas* (patas monkey). ♂. Captive. RCS Odonto. Mus., G 68.1484. Juvenile.
A: left side. M_1 empty socket.
The P^3 and P^4 are more buccally situated than normal and P^3 has begun to erupt through the outer alveolar plate above the crown of its deciduous predecessor. P^4 is misplaced, its crown being over the roots of m^1 instead of those of m^2. M^3 cannot be seen but it is situated at a high level on the posterior aspect of the maxilla, facing mesially and therefore impacted against the roots of M^2.
B: right side.
The disordered arrangement of the teeth is very similar. M^3 is not so high up as on the opposite side and is less tilted (Colyer, 1928).

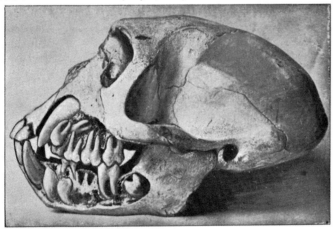

A

B

Section 2: Variations in Position

between these skulls and similar juvenile stages, preferably from the wild, would be needed. What does appear to be certain is that mesial tilting and other positional variations of M_3 are more common in captive patas monkeys. There were 12 (22%) examples among the 54 skulls of captive animals and only 2 (5%) among the 42 skulls from the wild.

Fig. 10.75. *Erythrocebus patas* (patas monkey). ♂. Captive RCS Odonto. Mus., G 68.1486. Juvenile. Windows have been cut in the bone to display the teeth. In the left mandible (above), M_3 is horizontal with the roots facing distally and extending to the ascending ramus. In the left mandible (below), the M_3 has erupted but it is tilted slightly lingually. The right M_3 and left M_2 have fused roots.
In the upper jaw, M^3 is impacted against the roots of M^2; a similar condition was present on the opposite side (Colyer, 1928).

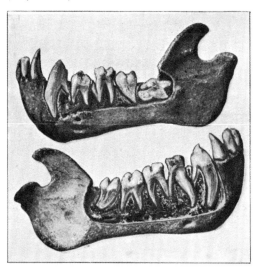

Genus *Cercocebus*, mangabeys

Table 10.3 shows that the incidence of positional variations is high in *Cercocebus*.

(a) Incisors

In mangabeys, the upper incisors vary in position more than the lower ones, the variation taking the form of a slight projection of I^1 or misplacement of I^2 lingually and mesially so that they are overlapped by I^1. The species showing the largest percentage of such irregularities was the agile mangabey (*C. galeritus agilis*) in which eight out of 48 specimens were affected. Two specimens showed incisor underjet.

The asymmetry of the tooth arches in Figure 10.77 is part of a general asymmetry of the cranium; measurements show that the left side is smaller than the right. The upper left arch is more curved than the right and is shorter, mainly because the pre-canine diastema is narrow, presumably because of lack of growth at the left premaxilla-maxillary suture. The disordered shape of the lower arch corresponds, and the lower left canine is probably more distally situated because it occludes with the disto-lingual aspect of the upper canine instead of its mesio-lingual aspect. The mandible itself is not asymmetrical although there is some evidence of injury to its left angle.

(b) Premolars

The most frequent positional anomaly of the premolars is rotation of P^4. Krapp and Lampel (1973) found rotation of one or both P^4 in five skulls (9.8%) of 51 white-collared mangabeys (*C. torquatus*) collected in Liberia. Less frequently, the P_3 may be placed obliquely to the arch and occasionally P^3 is rotated. Marked irregularity of the premolars is rare in wild

Fig. 10.76. *Erythrocebus patas* (patas monkey). ♀. Captive. RCS Odonto. Mus., G 68.1481. Sub-adult.
A and B: right and left side views. The lower left canine has been lost from the specimen. There is incisor underjet and, on the right, the lower cheek teeth are in slightly pre-normal relationship (note the interval between the occlusal surfaces of P^4 and P_4). On the left, m_3 is still present and P_4 is erupting to its buccal side. Both P_4 are rotated. The erupting right M^3 is just visible; the left M^3, a large tooth connate with an extra molar is unerupted high upon the posterior surface of the maxilla.
C: mandibles, lingual view. The M_3, displayed by removal of bone, are unerupted with roots not quite fully formed. They are slightly more mesially tilted than normal for this developmental stage and, as there seems to be plenty of room for them to erupt into (note the space between them and the M_2), it seems likely that they would have erupted.

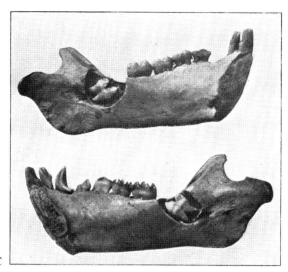

C

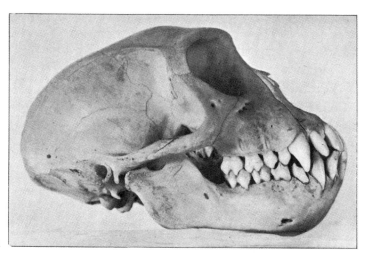

A

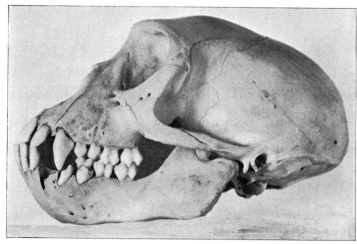

B

animals but the frequency shows a considerable difference between the two groups of *Cercocebus* (Table 10.15), being significantly greater in the albigena group than in the torquatus group.

Under captive conditions, irregularities of the premolars are more common.

Genus *Macaca*, macaques

The normal shape of the tooth arches in macaques is shown in Figure 10.78. The upper incisors form a smooth curve; the line of the upper cheek teeth curves buccally, the arch being wider between the M^1 and becoming distinctly narrower between M^3. There is considerable variation in the degree of curvature; when it is pronounced, the lingual surfaces of the three molars in particular may each project a little beyond the one in front, producing an echelon appearance. In the mandible, the line of the cheek teeth is much straighter; P_3 is usually placed a little obliquely to the

Table 10.15 Positional variations of premolars in various taxa of *CERCOCEBUS* from the wild

Species and sub-species	No. of specimens	Positional variations of premolars No.	%
Torquatus group			
Cercocebus torquatus and *C. galeritus*	91	15	16.5
Albigena group			
C. albigena albigena	21	16	76.2
C. a. zenkeri	32	8	25.0
C. a. johnstoni	106	72	67.9
C. aterrimus	21	13	61.9
Total Albigena group	**180**	**109**	**60.6**

Fig. 10.77. *Cercocebus torquatus torquatus* (white-collared mangabey). ♂. BMNH 1903.2.4.1.
A: The maxillary arch is asymmetrical, mainly because the pre-canine diastema is narrow and the left canine and cheek teeth are more mesial than on the left, but also because the cheek-tooth row is curved buccally whereas on the right it is straighter.

B: The lower arch corresponds in that the left cheek-tooth row is curved. However, the left canine is more distal than normal so that it occludes with the disto-lingual aspect of the upper canine instead of the mesio-lingual. The lower incisors incline to the left but the right I_2 is absent.

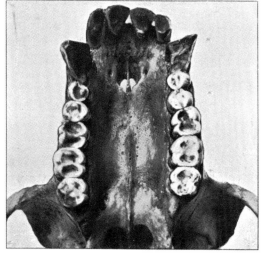

A

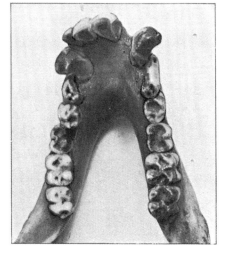

B

line of the arch with its buccal surface facing slightly distally. Positional variations in the genus *Macaca* are summarized in Table 10.3 and those in the various *Macaca* taxa in Table 10.16.

(a) Incisors
Anomalous positions of the upper incisors include a slight labial location of one or both I^1 and rotation of I^2. Irregularity of the premolars was present in about 30% of the specimens showing incisor irregularities. Extreme crowding, as in Figure 10.79, is rare.

A slightly irregular arrangement of the mandibular incisors occurs but is uncommon and grosser irregularities such as in Figure 10.80 are rare. One macaque skull had incisor underjet.

(b) Premolars
The majority of the positional irregularities were in the region of the premolars, nearly 25% of the specimens showing some degree; in about one-third of these, both upper and lower arches were affected. The most common variety in the upper arch was rotation of P^4 with

Table 10.16 Positional variations in various taxa of *MACACA* from the wild

Species	No. of specimens	Class of tooth					Total specimens affected			
		I	C	P	mod. rot. P4	M	with P4 column		without P4 column	
							No.	%	No.	%
Macaca sinica	17	1	0	3	1	0	5	29.4	4	23.5
M. radiata	34	1	0	5	0	0	9	26.5	9	26.5
M. cyclopis	20	2	0	1	5	0	7	35.0	2	10.0
M. nemestrina	35	1	0	2	6	1	9	25.7	3	8.6
M. silenus	6	0	0	0	0	0	0	0	0	0
M. fuscata	51	3	0	9	4	2	18	35.3	14	27.4
M. mulatta	88	12	0	19	7	1	34	38.6	27	30.7
M. fascicularis	195	18	0	31	13	6	60	30.8	47	24.1
M. f. philippinensis	31	3	0	4	4	0	8	25.8	4	12.9
M. assamensis	28	3	0	6	4	1	12	42.9	8	28.6
M. nigra	67	4	0	7	6	0	16	23.9	10	14.9
Macaca sp.	16	1	0	0	3	0	4	25.0	1	6.3
Total	**588**	**49**	**0**	**87**	**53**	**11**	**182**	**30.9**	**129**	**21.9**

In skulls in which two or more classes of teeth varied in position, each class is recorded individually; for example, if in a skull one or more of the incisors, premolars and molars varied, each appropriate class would be credited with one and the column 'total specimens affected' with one only. Rotation of P4 (by less than one sixteenth of a circle) has been excluded; moderate rotation (one sixteenth to one eighth of a circle) of P4 is listed in a column 'mod. rot. P4' separately from other positional variations of the premolars.

Fig. 10.78. *Macaca*. The normal tooth arches. The upper cheek-tooth rows are curved; the lower ones are nearly straight.

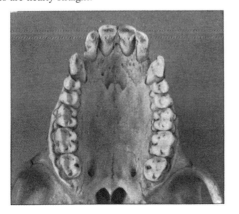
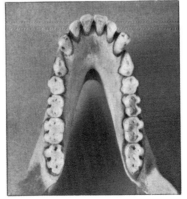

the lingual cusp facing mesially (Fig. 10.81). In the mandible, P_4 and less often P_3 may be rotated; P_4 may be squeezed out of the arch (Fig. 10.82). Although in *Macaca* there appears to be a tendency for rotation of the premolars, extreme misplacement is rare in the wild.

There was a considerable difference in the incidence of positional variations of the premolars between two of the sub-species of rhesus monkeys (*Macaca mulatta*). 70.6% of 17 *M.m. villosa* skulls showed variations in position compared with only 23.8% of 63 *M.m. mulatta* and the degree of variation was also more marked in *M.m. villosa*.

(c) Molars

It seems that, when the cheek teeth are crowded from lack of space or disproportion between the size of the teeth and that of the jaw, there is an accentuation of the echelon appearance due to slight rotation of the teeth. This may be accompanied by a further indication of lack of space, lingual misplacement of M^3 (Fig. 10.83) so that the arch is particularly narrow between the M^3. Lingual misplacement of M^3 may be so severe that only its buccal cusps occlude with M_3. In the mandible, M_1 or M_2 may be rotated (Figs. 10.84 and 10.85).

Fig. 10.79. *Macaca mulatta mulatta* (rhesus monkey). ♀. BMNH 1931.1.11.13. The upper incisors are crowded and overlapping. The rest of the arch is normal.

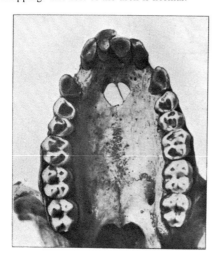

Fig. 10.81. *Macaca fascicularis philippinensis* (Philippine crab-eating macaque). ♂. BMNH 1895.8.2.1. Both P^4 slightly rotated with the lingual cusp turned mesially. Right I^2 slightly lingual to the arch.

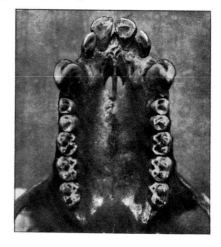

Fig. 10.80. *Macaca nigra* (black ape). BMNH 31*h*. A mandible in which both I_2 are tilted buccally so that their crowns project laterally beyond the canines and there are wide gaps between I_2 and I_1. The right canine is more mesially-situated than the left.

Fig. 10.82. *Macaca mulatta villosa* (rhesus monkey). ♂. BMNH 1914.7.10.4. Left P_4 misplaced buccally, partially overlapping P_3. The condition was bilaterally symmetrical.

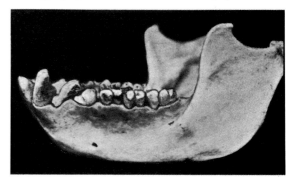

(d) Asymmetry of the arch

Colyer (1951b) described a toque monkey (*Macaca sinica* – RCS Odonto. Mus., G 8.2) in which asymmetry of the face and basal portions of the skull are associated with the early loss of the right M^1. The anterior parts of both maxillae are deflected to the right, so that the left upper canine and cheek teeth are more mesial than those on the right. The space originally occupied by the right M^1 has been filled by the mesial movement of M^2 and M^3 and distal movement of P^4 and P^3, the movement being more apparent in the crowns than in the roots. The upper molars on the right are in advance of the lower molars.

Fig. 10.83. *Macaca fascicularis sublimitus* (crab-eating macaque). ♀. BMNH 1864.4.12.1. The crowns of both M^3 are tilted distally. The curve of the cheek-tooth arch is exaggerated and the arch between the M^3 is particularly narrow. Both the buccal and lingual surfaces of the arch are uneven, giving an echelon appearance because all the teeth are slightly rotated.

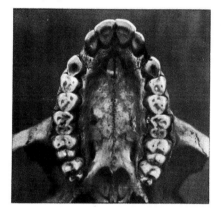

Fig. 10.84. *Macaca fascicularis aurea* (crab-eating macaque). ♀. BMNH 1914.12.8.11. Both M_1 are slightly rotated with the buccal surfaces turned mesially.

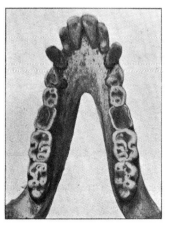

(e) The effect of captivity

Positional anomalies are more common and more marked in captive macaques than in those from the wild. Swindler and Sassouni (1962) found that the upper incisors did not reach the plane of the lower incisors (incisor open bite) in eight captive rhesus monkeys that habitually sucked their thumbs, fingers or toes. The one captive monkey that did not have this habit did not develop open bite and neither did any of 38 wild-born rhesus, which presumably did not thumb-suck.

Examples of generalized crowding of the post-canine dentition, probably due to insufficient growth of the bone, are shown in Figures 10.86 and 10.87.

Extreme displacement of P^4 has been referred to under the section dealing with the orang-utan (p. 173). The captive macaque affords another example. In the juvenile animal in Figure 10.88, the right P^4 has erupted mesially to its deciduous predecessor and, no doubt in consequence, P^3 is misplaced buccally.

Examples of extreme misplacement of P^4 with its deciduous predecessor still present are shown in Figures 10.89 and 10.90. In both, the crowns of the teeth are some distance from their normal situation, close to or even between the roots of the m^4. It is noteworthy that the m^4 show little or no resorption of the roots which should be quite advanced at this stage of development; indeed one specimen (Fig. 10.90) is adult. When permanent successors are absent, sometimes the roots of the deciduous teeth undergo normal resorption and sometimes not. On the whole, it seems more likely that these premolars are in these abnormal positions because they developed there *ab initio* and not because of failure of the tooth resorption process. It is noteworthy that the bone around P^4 in

Fig. 10.85. *Macaca mulatta mulatta* (rhesus monkey). ♀. BMNH 1931.1.11.18. Both M_2 slightly rotated with the buccal surfaces turned mesially.

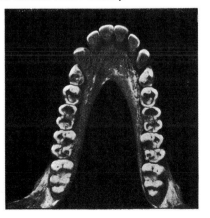

one specimen (Fig. 10.90) was abnormally dense (see Chapter 16, p. 246).

Among the captive macaques, there was one example of impaction of M_3. The crown was tilted mesially, the mesial margin impinging against the convex distal surface of the crown of M_2 (Fig. 10.91).

Of the 314 macaques from the captive state, 39.8% showed positional variations compared with only 30.9% of 588 from the wild.

In 1918, several hundred young rhesus monkeys (*M. mulatta mulatta*) were brought to the London Zoo for purposes connected with the Great War. They were transported in crowded conditions and most of them died of disease shortly after their arrival, but the skulls were kept and Colyer (1921) studied their dentitions. In 1934, 254 of the skulls were presented to the Odontological Museum of the Royal College of Surgeons. All have the deciduous dentition in place

Fig. 10.86. *Macaca nigra ochreata* (booted macaque). Captive. RCS Odonto. Mus., G 68.4. Symmetrical bilateral rotation of P^3 and P^4.

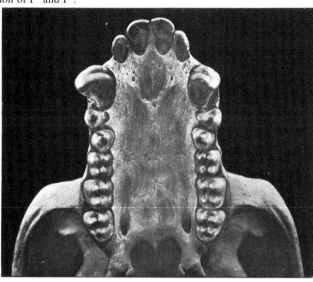

Fig. 10.87. *Macaca arctoides arctoides* (bear macaque). ♂. Captive. BMNH 1866.4.25.3. Generalized symmetrical bilateral crowding of the post-canine arch, with M_3 impacted beneath the distal convexities of the crowns of M_2.

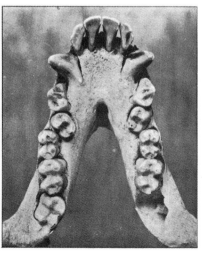

Fig. 10.88. *Macaca* sp. Captive. RCS Odonto. Mus., G 68.43. Juvenile. The right P^4 has erupted mesially to its deciduous predecessor m^4. The arch seems to be narrow (compare Fig. 10.78) but there is no lack of space for the teeth.

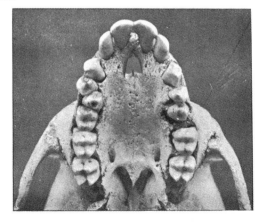

and some have M1 as well. Although 15 (5.9%) showed crowding of the deciduous incisors, the prevalence of positional abnormalities is not particularly high in these young animals.

However, of the eight juvenile skulls (those in which the permanent incisors, canines or premolars are beginning to erupt), five have some positional abnormality. In one (♂, G 68.4152), the developing lower incisors seem to be more crowded than usual; in another (G 68.4151), the lower incisors are crowded and the canines are erupting lingually to the line of the arch; in G 110.671 (♀), the right P_4 is beginning to erupt and is rotated through 180°; in A 84.271 (♂), the left m_3 is rotated through 90° with its buccal surface facing mesially, and P_3 is erupting into the large space between it and P_4. In G 68.411 (♀), both P^4 are

Fig. 10.89. *Macaca fascicularis* (crab-eating macaque). Captive. RCS Odonto. Mus., G 68.418. Juvenile. Windows have been cut to display the teeth.
On both sides, P^4 is not in its normal position with its crown close to and between the resorbing roots of m^4, but is high up in the maxilla and tilted distally; on the left, the tooth is nearly horizontal and its crown has pierced the facial surface of the bone.

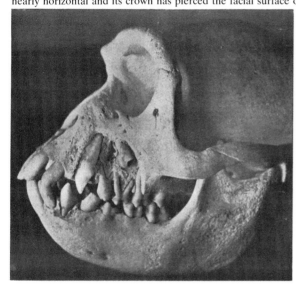
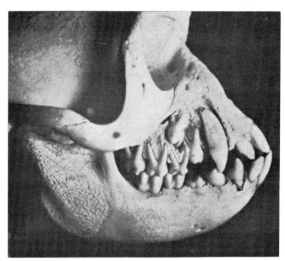

Fig. 10.90. *Macaca* sp. (macaque). Captive. RCS Odonto. Mus., G 68.415. Both P^4 are deeply embedded in the bone above the roots of m^4 and M^1. The roots of neither m^4 show any resorption although this is an adult. Incisors and canines with ring enamel hypoplasia. The bone removed to display the teeth was unusually dense.

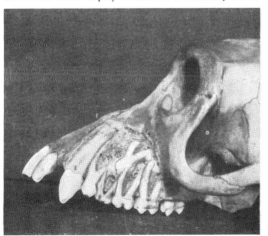
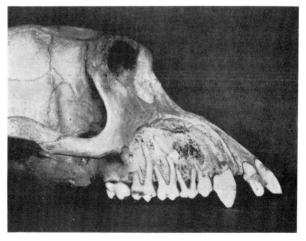

unerupted and misplaced lingually; the left one is lying horizontally and has a conical crown partially divided by a longitudinal groove; that is, shows features of connation. The left M^1 is misplaced lingually and rotated so that its buccal surface faces slightly distally.

Genus *Papio*, baboons

In the baboon, the upper incisors form a regular curve (Fig. 10.92) and the canines, which in males are powerful weapons, are set at the corners of the arch; the premolars and molars form straight rows converging only slightly in the region of M^3, which tend to be slightly rotated lingually. Positional variations in *Papio* are summarized in Table 10.3.

(a) Incisors

An I^1 may protrude slightly or an I^2 may be misplaced lingually. Figure 10.93 shows the skull of a female in which the anterior part of the upper arch is narrow and there is crowding of the incisors. I_2 may be misplaced lingually as a manifestation of lack of space.

(b) Canines

In one skull, an upper canine was over-erupted (Fig. 10.94) and inclined mesially; thus it passed mesially to the lower canine, barely occluding with it. The lower canine was inclined distally. It seems likely that one of these mutual inclinations and misplacements was secondary to the other; for instance, if the upper canine erupted mesially to the lower one, the lower ones would tend to be forced into distal inclination.

(c) Premolars

Slight rotation of P4 is common; in the maxilla, the tendency is for the lingual cusp to face mesially. In seven skulls, the maxillary tooth was rotated between one-eighth to three-sixteenths of a circle. Extreme rotation is rare in the wild state; in the example shown

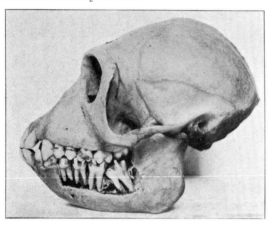

Fig. 10.91. *Macaca mulatta* (rhesus monkey). Captive. RCS Odonto. Mus., G 68.47. Bone has been removed to display the teeth. M_3 impacted under the distal convexity of the crown of M_2. × c. 0.75.

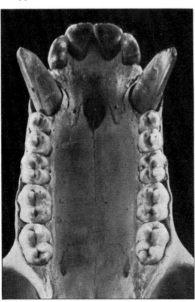

Fig. 10.92. *Papio hamadryas* (Arabian baboon). ♂. The normal upper tooth arch in baboons. × c. 0.75.

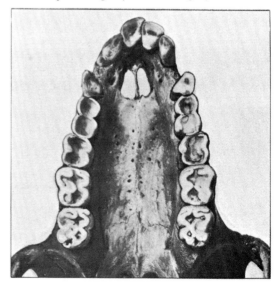

Fig. 10.93. *Papio anubis* (olive baboon). ♀. Wild. BMNH 1914.3.8.2. Upper incisors crowded with the I^1 overlapping. The anterior part of the arch is narrow and both P^4 are misplaced lingually. Reduced slightly.

in Figure 10.95, an upper premolar is rotated about 90° and it is interesting that this has occurred in spite of there being plenty of space for it. This is a young animal with the deciduous canines just being replaced and it is likely that the rotated P^4 has only just erupted into the space provided by the loss of its predecessor. In the pongids, as in man, the predecessors (m4) are distinctly larger than the premolars that succeed them; this provides some space in which adjustive movements of the teeth occur associated with the establishment of the occlusal relationships of the permanent teeth.

P^4 may be misplaced slightly buccally or lingually (Fig. 10.93) but gross misplacements as in Figure 10.96 are rare. Misplacements of P3 are less frequent than those of P4.

(d) Molars

Baboons show little variation in the position of molars. In three, the M^3 were slightly misplaced lingually, so affecting their occlusion with the mandibular teeth; in one specimen, an M^3 was rotated 90° (Fig. 10.97). The specimen shown in Figure 10.98 has a much disordered post-canine dentition on the right side due to the tilting

mesially of M^2 and its impaction against M^1; consequently M^2 is only partially erupted and only its distal cusps reach the occlusal plane. M^3 is also inclined

Fig. 10.95. *Papio ursinus* (chacma baboon). ♂. Wild. BMNH 1928.9.11.7. Juvenile. Left P^4 rotated with the buccal surface facing mesially. Note that there is ample space for it; the deciduous canines are still present and P^4 may have only just replaced the larger m^4. Right P^3 has not completely replaced m^3. Reduced slightly.

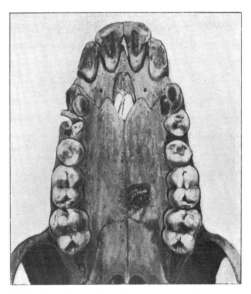

Fig. 10.94. *Papio ursinus* (chacma baboon). ♂. RCS Odonto. Mus., G 68.61. Left upper canine is misplaced mesially (note the space between it and P^3), is inclined mesially and is grossly over-erupted. It passes mesially to the lower canine and barely occludes with it. × c. 0.75.

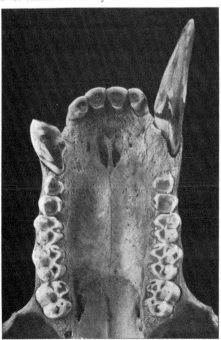

Fig. 10.96. *Papio hamadryas* (Arabian baboon). Wild, died soon after capture. RCS Odonto. Mus., G 68.73. Upper right P^3 lingual to the canine. × c. 0.75.

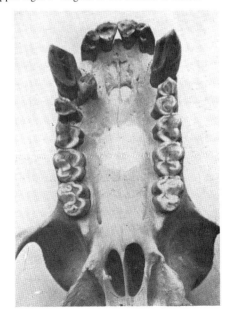

mesially. Unfortunately the mandible is missing so the nature of the occlusion can only be guessed at. Judging from the distance between roots of M^1 and those of M^2, there is basically no shortage of space. The orientation of M^2 is similar to that which pertains during development so perhaps, for some reason, the crown moved mesially during eruption and did not rotate downwards early enough and so became caught under the convexity of the distal surface of M^1.

(e) Effects of captivity

Baboons thrive under captive conditions and, as many are brought into menageries when quite young, the effects of captivity upon the development of irregular positions of the teeth can be studied. The following examples illustrate the types of irregularities seen in the captive animal. The jaws of one with crowded and irregular arches are shown in Figure 10.99 compared with a normal specimen. The ratio of the length of palate in relation to the cheek-tooth length in the normal specimen is 55.5%, in the abnormal one 59.3%.

In the yellow baboon in Figure 10.100, both P^4 were symmetrically rotated and both P^3 were misplaced buccally, the left P^3 so far that it was out of occlusion.

A rarer type of irregularity that may possibly be due to excessive crowding of the developing dentition is seen in Figure 10.101. A P^4 has erupted mesially to its deciduous predecessor and has displaced the P^3. In a juvenile skull (Fig. 10.102), all four P4 are embedded in the bone a slight distance away from the roots of their deciduous predecessors which show no sign of resorption. The rest of the dentition appears to be normal although the tooth arches may be narrower than usual. The cancellous bone of both upper jaws seems to be close-textured with small marrow spaces and the cortex is unusually thick. It is recorded that, when the windows in the jaws were cut to display the teeth, the bone cut easily with a knife so that it is possible that some disorder of the bone may be related to the misplacement of the P4.

An example of extreme irregularity of the teeth associated with inadequate growth of the jaws is shown in Figure 10.103. The incisors are crowded and the premolars are disordered although not particularly crowded. The right P^4, well spaced from its fellows, has

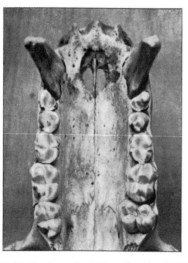

Fig. 10.97. *Papio ursinus* (chacma baboon). Wild. BMNH 1910.6.17.1f. The left M^3 rotated 90° so that its buccal surface faces mesially. ×*c.* 0.67.

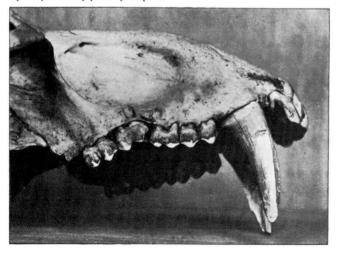

Fig. 10.98. *Papio anubis* (olive baboon). ♂. Wild. BMNH 1908.8.9.41. The post-canine occlusal plane is disordered because M^2 is tilted and impacted against the distal aspect of M^1; consequently it is only partially erupted.

erupted towards the palate, the root being situated in a direction distally and towards the mid-line. The left P^4 is embedded in the palate, the crown of the tooth almost reaching the mid-line, the root running obliquely upwards, distally and buccally. The view of the mandible shows that the enamel of all the permanent teeth is severely affected by enamel hypoplasia and M_2 is impacted against M_1.

A skull of an Arabian baboon showed asymmetry of the bones of the face. Viewed from the front, the left maxilla appears to be twisted downwards and to the left (Fig. 10.104). The effect of the distortion is seen from the palatal aspect. The left P^3 is misplaced buccally and the P^4 a little lingually. The left maxilla is smaller than the right and the space enclosed by the temporal fossa and the zygomatic arch is smaller on the left than on the right. Measurements show that the left side of the mandible is not as well developed as the right, there being a marked difference in the size and shape of the ascending rami. There are no signs of injury to the bones.

Of the 153 specimens from the captive state, 69 (45%) had positional irregularities, a significantly larger proportion than in animals from the wild.

Fig. 10.99. *Papio porcarius* (chacma baboon). Two captive specimens. The one on the right (RCS Odonto. Mus., A 88.3) has normal tooth arches. The one on the left (RCS Odonto. Mus., G 57.31) has a generally crowded and irregular post-canine dentition, especially in the upper jaw; both upper and lower arches are narrow in the premolar regions and also between the M3. The maxillary tuberosities are small, a further indication that the jaws are too small to accommodate the teeth. × c. 0.5.

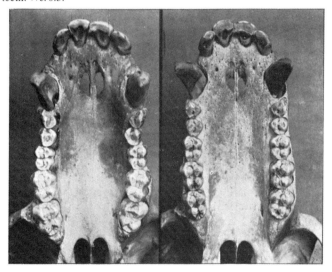

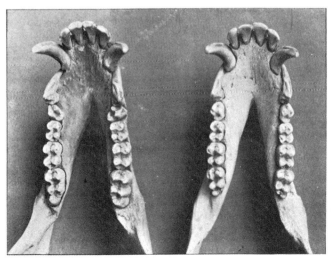

Section 2: Variations in Position

Genus *Mandrillus*, mandrills

Positional variations in *Mandrillus* are summarized in Table 10.3.

Positional irregularity of premolars seems to be a feature of mandrill. In the 35 specimens from the wild state, the following malpositions were seen:
 (i) Left P^3 misplaced buccally so that it did not occlude with the mandibular teeth. The left P_4 was rotated about 90°.
 (ii) P^4 rotated with the lingual cusps facing partially mesially.
 (iii) Both P^3 rotated so that the buccal surfaces faced towards the palate. This was not associated with shortage of space because there were wide gaps between the P^3 and the canines.
 (iv) In three specimens, both P^4 showed moderate rotation.
 (v) In two, both P^3 were rotated with the lingual cusps facing distally.

Among the 20 specimens from captive animals, there were the following malpositions:
 (i) The right P^4 was misplaced with its crown directed disto-buccally and the roots towards the palate.
 (ii) The left P^4 was rotated with the buccal surface in contact with the M^1.
 (iii) The right P^3 was rotated about 40°, the lingual surface facing mesially.
 (iv) A P_3 was rotated about 90°.
 (v) In two skulls, both P^4 were moderately rotated.

Fig. 10.100. *Papio cynocephalus* (yellow baboon). Captive. BMNH 1866.4.25.2. The P^3 are misplaced buccally and the P^4 are rotated with the buccal surfaces turned distally. Reduced slightly.

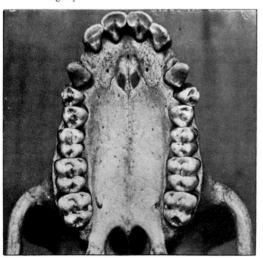

Genus *Theropithecus*, gelada baboons

Positional variations in *Theropithecus* are summarized in Table 10.3.

In four of the 21 specimens, the P^4 were a little

Fig. 10.101. *Papio ursinus* (chacma baboon). RCS Osteo. Series, 191b. The right P^4 (pointer) has erupted mesially to m^4 and P^3 is misplaced towards the buccal aspect of the canine. × *c.* 0.67.

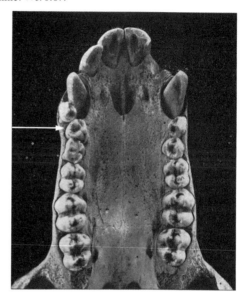

Fig. 10.102. *Papio* sp. (baboon). Captive. RCS Odonto. Mus., G 68.65. Juvenile. Windows have been cut to show that P^4 and P_4 are embedded in the bone just beyond the roots of their deciduous predecessors which show no resorption. The opposite jaws were similarly affected. × *c.* 0.67.

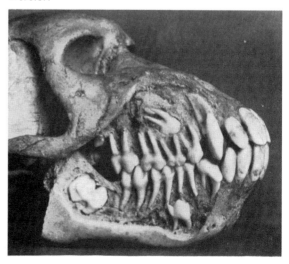

lingual to the arch or rotated. In one (Fig. 10.105), the partially-erupted maxillary canines were misplaced and occluded within the mandibular arch.

Fig. 10.103. *Papio* sp. (baboon). Captive. RCS Odonto. Mus., G 52.16. Juvenile. Upper arch narrow with crowding of the incisors and irregularity of the post-canine arches, the predominant feature being that the left P^4 is horizontal and buried in the palate near to the midline. Bone has been removed to reveal the root. Its predecessor is still present.

The lateral view of the right mandible shows that M_3 is tilted mesially and impacted against the distal surface of M_2. All the teeth show severe ring enamel hypoplasia. At least three rings can be identified on the crown of the canine, indicating three spaced episodes of systemic disorder (see Chapter 20).

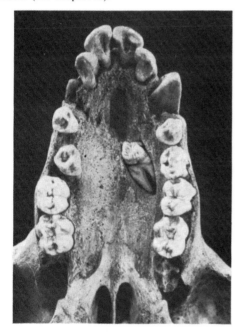

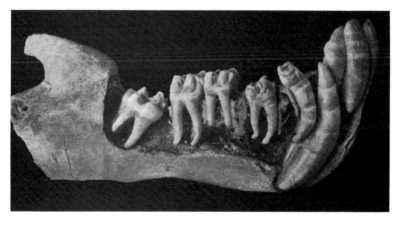

Family **Cebidae**. New World monkeys

Genus *Cebus*, capuchin monkeys

The normal type of upper arch in *Cebus* is shown in Figure 10.106. The bucco-lingual axes of the cheek teeth are slightly oblique to the line of the arch; they form a curve with the greatest convexity in the region of M^1; the mandibular arch is somewhat similar.

In plan view, M^1 is more or less square, the lingual border of M^2 is more rounded and M^3 is oval. The three premolars are flattened in the mesio-distal direction and, when there is insufficient room in the arch, they tend to be displaced in either a buccal or a lingual direction and not rotated as in Old World monkeys.

Positional variations were noted in about 20% of the specimens of *Cebus* and are summarized in Table 10.17. Table 10.18 shows that there is little difference between the various species of *Cebus* in the prevalence of positional variation.

(a) Incisor occlusion

The upper and lower incisors normally met edge-to-edge; underjet occurred in about 11.7% (Table 10.17). However, according to della Serra (1955a, p. 184) who examined 225 *Cebus* skulls, overjet was the most common type of occlusion, occurring in 86.9%, with 11.1% meeting edge-to-edge and only 0.4% with underjet. The differences between Colyer's and della Serra's findings cannot be explained, especially as della Serra seems to have been unaware of Colyer's work.

(b) Incisors

Sometimes the upper incisors were crowded with I^1 overlapping (Fig. 10.107) or I^2 was misplaced lingually (Figs. 10.108 and 10.109). Schultz (1935) noted crowd-

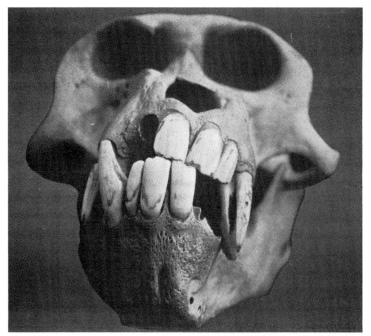

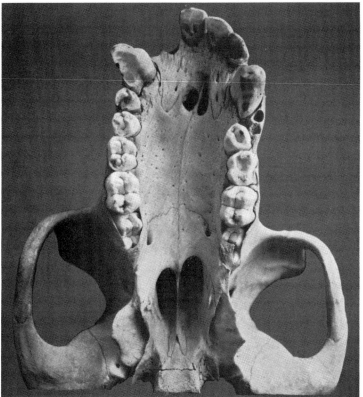

Fig. 10.104. *Papio hamadryas* (Arabian baboon). RCS Osteo. Series, 180. Marked asymmetry of the facial skeleton which seems to be predominantly due to the smallness of the left premaxilla and maxilla. The midline of the upper arch is about the width of I^1 to the left, the left canine is misplaced distally and the post-canine teeth misplaced distally to a less extent so that there is no space for P^3 which is now lost; its alveoli show it was misplaced buccally. There was some asymmetry of the mandible, most measurements being less on the left.

ing of the upper incisors in 1% and of the lower incisors in 1.9% of 105 *Cebus* skulls. A white-throated capuchin (*C. capucinus*, ♀, KU 111, 336) had the left I_1 tilted lingually (Smith, Genoways and Jones, 1977).

(c) Canines
In one brown pale-fronted capuchin (*Cebus albifrons*, ♂, BMNH 1844.1.18.22), the right upper canine was so misplaced buccally and distally that it made no contact with the lower teeth (Colyer, 1919).

Fig. 10.105. *Theropithecus obscurus* (gelada baboon). FMNH 27184. Both upper canines curve lingually so that they occluded within the mandibular arch.

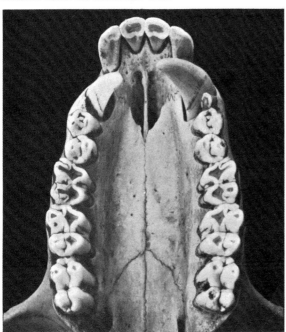

Table 10.17 Positional variations in CEBIDS from the wild.

Genus	No. of specimens	Class of tooth				Total specimens affected		Incisor underjet	
		I	C	P	M	No.	%	No.	%
Cebus	538	13	1	9	95	110	20.4	8	1.5
Lagothrix	82	6	0	8	2	16	19.5	22	26.8
Brachyteles	25	1	0	5	0	6	24.0	9	36.0
Ateles	186	6	2	7	15	29	15.6	4	2.2
Alouatta	781	25	3	128	21	166	21.3	191	24.5
Pithecia	148	7	0	8	2	16	10.3	0	0
Cacajao	23	0	0	0	0	0	0	0	0
Saimiri	100	3	1	4	1	9	9.0	0	0
Callicebus	122	5	1	6	2	13	10.7	1	0.8
Aotus	117	6	0	1	2	9	7.7	0	0
Total	**2122**	**72**	**9**	**176**	**140**	**375**	**17.7**	**235**	**11.7**

In skulls in which two or more classes of teeth varied in position, each class is recorded individually; for example, if in a skull one or more of the incisors, premolars and molars varied, each appropriate class would be credited with one and the column 'total specimens affected' with one only.

Fig. 10.106. *Cebus apella* (brown capuchin). ♂. BMNH 1897.10.3.5. The normal upper tooth arch in *Cebus*. ×*c*. 1.25.

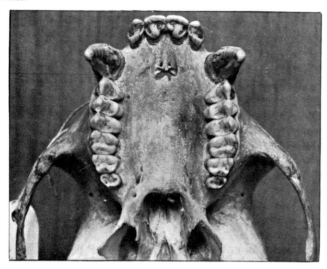

Fig. 10.107. *Cebus apella* (brown capuchin). ♀. BMNH 1900.11.5.2. The upper incisors are crowded with the I¹ overlapping each other (Colyer, 1919).

Fig. 10.108. *Cebus albifrons* (brown pale-fronted capuchin). RCS Odonto. Mus., G 68.882. Upper incisors crowded and left I² slightly rotated.

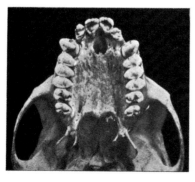

Fig. 10.109. *Cebus albifrons* (brown pale-fronted capuchin). ♂. BMNH 1928.5.2.24. Upper incisors crowded with both I² misplaced lingually; the left P³ is also slightly misplaced lingually. ×*c*. 1.33.

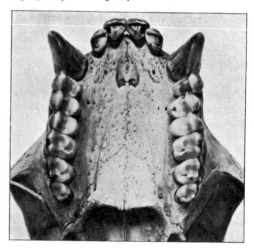

Table 10.18 Positional variations in some species of *CEBUS*

Species	No. of specimens	Total specimens affected	
		No.	%
C. albifrons	12	5	41.7
C. apella	180	42	23.3
C. nigrivittatus	16	4	25.0
C. capucinus	120	27	22.5

(d) Premolars

In *Cebus*, the P2 is usually placed slightly obliquely to the line of the arch. The obliquity is sometimes exaggerated; alternatively one or more premolar may be slightly misplaced buccally or lingually (Fig. 10.110). Schultz (1935) noted crowding of the upper premolars in 2.9% and of the lower premolars of 1.0% in 105 *Cebus* skulls, but there was only one skull with a misplaced premolar (right P^4) in his sample of 118 *Cebus capucinus* from Nicaragua and Mexico.

(e) Molars

In apes and monkeys, the buccal cusps of the M^3, in occlusion, overlap the buccal cusps of M_3. It is common in *Cebus* for the buccal cusps of M^3 and M_3 to meet edge-to-edge or for the buccal cusps of the M^3 to pass lingually to those of the M_3. These variations are not included in Table 10.17. The positions of the upper molars vary far more frequently than those of the lower molars. M^3 is often placed quite obliquely to the arch as in Figure 10.111; Smith, Genoways and Jones (1977) noted an example in a white-throated capuchin (*C. capucinus*, ♀, KU 108,384). The crown may be tilted mesially; this variation was present on one or both sides in six skulls.

In the mandible, the roots of M_3 usually incline distally; sometimes the whole tooth may be tilted mesially (Fig. 10.112). This was seen in four mandibles. In one mandible (Fig. 10.113), both M_3 were horizontal with the occlusal surfaces directed mesially, impinging on the roots of M_2. In another (Fig. 10.114), one M_3 and both M^3 were impacted. Smith, Genoways and Jones (1977) found three skulls of white-throated capuchins (*C. capucinus*) among 84 collected in Nicaragua with impacted molars. In two (♂, KU 108,387 and ♀, KU 104,359), the right M^3 and M_3 were impacted and, in the third (♀, AMNH 176,629), the left M^3.

In one brown capuchin (*Cebus apella*, AMNH 42318), M^1 and M^2 were rotated and M^3 were absent; in a mandible of the same species (♂, BMNH 1927.11.1.3), there was extreme rotation of M_2, the mesio-distal axes of the crowns being placed almost transversely to the line of the arch, the buccal surfaces facing mesially.

Fig. 10.111. *Cebus capucinus* (white-throated capuchin). ♀. BMNH 1868.7.9.1. The right M^2 is slightly misplaced lingually and both M^3 are rotated so that the buccal surfaces face slightly mesially and contact with M^2 is abnormal (Colyer, 1919).

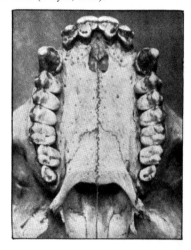

Fig. 10.112. *Cebus apella* (brown capuchin). ♂. BMNH 1845.6.17.2. Both M_3 are slightly rotated and tilted so that the buccal and occlusal surfaces face slightly mesially (Colyer, 1919).

Fig. 10.110. *Cebus capucinus* (white-throated capuchin). ♀. BMNH 1902.3.5.15. The arrangement of the premolars is irregular; the right P^3 and both P_3 are buccally misplaced (Colyer, 1919).

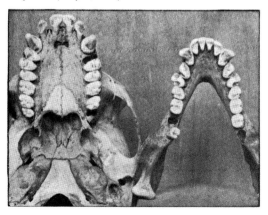

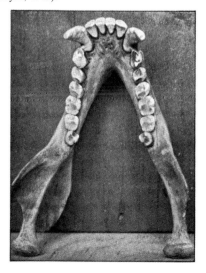

Lingual misplacement of the M^2 and M^3 on one or both sides (Figs. 10.111 and 10.115) was present in 14 capuchin skulls.

An example of slight rotation of all the post-canine teeth producing an echelon appearance is shown in Figure 10.116.

From the above description, it will be apparent that there is a wide range of variation in the position of the molars in *Cebus*.

Among the specimens of *C. apella*, there was a series of 22 from Kartabo, Guyana. Of these, 13 showed positional variations of the teeth, including three with irregularity of the incisors (an uncommon variation in *Cebus*), two in which both M^3 were tilted and one with rotation of both M^2. In two specimens, M^3 was absent, in one bilaterally and in the other the left.

(f) The effect of captivity

Capuchin monkeys are common in laboratories and menageries and it is therefore possible to observe the effects of captivity on the development of positional anomalies of the teeth. Strauss (1985) described three young sibling capuchins in which malposition of the incisors was associated with an expansile dystrophy of the jaws. Young animals seem liable to develop rickets and in these animals extreme irregularities of the premolars may follow, the teeth at times erupting through the buccal aspect of the alveolar process. Variations in position occurred in 39 (32.5%) of 120 captive capuchins compared with 20.4% of animals from the wild (Table 10.17). Positional irregularities of the premolars are more frequent in animals from the captive state than from the wild; in 120 captive specimens, there were 14 (11.7%); in 538 from the wild state, only nine (1.2%).

Figure 10.117 shows an unusual misplacement of a P^2 and upper canine, with a space between P^2 and P^3. There is no trace of a retained part of a deciduous tooth which would seem to have been a possible cause.

Captive conditions do not seem to increase the tendency for the molars of capuchins to vary in position, the percentage of specimens from the wild and captive states being approximately the same, but the

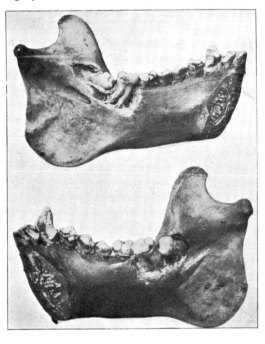

Fig. 10.113. *Cebus apella* (brown capuchin). ♀. BMNH 1925.2.1.3. Bone has been removed to display the teeth. Both M_3 lie more or less horizontally in the ascending rami with the occlusal surfaces impacted against the distal aspects of the crowns of M_2 (Colyer, 1947a). Enlarged slightly.

Fig. 10.114. *Cebus capucinus* (white-throated capuchin). ♀. BMNH 1904.7.6.2. Bone has been removed to display the teeth.
Left: In the right maxilla, M^3 is unerupted and impacted beneath the distal convexity of the crown of M^2.
Right: The right M_3 is tilted lingually and impacted under the distal convexity of the crown of M_2 (Colyer, 1919, 1947a). Enlarged slightly.

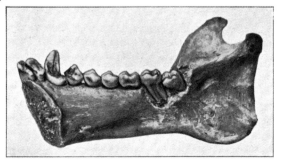

proportion showing tilting of the molars was larger in the captive state, being 5.8% compared with 2.1% in those from the wild state.

Genus *Lagothrix*, woolly monkeys

Positional variations were noted in about 20% of woolly monkeys and are summarized in Table 10.17.

(a) Incisor occlusion

There seems to be a definite tendency to incisor underjet; it occurred in about a quarter of the *Lagothrix* skulls examined. However, della Serra (1955a) said that none of his 28 *Lagothrix* skulls showed underjet; overjet was far more common, occurring in 44%, with edge-to-edge in 28%.

(b) Incisors

Irregularity of the upper incisors occurred in five skulls and was shown by overlapping of I^1 or a labial misplacement of these teeth. Slight rotation of I^2 was common and would seem to fall within the limits of the normal, but the rotation may be sufficiently marked to be counted as an abnormality. Irregularity of the lower incisors was seen in only one skull.

(c) Premolars

In about 10% of the specimens, the premolars were irregular in position, as shown by the following specimens:

(i) *Lagothrix lagothricha lugens* (common woolly monkey). ♀. BMNH 1895.9.1.4. (Fig. 10.118). There is a buccal misplacement of the left P^4. The degree of rotation of the I^2 may be regarded as within normal limits (Colyer, 1919).

(ii) *L. lagothricha lagothricha* (Humboldt's common woolly monkey). ♀. BMNH 1900.11.5.15. (Fig. 10.119). The left P^4 is rotated so that the buccal surface faces distally; on the opposite side, there is a wide gap between P^4 and M^1 (Colyer, 1919).

(d) Molars

Two specimens showed positional variations of the molars:

(i) *L. lagothricha*. ♂. BMNH 1857.10.17.1. The

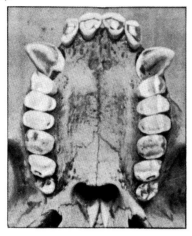

Fig. 10.115. *Cebus apella* (brown capuchin). ♂. BMNH 1903.9.4.19. Both M^3 tilted mesially and impacted against the distal aspects of M^2, particularly on the right (Colyer, 1919).

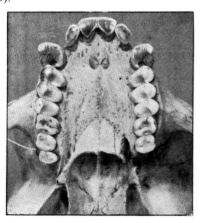

Fig. 10.116. *Cebus nigrivittatus* (wheeper capuchin). ♀. BMNH 1905.5.24.2. All the post-canine teeth are slightly rotated so that the buccal surfaces face slightly mesially, producing an echelon appearance. Left M^3 absent (Colyer, 1919).

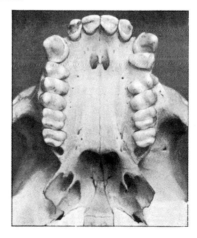

Fig. 10.117. *Cebus apella* (brown capuchin). ♀. Captive. RCS Odonto. Mus., G 68.883. The left canine and P^2 misplaced mesially and slightly buccally so that there is a space between P^2 and P^3. The mandibular arch is normal. normal.

roots of both M^3 slope unduly distally and the right M^3 is misplaced a little lingually (Colyer, 1919).

(ii) *L. lagothricha*. ♂. BMNH 1913.10.24.8. Both M^2 are rotated, the mesio-buccal borders being turned distally (Fig. 10.120). The rotation is more marked on the right than on the left.

(e) Asymmetry of the arch

In one specimen, there was marked asymmetry of the upper facial skeleton (Fig. 10.121); the mandible and the rest of the skull were unaffected so the cause was an interference with growth more or less limited to the right maxilla.

Genus **Brachyteles**, woolly spider monkeys

The chief feature of *Brachyteles* is the tendency to incisor underjet which occurred in 36% of the 25 skulls examined (Table 10.17). della Serra (1955a) found a

Fig. 10.118. *Lagothrix lagothricha lugens* (common woolly monkey). ♀. BMNH 1895.8.1.4. Left P^4 misplaced buccally.

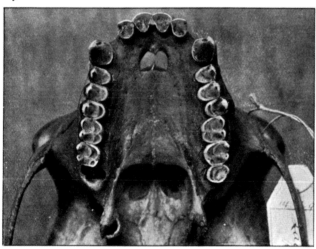

Fig. 10.119. *Lagothrix lagothricha lagothrica* (Humboldt's common woolly monkey). ♀. BMNH 1900.11.5.15. Left P^4 rotated so that its buccal surface faces mesially. A space between P^4 and M^1 on the right is associated, as is a similar space in the lower arch, with periodontal disease. The upper incisors have been lost *ante mortem* (Colyer, 1919).

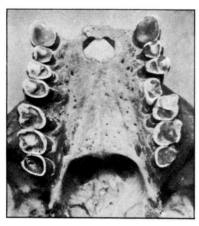

Fig. 10.120. *Lagothrix lagothrica* (common woolly monkey). ♂. BMNH 1913.10.24.8. Both M^2 slightly rotated so that their buccal surfaces face slightly distally.

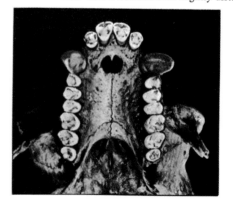

similar prevalence of underjet (33.3%) in the 18 skulls he studied, 33.2% being with overjet and 11.1% with the edge-to-edge condition.

In one specimen, there was a slightly crowded arrangement of the mandibular incisors.

An extra upper premolar which was present in one skull had caused P^2 and the canine to erupt mesially to their normal positions and this had led to malposition of the mandibular canine (Fig. 10.122). Five skulls showed slight variations in the positions of premolars.

Genus *Ateles*, spider monkeys

The chief interest in *Ateles* is the marked character of the positional anomalies that occurred in about one-fifth of the 29 skulls that varied (Table 10.17).

Underjet occurred in only 2.2% of the 186 *Ateles* skulls examined; della Serra (1955a) found none in the 40 skulls that he studied, of which 87.5% showed overjet and 2.5% the edge-to-edge condition.

In two skulls, the canines were irregular in position. In a white-whiskered spider monkey (*A. belzebuth marginatus*, ♀, BMNH 1853.3.19.73), the upper right canine was tilted buccally, but this was probably related to the presence of an extra premolar in each maxilla (Chapter 3, p. 54 and Fig. 3.62).

The asymmetry of the upper arch in a black spider monkey (Fig. 10.123) is of special interest because it hints at a relationship between the size of a jaw bone and the overall size of the teeth it bears. On the right, there are three incisors instead of the normal two; in addition, the canine and post-canine teeth are more distally situated than on the left.

The lower arch is normal and, bearing in mind that a pre-canine diastema separates the upper incisors from the canine, the presence of the extra incisor could not have prevented the canine and other teeth assuming their normal positions. Nevertheless, measurement shows that the distance between the alveolare (the bony crest between I^1) and the mesial margins of the canine sockets is 2 mm greater on the right than the left. The canines, of course, are situated immediately distal to the premaxillary-maxillary sutures; hence the measurements referred to are those of the premaxillae.

Fig. 10.121. *Lagothrix lagothrica poeppigii* (brown woolly monkey). ♀. BMNH 1866.3.28.7.
A: palatal view.
The upper arch is asymmetrical, principally because the whole upper jaw is at an angle to the base of the skull so that the centre of the incisors is some distance to the right. Note the angle of the line between the distal aspects of M^2 (the left M^3 has been lost and the right M^3 is tilted distally) and the median plane of the skull. In addition, the right post-canine arch is crowded and irregular, mainly because P^4 is misplaced lingually.
B: the right lateral view shows incisor underjet which is part of pre-normal occlusion on that side; note that P_2 can be seen mesial to the upper canine and all the lower cheek teeth meet the upper ones too mesially by about half the width of a premolar. This pre-normal occlusion is probably due to defective growth of the right upper jaw and is therefore an example of superior retrusion. The post-canine occlusion was normal on the left (Colyer, 1919).

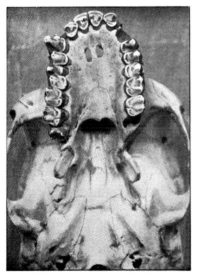

A

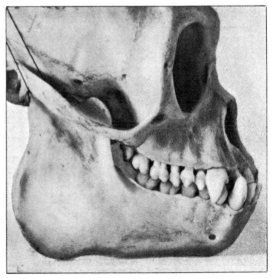

B

Unfortunately, in this specimen, these sutures are virtually obliterated, although what seems to be a trace can be seen immediately mesial to the socket of the right canine. Thus, the argument is that the right premaxilla, or at least its alveolar process, is large and that this is related to the presence of the supernumerary tooth.

In four out of the seven skulls showing variations of the premolars, there was considerable misplacement of the teeth (Fig. 10.124). Schultz (1935) figured a skull of a black-handed spider monkey (*A. geoffroyi*) in his own collection (no. 975) in which the right P^4 was lingually misplaced. He also noted crowding of the upper premolars in 3.4% and of the lower premolars in 1.5% of 203 adult black-handed spider monkey skulls from Nicaragua and Mexico (Schultz, 1960, p. 384).

The majority of variations in *Ateles* affected the molars (Table 10.17), the upper teeth being affected more often than the lower. The cheek teeth may be arranged in an echelon manner as described under *Macaca* (p. 202); the M^3 may be rotated obliquely or misplaced lingually (Fig. 10.125). Schultz (1960, p. 383) found M^2 and M^3 displaced lingually in three of his *A. geoffroyi* skulls. In three mandibles, the M_3 were tilted and in one of these skulls the left M^3 was impacted. Among 64 black-handed spider monkeys, Smith, Genoways and Jones (1977) noted impaction of M^3, bilaterally in two skulls and on the right in one. *Ateles* shows considerable variation in the number of the molars (p. 55, Table 3.6).

The specimen in Figure 10.126 shows asymmetry of both the upper and lower arches. In the upper jaw, the right canine and post-canine teeth are more mesial

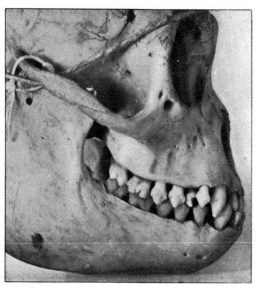

Fig. 10.122. *Brachyteles arachnoides* (woolly spider monkey). ♀. BMNH 1845.4.2.7. The occlusion between the right P_2 and P^2 disturbed by a supernumerary upper premolar. Note the incisor underjet (Colyer, 1919).

Fig. 10.123. *Ateles paniscus* (black spider monkey). ♀. BMNH 1872.4.11.5.
A: palatal view.
Upper arch narrower than normal between the premolars. An empty incisor socket indicates that there had been three incisors on the right instead of two. On the right, the canine and post-canine teeth are all more distal than on the left. The distance from the alveolar crest between I^1 (alveolare) and the mesial margin of the socket of C is greater on the right.
B: the lateral view shows that the right upper arch is post-normal. P_2 occludes mesially to the upper canine instead of distally to it and P_4 occludes between P^2 and P^3 instead of between P^3 and P^4. Occlusion on the left was normal.

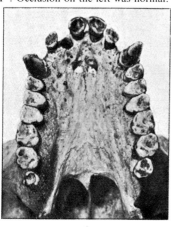

A

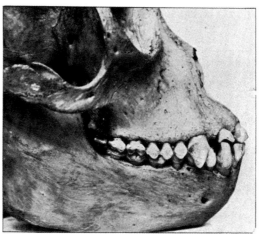

B

than on the left. The lower arch is similarly asymmetrical with some evidence that the abnormality is on the right side because there is a wide post-canine diastema into which the upper canine occludes and the right post-canine teeth are crowded. It is possible that the primary cause was defective growth of the right side of the mandible and the asymmetry of the upper arch secondary.

A remarkable condition is present in a black spider monkey (*Ateles paniscus*) (Fig. 10.127). The

Fig. 10.124. *Ateles geoffroyi vellerosus* (Mexican spider monkey). ♀. BMNH 1889.12.7.2. P^2 and P^3 on both sides are rotated so that their buccal surfaces face slightly mesially; they are also misplaced lingually, especially the P^2 (Colyer, 1919).

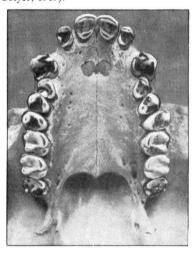

extreme variations in the positions of the teeth would appear to be due to abnormal development of the mandible.

Genus *Alouatta*, howler monkeys

Positional variations in *Alouatta* are summarized in Table 10.17.

In *Alouatta*, the incisors are small compared with the size of the skull and they are arranged in a curve; there is usually a well-marked pre-canine diastema; the premolars and molars are arranged in straight lines, the two sides gradually diverging as they approach the back of the mouth. The normal upper arch is shown in Figure 10.128.

(a) Spaces between the teeth

Spaces are present between the teeth in many skulls. I^1 may be separated from I^2, all the incisors may be spaced, or only the I^1 (Fig. 10.129), the interval being as much as 3 mm. Spaces occur between the mandibular canines and P_2; in a black-and-red howler monkey (*Alouatta belzebul*, ♀, BMNH 1904.7.4.6), in addition to the spaces between these teeth, there was a space between the left P_2 and P_3 as well as spaces between the upper incisors and the lower incisors. In a red howler (*Alouatta seniculus*, ♂, BMNH 1908.3.7.3), spaces were present between P_3 and P_4.

(b) Incisor occlusion

Incisor underjet was present in one-fifth of the specimens. In no instance was the occlusion of the

Fig. 10.125. *Ateles belzebuth belzebuth* (long-haired spider monkey). AMNH 30192. On both sides, M^2 and M^3 are misplaced symmetrically lingually and both M^3 are rotated so that the buccal surfaces face slightly mesially. Enlarged.

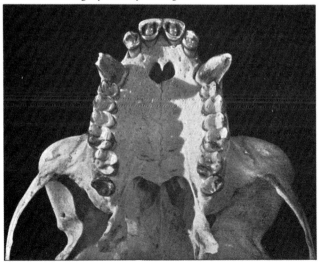

Section 2: Variations in Position

premolars and molars abnormal. Colyer found that the various species seemed to show a difference in the degree of tendency to incisor underjet (Table 10.19). However, della Serra (1951) also found differences between the species of *Alouatta*, but Table 10.19 shows that his findings are quite different from Colyer's, both between species and for the genus as a whole (compare Table 10.17).

(c) Incisors

Irregularity in position of the maxillary incisors, which occurred in about 3% of the *Alouatta* skulls examined, is shown by a slight labial misplacement of one or both I^1 or rotation of I^2. Smith, Genoways and Jones (1977) described a mantled howler monkey (*A. palliata*, ♀, KU 106,307) in which the space between the upper canines was narrower than usual; both I^2 were misplaced disto-lingually and were in abnormal occlusion with the lower canines. An irregular arrangement of the mandibular incisors is rare.

(d) Canines

Three skulls showed irregularity of canines. In one (Fig. 10.130), the mandibular right canine is deflected

Fig. 10.126. *Ateles geoffroyi* (black-handed spider monkey). BMNH 808a.
A: the upper arch is asymmetrical, the right canine and P^2 being more mesial than the left and its midline being to the right. Both M^3 are absent.
B: the lower arch is correspondingly asymmetrical; the midline of the incisors is misplaced to the left, and the right canine is misplaced mesially, so that the upper canine occludes into a postcanine diastema as can be seen in the lateral view.
C: side view. The lower right post-canine teeth are crowded and P_2 and P_3 occlude buccally to the upper arch instead of lingually (cross-bite) and are in post-normal occlusion by about one-half a premolar width. There is incisor overjet (Colyer 1919).

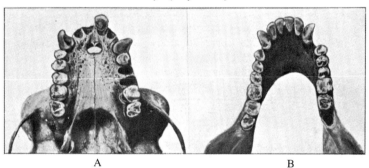

A   B

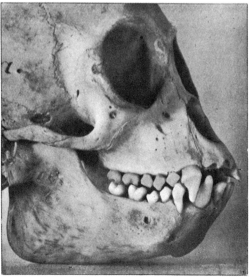

C

lingually to the lingual aspect of I^2 which is separated from I^1. I_2 is elongated, the crown, when the teeth are in occlusion, occupying the gap between the upper incisors. The mandibular left canine, partly erupted, is also misplaced, the root being directed towards the mid-line and the crown buccally. This type of abnormality is rare in monkeys but is not uncommon in the gorilla (see p. 164). Two other examples of irregularity

Fig. 10.127. *Ateles paniscus* (black spider monkey). ♂. BMNH 1927.3.6.1. There are general irregularities in both upper and lower arches; for instance, the left upper canine is misplaced buccally. Some teeth have been lost from disease during life. However, there is a grossly abnormal occlusion with generalized inferior protrusion (note that the right P_2 occludes mesially to the upper canine). The lower incisors are so far in front of the uppers that occlusion between them would have been impossible. The mandible seems to be slewed over to the left and, on that side, post-canine teeth pass to the buccal side of the upper ones; perhaps secondarily to this absence of occlusion, the molar-region height on the left is 0.5 mm greater than on the right. The overall length of the left mandible is 0.6 mm shorter on the left. It is likely that this asymmetry of the mandible is the primary cause of the disordered occlusion; most of the upper arch irregularities could be secondary.

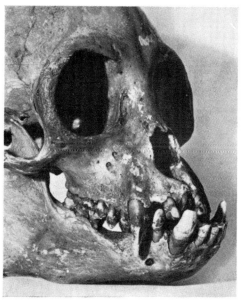

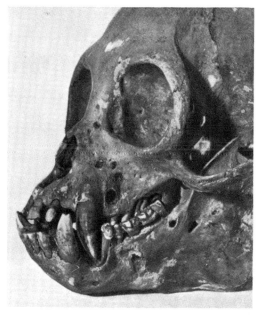

226 Section 2: Variations in Position

of the canines were traceable to the presence of extra teeth in the incisor region. Schultz (1960, fig. 6, p. 382) figured an *A. palliata* skull in which the upper right canine and P^2 are both misplaced labially and the corresponding deciduous teeth have been retained.

(e) Premolars

In *Alouatta*, one or more of the premolars may be rotated or displaced from the arch (Figs. 10.131 and 10.132). Rotation of the maxillary teeth is usually slight; there was, however, one skull in which the left P^4 was rotated through 180°. Schultz (1960) noted crowding of the upper premolars in 1.6%, but none of the lower premolars, among his 378 adult *Alouatta palliata* skulls from Nicaragua and Mexico. In one, the right P^3 was impacted and in another both P^4 were rotated.

Sometimes the three maxillary premolars are placed a little lingually to the molars (Fig. 10.133); this may be due to slight rotation of the teeth, the lingual cusps being turned mesially and the disto-buccal borders distally. In the case of P^4, this would bring the disto-buccal border lingual to the line of the buccal surface of the M^1 and so make P^4 appear to be lingual

Table 10.19 The incidence of incisor underjet in various species of *ALOUATTA*, contrasted with the data of della Serra (1951). See also Table 10.17

Species	No. of specimens	Incisor underjet No.	%
A. seniculus	265	59	22.3
A. caraya	40	17	42.5
A. belzebul	32	8	25.0
A. palliata	106	46	43.4
della Serra (1951)			
A. seniculus	47	21	44.7
A. caraya	43	12	27.9
A. belzebul	68	36	52.9
A. fusca	48	13	27.1
Total	**197**	**86**	**43.6**

Fig. 10.129. *Alouatta fusca* (brown howler). ♀. BMNH 1893.1.1.10. The incisors are spaced with a particularly wide space between the I^1.

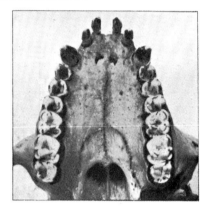

Fig. 10.128. *Alouatta seniculus* (red howler). ♂. BMNH 1903.9.1.1. Normal upper tooth arch in *Alouatta*.

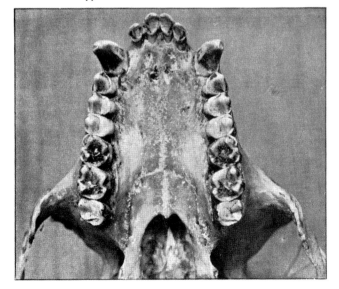

to the line of the buccal surfaces of the molars. Schultz informed Colyer that in his collection of 129 skulls of mantled howler monkeys (*A. palliata*) from Nicaragua the premolars were placed 'inside a line which continues the direction of the lateral aspects of the molars in at least 70 specimens'. Smith, Genoways and Jones (1977) found a similar anomaly in a male mantled howler (KU 107,171) from the same area.

The mandibular premolars do not differ in position as much as the maxillary teeth but rotation is sometimes extreme (Fig. 10.134).

(f) Molars
In four specimens, maxillary molars were rotated, the teeth involved being (i) right M^1; (ii) right M^2 (Fig. 10.135); (iii) right M^3; (iv) left M^2 and M^3. The most common irregularity of the mandibular molars was rotation of M_3 so that the mesial surface faced buccally or lingually. Buccal misplacement of the tooth occurs (Fig. 10.136). Among 137 skulls of mantled howler monkeys (*A. palliata*) from Nicaragua, there were none with impacted molars (Smith, Genoways and Jones, 1977).

Fig. 10.130. *Alouatta* sp. ZMB 23609. Lower right canine (pointer) inclined lingually and towards I_2 which occludes in a space between I^1 and I^2. Lower left canine only partially erupted and misplaced with its crown directed buccally mesial to the upper canine.

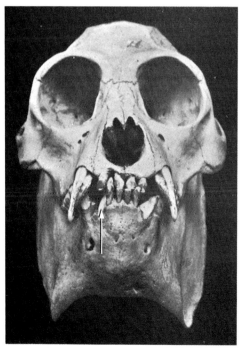

Genera ***Pithecia*** and ***Cacajao***: saki and uakari monkeys

The normal arrangement of the maxillary teeth in saki and uakari monkeys is shown in Figure 10.137. The incisor arrangement is slightly different from that of most of the other cebid genera; instead of being almost vertical in the jaws, the incisors, especially in the mandible, are inclined labially.

The canines form distinct corners to the arch and there is a wide pre-canine diastema in the maxilla. The line of the cheek teeth commences slightly lingually to the canines and follows a direction a little lingually and then buccally at M^1.

Fig. 10.131. *Alouatta* sp. ♂. BMNH 1851.4.23.1. Several premolars lingual to the line of the arch (Colyer, 1919). Note the relation of the left P^2 to the canine compared to the normal specimen in Figure 10.128.

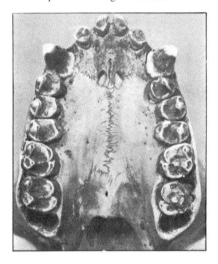

Fig. 10.132. *Alouatta fusca* (brown howler). USNM 113435. A disordered arrangement of the right premolars.

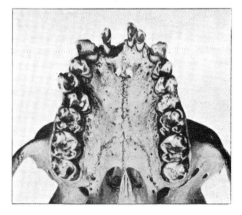

There is a tendency for the I^2 to be situated slightly lingually so that they occlude edge-to-edge with the lower incisors, whereas the I^1 slightly overlap the lowers labially. The lower incisors sometimes show slight labial or lingual misplacement of one or both I_2.

Positional variations are summarized in Table 10.17.

None of the 171 skulls of *Pithecia* and *Cacajao* examined had incisor underjet; della Serra (1955a) also found none in 10 *Cacajao* and 94 *Pithecia* skulls. Instead, there were differences in the mode of occlusion of I^1 and I^2, with I^1 overlapping I_1 labially, and I^2 meeting I_2 edge-to-edge; della Serra found this in 5.2% of his *Cacajao* skulls but not in *Pithecia*. The most common occlusion that he found was overjet, which occurred in 89.3% of his *Cacajao* and in 96.5% of his *Pithecia*.

Insufficient space in the region of the premolars results in the P^3 and P^4 being pushed lingually, and P^2 being rotated, with the lingual cusps turned a little mesially (Fig. 10.138). More rarely one of the premolars may be displaced from the line of the arch.

Fig. 10.133. *Alouatta caraya* (black howler). ♀. BMNH 1850.9.6.14. The arch is narrow between the premolars which are slightly lingual to the molars (Colyer, 1919).

Fig. 10.135. *Alouatta seniculus* (red howler). ♂. Juvenile. BMNH 1927.8.11.10. The premolars are erupting. The right M^2 is rotated so that its bucco-mesial angle is facing mesially.

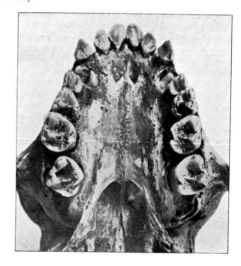

Fig. 10.134. *Alouatta fusca* (brown howler). ♀. BMNH 1852.12.9.9. Both P_4 (pointers) rotated through nearly 90° so that the buccal surfaces face distally.

Fig. 10.136. *Alouatta palliata* (mantled howler). ♀. BMNH 1903.3.1.6. Both M_3 symmetrically misplaced buccally.

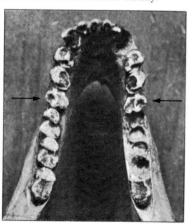

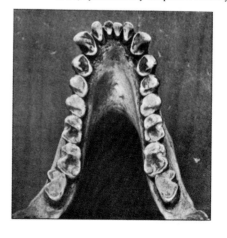

Variations in position of the molars are rare. In one skull, the left M^3 was tilted mesially and in another the right M^3 was slightly lingual to M^2.

Fig. 10.137. *Pithecia monachus* (monk saki monkey). ♂. BMNH 1927.8.11.23. The normal upper tooth arch in *Pithecia*. Enlarged slightly.

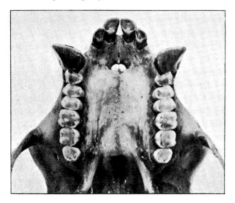

Fig. 10.138. *Pitheca pithecia* (white-faced saki monkey). ♂. BMNH 1842.4.29.8. P^3 and P^4 symmetrically misplaced lingually.

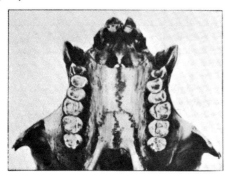

Genus *Saimiri*, squirrel monkeys

Positional variations in *Saimiri* are summarized in Table 10.17.

On the basis of a sample of 780 wild-caught squirrel monkeys, Beecher, Corruccini and Freeman (1983) concluded that incisor overjet is the prevalent form of occlusion in *Saimiri*. None of Colyer's sample of 100 skulls showed underjet (Table 10.17).

The squirrel monkeys showed considerable variation in the position of the upper incisors. The teeth in a few skulls were arranged with the labial surfaces forming a regular curve but, in the majority, I^2 were placed a little lingually to I^1 to the extent of about one-half the width of the crown of I^1. The lingual position of I^2 is usually associated with some rotation, with the mesial borders directed lingually, and may affect one side only. The three skulls in Figure 10.139 show the range of variation in the position of the upper incisors. Schultz (1935) noted crowding of the upper incisors in 2.1% of 48 *Saimiri* skulls and none in the lower incisors.

In one skull (Fig. 10.140), the right upper canine is misplaced distally and in occlusion covers the buccal surfaces of P_2 and P_3. Insufficient room for the maxillary premolars is shown by a lingual bend of the arch in the positions of P^3 and P^4 (Fig. 10.141). In a red-backed squirrel monkey (*Saimiri oerstedii*, ♀, BMNH 1903.3.1.20), the right M^3 was tilted.

Genus **Callicebus**, titi monkeys

The usual arrangement of the teeth in the titi monkey is shown in Figure 10.142. I^1 are a little in advance of I^2; in occlusion they overlap the lower teeth, I^2 meeting I_2 edge-to-edge. Sometimes there is a space between I^1 and I^2. The upper canines are not so well-developed as

Fig. 10.139. Three specimens of *Saimiri* to show the range of variation in the position of the maxillary incisors.
A: *S. sciureus* (common squirrel monkey). ♂. BMNH 1928.5.2.49.
B: *S. sciureus* (common squirrel monkey). ♂. BMNH 1928.5.2.29.
C: *S. oerstedii* (red-backed squirrel monkey). ♂. BMNH 1904.2.7.2. In (B) and (C), the I^2 are placed a little lingually to I^1.

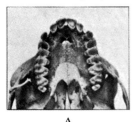
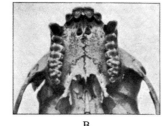
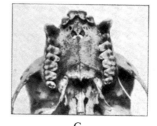

A B C

in many of the other genera of monkeys and in consequence there is only a small pre-canine diastema which is absent in about one-fifth of the animals. The upper cheek teeth form curves, the widest part of the arch being between the M^1. The buccal surface of the canine usually projects a little beyond that of the P^2 and the buccal surface of the M^1 beyond that of P^4. The mandibular teeth (Fig. 10.142) form a more regular

Fig. 10.140. *Saimiri sciureus* (common squirrel monkey). ♀. BMNH 1908.3.7.14. Right upper canine inclined distally and buccally (Colyer, 1919).

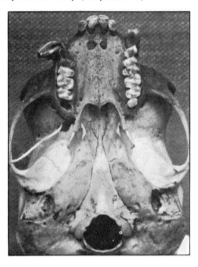
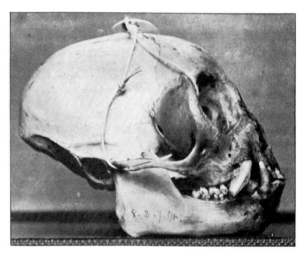

Fig. 10.141. *Saimiri sciureus* (common squirrel monkey). ♂. BMNH 1849.10.9.25. A symmetrical lingual curve in the arch in the region of P^3 and P^4.

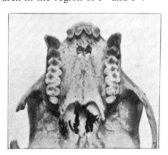

Fig. 10.142. *Callicebus moloch* (dusky titi monkey). ♀. BMNH 1908.5.9.9. The normal upper and lower tooth arches in *Callicebus*.

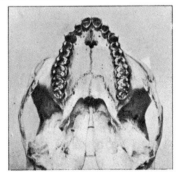

arch, the incisors being set in an even curve with sometimes a space between the I_1.

Positional variations in *Callicebus* are summarized in Table 10.17.

Incisor underjet was rare; in the only example, all the lower incisors occluded in advance of their normal positions (Fig. 10.143); the cause seemed to be defective growth of the upper jaw, so that it is an example of superior retrusion rather than inferior protrusion.

In three skulls, the labial misplacement of I^1 was marked; in two, the mandibular incisors were irregular in position.

Irregularity in the position of a mandibular canine was seen in a masked titi monkey (*Callicebus personatus*, NRM A620151); the tooth had erupted mesially to the normal position owing to the presence of an extra premolar on that side. P^2 is normally a little lingual to the line of the upper canine and other premolars (Fig. 10.144). In one skull, the left P^4 was misplaced lingually; in two skulls, there was more general irregularity of the premolars in both the upper and lower jaws (Fig. 10.145). In two others, one premolar was misplaced lingually.

Dr G. Hagmann gave Colyer the details of the 42 specimens of dusky titi monkeys (*Callicebus moloch*) in his collection from the neighbourhood of Taperinha, Santarèm, Brazil. They showed the high variability at times seen in animals from the same locality. In four,

Fig. 10.143. *Callicebus moloch* (dusky titi monkey). ♀. BMNH 1908.5.9.8.
Left lateral view. Apart from the incisor underjet, the lower cheek teeth on both sides are in pre-normal occlusion by the width of a premolar. Note that P_2 on the left occludes in front of the upper canine. × *c*. 1.25.
The view of the base of the skull shows slight asymmetry of the upper arch, with slight buccal bowing of the right post-canine row and lingual misplacement of both M^3.
The profile of the premaxilla is less sloping than normal and it is therefore possible that the basic defect is in the size of the upper jaw; that is, this is superior retrusion rather than inferior protrusion (Colyer, 1919).

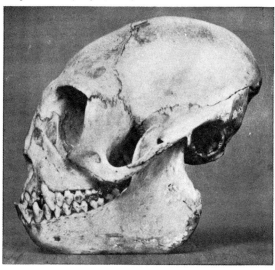
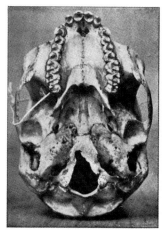

Fig. 10.144. *Callicebus torquatus* (widow monkey). ♀. BMNH 1925.12.11.8. The upper arch is narrow, mainly because of symmetrical lingual misplacement of P^3. The post-canine mandibular tooth rows are rather straight. Compare with Figure 10.142.

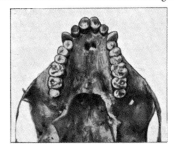
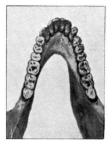

232 Section 2: Variations in Position

the incisors were in abnormal positions; two in the mandible, one in the maxilla. In one skull, there were positional abnormalities among both the upper and lower incisors and of the left P^2. In two specimens, the M_1 were a little beyond the line of the other teeth. The deciduous canine persisted in the mandible of an adult female, the crown of the permanent tooth having emerged to the lingual aspect of the deciduous tooth.

Genus *Aotus*, douroucouli monkeys

The genus *Aotus* shows little tendency to vary in tooth position (Table 10.17) and then mostly in the incisors. The upper incisors normally are arranged in a curve, but it is not unusual to find I^2 placed a little lingually to I^1; this misplacement may be marked. Overlapping or crowding of the upper or lower incisors occurs, and sometimes of both (Fig. 10.146).

The only irregularity in position of the premolars was rotation of P^2 with the lingual cusp turned distally. In one skull, both M^3 were placed obliquely to the line of the arch; in another, the right M^2 and M^3 were lingual to the arch.

Fig. 10.145. *Callicebus moloch* (dusky titi monkey). ♀. BMNH 1926.5.5.15. Upper tooth arch narrow and crowded especially mesially; also general asymmetry. Note the relative positions of the M^3 with similar, but less severe, narrowness and crowding of the lower arch.

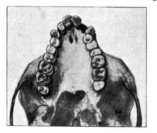
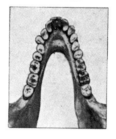

Family **Callitrichidae**. Marmosets

Positional variations in the *Callitrichidae* are summarized in Table 10.20.

I^1 are normally a little labial to I^2 and in occlusion overlap the mandibular teeth. There is considerable variation in the relative positions of I^1 and I^2; they may differ on the two sides or I^2 may be so lingual as to constitute an abnormality (Fig. 10.147).

In the mandible of a saddle-back tamarin (*Saguinus fuscicollis*) (Fig. 10.148), on the right, M_2 is a little more distal than the left M_2 and between the M_1 and M_2 there is a gap. M^1, the premolars and the upper canine are more mesial than the corresponding teeth on the left and occlude in advance of their normal positions. This mesial movement has caused the incisors to be deflected to the left and the right canine to strike the buccal aspect of I^2.

Irregularity in position of the premolars is uncommon. In one skull (Fig. 10.149), the left P^2 was rotated, the lingual cusp facing mesially; in another (Fig. 10.150), there was extreme rotation of the left P^3, the buccal surface facing directly distally; in three skulls, the premolars were crowded.

An example of positional anomaly of the molars is shown in Figure 10.151.

Sub-order **PROSIMII**

Family **Indriidae** (indris and sifakas); family **Lemuridae** (lemurs); family **Lorisidae** (lorises, galagos, pottos and bushbabies); family **Cheirogaleidae** (mouse lemurs); family **Daubentonidae** (aye-ayes); family **Tarsiidae** (tarsiers)

When three premolars are present, they will be designated as P2, P3 and P4; when two are present, as

Fig. 10.146. *Aotus trivirgatus* (douroucouli monkey). BMNH 1914.4.25.30. Slight crowding of the upper incisors, with some overlapping of the I^1, and more marked crowding of the lower incisors, the left I_2 being misplaced labially (Colyer, 1919).

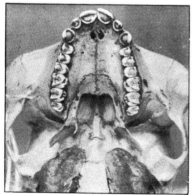
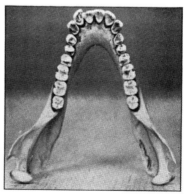

Table 10.20 Positional variations in the CALLITRICHIDAE

No. of specimens	Class of tooth				Total specimens affected		Incisor underjet	
	I	C	P	M	No.	%	No.	%
211	7	2	5	7	20	9.5	0	0

In skulls in which two or more classes of teeth varied in position, each class is recorded individually; for example, if in a skull one or more of the incisors, premolars and molars varied, each appropriate class would be credited with one and the column 'total specimens affected' with one only.

Fig. 10.147. Three specimens to show the range of variation in the position of the upper incisors in the *Callitrichidae*. In all three, the I^2 are slightly lingual to I^1 and in (C) the incisors are spaced. Enlarged slightly.

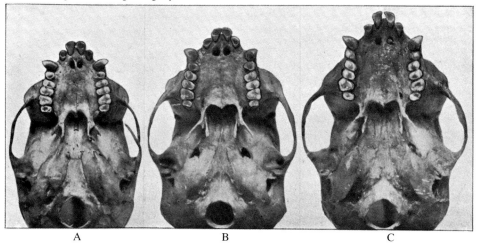

A B C

Fig. 10.148. *Saguinus fuscicollis leucogenys* (saddle-back tamarin). ♂. BMNH 1900.7.7.2. There is incisor underjet and pre-normal occlusion of the lower cheek-teeth on the right. Note from the lateral view that P^4 occludes with the middle of M_1 instead of between M_1 and P_4. The occlusal view of the lower arch shows some asymmetry with the right canine and all the cheek-teeth, apart from M_2 which has been lost from disease, are more mesial than the left. The left M_3 is absent. In occlusion, the midline of the lower arch was to the left (Colyer, 1919).

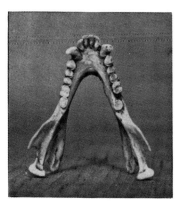
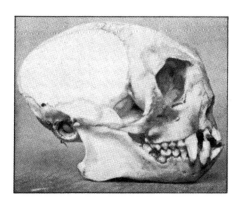

P3 and P4 (see Chapter 3, pp.59–61). The normal upper tooth arch in the Indriidae is shown in Figure 10.152. The buccal surfaces of the two premolars form a straight line but there is a sharp lingual bend in the position of M^3. In the mandible, both P_4 are usually slightly rotated so that they are slightly overlapped buccally by M_1.

Variations in the position of the teeth of pro-

Table 10.21 Positional variations in PROSIMIANS

Family	Sub-family	No. of specimens	Total specimens affected No.	%
Indriidae	—	90	17	18.9
Lemuridae	—	189	32	16.9
Lorisidae	Galaginae	241	10	4.1
Lorisidae	Lorisinae	112	17	15.2
Tarsiidae	—	9	1	11.1
Daubentoniidae	—	9	0	0.0
Total		**650**	**77**	**11.8**

In the Indriidae, skulls showing only rotation of P_4 have not been included because that condition appears to be normal.

Fig. 10.149. *Saguinus midas tamarin* (negro tamarin). ♂. BMNH 1904.7.4.17. Left P^2 slightly rotated clockwise.

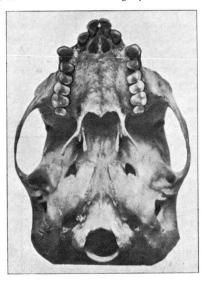

Fig. 10.150. *Cebuella pygmaea* (pygmy marmoset). USNM 20228. Left P^3 misplaced lingually and rotated by about 45° so that the buccal surface faces distally. Enlarged slightly.

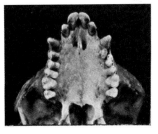

Fig. 10.151. *Saguinus fuscicollis leucogenys* (saddle-back tamarin). ♂. BMNH 1927.11.1.23. Upper arch more or less symmetrically irregular and narrow, particularly between the P^4. Both M^2 lingually misplaced. × c. 1.5.

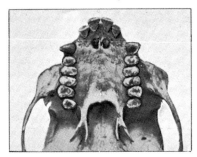

Fig. 10.152. *Propithecus diadema* (diadem sifaka). Captive. BMNH 1855.12.24.58. The normal upper tooth arch in the Indriidae. There are two premolars in each quadrant.

simians are summarized in Table 10.21. Most variations were slight and there were none with complete misplacement of any tooth.

Fig. 10.153. *Propithecus verreauxi coquereli* (Coquerel's sifaka). ♂. BMNH 1891.1.22.1. The buccal surfaces of P³ (pointer) are turned slightly distally.

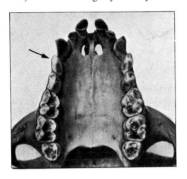

Fig. 10.154. *Lemur macaco mayottensis* (Mayotte Island lemur). ♂. BMNH 1906.6.3.12. Right P⁴ rotated with its buccal surface turned mesially.

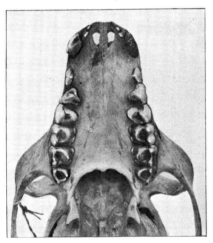

Fig. 10.155. *Galago crassicaudatus* (greater bushbaby). ♂. BMNH 1909.3.2.1. Upper premolars crowded and overlapping with some rotation of both P³ with buccal surfaces turned distally.

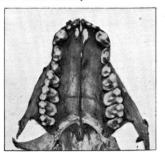

(a) Premolars

Variations in the positions of the maxillary premolars took the form of slight rotation of the teeth (Fig. 10.153). The Indriidae have short broad muzzles and, even though there are only two premolars in each quadrant and therefore crowding of the arches is uncommon, variations in the position of these teeth are frequent.

In the Lemuridae, slight rotation of the three premolars was present in about a quarter of the specimens of *Lepilemur* and *Haplemur* and in one-fifth of those of *Lemur*, but it may be marked (Fig. 10.154).

Fig. 10.156. *Nycticebus coucang borneanus* (slow loris). BMNH 1888.2.9.1. Right P⁴ misplaced lingually, and both M³ rotated with the buccal surfaces facing slightly mesially.

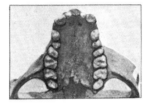

Fig. 10.157. *Perodictus potto* (potto). BMNH 1882.1.24.1. Both P⁴ misplaced slightly lingually with the buccal surfaces turned mesially. Right P³ misplaced buccally.

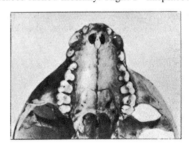

Fig. 10.158. *Propithecus verreauxi* (Verreaux's sifaka). ♀. BMNH 1925.4.10.2. The M³ are misplaced slightly lingually.

Anomalies in the positions of the premolars were present in about 20% of the Lorisidae; some examples are shown in Figures 10.155, 10.156 and 10.157.

(b) Molars

M^3 may be rotated (Fig. 10.156) or misplaced lingually (Fig. 10.158).

(c) The effects of captivity

Lemurs in the captive state seem particularly liable to develop bone disorders; these are often associated with irregularities in the position of the teeth:

(i) *Lemur mongoz cornatus* (crowned mongoose lemur). (Fig. 10.159). Both P^2 and both P^4 slightly rotated.

(ii) *L. catta* (ring-tailed lemur). (Fig. 10.160). Both P^3 are rotated. In the mandible, both m_2 are still present and P_2 have erupted distal to them; all the premolars are symmetrically rotated with buccal surfaces facing distally; they also overlap, an indication of overcrowding, perhaps secondary to the presence of additional teeth, namely the retained m_2 in the arch. The region of the angles of the mandible and the condyles is greatly thickened and distorted.

(iii) *L. catta*. (Fig. 10.161). The bones show thickening but are of dense texture. The left m^2 and both m_2 and m_3 are still present. The left P^2 is unerupted

Fig. 10.159. *Lemur mongoz coronatus* (crowned mongoose lemur). Captive. RCS Odonto. Mus., G 69.41. Both P^2 are slightly rotated; on the left with the buccal surface turned distally, on the right turned mesially. Both P^4 are symmetrically slightly rotated and misplaced lingually.

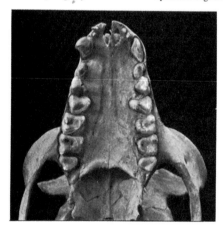

Fig. 10.160. *Lemur catta* (ring-tailed lemur). Captive. RCS Odonto. Mus., G 51.4731. Both P^3 are rotated with the buccal surfaces facing distally.

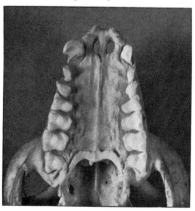

Fig. 10.161. *Lemur catta* (ring-tailed lemur). Captive. RCS Odonto. Mus., G 51.473.

A: The right P^3 slightly overlaps the buccal surface of P^2. The left m^2 is still present with the P^2 unerupted. The left P^3 is rotated with buccal surface facing mesially, P^4 is only partially erupted and M^3 is rotated with its buccal surface facing distally. The right M^3 is unerupted and embedded in what seems to be unusually dense bone.

B: The mandible. On the left, P_3 has erupted disto-lingually to m_3 which is still present; on the right, where bone has been removed to display the teeth, P_3 remains unerupted beneath m_3.

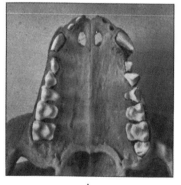

A

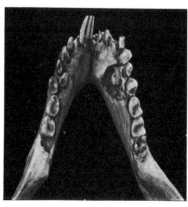

B

above m^2 and appears to be ankylosed; P^4 is only partially erupted and M^3 rotated; the right M^3 is unerupted. In the left mandible, P_3 has erupted distally to the retained m_3; on the right, P_3 remains unerupted beneath m_3. This appears to be mainly a disorder of eruption including shedding of the deciduous teeth at the proper time and may well be secondary to some bone disorder.

(iv) *Lemur macaco macaco* (black lemur). (Fig. 10.162). The bone, more particularly of the maxillae, is thickened and is very dense in character, which would be consistent with healed rickets. There is considerable crowding and misplacement of the upper premolars. In the mandible, the premolars are irregular in position and both M_1 are unerupted.

Fig. 10.162. *Lemur macaco macaco* (black lemur). Captive. RCS Odonto. Mus., G 69.43.
A: The upper arch is narrow and the arrangement of the post-canine teeth is disordered, predominantly by lingual misplacement of both P^4.
B: The lower tooth arch is also irregular, e.g. the left P_2 is rotated so that the lingual surface faces mesially, but it is more spaced than the upper arch because the M_1 are not erupted.
C: The right M_1 deeply embedded in the bone; a window has been cut to display the tooth. The enamel of several teeth, particularly P^4, is hypoplastic.

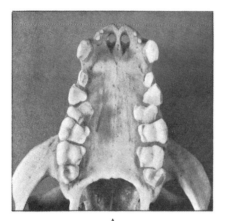

A

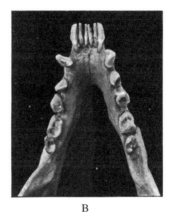

B

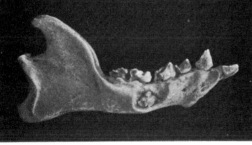

C

CHAPTER 11

Order **CARNIVORA**

In most of the carnivores, the snout is long and narrow but, in the Felidae, which is the most specialized of the carnivore families, the face, including the palate and the upper and lower jaws, is broad and short; the number of molars is reduced to one in each quadrant. Positional variations in the various carnivore families are summarized in Table 11.1.

Family **Felidae**. 1236 specimens examined

In the normal arrangement of the teeth in Felidae, the upper incisors are in a nearly straight line and the buccal surfaces of the four upper post-canine teeth form straight lines which diverge sharply; the minute M^1 (Fig. 11.1) is situated against the lingual surface of the distal end of P^4. The general shape of the mandibular arch is similar, but the mesial surface of the carnassial M_1 is oblique and is overlapped buccally by P_4.

The more common variations in position are (i) irregularity or crowding of the mandibular incisors, with I_3 misplaced labially; (ii) rotation of P^3 and less frequently of P^2; (iii) increase in the overlap of M_1 by P_4. The variations found in the various species of felids are summarized in Table 11.2.

Table 11.2 shows that irregularities in position are more common in lions (*Panthera leo*) than in other members of genus *Panthera*. A feature of lions is that the upper premolars may either be in contact or slightly spaced. The amount of spacing can be measured by calculating the length of the upper cheek teeth as a percentage of the length of the palate. The lower the percentage the greater the amount of space. Table 11.3 shows that the amount of spacing tends to be less in female lions than in males and that this is also true among lions from the same geographical area, but there is some overlap of the ranges of variation. Figure 11.1 shows different degrees of spacing in two females from the same area of Uganda.

Variations in tooth position are uncommon in the tiger (*Panthera tigris*) (Table 11.2). Figure 11.2 shows the mandible of a tigress, probably captive, in which the lower left canine is situated much more mesially than normal, displacing I_3 labially, so that it occludes mesially to I^3. The upper left canine is also situated more mesially than normal so that the pre-canine diastema is narrow. The primary cause of the disorder

Table 11.1 Positional variations in CARNIVORE families from the wild

Family	No. of specimens	Class of tooth				Total specimens affected	
		I	C	P	M	No.	%
Felidae	1236	39	0	40	0	79	6.4
Hyaenidae	133	0	0	20	0	20	15.0
Viverridae	1566	45	4	49	23	123	7.9
Mustelidae	1746	135	0	218	0	324	18.6
Procyonidae	355	6	1	60	5	67	18.9
Ursidae	322	12	0	19	9	38	11.4
Canidae	1478	14	0	91	10	106	7.2

In skulls in which two or more classes of teeth varied in position, each class is recorded individually; for example, if in a skull one or more of the incisors, premolars and molars varied, each appropriate class would be credited with one and the column 'total specimens affected' with one only.

Table 11.2 Positional variations in FELIDS from the wild

Genus and species	No. of specimens	Class of tooth				Total specimens affected	
		I	C	P	M	No.	%
Panthera leo	127	8	0	11	0	19	15.0
P. tigris	90	2	0	0	0	2	2.2
P. pardus	109	3	0	6	0	9	8.3
P. onca	22	0	0	0	0	0	0
P. nebulosa	9	2	0	0	0	2	22.2
Total *Panthera*	**357**	**15**	**0**	**17**	**0**	**32**	**9.0**
Acinonyx jubatus	17	0	0	0	0	0	0
Felis lynx	64	4	0	0	0	4	6.3
F. concolor	52	1	0	1	0	2	3.8
F. serval	41	0	0	0	0	0	0
F. bengalensis	81	0	0	1	0	1	1.2
F. marmorata	16	1	0	0	0	1	6.3
F. aurata	22	1	0	0	0	1	4.5
F. viverrinus	19	0	0	0	0	0	0
F. pardalis	89	6	0	8	0	14	15.7
F. yagaouarondi + *F. geoffroyi*	88	2	0	2	0	4	4.5
F. guigna	19	1	0	1	0	2	10.5
Felis (other spp.)	371	8	0	10	0	18	4.9
Total *Felis*	**862**	**24**	**0**	**23**	**0**	**47**	**5.4**
Total FELIDAE	**1236**	**39**	**0**	**40**	**0**	**79**	**6.4**

appears to have been that the lower canine erupted too far mesially and so was caught in front of the caniniform I^3. There was no evidence of injury.

A male tiger from the wild that had suffered a spectacular misplacement of a canine, described by Colyer (1951a) is shown in Figure 11.3. The upper right canine has erupted just to the right side of the midline of the palate and the crown lies obliquely across the left palate. The lateral surface of the crown would have been in contact with the dorsum of the tongue and, in consequence, is so much worn that the pulp cavity is exposed. The root apex is more or less in its normal position immediately beneath the facial surface of the maxilla; the main body of the root can be seen through the anterior nares crossing the nasal cavity just above its floor. P^2 is rotated through nearly 90° so that its buccal surface faces mesially. Mesially to it in the normal position of a canine is a shallow, but wide, empty socket. There has evidently been an abscess around the end of the root of the permanent canine with much reactive thickening of the bone on the facial aspect of the maxilla. The area of affected bone around the root extends to the empty socket with a probable sinus opening into it.

One interpretation is that the canine was misplaced lingually in development, as canines sometimes are, and consequently erupted obliquely into the palate; once its erupted tip came within the action of the tongue, the tongue would have tended to increase its obliquity, pressing it to the left. Lacking its permanent successor in the normal position, the deciduous tooth, it is supposed, remained long after the normal time of shedding until ultimately it became involved in

Fig. 11.1. *Panthera leo* (lion). Above: ♀. BMNH 1903.11.7.31. Normal upper tooth arch with the cusps of the teeth in more or less straight rows and slight spaces between the teeth. ×c.0.5.
Below: ♀. BMNH 1903.11.7.32. Upper jaw of another animal from the same region of Uganda. P^1 and P^2 are slightly rotated and P^1 slightly overlaps the buccal surface of P^2. ×c.0.5.

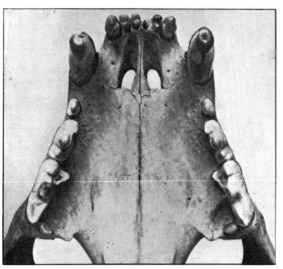

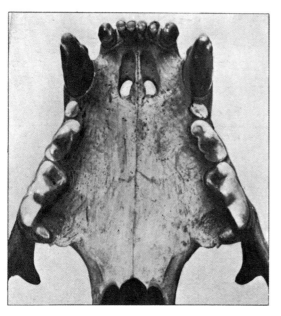

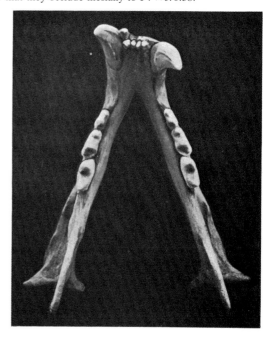

Fig. 11.2. *Panthera tigris* (tiger). ♀. RCS Odonto. Mus., G70.14. Lower left canine and I_3 misplaced mesio-labially so that they occlude mesially to I^3. ×c.0.38.

the abscess that developed around the root of the misplaced canine. The abscess would be an obvious consequence of the exposure of the pulp by wear. There are some features that this hypothesis does not readily account for. The misplaced canine is not fully erupted; its cervix can be seen about 1 cm above the floor of the nasal cavity. The root above that level is bare of bone; however, there may originally have been a thin layer of covering bone that was destroyed by the abscess. Another strange feature is that the pulp appears to have been exposed without the formation of reactionary dentine.

An alternative interpretation, which was favoured by Colyer (1951a), is that the canine had been displaced by trauma after it had erupted and was fully formed (the root is not distorted); the trauma had fractured the tip of the crown, exposing the pulp with the abscess as a consequence. The shallow socket being the one that was originally occupied by the canine, its healing and obliteration were perhaps delayed by the suppuration that involved it. P^2 was displaced at the same time.

In the felids, P^2 is liable to be rotated (Fig. 11.4).

It is difficult to assess the effects of captivity on felids for so much depends upon the age the animal is brought into captivity. Young lions and leopards are liable to develop skeletal disease which may affect growth of the jaws and thus the position of the teeth (Fig. 11.5) (pp. 333–334, 543, 603, 604). Abnormalities of tooth position are certainly more common in captivity than in the wild state (Table 11.4).

There was one example of irregularity of the teeth in a domestic cat (Fig. 11.6). The mandibular canines were inclined anteriorly so that they projected beyond the upper canines.

Family **Hyaenidae**. 133 specimens examined

The normal arrangement of the upper teeth in the brown hyaena (*Hyaena brunnea*) is shown in Figure

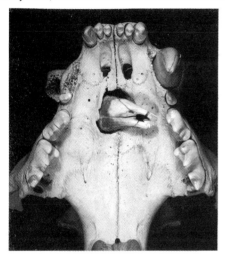

Fig. 11.3. *Panthera tigris* (tiger). ♂. RCS Odonto. Mus., G 98.111. The upper right canine has erupted transversely in the palate and has become worn by contact with the rough dorsum of the tongue so that the pulp is exposed. The empty socket in the normal position of the right canine is healing and probably is the socket of the recently-shed, retained deciduous canine. × c. 0.42.

Table 11.3 The amount of upper cheek-tooth spacing in male and female LIONS *(Panthera leo)* as shown by the relationship between the length of the teeth and the length of the palate

Sex	No.	Mean length of palate (A)	Sum of mean length of cheek teeth (B)	B × 100/A Mean	Range
male	21	133.8	73.5	54.9	46.5–58.8
female	21	113.7	67.4	59.3	53.8–64.3
Kenya					
Male	5	—	—	55.8	53.4–57.2
female	4	—	—	58.7	55.0–62.2
Mantemutonda river, Transvaal					
male	5	—	—	52.5	46.5–55.3
female	6	—	—	56.8	53.8–59.1

In felids, the palatal length was measured from the anterior surface of the left and right I^1 instead of the prosthion. All measurements in mm.

11.7. The positions of the premolars are variable; they may be spaced or in contact, but both conditions can be regarded as within the normal range. In the mandible, there is a space between the P_2 and the lower canine, the premolars are in contact and P_3 and P_4 are usually placed obliquely producing an overlapping echelon arrangement.

The only variations in position seen in the hyaenas were of the premolars (Table 11.1), most commonly rotation of P^2 and less often of P^3 (Fig. 11.8) and of P_2 and P_3, P_2 tending to be misplaced buccally or lingually. They were more common in *Hyaena* than in *Crocuta*.

A marked example of crowding and impaction of premolars was in a sub-adult spotted hyaena (Fig. 11.9).

Family **Viverridae**. 1566 specimens examined

In this family, the palate is narrower and longer than in the Felidae. The normal arrangements of the upper teeth are shown in Figures 11.10, 11.11 and 11.12. In the small Indian civet (*Viverricula*), the incisors form a slight curve. The six upper post-canine teeth as far as the distal margin of M^1 form a curve with the concavity buccally, but M^2 and M^3 form a line passing lingually at a sharp angle. There are spaces between the premolars (Fig. 11.10). In the palm civet (*Paradoxurus*), the incisors form a flatter curve; the premolars are closer together and the lingual bend of the arch is not so abrupt (Fig. 11.11). In the mongoose (*Herpestes*), the

Fig. 11.4. *Panthera pardus* (leopard). ♀. BMNH 1927.2.11.10. Left P^2 rotated about 90° with the buccal surface facing mesially. Reduced.

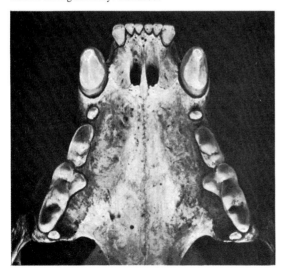

incisors are set in a straight line; the premolars are close together and there is little curvature in the line formed by them; the lingual bend of the upper arch is marked (Fig. 11.12).

In the mandible, the lines of the cheek teeth are straight and diverge buccally. However, there are a number of differences between the genera. In *Viverricula* and *Herpestes*, the M_3 are situated a little more lingually than the other cheek teeth.

The incidence of variations in the position of the teeth in the *Viverridae* is shown in Tables 11.1 and 11.5.

Fig. 11.5. *Panthera leo* (lion). Juvenile. Captive. RCS Odonto. Mus., G 46.1. The thickening of the jaws with poorly mineralized bone suggestive of rickets has led to the malposition of several teeth. However, the upper incisors are well spaced. The permanent canines have erupted alongside the deciduous ones. × c. 0.36.

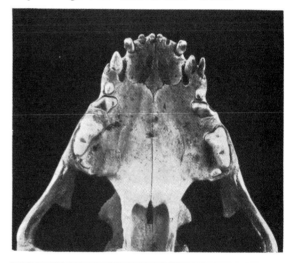

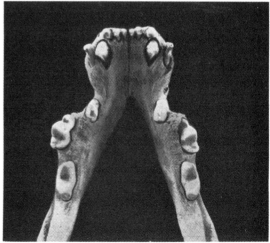

(a) Incisors

Incisor overjet, which is rare amongst mammals from the wild, was seen in a mongoose (Fig. 11.13). The overjet is marked and involves the left canines but the rest of the occlusion is normal. Not only does the premaxilla project beyond the mandible but it is asymmetrical and seems disproportionately large. This is borne out by the measurement of the cranium length which is 7% longer than the mean of 20 normal skulls whereas mandible length is only 0.4% longer.

Table 11.4 Positional variations in four species of FELIDS showing the differing incidence in animals from the wild and captive states

Genus and species	Wild state			Captive state		
	No. of specimens	Total specimens affected No.	%	No. of specimens	Total specimens affected No.	%
Panthera leo	127	19	15.0	23	5	21.7
P. pardus	109	9	8.3	27	6	22.2
Acinonyx jubatus	17	0	0.0	10	6	60.0
Felis pardalis	48	6	12.5	9	3	33.3

Specimens with gross bone changes have been omitted.

Table 11.5 Positional variations in the two main sub-families of VIVERRIDAE

Family and sub-family	No. of specimens	Class of tooth				Total specimens affected	
		I	C	P	M	No.	%
Viverrinae	640	8	—	17	21	45	7.0
Herpestinae	908	35	4	32	7	76	8.4

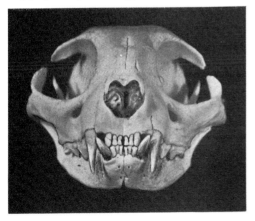

Fig. 11.6. *Felis catus* (domestic cat). RCS Odonto. Mus., G 70.4. Lower canines directed obliquely forwards so that they project beyond the upper canines. *c.* Natural size.

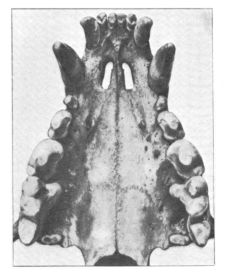

Fig. 11.7. *Hyaena brunnea* (brown hyaena). BMNH 1866.5.2.1. The normal upper tooth arch in the genus *Hyaena*. Minute M^1 are situated lingually to P^4. $\times c. 0.5$.

244 Section 2: Variations in Position

Irregularity in the position of the incisors was confined to the mandibular teeth. I_2 may be lingual to the line of the other teeth or I_3 may project labially.

(b) Canines

Positional abnormalities of the canines are rare. In one specimen, the right maxillary canine was misplaced distally and buccally; in three, the mandibular canines had erupted mesially to their normal positions so that in occlusion they passed between I^2 and I^3 (Fig. 11.14).

Fig. 11.8. *Hyaena hyaena* (striped hyaena). BMNH 1905.5.28.2. Both P^2 are rotated with the buccal surfaces facing mesially, the right more than the left. $\times c. 0.5$.

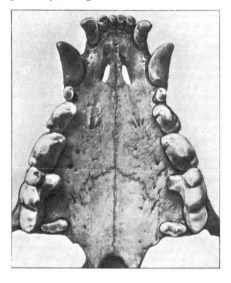

Fig. 11.9. *Crocuta crocuta* (spotted hyaena). Sub-adult. BMNH 1930.3.6.4. The right P^3 and P_3 are not fully erupted; P_3 is misplaced buccally. $\times c. 0.5$.

Mesial or labial misplacement of the canine usually leads to crowding of the incisors (Fig. 11.15).

(c) Premolars

The most common positional variation of the maxillary premolars is rotation (Fig. 11.16) which at times amounts to about 90° so that the buccal surface is brought against the mesial surface of P^4. More rarely,

Fig. 11.10. *Viverricula indica* (small Indian civet). ♂. BMNH 1907.7.20.8. The normal upper tooth arch in the genus *Viverricula*. $\times c. 1.33$.

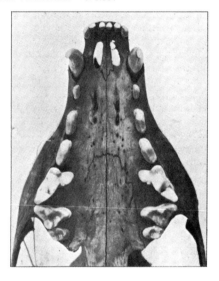

Fig. 11.11. *Paradoxurus hermaphroditus* (common palm civet). ♀. BMNH 1879.11.21.280. The normal upper tooth arch in the genus *Paradoxurus*. The incisors form a flatter curve than in *Viverricula*, the premolars are closer together and the molar rows do not bend lingually so abruptly.

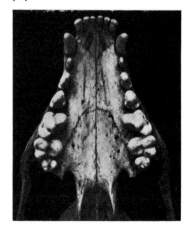

P^2 or even P^1 may be rotated. Other types of misplacement of premolars (Fig. 11.17) occur but are rare.

In the marsh mongoose (*Atilax*), a slight degree of crowding of the premolars may be regarded as normal, but was occasionally more marked though symmetrical (Fig. 11.18).

P_4 is usually slightly rotated; occasionally rotation is severe and unilateral (Fig. 11.19). Marked misplacement of P_4 is rare; an example is shown in Figure 11.20.

Fig. 11.12. *Herpestes ichneumon* (Egyptian mongoose). BMNH 1919.7.7.3492. The normal upper tooth arch in the genus *Herpestes*. The incisors form a straight line, the cheek teeth are close together and curve outwards only slightly; however, M^2 and M^3 are in a line directed abruptly lingually. Enlarged slightly.

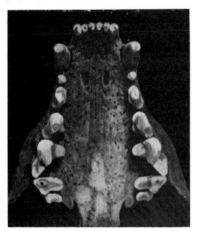

Fig. 11.13. *Ichneumia albicauda* (white-tailed mongoose). BMNH 1929.5.28.2. Marked incisor overjet and the lower left canine and P_1 are in post-normal occlusion. However, the rest of the occlusion is normal. The premaxilla is asymmetrical and seems unusually large so that, although there is crowding of the incisors in both jaws, the main cause of the malocclusion probably was in the upper jaw. × *c*. 0.66.

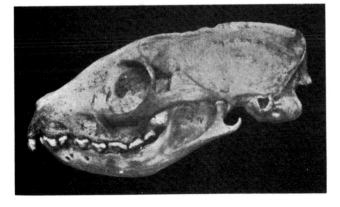

Fig. 11.14. *Herpestes sanguineus cauui* (slender mongoose). ♂. BMNH 1904.5.1.33. Lower right canine misplaced mesially so that it occludes mesially to I^3. × *c*. 1.33.

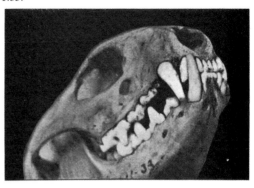

Fig. 11.15. *Herpestes ichneumon* (Egyptian mongoose). ♂. BMNH 1921.9.6.3. Incisor underjet and the lower left canine is misplaced so that it occludes mesially to I^3 which it has displaced into the pre-canine diastema. The left lower cheek teeth are also in slightly pre-normal occlusion. The whole of this cranium is deformed as if bent in the sagittal plane. Enlarged slightly.

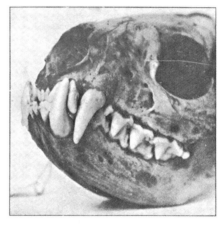

Fig. 11.16. *Herpestes ichneumon* (Egyptian mongoose). ♀. BMNH 1898.6.5.5. Symmetrical rotation of both P³, with the buccal surfaces facing mesially. *c*. Natural size.

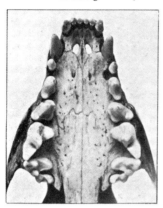

Fig. 11.17. *Genetta tigrina stuhlmanni* (Stuhlmann's large-spotted genet). ♂. BMNH 1907.4.6.12. Right P² misplaced distally so that it is partly lingual to P³. Natural size.

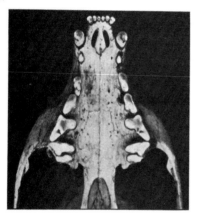

Fig. 11.18. *Atilax paludinosus* (marsh mongoose). ♂. BMNH 1909.3.2.14. Symmetrical crowding with overlapping of the upper premolars. The left canine has split following skull preparation.

In about 10% of the specimens of Herpestinae showing variations of the premolars, both the upper and the lower teeth were affected, sometimes asymmetrically so.

(d) Molars

The molars are sometimes rotated. In *Bdeogale*, some degree of rotation of M² was present in four out of 15

Fig. 11.19. *Viverricula indica* (small Indian civet). ♂. BMNH 1912.3.4.1. Lateral views of the hemi-mandibles. The left P₄ is misplaced buccally and distally and slightly overlaps M₁. Enlarged slightly.

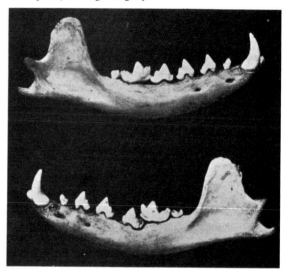

Fig. 11.20. *Paradoxurus jerdoni* (Jerdon's palm civet). ♀. BMNH 1913.8.2.24. Right P₄ misplaced buccally and out of occlusion.

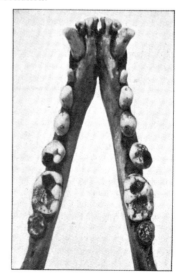

specimens; an example in *Bdeogale* is shown in Figure 11.21 and in *Rhynchogale* in Figure 11.22.

Rotation of M_2 was seen in *Paguma*, *Paradoxurus* (Fig. 11.23) and *Arctogale*. Figure 11.24 shows an example of bilateral misplacement of M_2 in *Paguma*.

(e) Spacing of cheek teeth

There are differences in the amount of spacing of the cheek teeth in three genera of viverrids (Table 11.6); it is least marked in *Herpestes*. The Table also shows the wide variation in the absolute length of the palate and the cheek teeth.

(f) Effects of captivity

In the Viverridae from the wild state, 7.9% showed variations in position (Table 11.1); of 91 specimens from the captive state, 23 (25.3%) varied.

There is a marked tendency to crowding of the mandibular incisors and tilting of P_4.

As defective growth of the jaws is probably the cause of the crowding, M2, being the last cheek tooth

Fig. 11.23. *Paradoxurus hermaphroditus kangeanus* (Kangean Island palm civet). ♂. BMNH 1910.4.6.37. Symmetrical rotation of M_2 with buccal surfaces facing mesially. Enlarged slightly.

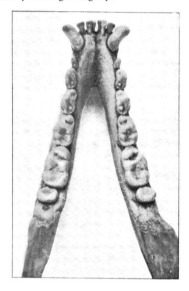

Fig. 11.21. *Bdeogale nigripes* (black-legged mongoose). ♂. BMNH 1897.7.1.5. Symmetrical rotation of both M^2 with buccal surfaces facing mesially.

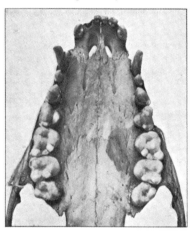

Fig. 11.24. *Paguma larvata ogilbyi* (Ogilby's masked palm civet). ♂. BMNH 1890.6.23.1. Both M_2 misplaced buccally.

Fig. 11.22. *Rhynchogale melleri* (Meller's mongoose). ♂. BMNH 1923.11.15.1. Rotation of both M^2 with buccal surfaces turned mesially.

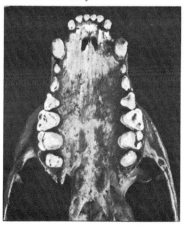

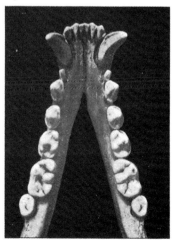

Fig. 11.25. *Viverra civetta* (African civet). Formerly RCS Osteo. Series, 542. Both M^2 tilted mesially and impacted under the distal convexity of M^1.
In the lateral view of the left mandible, M_2 is tilted lingually and has remained partially erupted. P_4 is rotated, tilted distally and impacted against M_1.

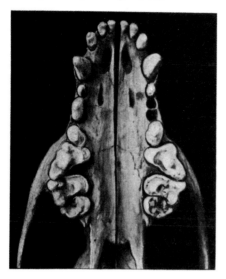

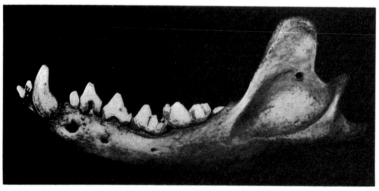

to erupt, is the tooth most commonly misplaced (Figs. 11.25 and 11.26).

In the white-tailed mongoose in Figure 11.27, there is a wide space between the upper left P^2 and P^3 which has not disturbed the occlusion a great deal because there is some more or less corresponding, and probably secondary, spacing between the lower premolars. There is no evident cause for this space; for instance no buried supernumerary tooth is present. It could be that there was an erupted supernumerary which was lost during life or there could have been a soft tissue tumour that held the teeth apart. The alveolar process in the space is a little pitted and rough.

Family **Mustelidae**. 1746 specimens examined

In most mustelid genera, the muzzle is shorter than in the Viverridae and there are marked differences between the genera in the arrangement of the premolars.

Irregularities of the lower incisors and variations in the position of the other teeth of mustelids are

Table 11.6 The amount of upper cheek-tooth spacing three genera of VIVERRIDS as shown by the relationship between the length of the palate (A) and the length of the cheek teeth (B).

Genus	No. of specimens	Length of palate (A)		Length of cheek teeth (B)		B×100A	
		Mean	Range of variation (%)	Mean	Range of variation (%)	Mean	Range of variation
Ichneumia	15	46.7	25.0	32.6	13.1	69.8	65.7–74.3
Atilax	15	44.9	21.2	32.6	18.6	73.3	65.6–79.1
Herpestes	15	41.1	15.6	32.6	12.2	79.3	71.8–84.7

All measurements in mm.

summarized in Tables 11.1, 11.7 and 11.8. Extreme malposition of individual teeth is rare.

(a) Incisors

An irregular arrangement of the lower incisors is common, the frequency differing in the various genera (Table 11.7) and in different species of the same genus

Fig. 11.26. *Paradoxurus hermaphroditus* (common palm civet). Captive? BMNH 1884.6.3.21. Both M_2 tilted mesially and impacted against M_1.

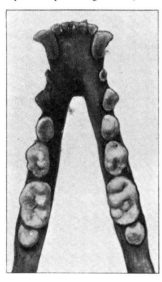

Fig. 11.27. *Ichneumia albicauda* (white-tailed mongoose). Captive. RCS Odonto. Mus., G 73.131. Left P^2 and P^3 rotated, P^3 with its buccal surface facing mesially, and there is a 3 mm space between P^3 and P^4; P^4 and the two left molars are situated more distally than on the right. There is slight corresponding spacing of the lower premolars so that the occlusion is more or less normal. Natural size.

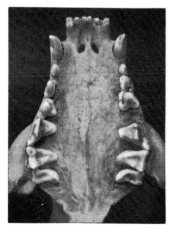

(Table 11.8). In some taxa, the irregularity is so common that it must be regarded as normal. Long and Long (1965) examined the dentitions of 900 American badgers (*Taxidea taxus*) from the wild, two of which had malpositioned lower incisors. In one (USNM 157833), the right I_3 was next to I_1 and I_2 was misplaced lingually to I_3; this may have been related to the partial eruption of the lower right canine. In the other (USNM 221782), both I_2 were lingual to I_1 and I_3.

Crowding of the incisors is often associated with crowded premolars (Fig. 11.28).

In a female hog-badger (*Arctonyx*. Fig 11.29), the mandible protruded well beyond the premaxillae. Measurements showed that the glenoid-incisor length was 9.7% less and the length of the mandible 3.5% less than the mean of three females with normal occlusion, so that the defect here seems to be in the growth of the premaxilla or maxillary complex.

Another example of underjet was in a ferret (Fig. 11.30). The lower post-canine teeth were also in slightly pre-normal occlusion, but the cause of the malocclusion is almost certainly the smallness of the

Fig. 11.28. *Lutra felina* (South American otter). Captive. BMNH 1856.11.10.5. In the upper jaw, the premolars are crowded and misplaced inwards so that both P^4 are overlapped slightly by M^1. The incisors are slightly crowded. The lower incisors are crowded with lingual misplacement of I_2.

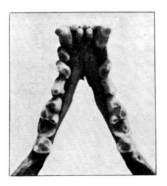

Table 11.7 The incidence of irregular arrangements of mandibular incisors and variation in the position of other classes of teeth in various MUSTELID genera

Sub-family	Genus	No. of specimens	Irregular lower incisors			Other classes of teeth		Total specimens affected		Total specimens affected excluding those with slight irregularity of the lower incisors only	
			Slight	Marked	%	Upper incisors	Premolars	No.	%	No.	%
Mustelinae	Martes	205	58	7	31.7	1	18	75	36.6	25	12.2
	Mustela	613	316	79	64.4	3	40	413	67.4	111	18.1
	Vormela	17	5	1	35.3	0	3	6	35.3	3	17.6
	Gulo	12	4	0	33.3	0	4	7	58.3	4	33.3
	Lyncodon	6	2	0	33.3	0	0	2	33.3	0	0
	Eira	60	0	0	0	0	6	6	10.0	6	10.0
	Galictis	59	18	6	40.7	0	4	25	42.4	9	15.2
Total Mustelinae		**972**	**403**	**93**	**51.0**	**4**	**75**	**534**	**54.9**	**158**	**16.3**
Melinae	Conepatus	167	9	0	5.4	0	22	28	16.8	22	13.2
	Mephitis	24	0	0	0	0	2	2	8.3	2	8.3
	Spilogale	9	1	0	11.1	1	2	3	33.3	3	33.3
	Poecilogale	21	2	0	9.5	0	3	5	23.8	3	14.3
	Ictonyx	73	5	3	11.0	0	6	13	17.8	9	12.3
	Poecilictis	22	0	1	4.5	0	6	7	31.8	7	31.8
	Taxidea	17	1	0	5.9	0	2	3	17.6	2	11.8
	Meles	85	6	0	7.1	0	21	23	27.1	21	24.7
	Arctonyx	18	0	1	5.6	0	0	1	5.6	1	5.6
	Mydaus	20	0	0	0	0	0	0	0	0	0
	Mellivora	50	6	0	12.0	0	6	10	20.0	6	12.0
	Melogale	46	1	0	2.2	0	4	4	8.7	4	8.7
Total Melinae		**552**	**31**	**5**	**6.5**	**1**	**74**	**99**	**17.9**	**80**	**14.5**
Lutrinae	Enhydra	21	0	0	0	0	0	0	0	0	0
	Lutra	161	47	36	51.6	3	59	105	65.2	75	46.6
	Pteronura	4	2	0	50.0	0	0	2	50.0	0	0
	Aonyx	36	3	1	11.1	0	10	12	33.3	11	30.5
Total Lutrinae		**222**	**52**	**37**	**40.1**	**3**	**69**	**119**	**53.6**	**86**	**38.7**
Total MUSTELIDAE		**1746**	**486**	**135**	**35.6**	**8**	**218**	**752**	**43.1**	**324**	**18.6**

The column headed 'Marked' under 'Irregular lower incisors' includes specimens where I_2 are misplaced lingually and are covered by less than one-third of the adjacent incisors; in many cases, both I_2 are completely lingual to the other incisors.
In skulls in which two or more classes of teeth varied in position, each class is recorded individually; for example, if in a skull one or more of the incisors, premolars and molars varied, each appropriate class would be credited with one and the column 'Total specimens affected' with one only.

premaxilla because the upper incisors are crowded and pre-canine diastemata are absent.

(b) Premolars

There was considerable variation in the positions of the premolars in the Mustelidae (Table 11.7); to illustrate the range of variation, details of several genera will be given separately.

Eira (tayras). – In about a third of the *Eira* skulls, the premolars were arranged in an even line (Fig. 11.31 A), the upper teeth being in contact or occasionally slightly spaced; in the remaining two-thirds, one or more of the upper premolars were oblique. P^3 was most often irregular in position and it may be rotated, sometimes to the extent of one-eighth of a circle which is sufficient to be recorded in Table 11.7. An example is shown in Figure 11.31 B.

Mephitis (striped skunks). – The premolars are arranged so that the buccal surfaces form an even line. The teeth are in contact. When there is insufficient room for the teeth, P^3 rotates and the buccal surface slides into the mesial concavity of the crown of the carnassial P^4 (Fig. 11.32); when space is more restricted, the rotation of P^3 is increased and P^2 is displaced lingually (Fig. 11.33). Irregularity to this extent is recorded in Table 11.7.

Martes (martens). – The upper premolars are set in an even line with the teeth in contact. In the mandible, the small P_1 was commonly rotated. In about a third of the specimens, many or all of the upper premolars were rotated, probably as a result of shortage of space. Where the crowding or rotation was severe, the distal surface of P^3 lay in the concavity of the mesial surface of P^4.

Heran (1970) described a skull of a male fisher (*M. pennanti*, National Museum of Canada 27567) in which the left P_3 and P_4 were misplaced; this may have been due to the retention of two of the deciduous molars.

Mellivora (honey badgers). – The maxillary premolars are usually set in a fairly straight line with slight spaces between the teeth. Insufficient room for the teeth is indicated when the premolars are in contact with one another or rotated (Figs. 11.34 and 11.35).

Poecilictis (Saharan striped weasels). – The

Table 11.8 The incidence of irregular arrangements of mandibular incisors in various MUSTELID species

Species	No. of specimens	Total specimens affected No.	%
Martes pennanti	8	6	75.0
M. americana	11	6	54.5
M. martes	36	11	30.6
M. foina	39	16	41.0
M. zibellina	16	4	25.0
M. melampus	15	5	33.3
Mustela altaica	10	8	80.0
M. nivalis	66	60	90.9
M. africana	19	10	52.6
M. erminea	158	108	68.4
M. kathiath	11	7	63.6
M. siberica	85	44	51.8
Galictis vittata	3	2	66.7
G. cuja	41	14	34.1
Lutra lutra	43	23	53.5
L. perspicillata	40	4	10.0
L. sumatrana	7	4	57.1
L. longicaudis	11	11	100.0
L. provocax	4	4	100.0
L. felina	16	15	93.7

Fig. 11.29. *Arctonyx collaris* (hog-badger). ♀. BMNH 1894.5.10.1. Marked incisor underjet with the lower canines occluding some distance in front of the upper ones. ×c. 0.66.

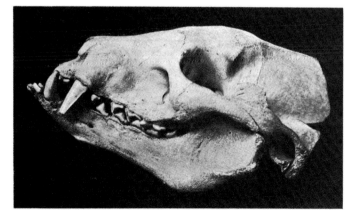

Fig. 11.30. *Mustela putorius furo* (ferret). RCS Odonto. Mus., G 73.23. Marked incisor and canine underjet; the lower post-canine teeth are in slightly pre-normal occlusion. × c. 0.8.

Fig. 11.31. *Eira barbara* (tayra).
A: ♂. BMNH 1905.2.5.12. Normal upper tooth arch in *Eira*.
B: ♂. BMNH 1903.9.4.43. Rotation of the premolars, particularly of P^3 (pointers) which are rotated with buccal surfaces turned mesially.

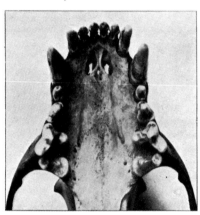

A

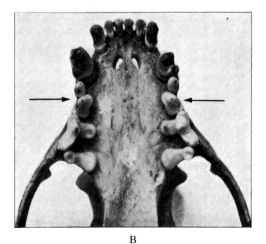

B

Fig. 11.32. *Mephitis mephitis* (striped skunk). ♂. BMNH 1894.5.9.7. Post-canine arch crowded and both P^3 symmetrically rotated with the buccal surfaces turned mesially.

Fig. 11.33. *Mephitis mephitis* (striped skunk). ♂. BMNH 1894.5.9.6. Post-canine arch crowded with P^3 more rotated than in Figure 11.32, but in this specimen P^2 are symmetrically misplaced lingually.

Fig. 11.34. *Mellivora capensis* (honey badger). ♂. BMNH 1914.7.23.13. The normal upper tooth arch. There are spaces between the post-canine teeth.

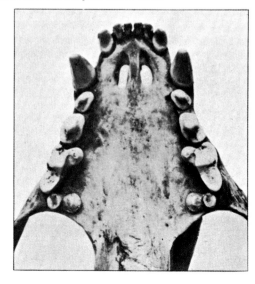

arrangement of the maxillary premolars varies. Normally a slight space is present between P^2 and P^3 with P^3 a little rotated (Fig. 11.36). Rotation of the P^2 sometimes occurs even when there is a space between it and P^3. Asymmetry in the arrangement of the premolars was seen in one skull; on one side, P^2 was rotated and, on the other side, P^3 (Fig. 11.37). In another example of asymmetry, there was considerable rotation of the right P^2 and P^3 whereas, on the left, the P^2 was rotated and P^3 placed rather straighter than in the normal (Fig. 11.38). The mandibular premolars were normally rotated slightly.

Ictonyx (striped polecats). – The maxillary premolars are usually slightly rotated; in about 40%, to the extent of one-eighth of a circle. An idea of the range of variation in the positions of the premolars can be gained from the Figures 11.39 to 11.42. In the first

Fig. 11.35. *Mellivora capensis* (honey badger). ♂. BMNH 1910.10.3.10. The upper arch is narrower than in Figure 11.34; there is slight rotation of the P^3 and carnassial P^4 and there are no spaces between the post-canine teeth.

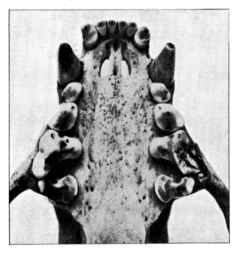

Fig. 11.36. *Poecilictis libyca* (Saharan striped weasel). BMNH 1898.4.3.1. The normal upper tooth arch in *Poecilictis*. ×c. 1.5.

Fig. 11.37. *Poecilictis libyca* (Saharan striped weasel). BMNH 1846.10.30.157. Right P^3 rotated with buccal surface turned mesially so that P^2 overlaps its buccal surface slightly; the left P^2 is slightly rotated with the buccal surface turned distally. ×c. 1.5.

Fig. 11.38. *Poecilictis libyca* (Saharan striped weasel). ♂. BMNH 1903.8.1.3. Right P^2 and P^3 rotated with buccal surfaces turned towards one another; the left P^2 is slightly rotated with its buccal surface turned distally. ×c. 1.5.

Fig. 11.39. *Ictonyx striatus* (striped polecat). ♂. BMNH 1897.11.5.12. In the majority of specimens, both the upper and lower premolars are to various degrees oblique to the line of the arch; that is rotated. Here the rotation is minimal and there are slight spaces between P^2 and P^3. ×c. 1.36.

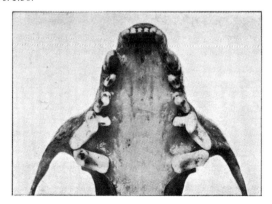

(Fig. 11.39), the rotation is slight and there are spaces between P^2 and P^3; in Figure 11.40, the rotation is more marked and spaces are present between P^3 and P^4; marked rotation (Fig. 11.41) is recorded in Table 11.7. The mandibular premolars are normally slightly rotated and this may be asymmetrical (Fig. 11.42).

Poecilogale (white-naped weasels). – There are only two maxillary premolars. The interesting feature is that P^3 is normally slightly rotated with a space between it and the upper canine (Fig. 11.43). More marked rotation (Fig. 11.44) is recorded in Table 11.7.

Spilogale (spotted skunks). – The maxillary premolars are in an even line with the distal end of P^3 slightly under cover of P^4. In the skull shown in Figure 11.45, the right P^2 is misplaced lingually.

Fig. 11.40. *Ictonyx striatus* (striped polecat). ♀. BMNH 1897.11.5.16. P^2 and P^3 are symmetrically rotated; spaces between P^3 and P^4. ×*c*. 1.25.

Fig. 11.43. *Poecilogale albinucha* (white-naped weasel). ♂. BMNH 1897.10.1.71. The normal upper tooth arch in *Poecilogale*. There are only two premolars, there is a space between P^3 and the canine but the P^3 are normally slightly oblique to the line of the arch. ×*c*. 1.5.

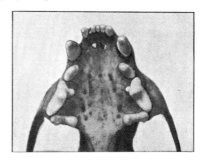

Fig. 11.41. *Ictonyx striatus* (striped polecat). ♂. BMNH 1905.5.1.10. P^2 and P^3 are more rotated than in Figure 11.40. ×*c*. 1.25.

Fig. 11.44. *Poecilogale albinucha* (white-naped weasel). ♀. BMNH 1910.7.28.6. P^3 are symmetrically more rotated than usual with the buccal surfaces turned mesially. ×*c*. 1.5.

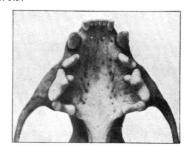

Fig. 11.42. *Ictonyx striatus* (striped polecat). ♂. BMNH 1904.9.1.35. Both P_3 rotated, the left one more than the right. Enlarged slightly.

Fig. 11.45. *Spilogale gracilis lucasana* (Alleghenian spotted skunk). ♂. BMNH 1898.3.1.54. Right P^2 slightly misplaced lingually. ×*c*. 1.33.

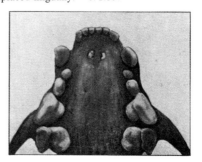

Conepatus (hog-nosed skunks). – The normal positions of the teeth are shown in Figure 11.46. In the maxilla, the buccal surfaces of the two premolars do not present an exactly even line, the distal end of P^3 being overlapped slightly by P^4; in the mandible, the premolars are definitely normally rotated. Deviations from this arrangement of the teeth are: (i) spaces between the maxillary premolars, (ii) rotation of P^3 (Fig. 11.47), and (iii) an increased rotation of the mandibular premolars to the extent of three-sixteenths of a circle.

Enhydra (sea otter). – The maxillary premolars are normally placed as seen in Figure 11.48, the only deviations from these positions being a decrease in the spaces between P^3 and the canines with the consequent misplacement of P^2 lingually.

Lutra (otters). – The specimen of a smooth-coated otter (*Lutra perspicillata*) may be taken as showing the normal arrangement of the maxillary teeth in otters (Fig. 11.49). P^1 lies to the lingual side of the canine and the distal end of P^3 is rotated a little lingually so that the buccal surfaces of the premolars present an uneven line. Decreased space for the teeth is shown by the rotation lingually of the distal end of P^3 along the mesial surface of P^4, the closure of the space between P^2 and the canine and lingual displacement of P^1 (Fig. 11.50).

Aonyx (clawless otters). – Variations in position of the teeth are less common in *Aonyx* than in *Lutra* (Table 11.7). In one specimen, the crown of the left P^2 was misplaced towards the palate so that in occlusion it passed to the lingual side of the lower teeth.

Family **Procyonidae**. 355 specimens examined

The genera of Procyonidae differ in the arrangement of the upper premolars and molars. The lower arches tend to be similar in the various genera, with straight rows of cheek teeth.

Coatis (*Nasua*) have long tapering jaws (Fig.

Fig. 11.46. *Conepatus semistriatus amazonicus* (Amazonian hog-nosed skunk). ♂. BMNH 1903.9.5.46. The normal tooth arches in *Conepatus*. P^3 slightly rotated so that the distal margins of the buccal surfaces are slightly overlapped by carnassial P^4.

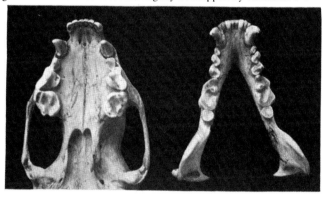

Fig. 11.47. *Conepatus chinga mendosus* (hog-nosed skunk). ♂. BMNH 1927.5.1.2. Marked symmetrical rotation of P^3 (pointers) with buccal surfaces turned distally. ×c. 1.33.

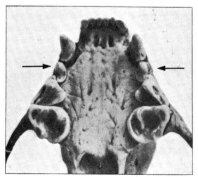

11.51). The double curves formed by the upper cheek teeth are very slight and flat compared with those in the Canidae. P^4 and M^2 are roughly triangular in plan view. P^1, P^2 and P^3 are sectorial and flattened bucco-lingually. The upper tooth-arch in the little coati mundi (*Nasuella*) is similar to that of *Nasua* but more slender and with spaces between the premolars.

In the racoon (*Procyon*), the jaws are slightly shorter and the premolars are usually in contact. The crown of P^3 is slightly triangular; the occlusal surface of M^2 faces slightly distally and there is no tuberosity. In the olingo (*Bassaricyon*), the upper cheek-teeth rows are nearly straight.

In the kinkajou (*Potos*), the jaws are shorter and broader and there are only three premolars. P^3 is distinctly triangular in plan view and is usually rotated slightly with its buccal surface turned mesially and slightly overlapped buccally by P^2 (Figs. 11.52 and 11.53); P_3 is usually similarly rotated.

In the ring-tailed cat (*Bassaricus*), the premolars may be slightly spaced or in contact.

Positional variations of the teeth of procyonids are summarized in Tables 11.1 and 11.9.

Genus ***Procyon***, racoons

The more common variations in *Procyon* are rotation of (i) P^3 and (ii) P_2 and P_3. In captive specimens, the rotation may be extreme (Figs. 11.54 and 11.55). Captive conditions seem to increase the liability to

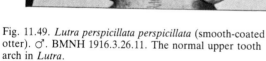

Fig. 11.48. *Enhydra lutris* (sea otter). BMNH 1928.4.21.23. The normal tooth arches in *Enhydra*.

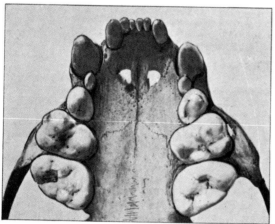

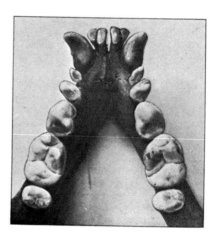

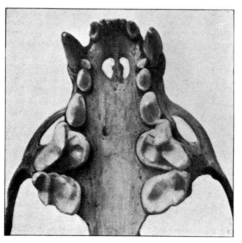

Fig. 11.49. *Lutra perspicillata perspicillata* (smooth-coated otter). ♂. BMNH 1916.3.26.11. The normal upper tooth arch in *Lutra*.

Fig. 11.50. *Lutra perspicillata perspicillata* (smooth-coated otter). ♀. BMNH 1895.5.20.1. Post-canine teeth crowded with no spaces between. The P^4 are slightly rotated and so slightly overlapped buccally by M^1.

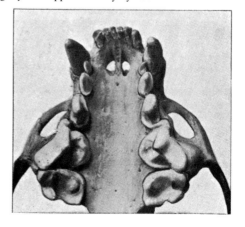

vary; of 17 captive specimens, 9 varied. Positional variations are more common in *Procyon cancrivorus* than in other species of *Procyon*. Of 32 specimens of the former, 14 were affected, and of 27 specimens of the latter only two.

Genus *Nasua*, coatis

The only example of irregularity of the incisors in the Procyonidae was in a skull of *Nasua*; the right I_3 was misplaced to the buccal side of the lower canine so that a gap remained between the two other incisors (Fig. 11.56).

Irregularities of the premolars are more common and, in *Nasua*, P^3 is frequently rotated (Fig. 11.57).

Rotation of the P^2 is less common. In the mandible on the other hand, irregularity in position seems more common in P_2 than in P_3.

In captive conditions, the prevalence of abnormalities of tooth position in *Nasua* is distinctly greater, more marked and more generalized (Fig. 11.58). Eleven (39.3%) out of the 28 captive specimens showed anomalies of position.

Genus *Potos*, kinkajous

The dentition of *Potos* affords good examples of the difficulties encountered in deciding between normal and abnormal arrangements of the teeth, especially if the apparent anomalies are symmetrical. The speci-

Table 11.9 Positional variations in the PROCYONIDAE from the wild

Genus	No. of specimens	Class of tooth				Total specimens affected	
		I	C	P	M	No.	%
Procyon	59	0	0	16	0	16	27.1
Nasua	197	1	0	30	0	29	14.7
Nasuella	17	0	0	0	0	0	0
Potos	41	1	1	11	5	15	36.6
Bassaricyon	10	0	0	1	0	1	10.0
Bassaricus	12	0	0	2	0	2	16.7

Rotation of P^3 to the extent of one-eighth of a circle can be regarded as normal in procyonids and has not been included in the Table.

Fig. 11.51. *Nasua* (coati). The normal upper tooth arch in the genus *Nasua*.

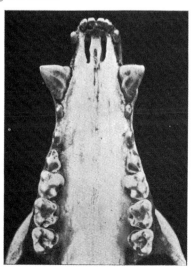

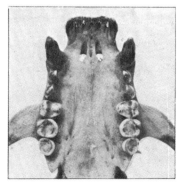

Fig. 11.52. *Potos flavus chiriquensis* (Panama kinkajou). ♀. BMNH 1903.3.3.24. Slight symmetrical rotation of P^3 (pointer) with buccal surfaces turned mesially; P^4 are rotated slightly in the opposite direction.

mens in Figures 11.52, 11.53 and 11.59 illustrate the range of variations in position of the maxillary premolars.

A moderately severe irregularity is shown in Figure 11.60. There is incisor underjet and the left lower canine is misplaced so that it occludes between I^2 and I^3. There are other minor irregularities and the left M^2 is absent.

Family **Ursidae**. 332 specimens examined

The cheek teeth of bears are set in straight divergent lines which turn slightly lingually at the level of the last molar. The length of the muzzle is very variable. Although bears have the full number of premolars, P1, P2 and P3 are small rudimentary teeth and, in the long-muzzled types, these premolars seldom persist through the whole life of the animal, P^2, P_2 and P_3 being usually lost at an early stage of life. The long-muzzled bears have spaces between the premolars, whereas in the short-muzzled bears the premolars are in contact and many specimens show a definitely crowded arrangement. Table 11.10 shows that the range of variation of the length of the palate, the length of the cheek teeth and the ratio between them varies greatly within each of the different species. This is even true of animals from the same geographical area, for example in 17 Asiatic black bears (*Selenarctos thibetanus*) from the same district the ratio varied from 45.2 to 57.0%.

Positional variations of the teeth of bears are summarized in Table 11.1, but do not present any features of especial interest. The mandibular incisors are more often irregular in position than the maxillary teeth, the irregularity seldom amounting to more than a slight displacement of I2 or I3. Two skulls showed incisor underjet.

P4, or more rarely P3, may rotate to the extent of a quarter of a circle. In a sloth bear (*Melursus ursinus*), the right P^3 was rotated, the buccal surface facing distally, m^4 was still present and P^4 had erupted mesially to its normal position; the left P^4 had the buccal surface directed lingually. The crown of the right P_1 was connate.

Molars are sometimes rotated; in polar bears (*Thalarctos*), slight rotation of M^1 seems to be a well-defined variation.

Family **Canidae**. 1478 specimens examined

The maxillary cheek-teeth are arranged with the buccal surfaces in a straight line; in the mandible, P_4 overlaps slightly the buccal surface of M_1. In *Canis* (wolves and jackals), the premolars are usually in

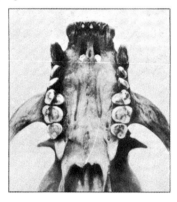

Fig. 11.53. *Potos flavus meridensis* (Venezuelan kinkajou). ♀. BMNH 1898.7.1.7. Symmetrical rotation of all the upper premolars with buccal surfaces turned mesially.

Fig. 11.54. *Procyon cancrivorus* (crab-eating racoon). BMNH 1884.2.8.6. Left P_4 incompletely erupted and impacted against the mesial surface of M_1.

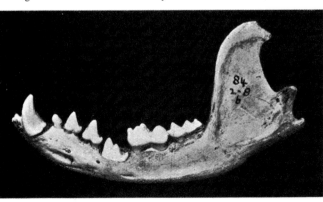

Table 11.10 Variation in the length of the palate and the length of the cheek teeth (the sum of the mesio-distal lengths of the upper molars plus P^4) in BEARS

Species and sub-species	No. of specimens	Length palate/length cheek teeth (%) Mean	Range	Range of variations (%) Length palate	Length teeth
Thalarctos maritimus	22	43.5	39.1–50.0	36.4	31.0
Ursus arctos arctos	21	51.1	43.1–57.8	40.9	32.2
U. arctos isabellinus	16	53.0	47.5–57.0	24.1	20.2
U. americanus	24	51.6	47.6–59.8	35.6	32.9
Selenarctos thibetanus	51	52.9		46.8	30.4
Helarctos malayanus	14	55.4	51.4–62.8	28.0	20.4
Tremarctos ornatus	9	55.0		25.1	16.1

All measurements in mm.

Fig. 11.55. *Procyon cancrivorus* (crab-eating racoon). ♂. BMNH 1900.7.24.1. Post-canine parts of upper and lower arches symmetrically crowded. Both P^2 and P^3 are rotated with buccal surfaces turned mesially. P_3 is incompletely erupted and impacted against the adjacent teeth.

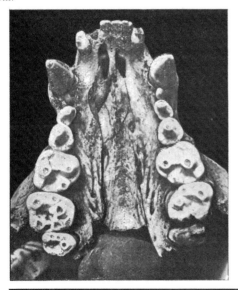

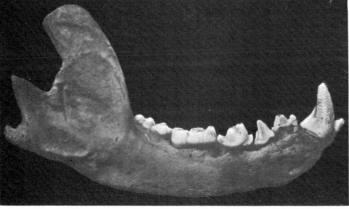

Fig. 11.56. *Nasua* sp. (coati). ♀. BMNH 1882.9.30.13. Right I₃ misplaced labially to the canine. Enlarged.

Fig. 11.57. *Nasua* sp. (coati). ♂. BMNH 1898.3.2.31. Both P³ rotated with lingual surfaces turned mesially.

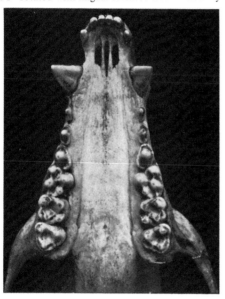

contact; in foxes (*Vulpes*), which have long and slender jaws, the teeth are spaced.

Positional variations in canids from the wild are summarized in Tables 11.1 (p. 239) and 11.11.

(a) Incisor occlusion

In normal occlusion, there is slight overjet, the lingual surfaces of the upper incisors occluding with the labial surfaces of the lower incisors. Both underjet and increased incisor overjet sometimes occur in wild canids and, when associated with equivalent abnormality of the cheek-tooth occlusion, are due to disproportion between the lengths of the upper and lower jaws. However, the usual difficulty exists of deciding, without craniometry, which jaw is at fault and of distinguishing, for example, superior protrusion from

Fig. 11.59. *Potos flavus megalotus* (Colombian kinkajou). BMNH 1865.5.18.22. Crowding of premolars, with symmetrical rotation of both P³ with buccal surfaces turned mesially. The lower premolars were also crowded.

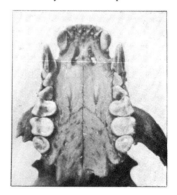

Fig. 11.58. *Nasua nasua* (southern coati). ♂. Captive. Formerly RCS Osteo. Series, 796. P³ symmetrically rotated with buccal surfaces facing distally (arrows). Compare with Figure 11.51. P₃ and P₄ symmetrically misplaced buccally.

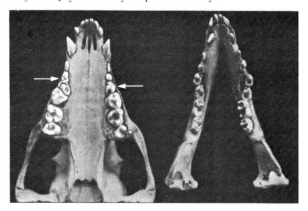

inferior retrusion and vice versa. This is illustrated by the following examples:

The skull shown in Figure 11.61, which was in the Museum of the Royal College of Surgeons, is that of a fox recorded as having come from near Stratford-on-Avon in the English Midlands; several foxes killed in the same neighbourhood were said to show similar peculiarities. The upper incisors are in marked overjet and the upper post-canine teeth are in slightly pre-normal occlusion. Although the disproportionately greater incisor overjet compared with the only slightly pre-normal cheek-tooth occlusion suggests that the upper jaw rather than the lower is at fault, strictly speaking there is the possibility that this is inferior retrusion and the mandible is shorter than normal. However, the length of the cranium was 7.2% more than the mean of 25 normal specimens and the length of the mandible only 0.6% more. This strengthens the

Table 11.11 Positional variations in CANIDS from the wild

Genus	No. of specimens	Class of tooth				Total specimens affected		Total specimens with asymmetry of premolars	
		I	C	P	M	No.	%	No.	%
Canis	407	1	0	24	6	29	7.1	23	5.6
Vulpes	544	2	0	11	2	15	2.8	59	10.8
Alopex	118	0	0	12	0	12	10.2	1	0.8
Cerdoyon	79	0	0	3	1	4	5.1	3	3.8
Dusicyon	136	3	0	11	1	14	10.3	21	15.4
Chrysocyon	15	0	0	0	0	0	0	1	6.7
Nyctereutes	29	1	0	2	0	2	6.9	0	0
Otocyon	30	0	0	1	0	1	3.3	0	0
Cuon	43	6	0	15	0	17	39.5	4	9.3
Speothos	14	0	0	3	0	3	21.4	0	0
Lycaon	63	1	0	9	0	9	14.3	0	0

In skulls in which two or more classes of teeth varied in position, each class is recorded individually; for example, if in a skull one or more of the incisors, premolars and molars varied, each appropriate class would be credited with one and the column 'total specimens affected' with one only.

Fig. 11.60. *Potos* sp. (kinkajou). ♂. BMNH 1905.7.5.6. The left lateral view shows incisor underjet and that the lower canine is misplaced mesially and occludes between I^2 and I^3, having displaced I^3 into the pre-canine diastema. The palatal view shows that P^3 are symmetrically rotated with buccal surfaces turned mesially. The right M^2 is absent (arrow).

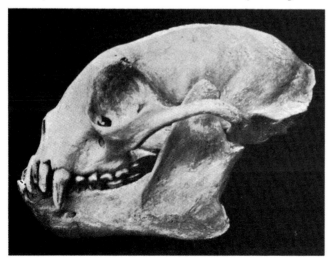
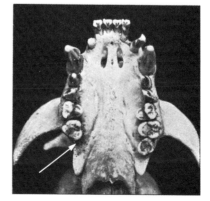

view that the disharmony of occlusion is basically due to an exceptionally long upper jaw.

The skull of a jackal in Figure 11.62 has incisor overjet with all the other upper teeth likewise in slightly pre-normal occlusion. However, measurement showed that almost certainly the lower jaw is at fault in being disproportionately short, and that this is an example of inferior retrusion, because the length of the cranium was only 0.2% less than the average of 13 normal specimens whereas the length of the mandible was 4.3% less.

In the fox skull in Figure 11.63, there is incisor underjet associated with a slightly pre-normal relationship of the lower post-canine teeth. This seems to have been due to inferior protrusion, that is the mandible being disproportionately long because, whereas the basal length of the cranium was 2.9% longer than the average of 25 specimens, the mandible was 8.0% longer. In another fox skull, there was incisor underjet which appears to have been due to superior retrusion rather than inferior protrusion because the cranium was 6.5% shorter and the mandible only 0.2% less than the average for 25 normal specimens.

These instances show that the interpretation of the cause of these occlusal disharmonies requires the assistance of craniometry, preferably more elaborate than the examples quoted here. Measurement at least of the lengths of cranium and of upper and lower jaw, but preferably of the tooth arches also, seems necessary.

Van Bree and Sinkeldam (1969) figured two red fox skulls, one with incisor underjet with the rest of the mandibular arch also in pre-normal occlusion, the other with incisor overjet with the rest of the mandibular arch in post-normal occlusion. Some measurements of jaw length made on the figures sug-

Fig. 11.61. *Vulpes vulpes crucigera* (British red fox). Formerly RCS Osteo. Series, 648. Marked incisor overjet. The upper post-canine teeth are also in slightly pre-normal occlusion; note that P^1 and P_1 are opposite one another instead of P^1 occluding between P_1 and P_2.

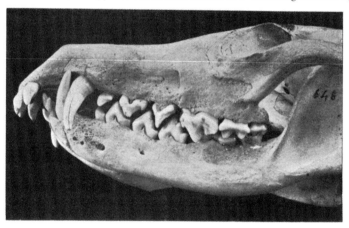

Fig. 11.62. *Canis mesomelas* (black-backed jackal). Incisor overjet; the remainder of the upper dentition also is in slightly pre-normal occlusion; P_1 has been lost. Note the lower canine occludes distally to the upper canine. $\times c. 0.66$.

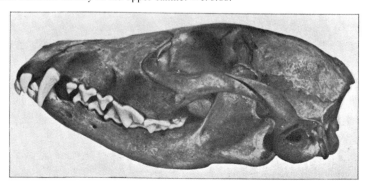

gest that in the first an unduly short upper jaw and in the second an unduly short mandible were the likely causes.

Marked shortening of the face occurs spontaneously in wild canids. Lawrence (1934) described the skull of a wild-shot coyote (*Canis latrans*, ♀, MCZ 29856) very similar in form to a bulldog, that is the facial part of the skull, including the upper jaw, was extremely shortened; as the lower jaw was of normal length, there was pronounced underjet. The teeth were normal in size and number but the upper premolars were very crowded and several were rotated.

A similar condition appears occasionally in red foxes (*Vulpes vulpes*) from the wild (Dönitz, 1868; Reinwaldt, 1962). Hauck (1942) showed that, although slight overjet is the normal type of incisor occlusion in wild canids, marked overjet and underjet occur occasionally (Table 11.12).

(b) Incisors
Irregular positions of the incisors are uncommon and are usually confined to the mandible.

(c) Premolars
Variations in the positions of the premolars take the form of rotation, the buccal surfaces of the teeth facing mesially (Fig. 11.64). The upper teeth vary more than the lower. The most variable tooth is P^3, the degree of rotation in some specimens bringing the tooth transverse to the line of the arch (Fig. 11.65).

An asymmetrical arrangement of the premolars is common in the canids; it is often limited to the first premolars. The percentage of specimens with asymmetrical premolars varies in the different genera, being particularly high in *Vulpes* and *Dusicyon* (Table 11.11). One sub-species of *Vulpes*, the Himalayan red fox (*Vulpes vulpes montana*), seems particularly liable to this variation; it was present in 7 out of 28 speci-

Table 11.12 The prevalence of malocclusion in various species of wild and captive CANIDS (from Hauck, 1942, p. 352)

Species	No. of specimens	No. of specimens affected		
		Marked incisor overjet	Slight incisor underjet	Inferior retrusion
Canis lupus	21	1	0	1
C. latrans	5	0	0	0
C. aureus	9	0	0	0
Vulpes vulpes	56	1	1	0
Cuon alpinus	1	0	0	0
Lycaon pictus	2	0	0	0

Fig. 11.63. *Vulpes vulpes vulpes* (Scandinavian red fox). BMNH 1911.6.3.1. Marked incisor underjet. The remainder of the mandibular dentition is in pre-normal occlusion; P_2 is opposite P^1 instead of occluding distally to it. ×c. 0.66.

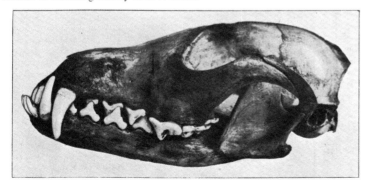

mens. The simian jackal (*Canis simensis*) shows a remarkable frequency of asymmetry; the variation was seen in 9 of the 10 skulls. An asymmetrical arrangement of the premolars is rare in *Alopex*, *Nycteuretes*, *Otocyon* and *Lycaon*. Examples are shown in Figures 11.66 and 11.67.

(d) Molars

Abnormality in position of the maxillary molars takes the form of tilting or rotation; most typically, the mesio-buccal border of M^2 moves into the concavity on the distal aspect of M^1; in a few specimens, M^2 was transverse to the line of the arch. Three specimens of *Cerdocyon thous* had spaces between the molars (Fig. 11.68).

Nellis (1972) figured the jaws of a wild coyote in which the crowns of both M_1 were tilted distally so that their distal margins were impacted beneath the mesial convexity of M_2, mentioning how similar it was to the condition described here in a dog (Fig. 11.90) and that he saw in two captive coyotes.

(e) Spacing of cheek teeth

Table 11.13 shows that, although there is considerable variation in the amount of cheek-tooth spacing in red foxes (*Vulpes vulpes*), the mean amount of spacing is similar in four of its sub-species. It is much less in the hunting dog (*Lycaon pictus*). In animals from the same geographical area, there is often a wide range of variation in spacing (Table 11.14).

In 489 coyote skulls from Central Alberta, Nellis (1972) found two with unusually wide spaces between P_2 and P_3 on both sides.

(f) The effect of captivity

Captivity has been shown to have a marked effect on the canid skull. In 1884, Nehring described the skulls of three sibling wolves in the Zoological Garden in Berlin bred from a pair of strongly-built Russian wolves captured in the wild. In all three siblings, the teeth were not reduced in size but, as both jaws were considerably shorter than in the parents, the teeth were crowded and overlapping. Klatt (1921) described a similar condition in a red fox born and bred in captivity. However, Fabian (1933) showed that captivity does not in itself lead to shortened jaws in canids; poor diet is a more likely explanation. He found that the conditions of captivity are critical; silver foxes that had been kept on a good diet and in good conditions for 10–15 generations showed no reduction in jaw size nor any abnormality of tooth position.

Canis familiaris (domestic dogs)

Domestic dogs are descended from wolves, *Canis lupus*, that were domesticated in prehistoric times (G.M. Allen, 1920; Clutton-Brock, 1981), although it

Fig. 11.65. *Canis mesomelas* (black-backed jackal). ♂. BMNH 1904.2.1.2. Both P^3 rotated through 90° so that the buccal surfaces face mesially. Enlarged slightly.

Fig. 11.64. *Speothos venaticus* (bush dog). BMNH 1846.2.13.10. Both P^3 symmetrically rotated with the buccal surfaces facing mesially. Enlarged slightly.

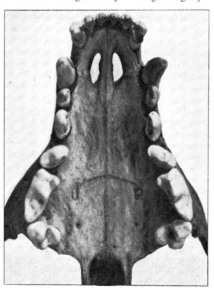

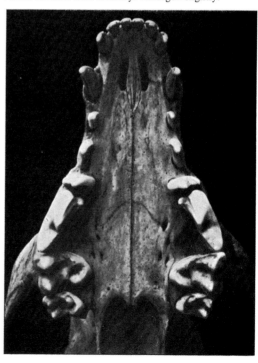

Table 11.13 The amount of upper cheek-tooth spacing in four sub-species of RED FOX *(Vulpes vulpes)* and in the HUNTING DOG *(Lycaon pictus)* as shown by the relationship of the length of the palate (A) to the length of the upper cheek teeth (B)

	Sex	No. of specimens	Length of palate (A)		Length of cheek teeth (B)		B×100/A	
			Mean	Range of variation (%)	Mean	Range of variation (%)	Mean	Range of variation
Vulpes vulpes	males	19	78.0	11.2	52.6	13.6	67.4	62.0–73.3
crucigera	females	11	73.4	17.7	50.4	17.6	68.7	61.0–72.3
V. v.	males	6	82.5	17.7	54.6	9.1	66.2	58.1–73.6
silaceous	females	3	73.0	16.2	49.5	5.9	67.8	64.6–71.0
V. v.	males	7	66.7	8.1	45.1	12.8	67.6	63.3–73.1
arabica	females	7	64.4	8.5	44.0	5.8	68.3	63.9–70.7
V. v.	males	11	49.5	12.8	33.4	13.1	67.5	63.0–71.4
pallida	females	4	47.7	7.0	32.8	11.1	68.8	67.5–71.2
Lycaon pictus	males	10	95.6	17.6	74.6	13.2	78.0	72.0–85.9
	females	17	96.3	12.8	72.6	18.6	75.4	71.7–81.4

All measurements in mm.

Table 11.14 The amount of upper cheek-tooth spacing in two CANID species from restricted localities

Species	District	No. of specimens	Range of variation of the mean length of the palate to the mean length of the cheek teeth (%)
Vulpes vulpes vulpes	Egersund, Norway	11	62.0–75.4
Lycaon pictus	M'toto Andei Railway Station Uganda	5	71.9–81.1

Fig. 11.66. *Cerdocyon thous jucundus* (crab-eating fox). ♂. BMNH 1921.1.1.2. In the upper jaw, all the teeth on the right are situated a little more mesially than those on the left. This is so in the lower jaw also, except for P_1 and P_2. ×c. 0.8.

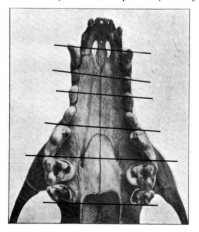
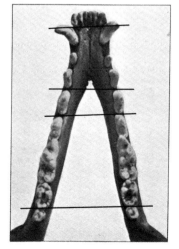

is possible that, in some areas, other wild canid stocks may have played a role.

The skulls of most breeds of dog are of roughly the same shape as those of wolves, but smaller. This reduction in size seems to have affected teeth less than the rest of the skeleton so the dentition tends to be slightly crowded (Fig. 11.69). Several breeds of dog, originating in different parts of the world, such as bulldogs, pugs, pekinese and S. American pug-like dogs (Allen, 1920), have been bred for their very short faces and this has led to gross malposition of teeth and malocclusion (Fig. 11.70). In a few breeds (greyhounds and borzois), the jaws are elongated compared with those of wolves and the teeth are even more spaced (Fig. 11.70).

An important contribution to the study of the dog, consisting principally of elaborate breeding experiments between breeds showing marked morphological differences, was made by Stockard and collaborators (1941). Although some of the ideas expressed, for example that endocrine imbalances play a large part in producing these morphological differences, would be challenged today, the work is of considerable value. Their book contains craniometrics and clear pictures of skulls of both pure-bred dogs and of hybrids, for example between bassett hound and saluki, bulldog and Alsatian, miniature griffon and dachshund (Stockard and Johnson, 1941, pp. 152–225).

(A) DENTAL ABNORMALITIES IN SHORT-FACED DOGS

The upper jaws are sometimes shortened in domestic dogs. The lengths of the upper and lower jaws appear to be under independent genetic control (Stockard *et al.*, 1941), so that, if the upper jaws are shortened, the mandibles may not be and may project beyond the premaxillae, curving upwards to meet them, thus giving rise to the short-faced skulls of bulldogs and pugs (Fig. 11.70). In domestic dogs, this shortening seems to be accompanied by an increase in the width of both jaws (St Clair and Jones, 1957), but this may be more apparent than real and needs confirmation by measurement. The maxillae and frontal bones are deformed, and so are the nasal and turbinate bones, which may result in respiratory dyspnoea; the tongue may be over-sized. The soft tissues may be of normal size, so the soft palate and the skin are corrugated and folded. Cleft palate is common in brachycephalic breeds (M. W. Fox, 1963).

The rounded brain-cases of brachycephalic dogs have features of neoteny, that is, the preservation of infantile characteristics.

Achondroplasia is a heritable condition in which there is defective development and growth of primordial cartilage. Consequently, as in the typical human achondroplasic dwarf, the head, although of normal overall size, is very brachycephalic, the nose is

Fig. 11.67. *Vulpes vulpes vulpes* (Scandinavian fox). BMNH 1925.11.3.17. (skull). RCS Odonto. Mus., G 8.22 (cast). Left P_2 and P_3 rotated with buccal surfaces turned distally; they are also misplaced distally so that P_3 is overlapped by P_2 and P_4. Right P_2 is rotated in the opposite direction and overlaps the buccal aspect of P_3. Reduced slightly.

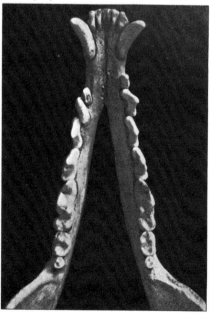

Fig. 11.68. *Cerdocyon thous* (crab-eating fox). ♀. BMNH 1925.2.1.80. Slight symmetrical spacing of the upper molars. × c. 0.8.

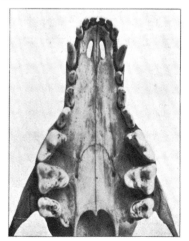

snubbed and the face flat; the limbs are very short but the trunk is little affected. Because the cranial base is small, to accommodate the brain there is compensatory increase in height and the brain-case tends to be globular. The face is flat and the upper jaw tends to be small because of involvement of the cartilaginous nasal capsule. Although cartilage in the condyles plays an important part in growth of the mandible, probably because that cartilage is secondary and not primordial, the mandible tends to be of normal size.

The short limbs of breeds of dog such as dachshund, Pekingese and basset hound greatly resemble

Fig. 11.69. Above: *Canis lupus* (wolf). RCS Odonto. Mus., A 149.21. × c. 0.36.
Below: *Canis familiaris* (domestic dog, Airedale). RCS Odonto. Mus., A 143.33. Showing the similarity of the skull of a dog with a muzzle of medium length like that of the wolf, which is the supposed ancestor of all domestic dogs. × c. 0.43. A palatal view of the wolf skull is shown in Figure 11.76.

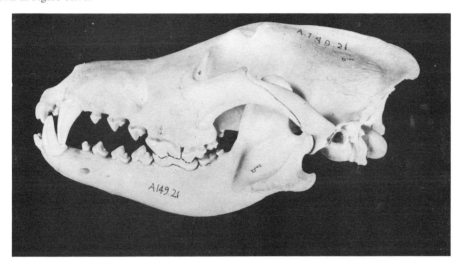

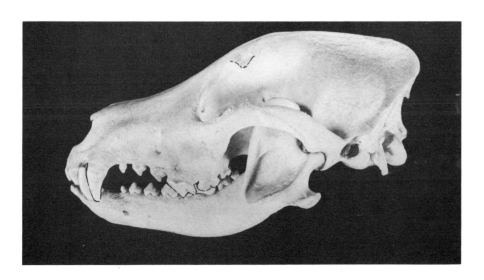

those of achondroplasia and it seems certain that achondroplasic dwarf mutants were incorporated in the ancestry of such breeds (Stockard, 1941, pp. 45–69). The fact that the skulls in these breeds are not short or achondroplasic need not detract from this view because achondroplasia occurs in partial forms in which the limbs only or the skull only are affected.

It seems likely that achondroplasic dwarf mutants have also contributed to the ancestry of many of the short-faced breeds of dogs. The skulls of the bulldog (Fig. 11.70), Boston terrier, Pekingese, and the miniature Brussels griffon show many of the features characteristic of the achondroplasic skull (Stockard and Johnson, 1941, pp. 272–288). They are particularly marked in the griffon which has a spherical brain case, below which is attached a tiny upper jaw crowded with

Fig. 11.70. Above: *Canis familiaris* (domestic dog, borzoi). RCS Odonto. Mus., G 71.2. × *c*. 0.44.

Below: *Canis familiaris* (domestic dog, bulldog). RCS Odonto. Mus., A 142.392. Showing the contrast between a dog with a proportionately long muzzle (borzoi) and one with a short muzzle (bulldog). × *c*. 0.51. Compare both with Figure 11.69. Palatal views of skulls of a borzoi and bulldogs are shown in Figures 11.71–11.76.

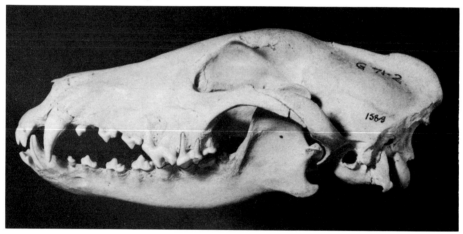

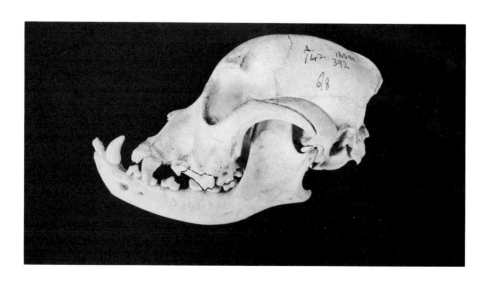

teeth, with a slender curved mandible projecting well beyond the upper jaw.

Short-faced toy dogs, such as the Pekingese, seem to have been selected for some unusual and attention-attracting characteristics. The evolution of toy dogs, some of which are skeletally bizarre, has been influenced by the standards for show dogs laid down by official judges in what many regard as an irresponsible manner (Patmore, 1984) because some carry genetic characters harmful to the animals.

When shortening of the jaw is marked, the general occlusion is grossly disturbed, and there is also a high rate of irregularity of position, of abnormality of number (Chapter 4, p. 86), and of retention of deciduous teeth (Chapter 16, pp. 350, 352). That such deformities should have been deliberately selected to create particular breeds of dogs is a sad reflection on human taste and consideration for animal welfare.

Five examples of the short palate and crowded arches typical of bulldogs are shown in Figures 11.71–11.75. The premolars in particular can only be accommodated by rotation and overlapping but the molars also are rotated in some degree. In all five instances, the mandible was not proportionately short and there was no crowding of the lower teeth; even some spacing of the premolars was present; however, in three of the skulls, one or both P_1 were absent, which provided some additional room. There was marked disproportion between the upper and lower jaws and gross superior retrusion with the lower incisors as much as 2.0 cm in front of the upper ones, so that any kind of incisor function was impossible. Tooth contact between the arches was limited to the two molars and part of P^4. It seems likely that these specimens were from animals bred and kept as pets, derived, of course,

Fig. 11.72. *Canis familiaris* (domestic dog, bulldog). RCS Odonto. Mus., A 142.391. The upper jaw is short and so all the post-canine teeth are crowded, with symmetrical rotation and overlap; in particular, the P^3 are rotated through 90° with the buccal surfaces facing mesially; both M^1 are slightly rotated with buccal surfaces turned distally and overlapped buccally by the P^4. Right M^2 absent. $\times c. 0.75$.

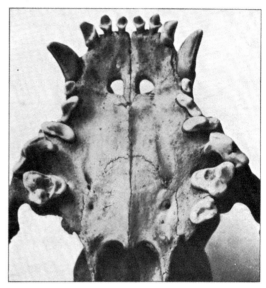

Fig. 11.71. *Canis familiaris* (domestic dog, bulldog). RCS Odonto. Mus., A 142.392. The upper jaw is short and in consequence the post-canine teeth are crowded. Both P^3 are misplaced lingually and are overlapped by P^2; the molars are rotated slightly with buccal surfaces turned distally. There was adequate occlusion with the lower arch because the sectorial lower teeth fitted between the overlapping upper premolars. $\times c. 0.63$.

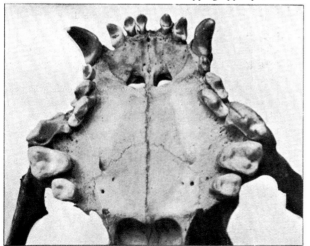

from bull-baiting ancestors; it is hard to believe that dogs that actually fought in the ring had cheek-tooth occlusions as poor as these.

(B) DENTAL ABNORMALITIES IN LONG-FACED DOGS

In long-muzzled breeds such as borzois and greyhounds, the teeth are placed in a regular manner with ample spaces between the premolars. The borzoi has a very long and narrow palate with the premolars well separated; in the greyhound, the palate is shorter and, although there are spaces between the premolars, the teeth are closer together than in the borzois (Fig. 11.76 and Table 11.15). In another borzoi skull (Fig. 11.77), the arrangement and size of the premolars are asymmetrical (Table 11.16).

Marked incisor overjet occurs occasionally in greyhounds (Fig. 11.78). The question arises whether this is due to the upper jaw being disproportionately long or the lower jaw disproportionately short. The comparison of the example shown in Figure 11.78 with the normal specimen, which is also shown, brings out the following points:

(i) The occlusion of P^4 and the upper molars with the mandibular teeth is similar in both specimens.
(ii) In the abnormal skull, the P^1 and P^2 are in advance of their normal relationships with the mandibular teeth.
(iii) In the abnormal skull, the mandibular canine occludes distally to the maxillary canine.
(iv) In the normal specimen, the mandibular premolars are spaced, in the abnormal one they are not.

The length of the skull was 1.5% more, and that of the mandible 6.9% less, than that of 15 normal specimens, suggesting that the disorder of occlusion was due to the lower jaw being disproportionately short.

(C) DENTAL ABNORMALITIES IN OTHER DOGS
(a) Disturbances of jaw length
As in wild canids, disturbances of jaw length can occur in dogs that characteristically have jaws of normal length. For example, Grüneberg and Lea (1940) showed that, in long-haired dachshunds, shortening of the lower jaw is a common inherited disorder in both sexes, and is exacerbated by slight lengthening of the

Fig. 11.73. *Canis familiaris* (domestic dog, bulldog). RCS Odonto. Mus., G17.114. Both P^1 are absent; the other three premolars on both sides are symmetrically rotated with buccal surfaces facing mesially. Both M^2 are rotated with buccal surfaces turned distally. The tooth in position of right I^2 belongs in the empty socket (arrow). $\times c.\,0.75$.

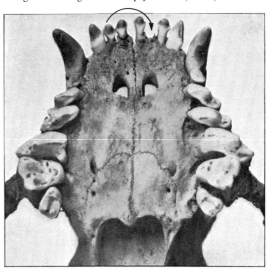

Table 11.15 The amount of upper cheek-tooth crowding in various breeds of DOG (*Canis familiaris*).

Length of muzzle	Breed	No. of specimens	Mean length of the palate (A)	Mean length of the cheek teeth (B)	B × 100 / A
Long	Borzois	6	117.9	74.8	63.4
	Greyhounds	15	107.9	69.3	64.2
Medium	Foxhounds	11	107.4	70.5	65.6
	Retrievers	13	96.8	68.1	70.4
	Fox terriers	14	78.2	57.7	73.8
	Mastiffs	6	100.6	70.7	70.3
Short	Bulldogs	23	63.7	70.5	110.7

Crowding is least in the long-jawed breeds and greatest in the short-jawed breeds.
All measurements in mm.

upper jaw which turns downwards anteriorly. There is marked incisor and canine overjet; as malocclusion is confined to the anterior teeth, it suggests that only the anterior parts of the jaws have been affected. They suggested that the elongation of the upper jaw was secondary to shortening of the lower jaw because, as the lower canines occlude distally to the upper canines, they exert pressure on the upper canines to move anteriorly. A similar condition also affected a strain of beagles (M. W. Fox, 1963).

Cranio-facial developmental anomalies occur in the domestic dog, of course; whether they are particularly common in those dogs that appear to have been bred for grotesqueness does not appear to have been established. Lantz and Cantwell (1984) described an example of hemifacial microsoma in a female toy poodle; the left mandible was poorly developed, some of its teeth absent and several other first and second branchial arch structures, zygomatic arch, muscles of mastication etc., were defective. The parents and three litter-mates were unaffected.

Marked incisor overjet occurs occasionally in many breeds, for example in foxhounds, dachshunds and terriers.

In a foxhound with incisor overjet and lower canines occluding lingually to the upper arch, the length of the skull was 2.2% less and that of the mandible 7.7% less than the average of seven normal specimens.

The skull of a sealyham (Fig. 11.79) shows considerable incisor overjet. The upper canines and the P^1 and P^2 are also in advance of their normal relationship with the mandibular teeth. However, P^3 and P^4, together with the molars, are in normal occlusion. There is marked crowding of the lower premolars. No measurements are available but the crowding of the lower arch compared to slight spacing of the upper premolars suggests that there has been defective growth of the mandible, presumably of the anterior part because the molar occlusion is normal.

Extreme incisor overjet is present in the skull of a feral dog of unknown breed in Figure 11.80. The

Table 11.16 The mesio-distal lengths of the premolars and of the spaces between them in a RUSSIAN GREYHOUND showing asymmetry

Upper teeth	Right	Left
Interval between the upper canine and P^1	8.1	6.0
Length of P^1	8.0	6.5
Interval between P^1 and P^2	2.3	7.0
Length of P^2	11.4	11.0
Interval between P^2 and P^3	3.6	2.4
Length of P^3	13.5	13.1
Width of P^4	20.4	20.4

Data supplied by Professor Magnus Degerbøl. All measurements in mm.

Fig. 11.74. *Canis familiaris* (domestic dog, bulldog). RCS Odonto. Mus., A142.38. On the left, P^2 and P^3 are side by side; on the right, they are accommodated in the arch by being rotated nearly 90° with buccal surfaces facing mesially. Both P^4 are rotated through 90°. × *c*. 0.75.

Fig. 11.75. *Canis familiaris* (domestic dog, bulldog). RCS Odonto. Mus., A142.39. The four premolars are accommodated within the small arches by being misplaced, but differently on the two sides. On the left, P^3 is rotated with buccal surface facing mesially and P^1 and P^2 overlap. On the right, P^3 is rotated with buccal surface turned distally, but P^4 is rotated so that its buccal surface is turned mesially. × *c*. 0.8.

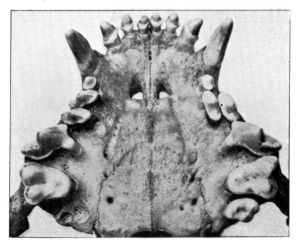

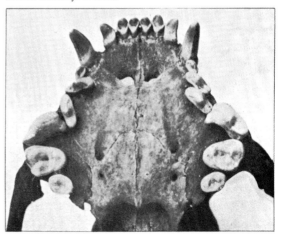

272 Section 2: Variations in Position

protrusion of the incisors is exaggerated by their anterior inclination, which tends to occur if the overjet is sufficient for the teeth to lie against the lower lip, thus exerting pressure on them. The whole of the lower arch is post-normal. There is little doubt that the basic disorder is shortness of the mandible.

The conditions present in all three skulls described above indicate that the overjet is due to subnormal development or growth of the mandible.

Figure 11.81 shows a skull of a fox terrier with marked incisor underjet compared with a normal skull. In the abnormal skull, in addition to the incisor underjet, the upper canine occludes some distance distally to the lower canine. P^1, P^2 and P^3 are also in slightly post-normal occlusion. On the other hand, P^4 and the molars are in normal occlusion. The basal length of the cranium was 13.1% less and that of the mandible 4.1% less than the average of nine normal specimens, including the one in Figure 11.81. The mandible and its tooth arch were similar to those of the normal specimen. Similarly, the fox terrier with incisor underjet has a shorter palate than normal. In the normal fox terrier, the length of the palate is 74.0 mm and there is ample room for the premolars; the length of the palate to

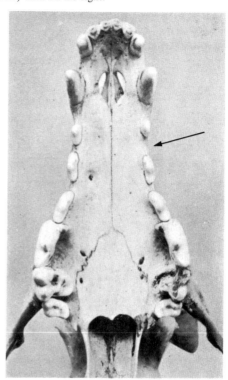

Fig. 11.77. *Canis familiaris* (domestic dog, borzoi). ZMUC 1506. The upper premolars are asymmetrically arranged so that the space between P^2 and P^3 is larger on the left (arrow) than on the right.

Fig. 11.76. Left: *Canis Lupus* (wolf). RCS Odonto. Mus., A 149.21. Centre: *Canis familiaris* (borzoi). RCS Odonto. Mus., G 71.2 (Lateral view Fig. 11.70). Right: *Canis familiaris* (greyhound). RCS Odonto. Mus., A 141.5.

In the borzoi and to a lesser extent in the greyhound, the jaws and palate are longer and the teeth are more spaced than in the wolf.

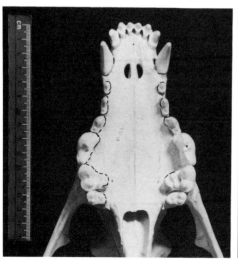
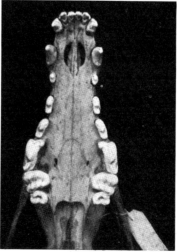
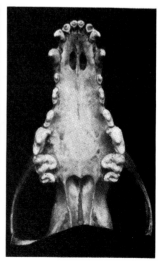

Fig. 11.78. *Canis familiaris* (domestic dog, greyhound).
Above: RCS Odonto. Mus., G 18.32. A skull showing normal occlusion. Note that the upper incisors occlude just anteriorly to the lower incisors and that the P^1 and P^2 do not actually meet P_2 and P_3. $\times c. 0.5$.
Below: RCS Odonto Mus., G 71.6. The mandible is short and the lower arch is in inferior retrusion. P_1 is distal to P^1 instead of mesial to it and the lower canine is distal to the upper canine. $\times c. 0.5$.

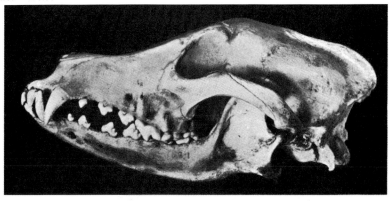

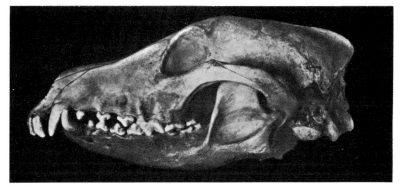

Fig. 11.79. *Canis familiaris* (domestic dog, sealyham). RCS Odonto. Mus., G 71.53. Incisor overjet with the lower canines occluding distally to the upper canines. No spacing of the lower premolars and hardly any of the upper premolars. $\times c. 0.5$.

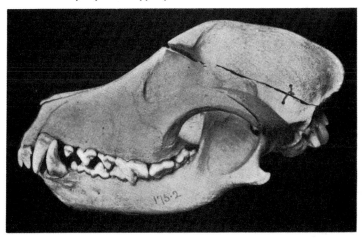

Fig. 11.80. *Canis familiaris* (domestic dog). BMNH 1919.7.7.2307. Extreme protrusion of the upper incisors and canines which are also inclined anteriorly. Both P_4 are tilted and rotated and the lower molars are more distal than normal so that M_2 is partly overlapped by the base of the coronoid process. Normally M_2 is anterior to the process.

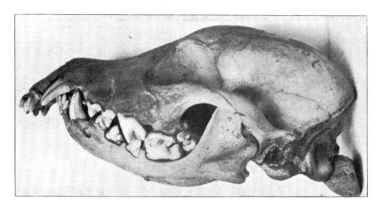

Fig. 11.81. *Canis familiaris* (domestic dog, fox terrier).
Above: RCS Odonto. Mus., A 143.6. A skull showing normal occlusion. Note that the premolars are slightly spaced and do not meet in occlusion.
Below: RCS Odonto. Mus., G 7.152. Gross disproportion in the length of the jaws; the upper arch mesial to P^4 is in post-normal occlusion, particularly mesially where I^3 and the upper canine occlude distally to the lower canine. $\times c.\,0.6$.

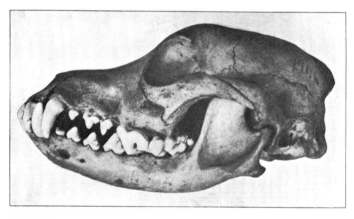

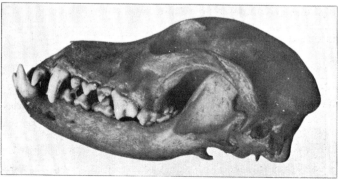

length of the cheek teeth is 74.5%; in the abnormal specimen, the length of the palate is 69.0mm, the premolars are crowded and the length of the palate to the length of the cheek teeth is 80.4%. These comparative measurements suggest that the underjet of the mandibular incisors and canines is due to sub-normal growth of the upper jaw, probably mostly of the premaxilla.

Another example of the same abnormality in a pariah dog is shown in Figure 11.82. P^1, P^2 and P^3 are distal to their normal relationships with the mandibular teeth and the mandibular canines are mesial to their normal positions; the occlusion of the fourth premolars and the molars is normal. The maxillary premolars are crowded. The term pariah is given to those nondescript dogs that haunt the villages of the Middle East, Asia and Africa (Lydekker, 1894).

The incidence of malocclusion in a large number of dogs of known breed has been recorded by Hauck (1942).

(b) Incisors

Irregularity in the position of the incisors is more common in short-muzzled than in long-muzzled breeds and is usually confined to the mandibular teeth, I_2 erupting lingually to the line of the other teeth. Rotation of the teeth is less common (Figs. 11.83 and 11.84).

Fig. 11.82. *Canis familiaris* (domestic dog, pariah). BMNH 166c. Disproportion between upper and lower jaws so that all the upper teeth mesial to P^4 are in post-normal occlusion; for instance, the lower canine occludes mesially to I^3. Nevertheless, the occlusal relationship between P_4 and the molars is normal. The mandible is probably of normal length but the anterior part of the upper jaw is under-developed. Reduced.

Fig. 11.83. *Canis familiaris* (domestic dog). Both I^2 are rotated symmetrically with the labial surfaces turned towards I^3.

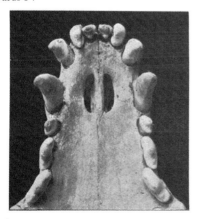

Section 2: Variations in Position

(c) Canines

In the wild canids, no positional variations of the canine were seen and, in the domestic dog, such variations are rare; four examples were noted. In three skulls, the right maxillary canine was misplaced as shown in Figure 11.85.

An extreme abnormality in the position of a maxillary canine was recorded by Cadéac (1910), the left upper canine having erupted in the palate near to the mid-line.

Fig. 11.84. *Canis familiaris* (domestic dog). RCS Odonto. Mus., G71.631. Right I_3 rotated with the labial surface (arrow) turned towards the canine. $\times c. 0.8$.

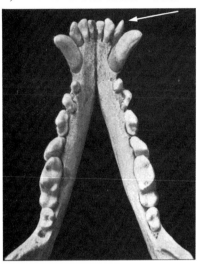

The head of a dog of unknown breed (RCS Odonto. Mus., G71.12) showed a remarkable condition. The right lower canine had been displaced lingually and over-erupted; it had pierced the palate so that there was a passage into the nasal fossa.

(d) Premolars

In some dogs with muzzles of normal length, there are spaces between the premolars (Table 11.16) but, in many, the premolars are in contact or even crowded. The principal tooth to be affected is P3 (Figs. 11.86–11.88).

P^3, although separated from P^4, may be placed obliquely to the line of the arch. In the skull of a collie (Fig. 11.89), both P^3 are placed to the lingual side of P^2 although there is ample room between P^2 and P^4 for the tooth. On the right, the crowding is exacerbated by the presence of an extra premolar.

An asymmetrical arrangement of the premolars is common in dogs and has already been noted in the skull of a borzoi (Fig. 11.77).

(e) Molars

In the section dealing with wild canids, attention was drawn to a curious displacement downwards of the distal portion of an M_1 in a coyote (p. 264). A similar irregularity is present in the skull of a foxhound (Fig. 11.90). The cause of the abnormality is traceable to the position taken by M^1. In the normal skull (Fig. 11.90C), the buccal surfaces of P^4 and M^1 form an even curve; the distal end of P^4 being in contact with the mesio-buccal portion of M^1. In the abnormal skull

Fig. 11.85. *Canis familiaris* (domestic dog, Pomeranian). RCS Odonto. Mus., G71.11. Right upper canine misplaced so that the lower canine occludes with its distal surface instead of mesially to it. This seems to be due simply to mischance. If the eruption of the lower canine slightly preceded that of the upper one which happened to engage mesially to the lower canine instead of distally, the inclined planes of the engaged surfaces might make it difficult for the teeth to disengage and the upper canine could be pressed mesially. $\times c. 0.83$.

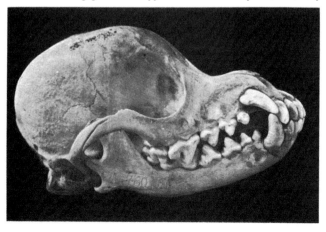

Fig. 11.86. *Canis familiaris* (domestic dog, retriever). Both P³ are symmetrically rotated through nearly 90° with the buccal surfaces turned mesially. ×c. 0.75.

Fig. 11.88. *Canis familiaris* (domestic dog). RCS Odonto. Mus., G71.32. Left P³ rotated with the buccal surface turned distally. ×c. 0.6.

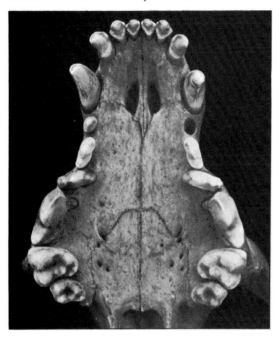

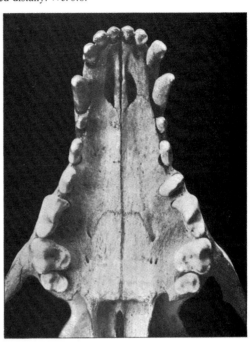

Fig. 11.87. *Canis familiaris* (domestic dog, Aberdeen terrier). RCS Odonto. Mus., A143.62. Symmetrical rotation of both P³ with buccal surfaces turned mesially. P² also symmetrically rotated but with buccal surfaces turned distally. ×c. 0.5.

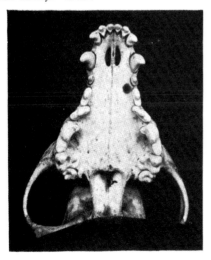

Fig. 11.89. *Canis familiaris* (domestic dog, collie). Both P³ symmetrically slightly rotated with their mesial margins overlapped buccally by P². A supplemental right P¹ is present. ×c. 0.6.

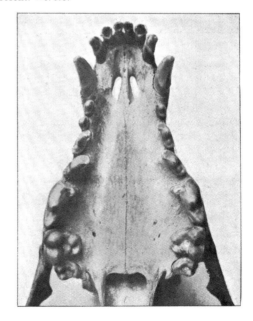

278 Section 2: Variations in Position

Fig. 11.90. *Canis familiaris* (domestic dog, foxhound). RCS Odonto. Mus., G 71.81. ×c. 0.5.
A: Right lateral view of skull. Right M_1 not fully erupted, being prevented by M_2 which has tilted mesially to lie partly over the distal part of M_1. This condition is probably secondary to the misplacement of the opposing P^4 and M^1.
B: Palatal view. Both P^4 are slightly rotated and their distal margins partially overlap the buccal surfaces of M^1 which are slightly rotated with their lingual surfaces turned mesially.
C: Normal upper tooth arch of a foxhound for comparison. ×c. 0.5.
D: Diagrams showing lingual views of the occlusion of left P4 and M1, on the left, in the normal skull; on the right, in the abnormal skull (G 71.81). In both diagrams, M^1 have been drawn as if tilted up lingually to display the occlusal surfaces.

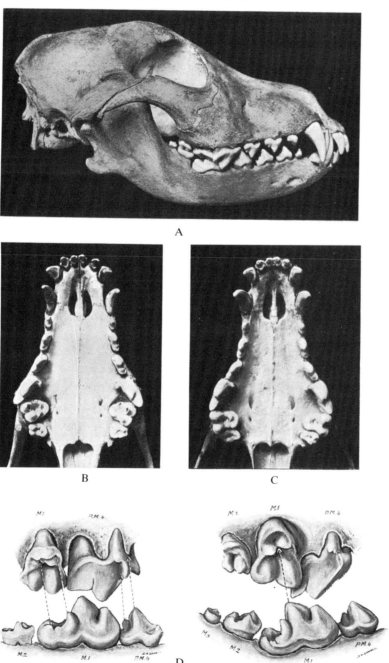

(Fig. 11.90 A, B), the distal end of P^4 overlaps the mesio-buccal portion of M^1 which is rotated and is, so to speak, under cover of P^4; the mesial portion of M^1 is depressed. The mesial two-thirds of M_1, which is blade-shaped, is covered by the distal three-fourths of P^4; the distal portion of P_4, which is at a lower level, meets the occlusal surface of M_1 (Fig. 11.90 D). In the abnormal skull, the rotation of the molar has caused the lingual cusp to be directed mesially; as a result of the alteration in the position of the tooth, the prominent cusp of M_1 strikes the occluding portion of M^1; the effect of this is to depress the lingual portion of M^1 and the distal part of M_1.

CHAPTER 12

Order **PINNIPEDIA**

Family **Otariidae**. 210 specimens examined (26 lacked mandibles)
Family **Phocidae**. 443 specimens examined (7 lacked mandibles)

There is a wide difference in the arrangement of the cheek teeth in the various species of the seals and amongst individuals of the same species.

In the Otariidae, the teeth are usually separated from one another by gaps which vary in size, the spaces between the maxillary molars being much larger than those between other teeth. The teeth, when in occlusion, interdigitate and the last tooth in the mandible

Fig. 12.1. *Zalophus californianus* (Californian sealion). BMNH 1873.11.5.8. The mandibular premolars are oblique to the line of the arch and in echelon (Chapter 13, p. 286) but the left P_4 (arrow) is more oblique than its isomere.

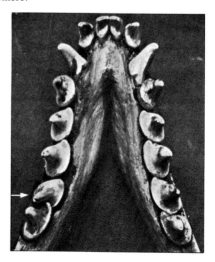

sometimes passes to the buccal side of the upper tooth.

Marked rotation of one or more of the premolars sometimes occurs (Fig. 12.1).

Amongst the Phocidae, certain species such as the Weddell seal (*Leptonychotes weddelli*) and the leopard seal (*Hydrurga leptonyx*) have the teeth spaced, whereas in others, for example the common seal (*Phoca vitulina*) and the monk seal (*Monachus monachus*), one or more of the cheek teeth are rotated. Ooë and Esaka (1981) have shown that in *P. vitulina* the amount of overlap of the cheek teeth decreases with advancing age.

CHAPTER 13

The Ungulates: Orders ARTIODACTYLA, PERISSODACTYLA, HYRACOIDEA AND PROBOSCIDEA

Order ARTIODACTYLA

Sub-order SUINA

Family Suidae

The muzzles of suids are elongated and the cheek teeth are set in straight, parallel lines. There is usually a pre-canine diastema in the upper jaw and a post-canine diastema in the mandible to accommodate the opposing canine tusk; there may also be spaces between the upper canine and P^1 and between the lower canine and I_3.

Genus *Sus*, wild Old World pigs. 191 specimens examined

In 33 skulls (17.3%) of wild pigs (*Sus scrofa*) in which the upper jaw was short compared with the sum of the mesio-distal lengths of the post-canine teeth, there was irregularity of the premolars, which was marked in eight (4.2%). Rotation of P_3 and P_4 or both occurred in ten (5%). In the bearded pig (*Sus barbatus*), which has a particularly long, slender muzzle, there were no irregularities of the premolars in the 19 skulls examined.

In three skulls of *Sus*, one or both M_3 were rotated

and, in one of them, a *S. barbatus* (formerly RCS Osteo. Series, 1760), the rotation was by nearly 90°.

Sus scrofa, domestic pigs

It is likely that pigs have been domesticated from local sub-species of the wild boar (*Sus scrofa*) in several different parts of the world. Western domestic pigs are descended from western sub-species of *Sus scrofa* and used to resemble them closely, except for being smaller in size. This was established on the basis of detailed craniological measurement by Nathusius (1864). However, in the nineteenth century, highly modified domestic Chinese pigs with high crania and short jaws were introduced into Europe and crossed with the indigenous, long-muzzled European breeds, so that now most pig breeds in Europe and North America have short muzzles and high crania (Groves, 1981; Nathusius, 1864). Darwin's account (1868) of the history of the domestic pig is still of much value.

In long-muzzled breeds, the incisor occlusion is usually edge-to-edge but, in short-muzzled pigs, the maxillae are often short relative to the mandibles, resulting in underjet. Table 13.1 shows that shortness of the jaws in these domesticated forms is not associated with shortening of the post-canine tooth rows, so these tend to be crowded. Henry (1935) found that the developing M_3 is placed higher in the ascending ramus in short-muzzled than in long-muzzled pigs and is more likely to be impacted (Chapter 16, p. 348).

According to Harvey and Penny (1985), deviation of the mandible to one side associated with malocclusion is common in tethered pigs and is probably due to them pulling on their tethers in one direction to reach food and water.

Deviation of the snout to one side can be the result of atrophic rhinitis (Penny and Mullen, 1975) even in wild boars (Lutz, 1988).

Genus ***Potamochoerus***, bush pigs. 60 specimens examined

The only positional anomalies were slight rotation of P^2 in five skulls and of P^4 in two, and more marked rotation of P^3 and P^4 in another (Fig. 13.1).

Genus ***Hylochoerus***, giant forest hogs. 15 specimens examined

In the specimens examined, there were no positional variations of the teeth.

Fig. 13.1. *Potamochoerus porcus koiropotamus* (bush pig). BMNH 1838.4.16.81. Left P^4 rotated with buccal surface facing distally; as it is much less worn than the right P^4, it appears to be much wider bucco-lingually. Right M^1 is missing $\times c. 0.66$.

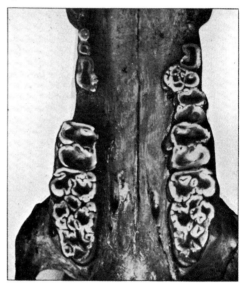

Table 13.1 The amount of upper cheek-tooth crowding in crania of various breeds of PIG *(Sus scrofa)* in the Odontological Museum of the Royal College of Surgeons

Length of muzzle	Breed	No. of specimens	Mean length of palate (A)	Mean length of cheek teeth (B)	$\frac{B \times 100}{A}$
Long	Large white	6	208.0	113.9	54.8
Short	Welsh	3	172.7	106.8	61.8
	Middle white	3	158.8	107.0	67.4

Crowding seems to be less in the long-jawed than in the short-jawed breeds, but the samples are small. Lengths in mm.

Genus ***Phacochoerus***, wart hogs. 79 specimens examined

The wart hog has powerful upper canines, curving upwards and outwards; the lower canine is smaller and occludes only with the lower part of the upper canine. M^3 is enormous, having about twice the mesio-distal length of M^2. M_1 is usually much worn before P3 and P4 erupt and is destroyed rapidly distally by approximal wear from M2 so that M2, in time, may come into contact with P4 (Fig. 13.2). Eventually, M2 also tends to break up and disappear so that M3 comes into contact with P4; there is therefore a continuous mesial drift of M3. I3 and P1 are lost early and, in old animals, the canines and M3 may alone be present as in a male wart-hog skull in the British Museum (BMNH 1921.4.23.1).

In one skull, the right P_4 was almost transverse to the arch (Fig. 13.3).

Genus ***Babyrousa***, babirusa. 58 specimens examined

A feature of babirusas is the curious arrangement of the upper canine in males. After emerging through the top of the snout, it takes an upward course and then curves backwards and eventually downwards and does not come in contact with the mandibular tooth as in the other suids. We have described this tooth in Chapter 17 (p. 366). There is considerable variation in the direction taken by the upper canines; usually they incline towards the mid-line and they may meet or even cross (Fig. 13.4).

There was slight rotation of P^3 in four skulls (6.9%).

Family **Tayassuidae**

Genus ***Tayassu***, peccaries. 45 specimens examined

In the Tayassuidae, the upper canines of males are directed downwards. The lower canines are directed upward, outward and backward, and are received into depressions in the bone anterior to the upper canines.

In a female collared peccary (*Tayassu tajacu*,

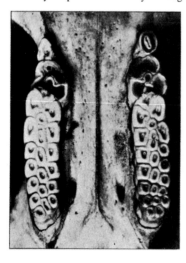

Fig. 13.2. *Phacochoerus aethiopicus* (wart hog). BMNH 1912.7.28.7. Wear has totally destroyed both M^1 so that the M^2 are in contact with the P^4 (no sign of P^3); the distal surfaces of the M^3 have been deeply cupped out by approximal wear by the convex mesial surface of the much more recently-erupted and mesially-moving M^3. $\times c.\,0.66$.

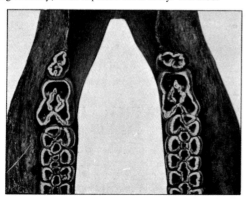

Fig. 13.3. *Phacochoerus aethiopicus* (wart hog). BMNH 1894.3.8.18. Right P_4 rotated with the buccal surface facing mesially; both M_1 have worn away. $\times c.\,0.66$.

Fig. 13.4. *Babyrousa babyrussa* (babirusa). BMNH 1900.3.30.19. The right upper canine is directed obliquely across the left upper canine. $\times c.\,0.2$.

BMNH 1903.7.7.128), I^1 overlapped each other and both I^2 were misplaced slightly lingually.

Irregularities in the positions of the premolars were present in 25%, usually in the form of slight rotation, but sometimes more marked (Fig. 13.5).

Dr G. Hagmann compared the dentitions of the two species of peccary from Taperinha, Brazil. He found that variations of the position of the premolars occurred in 22 (27.8%) of 79 *T. tajacu*, which has a narrow palate, and in only 21 (18.0%) of 115 *T. albirostris*, which has a broad palate. The opposite was true of the incisors, 3.8% varying in *T. tajacu* and 11.3% in *T. albirostris*.

In one captive collared peccary, the left I^2 was almost horizontal behind I^1 and there were slight irregularities of the post-canine teeth (Fig. 13.6).

Family **Hippopotamidae**. 30 specimens examined

In one skull, the lower canines curved out horizontally with their tips facing distally (Fig. 13.7). The cheek teeth, which normally incline lingually in a straight line, are arranged in a curve. Embedded in the bone beneath the left lower cheek teeth are four denticles (Chapter 25, p. 585 and Fig. 25.13).

Rotation or misplacement lingually of premolars was present in 10 specimens. An example is seen in Figure 13.8; the left P^4 has erupted towards the palate and is tilted. The premolars were asymmetrically arranged in two skulls.

In one hippopotamus skull (Fig. 6.6), the left M_1 is absent and the M_2 and M_3 have moved mesially.

Sub-order **TYLOPODA**

Family **Camelidae**

Camelids are in a separate sub-order from the ruminants because they have neither horns nor antlers, the cannon bones are incompletely fused and the stomach has only three chambers. Incisors and canines are present in the upper jaw, but the first two upper incisors are lost in adolescence and replaced by a fibrous pad, similar to that in the ruminants.

Genus ***Lama***, llamas, guanacos and alpacas

The upper cheek teeth of llamas are arranged in slight curves with their convexities outwards. There are only two premolars on each side in both jaws, P_3 being often lost quite early in life; P^3 is a small narrow tooth; P^4 is also small, with its buccal surface lingual to the line of the molars. The position of P^3 is variable (Figs. 13.9 and 13.10).

Sub-order **RUMINANTIA**

Ruminants have such a similar basic morphology, particularly of their dentitions, that it is only necessary to describe a single basic type of dentition and point out minor differences in some genera.

The upper cheek teeth are arranged in slightly curved rows with the convexities outwards. Ideally, the buccal surfaces of the tooth rows form an even line, apart from the prominent vertical ridges that the sur-

Fig. 13.5. *Tayassa tajacu* (collared peccary). BMNH 1921.1.1.78. Symmetrical rotation of P^3 with buccal surfaces facing mesially. Left P^2 missing. Right P_4 misplaced buccally. ×c. 0.5.

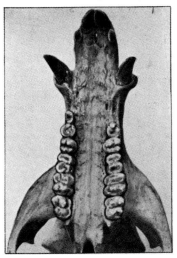
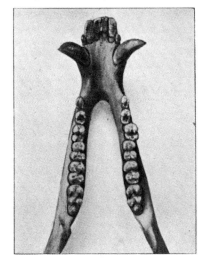

Fig. 13.6. *Tayassu tajacu* (collared peccary). ♀. Captive. Formerly RCS Osteo. Series, 1744. Left I^2 obliquely horizontal and only partially erupted. Both P^4 misplaced lingually and both M^3 tilted mesially against M^2. ×c. 0.6.

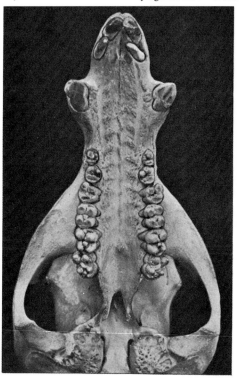

faces bear (Fig. 13.11). Being arranged in a curve, the more or less rectangular crowns have variable obliquity to the midline. However, it is common for the distal teeth to be slightly oblique to the arch, with their buccal surfaces facing slightly distally. This may be so even if the rows are nearly straight and may affect the mesial part of the row as well (Fig. 13.44). This obliquity of the teeth to the line of the arch, which has to be distinguished from the obliquity to the midline consequent on the curve of the arch, produces a step-like or echelon appearance to the buccal surface of the row, with each tooth being overlapped by the tooth behind (Figs. 13.35 and 13.44).

Minor degrees of obliquity are very variable in particular species and are so common that it is extremely difficult to decide what is the interface between normal and abnormal. For some ruminant genera, Colyer simply attempted to establish the range of variation. When the obliquity was unusually marked and affected single teeth, we refer to the increased obliquity as an abnormal rotation. Obliquity and rotation are usually associated with some buccal overlap. If the distal surface of a particular cheek tooth was overlapped by the tooth behind by a quarter of its buccal-lingual width or less, the amount of overlap was designated as one-quarter; if by one-third to one half, as one-third; if by one-half or more, as one-half.

In the mandible, the incisors are arranged in a curved fan-like fashion. The lower cheek-tooth rows

Fig. 13.7. *Hippopotamus amphibius* (hippopotamus). RCS Odonto. Mus., G 77.5. The lower canines have grown out horizontally and turned distally. ×c. 0.2.

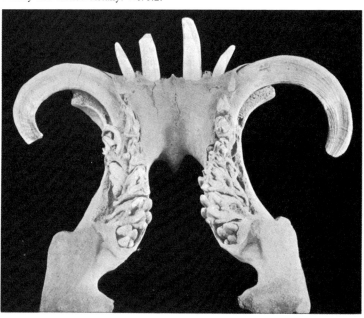

are similar to the upper ones but with less curvature. The morphology of the lower cheek teeth is such that it is their lingual surfaces that present an even line. Obliquities of the teeth corresponding to those of the upper arch are common, but are usually less marked than in the upper jaw. The lower arch is distinctly narrower than the upper; thus, disparity between the arches, anisognathism, means that in general, during mastication, only one side is in occlusion at a time.

Positional anomalies are more common among the premolars than the molars, perhaps because in this group of mammals the premolars erupt after the first

Fig. 13.8. *Hippopotamus amphibius* (hippopotamus). Formerly RCS Osteo. Series, 1864. The left P^4 is misplaced lingually and rotated with buccal surface turned mesially. $\times c. 0.3$.

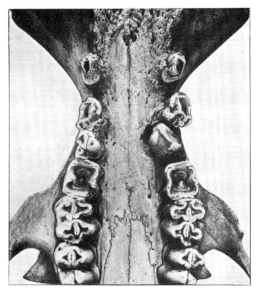

Fig. 13.9. *Lama guanicoe* (guanco). ♀. BMNH 1902.1.1.111. Both P^3 are rotated, the buccal surface facing mesially on the right and distally on the left. The mandibular arch was normal. $\times c. 0.66$.

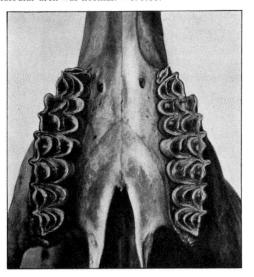

Fig. 13.10. *Lama guanicoe* (guanco). ♂. BMNH 1890.2.20.15. Both P^3 are rotated with buccal surfaces turned mesially and both P^4 are misplaced lingually. $\times c. 0.66$.

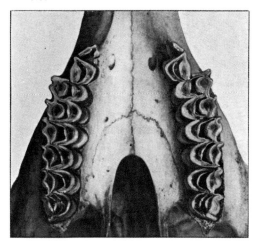

Fig. 13.11. *Synceros caffer nanus* (dwarf buffalo). ♀. BMNH 1930.11.11.424. The typical normal upper tooth arch in ruminants. $\times c. 0.5$.

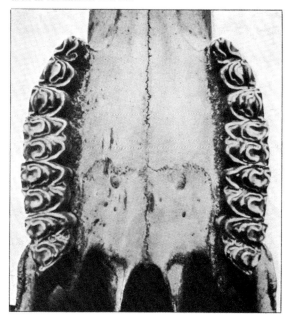

and second molars. Mesial drift of the molars, continuing while the deciduous molars are being replaced by the premolars which have shorter mesio-distal lengths, diminishes the space available for them (see Chapter 9, p. 157).

A total of 1995 ruminants were examined; 181 (9.1%) showed major irregularities in the position of the teeth.

Family **Tragulidae**

Canines are present in both jaws, the upper ones being particularly well developed.

Genus *Tragulus*, chevrotains. 103 specimens examined

The echelon arrangement of the molars and P^4 is marked. The overlap of P^4 is from one-quarter to one-third, smaller amounts being present in about 12% and more in 13%. In four, there was rotation of one or both P^3. In three of these, there was no associated rotation of P^4 and, in one, P^4 was slightly rotated.

A skull showing rotation of P^3 is seen in Figure 13.12.

Family **Cervidae**

An example of an achondroplasic or so-called bulldog skull in a one-year old roe buck (*Capreolus capreolus*) was described by von Braunschweig (1980).

According to Stieve (1941), in roe deer the incisors normally close just behind the anterior margins of the bony premaxillae. He found several skulls which he regarded as abnormal in which the incisors closed slightly in front of the premaxillae, and one that closed about two centimetres behind their anterior margin. This seems to amount to parrot mouth (see this Chapter, p. 294, 309) which has been described in several other species of deer. Three or four examples were found among 1000 white-tailed deer (*Odocoileus virginianus*) from the wild (Severinghaus and Cheatum, 1956). The literature was reviewed by Short (1964) who described two examples in mule deer (*O. hemionus*).

Jenks, Leslie and Gibbs (1986) described the skull of an infant white-tailed deer with the bull-dog deformity in which both the shortened upper jaw and the projecting mandible were markedly curved to one side.

Whitehead (1985) found an incidence of inferior retrusion of about 10% in a population of feral sika deer (*C. nippon*) in Lancashire, England. This is probably the result of founder effect (p. 10) as the deer are descended from a few which were introduced into the area for hunting in the early years of the present century.

Moschus moschiferus, musk deer. 28 specimens examined

Excessive rotation of P^4 was noted in three skulls (Fig. 13.13).

Fig. 13.13. *Moschus moschiferus* (Siberian musk deer). ♀. BMNH 1908.2.29.4. Symmetrical rotation of P^4 with buccal surfaces turned distally.

Fig. 13.12. *Tragulus javanicus affinis* (chevrotain). ♀. BMNH 1914.12.8.237. Right P^3 rotated with buccal surface turned distally. ×c. 1.5.

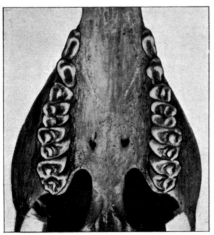
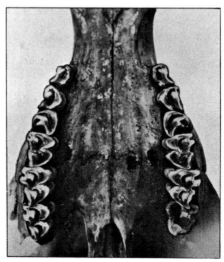

Genus *Muntiacus*, muntjac. 95 specimens examined

There is a slightly oblique arrangement of the cheek teeth with consequent echelon appearance of the buccal surfaces. In about 25%, the overlap of P^4 was one-quarter, in 10% about one-third and in three specimens (3.2%) one half (Fig. 13.14). More rotation of P^3 was present in four (Fig. 13.15). In two, there was an excessive symmetrical rotation of P^2 (Fig. 13.16).

Fig. 13.14. *Muntiacus muntjak vaginalis* (Indian muntjac). ♂. BMNH 1921.8.2.29. The rotation of the premolars, especially that of the P^4 with the buccal surfaces turned slightly distally, is greater than normal. ×c. 0.75.

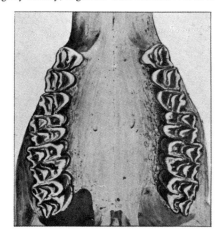

Fig. 13.15. *Muntiacus muntjak* (Indian muntjac). BMNH 1924.9.4.10. Premolars irregularly rotated; in particular, the buccal surfaces of P^3 face more distally than normal. ×c. 0.75.

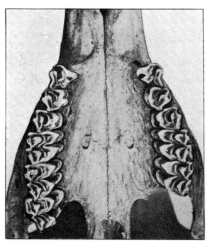

Genus *Odocoileus*, North American deer

Overlap of P^3 and P^4 ranges from one-quarter to one-third. Minor disorders in the arrangement, often different on the two sides, were present in seven specimens (Fig. 13.17).

Genus *Mazama*, brocket deer. 60 specimens examined

The usual overlap of P^4 was from one-third to one-half; in about 10%, the overlap was less. Greater overlap of P^3 or P^4 was present in eight; in one (*M. americana*, BMNH 1846.6.1.24), the right P^3 was rotated sufficiently to bring the disto-buccal border in contact with the mesio-lingual border of M^1; on the left, P^4 had erupted mesially to its normal position but, as the P^3 had been lost, it was not possible to determine the exact nature of the irregularity. One skull (*M. americana zetta*, BMNH 1885.4.22.7) had the right P^2 rotated so that the mesial portion faced lingually.

Genus *Hippocamelus*, guemal. 9 specimens examined

There was a tendency for more rotation of the P^3 compared with P^4 than in other Cervidae (Fig. 13.18).

Fig. 13.16. *Muntiacus muntjak rubidus* (Indian muntjac). ♂. BMNH 1899.12.9.83. The arch is unusually narrow between the P^2 which are slightly rotated with buccal surfaces turned distally.

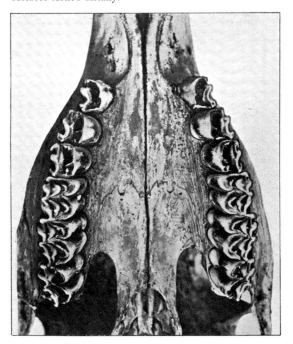

Genus ***Blastocerus***, marsh deer. 15 specimens examined

There was little rotation of the premolars; only in one specimen was the rotation marked.

Genus ***Capreolus***, roe deer. 52 specimens examined

The upper cheek-tooth arch bends sharply lingually at the junction of P^2 and P^3 but normally these teeth do not overlap. The overlap of P^4 was one-quarter, with less overlap of P^3, in about 40%. Greater overlap indicates insufficient room for the teeth (Fig. 13.19). The overlap of P^4 may be as much as one-half (Fig. 13.20). P^3 seems to be more liable to be rotated or misplaced than P^4; in three specimens, it was overlapped by P^4 by one-half (Fig. 13.21); in one (Fig. 13.22), extreme rotation of the tooth was associated with an

Fig. 13.17. *Odocoileus virginianus peruvianus* (white-tailed deer). ♂. BMNH 1894.11.20.3. Symmetrical slight rotation of P^3 and P^4 with buccal surfaces turned distally. ×c. 0.66.

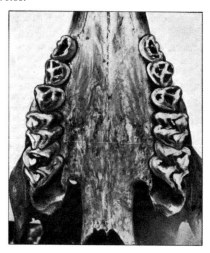

Fig. 13.18. *Hippocamelus bisculcus* (Chilean guemal). BMNH 1914.10.12.1. Symmetrical marked rotation of P^3 with buccal surfaces turned distally. ×c. 0.6.

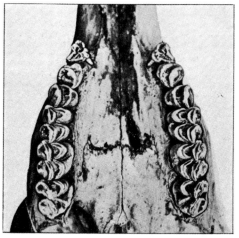

Fig. 13.19. *Capreolus capreolus pygargus* (Central Asian roe deer). ♂. BMNH 1905.3.21.2. Symmetrical crowding of the premolars with slight rotation of P^3 and P^4 with buccal surfaces turned distally. ×c. 0.85.

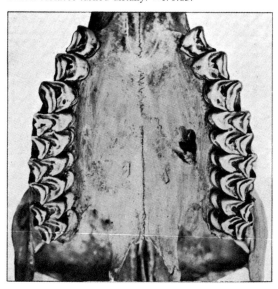

Fig. 13.20. *Capreolus capreolus capreolus* (roe deer). ♀. BMNH 1933.1.10.2. Symmetrical rotation of P^4 with buccal surfaces turned distally. ×c. 0.8.

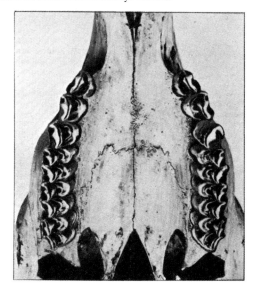

increased rotation of P^4. The position of P^2 differed from normal in about 40% of specimens; in one, the teeth were almost parallel with the mid-line of the palate (Fig. 13.23). Table 13.2 summarizes the amount of overlap of the upper premolars and shows that the normal overlap of P^3 by P^4 and of P^4 by M^1 is one-quarter.

Genus **Hydropotes**, Chinese water deer. 18 specimens examined

In males, the upper canines form well-developed tusks. Normally, the overlap of P^4 is about one-third, the overlap of P^3 is one-quarter, and the overlap of P^2 about one-eighth. The amount of overlap is variable and the degree of rotation may not be exactly the same on both sides. In one specimen, the left P^3 was rotated

Fig. 13.22. *Capreolus capreolus capreolus* (roe deer). ♂. BMNH 1911.12.5.1. Crowding of the premolars with overlapping and, on the right, rotation of P^3 with buccal surface turned distally. The arch is unusually narrow between the P^3. $\times c.\,0.75$.

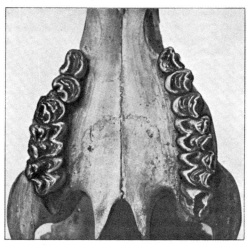

Table 13.2 The degree of overlap of the upper premolars in 50 skulls of ROE DEER *(Capreolus)*

No. of specimens	Degree of overlap by tooth behind		
	P^4	P^3	P^2
1	Nil	Nil	Nil
1	Nil	1/3	Nil
20	1/4	1/4	Nil
12	1/4	1/4	1/4
3	1/4	1/4	1/6
1	1/4	1/3	Nil
2	1/4	1/2	Nil
1	1/4	1/4	1/8
1	1/6	1/6	Nil
1	1/6	1/3	1/3
2	1/3	1/3	1/4
3	1/3	1/4	Nil
1	1/3	1/2	1/3
1	1/2	1/3	1/4

The specimens shown in Figures 13.20–13.22 are not included.

Fig. 13.21. *Capreolus capreolus capreolus* (roe deer). ♀. BMNH 1910.10.18.1. Symmetrical rotation of P^3 with buccal surfaces turned distally. $\times c.\,0.8$.

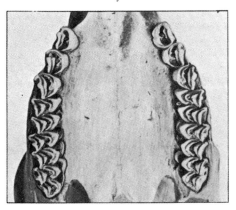

Fig. 13.23. *Capreolus capreolus bedfordi* (Chinese roe deer). ♀. BMNH 1911.2.1.263. Symmetrical slight rotation of the P^2 with buccal surfaces turned distally; they are normally turned slightly mesially.

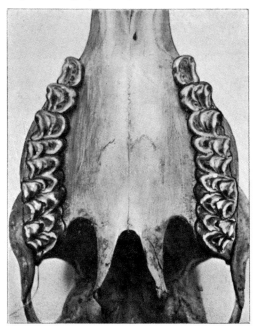

with the buccal surface facing distally (Fig. 13.24). In two, the overlap of P^4 was one-half and, in another, the right P^3 was overlapped to a similar extent.

Family **Bovidae**

Genus ***Bos*** (*senso lato*), gaur and banteng. 65 specimens examined

In the original edition, Colyer mentioned that he had examined 65 skulls of *Bos* from the wild. Strictly speaking, the only wild forms of the genus *Bos* are the extinct species *B. primigenius* and *B. namadicus*, and the very rare wild yak (*B. grunniens*). It seems likely that Colyer used the term *Bos* in the wide sense to include *Bibos*, the gaur and the banteng. In one skull, the left P^2 was rotated so that its buccal surface faced distally.

Bos taurus, domestic unhumped cattle, and *Bos indicus*, domestic humped cattle. 38 crania and 62 mandibles examined

In the domestic ox, slight overlap of the lower incisors is common in young animals but, as the teeth wear, the overlap tends to disappear. However, even in adults, crowding, overlap and rotation of the incisors sometimes occur (Figs. 13.25 and 13.26). Joest (1926, fig. 83) illustrated a mandible in which all the incisors are rotated, some by more than 180°. Grosser misplacement, as in Figure 13.27 where an incisor has erupted on the lingual aspect of the mandible, is more rare. A large abscess in connection with the retained, deciduous incisor may have contributed to this misplacement.

Increased rotation of the premolars occurred in one ox skull, P^4 being overlapped by M^1 to the extent of one-half; in one mandible, the left P_2 was rotated so that the buccal surface faced distally.

Niata cattle were a domestic breed that lived in South America near the river Plate in the nineteenth century; they are now extinct (Andresen, 1931). They were described by Charles Darwin (1868). The skull shown in Figure 13.28 was presented to the Royal College of Surgeons by Darwin in 1840 but was unfortunately destroyed in 1941. It was described by Richard Owen in the *Catalogue of Osteological Speci-*

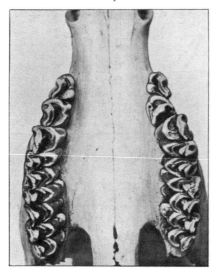

Fig. 13.24. *Hydropotes inermis* (Chinese water deer). ♀. BMNH 1908.11.14.11. Left P^3 rotated nearly 90° with buccal surface turned distally. $\times c. 0.75$.

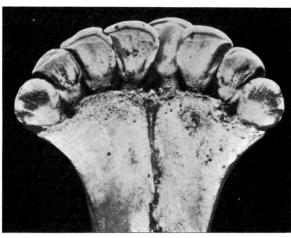

Fig. 13.25. *Bos taurus* (domestic ox). RCS Odonto. Mus., G 78.51. Lingual aspect of the anterior portion of the mandible; the incisors are irregular in arrangement and overlap, especially the I_1. $\times c. 0.9$.

Ch.13. The Ungulates 293

mens (1853) as follows: 'It is remarkable for the stunted development of the nasals, premaxillaries, and fore part of the lower jaw, which is unusually curved upwards to come in contact with the premaxillaries. The nasal bones are about one-third the ordinary length, but retain almost their normal breadth. A triangular vacuity is left between them, the frontal, and the lachrymal, which latter bone articulates with the premaxillary, and thus excludes the maxillary from any junction with the nasal.' Figure 13.28 shows these features, and the palatal view shows the exaggeration of the normal curve and narrowness of the arch in the premolar region.

The shape of the skull is suggestive of achondroplasia, although it does not seem to have the globular braincase which is one of the characteristics (see achondroplasia in dogs, Chapter 11, pp. 266–268). Whether the Niata cattle were deliberately bred for these, or some other characteristics, is a matter of

Fig. 13.26. *Bos indicus* (domestic humped ox). RCS Odonto. Mus., G 78.511. Lingual aspect of the anterior portion of the mandible; the left I_3 (arrow) with lingual surface turned mesially. Palatal mucosa intact $\times c.\,0.75$.

Fig. 13.27. *Bos taurus* (domestic ox). RCS Odonto. Mus., G 78.5. Lingual aspect of anterior portion of the mandible; the remains of the left i_2 are still in position; the I_2 is erupting at least 1 cm lingual to the arch. The mucosa is intact. Reduced slightly.

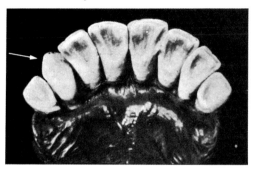
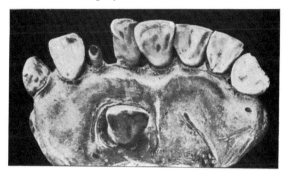

Fig. 13.28. *Bos taurus* (Niata domestic ox). Formerly RCS Osteo. Series, 1148.
Left: Lateral view. There is stunted development of the premaxillae in particular. The incisor region of the mandible is upturned. $\times c.\,0.2$.
Right: Palatal view. The curve of the cheek-tooth arch is exaggerated and the teeth are crowded so that their buccal surfaces form in irregular line (compare with Fig. 13.11). $\times c.\,0.4$.

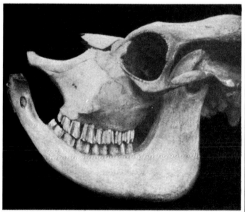
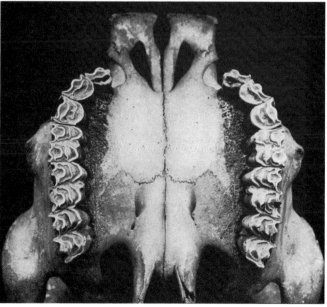

speculation; Darwin (1868) mentioned that they had unusually long hind legs. He pointed out that, although they could thrive on lush pasture, in times of drought without special care they tended to perish because the faulty jaw relationship, combined with their short lips, prevented them from browsing or grazing efficiently.

Achondroplasia, or a condition resembling it, occurs spontaneously in cattle, especially in the small Dexter breed (Crew, 1924), leading to shortened limbs and malformation of the skull. The term bulldog calf is commonly used because the head resembles that of the bulldog; an example is figured by Joest (1926, fig. 82). Usually the affected calves are either born dead or die before reaching maturity. Gilmore (1949) distinguished four forms of inherited achrondroplasia and discussed the mode of inheritance. It has been argued (Carroll, Gregory and Rollins, 1951; Gregory et al., 1953; Gregory, Koch and Swiger, 1962) that the condition is mediated by a deficiency of pituitary thyrotropic or thyroid hormone.

In a useful review, Brandt (1941) provided evidence that the condition in a lethal form occurs in a wide variety of breeds, for example, the Telemark cattle of Norway and in Nganda cattle of Uganda.

In herds where the gene for achondroplasia appears to be common, the detection of the carrier state, that is of animals in which the gene is carried but not manifest in the phenotype, is of importance so that carriers are not used for breeding. According to Emmerson and Hazel (1956), the thoracic and lumbar vertebrae in carriers show in a minor form the compression of the bodies and some other morphological features that occur in a gross form in full achondroplasic dwarfism; features can be detected radiographically in the week-old calf if it is a carrier of the gene. The more rare ateliotic or hypopituitary dwarfism, in which the animal is small but the body proportions are normal, also occurs in cattle (Emmerson and Hazel, 1956).

Bulldog head sometimes occurs in as non-lethal form. Becker and Arnold (1949) gave details of its inheritance in a Jersey herd in which many of the animals closely resembled Niata cattle and, like them, lived to maturity. The condition survived through three generations, apparently controlled by a single recessive gene.

Another heritable jaw defect found in cattle is known colloquially as parrot-beak, corresponding in our terminology with incisor overjet. The calf is born with a lower jaw that may be only half the normal length; the tongue is of normal length and therefore protrudes and hangs down over the lower jaw. Calves are either born dead or die after a few weeks. Although first described in English short-horn cattle, the condition seems to occur occasionally in most breeds in North America and New Zealand though it is usually traceable to a short-horn origin. It is sometimes associated with gross crowding and impaction of the cheek teeth, more severely in the mandible (Heizer and Hervey, 1937). In general, the anomaly appears to be transmitted as a simple recessive factor, perhaps male sex-linked (Annett, 1939). Good descriptions and reviews of the literature of lethal jaw anomalies in cattle are available in Ranstead (1946) and Grant (1956). A previously unrecorded type associated with generalized diarthrodial joint abnormalities was described by Jayo, Leipold, Dennis and Eldridge (1987).

Donald and Wiener (1954) studied occlusion in 459 cattle of various breeds and crosses and graded the degree of overjet and underjet on a numerical scale. They found no difference in the prevalence of malocclusion between young and mature animals, but underjet did appear to be more prevalent in shorthorns and beef crosses than in Ayrshires and Friesians. According to Andrews (1985a), slight underjet is usual in young calves.

In Great Britain, only licensed bulls may be used for breeding. The licensing authorities regard the normal incisor relationship as incisors against the anterior half of the upper gum pad and refuse a license if the incisors pass in front of the pad (undershot) or if the incisors occlude either with the posterior half of the pad or behind it (overshot). Using mainly the licensing data for bulls about 10 months of age for 1961–1967, Wiener and Gardner (1970) studied the prevalence of the various types of incisor occlusion in various breeds. Among 3289 bulls of several dairy breeds, the prevalences were: normal 96.6%, overshot 2.0%, undershot 1.4%. Among 2293 bulls of beef breeds, the prevalences were: normal 84%, overshot 13.7%, undershot 2.3%. These differences, especially those for overshot jaw, were statistically significant. The overshot condition was particularly common (20.1%, only 19 animals) in Jerseys (a dairy breed) and in Hereford (a beef breed) bulls (22.5%, 835 animals). Some more detailed data in respect of a Galloway herd supported the view that the incisor relationship is heritable but this material could not throw conclusive light on the mode of inheritance (because presumably only bulls with normal incisor relationship were bred from). However, Wiener and Gardner speculated whether the longer suckling period customary in dairy-calf management might account for the higher prevalence of a normal incisor relationship in dairy bulls; this suggestion accords well with the views of orthodontists in respect of human infants. Meyer and

Becker (1967) observed underjet in 31 calves but found that the anomaly disappeared at the age of 3–6 months.

Genus *Ovis*, sheep

Positional variations are rare in wild sheep. In skulls of domesticated animals, crowded incisors and rotation of the premolars are more common (Figs. 13.29 and 13.30).

The exact normal or optimal functional relationship between the incisors and upper gum pad in sheep remains uncertain. The classification of incisor occlusion by Hitchin (1948, 1957) into three types is commonly accepted. In Type 1, the incisors occlude on the pad behind its anterior margin; the incisor wear tables are approximately at right angles to the long axes of the teeth. In Type 2, the incisors occlude with the slightly sloping or rounded anterior edge of the pad and the wear tables are bevelled with a sharp chisel-like labial edge. In Type 3 the incisors pass completely in front of the upper pad and occlusion is between the upper and lower pads. In grazing, the grass is gripped between the pads and dragged across the incisors by a movement of the head. The grass is torn or cut through and the incisors become unevenly worn with an irregular labial bevel. It seems that Type 3 occlusion must be regarded as a malocclusion. It is common in some flocks, particularly among lowland Leicester sheep. Some experimental breeding experiments (Hitchin, 1948, 1957) suggested that the malocclusion is heritable. Affected sheep can thrive, except under severe hill-farming conditions, but the incisors are susceptible to periodontal disease leading to broken mouth (Chapter 24). The condition is known most commonly as undershot jaw (in New Zealand as overshot) but, in the terminology we have adopted (p. 106), it conforms to incisor underjet.

Additional consideration of the incisor apparatus is given in Chapter 24 (pp. 552–557).

Fig. 13.29. *Ovis aries* (domestic sheep). RCS Odonto. Mus., G78.2. Left I_2 rotated through nearly 180° so that its lingual surface faces that of I_3.

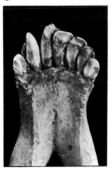

Richardson *et al.* (1979) provided some biometric information on the relationship of incisor-pad position to skull morphology.

The opposite condition, overshot jaw corresponding in our terminology to incisor overjet, where the mandible appears to be too short and, in severe examples, incisors occlude behind the upper pad, also occurs. It is known amongst breeders as 'rat-mouth' and an example is shown in Figure 13.31. Six specimens with this condition compared with 15 normal specimens of approximately the same age showed that the average basal length of the crania was 15.2% longer, and the mandibles were 2.2% longer than in the normal specimens. Nordby, Terrill, Hazel and Stoehr (1945) found that it occurred in 24 out of 1500 ewes of Rambouillet sheep (1.6%) in a flock in the USA. They showed from some breeding experiments that it was probably transmitted as a dominant with incomplete penetration, but the results were not completely conclusive because there is some possibility that the condition can be produced by non-genetic causes such as maternal malnutrition and ageing. For example, according to Suckling (1975) and Purser, Wiener and West (1982), in Type I incisor relationship there is slight forward movement of the incisors in relation to the upper pad as age advances.

It has to be borne in mind that slight degrees of incisor underjet, that is Hitchin's class 3, can develop during life by the incisors being tilted forwards on the anterior slope of the upper pad. This appears to be what Franklin (1950) referred to as the false undershot condition and he showed it can be produced by

Fig. 13.30. *Ovis aries* (domestic sheep). BMNH 1858.6.24.134. Right P^2 rotated, with the buccal surface turned distally, and misplaced lingually. Reduced.

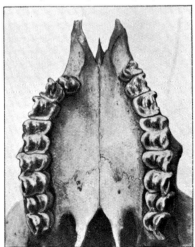

296 Section 2: Variations in Position

malnutrition. On the other hand, Wiener and Purser (1957) found that a low level of nutrition increased the tendency to overjet in Blackface, and Welsh sheep and various crosses, presumably because the mandibles were slightly underdeveloped.

A condition known as open mouth in which there is a gap between the incisors and the upper pad when the cheek teeth are in occlusion occurs in sheep. Nisbet, Butler, Robertson and Bonnatyne (1968) studied seven examples from a Scottish hill farm. All showed generalized osteoporosis and other evidence of rickets due to phosphorus and vitamin D deficiencies. The immediate cause of the anterior open bite appeared to be shortening over the vertical ramus of the mandible.

Genus *Capra*, goats. 50 specimens examined

Goats belonging to the Nubian group of breeds have strongly convex facial profiles and incisor underjet (Epstein, 1971). The skull of a Nubian goat (Fig. 13.32) shows that the nasal bone is short and convex, the premaxilla is turned downwards and the anterior part of the maxilla is very short; this has resulted in incisor underjet and post-normal occlusion of the upper cheek teeth. In another Nubian goat (Fig. 13.33), the shortness and narrowness of the anterior parts of the maxillae are associated with misplacement and rotation of the premolars; the mandible shows incisor underjet.

A West African dwarf goat with a short mandible

Fig. 13.31. *Ovis aries* (domestic sheep). The lower incisors have an abnormal overjet relation to the premaxillae. Overshot jaw. Note that the lower cheek teeth are not in post-normal occlusion. ×c. 0.33.

Fig. 13.32. *Capra hircus* (Nubian goat). ♂. BMNH G16. The facial profile is strongly convex, the maxillae are short and the premaxillae are turned downwards. There is post-normal occlusion of the upper cheek teeth and incisor underjet. Undershot jaw. ×c. 0.37.

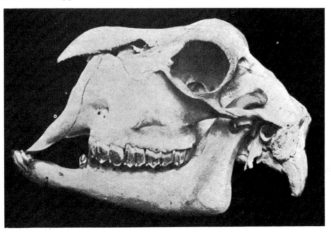

lacked the lower incisor and canines (Emele-Nwanbani and Ihemelandu, 1984).

In an important study of feral goats on islands in the Aldabra Atoll in the Indian Ocean, Van Vuren and Coblentz (1988) found evidence for founder effect (Chapter 1, p. 10). On Malabar Island, 17 out of 55 goats (31%) had one or more cheek teeth displaced either lingually or buccally; of the 25 malpositioned teeth, 14 were maxillary and 11 mandibular; 19 were premolars and 6 were molars. The malposition had resulted in severe malocclusion, food packing and loose teeth. However, none of the 292 goats collected on the neighbouring island of Grande Terre had misplacements. Although the assertion was not supported by skull measurements, the Malabar goats seemed to suffer a shortening of the skull in general and of the maxilla and mandible in particular; this had resulted in inadequate space for the eruption of the permanent teeth. The absence of malposition of the teeth amongst the Grande Terre goats appears to rule out any environmental cause, so it was concluded that the animals on Malabar were very inbred, due to founder effect; additional evidence for this view was provided by the lesser variation in coat colour among the goats on Malabar than on Grande Terre.

Genus *Alcelaphus*, hartebeests. 67 specimens examined

The skull is long and narrow, the upper cheek teeth rows are only slightly curved and there is hardly any echelon arrangement (Fig. 13.34). The only positional irregularities noted were:
(i) *Alcelaphus caama* (Cape hartebeest). BMNH 1931.2.1.59. Both P^2 rotated with buccal surfaces turned mesially.
(ii) *A. lichtensteini* (Lichtenstein's hartebeest). BMNH 1908.2.14.3. Both P^3 rotated.

Genera *Cephalophus* and *Sylvicapra*, duikers. 256 specimens examined

The cheek-teeth rows are more curved in *Cephalophus* (Fig. 13.35) and *Sylvicapra* than in *Alcelaphus* and the echelon arrangement is marked in the molar region. Slight symmetrical rotations of upper premolars are so common that they must be regarded as normal.

There was considerable range of variation in the positions of the teeth; in a few specimens, there was no rotation of the premolars; in many, there was a slight increase of rotation. The following anomalies were noted: marked rotation of the molars (Fig. 13.36); overlap of P^4 to the extent of one-half (Fig. 13.37); overlap of P^3 by one-third; rotation of P^2 (Fig. 13.38);

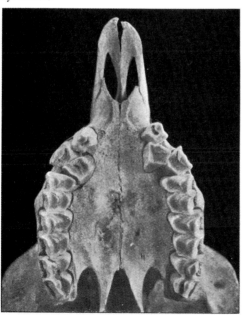

Fig. 13.33. *Capra hircus* (Nubian goat). RCS Odonto. Mus., G 78.3. Right P^4 and left P^3 misplaced lingually; right P^4 also rotated about 90° with buccal surface facing distally. × c. 0.8.

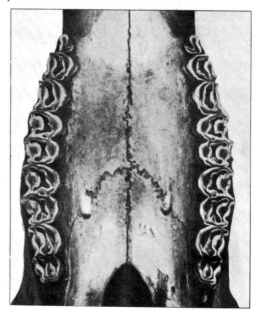

Fig. 13.34. *Alcelaphus caama* (Cape hartebeeste). ♀. BMNH 1931.2.1.58. The normal upper tooth arch in *Alcephalus*. × c. 0.66.

general irregularity of the premolars (Fig. 13.39); asymmetry of the premolars (Fig. 13.40). In addition, there were two examples of lingual misplacement of upper premolars presumably due to retention of parts of deciduous teeth (Fig. 13.41 and 13.42). P^3 was more often affected than P^4.

Table 13.3 shows the prevalence of anomalies of this kind among the 256 skulls of *Cephalophus* and *Sylvicapra*.

Table 13.3 Positional variations in the two genera of DUIKER

Genus	No. of specimens	Total specimens affected No.	%
Cephalophus	163	40	24.5
Sylvicapra	93	12	12.9

Fig. 13.35. *Cephalophus monticola maxwelli* (Maxwell's duiker). BMNH 1894.7.25.4. The normal upper tooth arch in *Cephalophus*. Note the echelon arrangement.

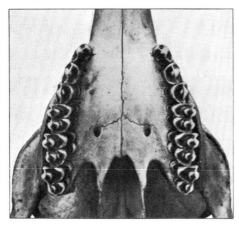

Fig. 13.36. *Sylvicapra grimmia leucoprosopus* (Angolan bush duiker). ♀. BMNH 1920.4.27.43. General crowding of the tooth arch with, in particular, symmetrical rotation of both M^1 (pointers), producing an exaggerated echelon appearance of the buccal surface of the arch.

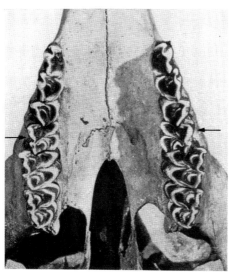

Fig. 13.37. *Cephalophus monticola nyasae* (Nyasa blue duiker). BMNH 1907.1.11.81. Both P^4 are symmetrically misplaced lingually and rotated about 45° with buccal surfaces turned distally.

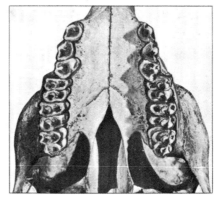

Fig. 13.38. *Cephalophus nigrifrons rubidus* (red mountain duiker). ♂. BMNH 1921.3.26.4. The buccal surface of the right P^2 faces distally. The empty socket of the left P^2 indicates that it was normally placed.

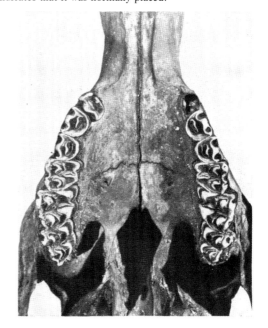

Genus *Ourebia*, oribis

The upper molars are normally slightly oblique, the overlap of P^4 is about one-quarter and the buccal surfaces of P^2 and P^3 form an even line. In about 25% of the specimens, there was slight irregularity of P^3. There was no rotation of P^4 in about a quarter of the skulls, but increased rotation was seen in about one-fifth; in only two did it amount to an overlap of one-half by M^1. In two specimens, there was an increased rotation of the M^1 and, in one, both P^3 and P^4 were misplaced (Fig. 13.43).

Fig. 13.39. *Cephalophus monticola monticola* (blue duiker). ♀. BMNH 1910.6.1.43. General irregularity of the premolars.

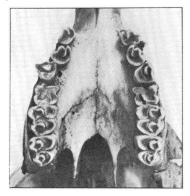

Fig. 13.40. *Cephalophus monticola anchietae* (Angolan blue duiker). BMNH 1927.4.27.1. Most of the premolars are slightly rotated, especially on the left. Enlarged slightly.

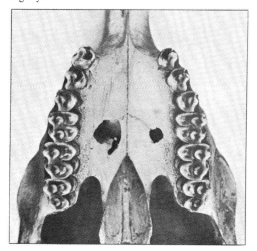

Genus *Raphicerus*, grysboks and steenboks. 50 specimens examined

The normal arrangement of the teeth is similar to that in *Ourebia* except that there is more overlap of P^4 by M^1 (Fig. 13.44). Echelon arrangement of the cheek teeth is a normal feature.

Fig. 13.41. *Cephalophus monticola maxwelli* (Maxwell's duiker). BMNH 1848.10.11.5. Between the left P^4 and M^1 is a retained fragment of m^4 (pointer). Presumably P^4 in consequence erupted more mesially than normal, perhaps m^3 was being shed at the time, and when P^3 was erupting no space was available for it in the arch. Right P^3 rotated so that its buccal surface is turned distally.

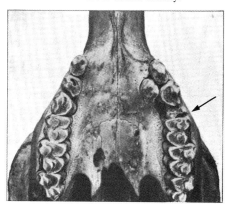

Fig. 13.42. *Cephalophus spadix* (Abbott's duiker). ♀. BMNH 1919.2.3.2. Left P^4 misplaced lingually and rotated so that its buccal surface is turned mesially. There is a gap between P^4 and M^1 which at present is empty but which probably was the site of retention of a part of m^4 which was the cause of the misplacement. Reduced slightly.

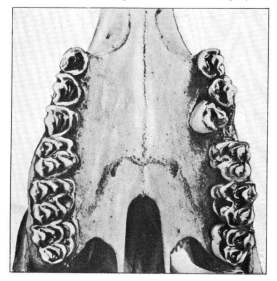

Greater than normal rotation of P^4 was present in about 14%, but seldom as much as in Figure 13.45 where the overlap is at least one-half.

In two skulls, P^3 was rotated with the buccal surface facing mesially. In one (Fig. 13.46), there was symmetrical rotation of both M^1.

Irregularity in position of the deciduous molars is uncommon in ungulates but two specimens of *Raphicerus* showed abnormal positions of m^3; in one, it was misplaced lingually (Fig. 13.47); in the other, the buccal surfaces were turned distally.

Fig. 13.43. *Ourebia ourebi aequatoria* (Sudan oribi). ♀. BMNH 1928.9.8.19. Left P^3 is rotated with buccal surface facing mesially and left P^4 rotated in the opposite direction.

Fig. 13.45. *Raphicerus campestris capricornis* (Transvaal steenbok). ♀. BMNH 1916.2.26.12. Symmetrical slight rotation of P^4 with buccal surfaces turned distally.

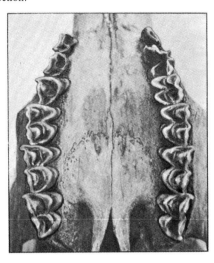

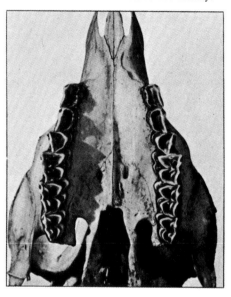

Fig. 13.44. *Raphicerus sharpei* (Sharpe's grysbok). ♂. BMNH 1901.6.26.4. The normal upper tooth arch in the genus *Raphicerus*. Note the echelon arrangement.

Fig. 13.46. *Raphicerus campestris neumanni* (East African steenbok). ♂. BMNH 1906.1.5.7. Crowded arches with symmetrical considerable rotation of the M^1 with buccal surfaces turned distally and slight rotation of P^4 in the opposite direction.

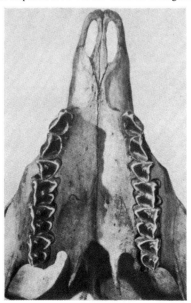

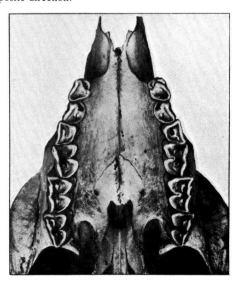

Genus *Nesotragus*, sunis

Overlap of P^4 was well-marked, amounting to one-third in the majority of the specimens. The two species of *Nesotragus*, *N. moschatus* and *N. livingstonianus* show a considerable difference in the prevalence of variation. In 21 specimens of *N. moschatus*, the overlap of P^4 was one-third or less in 17 and one-half in four; among 11 *N. livingstonianus*, there was overlap of P^4 of one-half in six, one being associated with misplacement of P^3 and one with misplacement of P^2 and P^3 (Fig. 13.48). Another (Fig. 13.49) showed rotation of both P^3.

Genus *Madoqua*, dik-diks. 168 specimens examined

The upper jaw is more slender than in *Nesotragus*. The upper cheek-tooth rows are but slightly curved; nevertheless, there is some obliquity in their arrangement, and the molars have an echelon arrangement, the amount of the overlap varying slightly. In a few specimens, there is scarcely any rotation of P^4 so that all the premolars are practically in an even line (Fig. 13.50).

The chief anomaly is greater overlap of P^4, sometimes by nearly one-half (Fig. 6.17), but extreme rotation of P^4 is rare. P^3 may be misplaced slightly lingually or rotated (Fig. 13.51 and 13.52). P^2 and P^3 may be lingual to P^4 which is usually not rotated.

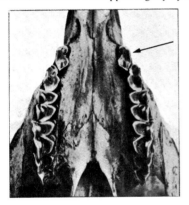

Fig. 13.47. *Raphicerus campestris campestris* (steenbok). ♀. BMNH 1902.12.1.33. Both m^3 (arrow) present with their buccal surfaces overlapped slightly by m^4.

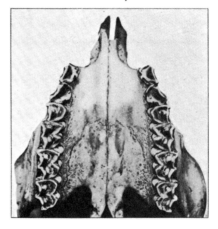

Fig. 13.49. *Nesotragus livingstonianus* (Livingstone's suni). BMNH 1913.7.23.2. Symmetrical rotation of P^3 with buccal surfaces turned mesially.

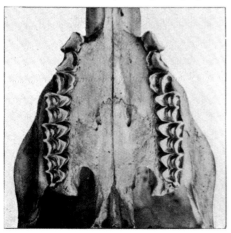

Fig. 13.48. *Nesotragus livingstonianus* (Livingstone's suni). ♂. BMNH 1903.1.21.1. Crowding and symmetrical overlapping of the premolars with rotation of both P^4 with buccal surfaces turned distally. Enlarged slightly.

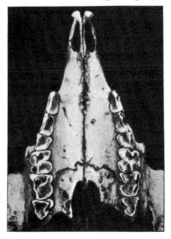

Fig. 13.50. *Madoqua saltiana phillipsi* (Phillips' dik-dik). ♂. BMNH 1911.7.2.28. The normal slight rotation of the premolars is absent. Enlarged slightly.

Section 2: Variations in Position

In *Madoqua guentheri*, rotation of P^4 is the rule rather than the exception; in 12 out of 15 specimens, it was overlapped by three-eighths or more and, in one of these, the rotation was extreme (Fig. 13.53).

Fig. 13.51. *Madoqua guentheri hodsoni* (Hodson's dik-dik). ♂. BMNH 1920.9.25.1. Symmetrical crowding and overlap of the premolars. P^4 and M^1 in particular are misplaced lingually.

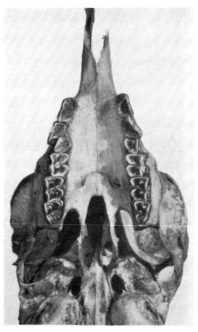

Fig. 13.52. *Madoqua kirkii cavendishi* (Cavendish's dik-dik). ♀. BMNH 1904.6.5.4. Symmetrical rotation of P^3 with buccal surfaces turned mesially, and partial overlap by P^2.

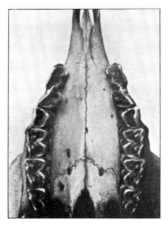

Genus ***Pelea***, Vaal rhebuck. 4 specimens examined

In one skull, the left P^3 was rotated by nearly one-quarter of a circle, the buccal surface being directed distally (*Pelea capreolus*, ♀, BMNH 1902.12.1.37).

Genus ***Redunca***, reed bucks. 47 specimens examined

The molars and P^4 form a fairly straight line with little echelon appearance but P^2 and P^3 are set at an angle to this line so that the arch is much narrower between the minute P^2. Nine skulls showed irregularities as follows:
  (i) One with rotation of both P^4.
  (ii) Three with rotation of one or both P^3 (Fig. 13.54).
  (iii) One with lingual misplacement of the right P^3 (Fig. 13.55).

Fig. 13.53. *Madoqua guentheri* (Guenther's dik-dik). ♀. BMNH BMNH 1904.5.9.33. Right P^4 rotated with buccal surface facing distally. × *c.* 0.75.

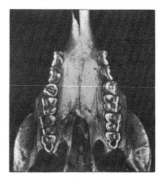

Fig. 13.54. *Redunca arundinum* (common reedbuck). ♀. BMNH 1926.11.1.93. Symmetrical rotation of P^3 with buccal surfaces turned mesially. × *c.* 0.66.

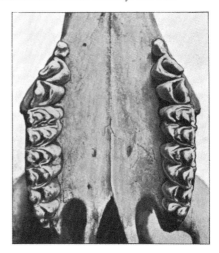

(iv) Four with rotation of P^2, one bilaterally (Fig. 13.56), the others unilaterally.

Genus ***Kobus***, water bucks. 57 specimens examined

The tooth arches are similar to those of *Redunca* and irregularities are uncommon, only two being noted; both were rotation of P^4, one unilaterally, one bilaterally.

Fig. 13.55. *Redunca arundinum* (common reedbuck). ♂. BMNH 1927.2.11.73. Right P^3 is misplaced lingually and rotated with buccal surface facing mesially. ×c. 0.66.

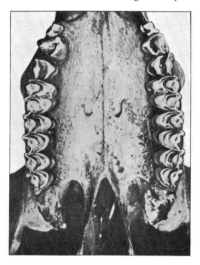

Fig. 13.56. *Redunca arundinum* (common reedbuck). ♂. BMNH 1924.8.3.22. Symmetrical rotation of P^2 with buccal surfaces facing distally. ×c. 0.66.

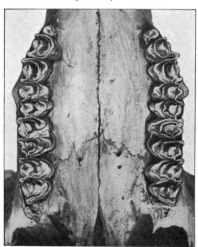

Genus ***Antilope***, blackbuck. 24 specimens examined

There is little rotation or echelon arrangement of upper cheek teeth. Two skulls showed variations. In one, the M^1 were rotated; in the other, the P^4 were overlapped by about one-half.

Genera ***Gazella*** and ***Procapra***, gazelles. 225 specimens examined

The echelon arrangement of the molars and P^4 is marked. Nineteen specimens (12.8%) showed major irregularities in the position of P^4, two of P^3, two of P^2 and, in six, P^2 and P^3 were lingual to the arch. In nine specimens, irregularities were confined to or more marked on one side.

The variations in position in different species of gazelle are summarized in Table 13.4, but the number of specimens of each species was too small to allow reliable inferences about inter-specific differences.

Gazella dorcas (dorcas gazelle). 32 specimens examined

P^4 is normally overlapped by about one-third; the buccal surfaces of P^2 and P^3 form an even line. Single instances of rotation of P^2, lingual misplacement of P^3 (Fig. 13.57) and rotation of M^1 were noted. In three skulls, P^3 was strongly rotated (Fig. 13.58), in four the overlap of P^4 was one-quarter and in five it was as much as one-half (Fig. 13.59).

Table 13.4 Positional variations in various species of GAZELLES

Species	No. of specimens	Total specimens affected No.	%
Procapra picticauda	9	0	0
P. gutturosa	9	1	11.1
Gazella sub-gutturosa	13	1	7.7
G. bennetti	18	3	16.7
G. thomsoni	15	2	13.3
G. spekei	6	2	33.3
G. rufifrons	23	0	0
G. granti	35	3	8.6
G. dama	9	1	11.1
G. soemmerringi	12	2	16.7
G. dorcas	43	5	11.6

Gazella rufifrons (red-fronted gazelle). 18 specimens examined

In this species, the rotation of P^3 is negligible, the normal overlap of P^4 is about one-quarter, but it was one-third in five and one-half in three specimens.

Gazella thomsoni (Thomson's gazelle). 12 specimens examined

This species seems to be one in which anomalous tooth position is particularly common. The following major variations were present:

Fig. 13.57. *Gazella dorcas* (dorcas gazelle). ♀. BMNH 1893.4.10.2. P^3 symmetrically misplaced slightly lingually.

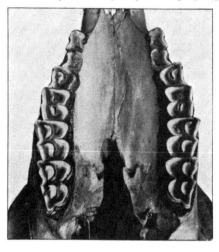

Fig. 13.58. *Gazella dorcas* (dorcas gazelle). ♂. BMNH 1925.5.12.98. Left P^3 rotated by nearly 90°, with buccal surface facing mesially.

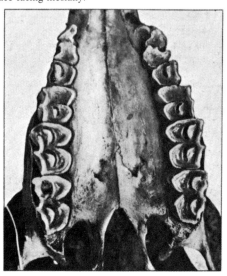

(i) P^3 rotated one-half, the buccal surface facing mesially instead of distally as is usual (♀, BMNH 1898.1.5.20).
(ii) P^4 rotated one-half. P^4 rotated with buccal surface in contact with the M^1 (Fig. 13.60).
(iii) Both P^3 displaced lingually. There was no rotation of P^4; the P^3 seems to have been driven along the convexity of the mesial surface of P^4 (♀, BMNH 1924.1.1.191).

Fig. 13.59. *Gazella dorcas* (dorcas gazelle). BMNH 1925.11.9.3. Symmetrical slight rotation of P^4 with buccal surfaces turned distally.

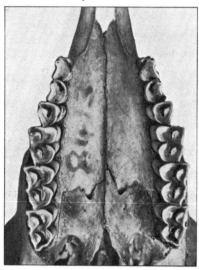

Fig. 13.60. *Gazella thomsoni* (Thomson's gazelle). ♀. BMNH 1907.5.28.6. Both P^4 rotated with buccal surfaces turned distally, the right only slightly, but the left by about 90°.

(iv) Right P_4 rotated one-half; left P_4 rotated more than three-quarters (♂, BMNH 1898.1.15.18).

Gazella spekei (Speke's gazelle). 5 specimens examined

All five specimens were affected with some kind of irregularity of P^4:
 (i) Both P^4 overlapped by one-quarter.
 (ii) Both P^4 overlapped by one-third.
 (iii) Both P^4 overlapped by one-half.
 (iv) Right P^4 misplaced buccally, the left overlapped by one-half.
 (v) Right P^4 overlapped by one-half, the left rotated with buccal surface facing almost distally (Fig. 13.61).

Gazella granti (Grant's gazelle). 26 specimens examined

Rotation of the premolars is not marked and irregularities of tooth position are uncommon. In the majority, P^4 was overlapped by about one-quarter and P^3 a little less. In six skulls, P^4 was overlapped by one-third and in two by one-half. In one third of the skulls, there was little rotation of P^3; in only one specimen was the overlap of P^3 as much as one-half.

Gazella bennetti (Indian gazelle). 11 specimens examined

The amount of overlap of P^4 is shown in Table 13.5. An example of extreme rotation of P^4 is illustrated in Figure 13.62.

Genus *Antidorcas*, springboks. 26 specimens examined

Considerable rotation of the premolars is frequently present; in eight specimens, one or both P^4 were overlapped by one-half and, in six, one or both P^3 were symmetrically rotated.

Genus *Oryx*, gemsbok and oryx. 23 specimens examined

The upper cheek teeth do not normally overlap, so their buccal surfaces form an even line. P^3 was misplaced in three skulls (Fig. 13.63).

Genus *Hippotragus*, roan antelope. 32 specimens examined

The upper cheek teeth do not normally overlap, so their buccal surfaces form an even line. Two skulls showed irregularities of position.

Table 13.5 The amount of overlap of P^4 by M^1 in 11 skulls of *GAZELLA BENNETTI*

| Degree of overlap | | No. of |
Right	Left	specimens
Nil	Nil	1
1/4	1/4	1
1/3	1/3	2
1/3	1/2	1
1/2	1/2	4
3/4	1/2	1
3/4	3/4	1 (Fig. 13.62)

Fig. 13.62. *Gazella bennetti* (Indian gazelle). BMNH 1923.10.13.14. Symmetrical rotation of P^4 with buccal surfaces turned distally.

Fig. 13.61. *Gazella spekei* (Speke's gazelle). ♂. BMNH 1891.6.20.5. Right P^4 slightly rotated with buccal surface turned distally; left P^4 slightly misplaced lingually and rotated so that the buccal surface faces distally. × c. 0.66.

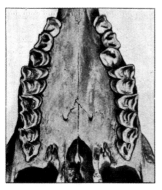
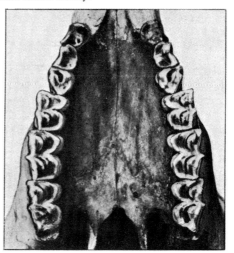

Genus **Tragelaphus**.

Tragelaphus scriptus (bushbuck). 90 specimens examined

Overlap of P^4 was usually about one-third and of the P^3

Fig. 13.63. *Oryx gazella* (gemsbok). ♀. BMNH 1931.2.1.43. Left P^3 misplaced lingually and slightly rotated with its buccal surface turned distally. × *c*. 0.5.

Fig. 13.64. *Tragelaphus scriptus massaicus* (eastern bushbuck). ♂. BMNH 1896.10.26.2. Both P^4 symmetrically rotated with the buccal surfaces turned slightly distally, whereas normally, like all the cheek teeth, they are turned in the opposite direction, slightly obliquely to the line of the arch (cf. Fig. 13.11). × *c*. 0.75.

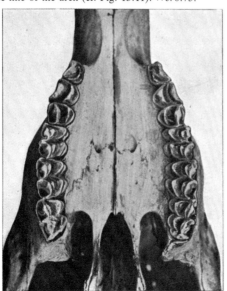

a little less. In about one-quarter of the specimens, P^4 and P^3 showed less overlap. Greater overlap of P^4 on one or both sides was present in 16, associated with less overlap of P^3 in seven and more in four. Greater overlap of P^3 to the extent of one-half was noted in seven skulls, three of which showed less overlap of P^4 than usual.

Fig. 13.65. *Tragelaphus scriptus* (bushbuck). BMNH 1893.7.9.28. Symmetrical crowding of the premolars generally and, in particular, rotation of P^3 with buccal surfaces turned distally. × *c*. 0.75.

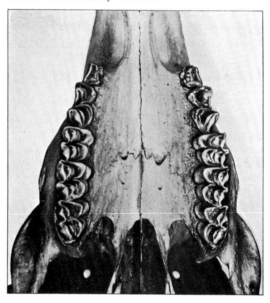

Fig. 13.66. *Tragelaphus scriptus* (bushbuck). ♀. BMNH 1924.8.3.57. Rotation of both P^3 with buccal surfaces facing distally, the left one much more than the right. × *c*. 0.75.

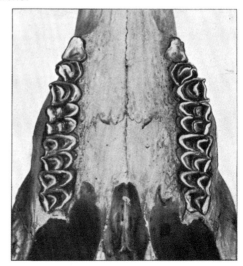

Differences in the amount of rotation of the right and left teeth were common. Figures 13.64 to 13.66 show the range of premolar rotation.

Tragelaphus spekei (sitatunga). 30 specimens examined

Overlap of the premolars was much less than in *T. scriptus*, that of P^4 being normally about one-quarter; rotation of P^3 was negligible. The specimen in Figure 13.67 shows slight rotation of P^3 with an increased lingual bend of the arch in the position of P^2. Overlap of P^4 or P^3 to the extent of one-half was seen in four skulls; in one of these (*T. spekei spekei*, ♀, BMNH 1900.10.3.4), both P^3 were rotated so that the buccal surfaces were in contact with the mesial surfaces of P^4, and in another the rotation of P^4 had brought the buccal surfaces against the distal surfaces of P^3 (Fig. 13.68).

Order **PERISSODACTYLA**

Family **Equidae**

Genus *Equus*, horses, zebras, donkeys and onagers. 65 specimens examined

In the skulls of three wild equids, there was slight crowding of the lower incisors and, in another, the right P^2 was rotated.

Equus caballus (domestic horses and ponies). 502 specimens examined

(a) Incisors

Positional anomalies of the incisors, including those associated with the persistence of the deciduous teeth are common, more so in the upper than in the lower jaw, ranging from slight crowding to a very complex arrangement of the teeth.

Extra teeth associated with retained deciduous teeth may cause a weird arrangement (Fig. 13.69). In

Fig. 13.67. *Tragelaphus spekei spekei* (Speke's sitatunga). ♂. BMNH 1929.11.3.40. The arch is unusually narrow between the P^2 and both P^4 are rotated with buccal surfaces turned distally, the right one more than the left. ×c. 0.66.

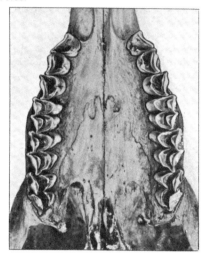

Fig. 13.68. *Tragelaphus spekei spekei* (Speke's sitatunga). BMNH 1926.9.5.11. Symmetrical rotation of P^4 with buccal surfaces turned mesially. ×c. 0.66.

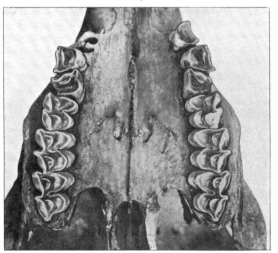

Fig. 13.69. *Equus caballus* (horse). RCS Odonto. Mus., G 76.21. Portion of upper jaw. Incisors crowded because of the persistence of three deciduous teeth slightly labial to the arch. ×c. 0.9.

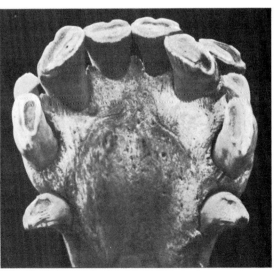

the upper jaw shown in Figure 13.70, there are two retained deciduous teeth and two extra permanent incisors. I^2 and I^3 have erupted, but the supernumerary teeth and the deciduous teeth have prevented the eruption of I^1.

Irregularity in the position of the incisors may occur independently of retained or extra teeth and this condition appears to be more common in the mandible than in the maxilla. In the portion of a mandible shown in Figure 13.71, there is a marked irregularity of the incisors and the canines are misplaced.

Cadéac (1910) illustrated a skull in which the left

Fig. 13.70. *Equus caballus* (horse). Portion of upper jaw; palatal mucosa intact. Gross crowding because, in addition to the six permanent incisors (*B*), two deciduous incisors (*C*) are still present. Two supplemental I^1 are completely unerupted in the positions (*A*). ×*c.* 0.66.

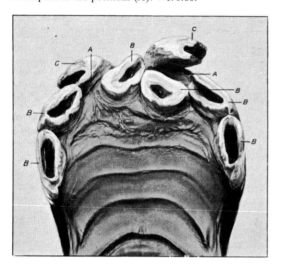

Fig. 13.71. *Equus caballus* (horse). Portion of mandible. Incisors crowded and overlapping, right I_1 misplaced lingually. The canines are misplaced and unerupted; bone has been removed to display them. ×*c.* 0.7.

Fig. 13.72. *Equus caballus* (horse). RCS Odonto. Mus., G 76.52. Gross incisor overjet (Colyer, 1914). ×*c.* 0.16.

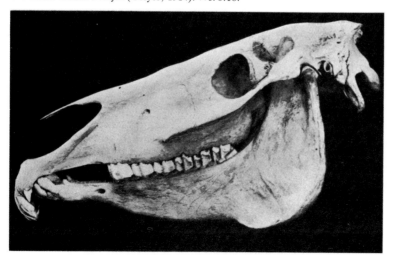

I$_3$ was buried horizontally in the jaw; the left I^3 was absent.

A midline gap between I1, upper or lower, sometimes occurs and might be due to some abnormality of the labial frenum which could not, of course, be detected in skeletal material.

Fig. 13.73. *Equus caballus* (horse). RCS Odonto. Mus., G 76.3. Portion of mandible. Right canine misplaced mesially and nearly in contact with I$_3$. Reduced slightly.

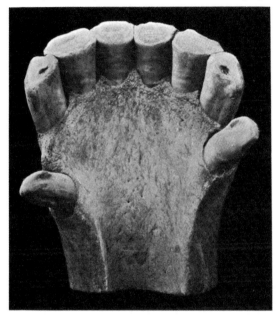

Fig. 13.74. *Equus caballus* (horse). RCS Odonto. Mus., G 76.41. Portion of upper jaw. P^3 symmetrically rotated with buccal surfaces turned distally. ×c. 0.5.

In veterinary practice, incisor overjet is known as parrot mouth; when the protrusion is slight, the lower incisors strike the lingual aspects of the upper ones; with more protrusion, the premaxillae tend to be bent downwards (Fig. 13.72) and to be associated with an abnormal occlusion of the cheek teeth; whether this is caused by inferior retrusion or generalized superior protrusion can only be determined by measurement of the jaws.

When the glenoid-incisor length was compared with that of the mandible (Chapter 9, p. 159) in the male skull shown in Figure 13.72, the mandible was found to be 9.3% longer and the glenoid-incisor length 18.1% longer, than in the average of seven normal specimens. In a female skull, the mandible was 1.5% longer and the glenoid-incisor length 11.15% longer than in eight normal specimens. These measurements suggest that in both instances the abnormality is due to overgrowth of the upper jaw rather than to undergrowth of the mandible.

Underjet is rare; Cadéac (1910, p. 129, Fig. 64) showed an example that appeared to be due to arrest of growth of the premaxillae. Kitt (1892) illustrated an example of a horse with overjet in which, on the basis of comparison with a normal skull, he concluded that the lower jaw was under-developed. Marked examples of under-development of the lower jaw are figured by Joest (1926, Fig. 81) and E. Becker (1970, Figs. 119 and 120).

Fig. 13.75. *Equus caballus* (horse). RCS Odonto. Mus., G 76.4. Portion of upper jaw bearing the premolars. P^3 and P^4 are symmetrically rotated with buccal surfaces turned distally (Colyer, 1906). ×c. 0.5.

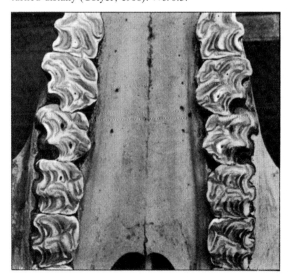

Section 2: Variations in Position

(b) Canines

The lower canine is normally separated by a space from I_3. The canine may erupt in contact with I_3, as shown in Figure 13.73, or the length of the space between I_3 and the canine may differ on the right and left sides.

(c) Premolars

Positional variations of the premolars are uncommon; only three examples were seen, one with rotation of the P^3 (Fig. 13.74), one with rotation of P^3 and P^4 (Fig. 13.75) and one with an irregular arrangement of the upper premolars. As the measurements indicated in Figure 13.76 emphasize, this was associated with interference in the growth of the right maxilla. The right half of the mandible is not as well developed as the left and the left incisors, P_4 and M_1 are irregular in position (Colyer, 1914).

An example of complete displacement of a premolar from the arch described by Cadéac (1910) is shown in Figure 13.77. The molars on the left are much more mesial than those of the right and it would seem that the mesial movement of the molars, perhaps P^4 which is misplaced lingually, pass to the buccal side of the lower cheek teeth and are consequently less worn than those in normal occlusion on the left side (C). $\times c.\,0.16$.

The sutures between the maxilla and adjacent bones have been defined in the lateral views and dimensions indicated (Colyer, 1914). $\times c.\,0.16$.

Fig. 13.76. *Equus caballus* (horse). RCS Odonto. Mus., G 76.42. Asymmetrical skull with defective growth of the right side of the cranium.
A: Palatal view showing upper jaw bent to the right and right cheek-tooth row crowded with P^4 misplaced lingually. $\times c.\,02$.
B: Right side showing that the upper cheek teeth, except

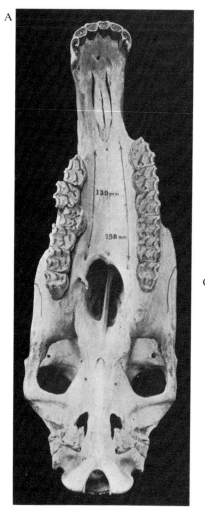
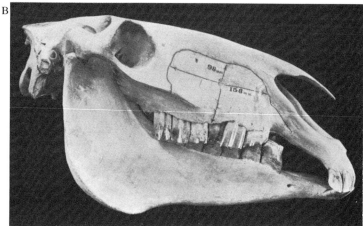
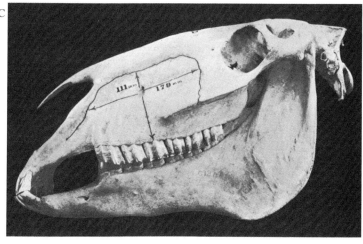

associated with premature loss of m^4, has eliminated the space for its successor P^4, which has erupted lingually to its normal position.

The small percentage of specimens showing variations in the positions of premolars is due, as pointed out previously, to the large amount of wear of the approximal surfaces of the teeth.

Fig. 13.77. *Equus caballus* (horse). Left P^4 misplaced lingually. (From Cadéac, 1910).

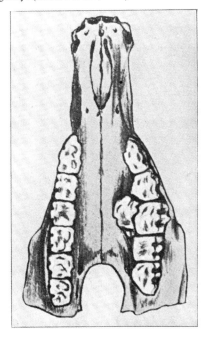

(d) Asymmetry of the skull

Although many examples have been noted in primates (Chapter 10), asymmetry of the skull seems to be rare in ungulates. Kitt (1892) figured the skull of a full-grown stallion in which the nasal bones and the maxilla were twisted slightly towards the right and then, with the premaxillae, markedly to the left, whereas the mandible remained straight. The upper right incisors occluded with the lower left incisors, resulting in extremely uneven wear, but cheek-tooth occlusion seems to have been unaffected.

Family **Tapiridae**

Genus *Tapirus*, tapirs. 30 specimens examined

Tapirs are of interest on account of the position assumed by the incisors in a large majority of the specimens; the premaxillae are bent down and the incisors do not seem to have been able to accommodate themselves to this position (Fig. 13.78). The root of I^2 is directed towards the mid-line and the slope of I^3 is similar though less marked. In a few cases, the incisors formed an even row.

Family **Rhinocerotidae**. Rhinoceroses. 45 specimens examined

No major irregularities were found. There was a tendency for the approximal surfaces of the teeth, as in other ungulates, to be worn away and, in some specimens, the amount of wear was extreme. It is interesting that the first illustration in the first volume of the

Fig. 13.78. *Tapirus terrestris* (South American tapir). BMNH 1847.4.6.1. Portion of upper jaw. The incisors are crowded and overlapping, with I^2 misplaced lingually.

312 Section 2: Variations in Position

Transactions of the Odontological Society (Bate, 1856–7) depicts a portion of the mandible of a fossil rhinoceros (*Rhinoceros hemitoechus*) showing irregular position of a premolar associated with retention of a deciduous tooth (Fig. 13.79).

Order **HYRACOIDEA**

Family **Procavidae**. 354 specimens examined (17 without mandibles)

Hyraxes have an unusual dentition. In the upper jaw, there is one pair of incisors, triangular in cross-section, growing from persistent pulps; in the mandible, there are four incisors of limited growth. The upper premolars and molars have a mesial extension of the buccal surface which slightly overlaps the buccal surface of the tooth in front. This gives the buccal surface of the tooth row an irregular outline.

The most common anomaly is a separation of the first incisors in the mandible varying from a slight space to a marked gap (Fig. 13.80). Table 13.6 shows that the anomaly is more common in the small-toothed rock hyrax (*Heterohyrax*) and the tree hyrax (*Dendrohyrax*) than in the large-toothed rock hyrax (*Procavia*).

Positional variations of the premolars were seen in about 4% of the specimens, usually taking the form of rotation of either P^1 or P^3 (Fig. 13.81).

As previously stated, the upper premolars and the molars overlap one another slightly; in the position of M^3, the overlap may be marked (Fig. 13.82).

Misplacement distally of M_3 is rare in hyraxes from the wild state; an example is shown in Figure 13.83.

Order **PROBOSCIDEA**

Family **Elephantidae**

The tusks form a gentle curve and are usually directed downwards and forwards and slightly towards the midline. Variations are not uncommon; the tusks may project horizontally with a slight curve upwards and outwards; the curvature on one side may differ from the other near the tips.

There was on view at the Empire Exhibition, London (1924), a skull of an Indian elephant in which the tusks were placed horizontally.

Misplacement of the teeth occurs in elephants that have been long in captivity. The famous African elephant Jumbo was brought to the London Zoo in 1865 and remained there until he was shipped to

Table 13.6 Separation of the first incisors in HYRAX

Sub-genus	No. of specimens	Slight separation	Marked separation	Total specimens affected No.	%
Procavia	170	33	9	42	24.7
Heterohyrax	53	20	6	26	49.1
Dendrohyrax	85	28	19	47	55.3

Fig. 13.79. *Rhinoceros hemitoechus* (fossil rhinoceros). Part of left mandible with the P_4 misplaced lingually to the retained m_4 (p. 352) (Bate, 1856–57; Colyer, 1914).

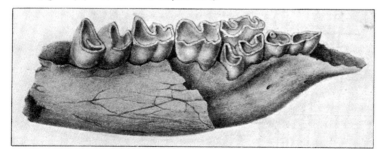

America in 1882, where he died in an accident. There was a great deal wrong with his dentition (Fig. 13.84). In elephants, the permanent molars normally succeed each other from behind; that is, as M1 becomes worn it is shed, M2 moves forward to succeed it and M3 emerges and gradually becomes functional also. On the right side of Jumbo's skull, presumably because the tooth-bearing part of the upper jaw has not grown sufficiently in length, M^3 has erupted on the lingual side of M^2. However, in addition, all the upper molars are misshapen, presumably also because of inadequate jaw growth. The right M^2 has a lingual curve and is rotated so that the convex buccal surface is turned mesially; the right M^3 has an opposite curve as if it had been compressed during growth against the slight convexity of the distal part of the lingual surface of M^2. On the left side, the relationship of the teeth is similar but M^2 has an angular bend buccally and M^3 is misshapen

Fig. 13.80. *Dendrohyrax dorsalis* (Beecroft's hyrax). ♂. BMNH 1903.1.6.3. An unusually wide median gap separates the lower incisors.

Fig. 13.81. *Procavia capensis* (large-toothed rock hyrax). ♂. BMNH 1904.9.1.82. Right P^1 rotated with its buccal surface facing mesially and its crown tilted mesially.

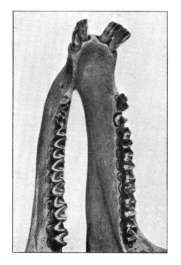

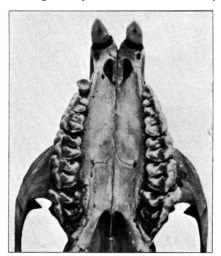

Fig. 13.82. *Procavia capensis* (large-toothed rock hyrax).
A: ♂. BMNH 1905.5.7.102. Normal upper tooth arch in the genus *Procavia*, showing the mesial extension of each tooth overlapping the tooth in front.
B: ♀. BMNH 1905.5.7.107. Symmetrical rotation of both M^3, which are also situated more lingually than usual so that the distal end of the arch is narrow. Right incisor region damaged.

A 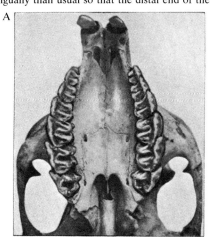 B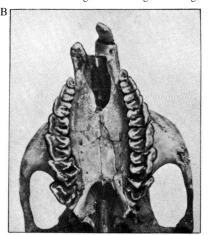

314 *Section 2: Variations in Position*

with a triangular occlusal surface, as if it had been compressed during growth against the distal part of the lingual surface of M^2.

The mandible in Figure 13.85 shows a similar crowding and misplacement of the molars. It is from an Indian elephant named Ellen who lived for several years in the Zoological Gardens in Copenhagen and who gave birth to three young during her captivity. The right M_2 is curved buccally and consists of about 12 plates and the left M_2 has only about 8, whereas normally there are 16. Both M_3 lie to the buccal side of M_2.

The left M^3 shown in Figure 13.86 has the distal part folded over on the buccal side. It consists of at least 26 plates, about 10 of which have come into use; the number is not exactly definable because of the

Fig. 13.83. *Procavia capensis matschiei* (Tanzanian large-toothed rock hyrax). BMNH 1910.12.16.1. Mandible. Left M_3 misplaced and tilted mesially. Bone has been removed to display its roots.

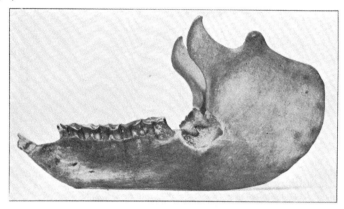

Fig. 13.84. *Loxodonta africana* (African elephant). "Jumbo". Captive for 17 years. Plaster cast of upper jaw. On the right, M^2 has a lingual curve and is rotated with its buccal surface turned mesially; M^3 has an opposite curve and is misplaced to the lingual side of M^2. On the left, the relationship of the teeth is similar, but M^2 has an angular bend buccally and M^3 has a triangular occlusal surface.

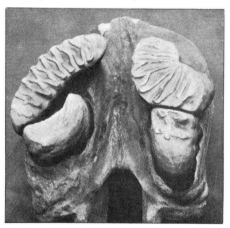

Fig. 13.85. *Elephas maximus* (Indian elephant). "Ellen". ♀. Captive for several years. ZMUC 1399. Lower tooth arch. On the left, M_3 is misplaced on the buccal side of M_2; on the right, M_2 has a sharp buccal curve around the M_3, which is unerupted.

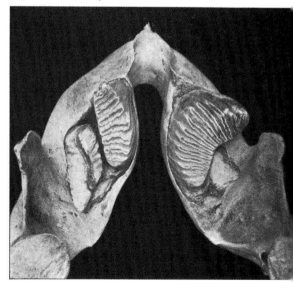

distorted character of the tooth. The specimen was in the collection of John Hunter. This deformity is not uncommon in fossil elephants and tends to be more common in recent forms. In tracing the elephant from its earliest appearance to modern times, there has been a progressive shortening of the jaw and a tendency towards an increase in the size of the molars. The effect of this would be to lessen the amount of room for the developing molars and so possibly lead to their distortion (A. T. Hopwood, personal communication to Colyer).

Fig. 13.86. *Elephas maximus* (Indian elephant). RCS Odonto. Mus., G 27.93. Left M^3 which has its distal part folded on the buccal surface. $\times c.\,0.32$.

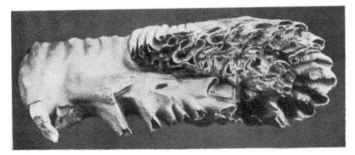

CHAPTER 14

Orders RODENTIA, LAGOMORPHA and INSECTIVORA

Order RODENTIA.
2554 specimens examined

It was not possible to undertake a complete survey of this order; the small size of many specimens rendered examination of the teeth too difficult. However, rodents seem to be remarkably free from positional irregularities of the teeth; only 11 (0.4%) were affected.

Family Sciuridae

In the genus *Marmota* (marmots), the P^3 are sometimes situated a little lingually to the other cheek teeth; in some specimens, this is accentuated and the mesial surface of P^4 slants mesially. This was seen in 14 of the 97 specimens examined, but in only one was the misplacement marked (Fig. 14.1).

In *Petaurista* (giant flying squirrels), the P^3 are small and situated a little lingual to P^4; the position of M^3 varies. Rotation of P_4 with the buccal surfaces facing mesially (Fig. 14.2) was seen in three out of 148 mandibles.

622 specimens of *Sciurus* (squirrels) were examined; there were no irregularities.

Family Ctenomyidae

395 skulls of *Ctenomys* (tuco-tuco) were examined; in one (*C. frater sylvanus*, BMNH 1921.1.2.10), both M^3 were more lingually situated than normal.

Family **Octodontidae**

In *Octodontomys gliroides* (mountain degu), the buccal surfaces of the upper cheek teeth are progressively turned more and more distally from the distal to the mesial ends of the arches, with P^4 being nearly transverse to the line of the arch (Fig. 14.3).

Family **Muridae**

Among 200 skulls of the red-backed vole (*Clethrionomys gapperi*) from Michigan in the Museum of Zoology, University of Michigan, Quay (1953) found 11 in which the arrangement of the cheek-tooth rows was irregular with consequential abnormal wear patterns. Most commonly, M^2 and M^3 were misplaced either labially or lingually; in the lower jaw, M_3 was more commonly misplaced lingually than labially, probably because the growing root of the incisor normally lies on the labial side of these teeth.

Family **Echimyidae**

Two specimens of *Proechimys* (terrestrial spiny rat) showed positional anomalies:
(i) *Proechimys guyannensis gularis*. ♀. BMNH 1903.4.5.4. (Fig. 14.4). The M^3 are slightly more lingual than the normal one on the left.

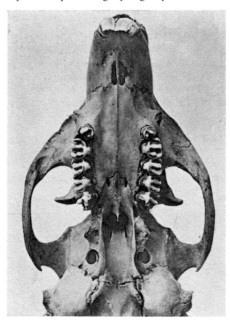

Fig. 14.1. *Marmota bobak* (Bobak marmot). ♂. BMNH 1909.4.3.110. Right P^3 rotated with buccal surface turned mesially and misplaced slightly lingually.

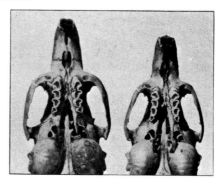

Fig. 14.3. *Octodontomys gliroides* (mountain degu).
Left: ♀. BMNH 1902.2.2.6. The normal upper tooth arch in this species. The widest diameter of the crown of M^3 is in the mesio-distal plane but that of M^2 is turned so that the buccal surface faces slightly distally; M^1 is turned still more and P^4 even more so that it is almost transverse to the line of the arch.
Right: ♂. BMNH 1902.2.2.8. The cheek-tooth rows are slightly irregular and in general the progressive rotation is increased.

Fig. 14.2. *Petaurista* sp. (giant flying squirrel). BMNH 1844.9.30.18. Symmetrical rotation of P_4 (pointer) with buccal surfaces turned mesially.

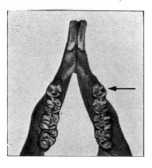

Fig. 14.4. *Proechimys guyannensis gularis* (spiny rat).
Left: The normal upper tooth arch.
Right: ♀. BMNH 1903.4.5.4. Bilateral lingual misplacement of M^3.

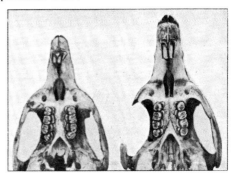

(ii) *P. guyannensis chrysaeolus*. ♂. BMNH 1911.6.5.4. (Fig. 14.5). The molars are irregular in position.

Family **Erithizontidae**

There were three anomalies of position in 98 specimens of the genus *Coendou* (tree porcupines):
 (i) *Coendou bicolor simonsi*. ♀. BMNH 1902.1.1.103. Both P_4 rotated.
 (ii) *Coendou vestitus pruinosus*. ♂. BMNH 1905.7.5.9.
  (Fig. 14.6). Right P_4 has the occlusal surface directed distally.
 (iii) *Coendou* sp. ♀. BMNH 1903.9.4.87. (Fig. 14.7). Both M_3 are rotated, the buccal surfaces facing mesially.

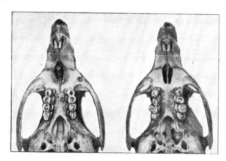

Fig. 14.5. *Proechimys guyannensis chrysaeolus* (spiny rat). Left: The normal upper tooth arch in this species. Right: ♂. BMNH 1911.6.5.4. Disordered, crowded tooth rows.

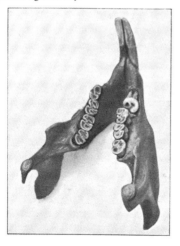

Fig. 14.6. *Coendou vestitus pruinosus* (tree porcupine). ♂. BMNH 1905.7.5.9. Right P_4 tilted distally with its crown impacted against M_1.

There were no anomalies in the families Castoridae, Caviidae, Ctenodactylidae, Dasyproctidae, Chinchillidae and Ochotonidae.

The rarity of positional anomalies and of crowding in the rodents may be partly due to the ample room for the teeth associated with the diastema between the incisors and cheek teeth and, in species in which the cheek teeth are of continuous growth, also to the wear of the approximal surfaces of the teeth.

Order **LAGOMORPHA**

Family **Leporidae**. Rabbits and hares. 1256 specimens examined

Four skulls with positional anomalies were found. In one (Fig. 14.8), the left I^1 was misplaced labially, slightly overlapping the right incisor. Slight variation in the position of the P^2 is common (Fig. 14.9). In one specimen, the left M_2 was rotated (Fig. 14.10).

Amongst 101 rabbit skulls of various domestic

Fig. 14.7. *Coendou* sp. (tree porcupine). ♀. BMNH 1903.9.4.87. Symmetrical rotation of M_3 with buccal surfaces turned mesially.

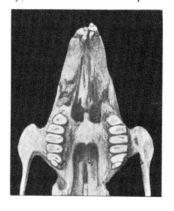

Fig. 14.8. *Lepus nigricollis* (Indian hare). ♀. BMNH 1909.6.10.7. The crowns of the left I^1 and I^2 are misplaced labially; that of the left I^1 overlaps the right I^1.

breeds, Nachtsheim (1936) found only one with any positional anomaly, a small chinchilla rabbit doe in which the right P^1 was rotated about 90°.

Ruprecht (1965d) described an anomalous skull of a hare (*Lepus capensis*) from the wild in Poland; it was greatly foreshortened and the braincase was globular. There was no record of the state of the post-cranial skeleton but the morphology of the skull suggested achondroplasia. An alternative discussed by Ruprecht is birth trauma; it seems unlikely that birth trauma of sufficient severity to damage the chondrocranium would occur in the wild, or, if it did, that the neonate would survive competition with the rest of the litter.

Chai and Degenhart (1962) found various skeletal anomalies in their stocks of inbred rabbits, including abnormally short upper or lower jaw. In some, the upper jaw was bent to one side and, as the lower jaw was normal, there was consequential overgrowth of the incisors. Many of the examples of overgrowth of the incisors of rodents and lagomorphs referred to in Chapter 17 were basically due to mismatch between the upper and lower jaws.

Both Fox and Crary (1971) and Huang, Mi and Vogt (1981) found that what they termed 'mandibular prognathism' in rabbits was inherited as a simple autosomal recessive with incomplete penetration. However, they did not distinguish between inferior protrusion, superior retrusion and incisor underjet, all of which can cause the lower incisors to occlude anteriorly to the upper incisors.

Fig. 14.9. *Lepus nigricollis dayanus* (Sind hare). ♀. BMNH 1913.8.8.135. Right P^2 slightly misplaced lingually.

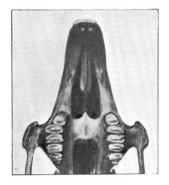

Chai (1970) bred rabbits between five inbred lines and found that malocclusions were mostly due to abnormal development of the upper jaw; in some cases, the I^1 crossed over one another and, in others, they erupted sideways so that they had no contact with the lower incisors.

As in rodents, the rarity of positional anomalies and crowding in lagomorphs may be partly due to the ample room for the teeth and to the approximal wear of the continuously-growing teeth.

Order **INSECTIVORA**

Colyer did not report positional anomalies of the teeth of the insectivores in the first edition of the book and we have found only one example in the literature. Harrison and Bates (1985) described a skull of an African hedgehog (*Erinaceus albiventris*) from the wild in which malposition of both P^3 was accompanied by absence of both P^2. Both P^3 were rotated in the same, anticlockwise, direction, the right P^3 by more than 90°, the left by more than 180°. In addition, the right P^3 was tilted disto-lingually to such an extent that it lay horizontally in a depression in the palate exposing the roots.

Fig. 14.10. *Lepus peguensis* (Burmese hare). ♀. BMNH 1910.9.15.1. Left M_2 (pointer) rotated and tilted mesially.

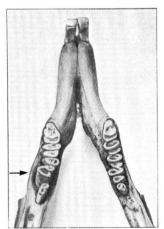

CHAPTER 15

Order MARSUPIALIA

Super-family **PHALANGEROIDEA**

Family **Macropodidae**

Positional variations in the Macropodidae are summarized in Table 15.1.

Sub-family **Macropodinae**. Kangaroos and wallabies. 856 specimens examined

The single lower incisor is procumbent; the upper incisors are arranged in an arch. The incisors are separated from the cheek teeth by a long diastema which, in the upper jaw, may contain a small canine. The cheek-tooth rows are curved with the convexity buccally. Figure 15.1 shows a notable feature of the macropods which is that P3 replaces not only m3 but also the tooth that is immediately mesial to it, m2, P2 or PC2, depending on the terminology employed (Chapter 1, p. 13).

In the sub-family Macropodinae, there is mesial drift of the cheek teeth throughout life; this is especially marked in kangaroos (*Macropus*). In young animals, M^1 is opposite the root of the malar process with the developing teeth projecting well back into the infra-temporal fossa; in old animals, M^4 may just be posterior to the malar process with a considerable interval between the tooth and the end of the maxilla. As a result of this mesial movement, the most anterior cheek teeth are progressively shed and, in old animals, the molars may be reduced to one in each quadrant. In macropods, M1 is always disproportionately worn, namely more than its eruption as the first in the permanent series of cheek teeth seems to warrant.

In animals in which there has been partial or

complete loss of teeth from fracture or excessive attrition, the mesial movement of the arch on the injured side is usually less than on the opposite side. In the skull shown in Figure 15.2, the right M_2 has been injured and part has been lost, the right M_1 and P_3 have not moved as far mesially as the corresponding teeth on the left. In the upper jaw, the mesial movement of the right and left teeth is equal.

That the developing teeth at the posterior end of the rows are important factors in mesial drift is shown by specimens in which a molar is absent. In the skull shown in Figure 15.3, the right M^4 is absent and the teeth on that side have not advanced to the same extent as those on the left. The unequal mesial movement of the cheek teeth, however, sometimes occurs with no discernible cause (Fig. 15.4).

(a) Incisors

Irregularities in the position of the incisors are rare. The labial surfaces in the majority of specimens do not form an even line, the teeth usually overlapping to some extent (Fig. 15.5).

(b) Premolars

The position of P3 is variable. In the upper jaw, it is in line with the molar series so that its buccal surface forms a continuous curve with the molars. The tooth may be placed with its mesio-distal axis almost parallel with the median line of the palate (Fig. 15.6) or its mesial surface may face lingually or buccally (Fig. 15.7). In a skull of *Lagochestes* (Fig. 15.8), both P^3 are misplaced lingually, the distal portions being overlapped by M^1.

In some species, this slight irregularity in the position of P^3 tends to be marked:

(i) *Macropus antelopinus* (antelope-kangaroo). In three specimens out of 11, P^3 was parallel with the median line of the palate; in two specimens, the mesial surface faced buccally and in four lingually; in two, the right P^3 was parallel with the median

Table 15.1 Positional variations in various MARSUPIAL families

Family	No. of specimens	Class of tooth			Total specimens affected	
		I	P	M	No.	%
Macropodidae	1050	5	33	5	43	4.1
Phalangeridae	853	2	4	2	8	0.9
Phasolarctidae	51	0	0	0	0	0
Peramelidae	213	0	1	0	1	0.5
Dasyuridae	331	1	4	0	5	1.5
Didelphidae	955	0	72	1	73	7.6
Total	**3453**	**8**	**114**	**8**	**130**	**3.8**

Fig. 15.1. *Macropus* sp. (kangaroo). Bone removed to show the normal relationship of the erupting P^3 to P^2 and the deciduous molar in the upper jaw. M^3 not yet erupted.

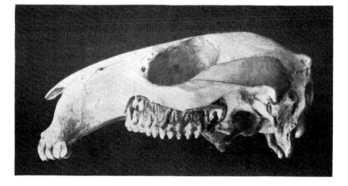

Fig. 15.2. *Macropus parryi* (wallaby whiptail). ♂. BMNH 1908.8.8.68. Probably because most of the crown of the right M_2 (pointer) was broken off in life (the fractured surfaces are now polished by wear), M_1 and P_3 have not undergone the normal mesial movement. The left lower and both upper cheek-tooth rows are normal but the right P^3 has no tooth to occlude with. Reduced slightly.

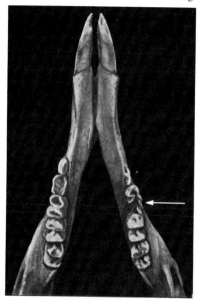

Fig. 15.3. *Macropus irma* (western brush wallaby). BMNH 1920.4.29.3. Right M^4 absent and, probably as a result, all the cheek teeth on that side are more distal than the left which have moved mesially in the normal way. There was no consequential effect on the lower tooth rows.

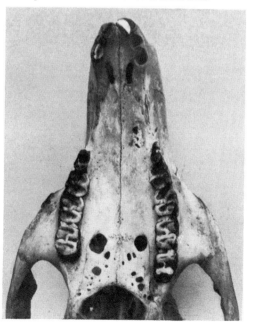

line of the palate, the left having the mesial surface facing lingually.

(ii) *Macropus parryi* (whiptail wallaby). In three out of 8 specimens, P^3 was parallel with the median line of the palate; in four skulls, P^3 was misplaced lingually and in one buccally.

In *Onychogalea lunata* (crescent nail-tailed wallaby), irregularities in the position of P^3 are common;

Fig. 15.4. *Lagorchestes conspicillatus conspicillatus* (spectacled hare-wallaby). ♀. BMNH 1913.1.28.6. The right cheek-tooth row is more distal than the left, for no apparent reason. The lower arch is similar, the right cheek teeth more distal than left.

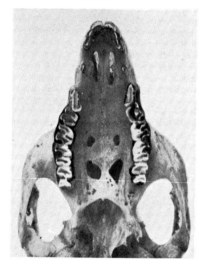

Fig. 15.5. *Macropus rufogriseus* (Bennett's wallaby). Left upper incisors are more lingually inclined and more overlapping than normal.

erupting as it does after M^1 has been worn down, it may become tilted over M^1 (Fig. 15.9). Sometimes, instead of replacing P^2 and m^3, P^3 may replace one only (Fig. 15.10).

Fig. 15.6. *Macropus bernadus* (Bernard's wallaroo). ♂. BMNH 1904.1.3.45. Both M^1 are, as is usual, more worn than the other teeth; their occlusal surfaces are in consequence overhung by the hardly worn P^3. Reduced slightly.

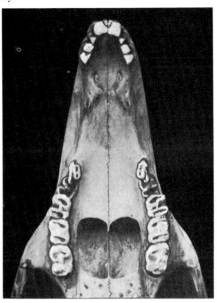

Fig. 15.7. *Macropus parryi* (whiptail wallaby). ♀. BMNH 1908.8.8.74. Both P^3 symmetrically rotated with mesial surfaces turned slightly buccally. Reduced slightly.

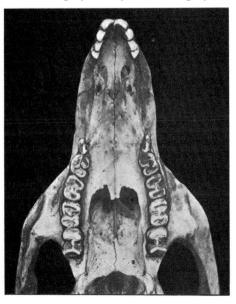

Marked irregularity of P^3 is seen in the genus *Macropus*. The tooth may have the crown tilted mesially or distally, it may be placed transversely to the line of the arch, or it may lie horizontally over the roots of m^3 (Figs. 15.11–15.14). There were nine such examples in 204 skulls of kangaroos and one in 174 of the small wallabies. In five skulls, the left P^3, in one the right P^3, and in four both P^3, were irregular in position.

Irregularities in position of P_3 were less common and less severe than those of P^3. P_3 may be placed obliquely; it may erupt before m_3, or it may replace it. Sometimes P_3 erupts lingually or buccally to m_3.

The variability in the position of P3 probably

Fig. 15.8. *Lagorchestes hirsutus hirsutus* (western hare-wallaby). ♀. BMNH 1906.10.5.20. Both P^3 misplaced lingually and partially overlapped by M^1.

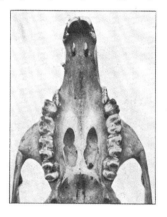

Fig. 15.9. *Lagorchestes conspicillatus conspicillatus* (spectacled hare-wallaby). ♀. BMNH 1901.8.2.5. The P^2 are present, unlike in Figure 15.8. Both P^3 rotated symmetrically with buccal surfaces turned distally; both teeth slightly overhang the much more worn M^1.

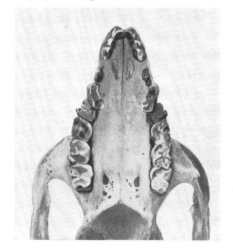

results from a combination of two factors: (i) the variable relationship of the developing P3 to m3 and to P2, (ii) the mesial movement of the molars. The developing P3 may be embraced by the roots of m3, may lie above the mesial root of m3 and the distal root of the P2, or be situated above P2. If P3 is to come into normal position, it is necessary that the time of eruption should synchronize with the time when the m3 and P2 have moved mesially under P3. If the times do not synchronize, abnormality in position will probably occur.

(c) Molars

Anomalies in the position of the molars are rare; the following were seen:

(i) *Lagorchestes hirsutus bernieri* (western hare-wallaby). ♂. BMNH 1906.10.5.16. Left M^1 rotated.
(ii) *Dendrolagus dorianus* (unicoloured tree kangaroo). ♀. BMNH 1899.9.7.1. (Fig. 15.15). Both M^4 rotated.
(iii) *Dorcopsis mysoiae* (forest wallaby). ♀. BMNH 1913.3.6.23. (Fig. 15.16). Left M_1, M_2 and M_3 irregular in position.

Fig. 15.10. *Lagostrophus fasciatus fasciatus* (banded hare-wallaby). ♂. BMNH 1906.10.5.12. Both P^3 have erupted mesially to the deciduous molars. Both M^4 are erupted; hence this is not a juvenile. The lower tooth rows were normal so neither P^3 was in functional occlusion.

Fig. 15.11. *Macropus giganteus giganteus* (eastern grey kangaroo). RCS Odonto. Mus., G 79.12. Right P^3 unerupted and misplaced with the crown tilted mesially. A similar, although less marked, condition was present on the left. Bone has been removed to display the roots.

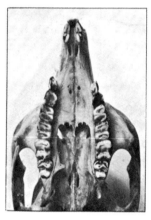
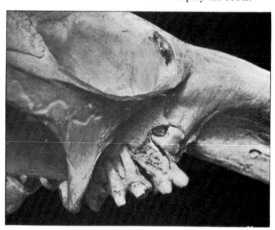

Fig. 15.12. *Macropus rufus rufus* (red kangaroo). Juvenile. Captive. BMNH 1862.12.26.3. Right M^4 not yet erupted. Some bone has been removed to display P^3 which is in process of replacing the deciduous molar and is tilted distally.

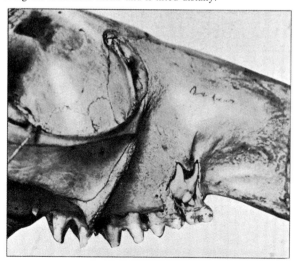

Sub-family **Potorinae**. Bettongs, potoroos and rat kangaroos. 194 specimens examined

Positional anomalies in this sub-family are rare; there was one skull in which both M_4 were tilted, the occlusal surfaces facing mesio-lingually.

Family **Phalangeridae**. Phalangers, ring-tails and brush-tailed possums. 853 specimens examined

Positional variations are uncommon; there were only eight (0.9%) (Table 15.1).

There are differences in the length of the muzzle in the various species of ring-tail *Pseudocheirus*; in those with long muzzles (Fig. 15.17A), the upper incisors form a small horse-shoe-shaped arch which occludes just labially to the procumbent I_1. The canine in the upper jaw is widely spaced between I^3 and the straight continuous row of cheek teeth. In the lower jaw, there is a long diastema between the procumbent I_1 and the compact row of P_3 together with the four molars, C, P_1 and P_2 being absent. Thus the upper canine and P^1 are not in functional occlusion. In those with short muzzles, the spaces between the upper teeth are considerably less and the teeth are sometimes in contact (Fig. 15.17B). In the mandible, there is no diastema and no C_1 or P_1, so the row of cheek teeth, beginning with P_2, is usually in contact with the recumbent I_1. With less space available for the premolars, the

Fig. 15.13. *Macropus bernardus* (Bernard's wallaroo). Juvenile. ♂. BMNH 1904.1.3.44. Left M^4 not yet erupted. Some bone has been removed to display P^3 which is erupting mesially to the deciduous molar and is rotated so that its buccal surface faces distally.

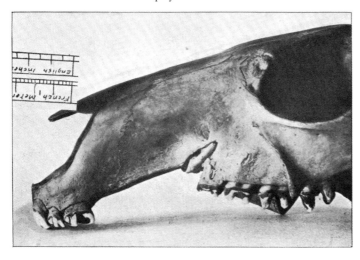

Fig. 15.14. *Macropus giganteus giganteus* (eastern grey kangaroo). Juvenile. RCS Odonto. Mus., G 79.1. Left P^3 misplaced horizontally above the roots of the deciduous molar. The outer plates of bone have been removed to display the roots (Colyer, 1907). The tilted most mesial lower cheek tooth is probably m_2. × c. 0.5.

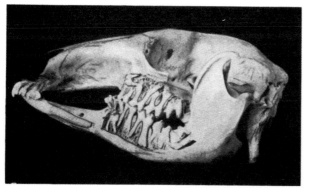

Fig. 15.15. *Dendrolagus dorianus* (unicoloured tree kangaroo). ♀. BMNH 1899.9.7.1. Both M⁴ symmetrically rotated with buccal surfaces turned slightly mesially.

Fig. 15.16. *Dorcopsis mysoiae* (forest-wallaby). ♀. BMNH 1913.3.6.23. The left mandibular molars form an irregular row, some being slightly misplaced lingually.

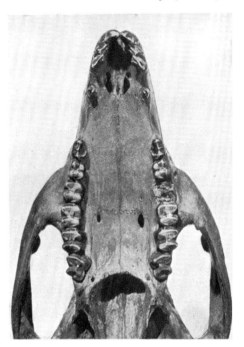

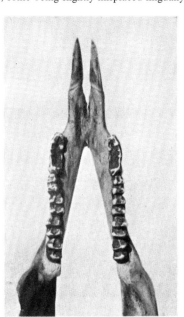

Fig. 15.17. Two specimens of *Pseudocheirus* to show the normal upper arches in a species with a short muzzle (A), and in a species with a long muzzle (B).
A: *P. archeri* (green ringtail). ♂. BMNH 1890.9.20.7. In this species, there is a slight gap between I¹ which are hook-shaped; otherwise all the other teeth in the arch are in contact.
B: *P. herbertensis* (mongan). ♀. BMNH 1890.9.20.5. There is a long space in the upper jaw between I³ and P¹ containing a small canine.

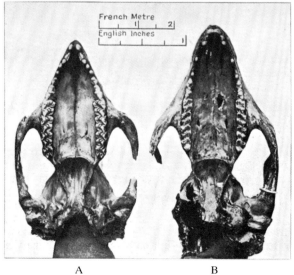

tendency is for P^1 to be squeezed out of the arch (Fig. 15.18); this was seen in two other specimens. Or P^2 may be set obliquely to the line of the arch as in a Tasmanian ring-tail *P. cooki* (formerly RCS Osteo. Series, 3688).

The specimens of phalangers and brush-tailed possums showing positional anomalies were:
(i) *Trichosurus vulpecula* (brush-tailed possum). Formerly RCS Osteo. Series, 3670. Both I^1 rotated.
(ii) *Phalanger orientalis carmelitae* (common phalanger). ♂. BMNH 1903.12.1.14. Both I^3 situated a little lingually.
(iii) *Trichosurus vulpecula* (brush-tailed possum). J 23080. Right P^4 erupting buccally to its normal position (Archer, 1975).
(iv) *Phalanger orientalis orientalis* (common phalanger). RCS Odonto. Mus., A 360.71. Right M_4 has the occlusal surface facing slightly lingually.
(v) *Phalanger maculatus* (spotted phalanger). RCS Odonto. Mus., G 79.5. (Fig. 15.19). Both M_4 tilted mesially.

Family **Phascolarctidae**. Koala. 51 specimens examined

No positional variations were noted in koalas (*Phascolarctos cinereus*) by Colyer, but Archer (1975) described a skull (J 5749) in which both mandibles were slightly twisted, resulting in general malocclusion.

Super-family **PERAMELOIDEA**

Family **Peramelidae**. Bandicoots. 213 specimens examined

The muzzle is long and narrow and the canines and premolars are widely spaced. Positional variations in the Peramelidae are summarized in Table 15.1.

One example of irregularity was seen in a female spiny bandicoot, (*Echymipera doreyana*). BMNH 1885.1.27.1. Both P^3 were lingually situated and in irregular occlusion with the mandibular teeth.

The upper premolars were asymmetrically arranged in one specimen (Fig. 15.20).

Super-family **DASYUROIDEA**

Family **Dasyuridae**. Marsupial cats, marsupial mice, etc. 331 specimens examined

The muzzle is fairly short but the canines and

Fig. 15.18. *Pseudocheirus cupreus* (coppery ringtail). Juvenile. BMNH 1905.11.28.25. The premolars on both sides are crowded and irregular. On the left, there is an empty socket where one has been lost. *P. cupreus* is one of the short-muzzled *Pseudocheirus* species.

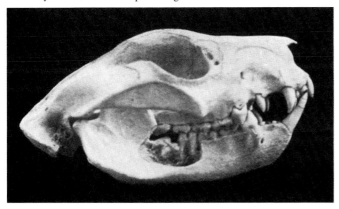

Fig. 15.19. *Phalanger maculatus* (spotted phalanger). RCS Odonto. Mus., G 79.5. Right M_4 partially erupted and tilted mesially with the crown impacted against the root of M_3. Bone has been removed to display the roots. The left M_4 was similarly affected. ×c. 0.8.

premolars are slightly spaced. Positional variations in Dasyuridae are summarized in Table 15.1.

A series of 30 skulls of *Dasyuroides byrnei* (kowari) in the Queensland Museum, all of which were bred in captivity, showed a higher incidence of brachycephaly and malocclusion than that in wild-caught animals (Archer, 1975). The series included:

(i) *Dasyuroides byrnei* (kowari). J11433. Captive. Right upper molars spaced and both lower molar rows crowded; general malocclusion.

(ii) *D. byrnei*. J11509. Captive. Left M^1 deflected disto-lingually.

(iii) *D. byrnei*. J10935. Right I^2 occludes lingually to the lower incisors.

Among 22 skulls of Tasmanian devil (*Sarcophilus harrisii*), two showed positional anomalies. In one, there was a crowded arrangement of the upper and the lower incisors; in the other, both P^3 were rotated, the buccal surfaces turned mesially.

In *Myoictis* and *Phascogale*, there were three examples of irregularity of P3:

(i) *Myoictis melas* (three-striped mouse). ♀. BMNH 191.1.11.104. Left P^3 wedged between P^2 and M^1.

Fig. 15.20. *Macrotis lagotis* (western rabbit-bandicoot). ♂. BMNH 1906.8.1.320. Right premolars more mesial in the arch than the contralaterals.

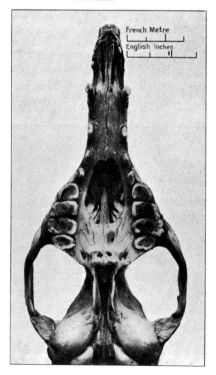

(ii) *Phascogale calura* (red-tailed wambenger). ♂. BMNH 1841.1240. Both P_3 wedged under M_1.

(iii) *Phascogale calura*. BMNH 1843.8.12.28. Both P_3 rotated with buccal surfaces turned slightly mesially.

(iv) *Phascogale calura*. WAM, M8069. Right M^1 rotated.

Fig. 15.21. *Didelphis azarae pernigra* (opossum). BMNH 1919.10.15.16. Both P^3 (pointer) tilted lingually so that they occlude inside the lower arch. They are also slightly rotated with buccal surfaces turned mesially.

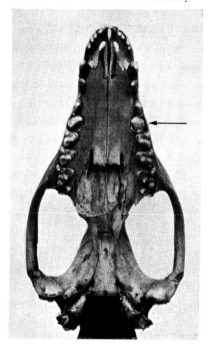

Table 15.2 Positional variations in some genera of the DIDELPHIDAE

Genus	No. of specimens	Total specimens affected	
		No.	%
Didelphys	177	19	10.7
Metachirus	123	9	7.3
Philander	71	1	1.5
Lutreolina	33	4	12.1
Monodelphis	76	0	0
Marmosa	460	39	8.5

The genera *Gliroma*, *Notodelphys* and *Chironectes* have been excluded because of the small numbers of skulls available for examination.

No positional anomalies were noted in other genera of dasyurids by Colyer, but two were listed by Archer (1975):
(i) *Sminthopsis crassicaudata* (fat-tailed dunnart). WAM, M4503. Left P_4 rotated through 180°.
(ii) *S. crassicaudata* WAM, M4497. Mandible very short, with open bite. On the right, the upper C occludes with I_3.

Superfamily **DIDELPHOIDEA**

Family **Didelphidae**. American opossums. 955 specimens examined

There is no diastema and adjacent teeth are in contact, the buccal surfaces of the cheek teeth forming a straight line. The prevalence of positional variations in the Didelphidae is summarized in Tables 15.1 and 15.2.

The position of P3 is variable in the Didelphidae. When there is insufficient room in the arch, P^3 tends to be rotated (Fig. 15.21). P^3 may be misplaced mesially or lingually but marked rotation is rare (Fig. 15.22).

Positional anomalies of P_3 take the form of tilting due to the overhanging portion of M_1 impeding the eruption of the distal part of P_3. Irregularity in position of P3 is a feature in the mouse opossums *Marmosa elegans* (Fig. 15.23) and *M. incania*.

In a few specimens, P1 was placed obliquely across the distal aspect of the canine.

Positional anomalies of the molars were rare; in the only one (Fig. 15.24), both M_4 were misplaced, the teeth being tilted and separated by a small space from M_3.

Fig. 15.22. *Didelphis* sp. (opossum). ♂. BMNH 1901.1.1.93. Rotation of both P^3 (pointer) with buccal surfaces turned mesially, the left more than the right.

Fig. 15.23. *Marmosa elegans pallidor* (mouse opossum). ♀. BMNH 1926.10.11.105. Right P^3 (arrow) slightly rotated with buccal surface turned mesially. Right P_3 crowded and appears to have been unable to complete its eruption.

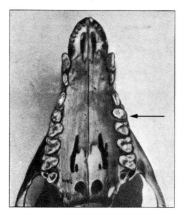

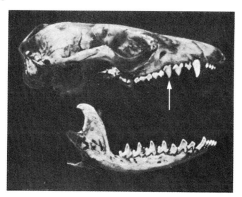

Fig. 15.24. *Didelphis marsupialis etensis* (Peruvian opossum). ♂. BMNH 1901.6.5.17. Right M_4 tilted mesially. Bone has been removed to display the tooth. The condition was symmetrical.

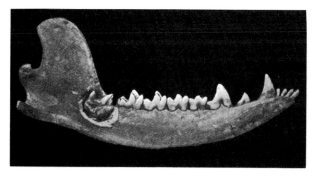

SECTION 3
Abnormalities of Eruption

CHAPTER 16

Variations and Disturbances of Eruption

Introduction

Eruption is the process whereby each tooth is carried from its developmental position within the jaw to its functional position in the mouth; movement is mainly axial in direction although rotation, tilting and lateral movement are also common components. Tooth emergence, the term used to distinguish the actual appearance of the tooth in the mouth from other phases of the process, occurs at particular times and in a particular sequence. Tooth emergence can be precocious or delayed, it may be incomplete or the tooth may not emerge at all, remaining buried in the jaw. The sequence and chronology of eruption differ to some extent from one taxon to another (Owen, 1866/1868).

In animals with high-crowned hypsodont teeth, such as horses and bovids, continuous eruption of the teeth as the crowns are gradually worn down is an integral feature of the masticatory apparatus (Ainamo, 1970; Ainamo and Talari, 1976). Even in non-hypsodont dentitions where tooth-wear is a less predominant feature, it is likely that a process known as continuous eruption occurs. This process has been most studied in man but some supportive observations have also been made on the cheek tooth dentitions of rats, mice and hamsters (reviewed by Miles, 1961), where there is strong evidence that, as age advances, recession of the gingival and alveolar crest margins (passive eruption) occurs leaving the crowns of the teeth more and more exposed. Whether this process is a compensation for reduction in the height of the crowns by occlusal wear is uncertain but, when such wear does occur, there is also some active eruption of the tooth produced by the deposition of bone on inner surfaces of the tooth sockets (Whittaker et al., 1985). The potential of the tooth to erupt further is not lost when the tooth comes into occlusion with its antagonists because, if at any time subsequently its antagonist is lost, its crown tends to erupt beyond the occlusal plane, as Anneroth and Ericsson (1967) and Moss and Picton (1967) found experimentally in macaques.

However, it is necessary to avoid thinking that all teeth erupt indefinitely if no tooth opposes them in occlusion. Some teeth or groups of teeth, although fully erupted in the sense that the whole crown emerges through the gingiva, never reach occlusion. Notable examples are the premolars of canids and of some marsupials. Thus it seems that, whereas the majority of teeth that normally reach occlusion given the appropriate circumstances, tend to over-erupt, in these other instances the eruptive force, whatever it is, is only sufficient to carry the teeth through the gingiva. Even though teeth may not actually occlude, they can still function; for instance, prey can be grasped or bones gnawed by canid premolars even though they do not meet in occlusion.

In many mammalian groups, some teeth, such as the incisors of rodents, or even the dentition as a whole, as in lagomorphs, have continuously-growing roots. It is well established experimentally that the rate of eruption of teeth of continuous growth is increased if their eruption is unimpeded by contact with their antagonists (see Chapter 17 p. 356).

Some of the examples described here are of teeth that have failed to emerge. Some are presumably due to a fundamental failure of the eruptive force which is ill-understood but which probably involves a complex of factors. Bone growth, involving both bone deposition and resorption, is associated with eruption, but whether this contributes to the eruptive force or whether the changes in the bone are secondary is uncertain.

Another simpler cause of failure of a tooth to emerge is the obstruction of its pathway by another tooth, leading to impaction. However, sometimes the erupting tooth is merely diverted by the obstacle and finally emerges in an abnormal position in the arch (Chapters 9–15).

As the shedding of deciduous teeth at the normal time can be regarded as part of the process of eruption of the succeeding permanent teeth, failure of the shedding process will be considered here. The retention of deciduous teeth beyond their normal time of shedding has not been studied systematically except in man but it seems safe to assume that what is known of the condition in man is largely applicable to other mammals. If a deciduous tooth is retained beyond its normal time, its successor sometimes remains quiescently buried in the bone or, especially if there is some resorption of the roots of the deciduous teeth, the permanent tooth may erupt beside its predecessor, in the line of the arch if there is room but often on the buccal side of the arch. Shedding is not induced by the presence or movement of the permanent successor because, if the successor is absent, the deciduous tooth may not be shed or, about equally commonly, may be shed at about the normal time.

Retention of deciduous teeth can be due to ankylosis of the resorbing tooth to the bone which, although difficult to demonstrate except by serial sectioning through the whole tooth, may be induced by infection or by heavy prolonged or intermittent stress or by more acute trauma.

A condition is well-documented in man in which a tooth, usually a deciduous molar but sometimes a permanent one, is partially erupted and overgrown by the eruption of the teeth adjacent to it so that it cannot erupt further; that is, it is impacted, but in many instances there is good evidence, such as the presence of occlusal wear, that before it became overgrown it was in occlusion and therefore fully erupted. Such teeth are known as submerged or, better still, as re-included because this implies that they have partially returned to their pre-eruptive condition. Darling and Levers (1973) have shown that such teeth, although at first erupting normally, later, unlike the teeth adjacent to them, are not carried axially by the growth of the alveolar process that occurs particularly during the replacement of the deciduous teeth and eruption of the permanent dentition. In other words, there is a localized failure of bone growth; many affected teeth have been shown to be ankylosed to the bone by a process of root resorption and replacement by bone.

It has been known since ancient times (Barrow, Simpson and Miller, 1974) that the consumption of the peas of the *Lathyrus* group, for instance the sweet pea, produces a condition in animals, cattle and horses in particular, and in man, known as lathyrism. Neurological symptoms and skeletal deformities are the commonest effect. It is now known that at least two different toxic compounds are responsible, one producing experimentally neuro-lathyrism and the other osteo-lathyrism (Selye, 1957). The compound responsible for osteo-lathyrism, beta-aminopropionitrile, appears to prevent the normal maturation of collagen and this has excited a considerable amount of experimental work, notably into the mechanism of tooth eruption because it is hypothesized that collagen contraction is the fundamental source of the force exerted during tooth eruption.

Disturbance of tooth formation appears not to have been noted in the natural disease but, in experimental lathyrism, a variety of effects on the teeth have been noted (Shore, Berkovitz and Moxham, 1984), including slowing of tooth eruption (Barrington and Meyer, 1966) which is why we have chosen this section to present this short account of lathyrism.

In lathyric rats, Sciaky and Ungar (1961) found a disturbance of the growth of incisors with accordion-like folding of the basal growing-end of the tooth indicative of a disturbance of the balance between the rate of growth of the tooth, the rate of its eruption and the rate of its wearing away in use. In some similar experiments in which sweet peas (*Lathyrus odoratus*) were fed to rats, Gardner, Dasler and Weinmann (1958) found that there was generalized stunted growth, that exostoses developed at areas of muscle and ligament attachment to the cranium and mandible, and that the mandibular condyle was enlarged and distorted. This raises the possibility that the effect on growing rat incisors could be secondary to changes in the morphology of the jaws and of the jaw joint. However, Sciaky and Ungar noted degenerative changes in the dental formative epithelium which suggests that the lathyritic toxin has a direct effect on the formative tissues of the tooth.

Transverse accordion-like folding of the basal ends of the incisors of rodents is met with in other conditions and appears to be due to a failure of the eruptive process. Obese-strain mice tend to lose their voracious appetites after a while and undergo a sudden decline in weight. Batt (1978) noted that, in this period of decline, the incisors often cease to erupt so that after a time no erupted part is visible. Radiography showed that the basal parts of the teeth were transversely folded, evidently due to the continuation of tooth formation even though the teeth had ceased to move axially. Sometimes only the upper teeth were affected; the lower incisors then became overgrown.

Retardation of eruption of both incisors and cheek teeth is also a feature of the screw-tail strain of mice (Bhaskar, Schour, MacDowell and Weinmann, 1951), in the incisors, associated with transverse folding of the basal ends.

Interference with tooth eruption is a feature of a variety of mutant strains of mice that have attracted the attention of geneticists. It is not possible to deal with them all here; Grüneberg (1963) has reviewed them fully. The basic skeletal disorder appears to be of the process of bone resorption and interference with tooth eruption is secondary to that. In the grey-lethal mouse (Grüneberg, 1936), continued growth of the incisors within restricted bony crypts leads to the formation of odontome-like (Chapter 25, p. 605) masses of tooth tissue which may protrude through the mental foramen.

Grüneberg (1963) referred to a strain of rabbit in which the absence of proper bone resorption leads to the formation of bones of great density (osteopetrosis) containing but little marrow. Both incisors and molars often fail to erupt. Osteopetrosis is also a predominant feature of the ia (incisor-absent) strain of rat, so-called because the incisors fail to erupt. Odontome-like masses at the basal ends of the incisors are a feature of this strain (Chapter 25, p. 605). Cotton and Gaines (1974) described a mutant strain of rat which seems similar to the ia strain but which they called tl (toothless) because both incisors and molars remain unerupted.

In man, tooth eruption is retarded by malnutrition; certainly such retardation was a notable feature of rickets (Colyer and Sprawson, 1931). It seems highly likely that malnutrition in animals delays tooth eruption also but only a little anecdotal evidence seems to be available, such as that derived from some white-tailed deer (*Odocoileus virginianus borealis*) maintained on over-browsed parkland where there was evidence that retarded premolar eruption was associated with undernourishment (Hamerstrom and Camburn, 1950).

In a number of specimens described by Colyer, there was evidently some widespread bone disorder which he categorized as rickets, usually associated with a disorder in the position or eruption of teeth. The bone was often thickened and of altered texture. Colyer referred to it as being softer than normal in the sense that it could be readily pared with a sharp knife or sometimes as being denser than normal. Sometimes the diagnosis of rickets was made by veterinary surgeons at the zoo where the animal died or the diagnosis was part of a museum catalogue entry without any supportive documentation. The works of Bland Sutton (1884b,c) and Christeller (1923) give insight into the state of knowledge at the time when many specimens were collected.

Over the past 60 years or so, rickets has become better understood, terms have been redefined and certain similar bone disorders have been more clearly distinguished.

Rickets affects bones during their growth period and the predominant cause is a deficiency of vitamin D. However, vitamin D is a complex of substances and deficiency of calcium or phosphorus or an imbalance of intake between the two also play an important part, as does secondary disturbance of parathyroid function.

The effects of the same sort of deficiency are different in the adult and are known as osteomalacia.

A bone disorder similar to vitamin-D deficiency rickets occurs in chronic kidney failure in young animals; it is mediated by secondary hyperparathyroidism and is known as renal rickets. An example in an adult dog, in which the jaws were both thickened and osteoporotic, was described by Kyle, Davis and Thompson (1985). Fibrous osteodystrophy is another related disorder due to over-action of the parathyroids and has features of both rickets and of osteomalacia.

It is obvious that a comprehensive account of these disorders cannot be given here. Good accounts are available in Resnick and Niwayama (1981, vol. 2) and by Jones and Hunt (1983, pp. 1167–1173), who figure examples of fibrous osteodystrophy affecting the jaws of a horse, a male spider monkey (*Ateles*) and a woolly monkey (*Lagothrix*) as well as the grossly distorted jaws of a 5½ month-old male terrier with renal rickets.

Joest (1926, figs. 102, 107) figured the mandible of a pig which was thickened by an osteodystrophy: there were many retained deciduous teeth and several permanent teeth remained deeply in the bone.

The differential diagnosis of these conditions in a skull with little known history is virtually impossible. Rickets, as the disorder of the growth period of bones, would be the one likely to affect the eruption of the teeth. It typically affects the cartilaginous growth plates, so signs of defective growth of the base of the skull would be expected. However, the trabeculae and even cortex of the bones of the jaws and of the calvarium are thinned and composed of not fully mineralized bone (osteoid) in rickets and there is a compensatory thickening of the bones by periosteal bone formation. The effects of renal rickets would be similar.

In osteomalacia, the replacement of fully mineralized bone by osteoid occurs on a large scale to an extent that the bones can be distorted by function. They also become thickened by periosteal bone.

In fibrous osteodystrophy, thickening of the bones is a notable feature to the extent that when the jaws are affected the teeth may become secondarily embedded in the periostally-formed new bony tissue which is largely osteoid. Jones and Hunt (1983) drew attention to species differences in the characteristics of fibrous osteodystrophy; for example, there is a hyperostotic type in horses. The condition is particularly common in pet and laboratory New World monkeys and the cause is still not understood in spite of some experimental work. In young cats, an almost exclusive diet of fresh beef hearts, which are low in calcium and high in phosphorus, causes the condition (Jowsey and Gershon-Cohen, 1964; Jowsey and Raisz, 1968).

When any of these disorders, which are essentially disturbances of calcium or mineral metabolism, occurred during the period of enamel formation, enamel hypoplasia (Chapter 20) would be expected to be present.

If any of these conditions becomes inactive, as they may do if the diet improves, the bone thickenings and distortions remain, of course, but the bone undergoes change towards normal which may be patchy and may be not to bone of normal texture but to bone of greater density than normal. This adds to the difficulty of diagnosis.

Most of the specimens Colyer referred to with some bone disorder are captive ones and it is a reasonable assumption that disorders of the rickets-type due to nutritional deficiencies are, or were, more common in captivity than in the wild.

We shall not, of course, attempt any re-diagnosis of these specimens even when we have been able to re-examine them. In a few instances, we have added additional comments.

Some additional discussion of rickets-like lesions and fibrous osteodystrophies in connection with so-called odontomes appears in Chapter 25 (pp. 601–605).

Eruption, like the rest of bodily development, is under endocrine control. The influence of the thyroid gland on eruption is moderately simple; thyroidectomy retards the process and the administration of thyroid extract accelerates it (Todd, Wharton and Todd, 1938, on sheep; Lusted *et al.*, 1953, on macaque monkeys). The influence of other components of the endocrine system is less marked and much more complex; Anderson, Seipel and Van Wagener (1953) showed that the administration of androsterone to two young male rhesus monkeys retarded the eruption of late-erupting teeth but this may have been secondary to the much more marked retardation of bone growth.

There are sometimes problems with determining the order of eruption, because two different criteria are employed for eruption. Most authors use the first appearance in the mouth (i.e. the emergence) of a particular tooth, whereas others use the final attainment of occlusal level (Bramblett, 1969). In practice, because the occlusal level is usually quickly reached, the criterion chosen has little effect on the sequence, except when the canine is large, as it is in many male primates, when the interval between emergence and arrival at the occlusal level may be great.

Variations in the sequence of eruption

(a) Family **Pongidae**. Great apes and gibbons

The usual order of eruption of the permanent teeth in all the pongids is:

M1, I1, I2, M2, P3 and P4, C, M3.

The lower tooth of each type erupts slightly in advance of the upper one. Although the canines begin to erupt before M3, the large canines of males may be the last teeth to take up their final position (Clements and Zuckerman, 1953; Schultz 1935, 1950; Randall, 1943/1944).

Gorilla gorilla (gorilla)

The usual order of eruption of the deciduous teeth is

i1, i2, m3, m4, c

(Selenka, 1899; Schultz, 1950). Eruption of the deciduous dentition begins at 3–4 months and is completed in the second year of life (Krogman 1930). Although i^1 usually erupts before i_1, and i^2 before i_2, the lower deciduous incisors may erupt before the upper; in six captive infant gorillas, Ussher-Smith, King, Pook and Redshaw (1976) found that i_1 preceded i^1, and that i^2 preceded i_2 in three animals and succeeded i_2 in the three others. The deciduous canines may erupt after m4. There was much variation in the times of the appearance of the deciduous teeth.

In the permanent dentition, I^2 may appear before I_2. The premolars appear in rapid and irregular succession, usually with P_2 emerging last (Schultz, 1950). In 17 out of 22 gorilla skulls, the canines emerged before M3; in two, they were emerging at the same time and, in three, M3 emerged before the canines. Occasionally M2 may erupt before I2.

Pan troglodytes (chimpanzee)

The usual order and chronology of eruption of the deciduous teeth is shown in Table 16.1. The order is the same as in the gorilla, except that the canines usually erupt after m4 (Nissen and Riesen, 1945). As in the gorilla, eruption of the deciduous teeth is usually completed at about 14 months of age (Zuckerman, 1928; Krogman, 1930; Schultz, 1940).

In chimpanzees, there is a wide range of variation in the order of eruption of the permanent incisors and, in more than half the skulls examined, I2 appeared after M2. According to Schultz (1940), the incisors may occasionally erupt even after the premolars. P3 erupts before P4 (Schultz, 1935). In 22 skulls, there was one in which eruption of M3 preceded that of the canines. Nissen and Riesen (1964) found that M3

Table 16.1 Eruption in the CHIMPANZEE *(Pan troglodytes)*

	Lower teeth No. of teeth	mean	range	Upper teeth No. of teeth	mean	range
Juveniles						
i1	32	84	53–126	32	95	40–161
i2	32	108	65–188	32	110	74–177
m3	32	143	72–210	32	120	74–180
m4	32	237	154–338	32	300	226–446
c	32	365	265–492	32	343	226–445
Adults						
M1	30	39.2	32–45	30	39.9	33–45
I1	30	68.9	60–84	30	67.5	54–81
I2	30	73.8	60–88	30	80.8	70–99
M2	30	77.5	67–88	30	81.5	68–94
P3	30	88.2	76–99	30	83.4	73–98
P4	30	89.7	73–109	30	88.2	75–100
C	29	107.7	95–121	28	107.9	91–121
M3	28	125.5	108–157	28	136.1	117–163

The age in days at the time of emergence of the deciduous teeth in 16 captive juveniles of both sexes (Nissen and Riesen, 1945), and in months for the permanent teeth in 15 captive adults of both sexes (Nissen and Riesen, 1964).

erupts slightly earlier in males than in females. A great deal of data on the eruption of teeth of the chimpanzee has been produced and analyzed by Nissen and Riesen (1945, 1964). Their results on the timing of the eruption are summarized in Table 16.1. They found no statistical differences in the eruption pattern of males and females except that, in males, M3 erupts rather earlier than in females and, as in other great apes, the eruption of the canines takes longer in males.

Pongo pygmaeus (orang-utan)

The usual order of eruption of the deciduous teeth is:

$i_1, i^1, m_3, m^3, i_2, i^2, m_4, m^4, c_1, c^1$

(Schultz, 1941), but Selenka (1898) found that m4 erupts after the canines. Eruption of the deciduous dentition is often completed by the age of one year.

In the permanent dentition, Selenka (1898) found much variation in the detailed order of appearance of I1, I2 and M2. Schultz (1941) noted that M2 frequently erupts before I2; occasionally, the canine emerges after M3.

Genera **Hylobates**, gibbons, and **Symphalangus**, siamangs

The usual order of eruption of the deciduous teeth in gibbons is

i1, i2, m3, c, m4

(Schultz, 1944). Eruption is usually complete within one year (Krogman, 1930).

The order of appearance of the permanent teeth in both the gibbon (Schultz, 1935, 1944) and the siamang (Mijsberg-van Rooijen and Mijsberg, 1931) is the same as in the great apes; the succession of premolars is variable.

Selenka (1899) found that M2 sometimes erupted before I2 and that occasionally the lower canine was the last tooth to erupt.

(b) Family **Cercopithecidae**. Old World monkeys

According to Schultz (1942), the order of appearance of the deciduous dentition in macaques (*Macaca* spp.) and the proboscis monkey (*Nasalis larvatus*) is:

$i^1, i_1, i_2, i^2, m_3, c_1, c^1, m_4, m^4$.

As these species belong to the different sub-families of the Cercopithecidae, it seems likely that this sequence holds good for the family as a whole.

The order of emergence of the permanent teeth in Old World monkeys is usually the same as in the Pongidae:

M1, I1, I2, M2, P3 and P4, C, M3.

I^1 may emerge before I_1, and I^2 before I_2; the position of the canine in the order is variable.

Sub-family **Cercopithecinae**

In 21 skulls of macaques (*Macaca* spp.), there were 10 in which the upper incisors appeared before the lower incisors and, in 11, the lower incisors appeared first. In two skulls, all four lower incisors had appeared before the upper incisors.

In *Cercopithecus*, the order of emergence of the premolars is usually P^4, P_4, P^3, P_3. In 13 skulls of *Macaca*, P3 erupted before P4. In male yellow and chacma baboons (*Papio cynocephalus* and *P. ursinus*), P^3 erupts before both P4 but P_3 appears after P^4 (Bramlett, 1969; Freedman, 1962; O. M. Reed, 1973).

Exceptions to the rule that canines emerge before M3 were seen in *Cercopithecus* but not in *Macaca*.

Hurme and Van Wagenen (1953, 1961) and Hurme (1960) established detailed statistical parameters for the eruption of the teeth in captive rhesus monkeys, *Macaca mulatta* (Table 16.2); they showed that in females the canines usually erupt before the premolars. Table 16.3 summarizes the time and order of eruption of the permanent teeth in a population of Japanese macaques (*Macaca fuscata*). Some of the teeth erupted slightly earlier in females than in males and sometimes lower teeth appeared slightly earlier than upper ones of the same type (Nass, 1977).

Table 16.2 Eruption in MACAQUES

	Females lower	upper	Males lower	upper
Juveniles				
i1	37	41	37	40
i2	44	66	48	66
c	110	103	99	95
m3	99	99	102	99
m4	179	201	190	212
Adults				
M1	18½	20½	21	22½
I1	36	33½	34½	36
I2	39½	39½	37	39
M2	44	48	44	45½
P3	50½	50½	55	52½
P4	54½	59	53½	55
C	48	55	57	61½
M3	87½	114½	83	92

The mean age in days at the time of emergence of the deciduous teeth, and in months of the permanent teeth, in the 98th percentile of a large group of captive macaques (*Macaca mulatta*). After Hurme (1960).

In females of *Cercopithecus ascanius*, *C. aethiops* and *Cercocebus albigena*, the canines tend to erupt after the premolars but before M3 (Wintheiser, Clauser and Tappen, 1977). According to Lampel (1963), the later eruption of the upper, but not necessarily the lower, canine in males is typical of the genus *Cercopithecus*.

Eruption of the deciduous and permanent teeth of the yellow baboon (*Papio cynocephalus*) is summarized in Figure 16.1. The canines appear at about the same time as the premolars, that is in advance of M3, but in males they achieve the occlusal plane well after M3 (Bramblett, 1969; Reed, 1973).

Table 16.3 Eruption in MACAQUES

	Females ($n=85$)		Males ($n=43$)	
	lower	upper	lower	upper
M1	1.75	2.0	2.0	2.0
I1	2.5	3.0	3.0	3.0
I2	3.0	3.0	3.0	3.0
M2	4.0 ▲	4.0 ▲	3.5 ▲	4.0 ▲
P3	4.0	4.5	4.0	4.0
P4	4.0	4.5	4.0	4.0
C	4.0	4.0	4.5	5.0
M3	6.0 ▼	6.0 ▼	7.0 ▼	7.0 ▼

The mean age in years at the time of emergence of the permanent teeth in a troop of wild-living macaques (*Macaca fuscata fuscata*). After Nass (1977). The arrows indicate when the eruption order may be reversed.

Sub-family Colobinae

In *Colobus* and the leaf-eating monkeys (*Presbytis*), the usual order of emergence of the premolars is

P^4, P_4, P^3, P_3

As in other cercopithecids, M2 usually emerges just after I2 but, in *Colobus* and *Presbytis*, M2 may appear before I2 or even before any of the permanent incisors, as in a banded leaf monkey (*P. melalophos femoralis*, ♀, BMNH 1908.7.20.4). In *Rhinopithecus* (snub-nosed monkey), M2 appears before I1, leading Schultz (1935, 1958) to infer that early eruption of M2 is a characteristic of the Colobinae.

In colobines, the canine usually appears before M3; for example, in 15 out of 21 skulls of red colobus (*C. badius*), the canines preceded M3 whereas, in 23 skulls of black-and-white colobus monkeys (*C. polykomos* and *C. guereza*), M3 preceded the canines in 20 specimens.

In *Presbytis*, M3 appeared before the canines in 27 out of 43 skulls but, in 11, the order was reversed and, in five, C and M3 were erupting at the same time. In 5 skulls, M3 appeared before any of the premolars and, in many others, M3 was emerging or functional before the premolars were functional. Such precocious eruption of M3 is shown in the following specimens:

(i) *P. melalophos femoralis* (banded leaf monkey). ♀. BMNH 1903.2.6.10. All M3 emerging; the deciduous molars are still present and I^2 is still unerupted.

(ii) *P. rubicunda* (maroon leaf monkey). BMNH

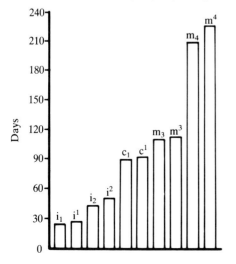
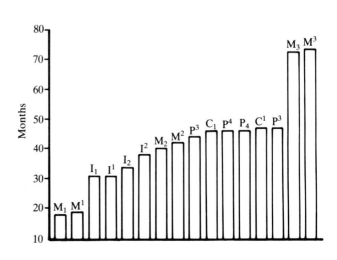

Fig. 16.1. Eruption in the yellow baboon (*Papio cynocephalus*). The timing and sequence of eruption of the deciduous teeth (A) and the permanent teeth (B). Eruption of the deciduous dentition is completed in 233 days and of the permanent dentition from 18 to 75 months. Redrawn from: Reed, O.M. (1973). By kind permission of the publisher.

338 Section 3: Abnormalities of Eruption

1908.7.17.3. All M3 emerging whilst the deciduous molars are still present.

In the proboscis monkey (*Nasalis larvatus*), the females have the usual eruption pattern but, in males, the lower canine and M3 appear simultaneously and the upper canine appears last (Schultz, 1942). In the snub-nosed monkey (*Rhinopithecus*), the canine usually appears last but it may erupt earlier in females (Schultz, 1935).

(c) Family **Cebidae**. New World monkeys

The sequence of eruption of the deciduous teeth varies between the genera:

Alouatta: i1, i2, m2, m3, m4 + c.

Cebus: i1, i2, m2, c, m3, m4 (Schultz, 1960)

Aotus: i1, i2, m2, m3, c, m4 (Hall, Beattie and Wykoff, 1979)

Saimiri: i1, i2, c, m2, m3, m4 (Long and Cooper, 1968).

The chronology of eruption of the deciduous teeth in *Aotus* and *Saimiri* is summarized in Table 16.4.

Table 16.4 Eruption in CEBID MONKEYS

	Aotus lower	*Aotus* upper	*Saimiri sciureus* lower	*Saimiri sciureus* upper
Juveniles				
i1	2.3	2.3	1.3 (–3)	1.3 (0–3)
i2	2.9	3.5	1.4 (1–3)	2.2 (1–3)
m2	3.4	3.5	4.8 (4–6)	4.9 (4–6)
m3	3.8	3.8	5.7 (5–7)	6.1 (5–7)
c	4.6	4.7	3.4 (3–4)	3.4 (2–4)
m4	5.5	5.6	8.3 (7–9)	8.9 (8–11)
Adults				
M1	4.3	4.9	5.1 (5–6)	5.5 (5–6)
M2	6.4	7.3	7.0 (7)	8.4 (7–9)
I1	9.6	9.4	9.2 (8–11)	9.7 (8–12)
M3	9.9	11.2	12.3 (11–14)	20.0 (19–22)
I2	10.6	10.8	9.6 (8–11)	12.0 (10–14)
P4	11.0	11.5	12.8 (12–15)	13.0 (12–15)
P3	12.3	11.9	15.0 (15)	13.0 (12–15)
P2	12.4	12.4	14.3 (12–16)	14.7 (14–15)
C	14.0	15.0	20.5 (19–21)	21.5 (21–22)

The mean age in weeks at the time of emergence of the deciduous and permanent dentition in 28 infant and 9 adult captive owl monkeys (*Aotus trivirgatus*), and in 10 infant and 2–8 adult captive squirrel monkeys (*Saimiri sciureus*). Values in parentheses indicate the range. (From Hall, Beattie and Wyckoff, 1979, and Long and Cooper, 1968.)

The order of appearance of the permanent teeth is usually the same as in the Cercopithecidae:

M1, I1, I2, M2, P2, P3, P4, C, M3,

with the additional premolar (P2) erupting just before P3. I^1 may precede I_1, and I^2 may precede I_2. The premolars appear in rapid and variable succession but most often in the order shown. In males, the last canine erupts slowly and may not reach the occlusal plane before M3 (Schultz, 1935, 1960).

In the douroucouli (*Aotus trivirgatus*), the order of eruption of the premolars is reversed and M_3 usually erupts before I_2 and M^3 before P^4, before any of the deciduous teeth have been shed. In one skull (BMNH 1925.12.11.10) in which both I^1 are just appearing, both M_3 have appeared and M^3 are erupting. The mean age of appearance of each of the deciduous and permanent teeth in *Aotus* is summarized in Table 16.4.

In *Cebus* (capuchins) and *Ateles* (spider monkeys), as in most New World monkeys, the canine usually precedes M_3, although the canine may be last in male capuchins (Schultz, 1935); in *Pithecia* (sakis) and *Alouatta* (howler monkeys), the canine usually erupts last in both sexes. In 27 skulls of *Alouatta*, the M3 preceded C in 17 skulls, C preceded M3 in seven and they emerged simultaneously in three.

Schultz (1935) found that in squirrel monkeys (*Saimiri*) M2 usually erupts before any incisors; although M^3 is the last tooth to erupt, M_3 usually appears before the premolars; however, Long and Cooper (1968) found that in living squirrel monkeys the canine is the last tooth to erupt, the order of eruption of the permanent dentition being:

$M^1, M^2, I^1, I^2, P^3, P^4, P^2, M^3$, C in the maxilla and

$M_1, M_2, I_1, I_2, M_3, P_4, P_2, P_3$, C in the mandible.

(d) Family **Callitrichidae**. Marmosets and tamarins

The sequence of eruption of the deciduous teeth in the common marmoset (*Callithrix jacchus*) is:

i1, i2, c + m2 + m3, m4

Eruption of the deciduous teeth begins in the first week of life and is completed in the fourth (Johnston, Dreizen and Levy, 1970). In the black and red tamarin (*Saguinus nigricollis*), Chase and Cooper (1969) found that the deciduous incisors and canines erupted before birth, followed by m2, m3 and m4 between birth and eight weeks of age.

The sequence of eruption of the permanent teeth in the common marmoset is:

M1, M2, I1, I2 + P4, P3 + P2, C

M1 erupts in the third or fourth month of life before any of the deciduous teeth have been replaced; the eruption of the permanent dentition is completed at 11–12 months (Johnston, Dreizen and Levy, 1970).

In the pygmy marmoset (*Cebuella pygmaea*), the sequence is similar except that I^2 erupts after P2. I^2 is caniniform in *Cebuella* and Byrd (1978) raised the question whether its late eruption, close to that of the canine, could indicate some influence by the canine morphogenetic field.

When completed eruption is used as the criterion, callitrichids fall into two groups (Byrd, 1981). In the marmosets (*Callithrix* and *Cebuella*), the molar-incisor sequence is M1, I1, M2, I2, but in the tamarins (*Leontopithecus* and *Saguinus*) it is M1, I1, I2, M2. Goeldi's marmoset (*Callimico*) resembles the tamarins, but has three molars in each quadrant and M3 is the last tooth to erupt. Byrd used these ontogenetic characters as a basis for separating callitrichids into three sub-families (Callitrichinae, Leontopithecinae and Callimiconinae) and pointed out that the eruption pattern in the Callitrichinae is the same as that in platyrrhine monkeys.

(e) Family **Lemuridae**. Lemurs

The sequence and timing of eruption of teeth of several lemur species has been tentatively established on the basis of very small samples (Eaglen, 1985). In the genus *Lemur*, the deciduous canines emerge first, followed by the deciduous molars in a front-to-back sequence, with the upper incisors emerging with, or soon after, the molars. The deciduous dentition is completed by six weeks in the ring-tailed lemur (*Lemur catta*) but not until ten weeks in the other species of *Lemur*.

The pattern of eruption of the permanent teeth seems to vary greatly between species; in *Lemur catta*, it is:

$$M1, M2, C_1, I^1 + I^2, P^4, P^3 + P_4 + P_3, P2 + M3, C^1$$

M1 emerges at the age of four months and eruption of the permanent dentition is completed at about 16 months.

(f) Order **CARNIVORA**

Table 16.5 gives the times and order of eruption of the deciduous teeth in various felids and shows that eruption occurs much earlier in domestic cats than in larger members of the family such as pumas, jaguars, leopards, lions and tigers.

In cats (*Felis* spp.), all the permanent incisors except I^3 erupt before the other teeth; P^2, I^3, M_1 and M^1 erupt before the canines and P3 and P4 erupt last. The detailed order of emergence is rather variable but can be:

$$I^1 + I_1 + I_2, I_3, I^2, P^2, I^3 + M_1, M^1, C, P^4, P^3, P_3, P_4$$

This is in close agreement with the order worked out by Hemmer (1966) for the cheek teeth of the domestic cat (*F. catus*) and by Pocock (1916b) for the snow leopard (*Panthera uncia*). Hemmer pointed out the similarity in the order of emergence of the cheek teeth in cats and snow leopards and contrasted it with that of lions (*Panthera leo*) in which Schneider (1959) found that M^1 is the last tooth to appear. However, Schneider's finding may be incorrect. In the skull of a juvenile lion from the wild (RCS Odonto. Mus., A 114.9), the order of eruption of the upper cheek teeth is P^2, $M^1 + P^4$, P^3 and, in a tiger (RCS Odonto. Mus., A 115.12), it is clearly P^2, M^1, P^4, P^3, as in cats.

According to Slaughter, Pine and Pine (1974), in felids and most other carnivore families, P^4 completes its eruption before P^3 and P_4 before P_3, enabling the upper carnassial (P^4) to shear against the lower carnassial (M_1) and P4 but, in viverrids and mustelids in which M_1 and P^4 are not carnassial, in the maxilla the order is still P^4, P^3, but in the mandible P_3 usually precedes P_4.

In the domestic cat, eruption of the permanent dentition begins at about $3\frac{1}{2}$ months and is completed by seven months (Habermehl, 1975).

In the spotted hyaena (*Crocuta crocuta*), the order of completion of eruption of the cheek teeth is:

$$P^1, P^4 + M^1, P^3, P^2$$

in the maxilla and

$$M_1, P_2, P_4, P_3$$

in the mandible (Slaughter, Pine and Pine, 1974).

In the Viverridae, the order of completion of eruption varies from one taxon to another. In *Paradoxurus hermaphroditus*, it is

$$P^1, M^1, P^4, P^2, P^3, M^2$$

in the maxilla, but, in other taxa, P^2 or M^2 or both may precede P^4. In the mandible, it is

$$P_1 M_1 P_2, P_3, P_4, M_2$$

In other taxa, M_2 may precede the premolars (Slaughter, Pine and Pine, 1974).

Berkovitz (1968b) found that in domestic albino ferrets (*Mustela putorius furo*) there are four upper and three lower deciduous incisors on each side (Chapter 4, p. 71); the deciduous incisors are functionless and only i^4 emerges into the mouth (at the age of

Table 16.5 Eruption in FELIDS

No.	Panthera leo lion 44		P. tigris tiger 27		P. pardus leopard 23		P. onca jaguar 2		Felis concolor puma 18		F. catus dom. cat 11	
	lower	upper	lower	upper	lower	upper	lower	upper	lower	upper	lower	upper
i1	22–23	23–24	20	12	26	26	17	<17	16–17	14–15	12–13	8–9
i2	25–26	29	23–24	20	26	27	20	25	19–20	17	13	11–12
i3	33–34	38–39	28–29	29	33	32	20	31–32	23	28	15	14–15
c	31–32	32	33–34	31–32	30	28	35	29	28	29	19	17–18
m4	48	85	52–53	64	40	69			37	47	24–25	31
m3	65	60–61	58–59	55	47	45			42	40	26	26–27
m2	—	117–152	—		—		—		—	61	—	37–93

The age in days at the time of emergence of the deciduous teeth in various captive and domestic felids (from Schneider, 1959, p.359).

Table 16.6 Eruption in MINK

	Lower teeth		Upper teeth	
	mean	range	mean	range
Juveniles				
m2	19.7	17–22	21.6	19–24
m3	21.4	19–24	23.5	22–25
c	22.4	21–25	22.7	20–26
m4	24.6	22–26	25.2	22–27
i2	39.3	33–46	36.2	28–42
i1	39.7	33–45	38.3	28–40
i3	40.2	35–49	18.7	16–21
Adults				
I1	46.7	46–49	45.8	44–47
I2	47.5	46–53	45.8	44–47
I3	50.8	48–56	47.9	44–51
M1	57.3	55–61	57.4	55–61
C	57.7	55–62	57.0	53–62
P2	60.8	58–63	60.5	58–62
M2	62.3	58–65	—	
P3	66.3	64–70	64.5	61–67
P4	68.3	64–71	60.9	58–64

The age in days at the time of emergence of the teeth in 65 ranch mink, *Mustela vison* (from Aulerich and Swindler

about 28 days). Like the other deciduous incisors, it may then either be resorbed completely or shed between 40 and 46 days. Upper and lower I1 and I2 appear at 46 days, I^3 at 54 days and I_3 at 68 days. This chronology is similar to those found by Aulerich and Swindler (1968) for ranch mink (*M. vison*) (Table 16.6) except that I_3 is much later.

Table 16.6 also gives times of eruption of the cheek teeth in *M. vison* and shows that P^2 erupts before M^2. However, Slaughter, Pine and Pine (1974) found that M^2 completes its eruption before P^2 so that the sequence of complete eruption was

M^1, P^2, P^4, P^3

in the maxilla and

M_1, M_2, P_2, P_3, P_4

in the mandible. The positions of P^2 and P_2 in the sequence differ from taxon to taxon.

The order of eruption of the teeth is the same in brown bears (*Ursus arctos*), polar bears (*Thalarctus maritimus*) and grizzlies (*U. horribilis*):

$P_1, I^1 + I^2 + P^1 + M^1 + I_1 + M_1, P^4 + I_2, I^3 + P_4 + M_2, P^3 + M^2 + I_3 + P_3, M_3 + P^2, P_2, C$

(Rausch, 1961; Pohle, 1923).

In the grizzly bear in the wild, P_1 appears in May (the third month of life) and the canines appear in October. Both I^3 and the canines are slow to erupt and the canines are not fully erupted until the animal's fifth summer (Rausch, 1961). Rausch found a four-month-old male grizzly from the wild with a precociously-erupted dentition similar to that of an animal of 16 months.

Slaughter, Pine and Pine (1974) found that the sequence of completion of eruption of the cheek teeth in the grizzly bear is the same as that for emergence given above, except that M_1 precedes P_1. In the American black bear (*Ursus americanus*), the sequence is the same as in the grizzly; in the Asiatic black bear (*Selenarctos thibetanus*), the order is

$P^1, M^1, P^4, M^2, P^2 + P^3$

in the maxilla and

$P_1 + M_1, P_4, M_2, P_2 + P_3, M_3$

in the mandible.

The times and order of emergence of the deciduous teeth in dogs are given in Tables 16.7 and 16.8. Kremenak (1969) found no statistically-significant differences in eruption times of the deciduous teeth in 40 male and 40 female puppies of various and mixed breeds (Table 16.7).

Mellanby's sample (Table 16.8) included dogs of large and small breeds. Eruption times were slightly earlier in large than in small dogs, suggesting that breed may influence the chronology of eruption. On the other hand, the times in the pure-bred beagles given in Table 16.7 are not identical with those found in 106 other pure-bred beagles (Table 16.8), showing that

Table 16.7 Eruption of the deciduous teeth in DOGS (*Canis familiaris*)

	Various and mixed breeds (40 males + 40 females)		Beagle (6 males + 10 females)	
	lower	upper	lower	upper
c	23	23	22	23
m3	23	28	25	30
i3	25	23	27	25
i2	26	23	28	24
m4	26	32	29	35
i1	31	24	31	25
m2	33	34	34	36

Mean times in days of emergence of the teeth (Kremenak, 1969).

Table 16.8 Eruption in the DOG

	Various breeds		Beagles	
	lower	upper	lower	upper
Juveniles				
c	20–26	20–28	22±3	22±2
i3	21–28	19–24	25±3	23±3
m3	21–28	25–35	26±4	29±4
i2	22–27	20.5–25	27±3	23±3
m4	24–32	31–37	28±4	34±4
i1	24–31	19–27	29±3	24±3
m2	28–39	30–39	31±4	35±4
Adults				
I1	107–121	113–126	117±5	115±6
P1	112–134	109–119	129±19	106±8
I2	113–129	124–136	123±5	123±6
M1	128–140	130–142	128±6	132±7
I3	131–148	113–160	134±6	133±7
C	148–154	153–172	145±5	146±7
M2	148–163	163–168	149±6	155±6
P2	154–164	150–163	149±10	149±8
P4	157–168	138–146	155±7	136±6
P3	158–170	156–164	156±6	153±6
M3			175±11	—

Ranges in the time in days after birth of the emergence of the teeth. Left, based on 17 puppies and four adults of various breeds (after Arnall, 1961 and Mellanby, 1929). Right, in 106 pure-bred beagles (Shabestari, Taylor and Angus, 1967).

there is quite wide variation amongst pure-bred animals even within the same breed, so it is not surprising that the sequence that Schultz (1935) established on 26 dogs of unknown breed differs slightly from those in Table 16.8. The order that Schultz established was:

$P^1, P_1, I^1 + I_1, I_2, I^2, I3, M_1, M^1, P^4, M_2, C_1, P^2, P_2, C^1, P^3, M^2, P_4, P_3, M_3$

Slaughter, Pine and Pine (1974) found that the usual sequence of completion of eruption of the cheek teeth in canids was

$P^1, M^1, M^2, P^4, P^2, P^3$

in the maxilla, and

$P_1, M_1 + M_2, P_2 + M_3, P_4, P_3$

in the mandible. As M_3 is usually the last tooth to emerge, it must erupt more quickly than P_3 and P_4. Slaughter et al. related the emergence of P_4 before P_3 to the fact that it has to shear against the carnassial P^4.

The eruption of the permanent dentition is complete by about six months after birth. Although the permanent canine erupts at about $5\frac{1}{2}$ months, it is common for the deciduous canines to be retained until the age of about seven months (Habermehl, 1975, p. 168).

In 489 coyotes (*Canis latrans*) from Central Alberta, Nellis (1972) found four skulls in which the upper deciduous canine was retained beyond the usual time of shedding (October); in one skull, the canine was still present in February of the following year.

(g) Order **PINNIPEDIA**. Seals

There is little information on the sequence of eruption in seals, but Loughlin (1982) found that P_3 erupts before P_4 both in phocids and otarids as in the mustelids. There is no close taxonomic relationship between seals and mustelids and Loughlin concluded that the eruption sequence of the lower premolars is of no value in the question whether phocids and otarids have substantially different phyletic origins.

(h) **UNGULATES** and other Mammals

In domestic pigs (*Sus scrofa*), the sequence of eruption of the deciduous teeth is:

$i3 + c, m^3 + m^4, i1, m_3 + m_4, m2, i_2, i^2$

i3 and c erupt before birth and i^2 at 12–13 weeks (Habermehl, 1975).

The sequence of eruption of the permanent teeth in both wild and domesticated pigs (*Sus scrofa*) is:

$M1 + P1, I3 + C, M2, I_1 + P2 + P3 + P4, I_2, I^2, M_3, M^3$

(Bull and Payne, 1982). A slightly different sequence was worked out by Silver (1969) from data derived from 18th century accounts of tooth eruption, but Bull and Payne (1982) showed that these data are unreliable and that the sequence in wild pigs, modern domestic pigs and domestic pigs from archaeological sites is that given above. The timing of eruption does seem to be affected by breed and particularly by whether members of the breed mature early or late. The teeth most affected are those that erupt early in the sequence (Table 16.9) (Reiland 1978).

The sequence of eruption of the deciduous teeth of domestic cattle (*Bos taurus*) is:

$i1 + i2, i3, c, m2 + m3 + m4$

(Brown, 1902; Grigson, 1982).

In early-maturing breeds, all the deciduous teeth may erupt before birth; in late-maturing breeds, i1 and i2 emerge before birth and the remaining teeth have erupted by 14–21 days (Zietzschmann, 1943).

The sequence of eruption of the permanent dentition of cattle is usually considered to be:

$M1, M2, I_1, P2 + P3 + M3, I_2, P4, I_3, C_1$

but some authors consider that P4 and M3 erupt at the same time and that I_2 erupts before the premolars (Grigson, 1982; Andrews, 1985a). The explanation for these differences is probably that the timing of eruption is variable both within and between breeds and is affected by whether the breeds are early or late maturing (Table 16.10). Another frequently-occurring problem is that authors do not usually give their criteria for considering that a tooth has erupted. Andrews (1982) gave details of the time of eruption of the permanent incisors, canines, M^1 and M^2 in a large sample of British cattle. He found that, as each successive pair of teeth erupted, the range of ages over which each stage of eruption occurred increased (Table 16.11).

Ratti and Habermehl (1977) used 328 ibex (*Capra ibex*) of roughly known age killed in the Kanton Graubünden in Switzerland to establish age changes. They found that the entire deciduous dentition except the canine had begun to emerge before birth; the canine appeared within a week of birth and, within a month, the eruption of the deciduous dentition was completed. The sequence of emergence of the permanent teeth was:

$M1, I_1 + M2, I_2 + P2 + P3 + P4 + M3, I_3 + C$

Table 16.9 Eruption in WILD AND DOMESTICATED PIGS (*Sus scrofa*)

	M1	P1	I3	C	M2	I1	P3	P2	P4	I2	M3
Domestic pigs											
Reiland 1978											
Improved Landrace, mandible	4	(5)	8	8	9	12	12	12	13	17	17
Habermehl, 1975											
early breeds	4	4	6	6	7	11	12	12	12	14	16
middle breeds	6	6	9	9	10	12	14	14	14	16	18
late breeds	8	8	12	12	13	14	16	16	16	18	20
Silver, 1969	4–6	3½–6½	8–12	8–12	7–13	12–17	12–15	12–16	12–16	17–20	17–22
Getty, 1975	4–6	5	8–10	9–10	8–12	12	12–15	12–15	12–15	16–20	18–20
Brown, 1902	5	(5)	7–8	9	10–12	12	12–15	12–15	12–15	17–18	17–18
Owen, 1866/8	6	6	9	9	10	12	12	12	15	18	18
Late 18th century data (Silver, 1969) (probably unreliable)	12	—	6–12	12	18–24	30–36	24	24	24	30–36	36
Wild boar and domestic pigs											
Lesbre, 1897/8	5	5	—	9–10	10	12–14	13–14	13–14	13–14	18–20	18–30
Wild boar											
Briedermann, 1965 mandible	4–5	—	10–12	10–12	(12)	14–16	14–16	14–16	14–16	18–20	21–24
Matschke, 1967											
mandible: average	6	7	9	9	13	14	16	16	16	20	25
range	5–6	5–8	7–9	7–11	12–14	13–14	14–16	15–17	14–18	18–22	23–26
maxilla: average	6	6	9	9	13	14	16	16	17	24	29
range	5–6	5–6	7–12	7–11	12–14	12–15	15–16	16–17	14–18	22–27	26–33

Teeth are listed in order of eruption; all ages are given in months. Most of the sources quoted do not distinguish between upper and lower teeth despite what appear to be significant differences in I2 and M3 (Matschke, 1967). (From Bull and Payne, 1982.)

Table 16.10 Eruption in DOMESTIC CATTLE (*Bos taurus*)

	Early maturing		Sisson & Gross.	Middle maturing	Late maturing		
	Ellen.-Baum early mat.	Silver modern	modern	Ellen.-Baum	Habermehl	Ellen.-Baum	Silver C19th
Deciduous							
i1	before birth	before birth	birth– 2 wks	before birth		before birth	before or some days after birth
i2	before birth	before birth	birth– 2 wks	before birth		before birth	before or some days after birth
i3	before birth	before birth/ 1st 2 weeks	birth– 2 wks	before birth		2–6 days	14 days
c	before birth	before birth/ 1st 2 weeks	birth– 2 wks	2–6 days		6–14 days	2–3 wks
m2	before birth	before birth/ 1st 2 weeks	birth– 2 wks	2–6 days		6–14 days	before or some days after birth
m3	before birth	birth– 3 wks	birth– few days	after a few days		14–21 days	before or some days after birth
m4	before birth	birth– 3 wks	birth– few days	after a few days		14–21 days	before or some days after birth
Permanent							
I1	17m	14–25m	18–24m	21m	22–24m	25m	18m
I2	22m	17–36m	24–30m	27m	28–31m	32m	30m
I3	32m	22–40m	36m	36m	36–40m	40m	42m
C	36m	32–48m	42–48m	45m	42–49m	52m	54m
P2	24m	24–30m	24–30m	26m	24–28m	28m	18m
P3	24m	18–30m	18–30m	26m	24–28m	28m	30m
P4	28m	28–36m	30–36m	31m	24–28m	34m	42m
M1	5m	5–6m	5–6m	5m	5–6m	6m	6–9m
M2	15m	15–18m	12–18m	16m	15–18m	18m	30m
M3	24m	24–30m	24–30m	26m	24–28m	28m	4–5y

m = months, y = years. (Data from Ellenberger-Baum (Zietzschmann, 1943); Silver, 1963; Habermehl, 1975; Sisson & Grossman, (Getty, 1975); Grigson, 1982.)

M1 emerges at about six months and eruption of the canine is complete at about four years of age.

In 29 feral goats (*Capra hircus*) from Scotland, the sequence and timing of eruption was similar to that of the Swiss ibex except that Bullock and Rackham (1982) showed that P4 and P3 erupted before M3 and C erupted after I3, at between 48.5 and 70 months of age.

The sequence found by Deniz and Payne (1982) for the mandibular teeth of Angora goats in Turkey was

$$M_1, M_2, I_1, P_2 + P_3 + P_4, I_2, M_3, I_3, C$$

In the females, eruption of the permanent dentition began at $3 \pm \frac{1}{2}$ months and was completed at $40 \pm c.6$ months. Eruption of some of the teeth was a little earlier in the males (Briedermann, 1965).

In sheep (*Ovis aries*), the deciduous incisors erupt before birth; the deciduous canines may erupt before birth or with the deciduous molars at about four weeks of age.

The sequence of eruption of the permanent teeth is:

$$M1, M2, I_1, M3 + I_2, P2 + P3 + P4, I_3, C$$

M1 erupts at about three months and eruption of the permanent dentition is completed at about 40 months (Habermehl, 1975). Although the sequence appears to be different from that in goats, in both sheep and goats, I_2, the premolars and M3 appear in such rapid succession in only 4–6 months that the differences may be due to individual variation.

In horses (*Equus caballus*), the deciduous molars erupt before birth, i1 at about six days, i2 at 3–8 weeks and i3 at 5–9 months. In general, the upper incisors erupt slightly earlier than the lower.

The sequence of eruption of the permanent teeth is:

$$M1, M2, I1 + P2 + P3, I2 + P4, M3 + C, I3$$

M1 erupts at about 1 year and I3 at about $4\frac{1}{2}$ years of age (Habermehl, 1975).

In elephants, the only incisors are the tusks (I^3), which are often absent in the female Indian elephant (*Elephas maximus*). They appear soon after birth and some authorities maintain that they occasionally have a deciduous predecessor (Ali, 1977). There are three milk molars and three permanent molars, only one or two of which are functional at a time, beginning with m2. As they are worn away and shed, they are replaced horizontally by a developing tooth from behind (Table 16.12) (Krumry and Buss, 1968).

In the manatees (Family Trichechidae), there is continuous replacement of the cheek teeth from behind. The process starts with the three deciduous molars which are present at birth; four or five permanent molars then erupt so that there are usually seven or eight functional cheek teeth in each quadrant. Continuous loss of the most mesial teeth then occurs through wear, the remnants being shed by resorption; in museum specimens, some of the partially-shed teeth at the mesial ends of the rows have often been lost. As teeth are shed, the rows move forward and new molars erupt from behind. As far as is known from the study of manatees in captivity, there seems to be a potential for the formation of unlimited numbers of molar replacements and it has been estimated that, during the life span of about 20 years, 30–40 molars may be formed in each quadrant (Domning and Hayek, 1984; see also Chapter 9, p. 155).

Information on tooth eruption in rodents and lagomorphs has been gathered by Habermehl (1975).

Table 16.11 Eruption in DOMESTIC CATTLE (*Bos taurus*)

	$\bar{X}$	Range	S.D.	V	No. of cattle
M^1	209	174–235	16.9	8.1	25
M^2	477	400–542	39.9	8.4	28
I_1	676	541–806	56.9	8.4	55
I_2	820	689–961	70.9	8.6	29
I_3	1095	902–1277	103.8	9.5	18
C	1343	1038–1742	146.4	10.2	41

The timing of emergence of some of the permanent teeth in cattle in Britain. $\bar{X}$ = mean, S.D. = standard deviation, V = coefficient of variation. (After Andrews, 1982.)

Table 16.12 Eruption in ELEPHANTS

m2	birth–2nd year
m3	1½–5th year
m4	2nd–11th year
M1	5th–19th year
M2	15th–60th year
M3	23rd–60+ years

The sequence and length of use of the cheek teeth in the African elephant (*Loxodonta africana*) (from Krumrey and Buss, 1968).

Laboratory guinea pigs (*Cavia aperea*) are monophyodont and by the first day of life the incisors, P4, M1 and M2 have emerged; M3 emerges at seven days.

Mice and rats are also monophyodont. In laboratory rats (*Rattus norvegicus*), I1 emerges at 8–10 days, M1 at 18 days, M2 at 21 days and M3 at 35 days.

Lagomorphs are diphyodont. In laboratory rabbits (*Oryctolagus cuniculus*), the deciduous dentition emerges before birth and, by the time of birth, I1 is already beginning to emerge. I^2 emerges before i^2 is shed so that, for a while, there are three incisors in each premaxilla. Eruption of the cheek teeth takes place from 3–5 weeks after birth. In the maxilla, the order is:

$$M^1 + M^2, P^2 + P^3 + P^4, M^3$$

In the mandible, the premolars precede all the molars:

$$P_3 + P_4, M_1 + M_2, M_3$$

The eruption sequence in the various taxa of Insectivora varies a great deal. As pointed out by Slaughter, Pine and Pine (1974), this is in accord with the phylogenetically loosely-knit nature of this order; that is, some members of the order may be more distantly related than being placed in the same order suggests. The sequence in *Setifer setosus* (the greater hedgehog tenrec),

m2, m3, m4, M1, M2, M3, P2, P3, P4

comes nearest to displaying what is regarded as the primitive mammalian sequence derived from the orderly mesial-distal sequence of the reptiles. In many other insectivores, however, P4 is the first premolar to erupt. Probably the front to back replacement of the deciduous cheek teeth broke down in the course of evolution, as the simple single-cusped P^3, and especially P^4, became progressively molarized and assumed more important functions in the increasingly heterodont dentition (Slaughter, Pine and Pine, 1974).

In marsupials, only P3 replaces a deciduous precursor; in macropods, P3 replaces not only m3 but also the tooth that is immediately mesial to m3 (m2, P2 or PC2 depending on the terminology employed. Chapter 1, p. 13). In the red kangaroo and the grey kangaroo (*Macropus rufus* and *M. giganteus*), M3 erupts at the same time as P3 but, in the wallaroo and the Bennett's wallaby (*M. robustus* and *M. rufogriseus*), P3 erupts between the early-erupting M4 and the later-erupting M3. The cheek teeth are progressively shed, mesio-distally, so that aged animals have only M3 and M4 or M4 alone (Kirkpatrick, 1964). P^3 sometimes fails to erupt at all; in 300 grey kangaroos, Kirkpatrick (1965) found 18 with one or both P^3 unerupted. The same was true of one amongst 84 red kangaroos and one amongst 106 Bennett's wallabies; none of the 71 wallaroos that he studied showed this abnormality. In macropods in which P^3 is unerupted, the deciduous molar persists.

Horton and Samuel (1978) found four mandibles among 34 of the Pleistocene fossil macropod *Macropus titan* in which the eruption of P_3 appeared to have been delayed. The teeth were more mesial in position than normal; it seemed that in the course of eruption, instead of erupting vertically, they had followed the course of the mandibular canal.

Non-eruption

There may be total or partial failure of the process of eruption. If a tooth does not erupt at all, it is usually because it is so misplaced in the tooth arch, or has developed so far from it (ectopically) that it is hardly likely to reach its destination; for example, a premolar lying horizontally in the palate of a baboon (Fig. 10.103) and two supernumerary teeth in the vertical ramus of the mandible of a gorilla (Fig. 3.3).

The term partial eruption is best reserved for instances where, in spite of malposition or some other cause, the tooth does emerge to some extent in the mouth. In the Section devoted to Abnormalities of Position (Chapters 9–15), there are many examples of captive animals in which there was not sufficient room for the teeth to erupt fully or to assume their normal relationships.

Teeth may also remain unerupted through what appears to be a fundamental failure of the eruptive process, as in the jaws of a captive mandrill (Fig. 16.2) where all four P2 remain unerupted. These premolars normally erupt before their roots are completed, and here the roots of at least the P^2 are complete. The cancellous bone of the jaws as a whole seemed unusually dense, which could affect eruption. However, only the P2 seems to be affected; most of the other teeth have already erupted, or are erupting normally.

Other examples of a disturbance of the eruptive process that seem to be related to abnormally dense or sometimes abnormally soft bone in captive animals are a chimpanzee (Fig. 16.3), two macaques (Fig. 16.4 and Chapter 10, p. 205, Fig. 10.90), a baboon (Chapter 10, p. 210, Fig. 10.102), a capuchin monkey (Fig. 16.5), a lion (p. 241, Fig. 11.5) and a polar bear (Fig. 16.6). An example in a horse was described by Joest (1926, fig. 86).

The skull of a domestic cat (Fig. 16.7) shows widespread failure of eruption. Only one upper incisor and the prominent cusps of P^4 and M^1 have pierced the bone and it seems unlikely that they had emerged through the mucosa. The majority of the teeth are buried in the jaws and invested by a layer of compact bone. The bone that was removed to display the teeth was unusually dense and some teeth seem to be ankylosed to it. It is hardly possible to speculate on the cause.

As a kitten, this animal was castrated and, starting at six weeks of age, suffered a series of convulsions over a period of a few months, which could be consistent with some hormonal disturbance, e.g. hypoparathyroidism. The animal was killed on account of an abdominal tumour.

Figure 16.8 shows an aged Asiatic wild dog (*Cuon alpinus dukhensis*) in which the P^2, P^3, P_3 and P_4 are only partially erupted; some have only just pierced the bone. The radiograph shows no abnormality of the roots or of the bone.

It is quite common in domestic dogs, especially in

Fig. 16.2. *Mandrillus sphinx* (mandrill). ♀. Captive. RCS Odonto. Mus., G 68.66. P^2 and P_2 unerupted on both sides. Bone has been removed to display the teeth. × c. 0.75.

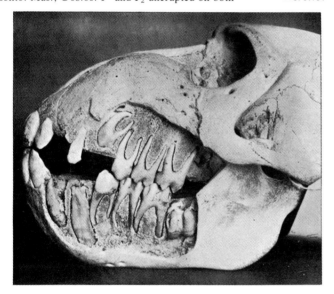

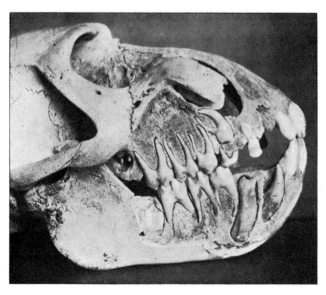

the short-muzzled breeds, for P_1 to remain only partially erupted, not necessarily in association with impaction or other evidence of shortage of space.

In a mandible of an adult tayra (*Eira barbara*) that was in the British Museum of Natural History, the crown of the right M_2 was partially erupted; the surrounding bone was unusually dense.

Impaction

The most common cause of impaction, that is when a tooth is unable to erupt because its pathway is obstructed by another tooth, is crowding of the arch; that is, the arch is too small to accommodate the teeth. This can affect the premolars when, owing to crowding, the molars move mesially whilst the deciduous molars are being replaced. Impaction of premolars was noted in an African civet (Fig. 11.25), two racoons (Chapter 11, Figs. 11.55, 11.54) and a tree porcupine (Chapter 14, Fig. 14.6). Gainer (1982) found three skulls in which one P^4 was impacted, and one with both P^4 impacted, amongst 1434 Nyasa wildebeest (*Connochaetes taurinus johnstoni*) in the Selous Game Reserve in Tanzania. However, impaction most commonly affects the last molars which are usually the last teeth to erupt so any failure of the growth process may result in insufficient space for them to erupt into. As already explained in Chapter 9, p. 155; Chapter 10, p. 171, they are then likely to undergo eruptive movements while still in a tilted position. Bodingbauer (1949*b*) found that, in three out of six pig skulls of short-muzzled breeds (Yorkshire and Berkshire), all the M3 were impacted (see Chapter 13, p. 283). Many examples of impacted last molars are to be found in Chapters 10–15; (e.g. Figs. 10.34, 10.70, 10.73, 10.74, 10.75, 10.76, 10.91, 10.113, 10.114, 11.25, 11.26, 13.83 and 15.19). In a few skulls, other teeth were impacted in addition to the last molars. Examples include a patas monkey (Chapter 10, p. 198, Fig. 10.73) in which both M_2 were impacted against M_1, and the right M_3 was impacted against M_2, and an African civet (Chapter 11, p. 248, Fig. 11.25) in which the left P_4 was impacted against M_1 and both M^2 were impacted under the distal convexity of M^1.

A rather different cause of impaction is when a deciduous tooth is lost from the arch through trauma or disease and the adjacent teeth move towards one another, usually by the mesial drift of the more distally situated teeth (Chapter 9, p. 157), so closing the space and preventing the eruption of the successional tooth. This seems likely to be the explanation of the impacted premolar of a horse in Figure 16.9.

Over-eruption

Although the evidence for the occurrence of over-eruption when a tooth loses its opposing tooth is overwhelming in respect of man and other primates (this chapter, p. 332), in dentitions in which continuous eruption and wear of high-crowned hypsodont teeth is an integral feature, as in horse and bovids (Ainamo and Talari, 1976), teeth may come to project beyond the occlusal plane simply because they have not become worn, as in the horse dentition in Figure 22.1. It is not always easy to distinguish this from over-eruption but in that example the overall height (from the apices of the roots to the occlusal surface) of the

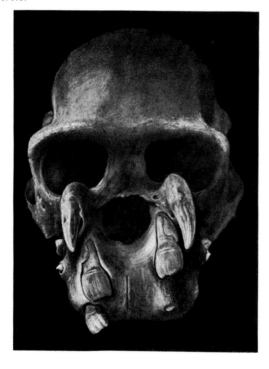

Fig. 16.3. *Pan troglodytes* (chimpanzee). Formerly RCS Special Pathology, 4125.1. Was estimated to be three years old when brought to the London Zoo and lived there for 27 years. Many teeth are grossly misplaced and unerupted or partially erupted. The upper canines are situated just below the orbits, their roots are sharply curved distally. The I^1 are partially erupted high up on the facial surface of the bone and other teeth can be seen projecting bizarrely from the sides of the maxillae. Only the right I^2 has erupted in approximately its normal position. Such enamel as can be seen shows no ring hypoplasia.

In the mandible, the I_1 were about 12 mm below the alveolar margin and several prominences on the surface of the bone seemed to indicate the position of several other unerupted teeth. There was a fundamental failure of the eruptive process probably due to a bone dystrophy. × c. 0.6.

projecting tooth was much greater than that of the adjacent teeth, showing that it had experienced less wear.

In the aged horse skull shown in Figure 16.10, the lower premolars project far above the original occlusal plane and are much taller than the other cheek teeth (Kitt, 1892). The bases of the crowns have been exposed to the level of the bifurcation of the roots. Three processes seem to have been involved: (1) a reduction in the usual amount of wear due to the loss or wearing away of the opposing upper premolars, (2) gingival and alveolar recession (passive eruption, see this Chapter, p. 331), which is common in elderly ungulates, and (3) some degree of over-eruption.

Retention of deciduous teeth

In primates, the retention of a deciduous tooth is occasionally caused by its malposition, as in the *Colobus* skull (P-C. Mus., Cam II 69) already mentioned (Chapter 3, p. 36) and illustrated in Figure 3.29, but it is more usually associated with the absence or malposition of the permanent successor. Both these basic conditions were present in a *Colobus* skull (USNM

Fig. 16.4. *Macaca* sp. (macaque). Juvenile. Bone has been removed to display the teeth. The animal was said to have been affected by rickets. At least one premolar is tilted and misplaced and it is doubtful whether it could have in due course moved into its proper position in the arch.

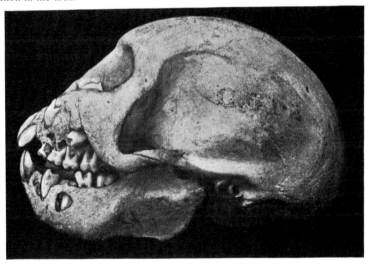

Fig. 16.5. *Cebus* sp. (capuchin monkey). Juvenile. Captive. Palatal view. The animal was said to have been affected by rickets after the permanent incisors, M1 and M2 had erupted and while the premolars were erupting. There is widespread thickening of the alveolar bone, which is seen particularly in relation to the crypts of both M^3, the eruption of which may be impeded.

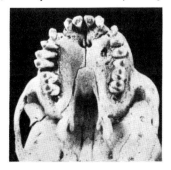

164584) described p. 38; the left P^3 was absent and m^3 had been retained, but the right m_3 was present and P_3 was erupting buccally beside it.

As positional abnormalities are more prevalent in captive than wild monkeys, it is not surprising that several instances of retained deciduous teeth were noted in captive macaques and patas monkeys (Figs. 10.69, 10.73, 10.76, 10.88, 10.89 and 10.90).

In primates, disturbances of the dentition, such as enamel hypoplasia, congenitally small or absent teeth and malformed teeth, are sometimes accompanied by the retention of deciduous teeth; this was noted in two *Colobus* skulls (Chapter 3, pp. 38–40) both of which were probably from captive animals. In a captive baboon (*Papio ursinus*) with jaws affected by a proliferative osteodystrophy, several deciduous teeth had been retained and their successors had erupted beside them (Bennejeant, 1936; Chapter 3, p. 49).

It is often difficult to distinguish between small extra teeth and retained deciduous teeth. A skull of *Colobus badius preussi* (P-C. Mus., M178) had an extra upper incisor on each side which Colyer (1940) thought might be retained deciduous incisors.

Among the prosimians, bone disorders are particularly common in captive lemurs. Two ring-tailed lemurs (*Lemur catta*) with bone disorders also had positional abnormalities of the premolars accompanied by retained deciduous molars (Chapter 10, pp. 236–237).

Small supernumerary teeth sometimes occur in the indrisoid lemurs (Indriidae) and in the sifaka (*Propithecus verreauxi*). Bennejeant (1936) showed them to be retained deciduous molars (Chapter 3, p. 60).

Most of the few instances of persistent deciduous teeth in carnivores are related to absence or malposition of the permanent successor. Teeth are frequently misplaced in short-faced breeds of dogs and retention of deciduous teeth is common; however, according to Catcott (1968), the malposition of the permanent

Fig. 16.6. *Thalarctos maritimus* (polar bear). RCS Odonto. Mus., A 170.12. A portion of a mandible with a small premolar embedded in the bone. $\times c. 0.5$.

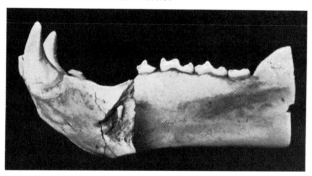

Fig. 16.7. *Felis catus* (domestic cat). ♂. BMNH RCS Odonto. Mus., G 47.1. Several of the teeth are embedded in the bone. The outer cortex of the bone has been removed. (Colyer, 1905.) Reduced slightly.

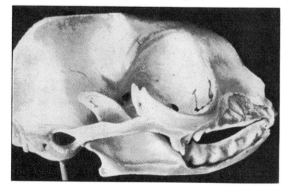

Ch.16. Variations and Disturbances of Eruption   351

Fig. 16.8. *Cuon alpinus dukhunensis* (Asiatic wild dog). ♀. BMNH 1907.11.14.2. A: maxilla. B: mandible. C and D: radiographs of mandible. Several of the premolars are only partly erupted.

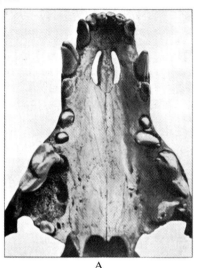

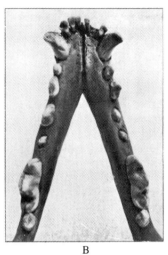

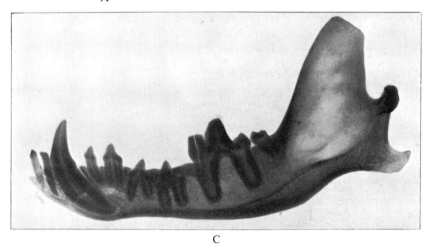

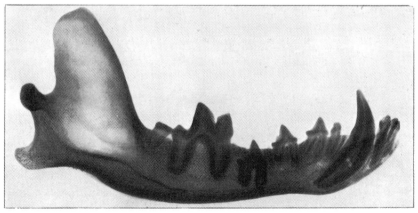

canines in these breeds is the result of the retention of the deciduous canines rather than vice versa. Becker (1970, figs. 180, 181) figured the skull of a poodle which appeared to have a double set of incisors in the upper and lower jaw due to the persistence of most of the deciduous canines and most of the deciduous incisors. A similar instance is depicted by Rossman, Garber and Harvey (1985, fig. 6–1).

In a duiker skull (*Cephalophus monticola*) (Fig. 13.41), a fragment of m^4 had been retained between P^4 and M^1; P^4 had erupted more mesially than usual. In perissodactyls, deciduous incisors are often persistent in horses and may be retained with the full complement of permanent incisors (as in Fig. 13.69) or even with extra permanent incisors (as in Fig. 13.70).

Joest (1926, fig. 90) figured a horse skull with five deciduous incisors retained beside the fully-erupted permanent incisors. Other examples are figured by E. Becker (1970, figs. 119, 120).

Although horses normally have three premolars in each quadrant, so-called wolf teeth (P^1) occurred in 16% of the skulls studied (Chapter 6, p. 121 and Table 6.2). Wolf teeth in horses vary considerably in size and morphology; Petit (1938) considered that they were sometimes retained deciduous molars (m^1) not P^1.

The mandible of a fossil rhinoceros (*Rhinoceros hemitoechus*) has a retained m_4 related to a misplaced P_4 (Fig. 13.79).

Few examples of retained deciduous teeth were seen in the other groups of eutherian mammals. A tooth which was either a supplemental I_2 or a persistent i_2 was noted in a rodent, an Indian gerbil (*Tatera*

Fig. 16.9. *Equus caballus* (horse). RCS Odonto. Mus., G 76.44. The eruption of P_4 has been impeded by approximating movement of P_3 and M_1. × *c.* 0.45.

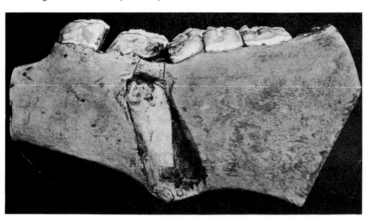

Fig. 16.10. *Equus caballus* (horse). Aged. The crowns of the lower premolars are much taller than usual and their bases are exposed. This is probably due to a reduction in wear, gingival recession and overeruption (from Kitt, 1892, fig. 5). Incisor underjet.

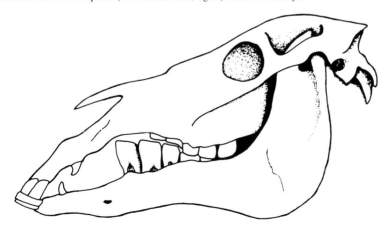

indica, HZM 10399), by Ingles, Bates and Harrison (1981) (see Chapter 7, p. 131).

The prairie-mole, *Scalopus aquaticus* (Insectivora), sometimes has rudimentary teeth in the diastema between I_2 and P_2. Microscopic examination suggests that these are retained deciduous teeth (Conaway and Landry, 1958).

In macropods, P2 and m3 are usually replaced by one tooth (P3) but occasionally P^3 replaces only one of them and the other is retained, as in the banded hare-wallaby (*Lagostrophus fasciatus*) shown in Figure 15.10.

Re-inclusion (see p. 332)

The female gorilla skull shown in Figure 16.11 was originally regarded as simply an example of four permanent molars in the left maxilla, where there are normally only three. However, we believe the evidence for the apparent supernumerary being a retained deciduous molar in an abnormal position is overwhelming and we have therefore dealt with it in the category of disturbed eruption. The process of re-interpretation of the specimen is of interest so we shall give it in full.

The first, third and fourth of the four molars correspond closely in shape to the contralateral M^1, M^2 and M^3 so that the tooth between M^1 and M^2 on the left, which in any case is about half their size, is the apparent supernumerary molar. It is more worn than the others and yet its occlusal surface is above the occlusal surfaces of the adjacent teeth so that it is no longer in occlusion. L. Bolk, who was able to examine the specimen carefully in Amsterdam, described it in a letter to Colyer as follows:

> It is a very interesting but not quite a normal case. Indeed there are four molars on the left side but the second molar in the row is the second deciduous molar which by some cause or other has been displaced backwards and found a place between the first and the second permanent molars. There were yet some remains of an abscess of the maxilla on this side and in relation with the first molar. Perhaps this may be the cause of the persisting of the second milk molar. The tooth must have been strongly worn down before it reached its place between the first and second molars, then there is no grinding facet corresponding with it in the row of the lower molars.

Fig. 16.11. *Gorilla gorilla* (gorilla). ♀. A.E.L. 1919/46. Four molars in the left maxilla. The first, third and fourth correspond in morphology to M^1, M^2 and M^3 in the right maxilla. The second of the left molars has the morphology of m^4 and is now overgrown by the adjacent molars. Its occlusal surface shows wear, so that at some time it was in occlusion. $\times c. 0.75$.

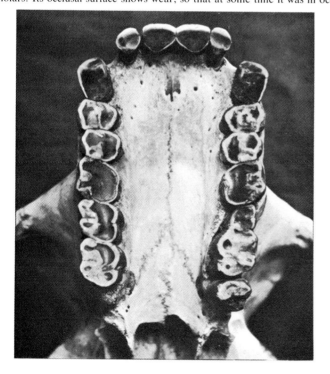

It is impossible to explain the presence of a deciduous molar in this situation except on the basis of an unusual disorder of the tooth band or of the tooth germs to which the band gave rise. However, it would seem that originally this tooth was fully erupted and in occlusion, but then was not carried down with the growth of the alveolar process that occurs in association with the eruption of the permanent molars and replacement of the deciduous molars by their successors.

It is thus an example of what we would today call a submerged or re-included deciduous molar (m^4), but it is particularly unusual because the tooth is situated distally to M^1 instead of mesially to it. Because m^4 erupts some time before M^2, it is necessary to postulate that the sequence of these teeth was disturbed and that M^1 erupted before m^4. The alternative, that m^4 erupted more distally than normal, leaving a space in the arch into which M^1 later erupted, seems less likely.

CHAPTER 17

Overgrowth of Teeth

Introduction

Teeth of continuous growth, such as rodent incisors, are kept to their appropriate functional length by wear; this implies an equilibrium between the rate of formation and the rate of wearing away. Occasionally, this equilibrium is not maintained and there is overgrowth of the teeth which, as in some examples described here, may reach grotesque proportions. It is not worthwhile to catalogue here all the descriptions that exist in the literature of single instances of this condition, some from the wild state, others from breeding colonies of captive animals, so we have restricted ourselves to descriptions that are usefully detailed, where previous literature is reviewed, or where the causation is discussed.

Teeth of limited growth can also to some extent appear to be overgrown beyond the level of the occlusal surfaces of the adjacent teeth when the teeth with which they normally occlude are lost, malposed or absent. This is often over-eruption (Chapter 16, pp. 348–349) but, in hypsodont teeth where wear is a predominant feature, over-eruption may be simulated by non-wear of the affected tooth; that is, it retains its normal overall height while adjacent teeth become shorter (Chapter 16, p. 348. Fig. 22.1).

Overgrowth of incisors in Rodentia and Lagomorpha

One of the most striking examples of overgrowth, seen in Figure 17.1, is of particular interest because it is a specimen collected by John Hunter, and is in the Hunterian Museum of the Royal College of Surgeons. The right I_1 of this American beaver must have

356 Section 3: Abnormalities of Eruption

emerged between the lips to curve beside and over the snout, and then, penetrating the skin, passed into the region of the mouth between the coronoid and articular processes. The articular process is atrophic, indicating the restriction of jaw movement associated with the incisor overgrowth.

Bland Sutton (1884*a*) recounted that the teeth of a beaver living in the Zoological Gardens, London, grew to such an extent that the animal could no longer open its mouth and was dying of starvation. The animal was restrained and about one inch of each lower incisor was excised. Recovery of well-being quickly followed. Nevertheless, several months later the condition recurred. The portions of teeth excised on that occasion (RCS Osteo. Series 2203) contained a pulp cavity obliterated with newly formed dentine (Bland Sutton, 1885*b*).

Figures 17.2, 17.3 and 17.4 show examples of overgrowth in the rabbit.

The many experimental studies that have been made on the growth of rodent and lagomorph incisors, directed mainly to throwing light on the general process of tooth eruption, are relevant to the processes involved in overgrowth.

Shadle, in a series of studies (Shadle, Ploss and Marks, 1944), recorded the weekly growth rates of the incisors of domestic rabbits, brown rats (*Rattus norvegicus*), guinea pigs and North American porcupines (*Erethizon dorsatum*) over periods of 24 to 54 weeks. There were species and sex differences but the rates were of the order of 100mm per year for the upper incisors of domestic rabbits, brown rats (*Rattus norvegicus*), guinea pigs and North American porcupines of eruption and therefore rate of formation is affected by the axial forces to which the incisor is subjected; if all axial force is removed, e.g. by removing the opposing incisor or repeated clipping of the cutting edge, the rate of eruption is almost immediately increased by as much as two-fold (Berkovitz, 1976).

Most tooth wear is brought about, not by tooth-to-tooth contact, but by abrasion against food, or against other materials when the incisors are used as tools for

Fig. 17.1. *Castor fiber canadensis* (American beaver). RCS Path. Series, 6. Right half of mandible with an overgrown I_1.

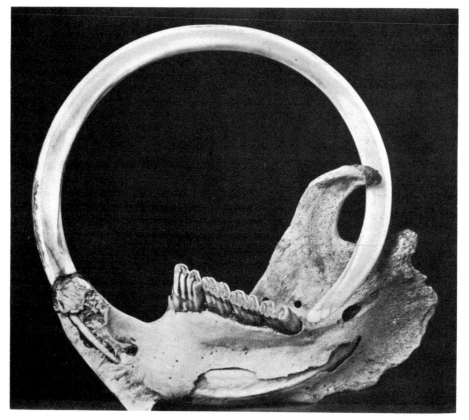

gnawing, as in nest building and so forth, the best-known example being the beaver's tree felling and dam building. However, Mills (1978) pointed out that such wear would not necessarily maintain sharpness but would tend to blunt the cutting edges in the same way as use blunts a chisel. According to Every (1970, 1975), many mammals deliberately sharpen their teeth by grinding them together, which Every calls thegosis (Every and Kühne, 1971). Rodents and lagomorphs, when they are sitting at rest, can commonly be seen to rub their teeth together, sometimes producing a chattering sound.

It may be that a process akin to thegosis helps to maintain the sharpness of the sectorial molars in carnivores upon which the efficiency of their scissor-action depends (Mellett, 1981). In the common vampire bat (*Desmodus rotundus*), the piercing function of the canines and upper incisors depends upon their sharpness. According to Vierhaus (1983), the sharpness of the canines is maintained by tooth-to-tooth thegosis and that of the upper incisors by what he calls tongue-to-tooth thegosis; that is rubbing of a specialized area of the tongue covered by horny papillae against the lingual surfaces of the incisors.

The wear surface of the rodent upper incisor, at least in the beaver, is more complex than is usually thought (J. W. Osborn, 1969). The worn dentine on the inner surface of the incisor above the sharp edge of enamel presents a sloping surface against which the lower incisor produces a scissor-like action when the mandible is in the protruded position. Above that level, however, as is depicted in Figure 17.5B, there is a step which provides an anvil-like surface against which the incisor bites in addition to its scissor action.

Fig. 17.2. *Oryctolagus cuniculus* (rabbit). RCS Odonto. Mus., G 37.391. The left I^1 has been removed in order to show that the right I^1 has crossed over and grown in the form of a circle, the point of the tooth finally entering the left maxilla. The overgrown I^2 can be seen on its lingual side.

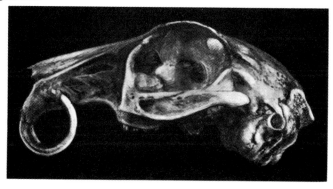

Fig. 17.3. *Oryctolagus cuniculus* (rabbit). RCS Odonto. Mus., G 37.39. All four incisors overgrown. Both I^2 had remained short. ×c. 0.95.

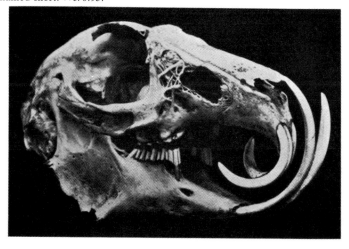

Sometimes, above the step there is a hollow produced by wear while the mandible is retruded in its molar-grinding position.

Rodent incisors are not simple curves but spirals (Poole and Miles, 1967), the lower incisor pursuing a more open curve than the upper. The geometry of the spiral differs in the various species (Landry, 1957) and becomes very evident if the incisor, especially an upper one, becomes overgrown, as is seen particularly well in the rat skull in Figure 17.6. A similar spiral upper incisor in a hoary marmot (*Marmota caligata*), dis-

Fig. 17.4. *Oryctolagus cuniculus* (rabbit). Formerly RCS Special Path. Series, 60.1. Two examples of overgrowth of all four I1. A. Anterior view. Both I^1 have grown towards the right side, the left one into the right palate and the right one to spiral outside the mouth. B. Lateral view. The I$_2$ project straight forward but the free end of the left one curves across to the right side.

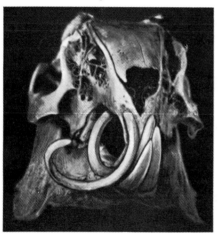

A

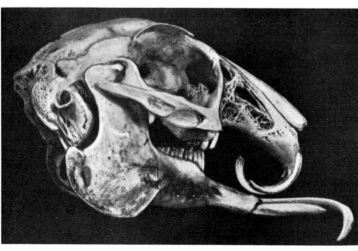

B

covered in the wild, was described by Pratt and Knight (1981).

When overgrown, upper incisors tend to recurve sharply backwards and to impinge upon the upper jaw. Lower incisors have a more open main curve which, in overgrowth, tends to carry the incisive edge in front of, or over, the snout.

Numerous single instances of overgrowth of rodent incisors occur in the literature. Only a few are worth quoting. Overgrowth of all four incisors was noted in a Townsend's ground squirrel (*Spermophilus townsendi*), caught in the wild, by Maser and Shaver (1976). The lower incisors were growing up to the right side of the snout; the upper incisors were curving backward, one into the posterior part of the palate, the other to the side of the tongue. The animal was not ill-nourished and was observed to feed by placing one side of its open mouth over long blades of grass and biting them off with its cheek teeth. A similar unilateral example in an American red squirrel (*Tamiasciurus hudsonicus*) was reported by H. C. Smith (1984).

Specimens have been described in which the lower incisors grow straight forward instead of curving upwards. The growing end of the incisor in rodents and lagomorphs is in close relationship to the roots of the cheek teeth; in an instance (Fig. 17.7) in a hare described in a paper by M'Intosh (1929), which is a source of a great deal of early literature and instances of tooth abnormalities, the straightness of the incisors was associated with inflammatory disease of the roots of the cheek teeth which may well have disturbed the differential growth at the bases of the incisors. A similar example in a rabbit is shown in Figure 17.8.

Rodent and lagomorph incisors can grow into

Fig. 17.5. A. Left: The skull of a wood chuck (*Marmot monax*) with the mandible in its posterior masticatory position. Note that the condyle projects posteriorly beyond the bony glenoid surface. The condyle is supported in this position by a fibrous roof or extension of the glenoid and free lateral movement is possible. In gnawing, the mandible is brought forward and the condyle lies in the groove-like bony glenoid which only permits antero-posterior movement.
Right: The cheek teeth are then out of contact, as they are here where the lower incisors are depicted even further forward in front of the uppers as they are during thegosis when the lower incisors are being sharpened. (Derived from Gehr, 1954).

B. The worn surfaces of the incisors of a beaver (*Castor fiber*). The upper incisor has a step which provides an anvil against which the lower incisor bites in gnawing. The arrow indicates where sometimes, in addition, there is a cupped-out hollow probably produced by contact with the lower incisor during molar grinding.
From: Osborn J.W. (1969). By kind permission of the publisher.

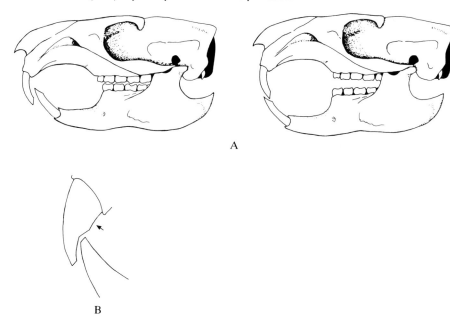

Fig. 17.6. *Rattus rattus* (black rat). Corkscrew-like overgrowth of I^1. The lower incisors have been deflected to the left, the upper to the right. The right I^1 has described one complete circle and three-quarters of another and, being well inclined outwards, has not encountered any obstacle. The left I^1, after describing a circle, penetrated the right maxilla to the depth of about 1.0 cm. The overgrown right I_1 has pierced the left maxilla and the parts above, its point projecting about 0.6 cm beyond the highest part of the skull. The overgrown I_1 caused the mandible to be quite fixed. The animal was caught in a starch works and it is probable that life was sustained by sucking flour. (From the *Transactions of the Odontological Society of Great Britain*, 1885, **17**, p. 119.)

Fig. 17.7. *Lepus capensis* (hare). Zoological Collection, Oxford University Museum, 1214a. Inflammatory disease around the roots of the cheek teeth has led to spacing and irregular wear. The overgrowth and absence of the normal curvature of the lower incisors is also a probable consequence of the inflammatory lesion affecting their basal growing ends (M'Intosh, 1929).

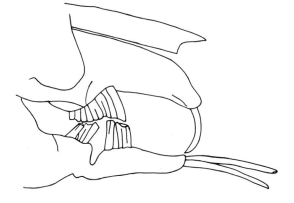

bizarre shapes (Fig. 17.9) which are hard to explain because, once formed, a length of tooth cannot be bent (M'Intosh, 1929). Perhaps when an incisor grows out of the mouth and has suffered loss of contact with the opposing tooth, it is subjected to forces which displace it in relation to the growing soft tissues at its base; it would then grow in a new curve; a further change in the direction of force exerted on the exposed tooth could induce a change in the curvature at the basal end.

At this point, it is necessary to consider the rodent-type of jaw mechanism as a whole (Fig. 17.5A). The mandibular condyles are cylindrical and orientated antero-posteriorly, resting in deep, groove-like glenoid cavities which are, of course, similarly orientated so that, when the mandible is brought forward during gnawing with the incisors, movement is largely in an antero-posterior axis with only a limited lateral component. However, in many species, to a varying extent, the posterior part of the glenoid groove is a fibrous roof which is flexible enough to permit a great deal of lateral movement when the condyle is in its posterior molar-mastication position. The mandible has to be brought back a considerable distance for all the lower cheek teeth to be brought into occlusion with the upper ones. In that position, the condyle is supported by the flexible glenoid roof and lateral and rotatory masticatory movements are free to occur.

In thegosis, it is highly likely that, not only are the upper incisors kept sharp by being scraped by the sharp edge of the lower incisors but, as suggested by Gehr (1954) and Osborn (1969), the lower incisors are brought in front of the upper so that they also can be sharpened by being scraped by the upper incisors.

According to Manville (1954), overgrowth of the incisors occurred in about 0.4% of a group of 9500 rats bred for a dental caries investigation. How common incisor overgrowth is in the wild, it is impossible to say; severe examples must be rare because an animal severely affected would be unlikely to survive for long, although Alexander and Dozier (1949) described a slightly emaciated sub-adult muskrat (*Ondatra zibethicus*) trapped in the wild which showed gross overgrowth of its upper incisors which formed a spiral curving into the mouth. However, the lower incisors were not much longer than normal and occluded against the anterior curve of the upper incisors. In this way, the animal maintained a just-adequate functional dentition. The authors calculated from the length of upper incisors that the malocclusion must have been present soon after birth.

Occasionally, the cause of incisor overgrowth is simple; one of the opposing teeth may be lost through injury or damaged in such a way that it no longer meets its fellow. An example in a female Polynesian rat (*Rattus exulans* USNM 359704) collected from the wild in Java was described by G. S. Jones (1979). One lower

Fig. 17.8. *Oryctolagus cuniculus* (rabbit), aged 4 months. An abscess of the tooth-bearing part of the mandible may have affected the growing end of the incisor and so may be responsible for a flattening of its normal curvature and consequent impossibility of occlusion with the upper tooth and overgrowth. The overgrown, normally inconspicuous, I^2 can be seen. (From Nachtsheim, 1936.)

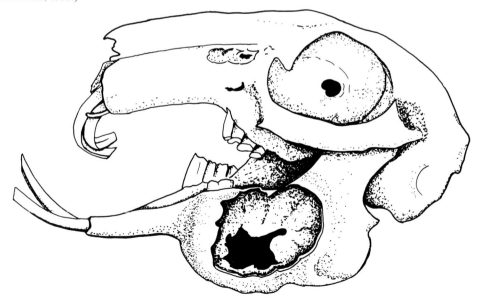

incisor was apparently congenitally absent; the incisor occlusion was in consequence disturbed so that the other lower incisor grew in a broad upward curve out of occlusion with the upper incisors which grew in backward and slightly outward curves to penetrate the outer surfaces of the upper jaw. Bland Sutton (1885c) described bilateral overgrowth of the incisors of a rabbit with a mobile fibrous union of a fracture of the mandible.

Incisor overgrowth associated with actinomycotic abscesses of the jaws was described by Dozier (1943) in muskrats (*Ondatra zibethicus*) raised in captivity, and by van Bree and Jansen (1962) in wild hamsters (*Cricetus cricetus*). Chaturvedi (1966) mentioned absence of tooth-pigmentation associated with incisor overgrowth in six specimens of bandicoot rats (*Bandicota bengalensis*) in the Zoological Survey of India in Calcutta. It seems likely that these were from the wild state. His paper is one of the few systematic studies where the skull and jaws were measured to discover whether the malocclusion of the incisors was associated with abnormal growth of the jaw. However, too few measurements of normal skulls were made for comparison for any convincing conclusions to be reached. Only a few other writers have mentioned disturbance of pigmentation. Pande (1945) found overgrowth of the incisors in six young rats in a colony of 80 maintained experimentally on a diet rich in fluorine; the normal pigmentation of the enamel was streaked with white patches which is a common effect of fluorosis as well as of dietary deficiencies. However, disturbance of pigment formation by the enamel organ is known to occur when eruption of the rat incisor is speeded by removing it from occlusion by clipping (Miles, 1963).

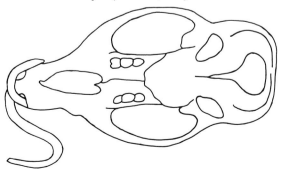

Fig. 17.9. *Rattus norvegicus norvegicus* (brown rat). BMNH. Ventral aspect of the skull. The left I^1 curves over the right side crossing the right I^1 which curves out of the mouth to the left; at an earlier stage, this tooth must have pursued a different curve and direction. The whole tooth is now S-shaped (M'Intosh, 1929).

In most instances of incisor overgrowth, the cause is far from clear. It is of particular interest to know whether the condition is inherited. It occasionally occurs in colonies of captive rodents, for example in chinchillas; as animals can only be kept alive with special care, the affected animals are usually destroyed so that they will not be bred from. Hence the question whether this disorder can be inherited goes unanswered. However, Nachtsheim (1936) described inheritance of the condition in a group of a Japanese breed of rabbits. Many animals were born with inferior protrusion; this presented no problem when the incisors were just erupting during the first week of suckling but, as soon as they were erupted, the lower incisors grew out in front of the upper jaw. Four affected females were kept alive by clipping the incisors at intervals and they were bred from. Seven out of 16 offspring were affected but the experiments were too incomplete to determine the exact mode of inheritance.

According to Günther (1957), a mutant producing a short bull-dog type of upper jaw tends to occur spontaneously in domestic breeds of rabbit and must be watched for because it is inherited as a dominant; because incisor occlusion is affected and therefore the normal wear and sharpening of the teeth do not occur, the animal will not thrive. In this instance, as in others where overgrowth of the incisors in lagomorphs is described, nothing was reported about the small I^2 that exist in lagomorphs immediately behind the main incisors (I^1) in the upper jaw. These I^2 are of continuous growth also and are presumably affected also by whatever causes the overgrowth of I^1. In a rabbit with the I^1 that curved in under the palate to form about three quarters of a circle, Friedlowsky (1869) showed quite clearly that the I^2 were also overgrown within the curve of the I^1 but were only about one third of their length. It seems likely that in many instances the slender overgrown I^2 would be broken off, especially if they grew out laterally from under the protection of the I^1.

Some experiments by Pastuszewska, Wyluda and Buraczewski (1979), who incorporated sea-krill protein in the pelletted diet of rats and mice, show how the balance of growth and wearing-away can be disturbed by slight changes of diet. The upper incisors became overgrown and both upper and lower ones became blunt. The animals lost weight probably because, having blunt teeth, they could not consume the pellets properly. The pellets of krill seemed to increase the hardness of the enamel and this could be the key factor that started a complex series of events (Mróz, Gasiorowska, Wyluda and Pastuszewska, 1979).

Caged guinea pigs (*Cavia porcellus*) seem to require in their diet something on which they can nibble. If they are deprived of hay, for instance, they tend to nibble their fur or that of their cage mates and produce large areas of baldness. Furthermore, their teeth, both incisors and cheek teeth, may become severely and rapidly overgrown with consequent rapid loss of condition and ultimately death from scurvy because of inability to consume green food (Paterson, 1967).

In most instances, the cause and the mechanism involved in overgrowth of incisors can only be speculated upon. The spiral geometry of the incisor could presumably arise from differential rates of cell proliferation in various parts of the growing basal end of the tooth or, alternatively, from some force in the environment of the tooth, acting on it not far from its growing end and tending to deflect it slightly. This could be a finely-balanced system, a minute disturbance of which might have an exaggerated effect at the occlusal end of the tooth where there may be another finely-balanced system upon which the maintenance of the length of tooth depends, namely the tooth-sharpening mechanism or thegosis. If that failed, the teeth could grow slightly out of contact with one another. The jaw joint in rodents in the forward gnawing position certainly allows free antero-posterior movement but may not allow sufficient lateral movement for the animal to be able to compensate for a minor malocclusion of the incisors, and, if it did not, the point would quickly, almost suddenly when the rapid growth of the incisors is considered, be reached when the opposing incisive edges could no longer be brought together. One or both opposing teeth would then be free to grow continuously longer.

Overgrowth of cheek teeth in Rodentia and Lagomorpha

The cheek teeth of lagomorphs and some true rodents, for instance the chinchilla and guinea pig, are of continuous growth. A condition, known to breeders of rabbits, guinea pigs and chinchillas (Miles, 1964) as 'slobbers' (Pollock, 1951), is due to overgrowth of the cheek teeth associated with loss of appetite, excessive salivation and paralysis, perhaps due to dehydration. The animals finally die of starvation (Zeman and Fielder, 1969). At autopsy, the overgrown, slightly medially inclined teeth are commonly found to have worn with razor-sharp medial edges and to impinge on the tongue or even almost meet in the midline over it. Beemer and Kuttin (1969) attributed the condition to failure to give the animals a diet that requires mastication or, alternatively, to provide them with wood to gnaw.

Hard and Atkinson (1967a) described an epidemic of slobbers affecting nine major guinea-pig breeding centres in Australia that turned out to be due to fluorosis (see Chapter 20). The commercial company that manufactured the pelleted food used at all the affected centres had substituted cheaper finely-ground rock-phosphate for the bone flour that had hitherto been used in the pellets. This mineral has a high content of fluoride. The animals showed a disinclination to eat and sat for long periods hunched and dribbling from their mouths. They lost weight and finally died with signs at necropsy of kidney failure and digestive tract infection. The striking thing was that the lower cheek teeth were much overgrown forming a partial arch over the tongue. The corresponding upper teeth were usually worn down to the gum although the anterior pair were sometimes overgrown and curved buccally as long tusk-like structures. The incisors were to some extent overgrown also. The sides of the cheek teeth were encrusted with rough dental calculus.

The condition seems certainly to have arisen from chronic fluoride poisoning; the kidney changes were in accord with this (Hard and Atkinson, 1967b). Fluoride incorporated into the growing teeth would affect the wear qualities of the tooth substance and lead to the irregular wear and malocclusion that is characteristic of chronic fluorosis in many animals. Indeed, there are examples of dairy cattle suffering similarly from fluorosis induced by the substitution in cattle feed of rock phosphate for bone meal (Chapter 20, p. 450). It is estimated (Fish and Harris, 1934) that normally in the guinea pig the continuously growing molars are worn away and completely replaced in about 40 days. The malocclusion of the overgrown teeth would not only have prevented the animals from masticating food properly but the arch formed by the teeth over the tongue would have interfered with the swallowing mechanism which depends upon the raising of the tongue. Thus, inability to swallow saliva was the probable cause of the dribbling from the mouth. Guinea pigs are highly sensitive to vitamin C deficiency and are therefore always given green vegetable as well as pelleted food. These animals would have been unable to consume the green food. It seems likely that death was due to a combination of fluoride toxicity, starvation and scurvy.

Overgrowth of teeth of continuous growth in other mammals

Many other groups of mammals have continuously-growing teeth; these may be incisors (Daubentiniidae, Proboscidae and Hyracoidea), canines (Suiidae) or both incisors and canines (Hippopotamidae).

The aye-aye (*Daubentonia madagascariensis*) is a primate but has a rodent-like dentition with one curved chisel-like incisor in each quadrant growing from a persistent pulp (Fig. 17.10). The incisors are used for scratching away the barks of trees in search of insects. In a specimen (Fig. 17.11) described by J. Murie (1867) from an animal maintained at the London Zoological Gardens for about five years, the two upper incisors extend downwards over the lip, diverge widely towards the cutting edges and pass in front of and to the outer side of the mandibular incisors. The upper incisors have marks of attrition on their mesial surfaces, the amount of which decreases towards where they emerge from the bone. The lower incisors show wear on their labial surfaces but curve backwards (Fig. 17.11B) so that, when the jaws were closed, the points of the teeth impinged on the palate; infection of the tissues followed and most of the palate was destroyed. There is nothing about the specimen to account for this excessive growth, but it may be that the animal, being in captivity, did not have the opportunity to use its incisors for its specialized food-gathering function.

Figure 17.12 shows overgrowth of an upper incisor (I^1) in a hyrax (*Procavia capensis*) resulting from misplacement distally of I_2 with which it normally articulates. Harrison (1982) described the overgrown continuously-growing left upper incisor of an adult male hyrax. The tooth had grown into the form of a half-circle curving slightly laterally with its tip just below the buccal surface of the left P^4. Functional occlusion of the rest of the dentition was possible but, perhaps because of the abnormal position of the upper incisor, the left P_1 had been lost during life. Overgrowth of the upper incisors appears to be common in the male hyrax. Stuart (1984) found such overgrowth in twelve out of 244 males but none in 291 females. The skulls of the affected animals were consistently larger than the unaffected ones, probably indicating that the affected animals were of advanced age.

The incisor of the hippopotamus shown in Figure 17.13 provides one of the most remarkable examples of overgrowth. The tooth grew in a perfect circle and the point must have re-entered the jaw, reaching the growing basal end at its centre. Busch (1891) described an overgrown hippopotamus lower canine which formed a spiral and was 90 cm long along its convexity. Friedlowsky (1869) described the skull of a hippopotamus in which a mandibular canine, of continuous growth, was malposed and grew to a considerable length in a backward curve out of the mouth (Fig. 17.14). Busch (1891) added some more information about this specimen. The animal was not fully adult; some deciduous molars were still unshed and the last permanent molars were unerupted. For some unapparent reason, the upper and lower canines on the left erupted in such a way that they did not meet and become mutually worn. Both grew unhindered with the preservation of their pointed ends in an unworn state (Fig. 17.14). The lower one was 35 cm long and the upper one 16 cm long. Busch pointed out that this difference in length suggests that the lower canine may normally grow more rapidly than the upper; the lower canine is the predominantly functional tusk, as it were, the upper one acting like a

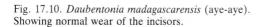

Fig. 17.10. *Daubentonia madagascarensis* (aye-aye). Showing normal wear of the incisors.

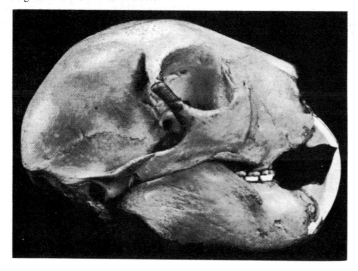

grind-stone against which the lower one is kept sharp. It is interesting that, in rodents generally, the lower incisors grow more rapidly than the upper (Berkovitz, 1976).

Overgrown boar's tusks, sometimes forming complete circles of about 60 mm inside diameter, are not uncommon in museum collections of all kinds. These specimens almost invariably come from the South Pacific islands where they were used as body ornaments reputed to confer fertility or to possess magical powers and often designating social rank. They were of such symbolic value that they were sometimes imitated in carved sea shell or by fitting two normal boar's tusks together (Jöchle, 1958). These circular tusks were deliberately manufactured by knocking out the upper tusks of young domesticated boars (Finsch, 1887). The lower tusks, which are of continuous growth, deprived of the normal wear mechanism, were free to grow and in due course the natural curve of the tooth brought the tip of the tusk against the skin of the cheek which it penetrated. The tip then recurved forwards, penetrating the bone, and eventually approached the vicinity of the growing base of the tusk. At the appropriate time, the boar was killed in order to harvest the cultivated tusks. A. B. Meyer (1896/1897) added a great deal of interest about these circular tusks and the way they were made. The nature of the Papuan economy and way of life, which were centred largely on the pig, were such that only the most rich and powerful could have much chance of keeping a pig for the long time needed for the tusk to grow to a full circle. When the piglet was about one-year old, both upper canines were smashed out with a wooden mallet. The pig was then carefully nurtured and often the tusk was filed as it grew to help secure the right curvature. Figure 17.15 shows a bilateral example, with the tusks *in situ* in the jaw,

Fig. 17.11. *Daubentonia madagascarensis* (aye-aye). Formerly RCS Osteo. Series, 302. Anterior (A) and side-view (B) showing overgrowth of the incisors.

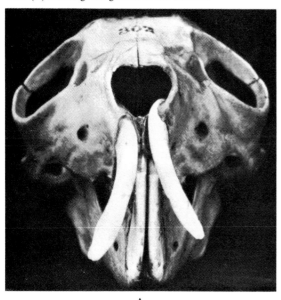

A

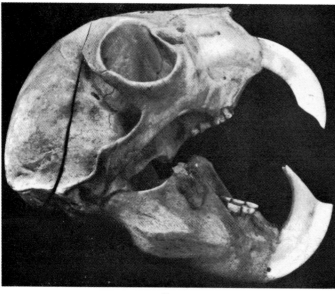

B

Fig. 17.12. *Procavia capensis* (Cape hyrax). RCS Odonto. Mus., G 43.1. Right I^1 overgrown. $\times c. 0.75$.

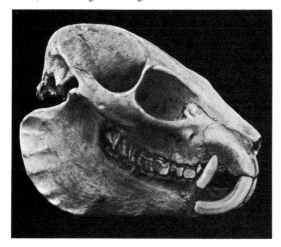

which is of special interest because it was figured by Cheselden (1733) in his *Osteographia*; it passed into the collection of John Hunter and later became part of the Hunterian Collection of the Royal College of Surgeons. The points of the teeth, following their normal curve, re-entered the mouth through the cheeks; the right tusk penetrated the bone on the outer aspect of the molars and made its way obliquely through the bone beneath the roots of the molars and emerged on the inner aspect of the mandible close to its own growing end, so forming a complete ring; the left tusk took a slightly different course and penetrated the body of the mandible behind the molar teeth. Unfortunately, this specimen did not survive the wartime damage to the Royal College of Surgeons in 1941. However, Figure 17.16 shows a similar specimen.

In the wart hog (*Phacochoerus aethiopicus*), the upper canine normally turns upwards so that only part of its undersurface occludes with the lower canine; marked overgrowth has been observed in a captive specimen (Fig. 17.17).

Fig. 17.13. *Hippopotamus amphibius* (hippopotamus). RCS Odonto. Mus., G 41.1. An overgrown tusk which has taken the form of a circle. $\times c. 0.33$.

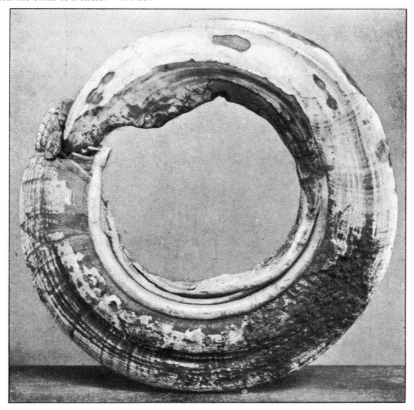

In the male wild pig or babirusa from Celebes, *Babyrousa babyrussa*, the upper canines grow upwards and erupt on top of the snout, curving backwards as a true tusk towards the forehead; the lower canines grow up beside the snout as in the other Suidae but, being directed slightly outwards and not being in occlusion with the upper canine, they are free to grow and project well up above the snout. The upper tusks turn forwards and a little outwards at their extremities so that normally, however long they become, they do not quite come into contact with the skin of the forehead. However, instances have occurred where the tusks grew backwards and penetrated the skull. M'Intosh (1929) figured an example from the Royal Scottish Museum, Edinburgh (1878-1-4), in which the right tusk had penetrated the bone but not pierced it because there had been a deposition of subperiosteal bone between the ingrowing tip of tusk and the brain. This appears to be a hazard associated with captivity; captive animals often rub their tusks against their prison walls, and even in one case against their keeper's legs (Bland-Sutton, 1884a, 1885b), and so modify the direction of growth of the tusks.

Mohr (1958) mentioned that the upper tusks in the babirusa have no enamel and are very brittle so that many museum specimens are imperfect. She also stated that the vast majority of babirusa skulls in collections are of males, perhaps because the females, which have short tusks which project only a little above the snout, are less spectacular to collect, or it may be that the females are so well protected by the aggressive males in the field that hunters rarely get the opportunity to kill females. Mohr also described a male babirusa in the Brookfield Zoo in Chicago in which the upper tusks pointed forward instead of curving back over the snout.

The lower canines in the male babirusa, which project backwards and outside the curve of the upper ones, are its most effective aggressive weapon; however, according to MacKinnon (1981), in fights between males the head is sometimes manipulated so as to hook a curved upper tusk on the opponent's lower tusk and so disable him, yet still to be able to inflict damage to him with the lower tusk.

According to A. B. Meyer (1896/1897), some artificially-overgrown boar's lower canines have been mistaken for babirusa upper tusks. The two can be distinguished because, apart from the presence of enamel on the boar's tusk, it is roughly triangular in cross-section, whereas the babirusa tusk is without enamel and is elliptical in section.

Meyer described a lower jaw of a babirusa in which the right canine had grown round in a complete circle and had re-entered the jaw, very much like the teeth in Figure 17.15, and within the jaw had overlapped the growing end. The tooth must have had an abnormal direction of growth because the babirusa lower canine is normally inclined outwards as well as backwards so that unhindered growth would carry the point of the canine well away from the jaw bone. This abnormal direction of growth may have been connected with a small supernumerary canine which Meyer says was present on the inner aspect of the overgrown canine. However, as the lower canines in babirusa,

Fig. 17.14. *Hippopotamus amphibius* (hippopotamus). A greatly overgrown left mandibular canine. The tooth, of continuous growth, curves outwards and then sharply backwards. (From Friedlowsky, 1869).

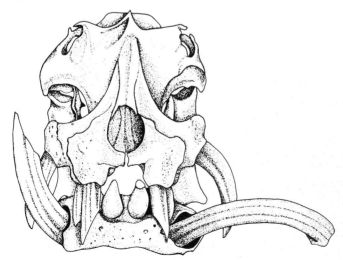

Fig. 17.15. *Sus scrofa* (wild boar). ♂. Formerly RCS Osteo. Series, 1798. The mandibular tusks have become overgrown; their points have penetrated the body of the jaw forming nearly complete circles. The engraving is from Cheselden (1733), 'The end of a boar's jaw in which one of the teeth is grown quite round and through the jaw again'. In printing, of course, the engraved image has been reversed.

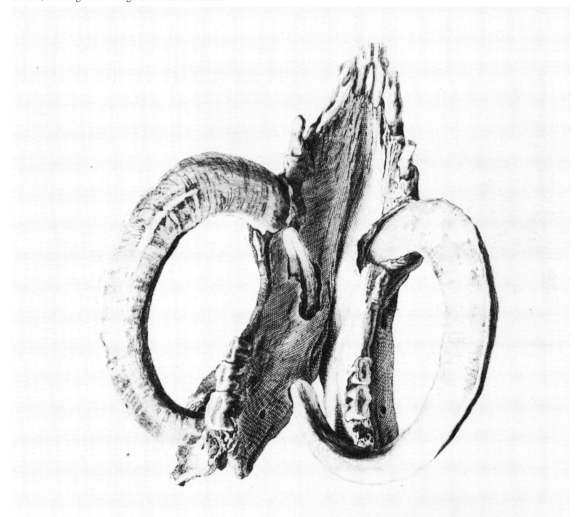

Fig. 17.16. *Sus scrofa* (wild boar). ♂. RCS Odonto. Mus., G 43.21. A mandible in which, very symmetrically, the normally curved tusks have grown in a spiral, re-entered the mouth, probably by penetrating the cheeks and, by pressure resorption of bone and the roots of teeth, have emerged on the medial aspects of the mandible. Pointer indicates an abscess cavity which contains a loose sequestrum. × c. 0.27. (From Miles, 1973.)

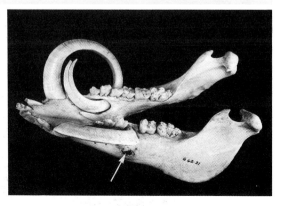

unlike those of pigs, are not in contact with the upper canines, the question arises why the lower canines do not usually grow to a greater length. The reason is that they are normally kept worn down by use as a digging tool. Meyer suggested that this specimen was perhaps a captive animal deprived of the opportunity to use its tusks and wear them away.

In whales of the family Ziphiidae, the dentition is reduced to a single flat tooth on each side of the mandible, which on closure remains outside the mouth

Fig. 17.17. *Phacochoerus aethiopicus* (wart-hog). ♂. Captive. Museum of the Dental School, Cairo. The upper canines are greatly overgrown and cross each other over the snout.

Fig. 17.18. *Mesoplodon layardi* (strap-toothed whale). ♂. Normally the teeth grow vertically to about this length but, in this specimen, they have curved inwards to meet in the midline. These are the only teeth these whales possess and it may be that the slight restriction of jaw opening produced here would not greatly interfere with feeding on squid. The specimen is viewed obliquely from in front so that the elongated jaws are fore-shortened.
From: Morzer Bruyns, W.F.J. (1973).

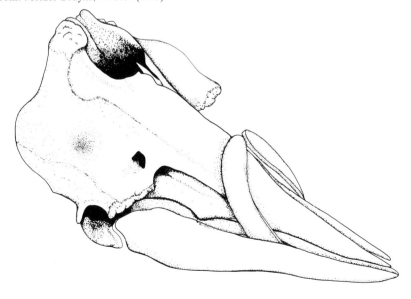

(and is therefore by definition a tusk). In the male strap-tooth whale (*Mesoplodon layardi*), the teeth are large and flat, are sited near the middle of the jaw, and grow upwards and backwards and slightly over the rostrum looking like a pair of inverted ribs. Sometimes the teeth become overgrown (Fig. 17.18) and form an arch over the rostrum which limits to a centimetre or two the amount by which the mouth can be opened (Morzer Bruyns, 1973). However, it is said that the diet of the strap-toothed whale consists of squid which it seems likely they suck into the mouth like a vacuum-cleaner (L. Watson, 1981) from the sea bottom or among rocks.

SECTION 4
Other Disorders of Teeth and Jaws

CHAPTER 18

Injuries of the Jaws

Introduction

A chapter on injuries of the jaws is included because the teeth are usually involved and play an important part in the way in which the fractured parts are displaced.

Fractures of the jaw and other injuries are common in animals living in the wild. The true prevalence is unknown because the evidence is based almost exclusively on those in which healing sufficient for the animal to survive occurs and on the few whose skeletons ultimately happen to be examined.

A great deal of old skeletal injury, especially fracture, is found in non-human primates in the wild (Schultz, 1939, 1956) perhaps because many sooner or later meet with accidents in the course of their remarkable leaps in the tree-tops. Other injuries derive from human and other predators. Orang-utans (Duckworth, 1912) and baboons (Bramblett, 1967) seem to be particularly susceptible; the heavy-bodied baboon, although largely terrestrial, takes refuge in trees and is a clumsy climber. Many examples are reviewed by Schultz of survival and recovery from, or adaptation to, severe multiple injuries. A capuchin for example was shot dead while apparently feeding in the trees as actively as its fellows. It was found to have previous fractures of both femora, one repaired with much shortening, much of its mandible had been shot or torn away, and there were fractures of the facial bones and sternum.

Colyer (1921) found three instances of healed severe injury to the premaxillary region among the skulls of 300 young rhesus monkeys which died soon after being brought to the UK for medical research during the First World War. These injuries seem likely to have been due to fighting induced by stressful crowded conditions of captivity.

Howell (1925) described four skulls which showed marked asymmetry, one a gorilla (Chapter 10, p. 167), two sea-lions (*Eumetopias jubatus*), and a grivet monkey (*Cercopithecus aethiops aethiops*, formerly *Lasiopyga griseoviridis*), and reviewed others from the literature. The cause in most instances was uncertain though assumed to be probably due to injury in infancy to the skull which affected its growth. The skull of the monkey had almost certainly suffered severe trauma probably from a bullet which destroyed or carried away the greater part of the vertical ramus of the left mandible. The whole of the facial skeleton is twisted and bent to the left side. It was impossible to determine how much this deformity was due to loss of bone and damage to the growth mechanism of what remained and how much was due to lack of muscle activity; there would have been loss of most of the masticatory muscle on the left as well as loss of the part of the bone on which those muscles acted.

An asymmetrical skull of a sub-adult female leopard seal (*Hydrurga leptonyx*, RCS Odonto. Mus., A 175.21) was described by Cave and Bonner (1987). There is evidence of severe crushing damage to the bone on the left side probably during infancy and perhaps by entrapment between ice floes. The left orbit, zygomatic arch and external auditory meatus had suffered the most severe direct damage, but all the bones of the face on that side are smaller than on the right and the muzzle is deflected to the left. Cave and Bonner attribute this asymmetry to the effect on the growth of the bones of damage to the muscles of mastication; the coronoid process of the left mandible is small. Probably in consequence of the smallness of the upper jaw, the mandibular teeth on the left are in pre-normal occlusion. However, the left mandibular condyle is slightly deformed so that the growth or the position of the mandible may have been affected by the initial trauma.

Injury to the jaws during their growth, apart from fracture, can have a large effect on the morphology of the jaws and the relation between the tooth arches. Spinage (1971) described the skull of an adult impala (*Aepyceros melampus*) which was deformed as result of an injury when young; the principal effect had been the loss of M_2 and M_3 on the left. The whole mandible on that side was much more slender than on the right and there was no projecting angle. Spinage had seen calves kicked by inexperienced dams and suggested that something of the kind could have been the cause of this deformed mandible.

Bland Sutton (1885c) quoted an account of an unusual injury in a horse which had lived in the wild for three years with a rope halter over its head and round its jaws. The rope had cut a deep groove in the lower border of the mandible.

The variety of possible injuries to the jaws is wide; instances are recorded of sticks or bones being wedged across one or other of the tooth arches, in a wolf (Chapter 23, p. 511), in a dog by Catcott (1968) and in a cat by Reed (1975). A bizarre example described by E. Becker (1970, p. 274) concerned an emaciated cow that had part of the mandible of a roe deer wedged in its lower tooth arch.

Fracture of the maxilla

Displacement of the fragments associated with fractures of the upper jaw depends mainly on the amount and direction of the force that produces the fracture; as there are no powerful muscles attached to the upper jaw, the pull of muscles contributes little to the displacement. For reasons that are still not well-understood, union of fractures of the upper jaw is frequently not true union by bone, but by fibrous tissue. This is well-established in respect of the human upper jaw and is also well illustrated by a number of animal specimens preserved in museums.

The united fracture of the maxilla in a lion (Fig. 18.1) is a good example. The fracture, probably produced by a wrenching force applied to the canine in combat, passed through the anterior palatine foramen, obliquely across the palatal process, through the middle of the post-canine alveolar process, through the infra-orbital foramen and up to the fronto-maxillary suture medial to the orbit. Union has occurred with slight upward displacement of the canine-bearing part of the bone. Only in the palate is there any true bony union (Fig. 18.1B). On the facial aspect (Fig. 18.1A) below the level of the infra-orbital foramen, the fractured surfaces are separated by a slight gap, probably filled in life by fibrous tissue; above the foramen, union has occurred with the formation of a dentate interlocking suture.

The skull of a colobus monkey (*Colobus polykomos palliatus*) (Fig. 18.2) has a similar fracture (Colyer, 1943b). The injury caused a gap behind the canine in the facial surface of the maxilla; the line of fracture passed upwards along the posterior border of the canine and then forwards across the nasal process. In the palate, the fracture crossed the premaxilla to the interval between the first incisors. The margins of the bone bordering the gap on the facial surface were rounded and presumably filled with fibrous tissue;

across the nasal process, there was a tendency to the formation of a dentate suture. There was no sign of callus anywhere but, stretching across the anterior aspect of the junction of the two premaxillae, there was a bridge of osseous tissue.

A fracture of the upper jaw in an American badger (*Taxidea taxus*) (USNM 248422) in which bony union appears to have occurred was described by Long and Long (1965). A lower canine was fractured and the upper incisors displaced backwards. There were fine vertical fracture lines on each side of the facial surface of the premaxilla, extending on to the floor of the nose, and a transverse one behind the incisors. The trauma had separated most of the premaxilla and displaced it slightly.

According to Pycraft (1913), among the large

Fig. 18.1. *Panthera leo* (lion). RCS Odonto. Mus., G 80.1.
A: A fracture of the maxilla through the infra-orbital foramen. Union is fibrous with, above the foramen, formation of a dentate suture.
B: Palatal view showing some true bony union just behind the incisive foramen. (From Colyer, 1913.) ×c. 0.42.

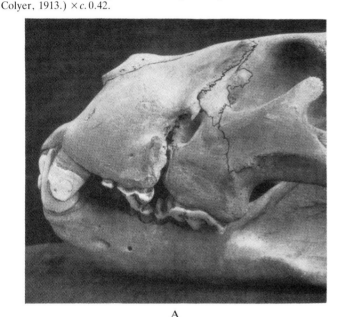

A

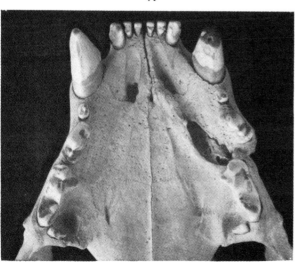

B

Fig. 18.2. *Colobus polykomos palliatus* (mantled colobus). ZMB 18798.
A: A fracture behind the right canine. Below, there is a gap with rounded edges but above, where the fracture line crosses the nasal process, there is a dentate suture.
B: Palatal view. The fracture crosses the premaxilla with a slight gap and passes between the two I^1 where there is a bridge of bone. (From Colyer, 1943b.).

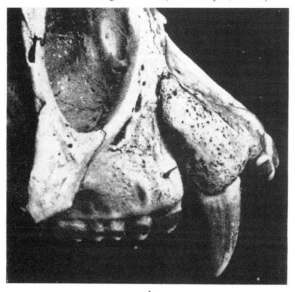

A

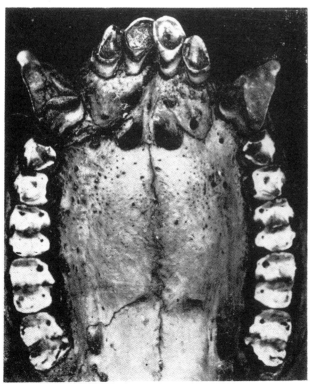

B

apes, the male, when endeavouring to discover the whereabouts of the female, utters loud cries to attract her, but this may attract another male on a similar quest and lead to a serious conflict, the powerful canines being used as weapons. The injury seen in the gorilla skull in Figure 18.3 is probably the result of such combat. The upper incisors have been torn away and there has been considerable loss of bone causing an opening into the right maxillary air sinus.

A fracture of the maxilla and premaxilla in a gorilla was mentioned by Randall (1943/1944).

There may be failure of union of fractures of the upper jaw, in which case, if the animal survives, the bony parts are separated by such an amount of fibrous tissue that movement of various degrees is possible. Alternatively, the bony surfaces may remain close and, under the influence of continuous movement, become congruous with the formation of a false joint or pseudo-arthrosis. Ruprecht (1965a) described this condition in the skull of a red deer stag (*Cervus elaphus*) which had exceptionally massive antlers and which had been the leader of its group. There had been a fracture across the front of the upper jaw producing a separation of the premaxillary region from the maxilla. Ruprecht suggested that the fracture probably occurred in combat during the mating season. The fractured surfaces were complexly congruent, with some lateral displacement, but a considerable amount of movement must have been possible. That this would be consistent with survival is less surprising if it is borne in mind that the bones of the skull in cervids, and in ruminants generally, are not so rigidly fixed together as they are for example in the carnivores, and very slight movement between many of the bones is normally possible; that is the skull has kinetic characters.

Other examples of pseudo-arthrosis are described in the next section dealing with the mandible.

Pilleri and Gihr (1969b) described a sub-adult male Amazon porpoise (*Inia geoffrensis*) in which the rostral end of the upper jaw had been destroyed on one side, by osteomyelitis or trauma or a combination of both. The bony surfaces were healed.

Fracture of the mandible

In fractures of the mandible, the fragments tend to be displaced by the action of the powerful muscles of mastication attached to the bone. Displacement associated with fracture through the horizontal tooth-bearing ramus depends greatly upon the position of the fracture in relation to the depressor muscles attached to the anterior fragment and to the mainly elevator muscles attached to the posterior fragment. The amount of displacement also depends on the angle of the fracture through the ramus. Some oblique fractures are unfavourable and leave the muscles free to separate the fragments; other obliquities are favourable and tend to restrict muscle action. Displacement is affected also by the nature of the dentition. If the crowns of the teeth are broad, contact of the teeth with those of the opposing jaw will tend to splint the jaws together. Where the teeth are narrow from side to side and occlude with a slicing action, as in the Felidae, displacement of fragments tends to be greater.

The skull of a cheetah (Fig. 18.4) is a good example. The impact of the blow causing the injury would appear to have been received on the lower carnassial (M_1), as the mesial cusp is broken and the distal cusp is slightly splintered. The posterior fragment of the mandible has swung upwards, a large amount of callus has been formed especially on the outer aspect, and it is evident that healing was accompanied by considerable suppuration but, as will be seen from the radiograph (Fig. 18.4B), union is not complete between the fractured ends of the bone. The mesial root of the molar is completely denuded of bone and a large portion has been resorbed. The occlusion of the teeth is good and there has been no movement of the major fragments towards the injured side, because it was held in position by the interlocking of the canines. This specimen shows upward and forward movement of the posterior fragment of the mandible, and also the disadvantage of teeth being involved in the line of fracture.

In the Felidae, the upward movement of the

Fig. 18.3. *Gorilla gorilla* (gorilla). ♂. RCS Odonto. Mus., G 82.11. Upper incisors torn away and the right maxillary air sinus exposed. × c. 5.58.

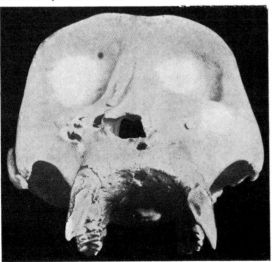

posterior fragment of the mandible may bring the M_1 against the palate and the pressure thus exerted on the tissues may lead to the formation of an opening through the palate into the nasal fossa as has happened in the leopard shown in Figure 18.5. The mandible was fractured through the region of P_4 and fragments of the tooth remained embedded in the bone (Fig. 18.5A). Nevertheless, there was good union with strong continuity of the bone and only slight deformity compared with the opposite side, as can be seen in the radiographs (Fig. 18.5C). However, upward movement of the posterior fragment of the mandible and consequent pressure of the broken M_1 caused a perforation of the palate (Fig. 18.5A).

Another example of this type of fracture is shown in Figure 18.6. The mandible on the right side has been fractured and the skull in the region of the zygomatic fossa severely injured. The line of the fracture passed through the position of P^4, this tooth and P^3 being probably lost at the time of the injury. The posterior fragment of the mandible has been drawn upwards and slightly inwards, the anterior extremity overlapping the end of the anterior fragment which has moved backwards slightly. In the Felidae, when the mandible is fractured in the region of the molars or premolars, the anterior fragment is usually kept from moving towards the injured side by the interlocking action of the canines. The movement of the anterior fragment in this particular case is due to the absence of the interlocking action of the canines owing to the loss of the greater part of the crown of the lower canine. The union of the fracture, as shown by the radiograph (Fig. 18.6B), is almost perfect. The injury carried away a considerable portion of the zygomatic arch and the posterior part of the maxilla. The palate was involved, the posterior part being destroyed and the remaining portion split, the line of fracture extending as far forwards as P^2. There are signs of bony union of the palate in two places. The fractured zygomatic bone also shows indications of repair. The upward movement of the posterior fragment brought M_1 into the cavity at the posterior part of the palate.

An instance of non-union of a fracture through the diastema-region of the mandible with the formation of a false joint (pseudo-arthrosis) in a fossil sabre-tooth tiger (*Smilodon californicus*) was described by Shermis (1985–1986) together with two instances of united fractures.

The fractured mandible of the leopard in Figure 18.7 presents several features of interest. On the right side (Fig. 18.7B), the bone was broken obliquely; the line of fracture extended from the outer side below the level of the M_1 upwards and inwards between M_1 and P_4 to the inner aspect of the bone. The posterior fragment moved upwards and inwards; the effect of this movement was to displace the anterior fragment outwards and forwards leaving a space of about 12 mm between M_1 and P_4. The upward movement of the posterior fragment was limited by the palate but the pressure of M_1 caused some resorption and in one place a small perforation of the bone.

On the left side (Fig. 18.7B and C), it seems that

Fig. 18.4. *Acinonyx jubatus* (cheetah). Captive. RCS Odonto. Mus., G 82.12.
A: A healed fracture of the right side of the mandible.
B: Radiograph of the mandible. The gap between the fractured surfaces is occupied by callus of much less density than the original bone. Union is not complete. ×c. 0.5.

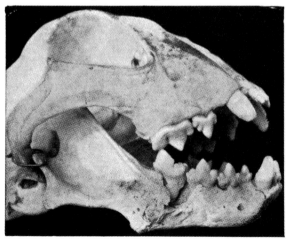

A

B

Ch.18. Injuries of the Jaws 377

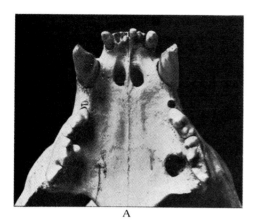

Fig. 18.5. *Panthera pardus* (leopard). BMNH 1929.12.23.1.
A: Palatal view of the upper jaw showing a perforation of the palate on the left by M_1. $\times c.0.66$.
B: The left side of the mandible showing a fracture through the region of P_4 which is well-united though the fragment bearing the carnassial tooth (M_1) is displaced upwards.
C: Radiographs of the mandible; the fractured left side is above, the uninjured right side below. Reduced slightly.

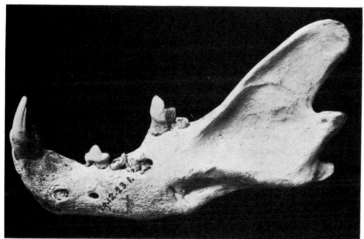

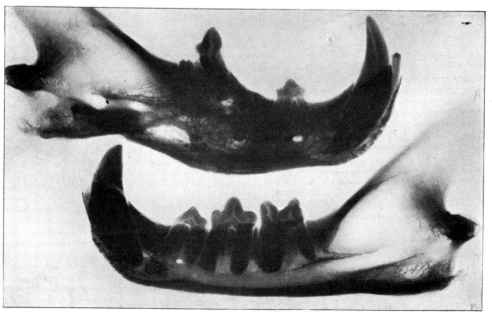

378 Section 4: Other Disorders of Teeth and Jaws

the line of the fracture started from the lower border and passed upwards towards the position of the canine. The lower left canine and incisors were shot away and the injury caused a fair amount of comminution of the bone forming the inner wall of the anterior fragment. The posterior fragment moved upwards and inwards, but not so much inwards as on the right, with the result that function was maintained on the left side and so probably prolonged the life of the animal. It is of interest to note that, on the side where function was maintained, the alveolar process covering the teeth appears to be quite normal, but on the right side there is a considerable degree of rarefaction of the bone. The anterior fragment is depressed and rotated, the tip of the lower right canine is below and opposite I^2 and I^3, and P_4 is to the buccal side of P^3 as shown in Figure 18.7A. The rotation of the anterior fragment was probably due to the oblique direction of the fracture on the right side. The fractures are well-united but, as the radiograph (Fig. 18.7D) shows, with more thickening by persistent callus on the right side where the displacement was greater.

Figure 18.8 shows an example of bony union in a double fracture of the mandible of a wild leopard, probably caused by a bullet as radiographs show fragments of metal in the bone. Most of the teeth, upper

Fig.18.6. *Panthera pardus* (leopard). RCS Odonto. Mus., G 82.121.
A: The skull has been severely injured and the mandible on the right side fractured with damage to several teeth. × c. 0.5.
B: Radiograph of the fractured right mandible.

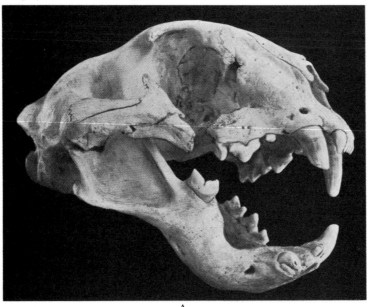

A

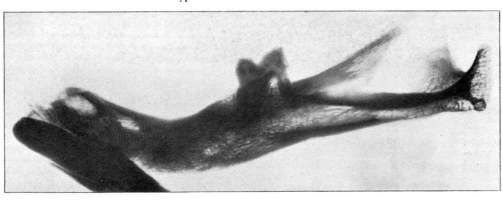

B

and lower, on the right have been shattered, though the surviving roots have been healed over. There are two united fractures of the horizontal ramus of the right mandible, one just anterior to the vertical ramus and the other a little posterior to the canine and mental foramen. Bony union is complete and radiographs show total obliteration of the line of fracture by remodelling (Fig. 18.8B). The only remaining evidence of the fractures are the two disturbances of the external contour of the bone, showing that the main displacement was an upward swing of the posterior fragment and some downward displacement of the intermediate fragment.

An additional point of interest is that, presumably as a result of loss of the medial wall of the inferior alveolar canal at the site of the posterior fracture, there is a large opening with rounded margins into the canal 1.5 cm in front of the right mandibular foramen.

In this specimen, although the line of fracture has passed through the sockets of the teeth, healing has been accompanied by but little suppuration, in marked contrast to most fractures in animals living in the captive state. Nevertheless, fractures occurring in well-conducted zoos may receive sophisticated treatment such as wiring the fragments together under protective antibiotic cover. Milton, Silberman and Hankes (1980) gave details of a double comminuted fracture through both ascending rami of the mandible in a female Bengal tiger sustained in a fight with a male tiger.

A fracture in a leopard through the mandible distal to the right canine, still in the stage of healing with ample callus, was described by Hemmer (1964).

The superior constrictor muscle of the pharynx is attached to the medial aspect of the mandible towards the posterior end of the horizontal ramus. Although its exact insertion differs in various species, when there is a fracture of the ramus in front of the muscle insertion, the muscle during swallowing produces an upward and inward pull on the posterior fragment which contributes significantly to the degree of displacement. If, however, the fragments override each other with the posterior fragment lying on the lateral side, inward movement of the posterior fragment is prevented and bony union is favoured. Such is the case of the fractured mandible of a hartebeeste with good union in Figure 18.9.

In the satanellus depicted in Figure 18.10, however, there was no favourable overlap and the posterior fragment was swung upwards and inwards by muscle action producing a considerable gap between the fractured surfaces. Nevertheless, union has occurred but with the formation of much callus. Possibly, as in the cheetah already referred to, interlocking of the large canines held the anterior part of the mandible in place. The end-result is a perfectly functional mandible with good occlusion.

Fractures in the region of large canines are often oblique, with overriding of the fragments. Figure 18.11 shows how such fractures heal. It is the skull of a chimpanzee which has a fracture of the mandible in the region of the left canine and P_3; these teeth, together with P_4, were lost as a result of the injury. Union has followed, with the formation of a large amount of callus. The major fragment (on the right) has swung over to the left (Fig. 18.11A) and has brought the lower premolars and molars forward to the extent of about half a premolar. On the left side, the forward movement of the smaller fragment is more marked and M_1 is in contact with the whole of P_4 and with part of P_3.

The radiograph (Fig. 18.11C) shows the overriding of the fragments, the larger right fragment having passed over the smaller fragment. The illustrations show that the upper and lower teeth do not occlude in an even manner. This irregularity of the occluding surfaces suggests that the injury occurred not long before the death of the animal. There is no question as to the relative positions of the mandible and maxilla at death because, when the skull was first seen, it had not been properly cleaned and the ligaments were holding the mandibular condyles in the correct position.

A remarkable injury to the mandible of a fox reported by Breuer (1932) is shown in Figure 18.12. The anterior part of the mandible bearing the incisors and canines has been split off and displaced to a position beneath the remainder of the mandible where it became firmly united with the undersurfaces of the rami (Fig. 18.12B). Masses of callus were formed at the fractured ends of the post-canine parts of the mandible which in consequence are almost in contact across the midline (Fig. 18.12C). On the right, a broken-off fragment of bone has united to the ramus and there is a deep cleft (Figs. 18.12B and C) which Breuer suggested is the result of a penetrating injury made by the canines of the attacker.

Lensink (1954) described a brown bear (*Ursus arctos*) in which the anterior end of the mandible had been shattered by a bullet but the fractured surfaces had come together and united, with consequential deformity of the post-canine occlusion and shortening the jaw by about 2.5 cm. The anterior shattered fragments, including the lower canine teeth, had become embedded in the lower lip. The lingual surfaces of the upper canines were deeply abraded, it was suggested by attempts to prehend food in the absence of incisor and canine occlusion. In spite of what would seem to be a considerable handicap, the bear appeared to be well-nourished with a 5 cm adipose covering.

A partially healed gunshot fracture of the mandible in a black bear (*Ursus americanus*) that appeared to have survived for a few months was described by Dyer (1981).

The skull of a three-year old female wolf (*Canis lupus*) described by Barrette (1986) is a further illustration of how an animal with an impaired jaw mechanism can nevertheless survive in the wild. Both mandibular joints were grossly malformed, presumably from birth and not from injury. The degree of functional impairment was uncertain but it could be deduced that the joints could not have provided the stable hinges upon which efficient scissor-action of the carnassials depend.

Union does not always follow fracture of the jaw in wild animals. The skull of a male gorilla in Figure 18.13 has an un-united fracture of the lower jaw. In the

Fig. 18.7. *Panthera pardus* (leopard). RCS Odonto. Mus., G 82.1212. With bilateral fractures of the mandible.
A: Right lateral view. An oblique fracture of the mandible in the region of M_1 with good strong union but with displacement upwards of the posterior fragment. $\times c.\,0.66$.
B: Left lateral view. A similar fracture of the mandible but further forward in the region of the canine tooth which was lost. On this side, the molars remain in functional relationship. $\times c.\,0.66$.
C: View of mandible from above. $\times c.\,0.75$.
D: Radiograph corresponding to the view of the mandible in C. $\times c.\,0.75$.

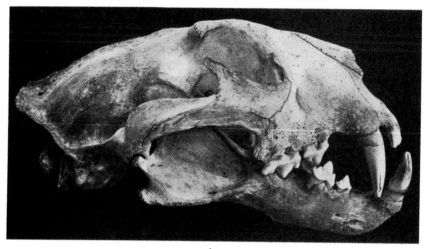

A

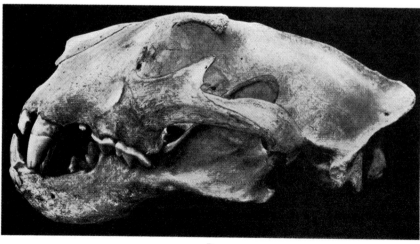

B

upper jaw, both I^1 have been lost, and I^2 and the upper canines have moved forward towards the mid-line. In the mandible (Fig. 18.13B), the incisors, canines and P_3 have disappeared as a result of the injury. The bone was fractured through the region of the left canine with loss of some of the substance of the bone. However, the fractured surfaces must have come into apposition and, with changes in those surfaces and those of the associated callus, there are now two pitted circular surfaces about 2.5 cm in diameter which fit each other very closely and were probably firmly united with fibrous tissue.

It is worth noting the changes that have occurred in the mandibular molar dentition (Fig. 18.13). On the left side, M_3 is in its normal position with its distal surface about level with the anterior border of the vertical ramus. On the right side, however, the molars must have migrated mesially about 10 mm as part of the adaptation to the disturbance of the occlusion resulting from the injury, because the distal surface of M_3 is about that distance in front of the anterior border of the vertical ramus.

Sometimes in fractures of the mandible there is so little displacement that union occurs with hardly any

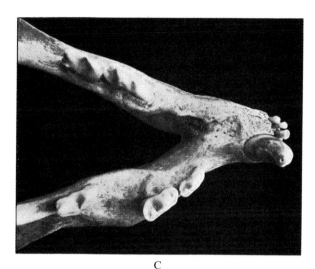

C

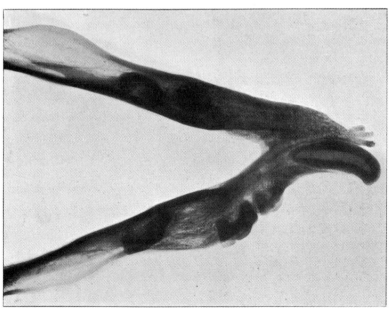

D

deformity in the shape of the bone. Often this lack of displacement is because the line of fracture is through a muscle attachment and the muscle tends to hold the parts together. This is probably the partial explanation of a double fracture of the mandible in a wild male bay duiker *Cephalophus dorsalis* (RCS Odonto. Mus., G 81.32) described by Colyer (1945a). One fracture passed obliquely through the region of the symphysis; there was bony union with little displacement, perhaps because the fracture line crossed the attachment of the anterior belly of the digastric muscle. A second fracture which was comminuted, that is so complex that fragments of bone were separated, was through the upper part of the vertical ramus just below the cor-

Fig. 18.8. *Panthera pardus* (leopard). RCS Odonto. Mus., G 82.1211. A double fracture of the right side of the mandible.
A: Right lateral view of the skull. Disturbances in contour of the mandible indicate a well-united fracture, with slight upward displacement of the posterior fragment, just in front of the vertical ramus and another, with more displacement, a little posterior to the mental foramen. ×c.0.6.
B: Radiograph of the right mandible. Union of the fractures is good, no sign remaining apart from the abnormal external contour. Some of the dark objects in the substance of the bone are metallic fragments. The broken-off roots are buried in the bone.

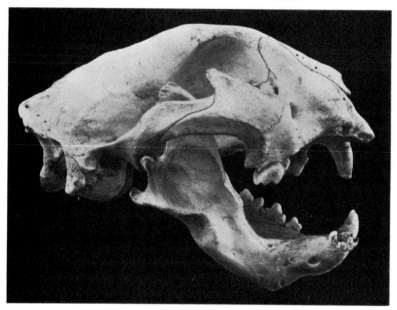

A

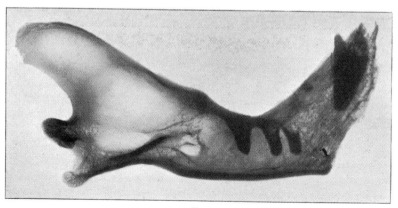

B

onoid and condylar processes. There was an extension of the fracture between the processes which separated them from each other as well as from the ramus. Nevertheless there was little displacement of the processes and union was proceeding when the animal was shot by a hunter. The anterior part of the fracture would have passed through the lower part of the insertion of the temporal muscle into the coronoid process and this may have prevented the main body of the muscle from pulling the process away from the ramus. More posteriorly, the fracture line may have passed through the upper part of the attachment of the masseter muscle to the bone; this could explain why the condylar process was not displaced, as would have been expected, by the pull of the lateral pterygoid muscle inserted into its anterior surface.

Another instance of survival of an animal in spite of severe limitation of masticatory function by injury is provided by a male red deer (*Cervus elaphus*) described by von Braunschweig (1974). Six weeks before

Fig. 18.9. *Alcelaphus buselaphus tschadensis* (Tchad hartebeeste). ♂. BMNH 1923.1.1.500.
A: Inferior surface of the mandible showing a healed fracture on the right side with slight overlapping of the fragments. ×c.0.25.
B: Lateral views. Above, right side showing the fracture just behind the cheek teeth and with little displacement. Below, left uninjured side for comparison. ×c.0.25.

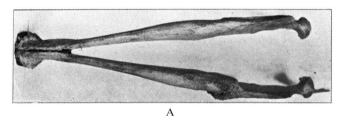

A

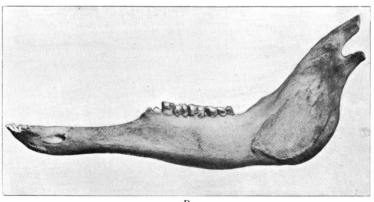

B

it was killed, the animal was seen in distress with its tongue hanging forward. The mandible was found to be fractured bilaterally behind the incisors, with such free movement that union was not possible. There was no breach either of the skin or the oral mucosa so that the fractures remained uninfected. The contents of the stomach consisted mainly of pine needles and twigs, from which it may be deduced that, being unable to graze with its incisors, the creature had browsed on tree foliage using its cheek teeth.

Although it is not firmly established that brittleness of the bones is a feature of fluorosis (Shupe and Olsson, 1980), it may be significant that Karstad (1967) found an animal with two fractures of the mandible, one healed and the other unhealed and recent, among white-tailed deer (*Odocoileus virginianus*) with high bone-F content and fluorosed pitted enamel (Chapter 20, p. 454). The animals were from the vicinity of an industrial complex.

In the mandible of an oryx shown in Figure 18.14, there was a fracture across the anterior end just behind the canines. It seems likely that there was suppuration; at any rate, osseous union did not follow, but there are a number of peg-like projections from the anterior fragment which fit into socket-like cavities in the opposite fractured surface so closely that the surfaces are firmly keyed together, even though there is no actual bony union.

Figure 18.15 shows a comminuted fracture in a wild reedbuck which has not united. There is considerable formation of bony callus, though with a cleft running through it, suggesting union largely by fibrous tissue but with some interlocking of opposing bony irregularities. The radiographs show, however, that embedded in the middle of this bony mass are large splinters of the original mandible which contribute to the interlocking union of the parts (Fig. 18.15B). It seems certain that this lesion was of at least some months' standing and that sufficient function persisted to enable the animal to sustain itself. In fact, had the animal not fallen to a rifle bullet, it seems likely that much more complete resolution would have occurred.

Bland Sutton (1885c) described the skull of a rabbit in which there was a pseudo-arthrosis between the un-united surfaces of a fracture of the mandible. The consequent malocclusion had led to overgrowth of the lower incisors and some of the molars.

Van Bree and Duguy (1970) described a bottle-nosed dolphin (*Tursiops truncatus*) with a pseudo-arthrosis that developed at the site of a fracture across the tooth-bearing part of the mandible on one side. The posterior fragment appeared to be partly enveloped by a new formation of bone, that is by an involucrum of the kind that occurs in osteomyelitis. They also described a male porpoise (*Phocoena phocoena*) which was found stranded and had a fracture across the anterior end of the mandible on one side. There was but little displacement and the fragments were held together by callus.

Fractures are said to be particularly common in the orang-utan when these heavy animals fall from trees. In the orang shown in Figure 18.16, the mandible has been fractured in three places; there is an oblique fracture through the anterior part of the body and fractures through both rami. The facial portion of the cranium has been completely separated from the brain case, the type of injury being similar to that seen at times in human beings as the result of lift accidents. The peculiar features of this specimen is the formation of new bone along the fractured edges of the mandible

Fig. 18.10. *Dasyurus hallucatus* (satanellus). ♂. BMNH 1904.1.3.94. Mandible with a slightly malunited fracture of the right side. ×c. 0.5.

and the total absence of a similar formation in the facial portion of the skull.

Fraser (1946) mentioned a healed fracture of the mandible of a white-beaked dolphin (*Lagenorhynchus albirostris*) close to the articular condyle, in consequence of which the affected side of the jaw was shorter than the uninjured side and the teeth did not occlude properly. An incompletely repaired fracture of the mandibular ramus of a bottle-nosed dolphin (*Tursiops truncatus*), just behind the tooth row, was referred to by Fraser (1953).

According to Kumar, Singh, Manohar and Nigam (1977), an inherent weakness of the mandible in the camel renders it liable to fracture, either during violent behaviour of the male during rut, when the animals are given to biting inanimate objects violently, or through

Fig. 18.11. *Pan troglodytes* (chimpanzee). P-C. Mus. Cam I 228.
A: Anterior view (left) and lateral view (right) showing a slightly malunited fracture of the mandible in the region of the left canine tooth with much callus formation. The anterior view shows that the fracture has united with the greater part of the mandible deviated to the left.
B: The mandible from above. The left canine and premolars were lost in the injury.
C: Radiograph of the mandible corresponding to B and showing the over-riding of the healed fragments.

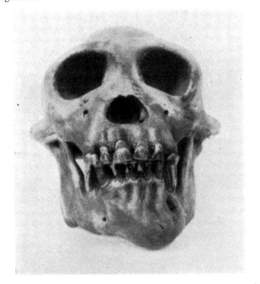
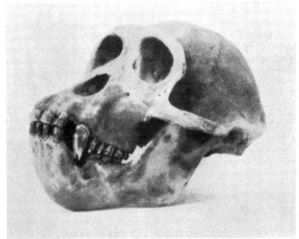

A

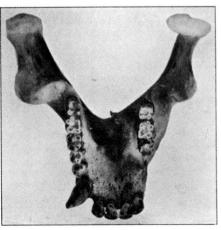

B

C

386 Section 4: Other Disorders of Teeth and Jaws

traumatic ill-treatment by their owners. Non-union of these fractures is common and can be dealt with, as shown by Purohit, Dudi, Chouhan and Sharma (1984), by excision of the anterior fragment. The exact post-operative functional state is not clear but the animals appear to be able to browse effectively with lips alone.

Fig. 18.12. *Vulpes vulpes* (red fox).
A: Front part of the mandible bearing the incisors and canines has been broken and displaced downwards and backwards beneath the rest of the mandible and united there. A mass of callus has formed over the free fractured ends (Breuer, 1932).
B: Right side of the mandible. (1) Callus in the line of fracture running to the end of the canal made by the bite of the opponent. (2) Posterior mental foramen. (3) End of the canal caused by the bite of the opponent. (4) Anterior mental foramen, with socket of P_1 directly over it. (5) Mass of callus over the fractured end.
C: The mandible from above. The masses of callus are seen in front. (1) Socket of the left P_3. (2) Cleft of the fracture split off from the inner side of the left ramus. (4) Fractured end of the symphysis. (5) Socket of the right P_3.

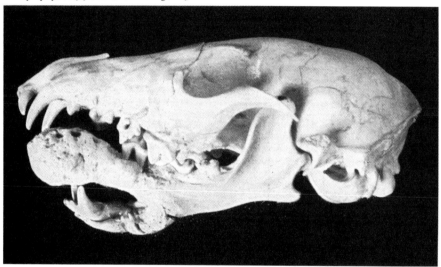

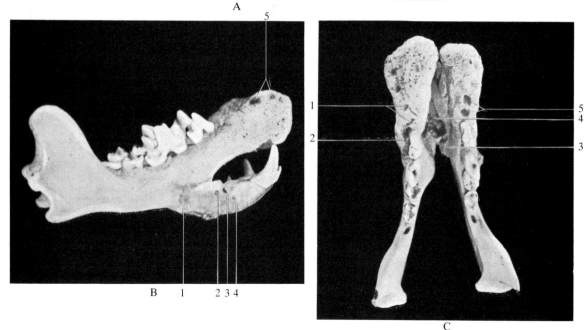

Doerr and Dieterich (1979) mentioned finding one healed fracture among 2440 mandibles of Alaskan caribou (*Rangifer tarandus*).

Hickman (1979) described the mandible of a male white rhinoceros (*Diceros simum*) in which, in the line of an old healed fracture, the P_2 was completely inverted so that the ends of the roots were in occlusion with the apposing teeth and had been so for long enough to produce considerable wear of the roots.

Schultz (1941) referred to a healed fracture of the mandible in an orang-utan; he also (1944) figured an example of good union of a fracture through the mandibular symphysis of a gibbon and referred to two other healed fractures of the mandible, both associated

Fig. 18.13. *Gorilla gorilla* (gorilla). ♂. RCS Odonto. Mus., G 82.1.
A: Lateral views of the jaws. There is non-union of a fracture of the left side of the mandible in the region of the canine. The molar occlusal relationship on that side (right Fig.) is normal but, on the other side (left Fig.), the relationship is abnormal; the lower molars have migrated mesially. There is a gap between the vertical ramus and the distal aspect of M_3.
× c. 0.75.
B: Upper surface of the lower jaw. The site of the fracture, the loss of incisors, canines and premolars, and the mesial position of the molars on the right side can be seen. × c. 0.36.

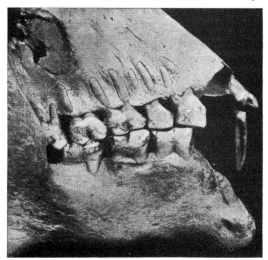
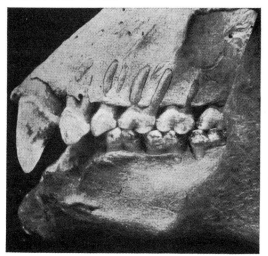

A

B

with multiple fractures of ribs and long bones. Examples from the wild of repaired fractures of mandibles and zygomatic arches in the New World monkeys *Alouatta*, *Cebus* and *Ateles* were described by Schultz (1960), the majority in males. Hershkovitz (1970) briefly described a male (*Saguinus oedipus geoffroyi*) tamarin in which there was apparently good resolution of a gunshot wound which destroyed the coronoid process and part of the condylar process on one side of the mandible.

The examples described here do not show, of course, the usual outcome of untreated fractures; they merely illustrate what the tissues are capable of.

Because of the firm attachment of the alveolar mucosa to the jaws, fractures through the tooth-bearing parts of the jaws are usually compound; that is the mucosa is torn so that there is ingress of bacteria from the mouth. Hence, and also in gunshot fractures, healing may be interrupted by infection. Osteomyelitis may occur as a complication of severe injuries but it seems that often the animal eventually recovers.

An example of survival following severe injury combined with infection is the mandible of a peccary (*Tayassu*) described by della Serra (1955b). A bullet had caused a fracture through the left horizontal ramus and had also injured and perhaps perforated the lower border of the right ramus. There was partial union of the fracture without much displacement but with the formation of much irregular callus in which there were a number of holes indicative of suppuration. The fracture site and the site of injury on the opposite side were joined by a bridge of bone below the level of the floor of the mouth, formed it may be supposed in the track of the bullet and in relation to the displaced splinters of bone.

E. Becker (1970, p. 268) figured some treated fractures of the mandible in horses and dogs.

A remarkable deformity of the mandible occurs in the sperm whale (*Physeter catodon*). The underslung mandible is long and slender and bears up to about 30 conical slightly-recurved teeth on each side. The anterior extremities of the two halves of the mandible are particularly slender and are joined over the distance corresponding to about two-thirds of the row of teeth by a greatly elongated symphysis. There are no functional teeth in the upper jaw so that the lower teeth close into shallow pits or sockets in the mucosa around the margins of the upper jaw. The deformity consists of a sharp lateral bending, or even spiral twisting, of the anterior end (Fig. 18.17). The condition was first described by Beale (1839) who had seen two examples and who reported that the condition was well-known to whalers who regarded it as the result of fights in the rutting season between adult males and younger male challengers in which the whales 'rush head first one upon the other, their mouths at the same time wide open, their object appearing to be the seizing of their opponent by the lower jaw, for which purpose they frequently turn themselves on the side; they become as

Fig. 18.14. *Oryx beisa* (Beisa oryx). RCS Odonto. Mus., G 81.3. Fracture of the mandible in which, although union is fibrous because the parts can be separated, as in the right view, peg-like projections of bone on the anterior fragment fit into socket-like cavities on the opposite surfaces (Colyer 1907). ×c. 0.65.

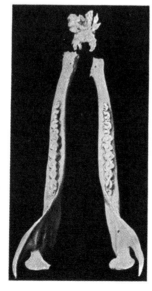

it were locked together, their jaws crossing, and in this manner they strive vehemently for mastery'. Another eye-witness account by a whaling-master, given to R.C. Murphy (1947), confirms the locking of the jaws in combat. According to L. Watson (1981), breaking off of teeth at the gum margin is one of the lesser consequences of these combats. Shaler (1873) described a dying sperm whale in which the lower jaw had been almost completely torn away and Casinos, Fitella and Grau (1977) described another which, at least for a time, had survived the tearing off of the rostral end of its mandible. Yet another, in which about 30 cm of the rostral end of the mandible had been torn off and healed over, was referred to by Murphy (1947). All three of these accounts concern males, which are always twice as large as the females.

Two more anecdotes from the literature of commercial whaling seem reliable enough to quote. Bullen (1898) gave a graphic account of a small whaling boat that was attacked by a sperm whale which seems likely to have crunched the boat in its jaws had its lower jaw not been deformed. 'At a short distance from the throat it turned off nearly at right angles, the part that thus protruded being deeply fringed with barnacles and plated with big limpets.' Bennett (1931) referred to a solitary old bull sperm whale that was found to have a fractured lower jaw which had re-knit but not in a straight line. Some teeth at the site of the fracture had

Fig. 18.15. *Redunca arundinum* (reed-buck). ♂. RCS Odonto. Mus., G 82.122.
A: Partially united fracture of the mandible. There is a great deal of dense callus through which there is a distinct vertical cleft. Nevertheless there is rigid union of the parts. × c. 0.68.
B: Radiograph showing, embedded in the callus, a large splinter of the original bone which appears to contribute to the interlocking of the parts. × c. 0.75.

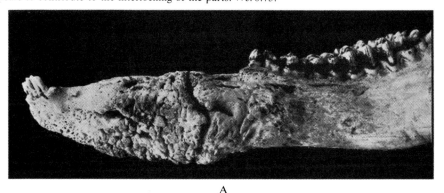

A

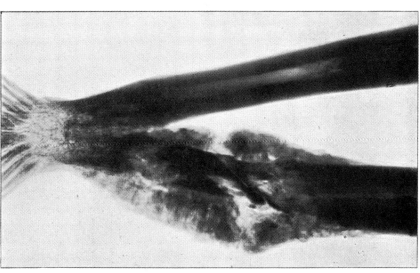

B

been broken across but were partly re-united. One had formed 'an independent wall of bone around the break, acting as a kind of splint'.

The manner in which sperm whales feed (Nakamura, 1968) offers another possible way in which the lower jaw could be injured. According to Scammon (1874), the whales seeking food swim with the mouth open and the lower jaw almost vertically downwards and may skim along the sea bed disturbing sediments and thereby may injure the jaw. Slijper (1979) added the colourful suggestion that the cuttle-fish, which are the principal prey, are attracted by the contrast between the whale's purple tongue and its white rows of teeth.

There seems little doubt that another way in which sperm whales can be injured is by encounters with submarine cables. Heezen (1957) described in detail a sperm whale found with two turns of cable around its

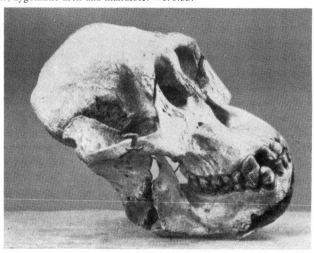

Fig. 18.16. *Pongo pygmaeus* (orang-utan). ♂. Formerly RCS Osteo. Series, 57. Multiple fractures through the orbits, zygomatic arch and mandible. × c. 0.33.

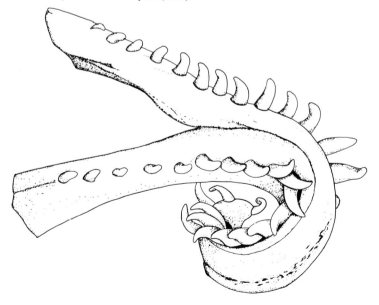

Fig. 18.17. *Physeter catodon* (sperm whale). Adult. ♂. Deformed lower jaw. The rostral extremity is coiled round beneath the left side of the jaw. A barnacle was attached to a tooth on the extremity. Drawn from Spaul (1964).

lower jaw and another around its head when a submarine cable was pulled up by a cable ship searching for a defect. He collected records of a further 13 instances. It seems that, especially if cables have been repaired, they may be loosely coiled on the sea bed and sperm whale swimming with open mouth over the sea bed can get first its jaw caught and then, in its struggles, may become more grossly entangled and, of course, drown; some might injure the jaw in this way and yet survive. Heezen mentioned one instance of a humpback whale being found entangled in a cable.

Nearly forty examples of sperm whales with this deformity have been recorded, most were listed by Clarke (1956) and Nakamura (1968). Some are preserved at the British Museum (Natural History), including one described by Spaul (1964, – BMNH 1965.3.10.1) (Fig. 18.17) and one by J. Murie (1865, 1867), who also described one that was in the possession of the Royal College of Surgeons (Physiological Series, 2452) but suffered destruction in 1941. Others (Thomson, 1867; Pouchet and Beauregard, 1889) probably still exist in the museums of the whaling areas of New England. A specimen described by Fischer (1867) is in the Muséum national d'Histoire Naturelle, Paris (A 3237).

It would be supposed that, if the deformity were produced by injury in the way described, the specimens would bear evidence of healed fractures because such acute bends or twists could not occur without splitting the bone. Several writers have failed to find convincing evidence of fracture. Many of the specimens are recorded as being males and some were diagnosed as male on the basis of the number of teeth borne by the jaw, which cannot be a completely reliable criterion; one of those in the British Museum (Natural History) is identified on its tooth numbers as female (J. Murie, 1867) and Nakamura (1968) listed seven females among two examples sexed on general body features. Whales with deformed jaws are said to be usually of normal size and weight so that the deformity appears to be no handicap in their search for food (Ash, 1954). However, it is said that the seas are so rich in squid that the whale normally depends but little upon its dentition, and indeed the teeth of the sperm whale do not erupt until several years after weaning.

J. Murie (1867) described the tooth of a sperm whale which was in the Staatliches Museum für Naturkunde in Stuttgart which was 'curved like a shepherd's crook, the tip of the crown and the end of the fang almost meeting'. It seems likely that it, and a similar one figured by Gaskin (1972), came from a deformed jaw.

Barnacles attached to a deformed jaw of a sperm whale have already been referred to. It is well known that barnacles commonly fasten themselves to some part of the surface of whales, including the baleen plates in whalebone whales (Clarke, 1966) and that the bases of the teeth in toothed whales are commonly parasitized in this way, usually by the stalked barnacle, *Conchoderma auritum*, which has a preference for hard surfaces such as shells, including those of the sessile barnacle, *Coronnula*, which does colonize the skin of whales. The areas of teeth affected are usually those where the parasites remain more or less undisturbed during closure of the jaws. For instance, Rice (1963) found *C. auritum* attached to the anterior teeth of the lower jaw in three out of the eight specimens of beaked whales (*Berardius bairdii*) that he examined. In this species, the dentition consists only of four teeth situated on the anterior tip of the lower jaw which normally projects beyond the upper, leaving the teeth exposed and out of occlusion. Nansen (1925) referred to receiving the tooth of a bottlenosed whale (*Hyperoödon ampullatus*) which had several large stalked *Conchoderma auritum* attached to its tip.

Bottlenose whales have dentitions similar to *Berardius* but there are only two functionless teeth at the end of the lower jaw. Many of the examples of heavy infestation of teeth with *C. auritum* are in the sperm whales with deformed jaws described above where the teeth are out of occlusion (reviewed by Clarke, 1966) and it seems that barnacles particularly colonize the region of the angle of bent jaws. Ash (1954) suggested that the bend creates a patch of dead water where young barnacles are less likely to be swept off. Once firmly attached and grown large, they are resistant to detachment. According to Ash, sperm whales are slower swimmers than many other whales and this may increase the chance of barnacles settling.

The susceptibility of teeth out of proper occlusion in odontocetes to be colonized by barnacles is illustrated by a spotted dolphin (*Stenella graffmani*) examined by Perrin (1969). Clumps of barnacles (*C. auritum*) were attached to the teeth in two areas. There was a malunion of a fracture of the mandible so that many teeth, including those colonized by barnacles, were out of occlusion.

In view of the bizarre curvatures of the jaws described in the sperm whale, it is interesting that, according to Busch (1892), the snout of the odontocete, *Inia geoffrensis*, is normally distinctly upcurved and is commonly also asymmetrically curved to left or right.

Fracture of the zygomatic arch

This class of fracture is of interest because it illustrates the effect of uncounteracted muscular action on the shape of the cranium and the jaws. An example of this type of injury is shown in Figure 18.18, the skull of a bear. The injury seems to have been a gun shot which fractured the palate and swept away the entire zygomatic arch on the right side. In the mandible, the posterior portion of the right coronoid process and the outer two-thirds of the condyle were destroyed. There was a fracture through the base of the anterior main part of the coronoid process but with good union and but little displacement. The effect on the skull and palate is instructive. The portion of bone shot away includes the whole of the origin of the masseter and part of the insertion of the lateral pterygoid muscle. The loss of these muscles has caused an unequal balance of muscular action which has resulted in a definite bend of the cranium. The line of flexion is about 25 mm posterior to a line connecting the two supra-orbital processes of the frontal bone. In the portion of the skull anterior to this line, the junction of the bones is not disturbed, but posterior to this line, the suture between the two parietal bones shows a definite curve towards the injured side. The bend of the cranium is well seen in the basal view (Fig. 18.18). The bend suggests that the injury restricted growth on the right side. A remarkable feature of the specimen is the formation of a new joint. The original glenoid cavity has been destroyed and the medial part of the condyle, which survived the injury, was displaced medially between the post-glenoid process of the squamosal bone and the pterygoid plate where the new concave joint surface has been formed.

There is also a healed perforating injury in the right parietal region. A hinged bone flap enables the intracranial surface of the bone to be seen; a cluster of stalactite-like spikes of bone projects from the healed inner surface.

Fractures of the zygomatic arch are among the commonest of 33 fractures in gorillas listed by Randall (1943/1944). Schultz (1958) mentioned repaired fractures of the zygomatic arch in two male colobus monkeys in the wild.

Other jaw conditions

Arthritis of the mandibular joint is sometimes connected with injuries but it is beyond the scope of this chapter, or indeed of this book, to deal systematically with this condition, or with the various disorders that produce bony changes in the jaws, some of which may have connections with injury. However, some reference will be found to the bone dysplasias and osteodystrophies (Chapter 16, p. 333–334; Chapter 25, p. 601 *et sequ.*) and so-called lumpy jaw (Chapter 23, p. 515–516, 518–520).

The papers of Schultz (1944, 1956, in particular) contain important examples of bone disorders in

Fig. 18.18. Probably *Thalarctos maritimus* (polar bear). RCS Odonto. Mus., G 81.1. Left: view of skull from above. Right: view from below. Right zygomatic arch lost from injury. × c. 0.2.

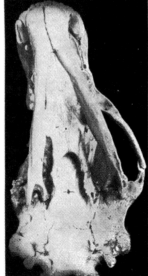

primates, even of yaws, a syphilis-like specific infection that produces destruction of bone. The importance of this condition, for instance, in respect of the main themes of this book, is that, when the jaws are affected, the changes have to be distinguished from some of those dealt with here. This is exemplified by the mandible of a female gorilla. The anterior part of the alveolar process has been destroyed, according to Wallis (1934), by a gunshot wound. However, Schultz (1956) thought that yaws could have been the cause of the bone destruction.

Jaw injuries in non-mammalian vertebrates

The powers of tissue regeneration are much greater in reptiles than in mammals; many are able to generate digits, limbs and even parts of the rest of the body. Regeneration of the tail in lizards is particularly well-known.

The crocodilians are not among those that regenerate limbs but they do survive severe injuries and regenerate tissue to an exceptional extent. This is illustrated by a captive adult male marsh crocodile (*Crocodylus palustris*) described by Brazaitis (1981) which suffered a bite from another crocodile which excised a large part of its upper jaw bearing five teeth. Within two years, there was almost complete restoration of the lost part with five teeth in function though slightly misplaced. The regeneration of so much lost jaw is even more remarkable than the regeneration of the teeth because the reptiles are polyphyodont and no doubt the dental lamina responsible for producing a continuous succession of teeth survived the injury. Incidentally, it cannot be assumed that in reptiles the cellular responses in healing and inflammation are the same as in mammals (Elkan, 1981). Furthermore, it seems that the formation of secondary cartilage does not occur in vertebrates other than birds and mammals (Hall and Hanken, 1985).

Among the many examples of recovery from injury, ranging from lost limbs to scars on head or trunk, found by Webb and Messell (1977) in 1345 estuarine crocodiles (*Crocodylus porosus*) caught, examined and released, was an adult which had lost part of the anterior end of the snout. Less dramatic were several examples of jaw misalignment; in three, the lower jaw was skewed to the left; in four, the lower jaw was short so that the anterior teeth did not occlude.

Biddulph (1937) described an adult estuarine crocodile with a partially healed fracture of the tooth-bearing part of the lower jaw; there was slight ventral displacement of the anterior fragment.

Many species of lizard are very aggressive and consequently tend to suffer injuries. Evans (1983) found signs of healed or healing fractures in about 7% of 300 fragmentary fossil lower jaws of the lower Jurassic lizard-like reptile, *Gephyrosaurus bridensis*.

The capacity of amphibia to regenerate lost parts is considerable, as is illustrated by the following:

A leopard toad (*Bufo pardalis*) which had survived severe injuries to the head was observed living, apparently successfully, in the wild despite being blind (Lambiris, 1978). The injury had destroyed one eye and all the face anterior to the eyes, except the lower jaw. Skin from the sides of the head had grown forward to cover the wounded tissues leaving a rigid-walled opening in place of a mouth through which the toad must somehow have gathered food.

It goes without saying that jaw injuries from fish hooks are common in game fishes such as salmon; healing can occur even though part of the upper or lower jaw is torn away, though such fish are usually in poor condition (Fulmer and Ridenhour, 1967).

Many fish, including the carp, have protrusible mouth parts which may be susceptible to injury. Menzel (1974) described a carp (*Cyprinus carpio*) that had survived total loss of the mouth mechanism and one eye from injury. No mouth opening could be found so it was assumed that the fish, normally a detritus feeder, had survived on food detritus that had crossed the gills during respiration.

CHAPTER 19

Injuries of the Teeth

Introduction
Injuries of the teeth are common in animals, and their study is both instructive and interesting. To facilitate description, these injuries will be divided into two groups; injuries to developing teeth and injuries to teeth that have erupted, further divided into (a) teeth of continuous growth and (b) teeth of limited growth.

Injuries to developing teeth
The effect of injury on a tooth in its crypt depends upon the stage of growth and the character of the injury. If the tooth is partly formed, for instance when the crown is formed and mineralized and the root is about to be formed, the crown may be displaced by an injury leaving the formative soft parts still in position but, because the formed and unformed parts are still connected, a bend is introduced between the two which may be quite acute. When recovery occurs and formation recommences, the bend and any associated deformity of the soft parts will be reproduced in the completed tooth. Such specimens often look as if they have been divided and then stuck together; hence this condition is often categorized as dilaceration.

Figure 19.1 shows an example of this condition. The left I^3 of a captive, sub-adult polar bear is lying horizontally along the floor of the nasal fossa covered only by a thin layer of bone. About one-third of the crown of the tooth had been mineralized at the time of injury and had been bent backwards, forcing it slightly into the as yet unformed portion (Fig. 19.1B). Although the injury was severe enough to disorganize a zone of the formative tissues, complete recovery seems eventually to have occurred because the shape of the greater part of the later-formed tooth is normal and its overall length is about the same as that of the corresponding tooth in the opposite jaw. The right maxillary canine is affected slightly by a similar disorder; whether by the same injury or another is impossible to say. This canine has not completely erupted and there is a zone of imperfect enamel and narrowing of the crown at about its middle (Fig. 19.1C). There is a slight enhancement of the normal curve of the root with a depression on its mesial surface and an elevation on its distal surface.

The absence of the upper incisor from the arch and the partial eruption of the canine have resulted in a modification of the normal relationship of the teeth. The incisors have shifted towards the left and the pulp cavity of the outermost incisor on the right side has been exposed owing to excessive attrition from abnormal contact with the mandibular canine. This specimen was first described by Colyer (1913).

A similar condition affecting the right I^2 of a baboon is depicted in Figure 19.2. The tooth was unerupted and lying horizontally in the palate with its crown facing forward. It must have been injured and slightly displaced when about one-half of the crown was formed. Recovery occurred with the completion of the rest of the tooth, which is only slightly shorter than

Fig. 19.1. *Thalarctos maritimus* (polar bear). ♂. Captive. RCS Odonto. Mus., G 98.2.
A: A left deformed I^3 lies deeply in the palate with crown pointing forward and has been exposed by the removal of a window of bone. The right canine shows a slight deformity and has not completed its eruption. × *c*. 0.5.
B: The incisor removed from the jaw. The crown is above.
C: The injured right canine is on the left and the normal one on the right. (Colyer, 1913.).

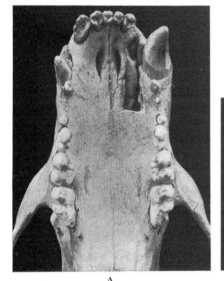

A   B

C

Fig. 19.2. *Papio ursinus* (Chacma baboon). ♂. Juvenile. Captive. RCS Odonto. Mus., G 96.5.
A: A deformed I^2 lies horizontally in the palate with crown facing forward and has been exposed to view by the removal of a window of bone. × *c*. 0.66.
B: The incisor removed from the jaw. The crown, which shows a transverse zone of injury, is above.

The molars show various stages of caries which has commenced in the occlusal surfaces. Primary caries of enamel. (See Chapter 21, p. 472.)

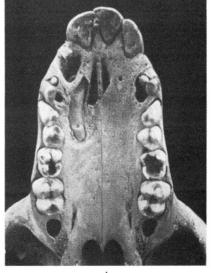

A   B

its fellow of the opposite side, though the later-formed part shows some irregularity in shape and there are patches of enamel hypoplasia. The apical part of the root is of more normal morphology so that recovery of the formative tissues from the injury must eventually have been nearly complete.

Injury may result in separation of some of the formative cells from the main portion; a separate denticle or one attached to the main tooth may then be formed. An example of this condition in a tantalus monkey is shown in Figure 19.3. The first-formed, incisive third of the crown of this incisor is irregular in shape and attached to it is a small denticle covered with enamel. The rest of the crown is more normal in shape but the root is stunted. Direct evidence of trauma is not present in this instance, nor in many others, but it seems a reasonable deduction to make.

It appears that trauma can sometimes produce supernumerary roots. Figure 19.4 shows a deciduous

Fig. 19.3. *Cercopithecus aethiops tantalus* (tantalus monkey). ♀. Juvenile. Captive. RCS Odonto. Mus., G 96.61. The right I^1 (exposed by bone removal) is malformed (the left tooth has been lost from the specimen). Enlarged.

Fig. 19.4. *Macaca mulatta mulatta* (rhesus monkey). A: Bone has been removed to display an unerupted m^4 which shows deformity possibly due to trauma. B: The deciduous molar removed from the jaw. Its components are interpreted as follows: (a) the main body

upper molar in a rhesus monkey in which a supernumerary root projects at an angle of about 90° to the main axis of the tooth, from what seems to be an enamel-covered displaced part of the crown. Such specimens have to be distinguished from those in which anomalous development produces grotesque forms of teeth, many of which, especially if they show signs of excessive uncontrolled growth, are categorized as odontomes (Chapter 25). Often the decision that the anomaly is of traumatic origin is made on the basis that the first-formed part of the tooth is normal and the later-formed part appears to be joined on at a faulty angle, with a zone of disturbed growth and defective formation at what can be called the join, as in this specimen.

Traumatic injury to a developing tooth can simply arrest its growth, though usually it produces some distortion as well, as in the unerupted canine in the mandible of a glutton (Fig. 19.5). The crown is normal but the root is short and globular; the abruptness of the change to globular form suggests injury. A radiograph (Fig. 19.5 B) shows that there is no well-defined pulp cavity. There has been some resorption of the surface of the crown with deposition of new bone fusing the tooth to the jaw (ankylosis) (Colyer, 1915).

The irregular mass of tooth between the right orbit and nares of a chimpanzee in Figure 19.6 has a recognizable crown and root and resembles an ill-formed canine tooth. Springing from its lower border, there is a curved tapering projection of tooth substance. The canine is missing from the dental arch and it is possible that the mass represents that tooth driven up by some injury when it was still a tooth germ.

The incisor of a horse in Figure 19.7 shows

of the crown; (b) the mesio-buccal root; (c) the disto-buccal root; (d) the palatal root; (e) a supernumerary root derived from what appears to be a displaced portion of the crown. ×c. 2.

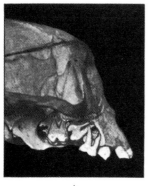

A

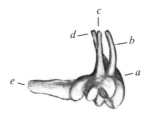

B

arrested growth which began when more than half the length of the tooth had been formed and may have been caused by trauma to it soon after it erupted. The tooth, instead of possessing a normal tapering root, is diminished in length and has a bulbous termination. The pulp chamber is open on the incisive edge. The incisors of horses are subjected to considerable attrition, the pulp progressively forming reactionary dentine. In this case, with the disappearance of the mark from attrition, the pulp chamber was exposed but infection of the bone did not follow because the apical foramen had become occluded by the irregular formation of the root.

When the injury is more severe, the tissues may become disorganized and yet recovery often occurs. The leopard shown in Figure 19.8 was injured when the deciduous canines were still in position; the tips of their crowns were broken off and their pulp cavities exposed. The unerupted permanent canines show signs of being injured by being driven up by the blow that had broken their deciduous predecessors (Fig. 19.8B). A ground section of the left tooth (Fig. 19.8C) showed evidence of the formed crown having been displaced upwards into the soft formative tissues which it seems were compressed concertina-fashion, because that is the pattern of the deformity it now shows.

Fig. 19.5. *Gulo gulo* (glutton). RCS Odonto. Mus., G 97.1.
A: Mandible with an unerupted malformed canine exposed by the removal of a window of bone.

B: Radiograph. The light zone around the unerupted canine is where the window of bone was removed. (Colyer, 1915.)

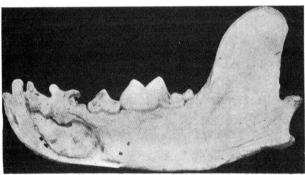

A

B

Fig. 19.6. *Pan troglodytes* (chimpanzee). ZMB 16969. There is an irregular mass of tooth tissue between the orbit and the anterior nares.

Fig. 19.7. *Equus caballus* (horse). RCS Odonto. Mus., G 96.4. Portion of the mandible from which bone has been removed to show a stunted incisor. (Colyer, 1915.) × c. 0.66.

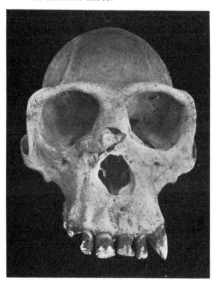
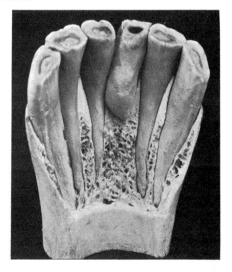

398 Section 4: Other Disorders of Teeth and Jaws

Figure 19.9 shows three views of a lion's maxillary canine with a deformity which is also likely to be due to a compression injury to the developing tissues above the partly-formed tooth. The crown is of normal morphology but, above that level, there is an abrupt change to an irregular mass of tooth tissue. A section through the tooth (Fig. 19.9B) shows a trabeculated mass of irregular dentine in what was the pulp cavity within the formed crown at the time of injury. With further recovery from the injury, a more compact layer of dentine was laid down which seems to have walled off the less-organized irregular dentine from the present pulp cavity in the root which has a thin wall of much-folded dentine. The deformity of the developing tissues seems to have been similar to that deduced from the leopard canine (Fig. 19.8). Embedded in the bone

Fig. 19.8. *Panthera pardus* (leopard). Probably captive. RCS Odonto. Mus., G 63.2.
A: The left side of the skull showing the injured upper deciduous canine. Bone has been removed to show the permanent successor with a deformed root. The condition on the opposite side was similar but the trauma appears to have been slightly less. × c. 0.5.
B: Two halves of the left permanent canine divided longitudinally. The deciduous predecessor is above with the crown facing left. Natural size.
C: A ground section of the permanent canine in B showing the concertina-like deformity of the tissues (Colyer, 1913, 1915.)

A

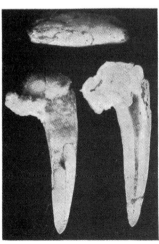

B

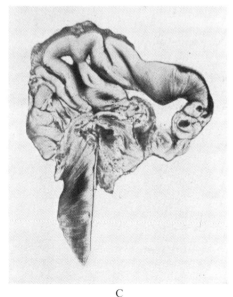

C

that surrounds this tooth, but separate from it, is a small mass of tooth tissue which is evidence that the pulp of the developing tooth can become isolated by trauma. Another feature of this specimen is the tapering projection arising from the base of the tooth; this tendency for the disorganized pulp to throw out branchlets is discussed in detail in connection with injuries to elephant tusks (this Chapter, p. 406 and following).

A remarkable instance in a male lion of deformity

Fig. 19.9. *Panthera leo* (lion). ♀. Probably captive. RCS Odonto. Mus., G 63.3.
A: Three views of a maxillary canine deformed by injury. ×*c*. 0.75.
B: The tooth has been divided longitudinally. (a) Irregular trabeculated dentine in the pulp cavity of the crown; (b) a wall of compact dentine separating (a) from the pulp cavity of the deformed root which has a thin wall of folded dentine (c).
The disorganized growth of dental tissues resembles that of a radicular odontome (Chapter 25, p. 595) (Colyer, 1915).

A

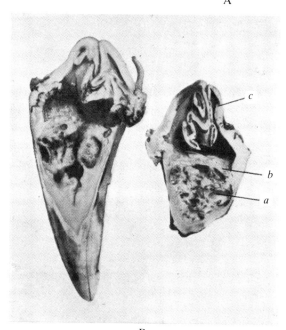

B

of all four permanent canines, as well as some incisors, possibly due to injury to their predecessors, was described by Colyer (1942a). The animal had lived at the Gardens of the Zoological Society of London for almost 13 years from the age of about six months. Death was ascribed to gangrene of the lung. In the upper incisor region on the left side (Fig. 19.10), there is a cluster of small toothlets many of which, on closer examination, are seen to be joined together; for example, one tooth unit consists of a double element and another consists of four elements. I^3 is more normal in size but most of the crown has been destroyed either by wear or fracture, and there is a widely-open pulp cavity. The left canine is unerupted; the tip of the crown is normal but the rest of the crown and the root are wider than normal. On the labial aspect of that part of the crown, there are three longitudinal grooves, two of which continue for the whole length of the root; these grooves give the tooth the appearance of being composed of two or even three elements (see connate

Fig. 19.10. *Panthera leo* (lion). ♂. Captive. RCS Odonto. Mus., G 98.11. The permanent canines are grossly deformed and unerupted, apparently due to interruption of their formation when about 1.0 cm of the tips of the crowns had been formed. Bone has been removed to expose the canines. Other teeth are abnormal.
A: Right upper jaw. At the site of interruption of growth, there is a spiky excrescence of dentine and the part of the tooth above shows two longitudinal grooves. ×c. 0.75.
B: Left upper jaw. The canine is similarly deformed; a longitudinal groove is better seen. ×c. 0.75.
C: Right lower jaw. The condition of the canine is similar but there are two ragged holes produced by resorption; one in the crown below the excrescence and the other in the root. The grooving of the root is obscured by cementum. ×c. 0.75.
D: Left lower jaw. The canine resembles that in the opposite jaw but there is much more resorption and much of the root surface is ragged with deep burrows extending to the pulp cavity (Colyer, 1942a). ×c. 0.75.

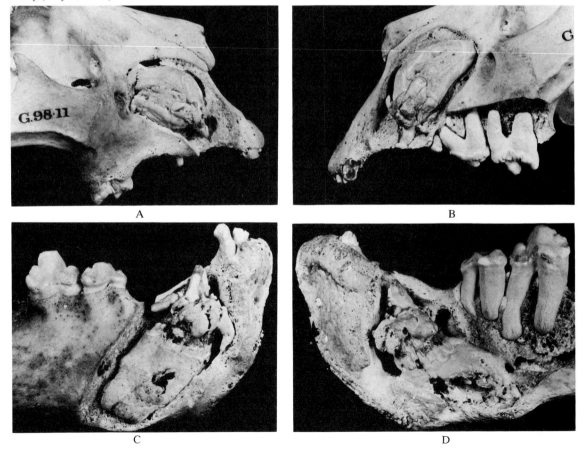

teeth, Chapter 1, p. 7). At the junction of the tip of this canine and the abnormal part just described, a spiky mass of tooth tissue, probably dentine, protrudes. P^2 and M^1 are not of completely normal morphology. They have the appearance of the mesial and distal halves having been slightly bent inwards towards each other.

The state of affairs in the right upper jaw is similar but the toothlets are worn down to the level of the bone and it is impossible to determine how many were joined to form larger tooth units. The right I^3 resembles that on the other side in being broken or worn away with a large pulp cavity exposed. The canine is similar to the one on the left; the tip is normal, there is a massive spiky excrescence and the longitudinal grooving is more marked than on the left side. The right P^3 is missing, though there is a shallow simple socket, and the right M^1 shows the bent morphology to a slight extent.

In the mandible, there are four stumps of what could have been normal incisors. The right canine is unerupted, the tip of the crown is normal and the condition of the rest of the tooth closely resembles the upper canines; there is a large spiky excrescence from the labial aspect of the crown, the root is short and has such a thick covering of cementum that its basic morphology is hidden although three vertical grooves can be discerned. Some sort of burrowing resorptive process, perhaps from within outwards, has affected the tooth and formed two ragged holes, one in a crown and one in the root.

The left mandibular canine is of similar morphology to the right and is unerupted. There has, however, been much more resorption which has produced a hole in the crown and led to a deeply-scalloped surface extending over the greater part of the labial surface of the root. The root appears in several places to be ankylosed to the bone. The spikes on the excrescent masses of all four canines point rootwards, suggesting that the teeth (only the tips of which would have been formed) were moving axially towards the mouth while the excrescences were being formed. There are two premolars only on each side of the mandible. They are of normal morphology. There are

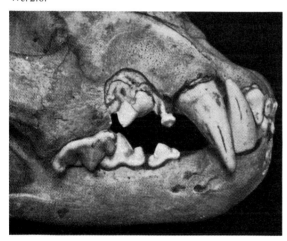

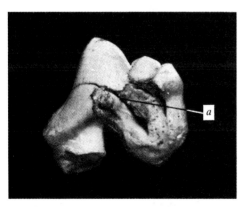

Fig. 19.11. *Panthera leo* (lion). ♀. RCS Odonto. Mus., G 102.11.
A: Bone has been removed to show a partly-erupted premolar which has been injured. Originally the only part of the tooth visible was the tip of the central cusp. ×c. 0.5.
B: The injured tooth shows signs of an oblique fracture at (a) where there is an enamel-covered area which seems to be a splinter of the tooth which has become re-attached. ×c. 2.6.

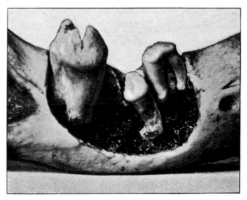

Fig. 19.12. *Panthera onca* (jaguar). Captive. BMNH 1858.5.26.9. Portion of the right mandible. Unerupted P_4 lacks the mesial and distal cusps and M_1 is malformed. An injury had broken off the lingual surface of P_3.

shallow sockets in the region of M_1 on each side; it seems likely that those teeth were lost only a few months before death.

Only one other feature deserves mention; the alveolar bone in the incisor regions of both upper and lower jaws is thickened but this probably has no connection with the disorder of tooth formation. A point in favour of trauma as the cause is that something appears to have happened suddenly to all four canines at the same time when about 1.0 cm of the tip of the crown was formed, which would be at about the end of the first year of infancy. The excrescences could have been formed by dental formative tissue displaced, or squashed, by the formed parts of the teeth being driven forcibly upon them. The deformity of the rest of the teeth could have been due to the trauma having permanently deformed the formative tissues. On the other hand, the widespread nature of the deformity, with the possibility that the cheek teeth are affected also and the almost strict symmetry of the condition of the canines, does not seem entirely in accord with trauma. However, all other possibilities that could be speculated upon, such as some genetic condition, seem even more unlikely.

In the skull of a lioness in Figure 19.11, the right P^3 was fractured at about the time that eruption was beginning. It is partly erupted and lying in a tilted position facing slightly distally. The injured tooth (Fig. 19.11B) shows a fracture starting from the mesial aspect of the main cusp and extending transversely towards the distal root. The enamel and part of the dentine covering the distal part of the tooth have been split off, disclosing two fractures which run in a direction oblique to the transverse fracture. The portion marked (a) is covered with enamel and is probably a small splinter of the outer part of the tooth which has been driven inwards and become re-attached. The growth of the pulp has been arrested and its surface covered with a layer of dentine and cementum. The splintered surface is rough and bears signs of resorption; the fractured surfaces show no signs of repair. There is no sign of inflammatory change or suppuration in the alveolar bone.

Another example of malformation of the teeth that Colyer believed to be due to trauma is in the jaguar shown in Figure 19.12. The lingual surface of the P_3 had been broken and the P_4 and M_1 are malformed. Normally, the P_4 has a large central cusp (protoconid) between two smaller cusps (paraconid and metaconid); in this skull, the right P_4, which is unerupted, has a well-developed protoconid, but the paraconid and metaconid are absent. The tooth has only one root instead of two, the end of which is thickened and ankylosed to the bone. The right M_1 is also malformed, the mesial and distal parts of its crown are as it were folded towards one another, the enamel covering the upper part of the crown is hypoplastic and the roots are fused together. Evidently, if trauma was the cause of these disorders, it must have occurred at the stage of development of the crowns of the teeth.

The crowns of premolar teeth develop in particularly close relationship to the roots of their deciduous predecessors; actually between the roots when the predecessors are multirooted, which is usually the case. As inflammatory conditions can develop on the roots of deciduous teeth, the question arises whether that can affect the development of the permanent successors. In man, where dental caries with consequent peri-radicular inflammation is common in the deciduous dentition, it is well recorded that such inflammation may produce localized defects (hypoplasia) of the enamel of some part of the permanent successor. Furthermore, trauma to the deciduous molar can displace the developing permanent tooth crown and produce dilaceration (see this Chapter, p. 397, Fig. 19.8). It might be supposed that, if the inflammation in connection with the deciduous tooth occurs at the early stage when the permanent tooth is a differentiating tooth germ, the future shape of its crown could be altered by the influence of toxins of infection, by inflammatory change reducing the blood supply to the tooth germ, or, for instance, replacement of tooth formative tissue by fibrous repair tissue.

In addition to the jaguar just described, the maxillae of a young rhesus monkey (RCS Odonto. Mus., G 68.411) provide another possible example. Both m^4 had been lost from disease and both M^1 had moved forwards. The unerupted P^4 were in an abnormal position and the crown of the right tooth was a simple cone. There was no proof that this malformation of the right P^4 related to the disease affecting the deciduous molar and it may be that it was a variation, that is a purely developmental defect.

The interpretation of museum specimens, usually without any helpful history, is highly subjective and largely dependent on comparison with other specimens where the evidence for a particular interpretation is stronger. This is illustrated by the following example which Sir Frank Colyer interpreted in the first edition of this book in 1936 as a malformation of probable genetic origin and placed it in what is now Chapter 3 but, in 1943 (b), he re-interpreted it as probably due to trauma, perhaps affecting the deciduous predecessors as described above. The specimen is that of a young adult, female colobus monkey (Fig. 19.13). The four upper incisors are abnormal in

Fig. 19.13. *Colobus guereza kikuyuensis* (eastern black-and-white colobus). BMNH 1912.7.1.1.
A: The upper tooth row; the incisors are abnormal in shape.
B: The labial aspects of the four incisors. The crowns are misshapen and the enamel surfaces are uneven and grooved.
C: Other aspects of the incisors. (a) Mesial surface of the right I^2; (b) lingual surface of the right I^1; (c) lingual surface of the left I^1; (d) lingual surface of the left I^2.
D: The lingual aspects of two normal I^1 from another skull for comparison.

shape but there are notable differences between the two sides of the body in the details of the abnormality. The labial surface of the right I^1 has an area of defective enamel near the incisive edge (Fig. 19.13B); the left I^1 has a groove running vertically from near the cervix to the incisive edge. The lingual surfaces of these teeth (Fig. 19.13C), but especially the left one, are very abnormal as can be seen by comparing them with the corresponding surfaces of two normal teeth in Figure 19.13D. The labial surfaces of both I_2 are of very abnormal shape (Fig. 19.13B), the predominant feature of the right being a vertical ridge and that of the left a transverse furrow. One of the principal features suggestive of trauma is the localized area of enamel hypoplasia on the labial surface of the right I^1 (cf. Chapter 20).

Colyer (1943b) described an I^1 in a Preuss's colobus (*Colobus badius preussi*. RCS Odonto. Mus., A 68.79) in which the mesial corner of the incisive edge appears to be missing. A longitudinal groove extends on the labial surface of the tooth from this defective corner up the whole length of the root. Colyer suggested that a part of the incisive edge may have been displaced slightly by trauma when no more than the incisive edge was mineralized. Associated displacement or damage to the formative tissues could persist and produce a groove.

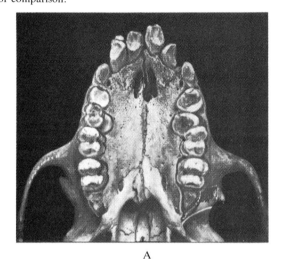

A

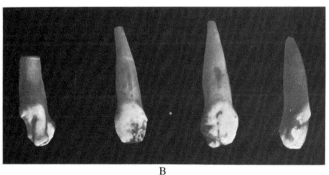

B

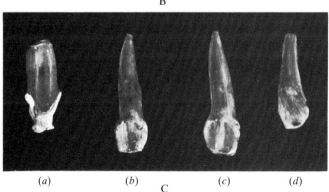

(a)   (b)   (c)   (d)
C

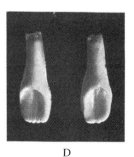

D

Section 4: Other Disorders of Teeth and Jaws

In Chapter 20, the effects are referred to of experimental mechanical damage to tooth germs in sheep, rats, guinea pigs and kittens. B. A. Levy (1968) described an irregular mass of dental tissues apparently formed from tooth-formative tissues detached by needle trauma in some experiments on rats. Following damage to tooth germs in neonatal guinea pigs and kittens by passing a needle through the skin over the jaw, Santoné (1937) found small masses of dental tissues, and on more than one occasion a small separate toothlet, adjacent to the teeth.

Injuries to erupted teeth

The effects of injury to teeth that have erupted depends upon the character of the injury and the type of tooth, in particular whether of continuous or limited growth. Teeth of continuous growth repair more actively and effectively than teeth of limited growth. There are several likely reasons why this is so: the pulps of teeth of continuous growth have a rich blood supply through the funnel-shaped growing ends of the roots; much of the pulp and tissues adjacent to the root consist of cells still actively reproducing themselves; active formative cells are constantly available to repair or replace damaged tissue.

(a) Injury to teeth of continuous growth

The injury may be insufficient to disorganize completely the formative tissue, but sufficient to alter the regular formation of the tooth and in some cases to cause the arrest of its growth. The portion of the tooth being developed at the time of injury usually has a crinkled appearance, the irregular formation increasing towards the growing end and the tissues showing signs of defective mineralization, as shown in the hippopotamus tusk in Figure 19.14. A similar example was described by Cave and Blackwood (1965).

The injury may be of sufficient severity to cause fracture of the tooth and in some cases repair is brought about by the cementing together of the fragments by a trabecular type of dentine which resembles the osteo-dentine found normally in some teeth, especially those of non-mammalian vertebrates. As an illustration, we may take the tusk of the hippopotamus (Fig. 19.15) described by Owen in his *Odontography* (1840–1845), as follows:

> The injury was indicated externally by the sudden constriction of the tusk at x, with an interruption in the enamel at that part, and irregular deposits of dentine both there and at the adjoining concavity of the tusk. A longitudinal section of the tusk showed the pulp cavity obliterated at the fractured part and for some distance below it, towards the base of the tusk, by a mass of osteo-dentine, deposited principally in the form of nodules closely impacted together, with their convex sides projecting into the re-established pulp cavity next the base; the general disposition of the osteo-dentine being very like that in the centre of the tooth of the great sperm whale (*Physeter*

Fig. 19.14. *Hippopotamus amphibius* (hippopotamus). RCS Odonto. Mus., G134.5. Portion of a mandible with an injured tusk in position. ×c. 0.2.

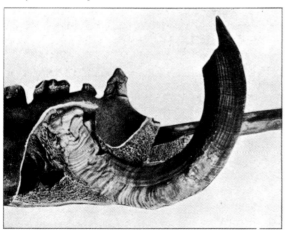

Fig. 19.15. *Hippopotamus amphibius* (hippopotamus). Formerly RCS Osteo. Series, 1903. A: Mandibular left canine showing dilaceration at x which demarcates the amount of the tooth formed at the time of the injury. B: longitudinal section through the tooth. (Owen, 1840–1845.)

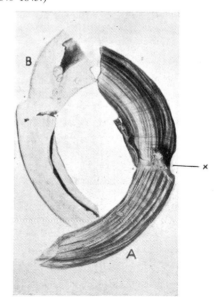

macrocephalus). The remains of the pulp cavity in the protruded part or crown of the tusk were unusually conspicuous in the form of a narrow canal near the concave side of the tusk, and opening, like a fistula, upon that surface beyond the fracture; another irregular slender canal extended transversely through part of the uniting substance and opened upon the concave side of the tusk.

Another example showing the unusual power of repair associated with the teeth of the hippopotamus is shown in Figure 19.16. This tusk was shattered near its growing end into several fragments which have become welded together by what may be cementum but seems more likely to be a new formation of dentine produced by a grossly damaged pulp.

It seems likely that such union of a fracture of a tusk could only occur if the parts were splinted, as they would be if the fracture affected the socketted part of the tusk.

Robins and Rowlatt (1971) found that broken incisors were common in a group of 600 aged mice, that is 22–33 months of age, compared with younger groups. There were broken incisors in 127 animals; in three instances, a newly-formed part of the tooth had erupted alongside, or among, the shattered remains of the original incisor. Injury is likely to have led to a period of cessation of growth of the tooth and, when formation recommenced, the newly-formed part of the tooth was not in continuity with the old and thus was free to erupt beside it (cf. Chapter 1, p. 7).

The tusks of elephants show some of the most remarkable and instructive examples of recovery of teeth of continuous growth from injury. The tusks are socketted into the upper jaw as shown in Figure 19.17.

The socketted portion is almost vertical and the growing end is situated just below the floor of the nasal cavity. This portion of the tusk contains a funnel-shaped pulp cavity the extremity of which in the adult animal reaches almost to the level of the margins of the bony socket. The tusk seems to be liable to two principal kinds of injury. Firstly, breaking off of the tusk, perhaps during combat or in falls, sometimes no doubt into concealed pits made to entrap the animals, and secondly damage from bullets and spears. When the tusk is broken off, the pulp may become exposed because in some cases the living pulp extends into the erupted part of the tusk.

Many specimens are recorded which show the remarkable powers of the enormous pulp to recover from such injuries. They attracted the interest of enquiring dentists, early in the history of science, before the microscope was in common use, because they presented examples of the behaviour of tissues that could be studied, interpreted and speculated upon with the unaided, naked eye (Miles, 1973). The earlier literature has been reviewed by W. D. Miller (1890–1891). Other specimens have been described by Busch (1890*a,b,c*), von Metnitz (1903), and Stannus (1911). There have also been recorded a number of instances in which tusks snapped off leaving the socketted portion *in situ* and in which a mass of ivory has been deposited in the pulp cavity sealing off the pulp from the external environment so that the formation of the tusk has continued, though in a disturbed manner (Miller, 1890–1891, 1899; Mitchell, 1903; Humphreys, 1926; Gainer, 1986). In many instances, the exposed part of the pulp must have died and abscesses may have arisen in connexion with the remainder; nevertheless the pulp has survived to go on forming the tusk which, although deformed, has continued to function. In such

Fig. 19.16. *Hippopotamus amphibius* (hippopotamus). RCS Odonto. Mus., G 134.1. An example of a comminuted fracture of a tusk which has healed. (Colyer, 1915.) × c. 0.4.

Section 4: *Other Disorders of Teeth and Jaws*

specimens, there is a tendency for longitudinal splitting or folding of the pulp to occur to form partially, or even entirely, separate small tusks.

Schulze (1957) described the sufferings of a 30-year-old Indian bull elephant in Leipzig Zoo from the stump of a tusk which had been split longitudinally; after several years, the tusk drooped at an angle because it had also been fractured transversely through the implanted portion. The tusk fell out the next day with a discharge of pus. Schulze also mentioned an elephant in Munster Zoo which broke its tusk, leaving the whole length of the pulp hanging free. The animal suffered so greatly that it had to be killed.

A specimen resulting from the fracture type of injury is shown in Figure 19.18. It is part of the tusk of an African elephant that remained in the jaw after the rest was broken off by a fall into a ravine, together with a portion added after the injury. The specimen is of particular importance because the interval between the fracture of the tusk and death was thought to be two years (Colyer and Miles, 1957). Thus, from the length of tusk formed after the injury, 340 mm, it was possible to calculate that the rate of growth of the tusk was approximately 3.3 mm per week which corresponds closely with 3.1 mm derived from a similar specimen (Humphreys, 1926) in the Odontological Museum of the Medical School, University of Birmingham, and can be compared with rates of growth of 1.3–4.3 mm per week recorded for the incisors of various rodents.

The specimen consists of two parts, distal and proximal, each with the form of a hollow cylinder and each with a proximal end with the funnel morphology of the growing end of the tusk. The distal cylinder overlaps the proximal one slightly and they are joined on one side by a mass of new irregular ivory which also closes the lumen of the pulp cavity in the proximal part from that of the fractured distal part.

There can be little doubt that the distal part **A** (Fig. 19.18) was the fragment of tusk originally left behind in the jaw after the rest had been broken off, and that the jagged distal end of this part was the site of

Fig. 19.17. *Elephas maximus* (Indian elephant). Longitudinal section through skull and part of tusk. After Owen (1840–1845). The inset diagrams show the probable condition of the tusk of the African elephant in Fig. 19.18 (a) immediately after the injury and (b) two years later.

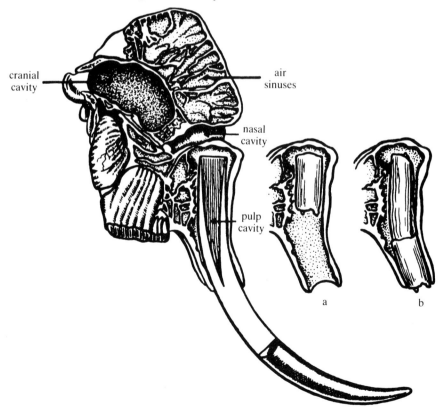

the fracture. The proximal part **B** must have been formed after the injury, gradually pushing **A** downwards until, to judge from the worn and stained appearance of the fractured edge, it projected slightly from the jaw as depicted in b of Figure 19.17b.

It may be supposed that the greater part of the pulp in the original tusk fragment became necrotic as a result of the injury but that the formative tissues in the opening of the funnel-shaped extremity of the tusk survived. At first a mass of unorganized dentine was formed, nearly filling in the pulp cavity and bridging over the raw surface of the pulp. This is the mass of ivory joining the two portions; in section it had an irregular gnarled structure. Subsequently, the formative tissues recovered and went on to form the part B which, as the other transverse sections showed, had a structure which became progressively more normal. The abnormal longitudinal fissure shown in Figure 19.18 can be attributed to the non-recovery of a small region of the formative tissue.

The wide funnel-shaped base of the implanted portion of the tusk, containing a pulp with a rich blood supply (Fig. 19.17), has to be borne in mind in interpreting such specimens. The portion of the tooth already formed may be dislocated on the growing portion, as in dilaceration, the injury being insufficient to destroy the existing formative cells, but sufficiently severe to arrest temporarily the growth of the tooth and perhaps to disorganize further growth.

Another tusk injured in this way is shown in Figure 19.19. There is a right-angle bend in the tusk and also crumpling of the dentine above the level of the arrow; the furrows between the corresponding ridges on the surface are filled with cementum. It is impossible to interpret this specimen reliably without knowing the orientation its principal axis had in the skull from which it came, but it is reasonable to suggest that the concave curved surface to which the arrow points is the anterior surface of the tusk and that the developing extremity, that is the short upper arm, would have been least likely to have been displaced. It follows that the part below the arrow was formed at the time of an injury which forced it sharply upwards together with some of the formative tissues attached to it, but leaving the primordial formative cells at the basal extremity of those tissues undisplaced; in other words, created a right-angle bend in the formative tissues. It must also be assumed that the direction of force also pushed the tusk upwards and compressed the formative tissues so that, when they recovered, they produced a length of crumpled tusk. However, when in due course it was, as it were, the turn of the undisplaced primordial cells to differentiate and produce the part of the tusk that is above the bend, they did so in the original growth axis of the tusk.

Tusks with double bends occur. The tusk shown in Figure 19.20 is normal in shape for the first 3 cm, then it gradually increases in girth for a distance of about

Fig. 19.18. *Loxodonta africana* (African elephant). RCS Odonto. Mus., G 122.71. The anterior surface of the tusk, showing (a) jagged fractured end; (b) fragment of distal part of tusk that has been turned down to show inner surface and irregular mass of ivory; (c) dotted line indicating original position of fragment; (d,e,f) positions of transverse sawcuts. (Colyer and Miles, 1957.)

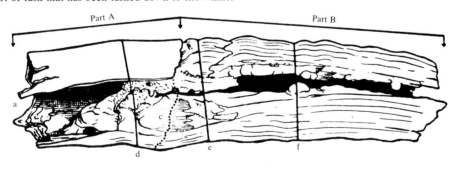

40 cm; next it takes a course at right angles for about 12 cm, still increasing in girth, and is then bent abruptly upwards. A longitudinal section showed that the portion of the tusk formed at the time of injury was dislocated almost at a right angle on the soft papilla. A noteworthy feature of the specimen is the splitting off

Fig. 19.19. *Loxodonta africana* (African elephant). RCS Odonto. Mus., G 121.1.
A: A tusk that was injured at an early period of growth. × *c*. 0.11.
B: Drawing of a longitudinal section of the tusk. There is crumpling of its structure above the level of the arrows.

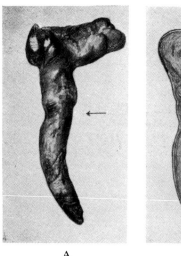
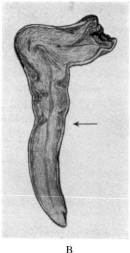

A   B

Fig. 19.20. Elephant. RCS Odonto. Mus., G 120.2. An injured tusk showing a double bend. A root-like branch of independent growth of tusk can be seen arising from the second bend on the left. (Colyer, 1915.) × *c*. 0.25.

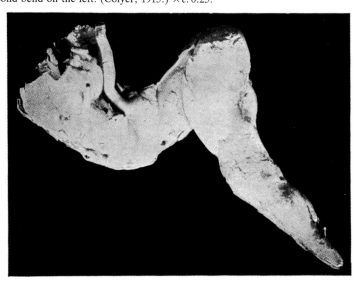

of what can be called branch roots from the main portion of the tusk.

The deformed tusk in Figure 19.21 is a good example of the power of repair of injured pulp tissues. When the tusk was broken off, the implanted part must have been split longitudinally with the pulp exposed beside it. The specimen is composed of a long splinter of the tusk (a) already formed at the time of the injury, and a large mass of tissue (b) formed after the injury. At the position (o) in the portion (b), there is an irregular opening leading to a cavity in the interior of the tusk. To determine the nature of the tusk it was divided into five pieces numbered 1–5.

The sectioned surfaces at the level between 1 and 2 (Fig. 19.21 B) not far from the worn and stained extremity of the tusk are composed partly of the normal cementum-covered splinter of original tusk which has an abrupt junction with a mass of irregular dentine containing many small circular patches of tissue which are known to workers in ivory as seeds. There is a slit at the junction between the two types of dentine which, in the section at the level between 2 and 3 (Fig. 19.21 C), leads into a large cavity which opens at the side of the tusk. This must be an abscess cavity which, as is seen in the section between 3 and 4 (Fig. 19.21 D), is completely shut off from the pulp by a solid mass of irregular dentine. The splinter of original tusk can still be seen at this level. Higher up the tusk, at the level between 4 and 5 (Fig. 19.21 E), much more of the surface consists of the original dentine of orderly structure in which the lozenge-pattern character of elephant ivory is seen. There is, however, newly-formed irregular seed-containing dentine at the periphery and also within the lumen of a central cavity which represents the original pulp cavity. The surface of the tusk at 5 in Figure 19.21 A has a slightly distorted funnel morphology of the developing end of a tusk with some dentine being formed on the surface of the splinter that extends up beyond the funnel. It seems likely that, at the time the tusk was broken, the splinter was forced up beside the formative tissues.

In this tusk, the injury to the pulp was followed by extensive suppuration, the pus in time was surrounded completely by a wall of reparative dentine, after which the pulp readily regained its power to form normal dentine.

This power of recovery of function on the part of the odontogenic cells is still more strikingly shown in the specimen in Figure 19.22. The tusk is sickle-

Fig. 19.21. Elephant. RCS Odonto. Mus., G 122.7.
A: An injured tusk. The proximal growing-end is to the right. (a) Indicates the ventrally-situated portion of the tusk already formed at the time of the injury; (b) portion of the tusk formed subsequent to the injury; (o) opening leading to a cavity in the tusk. The numbered lines above indicate the positions at which the tusk was divided. × c. 0.2.
B–E: Drawings of proximal surfaces of the numbered segments of tusk 1 to 4 respectively. c = cementum; d = dentine of normal structure; s = seed-containing reparative dentine; o = opening into the central abscess cavity.

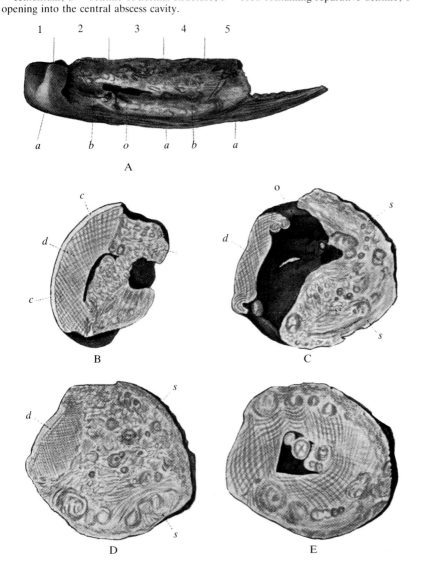

Fig. 19.22. Elephant. RCS Odonto. Mus., G 121.33.
A: Deformed tusk. The formative end is above and a stained worn surface to the left.
Numbers 1–5 indicate the segments into which the tusk was divided for examination. ×c. 0.2.
B: The surface of the section at the junction of the portions 1 and 2. ×c. 0.75.
C: The surface at the junction of the portions 2 and 3. ×c. 0.75.

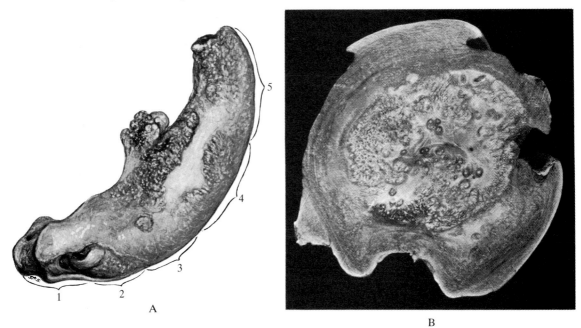

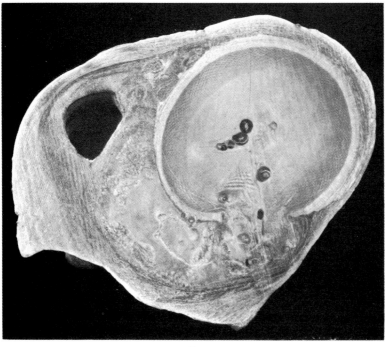

D: The surface at the junction of the portions 3 and 4. The circular area at the exact centre of the tusk is a large seed. There are two similar ones more peripherally. ×c. 0.75.
E: The surface at the junction of the portions 4 and 5 is composed almost entirely of ivory of normal structure and invested with thick cementum. (The sectioned surfaces are reduced about one-quarter.)

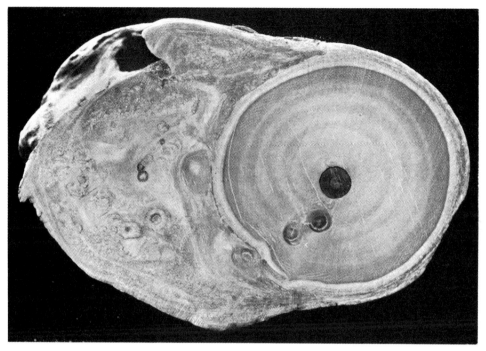

D

E

shaped. The greater curvature is fairly regular in outline except in places where there are irregular patches of cementum. On the lesser curvature, there is an outgrowth of tissue which has an irregular surface and ends abruptly about two-thirds up the tusk in a collection of nodular masses. The tusk has been divided into five numbered portions. The section at the level 1/2 not far from the worn, stained and furrowed extremity (Fig. 19.22B) suggests that this part of the tusk is composed of a solid mass of irregular dentine with a nodular structure with, at its periphery, a thick covering of cementum. At the level between the segments 2 and 3, which includes part of the outgrowth (Fig. 19.22C) on the surface of the tusk, there is a section through tusk that is nearly normal, apart from included laminated nodules near the centre and irregular dentine corresponding with what seems to be a gap in the cementum covering. The mineralized tissue of irregular structure in this gap is continuous with the mass of similar tissue on the surface of the tusk. This mass contains a cavity, an abscess cavity which, in the section higher up (Fig. 19.22D), opens on the external surface. Although some of this mass is probably cementum, the general appearance suggests the possibility that some of the mass derives from pulp tissue extruded through what we have referred to as a gap in the cementum of the original tusk.

The opposite extremity of the specimen, segment 5, has a distorted funnel-shape where the tusk continues to grow. That segment at the junction between 4 and 5 shows that recovery of the pulp has been complete and dentine of normal structure has been formed (Fig. 19.22E).

Fractures of the tusk in a longitudinal direction are common and, even after severe injury, good repair often follows. The tusk shown in two planes in Figure 19.23 was fractured and separated longitudinally into two portions (a) and (b) which have been joined together by the formation of fresh tissue. The portion (a) is regular in shape, although there is a longitudinal groove at (c) which probably arises from a subsidiary split with some over-riding of the edges which is now rounded by the addition of cementum. The proximal end of the portion (a), where there is a wide transverse slit, has the funnel-shape of a growing end and it seems likely that most of portion (a) up to about the level of the slit is tusk formed before the injury. The slit opens into the pulp cavity which a transverse section (Fig. 19.23B) through the tusk at level (1) shows contains trabeculae of the irregular dentine which also joins the two portions (a) and (b) together. This section shows that much of (a) is composed of orderly ivory, so confirming that it is part of the original tusk, and also shows that portion (b) is composed partly of a length of original tusk joined to (a) by a new formation of more irregular dentine.

Part (b) contains a pulp cavity continuous with that of (a) and ends, at a higher level than the transverse slit above (a), in a small funnel-shaped opening. It seems likely that, had the animal survived, something resembling a small tusk would have continued to be formed there. What would have happened at the transverse slit on the side (a), it is impossible to be sure. The formation process may have virtually ceased there which would explain why the funnel end above (b) is at a higher level. However, the proximal end of the specimen consists of a club-shaped mass without external openings and a section through it at level 3 (Fig. 19.23C) shows that it is composed mostly of irregular nodular dentine but with some areas of more orderly structure in which there are traces of the stripes and lozenges of normal ivory.

The specimen in Figure 19.24 shows not only the power of the pulp to recover from injury, but also indicates how subsidiary tusks may be formed. The tusk, at an early period of its growth, must have received a severe injury which split it longitudinally. At a later period, the tusk was broken transversely a second time. The upper end (Fig. 19.24A) is irregularly fractured with much staining and some evidence of wear on some of the prominences. There is no opening into the pulp cavity.

The piece I, with a similar smaller piece IV attached to it, runs the whole length (45cm) of the specimen and at the proximal end has the funnel-shape characteristic of the forming-end of a tusk; this part seems to be the product of orderly tusk formation in spite of an injury that disorganized the growth of the rest.

Within the concavity of this part of the tusk, there is what seems to be a new formation, consisting of two folded columns joined firmly with the more solid distal mass and in the process of being gradually united with portion I by deposition of more dentine within it. There is at K a slender column or tusklet which is entirely free, apart from its incorporation in the distal mass about 8.0cm from the fractured surface.

The sectioned surface at level A (Fig. 19.24B) shows a structure suggestive of folding similar to that seen on a smaller scale in the previous specimen (Fig. 19.23B).

The relationship of the various elements of which this tusk is composed can be appreciated on the distal aspect of the division at level (B) (Fig. 19.24C). The main outer plate I is separated from the column III by a slit-like pulp cavity whereas, more distally (Fig.

Fig. 19.23. *Loxodonta africana* (African elephant). RCS Odonto. Mus., G 122.3.
A: Two views of a tusk which has been split longitudinally into two portions, (a) and (b), with a groove (c) between. The numbers 1–3 indicate the levels at which the tusk was divided. The worn exposed end of the tusk is below. ×c.0.16.
B: Drawing of the proximal surface at level 1. (d) is dentine with normal ivory patterns, indicating that both main portions (a) and (b) incorporate a longitudinal splinter of the original tusk. (e) indicates two possibly additional small splinters, (f) areas of folded structure corresponding with the longitudinal grooves on the surface of the tusk, (g) post-injury dentine of compact structure, (h) irregular nodular dentine joining the main portions (a) and (b), (p) pulp cavity, (c) cementum. ×c.0.66.
C: Drawing of the transverse section made at level 3. (d) areas showing a trace of ivory structure; (c) cementum; (g) compact dentine; (s) seed-containing reparative tissue. ×c.0.5.

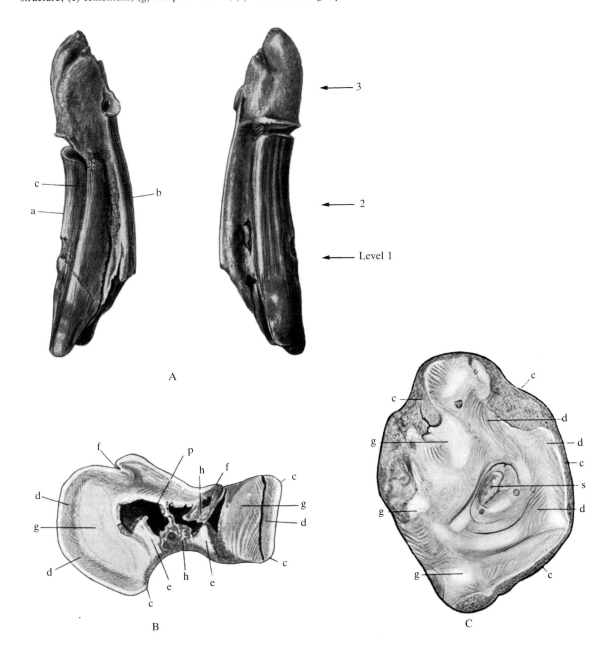

414 Section 4: Other Disorders of Teeth and Jaws

Fig. 19.24. *Loxodonta africana* (African elephant). RCS Odonto. Mus., G 122.4.
A: Portion of a fractured tusk. The fractured end is above, the formative end below. I and IV are splinters of old tusk. II and III are joined columns of new formation. (A) and (B) indicate the levels at which the tusk was sectioned transversely. (K) is a detached tusklet. $\times c.\,0.13$.
B: Drawing of the distal surface of the section at level (A). The main structures are labelled as in A. The outer plates I and IV, corresponding with the splinters of the old tusk in A, are composed of ivory of orderly structure. The column II is joined to I by dentine at (G); column III is joined by nodular dentine. (A), (B), (C), and (C_1) indicate a thick layer of cementum; (D), (D_1), (E), (F), and (F_1) indicate cementum of less-orderly structure. $\times c.\,0.5$.
C: Drawing of the distal surface of the section at level (B). The column III is now separated from I by a split-like cavity filled in life with tooth pulp; it communicates with the area between the main mass and IV in which the main pulp was situated. $\times c.\,0.5$.

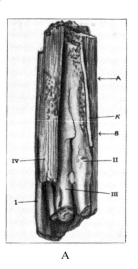

A

19.24B), the two were joined by irregular nodular dentine. The part of outer plate IV and, of course, the tusklet K, are also separated. Nearer the proximal end of the specimen, most of the elements in (Fig. 19.24C), apart from I and IV, would lie separately in the pulp.

It seems almost certain that there was an initial injury which avulsed the tusk, leaving behind only the formative tissues of the pulp and perhaps part of the funnel-shaped end of the partly-formed tusk. The pulp recovered and continued forming tusk in an orderly fashion on that side, but the rest of the pulp was disorganized, perhaps by scar tissue within its substance, and began forming a number of imperfectly organized columns or tusklets. As these columns were close together, the formation of more dentine by the pulp tissue between them brought about their fusion into a solid mass which is longitudinally-grooved on its outer surface. Presumably in due course this new tusk extruded beyond its socket and became functional and then was fractured a second time.

Another specimen which seems to throw light on the formation of secondary tusks is shown in Figure 19.25. There is a main tusk (A) about 58 cm in length which is split into three longitudinal elements; attached to it are two smaller tusks (B) and (C), which are placed obliquely across the main tusk. The compound tusk had been fractured some time before death because the fractured surface was worn to a blunt rounded surface. Probably at the time of death, the tusk had been injured by a bullet and one of the smaller tusks (C) had been broken off. Examination of the surfaces of cross-sections of the main tusk at six levels showed that the three elements of which it was composed fitted closely together as if they had originally been one and were then split asunder (Fig. 19.25B). Only their external surfaces were covered with a layer

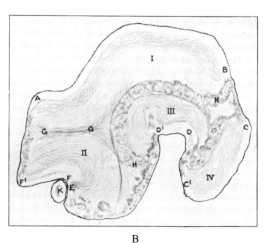

B

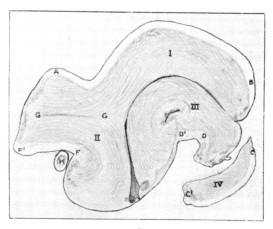

C

Fig. 19.25. *Loxodonta africana* (African elephant). RCS Odonto. Mus., G132.3.
A: An injured tusk. Inset above: view of the erupted worn distal end; Inset below: view of the growing end. (A) consists of a main tusk split into three (a,b,c) longitudinally at the distal end. (B) and (C) are separate smaller tusks. (D) the site of a bullet injury which broke (C) off short. Arrows 1 to 6 indicate the levels of cross-section. ×c.0.16.
B: Drawings of the surfaces of sections at levels 1, 2 and 6. a, b, and c indicate the three parts of the main tusk in the upper inset in A. (s) irregular nodular dentine; (ce) cementum; (e) poorly-formed cementum; (m) metallic fragments of a missile within the surface 2. ×c.0.66.

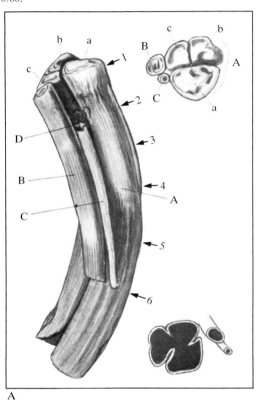

A

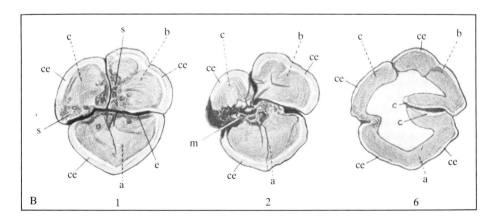

B

of regular cementum; the mutual surfaces were composed of nodular dentine and, only higher up at level 6, are partially covered by cementum. The section at level 6 close to the forming end showed that the pulp was common to all three elements and at that level the split was largely healed although deeply-grooved on its surface. However, it seems reasonable to believe that the two additional tusklets were the result of the same

Fig. 19.26. *Elephas maximus* (Indian elephant). RCS Odonto. Mus., G 132.4.
A: Two views of a fractured tusk. The distal extremity (below) presents a worn, fractured surface. There is a zone of interrupted growth at (a) indicating that the greater part of the specimen was formed after the injury; in the part above a, there is a system of longitudinal grooves indicative of division of the tusk into columns. At the proximal end formed towards the end of the animal's life, this process of division into columns is complete enough to form several slender tusklets. The columns have been serially numbered up to 13. The letters on the left indicate the segments into which the specimen was divided for examination. × c. 0.16.
B–E: Drawings of the surfaces between the five segments. × c. 0.5.
B: The level between segments A and B. (p) the pulp cavity containing a pulp nodule and walled off by solid dentine above; (d) dentine of normal structure formed before the injury. Covering these parts is a layer of cementum 0.8 cm thick. (f) indicates areas of infolding; (n) an area of nodular irregular dentine. (5), (7), (8), (10) and (11) correspond to columns seen on the surface and numbered in A.
C: The level between segments B and C. (s) indicates a probable abscess cavity. It does not communicate with the cavity (p) in B. Other lettered and numbered areas are continuous with those in B.
D: The level between segments C and D. The disorganization of growth with formation of separate columns, some joined by cementum, is here more evident. The lettered areas correspond with those in the other sections and the numbered areas correspond with numbered columns in A.
E: The level between segments D and E. The central area (d) as well as some of the peripheral columns show the lozenge ivory pattern. An additional column (13) has come into existence. Column (12) is now separate and column (5) nearly so.

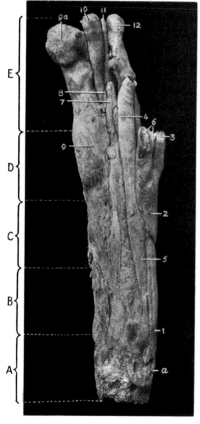 

A

Ch.19. Injuries of the Teeth 417

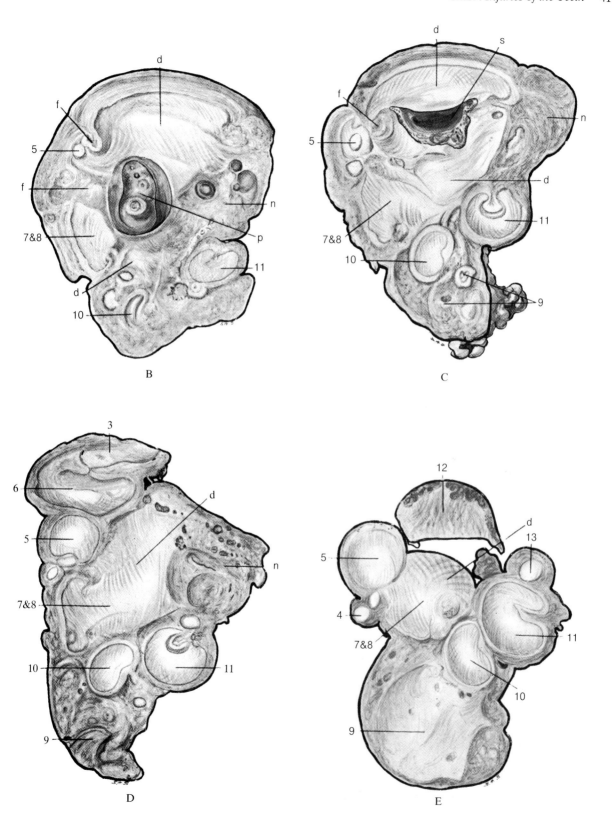

B

C

D

E

418 Section 4: Other Disorders of Teeth and Jaws

injury that split what now is the main tusk, although it is impossible to deny that it may have originally developed as a triple tusk (Chapter 6, p. 126).

A specimen typical of the tendency for longitudinal splitting or folding of the pulp to occur is shown in Figure 19.26. The distal end of the tusk presents a worn, fractured surface and a zone of interrupted growth shows that the greater part of the specimen was formed after the tusk was fractured. The outer surface of this part shows longitudinal grooving which becomes progressively deeper towards the proximal end and which, at the time of death and removal of the tusk, had led to the emergence of several entirely separate growing elements.

The surface of the section at AB (Fig. 19.26B) a little above the zone of injury at (a) shows the central cavity of the original pulp cavity containing a pulp nodule of irregular dentine. This cavity is open below to the outside environment and the pulp in it must have become necrotic because the cavity is closed off above this level by a bridge of dentine.

The cut surfaces at higher levels (Figs. 19.26C and D) show increasing separation into separate columns or tusk elements, a process which becomes complete for several elements above the level of Figure 19.26E.

The formation of subsidiary tusks after injury is probably brought about as follows: the injury disorganizes the soft pulp, or formative part of the tooth, and peri-cementum dips into the pulp thus separating a portion or portions from the main body. The fold gradually extends around the portion thus separated, and eventually envelops it completely.

Whether instances of double or multiple tusks (Chapter 6, p. 128) could be due to trauma even though direct evidence of trauma is absent is a matter of speculation. A few specimens which Sir Frank

Fig. 19.27. *Elephas maximus* (Indian elephant). RCS Odonto. Mus., G 120.3.
A: A deformed tusk. The main mass (B), which is 34 cm in greatest diameter, is smoothly rounded but the surface (A), from which the nodular mass (D) projects, is irregular. From the opposite surface, the smoothly-rounded cylindrical mass (C) projects. ×c. 0.05.
B: A longitudinally-sectioned surface. The main masses are labelled as in A. (d) an area of dentine marked by parallel streaks; (s) seed-containing irregular dentine; (a) a large such area from which pass two white streaks (b); (ce) cementum. ×c. 0.5.

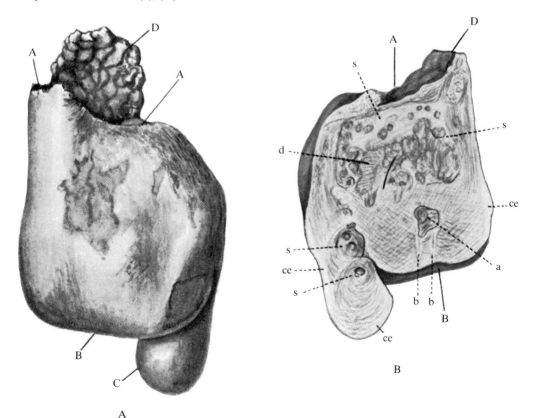

Colyer chose to place in this chapter in the original version of this book because of the possibility that they were caused by trauma, we have chosen to place among the developmental variations (Chapter 6).

The mass of the tooth tissue in Figure 19.27 was removed from the tuskless side of a rogue elephant. The animal measured 3.05 m at the shoulders and had only one tusk exposed. The orientation in the jaw of this deformed tusk is unknown but it was encased in a thin covering of bone from which the nodular mass of tissue (D) projected and was embedded in dense soft tissue which contained six small oval pieces of mineralized tissue. The surface from which the mass D projects is irregular; the surface of most of the rest is smooth and globular and there is a smoothly rounded cylindrical projection which appears bent towards the main mass.

A longitudinal section shows that the lower half of the tusk (Fig. 19.27B) is of fairly regular ivory and contains a small area of irregular dentine near the centre. The cylindrical projection is formed of dentine with a laminated structure surrounding two islands of irregular dentine. The centre of the upper part of the main mass consists of a roughly rectangular mass of heterogeneous structure; parts of it (d) are of near normal ivory; there is a large central area of compact dentine containing seeds and a large part is nodular irregular dentine. Most of the specimen is invested by an ill-defined layer of cementum.

The disorganization and types of tissue formed in this specimen closely resemble those found in specimens already described which can reliably be attributed to injury. This lends to the suggestion that this mass of tissue is something formed by a dentine papilla left behind when its tusk was totally avulsed by an injury. The disorganized growth of dental tissues in this and other specimens described here resembles that found in odontomes (Chapter 25).

Elephant tusks with a spiral morphology occur and there is reason to believe that they are the result of injury to the developing part of the tusk. An example which is of particular interest because of its historical associations is shown in Figure 19.28. It originally belonged to the Royal Society and was described by the secretary of the Society, Nathaniel Grew (1681), in the following terms:

> A spiral or wreathed tusk of an elephant. Presented from the Royal African Company by Thomas Crispe, Esq. It is twisted and wreathed from the bottom to the top with three circumvolutions standing between two straight lines. 'Tis also furrowed by the length. Yet the furrows surround it not as in the horn of the Sea Unicorn, but run parallel therewith. Neither is it round as the said horn, but somewhat flat. The tip very blunt.

In 1781, at the time when the Royal Society was moving to new premises in Somerset House, the Society transferred its collection of rarities to the British Museum, then only in its infancy (Lyons, 1944). In 1809, the British Museum disposed of a number of anatomical preparations to the Royal College of Surgeons for the sum of £175.10s. and this original Royal Society specimen was included (Cope, 1959).

The tusk measures about 140 cm from the tip to the base along the outer curvature. At the growing end, the ivory does not completely encircle the pulp cavity, there being a slight gap extending down the tusk for about 20 cm; beyond this is a shallow depression, gradually decreasing in width, which continues for half the length of the tusk. The remaining portion of the tusk appears to be well formed. The curvature is towards the grooved side (Fig. 19.28).

A lesser degree of spiral curvature is seen in the specimen in Figure 19.29. The growing end of the tusk presents a cleft which runs about 16 cm along the tusk, is followed by a groove which rapidly widens to 25 mm, and then gradually narrows until 20 cm from the tip where it is represented by a simple fissure; after a

Fig. 19.28. *Loxodonta african* (African elephant). RCS Odonto. Mus., G 122.8. A spiral tusk. × c. 0.09.

course of 10 cm, the fissure splits to enclose an island of dentine; further on, another island of dentine appears surrounded by a fissure. The arrangement of the tissues forming the terminal part of the tusk suggests that the tusk was injured at an early stage of its growth. The tusk curves towards the grooved side.

The tusk in Figure 19.30 is about 75 cm long. The distal third is more distorted than the photograph suggests, the final twist taking a direction almost horizontal to the long axis of the tusk. A deep groove, which starts near the tip, transverses the tusk for about two-thirds of its length. One border of the groove is flat, the other shelving as shown in the outline of its growing end. The first half of the groove has an irregular surface; the second half has a smoother surface as if covered with a layer of freshly-formed tissue. In this specimen, we have evidence of an injury sufficiently severe to break off a flake of the tooth, including both the exposed and the implanted parts. The fractured surface of the implanted portion healed over, but the injury was sufficient to cause complete disorganization of the regular formative activity of the pulp.

The superficial portion of the thin-walled socketed portion of the tusk on the face of the elephant exposes it to injuries of the second type mentioned on p. 405 of this Chapter; namely, injuries from penetrating missiles, such as bullets and spears, aimed at the head from in front.

In earlier times, the trade in elephant ivory was considerable and, in sawing the tusks into suitable lengths for the great variety of purposes for which ivory was then used, it was common to discover segments of the tusk spoiled, as far as industrial purposes were concerned, by past injuries, the most common and fascinating of which was injury from musket-balls and bullets found embedded in the ivory (Figs. 19.31,

Fig. 19.29. *Loxodonta africana* (African elephant). RCS Odonto. Mus., G 122.81. A tusk with a spiral curve. A cleft seen in the profile of the growing end extends some distance along the inside of the curve of the tusk. ×c. 0.11.

Fig. 19.30. *Loxodonta africana* (African elephant). RCS Odonto. Mus., G 122.82. A twisted tusk with a deep groove which is seen in the diagrammatic profile of the growing end. ×c. 0.2.

19.32). Specimens of this kind attracted the attention of many early biological scientists, the first, according to Busch (1890b) being Ruysch (1719). However, among those who described specimens, one of the most notable is Goethe, the poet and natural philosopher who, some time before 1798, described with remarkable precision thirteen preparations of abnormal ivory, including seven containing musket-balls. His specimens were obtained from the makers of ivory combs who, discovering these faulty segments of tusk, would 'leave them aside to sell to interested scientists'. In a passage which appears to date from 1823, he mentions that such specimens had become less easy to obtain because the trade in ivory was diminished and also because, perhaps with improvements in firearms and the mode of hunting elephants, fewer animals suffered sub-lethal injuries. In case there could be any chance of Goethe's specimens being still traceable, it is worth mentioning that he presented them to his former teacher, J.C. Loder, then of Moscow (Goethe, 1823, 1949). Several specimens of bullets incorporated in ivory were collected by John Hunter and are still to be seen in the Hunterian Museum of the Royal College of Surgeons of England (Figs. 19.31, 19.33).

The earliest writers on the subject were puzzled to explain how the ivory could contain a musket-ball without there being any evident point of entry; they were particularly intrigued when the ball of soft lead showed no sign of flattening from impact with hard ivory. Combe (1801), however, seemed to understand the essentials of the growth mechanism of the tusk and that a missile could, by penetrating only a thin layer of bone, enter the soft tissue of the growing end of the tusk and be incorporated into its substance by subsequent growth. Certainly by the time Owen wrote his *Odontography* in 1840–1845, this mechanism was well understood (Fig. 19.17) and it was realized that growth of cementum over the surface of the tusk could conceal a point of entry of a missile through the tusk itself.

Fig. 19.31. Elephant. RCS Pathology Series, T117. An iron bullet embedded in a tusk. (Colyer, 1915.)

Fig. 19.32. Elephant. RCS Odonto. Mus., G128.3. An iron bullet embedded in a tusk. *c*. Natural size.

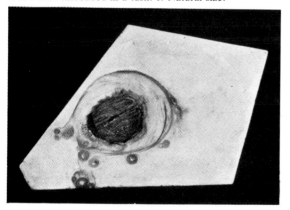

Fig. 19.33. Elephant. RCS Pathology Series, T115. A lead bullet embedded in a tusk. (Colyer, 1915.)

When the pulp is injured by a penetrating missile, in favourable circumstances, the reaction of the pulp is two-fold: (i) the disorganized pulp is in time mineralized into an irregular dentine; (ii) the missile is cut off from the pulp by a layer of hard tissue.

The specimen in Figure 19.34 is a good example of the repair after injury to the pulp by a bullet which passed through the base of the tooth; the soft tissues have been replaced by a bar of reparative dentine. There is a circumferential groove or zone of injury a little proximal to the point of entry which suggests that the missile penetrated the thin extremity of the basal end of the tusk. About 40mm of the tusk was formed after this zone of injury, from which it is possible to estimate, allowing a 2-week period for the odontoblasts to recover, that the injury occurred about 14 weeks before death. Wedl (1870) described a similar specimen.

When the bullet penetrates the alveolar process and the wall of the tooth but fails to escape from the tooth-wall, the wound on the outer surface is closed by cementum and the pulp coats the bullet with a layer of irregular dentine; when this is completed, ivory of normal structure is formed. In the case of an iron bullet, the metal oxidizes and the ivory enclosing it is often irregular in type, taking on the character of an osteo-dentine. The ivory does not form in close contact with the metal, so that the bullet usually lies freely in a cavity (Figs. 19.31, 19.32). With a lead bullet, the layer of primary encasing-ivory is often quite thin; it is less irregular in character and may be formed in the closest contact with the metal as in the examples shown in Figures 19.33 and 19.35. In one (Fig. 19.35), the force of impact has splayed the lead but the ivory has nevertheless formed in close contact with the metal.

The type of metal of which the missile is composed seems to have a considerable influence on the healing power of the pulp. According to W.D. Miller (1890–1891, 1899), who examined at least 58 instances of metallic foreign bodies within tusks, lead bullets appear to produce the least irritation. He goes on to say that this observation led Dr Cunningham to use lead in his experiments on the implantation of substitutes into the jaws of dogs and monkeys. It seems possible that this refers to George Cunningham, the widely-active Cambridge dental practitioner (1851–1919) who was a strong protagonist of dental care as part of public health in this country (Cunningham, 1911; Miles, 1972). He is known to have practised tooth implantation (Cunningham, 1889) but no published account has been traced of the experimental work referred to by Miller.

Where the bullet passes through the wall of the tooth and lodges in the substance of the pulp, the nature of the repair depends largely on the amount of damage to the soft tissues. In the specimen in Figure 19.36, the bullet came to rest on the inner aspect of the

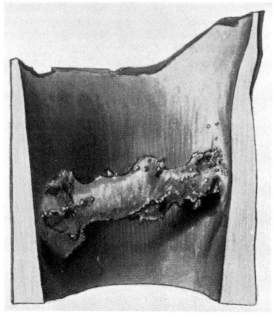

Fig. 19.34. Elephant. RCS Odonto. Mus., G 129.2. Portion of tusk, basal extremity above, with a bar of reactionary dentine at the site of a transfixing injury. × c. 0.66.

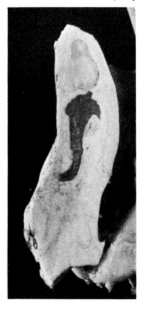

Fig. 19.35. Elephant. RCS Odonto. Mus., G 129.1. A lead bullet embedded in a tusk. (Colyer, 1915.) c. Natural size.

Fig. 19.36. Elephant. RCS Odonto. Mus., G 123.11. Portion of a tusk; a mass of reactionary dentine, enclosing a bullet, is attached to the inner wall of the pulp cavity. There is an abscess cavity in the tusk wall. × c. 0.43.

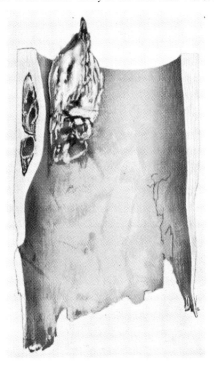

tooth-wall and so caused but little injury to the pulp. The entry wound is shown by an irregular opening which leads into a cavity in which an iron bullet is lodged. The walls of the cavity, formed by an irregular deposit of dentine, are incomplete and the cavity is not therefore cut off completely from the pulp tissue. It is possible that this incomplete isolation of the bullet was due to the life of the animal having been cut short.

Figure 19.37 shows another specimen with only localized disorganization of the pulp. The external surface of the tusk shows the point of impact of the bullet from which extends a longitudinal split. The margins of the split are rounded by the irregular deposition of new ivory. The inner surface of the tusk at the point of impact presents an elongated mass of ivory, the extremity of which is attenuated and ends in a cup-shaped expansion which partially surrounds the bullet, now lost, which evidently came to rest in the substance of the pulp.

Sometimes, however, a large area of the pulp tissue is involved in the injury as shown in Figure 19.38. The section has been made a little beyond the site of entry of a bullet which is embedded in a mass of irregular reparative dentine laid down by the pulp, much of which must have survived the injury. The mass contains a few fragments of ivory carried into the pulp by the bullet. The opposite side of the tusk is fissured which was presumably produced by the impact of the

Fig. 19.37. Elephant. RCS Odonto. Mus., G 123.1.
A: Portion of tusk with a longitudinal fissure extending from an impact injury not far from the present basal extremity above. × c. 0.5.
B: Section through the portion of the tusk. The margins of the fissure are rounded by deposition of new ivory and a mass extends from the point of impact and ends in a cup-shaped depression (A) in which the bullet rested. (Colyer, 1915.)

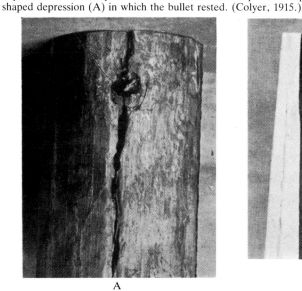

A

B

bullet which then fell and bounced back into its track through the pulp. Much of the surface formative tissue of the pulp must have survived because there is a layer of ivory of more or less normal structure which lines much of the original tusk. However, at the site of the cavity (Fig. 19.38), there is no such layer of new ivory. That part of the pulp must have died following the injury.

In the specimen depicted in Figure 19.39, there is a solid bar of hard tissue stretching across the pulp cavity. The specimen has been divided in a vertical plane and each half again divided at right angles to the vertical cut. Unfortunately, these operations were carried out by one not very skilled in such work and the results have detracted somewhat from the value of the specimen. The right half contains a fragment of iron and the left a brass bullet. In separating the right half, saw-cuts were made in three directions and the portions were forced apart leaving a rough irregular surface beyond the iron fragment.

A composite drawing prepared from the longitudinal-sectioned surfaces is shown in Figure 19.39B. The tissue around the iron missile is dense and fits more closely around it than is usually the case with iron missiles. Above this missile is a space which, in the opposing sectioned surface, is continuous with a channel (e) which opens on the surface of the tusk and is part of the opening of entry of the missiles into the tusk. This opening has been partially closed by irregular dentine. Much further into the bar of tissue, on the left, there is an irregular elongated cavity, at the far end of which the brass bullet lies loosely. This cavity must, like the space related to the iron missile, be part of the original path of destruction of the missiles through the pulp. The two cavities are made discontinuous by a mass of reparative dentine of irregular structure containing some seed-like bodies.

The spindle-shaped pieces of tissue (sp) (Fig. 19.39A) which project from the surfaces of the bar are seen in section in two places in Figure 19.39B and are dense, structureless and translucent. It is hard to account for them but it seems possible that they were so-called pulp stones already formed within the pulp at the time of the injury and were then incorporated in the reparative dentine.

The thickness of the walls of the tusk at the time of injury was about 0.5 cm (k in Fig. 19.39B) which means that the area struck by the missiles must at the time have been close to the funnel-shaped formative end of the tusk. Subsequently, a thickness of about 1.2 cm was laid down.

It seems unlikely that the two bullets, one brass and the other iron, entered this area on different occasions; it is possible that both were part of the discharge of a blunderbuss type of firearm which had been loaded with various bits of available metal and

Fig. 19.38. Elephant. RCS Odonto. Mus., G 128.23. Section through a tusk showing a lead bullet embedded in a mass of reparative dentine. (a) the site of entry of the bullet; (b) a fissure in the wall of the tusk, presumably due to the bullet hitting it and then falling back in its path of damage; (d) ivory of more or less normal structure formed subsequent to the injury. No such ivory has formed on the part of the pulp cavity to the right where presumably the pulp died following the injury. ×c. 0.4.

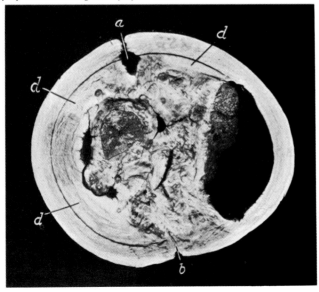

both missiles entered the tusk at the point of entry (a). An important feature of the sections represented in the diagram (Fig. 19.39B) is the various types of tissue formed by the pulp tissues in response to the injury which must have actually destroyed the tissue in the pathway of the missiles. In the supposed pathway itself, the dentine formed is of maximal irregularity whereas, at the periphery of the bar, it is more dense and orderly to the extent that some of it shows a lamellar pattern with traces of the lozenge pattern characteristic of properly-formed ivory.

The pulp cavity or pulp of the elephant tusk may

Fig. 19.39. *Elephas maximus* (African elephant). RCS Odonto. Mus., G 128.22.
A: Portion of a tusk with a bar of tissue stretching across the pulp cavity. (a) point of entry of bullets; (b) slight thickening of cementum at the opposite end of the pathway of the missiles; (sp) end of a spindle-shaped spike of displaced dentine; (d–d) one plane of section of the specimen. ×c. 0.66.
B: Diagrammatic view of a longitudinal section through the specimen. (a) marks the point of entry now filled in with irregular dentine; (e) a channel, the remaining path of entry; (f) a continuation of the channel in the tissue leading to the iron fragment at (i); (k) thickness of ivory formed at the time of injury; (g) brass bullet above which is the original path; (n) new ivory of normal structure; (h) areas of dense structureless tissue; (sp) spindle-shaped bodies; (o) areas of irregular structure containing in places seed-like bodies; (c) probably an abscess cavity filled with irregular dentine. ×c. 0.66.

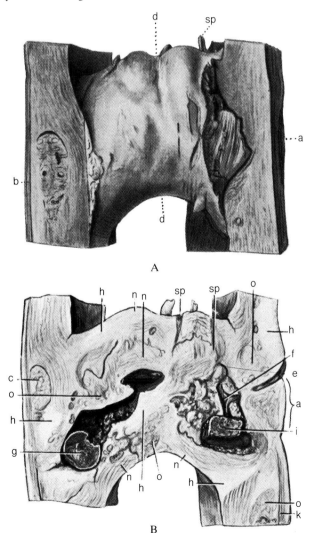

contain pulp stones of great size. There is one (RCS Odonto. Mus., G130.1) which is 28 cm long. Some may enclose bullets (Fig. 19.40). A pebble, probably derived from an old muzzle-loader of a native hunter, was found in the diseased flesh of a suppurating tusk (RCS Odonto. Mus., G131.11).

Figure 19.41 shows a specimen in which an iron musket-ball has penetrated the wall of the tusk, traversed the pulp, and come to rest in the opposite wall. A mass of irregular nodular dentine has formed in the path of the ball which now lies loosely in a cavity.

A remarkable specimen, described first by J. Murie (1870) and later by Sir John Bland Sutton (1910), is shown in Figure 19.42. It is part of a tusk in the pulp cavity of which the iron head of a spear is lying. The spear must have entered the open developing end of the tusk from above and almost certainly is part of a weighted spear dropped from a tree. The hunter lies in wait on a branch overhanging a waterhole where elephants are likely to come to drink and, if an elephant moves into a favourable position, the weighted spear is dropped (Christy, 1924; Miles and White, 1960). Combe (1801) described a closely-similar specimen and accurately interpreted it as being due to the entry of the spear from above through the thin plate of bone that separates the funnel-shaped end of the tusk from the nasal cavity, further growth of the tusk then carrying the spear-head more deeply into the tusk (Fig. 19.42). W. D. Miller (1899) illustrated a similar specimen that he saw in London in the possession of an ivory dealer who refused to part with it for less than 500 dollars.

In describing various specimens, we have referred to abscess cavities in ivory. Often the injury is associated with necrosis of some of the pulp followed by infection and suppuration, in spite of which, as we have seen, the pulp may recover its functional activity. This recovery may be associated with the shutting off of the suppurating area or collection of pus by a new formation of dentine. Figure 19.43 shows such an abscess cavity in the wall of a tusk (J. Tomes, 1848). The diagram (Fig. 19.43B) of the sectioned surface shows that the suppuration occurred on the deep surface of the thin funnel-wall of the growing end. Surviving pulp adjacent to suppuration laid down new dentine on the wall and within the substance of the pulp itself around the suppurating area, so encapsulating it completely. A formation of irregular dentine has encroached upon the lumen of the abscess at (r). The abscess cavity is lined in part by a thin structureless semi-transparent mineralized tissue which W. D. Miller (1890–1891) described in similar situations and called the 'stratum primitivum'.

In the specimen in Figure 19.44, the abscess cavity occupies the whole width of the pulp chamber. The outer wall of the cavity is formed by a stratum primitivum, which sends numerous processes towards the centre of the cavity and so divides it into partitions or loculi. The mineralized deposits give an arborescent appearance to the specimen.

The complete encapsulation of abscesses seen in the instances described above does not always occur. Figure 19.45 shows the basal portion of a tusk which has been divided in a longitudinal direction. The pulp,

Fig. 19.40. Elephant. RCS Odonto. Mus., G128.4. A mass of ivory loose in the pulp cavity of a tusk. It encases an iron musket-ball which remains exposed at lower left. (Colyer, 1915.) ×c. 1.3.

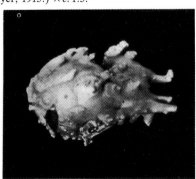

Fig. 19.41. Elephant. RCS Odonto. Mus., G128.2. Transverse section through a tusk showing a repair of the tissue by irregular reactionary dentine (B), known to ivory-dealers as 'pith', following injury by an iron musket-ball which lies loosely in a cavity (A). The lozenge-pattern characteristic of elephant ivory in cross-section is discernible. The outer layer of cementum, which is sometimes thick, is known in the trade as the 'bark'. (Colyer, 1915.) Natural size.

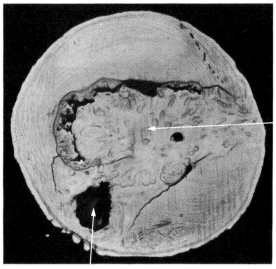

for the greater part, has been replaced by a mass of reparative dentine enclosing an abscess cavity which may be likened to a cavern in a limestone rock with stalactites hanging from the roof. The abscess cavity communicates in one place with a longitudinal cleft on the external surface of the specimen.

Bricknell (1987), among 178 tusks of mammoths, found evidence of fracture in one; because these enor-

Fig. 19.42. Elephant. RCS Odonto. Mus., G 126.1. Spearhead 180 mm long lying free in the pulp cavity of a tusk but partially encapsulated by a blister-like new formation of ivory. (Colyer, 1915.) ×c. 0.29.

Fig. 19.43. Elephant. RCS Odonto. Mus., G 125.1.
A: A specimen described by J. Tomes (1848) as an abscess cavity on the surface of the pulp which became encapsulated completely by the formation of new dentine. A V-shaped section has been sawn out. The thin dentine (pointer) near the base of the tusk (left) already formed at the time of formation of the abscess can be recognized. ×c. 0.5.
B: A drawing of one of the sectioned surfaces. (ce) cementum; (d) dentine of normal structure; (b) dark layer at the dentine-cementum junction; (s.p.) thin layer of structureless tissue (stratum primitivum) which, between the points (c), lines the abscess cavity; (r) reparative dentine that has partly filled the abscess cavities; (n.d.) new dentine formed subsequent to the encapsulation of the abscess; (e) dense dentine, the last formed by the pulp in this specimen.

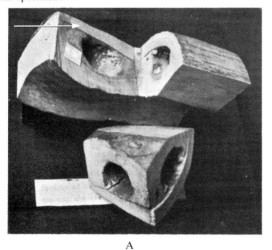
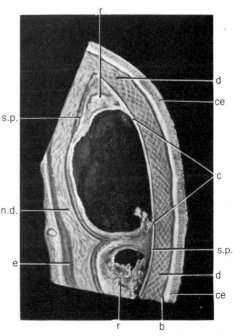

mous tusks are said to have been used to sweep away snow and ice in order to reach food, he had expected to find more.

(b) Injury to teeth of limited growth

Teeth of limited growth do not have the same power of recovery to injury as those of continuous growth, although, provided that they are not yet fully developed, so that their pulps are large and have a good blood supply through the large opening of the still-growing root, the power of recovery may be considerable; however, injury may arrest or distort their growth. When the tooth is fully grown, the pulp has only a limited blood supply through narrow apical canals and, if the pulp is directly involved in the injury, it is likely to become necrotic, the more so if infection gains access to the injured tissues.

In young children, a fall with the mouth open may drive a deciduous upper incisor up into the potential space beside the crown of the developing permanent successor. Garlick (1954b) referred to a similar displacement of lower incisors in cattle. The teeth may retain or regain their pulp circulation, the fracture of the bony socket heals and the tooth may remain in its depressed position.

According to Wegner (1962), injuries to the canines in particular are common in the wild in anthropoids that live on hard-shelled fruits as they tend to break their teeth on the hard shells. He described in well-illustrated detail seven such instances, four in the gorilla and three in the orang-utan, in which the fracture of the canines laid open the pulp which became necrotic, and thus provided a portal of entry for infection deep in the jaw leading to fistulous abscesses.

Hill (1953a) described the skull of a male mandrill (*Mandrillus sphinx*), zoo-maintained from three years of age, in which the right upper canine was misplaced palatally so that it crossed the mouth obliquely. There was a history of angry cage-biting and it is assumed that this played a part in displacing the canine. In closing the mouth, the point of the canine came to rest in an impression in the gum and underlying bone on the lingual side of the lower canine of the opposite side. The displacement of the canine probably occurred soon after it emerged into the mouth because the fully-erupted upper canines in the male are very long and project well down outside the lower arch when dis-

Fig. 19.44. Elephant. RCS Odonto. Mus., G 125.5. Portion of a tusk showing a large abscess cavity filled with a formation of reparative dentine with trabecular structure, in one place of arborescent appearance. × c. 0.05.

Fig. 19.45. Elephant. RCS Odonto. Mus., G 125.6. Sectioned surface of one-half of the basal portion of a tusk. The growing end is above and has only a slight hollow because the funnel-shaped cavity has been largely filled with irregular seed-containing reparative dentine near the centre of which there is an irregular abscess cavity lined with stalactite-like projections. The cavity communicates with a deep longitudinal slit on the surface of the specimen. Part of the slit can be seen in section to the left of the abscess cavity (arrow). × c. 0.5.

placement into the palate would be unlikely even if the mouth was open.

Three of the canines of a captive sun bear (Fig. 19.46) show abnormalities probably caused by trauma; only the lower right canine is unaffected (Fig. 19.46D). The common feature of the other three is a ring-like groove in the enamel a short distance from the cervix of the tooth. The left upper canine is less abnormal, with the root complete, apart from a general thinness and irregularity of the surface of most of the enamel. The lower left canine not only shows a more marked ring-like groove, but the crown and root appear to have been squashed on one another; in other words, they show the features of dilaceration; the crown was nearly completed at the time of an injury which slightly displaced it in relation to the formative pulp. The pulp had recovered because about half of the normal length of root had been formed after the injury though the root itself is a bit mis-shapen. However, the root is discoloured and it lies in a large abscess cavity in the bone with several sinus openings on the outer surface. That this suppuration is associated with death of the pulp is shown by the wide funnel-opening typical of a growing root.

The upper right canine has a morphology similar to the tooth just described though it is even more malformed. Furthermore, much of the crown appears to have been worn away or broken off because it presents a large opening into the pulp cavity which is very large and has thin walls. A longitudinal fissure runs the length of the root; at the coronal end of the fissure a spine-shaped projection of enamel is directed root-wards (Fig. 19.46E). A cross-section through the root shows that the fissure is formed by an infolding of the tissues. Drawing upon larger-scale examples of injuries to developing elephant tusks, this specimen may be interpreted as the result of trauma which produced the vertical fracture of the formed part of the

Fig. 19.46. *Helarctos malayanus* (sun bear). Captive. RCS Odonto. Mus., G 98.1. Three of the permanent canines have suffered injury. A: Palatal view; both canines show deformity from injury. The right canine has a widely-open pulp cavity. $\times c. 0.75$.
B: Medial aspect of the right maxilla, showing that the canine has been driven up to produce an elevation on the wall of the nose.
C: The left side of the mandible; the canine is injured and abscessed. There is the opening of a suppuration sinus opposite its root. $\times c. 0.5$.
D: The four canines. (a) the right mandibular canine which is normal; (b) the left maxillary canine showing the slenderness of its crown, principally because the enamel-covering is thin; (c) the short, stunted left mandibular canine which has a funnel-shaped end to the root because formation ceased before it was complete; (d) the right maxillary canine, similarly stunted; the end of the root is funnel-shaped. Note the spike of tissue that projects rootwards from the cervical enamel.
E: A larger view of the opposite side of the right maxillary canine showing that the projection is associated with a longitudinal fissure. Enlarged.
F: A transverse section through the right maxillary canine showing the infolding at the site of the longitudinal fissure. The pulp cavity is partly filled with trabecular reparative dentine. (Colyer, 1915.)

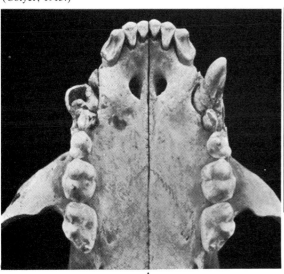

A

B

continued over

Figure 19.46 continued

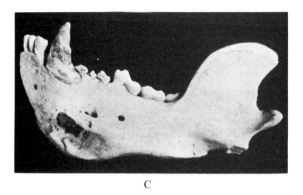

C

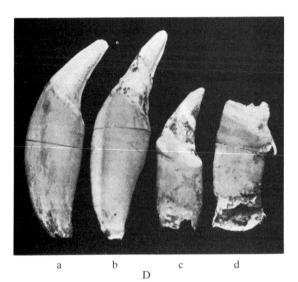

a b c d
D

E

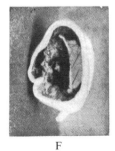

F

crown which was the beginning of the fissure; the developing soft tissues were also torn and compressed so that, when formation of the tooth recommenced, the distortion was perpetuated. The projection seems likely to have arisen from a herniation of formative tissues brought about by compression. Apparent extrusions of dentine, covered sometimes with enamel, and known as odontoceles, projecting from the surfaces of human teeth have been described (Miles, 1954).

The transverse section (Fig. 19.46F) shows that the pulp cavity in the root portion is partially filled with sponge-like trabeculae of what may be presumed to be reactionary dentine.

The tooth shows some evidence of having been driven backwards because the root portion, covered by a thin layer of bone, projects into the nasal fossa. It is associated with suppuration and there is a large opening between the nasal fossa and the infratemporal fossa which is almost certainly fistulous and leads from the area of suppuration around the end of root. This tooth, like the lower left canine, is discoloured by suppuration. Both these teeth must have survived long enough for a considerable part of the root to be formed after the injury and then the pulp must have succumbed to infection. Perhaps the original injury was accompanied by infection and suppuration, which continued at a low level while much of the root was formed, but finally progressed and destroyed the pulp.

The position of the zone of injury is the same in all three of these canine teeth, so it can be assumed that they were all injured by the same traumatic event. What this event might have been in a captive animal with one canine escaping, is a matter for speculation, although the canines are obviously vulnerable and could be injured in attacking another animal.

The pulp can sometimes permanently survive quite severe injury. The left maxillary canine in a mona monkey which died in captivity (Fig. 19.47) can be interpreted as having been fractured through the crown, widely exposing the pulp. Probably at the time the root was incomplete so that the pulp had a rich blood supply through a funnel-shaped end of root; otherwise it is difficult to believe that the pulp could have survived as it evidently did. The longitudinal section shows that on the labial aspect a fragment of the crown (a) was detached but then became re-attached by being incorporated in a mass of reparative dentine. The root of the tooth was completed.

Figures 19.48 and 19.49 show two molars with roots projecting at abnormal angles; in one (Fig. 19.48), in a rhesus monkey, the root projects from the crown and in the other (Fig. 19.49), in a baboon, from the root. Details of the morphology of these roots and of the areas from which they project are strongly suggestive of distortion by trauma.

On the other hand, there are instances where the morphological appearances are suggestive of trauma but the bilateral distribution of the malformation makes a systematic cause affecting development much more likely. Figure 19.50 shows both M^2 and the right M_2 from a mangabey monkey which have deformities of the roots together with the unaffected left M_2 for comparison. Each of the three deformed molars has an enamel defect on the buccal side close to the root. In the left M^2, the disto-buccal root is absent; in the right M^2, the disto-buccal root is diminutive and abruptly bent mesially so that it lies lingual to the mesio-buccal root. The two roots of the right M_2 are fused.

Fig. 19.47. *Cercopithecus mona* (mona monkey). ♂. Captive. RCS Odonto. Mus., G 98.81.
A: Palatal view. The crown of the left canine has been broken off but there is a surface of hard tissue and the pulp cavity is not exposed.
B: Drawing of a longitudinal section through the left canine. The root is below. (d) dentine; (e) enamel; (a) a fragment of the tooth separated by the injury; (r) reparative dentine, some of which joins the fragment (a) to the tooth. × c. 2.

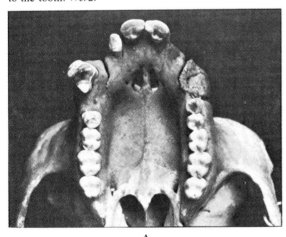

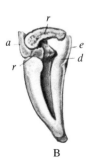

Whatever caused these three disturbances of development acted at the time when the teeth were developing in the jaws and root formation was just about to commence. It is difficult to envisage a traumatic incident that could be localized in three such areas. Figure 19.51 shows some similar distortions of

Fig. 19.48. *Macaca* sp. (macaque). RCS Odonto. Mus., G 97.11. Distal aspect of a mandibular molar. Instead of a distal root which should resemble the mesial one, there is a small root which projects from the distal aspect of the crown.

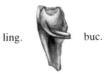

Fig. 19.49. *Papio* sp. (baboon). RCS Odonto. Mus., G 97.12. Mesial surface of a maxillary molar. The mesio-buccal root arises closer than normal to the crown and projects mesially.

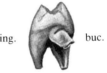

Fig. 19.50. *Cercocebus atys atys* (sooty mangabey). ♂. Captive. RCS Odonto. Mus., G 97.14. Three of the second permanent molars show deformities of the root morphologically suggestive of traumatic injury. (a) left upper molar; (b) right upper molar; (c) right lower molar; (d) the normal left lower molar.

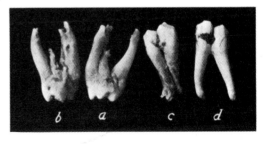

Fig. 19.51. *Macaca sinica* (toque monkey). ♀. Captive. RCS Odonto. Mus., G 97.16. Left and right M^2; the roots are distorted in their direction. (a) left tooth, buccal surface; (b) left tooth, lingual surface; (c) right tooth, buccal surface; (d) right tooth, lingual surface.

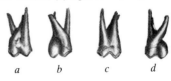

432 Section 4: Other Disorders of Teeth and Jaws

roots of both M^2 of a toque monkey which, although suggestive of trauma, because of the bilateral symmetry, are more likely to be an example of abnormal development due to ill-understood genetic causes. The interpretation of such specimens is subjective and speculative. Those interested in the specimens depicted in Figures 19.48–19.51, all of which died in captivity, should consult Colyer's interpretation in the First Edition (p. 567).

Figure 19.52 shows an unusual injured I^1 in a wild sub-adult gorilla. No history is available but there is a healed depressed fracture of the left parietal bone. The root of the incisor has a cupped-out surface from the edge of which a peg-like spike projects. The alveolar bone presents a congruous surface with a socket into which the spike fits and an adjacent concavity which is more compact and lighter in colour than the rest of the bone; the question arises whether it is a dentine surface. Only microscopic examination of a section of the tissue could determine this.

James and Barnicot (1949) noted that one or both P^3 were missing in 8 out of 16 zoo-maintained adult baboons; radiographs in many cases showed fragments of the roots of these teeth within the bone. This tooth is normally small with short roots. They suggest that the high rate of loss of the tooth is because it is situated in a region of the mouth which is subject to stress in tearing food and this renders it liable to be wrenched out or fractured.

Fig. 19.52. *Gorilla gorilla* (gorilla). ♀. Wild. P-C. Mus., M.180. The left I^1 had been injured. ×c. 0.44.

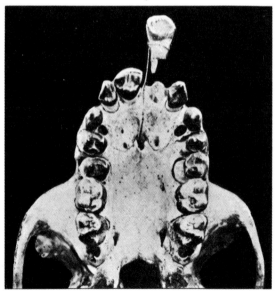

According to Garlick (1954b), fracture of the incisors in cattle is less common than otherwise would be the case because the teeth normally have some mobility so that, if subjected to sudden forces, they tend to 'duck away' rather than fracture. E. Becker (1970, p. 261) described several instances of fractures of the incisors of horses.

The molar teeth of elephants are sometimes injured by bullets. The one shown in Figure 19.53 was described by C. S. Tomes (1877). The animal was an elderly male which had frequented a coffee-growing district in Ceylon and was known locally as 'an old rogue'. When killed by a rifle shot, evidence was found of at least six other previous bullet wounds, some suppurating. The following is Tomes' description which is accompanied by some fine photogravure pictures of the intact tooth:

> The tooth is believed to have been the left lower molar, but owing to the condition of the carcase this was not positively ascertained. The bullet apparently struck the side of the tooth, somewhat below and considerably behind its middle, and until the tooth was cut through, remained loosely enclosed within an irregular cavity capable of holding about two fluid ounces. To what extent it broke up and shattered the dental tissues cannot now be seen, for the cavity in which the bullet lies has obviously been excavated by subsequent inflammatory processes, and its walls are formed by uninjured tissues, which, from the regularity with which the plates of dentine and enamel are disposed, were obviously formed prior to the infliction of the injury (Fig. 19.53). The arrangement of the enamel, dentine and cement is perfectly normal in the whole of that part of the tooth which lies in front of the bullet hole and also in that part which lies immediately behind it. At a distance of about an inch and a half behind the bullet hole the plates of dentine and enamel become stunted and irregular in form and in the remains of the last plate osteo-dentine has taken the place of true dentine.
>
> The main interest of the specimen lies not so much in the modifications brought about in the development of its posterior portions, which are surprisingly slight, as in the extraordinary extent to which dentine, enamel and cement previously formed have been excavated and removed by the morbid processes set up in the pulp.
>
> It is difficult, and indeed impossible, to decide at what precise point the bullet had struck; but judging from the few data remaining, it would

appear to have been above the level of the common pulp-chamber, a little below the point indicated at (d) in Figure 19.53 A. But the exact original point of entry is obscured by subsequent processes of absorption, and redeposition of calcified material, by which all the edges of the hole have been rounded down and smoothed.

Intense inflammation of the pulp seems to have ensued, and the existing walls of the pulp cavity have been attacked by absorptive action, even at a distance from the point of injury. Thus at (h) and (h_1) (Fig. 19.53 B), the dentine which must once have been solid in these plates, just as it yet remains in the younger plates behind them, has been removed, so that hardly a vestige of dentine remains in them, though the enamel is intact. In front of this, the absorption has gone on to such an extent as to weaken the tooth, until all that lies above the pulp chamber has crushed down under the force of mastication; hence at (e) in Figure 19.53 A there is a large hole leading straight into the pulp cavity, and the plates of dentine and enamel above it are gone. This clearly shows that the Elephant was making use of the tooth as a grinder even after all these inflammatory processes were in full action.

The tooth is chiefly remarkable as an example of the extraordinary extent to which an inflamed

Fig. 19.53. *Elephas maximus* (Indian elephant). RCS Odonto. Mus., G 127.1.
A: A molar tooth which has been injured by a bullet. (a) front part of the tooth broken down to the level of the pulp cavity; (b) front surface of the anterior plate which remains standing; (c) posterior part of the tooth; (d) irregular opening with rounded edges leading to the cavity in which the bullet rests; (e) opening into the pulp cavity; (f) opening leading to the cavity in which the bullet rests. The plane of section can be seen.
B: Horizontal section through the tooth. (a) front part; (e) posterior part; (e) opening into pulp cavity; (g) iron bullet lying in cavity; (h), (h_1), (h_2), (h_4) plates from which the dentine has been removed by resorption; (k) plate but little affected by resorption. (l) plate formed prior to injury, but almost all resorbed; (l_2) plate almost normal; (i_2), (i_3), (i_4) distorted plates formed subsequently to injury. (Colyer, 1915.) $\times c. 0.3$.

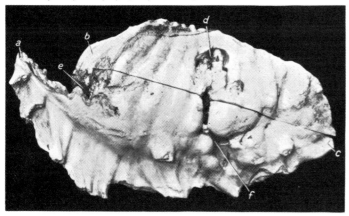

A

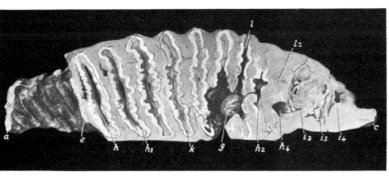

B

pulp has burrowed about, and eaten its way through pre-existent hard tissues of the tooth; and what is rather remarkable is, that this absorbent action has gone on with the most activity at some distance in front of the injury. Thus the greatest amount of absorption seems to have happened in front of (e); the dentine has been most completely removed from the plate (h), less so from (h_1) and still less from (k). The absorption in this front part of the tooth is curiously regular, and is limited to the removal of the dentine from inside the enamel, as is the case also in the plate (h_3) behind the injury.

But behind this point absorption has drilled out holes of irregular form, and has eaten away any tissue which happened to come its way indiscriminately.

A similar specimen was described by W. D. Miller (1899). A leaden bullet passed through the alveolar process and slightly penetrated a mandibular third molar which was still unerupted. The force of the blow ruptured the tooth and completely distorted the plates, which at the time were in the process of formation. A cross-section of the tooth is shown in Fig. 19.54. The injury was repaired without suppuration, the leaden bullet being enclosed in cementum of normal structure.

The difference in the type of repair in these two specimens is due to septic infection in the first instance, but not in the second. In the specimen recorded by C. S. Tomes (1877) the bullet was of iron, a metal liable to oxidize and so set up more irritation, the wound was made in an erupted tooth, and there was probably direct infection from the mouth. In the one described by Miller, the bullet was of lead, a metal that is easily tolerated by the tissues; the track made by the bullet would have been rapidly filled with blood, and so infection of the injured tooth from the mouth was prevented.

The teeth, particularly of carnivores, may be injured when animals are caught in traps during their efforts to free themselves. Kuehn *et al.* (1986) found injured teeth, mostly not severely so, in over half the wolves caught in traps for marking and release.

The carnassial teeth of carnivores, especially those of felids, are subject to great strains in tearing flesh and crunching bones; it is perhaps not surprising that the teeth occasionally fracture. The right M_1 of a lion in Figure 19.55 has been split obliquely through

Fig. 19.54. *Elephas maximus* (Indian elephant). Section through a mandibular third molar which has been injured by a lead bullet. (Miller, 1899.).

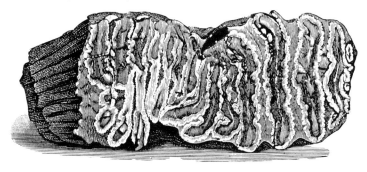

Fig. 19.55. *Panthera leo* (lion). Captive. RCS Odonto. Mus., G 102.1. Right half of the mandible with an oblique fracture through the crown of the first molar. (Colyer, 1915.) $\times c. 0.55$.

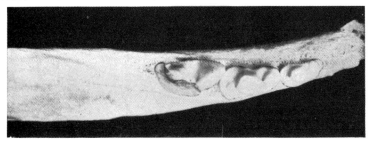

the pulp cavity. The two parts of the split tooth are firmly implanted in the jaw and the bone shows only slight inflammatory change and no signs of suppuration.

Even in domestication, the slender canines of the cat appear to be liable to injury. Schlup (1982) found various degrees of fracture of one or more canines, usually the upper ones, in 14% of 200 cats brought to a clinic for other reasons. Fractures were much more common, as would be expected, in house cats living partly out of doors than in more pampered pure-bred cats. In some cases, merely the tip of the canine was broken off but in about 38% of the fractures the pulp was exposed and 10% were abscessed. It is interesting to compare these minor injuries to the avulsion of the massive canines of the relatives of the felids, the sabre-toothed tiger (*Smilodon californicus*) described by Shermis (1985).

Among the seals, injury, including wear, of the teeth seems only to be common in the leopard seal (*Hydrurga leptonyx*), probably because it is more aggressive than other seals and tends to attack and feed on them. Hamilton (1939) described the skull of a particularly large female leopard seal (formerly RCS Osteo. Series, 1902) in which many teeth were broken, including a canine and most of the upper cheek teeth. Several teeth had been lost, presumably by becoming abscessed following fracture.

Neuville (1932) described several examples of fracture of the tips of sperm whale teeth; several showed evidence of wear indicating considerable use after the fracture. According to L. Watson (1981), teeth become broken during fights between bull sperm whales when they lock jaws with each other (see Chapter 18, p. 388–389).

Occasionally the cheek teeth of equids may split longitudinally in the plane of the columns but, in domesticated horses, this is usually in teeth weakened by caries. E. Becker (1970, Fig. 261) showed several non-carious instances (see also Chapter 21, p. 438). Penzhorn (1984) mentioned an instance in a zebra (*Equus zebra zebra*) in the wild. There were irregular wear of the opposing teeth and inflammatory alveolar changes but these were probably secondary to the fracture of the tooth.

Adaptation of the dentition after injury

Injuries to jaws and teeth and loss of teeth can be a severe handicap to the animal and, in the wild, can lead to death from starvation. Examples have been given in Chapter 18 of the maintenance of a functional dentition in quite severe jaw injuries. The two following examples illustrate adaptation to less severe injuries. Figure 19.56 shows a male coypu (Bland Sutton, 1888a) which died in the London Zoo of suffocation from a large cystic bronchocele. About three weeks before death, it fractured the right I^1 and the left I_1.

Fig. 19.56. *Myocaster coypus* (coypu). ♂. Captive. RCS Odonto. Mus., G 38.2. Adaptation of the incisors following injury (Colyer, 1915.)

Fig. 19.57. *Presbytis vetulus monticola* (purple-faced langur). ♀. BMNH 1920.9.26.1. Adaptation of the incisors across the midline following loss of teeth.

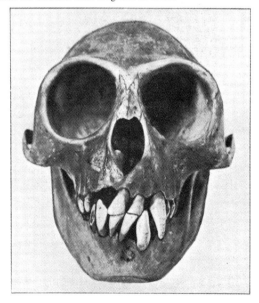

Within that short time before death, the animal had established a new relation between the fractured teeth and the left I^1 is in good gnawing relationship with I_1 of the opposite side. Had the animal survived, no doubt the fractured incisors, which would have quickly grown to the occlusal plane, would have reestablished the former relationship.

The leaf monkey whose skull is shown in Figure 19.57 lost the right upper incisors and canine and the left lower incisors from injury and was thus left with no functional incisors. There has been considerable drift of the remaining incisors across the midline to take the places of the missing teeth, so leading to the restoration of a good functional incisor occlusion.

CHAPTER 20

Enamel Hypoplasia

Introduction
Enamel hypoplasia is the term used for conditions where the formation of enamel is defective, usually in quantity, although there is a condition, usually called enamel hypocalcification, in which the enamel is of normal thickness but is not fully mineralized; therefore it does not have the characteristic translucency of enamel but is white and opaque and may be grossly softer than normal. Enamel hypocalcification is little referred to in species other than man.

There are varieties of enamel hypoplasia which are of genetic origin; these would naturally tend to affect enamel formation over the whole tooth surface and may affect both deciduous and permanent dentitions. Enamel may be virtually totally absent but, more commonly, the crown has a thin partial covering of enamel which has an irregular and uneven surface. Proof of genetic origin is heritability and this is impossible to establish in museum specimens and where the condition is only observed at death.

Statistics on the prevalence of enamel hypoplasia in various species and under various conditions, for instance captive compared with wild state, are scarce.

Systemic causes
Many systemic environmental conditions can affect enamel formation, for instance, malnutrition of non-specific or specific character and systemic infections. Disorders of the local environment of the tooth, such as local infection and trauma, can also produce lesions of the enamel.

Many systemic conditions exist only for limited periods and therefore they produce zones of defective formation in the incremental pattern of enamel formation, that is produce ring-like zones around the crown of the tooth. Because the environmental condition nearly always affects dentine formation also, these zones of defective enamel formation are associated with corresponding zones in the dentine, discovered, of course, only when the teeth are sectioned. The defective dentine takes the form of so-called interglobular dentine in which there is incomplete fusion between the globular calcospherites characteristic of dentine formation.

The nature of the zones of defective enamel formation varies a great deal and depends on the severity of the cause and the length of time it acts. If not severe and acting for only a short time, perhaps up to a few weeks, there may be merely a ring of shallow pits with some areas of enamel opacity or whiteness. If the severity is greater and the condition lasts longer, the pits would be deeper and confluent, producing ring-like grooves around the teeth. If enamel formation is severely affected for periods lasting for months, there may be a band of defective enamel, or the enamel may be totally absent, and the dentine at the base of the band-like lesions may be stained brown or black from food, or from caries.

The enamel hypoplasia, of course, only affects the tooth crowns being formed at the time when the environmental cause is acting; those with crowns already formed and those whose formation has not yet started are unaffected. Thus, from the pattern of which teeth and which parts of teeth are affected, provided that the chronology of tooth development for the species is known, as in the dog (see this Chapter, p. 447) and in man, it is possible to deduce the age at which the cause acted.

Experimentally, zones of hypoplasia can be produced in enamel and dentine in a variety of ways, for instance by keeping animals on diets deficient in vitamin D or in calcium, or in other ways which interfere with the mineralizing mechanism. Kreshover (1944) showed the effect on ameloblasts of tuberculosis in the mouse or at least of the systemic debility induced by tuberculosis. The most certain and simple way of inducing a well-defined zone of hypoplasia in the teeth, as shown by Erdheim (1906) in rats, is to remove the parathyroid glands and so induce tetany. It is possible that the effects of a number of disturbances of mineral metabolism, including vitamin D deficiency, are mediated by the parathyroid glands.

The causes mentioned only produce enamel hypoplasia if they act during the process of tooth formation. Rickets, for example, if it occurs after tooth formation is complete leaves no mark on the teeth. It perhaps goes without saying that enamel hypoplasia due to systemic causes must be expected to show a general symmetry between the two sides of the dentition though, as some of the following examples show, not always in every detail.

A type of enamel hypoplasia, endemic enamel fluorosis, with special characteristics arises from the ingestion of fluorides. In animals, the fluoride is largely derived from vegetation and therefore herbivores are mainly affected. However, the condition also affects man, where drinking water seems to be the main source of the fluoride. Endemic fluorosis is dealt with separately (p. 450) at the end of this Chapter.

The tetracycline compounds, administered to combat infection, have an affinity for mineralizing tissues and, if enamel and dentine are being formed at the time, are deposited as yellowish lines or bands which become visible when the teeth erupt. Because of the obliquity to the surface of the lines of growth in enamel, such lines, even if narrow, are seen as a broader, diffuse yellow or brown, band on the surface of the tooth. The colour deepens with time after eruption because of a phototropic effect. The dentine is always stained more heavily than the enamel and, because the enamel is translucent, must make some contribution to the appearance of the tooth. The tetracyclines fluoresce under ultraviolet light which facilitates the study of teeth both in the living and in histological sections. The condition has been most fully studied in children (Pindborg, 1970) and it is known that the depth of staining depends, not only on the dosage and length of the period of administration, but also on the actual compound used. Oxytetracycline stains with less intensity than tetracycline itself and dioxytetracycline is said not to produce staining. Intermittent dosing with intervals as short as a few days produces a series of sharply-defined lines in the dental tissues. The drugs cross the human placental barrier so that the drug administered in the latter half of pregnancy can affect the infant's teeth.

These drugs are widely used in veterinary practice and in the dog at least are known to produce the appearances described above (Catcott, 1968).

Tetracycline is known to have a harmful effect on enamel formation in children, producing defective enamel as well as staining (Witkop and Wolf, 1963); this was observed by Bennett and Law (1965) in their experimental administration of tetracycline to dogs.

The quality of structure of enamel is relevant to its susceptibility to dental caries; this is discussed in particular on p. 458–459 of Chapter 21.

Local causes

Enamel hypoplasia may be produced by localized conditions that affect single teeth or single areas of the crown; for instance abscesses and trauma. The crown of a permanent tooth that succeeds a deciduous one is in such close relationship to the root of the deciduous tooth that an abscess on the root of the latter can produce a localized effect on the enamel organ of its successor and can cause a patch of defective enamel. Binns and Escobar (1967), for example, produced areas of defective enamel experimentally on premolars in dogs by creating abscesses on the deciduous predecessors and were able to study the resulting damage to the ameloblasts.

Large-scale examples of the effect of trauma on tooth development are described in Chapter 19. Trauma that does not lead to actual displacement of the dental tissues is hard to diagnose. Some possible examples are referred to later in this chapter.

The range of enamel defects that can be produced by trauma in sheep incisors was shown by Suckling (1980); she produced localized defects ranging from enamel opacities, considered to be due to interference with the mineralization phase of amelogenesis, to absence of enamel, considered to be due to interference with the secretory phase. The defects were produced by inserting an instrument close to the tooth germ but avoiding direct damage to the already-formed enamel surface. Deliberate damage of tooth germs was produced in neonatal guinea pigs and kittens by Santoné (1937) and in neonatal rats by Levy (1968) by passing a needle through the skin over the jaw. Localized areas of enamel hypoplasia were the most common results but often larger-scale damage to the dental formative tissues was produced (see Chapter 19).

Anthropoid Apes

In the wild state, slight degrees of what must be called enamel hypoplasia are common in both the apes and monkeys, though less so than in animals in captivity, and really severe forms are rare. The most common variety consists of slight transverse, but sometimes longitudinal, grooves and shallow depressions. Frequently only the canine teeth are affected, usually symmetrically, the mandibular ones more than the maxillary (Figs. 20.1, 20.2); one gorilla from the wild had nine grooves on each mandibular canine. The defects are usually on the cervical parts of the crown; the occlusal halves of the crowns tend to be less severely affected, but this may be because minor grooves are worn away during function. This may well be true also in the gibbon (Fig. 20.2) where the tips of the mandibular canines lack the grooves which are very evident on the cervical parts of the crowns. The enamel of the maxillary canines is smooth from wear and could originally have borne shallow grooves.

Schuman and Sognnaes (1956) found that slight transverse grooves were common on the surface of the labial enamel in a large collection of skulls of chimpanzees from the wild in the Peabody Museum, Harvard University. The enamel of the cheek teeth was commonly grooved also or showed shallow irregular depressions. The enamel defects were similar

Fig. 20.1. *Gorilla gorilla* (gorilla). Wild state. P-C. Mus. FC 133. Grooves typical of enamel hypoplasia of systemic metabolic origin across the incisors and canines.

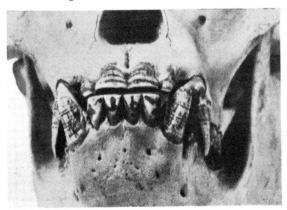

Fig. 20.2. *Hylobates concolor gabriellae* (Gabriel's gibbon). Wild. BMNH 1908.11.1.1. A series of spaced grooves on the mandibular canines and first premolars.

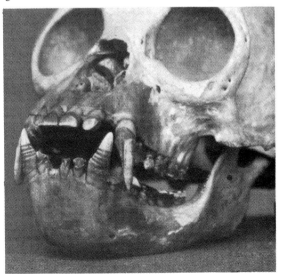

to Mellanby's category of 'slight enamel hypoplasia' (p. 445 of this Chapter). Sections of the teeth showed that the enamel defects were not associated regularly with any corresponding defects in the incremental pattern of the dentine, although zones of defectively mineralized interglobular dentine were present in 17% of the 42 teeth that were sectioned. This lack of correspondence between enamel lesions and the dentine judged to have been formed at the same time was also found in a few teeth examined of gorilla, gibbon and orang-utan, in contrast to the rhesus monkey where such defects were notably absent. The findings of Molnar and Ward (1975) on the microstructure of the enamel and dentine in five anthropoid apes, four baboons and seven macaques, some captive and some from the wild, were similar. Exaggerated incremental lines in enamel and dentine, interpreted as indicative of defective mineralization, were common in both apes and cercopithecoids but tended to be more severe in the apes. As the numbers were so small, no significance can be attached to the slight differences between wild and captive.

An example of enamel hypoplasia in a captive chimpanzee is shown in Figure 10.20.

Some observations by Jones and Cave (1960) suggest that enamel hypoplasia is more common in the wild than is generally supposed. They found six examples among twelve adult chimpanzees wild-shot in Sierra Leone. One example, in a young adult female,

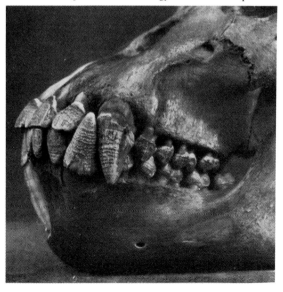

Fig. 20.3. *Pan troglodytes* (chimpanzee). Probably wild. Formerly RCS Osteo. Series, 7. Widespread enamel hypoplasia suggestive of a genetic origin because not in accord with a particular chronology of tooth development.

was severe; there were ring-like grooves, involving the whole thickness of the enamel, across the middle of the crowns of the first incisors (the second incisors were lost post mortem); a little beyond the tips of the canines, there were grooves which were deep enough to weaken the teeth so that one tip had broken off. There were also grooves close to the tips of the cusps of the first premolars. From the little exact information there is about the chronology of tooth development in the chimpanzee, these lesions would accord with a short severe metabolic disturbance at about birth. The other five examples showed much shallower transverse grooves with a similar distribution.

The chimpanzee shown in Figure 20.3 was in the Osteological Collection of the Royal College of Surgeons, having been presented in 1851 as probably from the wild. The specimen was unfortunately destroyed during the war in 1941 but the enamel was described as being of milky appearance. All the enamel surfaces of all the teeth showed a pattern of closely-set grooves, predominantly transverse but with longitudinal elements, and also many minute pits. The enamel hypoplasia is rather different from most hypoplasia of probable systemic origin, partly because it involves all the enamel of all the teeth, suggesting the possibility of a genetic origin, especially as it resembles some instances of known heritable enamel hypoplasia in man (Schultze, 1970). A similar condition occurred in two other chimpanzees, one wild and one captive, and in three captive orang-utans. Jones and Cave (1960) described similar enamel hypoplasia in a chimpanzee that had lived in the London Zoo for 24 years. Most of the enamel surface of all the incisors and canines, and to a less extent of the post-canine teeth, was uneven. This animal would have been born in the wild, and, whatever the cause, it would have acted over a long period beginning before birth. In the absence of captive pairs that might be bred from, it is impossible to do more than speculate whether some of the enamel hypoplasias described here are of genetic origin.

No doubt enamel hypoplasia occurs in the orang-utan although we know of no example to refer to, but, if met with, it has to be borne in mind that a wrinkled surface is a normal feature of enamel in this species, especially of the cheek teeth (Selenka, 1898).

A rare example of severe enamel hypoplasia in a primate in the wild is shown in the gibbon in Figure 20.4. Most of the enamel over the incisive halves of the incisors is missing; there is then an abrupt change to enamel of normal thickness over the cervical halves of the teeth. The other teeth were unaffected, perhaps because only the occlusal surface of the first permanent molars and tips of the canines would have been forming at the time of the systemic cause of the lesions. The cause here, as in the other instances described, can only be speculated on; perhaps a period of malnutrition during a prolonged drought.

Monkeys

Mild enamel hypoplasia, in the form of slight transverse and longitudinal grooves and shallow depressions, is common in certain genera of monkeys, especially in baboons (*Papio* spp.) and mangabey monkeys (*Cercocebus* spp.) (Colyer, 1947b).

Colyer in the first edition of this book referred to having observed enamel hypoplasia in 7 out of 288 baboon skulls from the wild. The particulars were as follows:

(i) In two, the incisors and canines were affected.
(ii) A row of pits across the labial surfaces of the upper first incisors.
(iii) Two well-defined grooves on the I^1, slight pitting on the second incisors, a patchy appearance of the enamel of the lower incisors.
(iv) In two specimens, the defect was limited to one of the I^1.
(v) In one specimen, the lower incisors and the right I^1 were affected, the other incisors being free from any defects.

In baboons kept in zoos, enamel hypoplasia is particularly common and well-studied, perhaps because they are often brought into captivity at an early age and are often available for study in large numbers. Even in normal baboons (Fig. 20.5), the crowns of the upper first incisors do not have the even surface found in most normal teeth and may show slight transverse grooving which appears to be an exaggeration of the normal imbrication lines of enamel and is part of the generalized unevenness of the enamel surface in orang (Selenka, 1898).

Figure 20.6 shows the incisor teeth of an adult captive baboon. The enamel of the incisive third of the crowns of the two upper incisors is irregular, pitted and defective; the corresponding lower incisors are not affected. It is impossible to say that the cause here was not some infection or trauma to the deciduous predecessors (this Chapter, p. 439) but a point in favour of the cause being some sort of systemic rickets-like condition is that the bones of the skull appeared to be friable and imperfectly mineralized.

The incisor teeth in Figure 20.7 are from another baboon, with skull bones that appear to be poorly mineralized. Both I^1 show widespread enamel defects on the incisive halves of the crowns; there are lesser defects a little lower down on the I^2 and zones of irregular rough enamel close to the necks of the lower incisors. There is by no means perfect symmetry between the two sides. It must be assumed that the crowns of the lower incisors were already formed at the

Fig. 20.5. *Papio* sp. (baboon). Labial aspects of the normal incisors. However, note the axial groove on the root of the left I^1 (Chapter 3, p. 19).

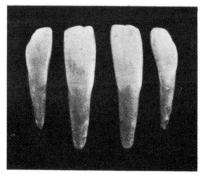

Fig. 20.4. *Hylobates lar* (common gibbon). Wild state. AMNH 54660. Enamel hypoplasia of the upper incisors.

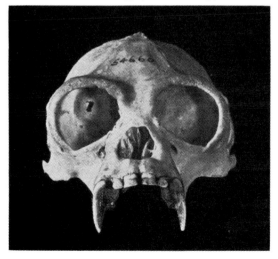

time when the systemic cause, whatever it was, acted and were certainly complete, with root formation proceeding, by the time the cause ceased. The condition of the upper incisors suggests that the cause started to act when the formation of their tips had just begun, and that it continued to act for several months until about one-half of each crown had been formed.

A young captive baboon with ring-type enamel hypoplasia indicative of three episodes of systemic disorder is referred to on p. 210 (Figure 10.103).

The mandible of a baboon in Figure 20.8 shows enamel hypoplasia of the groove type confined to the labial surfaces of the incisors and canines; the pits in the surface of the enamel are shallow; there is a deficiency of tissue alternating with an approximately normal amount; the teeth show longitudinal grooves and the enamel appears to be fairly well-formed. The crowns of the premolars are pitted and there are shallow grooves across the first and second molars; the upper halves of the crowns of the third molars are affected. One M^3 is shown in Figure 20.8B; a microscopic section of the same tooth (Fig. 20.8C) shows a typical zone of interglobular, poorly-mineralized dentine corresponding with one of the periods of defecive enamel formation.

Colyer (1921) found three examples of enamel hypoplasia of single teeth among wild-caught rhesus monkeys which he attributed to sepsis or trauma during development, but Schuman and Sognnaes (1956) found a notable absence of hypoplastic defects in the enamel and dentine in 103 skulls of rhesus monkeys from the wild.

The teeth in Figure 20.9 are from a captive black ape (*Macaca nigra*). They show many closely-set zones of defect across or around the tooth, too closely-set for there to be any really normal enamel in between. This indicates a long period of malnutrition or general debility for the whole period when the crowns were being formed, with many periods of partial remission. The skull bones were without any gross evidence of disorder.

An example of ring-type enamel hypoplasia in a captive macaque is shown in Figure 10.90.

What appeared to have been a generalized softness and opacity of the enamel, and stunting of some teeth, was found in young rhesus monkeys (*Macaca mulatta*) maintained by Berdjis *et al.* (1960) on a diet deficient in vitamin B_6 (pyridoxine) for periods up to 16 months. Unfortunately, it is unclear whether only enamel being formed during the experimental period was affected so that the experiment is impossible to evaluate.

Figure 20.10 shows the skull of an adult mangabey (*Cercocebus*) from the wild in which there are two well-spaced ring lesions in the enamel of the first incisors in both jaws and only one groove in the I^2 which, it must be assumed, corresponds to the cervically-placed groove in the I^1. It is clear that the formation of I^2 began after the animal recovered from the systemic disorder that caused the first ring lesion near the incisive edge of I^1.

Enamel hypoplasia may occur in colobus monkeys in the wild. In an Abyssinian guereza (Fig. 20.11), it was so severe that the dentine of the incisors seems to

Fig. 20.6. *Papio* sp. (baboon). Captive. RCS Odonto. Mus., G 52.15. Labial aspects of the incisors. Both I^1 show enamel hypoplasia, slightly more severe on the right side.

Fig. 20.7. *Papio anubis* (olive baboon). Juvenile. Captive. RCS Odonto. Mus., G 52.13. Labial aspects of the incisors showing enamel hypoplasia.

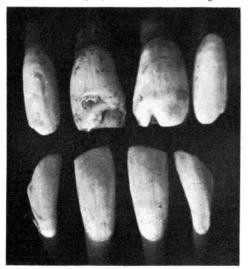

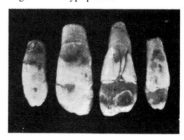

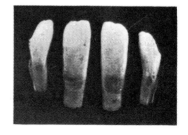

be devoid of enamel except for a few patches and at the cervix. In another, a *Colobus guereza caudatus* (Fig. 3.32), originally described by C. S. Tomes (1898) and described in Chapter 3 (p. 38), many of the teeth, as well as having deficient enamel, are unusually small and some appear to have been congenitally absent.

However, microscopy showed that what enamel was present was of normal structure and properly mineralized. Another colobus monkey (Fig. 3.33) showing similar features was described by Remane (1926). Several teeth were absent, though it is impossible to be certain that they were not lost from accident

Fig. 20.8. *Papio* sp. (baboon). ♂. RCS Odonto. Mus., G 52.16.
A: Lateral view of mandible showing hypoplasia of the teeth. A window has been cut to display the teeth. ×c. 0.85.
B: An M^3 from the same animal.
C: Part of a longitudinal ground section through the molar in B, showing the grooves in the enamel on the buccal surface, the disturbance of the growth lines in the enamel and a zone of interglobular dentine.

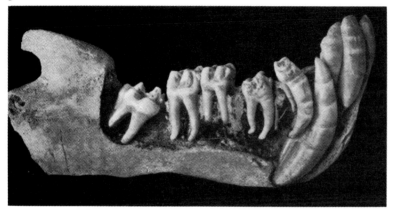

A

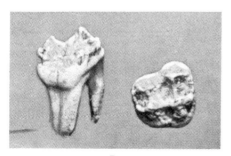

B

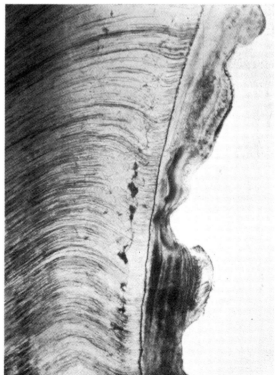

C

444 Section 4: Other Disorders of Teeth and Jaws

or disease, and several were of dwarf-like proportions. Although the animal was fully adult, several deciduous teeth were still functional and their successors presumably absent.

The fourth example (RCS Odonto. Mus., G 54.4)

Fig. 20.9. *Macaca nigra* (black ape). ♀. Juvenile. Captive. RCS Odonto. Mus., G 54.22. Labial aspects of the incisors. I1 show many closely-set zones of defect across the teeth.

Fig. 20.10. *Cercocebus albigena ituricus* (grey-cheeked mangabey). Wild. P-C. Mus. C 264P. Two well-marked grooves across each first I1 and a single line on each I^2, chronologically corresponding to the upper line in the I1.

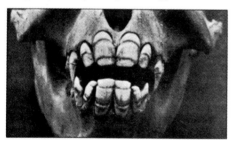

Fig. 20.11. *Colobus guereza gallarum* (Abyssinian guereza). Wild. FMNH 27055. The incisal halves of the

is an adult female *Colobus polykomos palliatus* shot in Tanganyika in which the enamel, especially of the M3, appears to be defectively formed and friable (Colyer, 1943b). There are yellow-brown patches and elsewhere most of the enamel is milky-white.

Only two of the specimens referred to are of the same *Colobus* species and at least three of them came from widely separated parts of Africa. Hence, it is doubtful whether the opinion Colyer expressed in the first edition of this book, that colobus monkeys are particularly susceptible to enamel hypoplasia, is sustainable.

Ring-like enamel hypoplasia is met with in the New World capuchin monkey (*Cebus*, Fig. 20.12). Colyer in the first edition of this book reported finding 11 examples in 538 specimens in various museums. The defect was limited to the canines, often to the mandibular ones.

Lemurs

An example of enamel hypoplasia in a captive lemur is shown in Figure 10.162.

Carnivores

A crinkled appearance of the enamel resembling hereditary enamel hypoplasia in man (Schultze, 1970) is not uncommon in civet cats (*Viverridae*) in the wild, for example in the young teeth of palm civets (*Paradoxurus*, *Paguma* and *Arctogalidea*), binturongs (*Arctictis*) and otter-civets (*Cynogale*). This wrinkled enamel is marked in two specimens of a rare civet from the Celebes in the British Museum (Natural History) one of which is depicted in Figure 20.13.

Colyer (1907) described widespread patches of

crowns of the incisors are devoid of enamel apart from a few islands.

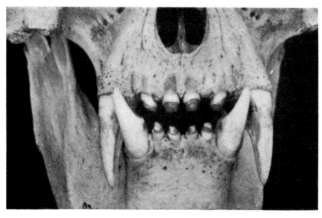

enamel hypoplasia in the teeth of a captive lion (RCS Odonto. Mus. G 52.1) aged about five years which at necropsy showed evidence of acute polyarticular arthritris.

In experimental studies of mineral metabolism in general and of enamel hypoplasia in particular, the dog has been much used. In this area, the work of Mellanby (1929) is notable. She induced enamel hypoplasia and the associated more occult hypoplasia of dentine with a variety of diets deficient in a number of mineralizing factors, particularly vitamin D. However, as can be seen in Figure 20.14, the enamel hypoplasia produced did not match exactly the ring-like horizontal bandings characteristic of spontaneous enamel hypoplasia, perhaps because the deficient diets covered the whole period of formation of the tooth crowns and the deficiency level was maintained so strictly constant that there were none of the intermissions seen in natural disease and malnutrition. There was even wrinkling of the enamel in the vertical axes of the teeth (Fig. 20.14C).

Enamel hypoplasia is a common sequel of canine distemper, affecting as it does young puppies in the tooth-formative period (Catcott, 1968; E. Becker, 1970, pp. 229–31) and is known among veterinarians as distemper tooth although other causes can less commonly be responsible (Kuiper and van der Gaag, 1982). According to Bodingbauer (1946, 1949a, 1955b), who made an impressive study of this condition, the deciduous dentition appears to escape, although it is occasionally affected in the offspring of bitches that suffered from distemper during pregnancy; abortion is the more common outcome.

In distemper enamel hypoplasia, the lesions are of the ring type and resemble those already described; they vary from rows of shallow pits to broad bands where enamel is missing and dentine is exposed (Rossman, Garber and Harvey, 1985, fig. 6–5), depending on the severity of the initial attack by the distemper virus, and perhaps even more on the severity and length of the period of general debility associated with the infection. Bodingbauer, however, did not find a close connection between the severity of the attack of distemper, the occurrence of enamel hypoplasia and the severity of the enamel lesions. Dubielzig (1979, 1980) described changes in the ameloblasts of puppies

Fig. 20.13. *Macrogalidea musschenbroekii* (brown palm civet). ♂. BMNH 1898.11.10.1. The enamel of the teeth shows a condition resembling hereditary hypoplasia.

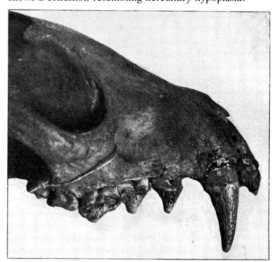

Fig. 20.12. *Cebus apella* (brown capuchin). Wild. BMNH 1911.6.7.1. Transverse grooves on the canines.

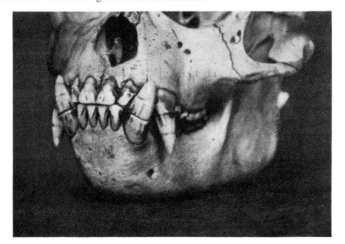

446 Section 4: Other Disorders of Teeth and Jaws

with distemper that are suggestive of virus infection, for instance the presence of intra-cellular inclusion bodies. Furthermore, Dubielzig, Higgins and Krakowka (1981) reported similar changes in the ameloblasts of gnotobiotic dogs by infecting them with distemper-virus particles obtained from the spleens of dogs suffering from distemper. Nevertheless, the possibility still remains that some associated or secondary nutritional influences play a part in the production of distemper enamel hypoplasia.

Fig. 20.14. *Canis familiaris* (domestic dog). The buccal surfaces of the mandibular carnassial teeth of four puppies. (A) on a normal diet, (B)–(D) maintained on diets to progressive degrees deficient in vitamin D. (B), slightly D-deficient producing slight irregularity of the enamel surface. (C), more severe D-deficiency producing enamel hypoplasia in the form of wrinkling of the surface. The relationship of the defects to the incremental pattern is not marked but there is some faint horizontal grooving in the lower half. (D), severe D-deficiency producing severe enamel hypoplasia with bare dentine close to the tip of the cusp. (E), longitudinal ground section through the tooth in (D), showing the thinness of the enamel layer and its absence over large areas (Mellanby, 1929).

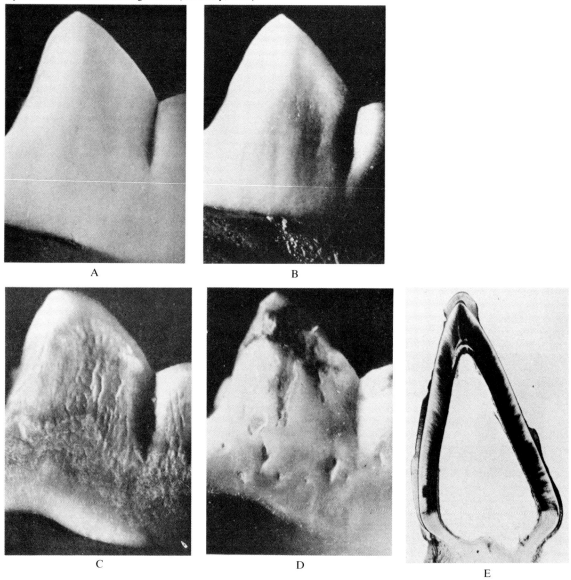

Figure 20.15 shows the distribution and extent of hypoplasia lesions in the dentition of a dog affected by distemper. The distribution of the lesions is similar to those depicted in Figure 20.16A from another dog (Bodingbauer, 1949a) known to have suffered an attack of distemper at four months of age. The undulating line indicates that the amount of each tooth crown that had been formed at the time of the illness corresponds with the chronological chart for tooth formation stages in the dog (Fig. 20.16B). From the width of the bands of hypoplastic enamel and further reference to the chart, it can be inferred that the period of disturbed metabolism lasted for about six weeks.

Eight instances were described by Arnbjerg (1986) in which enamel hypoplasia apparently due to distemper-like infections was accompanied by stunting of the roots of the teeth.

Bodingbauer believed that, in some of the experiments in which dogs have been used to investigate the relationship between rickets and enamel hypoplasia, the possibility of the animals being infected by the distemper virus has been insufficiently guarded against. For instance, he claimed to find evidence in the experimental regime and history of the animals in some such experiments by Bauer (1925) that several of the animals suffered from distemper during the prolonged nutritional experimental period. It is interesting to note that, having scrutinized her publications, Bodingbauer excluded the experiments of Mellanby (1937) from this stricture.

An adult dog in which the whole of the crowns of the teeth were brown and translucent was described by Meyer, Suter and Triadan (1980). The dentine structure was grossly dysplastic and, on the basis of the similarity of the dentine structure to well-authenticated instances in man, the diagnosis was made of odontogenesis imperfecta of genetic origin, although the heritability could not be established.

Nellis (1972) found enamel hypoplasia in three skulls of wild coyote (*Canis latrans*) out of a group of 489; he stated that, in the most severe one, the lesions were similar to those depicted here in a dog (Fig. 20.15).

Ungulates

An example of enamel hypoplasia in a foal, associated with congenital absence of teeth, is shown in Figure 20.17. The jaws are slender, particularly the tooth-bearing ramus of the mandible. The teeth present are deciduous. In the upper jaw on each side, there are only two incisors instead of three and two erupted cheek teeth; two others are still in their crypts. In the lower jaw on each side, there are two incisors, one erupted cheek tooth and another unerupted in its crypt. Further back in the jaw, there is on each side a crypt containing a small toothlet and more distally still several other small crypts only one of which contains a minute toothlet. The other crypts may have contained tooth germs which had not begun to mineralize (Colyer, 1913). Radiographs (Fig. 20.17B) show total absence of permanent successors. The occlusal parts of

Fig. 20.15. *Canis familiaris* (dog). RCS Odonto. Mus., G 139.2. The animal had suffered from distemper. The bands of affected enamel on the incisors and canines correspond approximately with the enamel affected in Fig. 20.16A; hence the attack of distemper was probably at about 4 months of age. × c. 0.68.

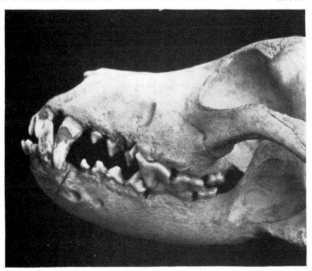

448   Section 4: Other Disorders of Teeth and Jaws

Fig. 20.16. *Canis familiaris* (domestic dog).
A: Diagrammatic representation of the right half of the permanent dentition of a bull terrier with enamel hypoplasia associated with an attack of distemper at 4 months of age. The shaded areas indicate affected enamel; the black areas indicate where enamel was totally absent. The other half of the dentition was similarly affected. Asterisks indicate the position of the teeth that were congenitally absent. The undulating line indicates the amount of each tooth crown formed at the onset of the illness and corresponds with that formed at 18 weeks in the chart B below. The margins of the defective bands of enamel are associated with recovery from the illness and would correspond approximately with the 23-week line in chart B (Bodingbauer, 1949a).
B: Chronological chart of formation (mineralization) of the permanent dentition in the dog based upon the clinical and radiographic examination of medium-sized orthognathic varieties (e.g. terriers, Dobermann). Line 1, 15 weeks; Line 2, 17 weeks; Line 3, 18 weeks; Line 4, 21 weeks; Line 5, 23 weeks (Bodingbauer, 1946).

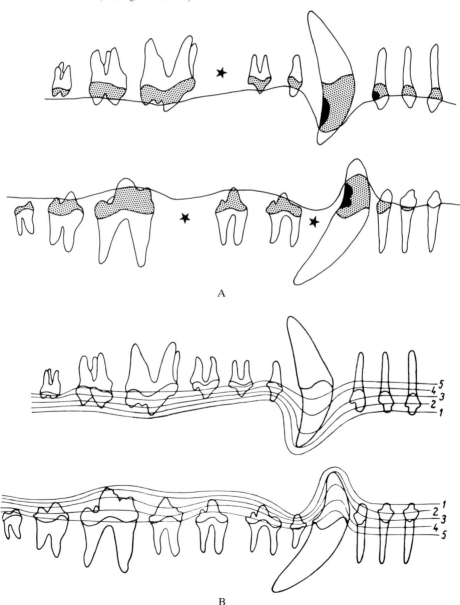

Fig. 20.17. *Equus caballus* (horse). RCS Odonto. Mus., G 52.11. The skull of a foal with congenital absence of teeth; enamel absent or very thin over many teeth.
A: The jaws, especially the tooth-bearing part of the mandible, are unusually slender. The erupted teeth are all deciduous; they are small and deformed principally because they lack enamel. A mandibular incisor was lost in preparation of the specimen. ×c. 0.33.
B: Radiograph of the upper incisors. The shadow normally cast by enamel is absent and there are no permanent successors.
C: Radiograph of the mandible. There are two erupted cheek teeth and some crypts containing unerupted deformed toothlets. ×c. 0.33.

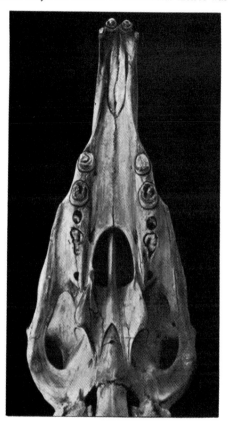
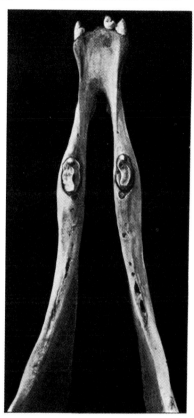

A

B

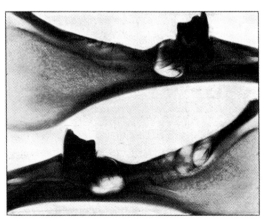

C

the incisors and the cheek teeth show, as far as can be determined with the naked-eye, total absence of enamel. The more distal teeth in the upper jaw are quite unworn because of the absence of their antagonists. The enamel over the remainder of the teeth is of normal appearance and thickness. We believe that, if this enamel hypoplasia was part of the genetic disturbance responsible for the absence of teeth, all the enamel would be defective. The parts of the teeth affected were formed before birth; gross nutritional or other systemic disturbances in the fetus seem to be rare, perhaps because, if they do occur, they lead to abortion. Whatever maternal or fetal disturbance was responsible for this enamel hypoplasia, there must have been some recovery which, from what is known of the chronology of tooth development in the horse, would have been soon after birth. Unfortunately, nothing is known of the history of this animal.

As tooth enamel is the product of ectoderm, it tends to be affected in genetic ectodermal dysplasia (see p. 40, 88 and 115). A gross example of this relationship was described by Dubielzig, Wilson, Beck and Robbins (1986) in a foal which only survived for two days. There was almost complete absence of epithelium from the surface of both skin and oral mucosa. The dentine parts of the teeth were of normal morphology covered by cementum with, in the crown parts, only a thin irregular layer of enamel matrix intervening.

Garlick (1954b), in an account of dental disorders in a group of 7480 cattle slaughtered for beef, reported a total of about 19% with various types of enamel hypoplasia. Fourteen per cent had simple flecks of opaque enamel and 3% had what appears to have been pitted hypoplasia. Garlick mentioned a type of hypoplasia which tends to affect symmetrical pairs of teeth in which the enamel is brown and less resistant to wear than normal. This appears to be an enamel hypocalcification that could be of genetic origin and heritable. Little appears to be known about the cause of any of the other types described.

Enamel defects in the form of areas or transverse zones of opacity or of defective enamel thickness occur in sheep (Suckling, 1979) and from their distribution appear to be due to systemic causes; for example, periods of drought or, as suggested by some experiments on young sheep by Suckling, Elliott and Thurley (1983, 1986), infection with nematode intestinal parasitic worms.

Henrichsen (1982) noted thinness of the enamel layer of certain teeth, predominantly M^3, among musk ox (*Ovibos moschatus*) in Greenland and various Alaskan islands. In an area of E. Greenland, the prevalence of this enamel hypoplasia was higher than 50%; only half of the lesions were bilateral. Henrichsen appeared to have in mind that this hypoplasia is of genetic origin because the musk ox was introduced into these areas at various times since about 1935 and he lists enamel hypoplasia together with congenital absence of certain teeth and abnormalities of shape of molars, which are also common (see Chapter 6, p. 114), as possibly indicative of genetic differences between the Canadian and European populations from which the Greenland and Alaskan animals were derived.

Enamel fluorosis

A condition known as mottling of the enamel or endemic enamel fluorosis has been recognized in animals and man in many parts of the world where the soil, and thus the drinking water and vegetation, has a high content of fluoride (Shupe and Olson, 1980). In addition, the condition is met with in animals grazing in the vicinity of factories, usually aluminium smelters, where the atmosphere is contaminated by the emission of fluoride-laden smoke. Dust particles contaminate the soil and settle on the growing vegetation. Another potential source of fluoride is the rock phosphate sometimes provided as a lick for the animals to provide a supplementary source of calcium and phosphate (Seddon, 1945) or added in a ground-up form to artificial animal feeds. Hillman (1979) found an association between enamel fluorosis and acute gastric dilatation in dairy cattle; there was evidence that this was mediated by acute hypocalcaemia induced by commercial cattle-feed containing fluoride-rich ground rock-phosphate substituted for bone meal. Severe fluorosis was reported in cattle fed with rock-phosphate supplements which, by accident, had not been subjected to the usual defluorinating treatment (Shlosberg, Bartana and Egyed, 1980). Fluorosis in guinea pigs that was due similarly to rock phosphate in pelletted food is referred to in Chapter 17, p. 362.

Enamel fluorosis can result from the consumption of fluoro-organic compounds. Loeffler *et al.* (1979) described discrete brown patches and broad yellowish bands on many teeth in three dogs whose food was accidentally contaminated with a fluoride-containing wood preservative. Both teeth and bone had a high fluoride content.

Endemic enamel fluorosis differs substantially from the enamel hypoplasia described; because the cause tends to act continuously during the lifetime of the animal, all the enamel of all teeth tends to be affected. There is some evidence that a high calcium diet tends to protect against the effect on teeth and skeleton of high fluorine intake (Franklin, 1950; reviewed by Shupe, Ammerman and Peeler, 1974). Selenium is another trace element in the soil that, in excess, can have a harmful effect on the well-being of grazing animals and thus, at least hypothetically, on their teeth (Hadjimarkos, 1970); manganese and molybdenum are also suspect in this respect (C.F. Mills, 1970).

The literature on endemic fluorosis seems to relate, apart from man, almost exclusively to domesticated grazing animals. It may be that grazers consume more fluoride from soil-dust deposited on grasses and plants close to the soil than leaf-eaters and carnivorous animals. Sheep tend to ingest considerable quantities of soil during grazing (Healy, 1968). Figure 20.18, however, shows a guenon monkey, a leaf eater, with patches of pigmented enamel on the incisors and premolars similar to those of fluorotic mottled enamel. There were also transverse grooves on the canines lined with thin pigmented enamel.

Although the deciduous dentition tends not to be affected (Franklin, 1950, p. 25), even though the bone ash of fetal calves may be rich in fluorine, occasionally it is severely affected. Crissman, Maylin and Krook (1980) illustrated severe fluorosis of all the deciduous dentition in a 6½-month-old heifer in an area of industrial-fluoride atmospheric pollution. According to Velu (1932) and Seddon (1945), the I_1 often escape. There is no simple reason for this; I_1 are the first of the anterior teeth to begin to be formed and mineralized but the animal is about 6 months old when this happens (Brown et al., 1960) and so is well past weaning.

In the mildest forms of enamel fluorosis, there are flecks of opaque or white enamel, particularly on the labial aspects of the teeth but they are not easily seen. If the patches are large, they may become brown after the teeth have been erupted for several years. There may be opaque stripes across the teeth in the incremental pattern. In the most severe form, much of the surface of the teeth may be whitish or brown in patches (mottled enamel) and marked by shallow pits which, if confluent, may form shallow areas of rough enamel that become deeply stained by foodstuffs. It seems that, even in severe fluorosis, the full severity only develops after the teeth erupt and, when the teeth first erupt, the enamel is only patchily opaque. It is only when the teeth have been erupted for a while that shallow pits and staining occur on the surface of the enamel; pitting continues and the pits become confluent so that in the course of time most or all of the enamel surface is irregular, rough and stained. It seems well established in man, and only less so in animals, that the labial aspects of the incisors are the most affected. This, and what process is responsible for the loss of the surface enamel after the teeth erupt, has not been adequately explained. It may be that the continued ingestion of fluoride has some local effect on enamel already affected by the fluoride incorporated in it during formation. In particular, fluoride in the water the animal drinks may be responsible; this could explain why the post-eruptive changes are most marked in the incisors. Exposure to the atmosphere or to light are potential influences that have also been considered.

More is known about the characteristics of endemic enamel fluorosis in man and the above description derives in part from that literature. It also derives from the account by Velu (1932) which contains as yet unmatched descriptive detail, relating mainly to sheep, though also to horses and cattle, concerning both spontaneous and experimental fluorosis. The description accords with those given by Seddon (1945) for sheep, by Garlick (1955) for cattle, and by Shupe, Olson and Sharma (1972) for a variety of herbivores, as well as that by Milhaud, Zundel and Crombet (1980) concerning experimental fluorosis in sheep and goats and of Maylin, Eckerlin and Krook (1987) on dairy calves and of Suttie et al., (1985) on white-tailed deer.

It is difficult to examine the lesions closely in living animals and thus to make longitudinal studies to see what post-eruptive changes occur in the lesions. A major difference between the human dentition and that of herbivores is that, in herbivores, the enamel is covered by a layer of coronal cementum which, although thin, may prevent the erosive process responsible for the post-eruptive pitting of the enamel, or at least obscure it. Incidentally, in fluorotic teeth, the coronal cementum has a high content of fluoride which is presumed to enter it after the teeth have erupted (Shearer, Britton, Desart and Sultie, 1980).

Fig. 20.18. *Cercopithecus mitis stuhlmanni* (Stuhlmann's guenon). Wild. USNM 236990. There are patches of pigmented enamel on the incisors resembling enamel fluorosis.

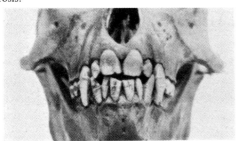

Many accounts are so lacking in description of the enamel lesions which they record as fluorosis that it is impossible to be sure that this truly is the disorder they deal with; for example, Franklin (1950) in his account of sheep maintained experimentally on a cereal diet deficient in calcium, referred merely to pitted and discoloured enamel as 'similar to enamel fluorosis' with no further detail. Certainly, however, the irregular wear of the cheek teeth in the animals is described in sufficient detail to convince that it closely resembled that found in endemic fluorosis. At least one important record is out of accord with our description of enamel fluorosis: Krook and Maylin (1979) categorized various degrees of brown discolouration as less severe forms of enamel fluorosis than what they call mottled enamel.

Ranking scales, usually numbered 0–5, have been used by various workers but there is little conformity between the systems so that comparison between various epidemiological studies is almost impossible. Study of the condition in animals would be helped if a scale could be devised as detailed as that of Dean (1934), subsequently modified by Møller (1982), which is used as a basis for studies of human enamel fluorosis on a world-wide scale. Neeley and Harbaugh (1954) have published a large series of coloured photographs of incisors of cows affected by fluorosis that come closer to what seems to be required. The rankings and photographs of fluorotic incisors given by Shupe, Ammerman and Peeler (1974) are also helpful. What many accounts fail to make clear is that enamel fluorosis of the incisors, apart from the excessive wear, appears to be a progressive condition.

The most notable feature of dental fluorosis in ungulates is the rapid and irregular wear of the cheek teeth which must be due to the softness of fluorosed enamel; however, it is said that the teeth, including the roots, are unusually brittle. Excessive wear and the abnormal wear pattern of the cheek teeth produce dramatic effects on the dentition that are harmful to the animals' well-being or thriftiness. The wear on the cheek teeth is often very uneven and does not always affect the corresponding upper and lower teeth equally. There is a tendency for the irregular wear to be greatest at the middle of the arches. Eventually not only may certain teeth be worn down almost to the gum but some teeth that oppose them tend to be left unduly long (Fig. 20.19). Furthermore, only parts of the occlusal tables of teeth may be worn excessively, for example the distal part, leaving the mesial part high and jagged. These jagged prominences may penetrate, in occlusion, approximately between adjacent teeth in the opposite arch, impinging on the gum and producing periodontal disease, which may lead to loss of teeth and consequent gross aggravation of the disorganization of occlusion. These changes naturally adversely affect the animal's food consumption and in particular affect the chewing of the cud which is so essential to digestion in ruminants. In consequence, the animals lose weight and condition. Several workers have com-

Fig. 20.19. *Ovis aries* (sheep). RCS Odonto. Mus., G 54.01. One of several animals affected by fluorosis due to grazing on pastures heavily contaminated with fluoride from the chimneys of a nearby aluminium-smelting works (Boddie, 1949). The buccal aspects of the teeth of the left side are depicted (mesial to the left) to show the irregular wear of the occlusal surfaces. The lower teeth are depicted tilted a little in order to show better the worn surfaces, including the deep crescentic infoldings of enamel.

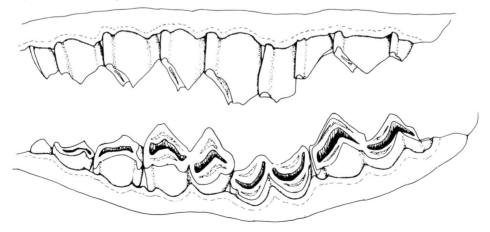

mented that affected animals seem to experience pain on drinking cold water (Boddie, 1949; Krook and Maylin, 1979; Shupe et al., 1987).

Fluorotic enamel is softer than normal as has been demonstrated by micro-hardness measurements by Shearer, Britton, Desart and Suttie (1980); they applied the tests to longitudinal sections of bovine teeth to avoid the coronal cementum and found that the outer layer of fluorotic teeth was about 40% less hard than normal, with much less change in the deeper layers; in fluorosis, there was reversal of the gradient which normally is hardest on the surface and softest close to the dentine. The tests suggested that fluorotic dentine might be harder than normal, but the difference was not statistically significant. Studying experimentally-induced fluorosis in sheep, Suckling and Purdell-Lewis (1982) confirmed that fluorotic enamel is softer than normal.

The softness of fluorotic enamel alone, if all teeth were affected equally, would not produce such gross disturbance of occlusion. Almost certainly, the uneven wear occurs because not all the teeth are equally affected by the fluorosis. The first-formed parts of the M1 may escape the full effect of the fluoride because they are formed before birth and during the pre-weaning period. In endemic fluorosis, there appears to be some trans-placental transfer of fluoride (Crissman, Maylin and Krook, 1980) and the fluoride content of milk seems to be increased (Sinclair, 1949), although not to the same extent as that of vegetation. It seems unlikely that the amount of fluoride incorporated in the teeth at those stages is high enough to alter the properties of the enamel much. Furthermore, even during the post-weaning period, the fluoride intake probably varies seasonally so that not all the teeth in the series or all parts of individual teeth are affected equally. Electron probe micro-analysis has shown that the fluoride content of the enamel of the teeth of animals given fluoride in the diet was very patchy (Shearer, Kolstad and Suttie, 1978). These inequalities of hardness could account for the unevenness in the occlusal wear plane already referred to. A factor that enhances this uneven wear is that each tooth in the arch, except those at the ends of the cheek-tooth arch, occludes with two teeth in the opposing arch. If the resistance to wear of these two is unequal, each half of each tooth would wear at a different rate with the consequence on the occlusal surfaces that is indicated diagrammatically in Figure 20.20. It is interesting to note that similar gross disturbances of occlusion developed in sheep maintained on varying levels of calcium-deficient cereal diets (Franklin, 1950) which would be expected to produce teeth of varying hardness. Shupe et al., (1987) compared the pattern of fluorosis lesions on the incisors with the pattern of uneven cheek-tooth wear in large numbers of cattle, including 176 in which the fluorosis was experimentally induced. There was some evidence that the pattern of wear was related to the timing of periods of maximal ingestion of fluoride as deduced from the incisor lesions.

Stiffness of movement and lameness are common in fluorotic animals and contribute to their poor condition because they are disinclined to graze over a large area. A lengthy account of the skeletal changes is inappropriate here but fluoride is deposited freely in the bones and, in severe fluorosis, the bones are thickened and are of increased density (sclerosis). This is thought to be due to an inhibitory effect on the bone resorptive process, and the periosteal surfaces of the bones tend to be uneven, even to the extent of formation of hyperostoses (exostoses). These changes are particularly marked in the metapodials and, in addition to urine analysis, provide clinical and radiographic features of diagnostic value (Shupe, Olson and Sharma, 1972). In very severe long-standing fluorosis, osteo-arthritic changes may cripple the animals.

The fluorosis that occurs in animals grazing in areas where herbage is contaminated by industrial atmospheric pollution is similar in respect of both tooth and bone effects. Particularly well-recorded

Fig. 20.20. Ungulate cheek-tooth rows, showing diagrammatically what would be expected to follow if teeth in the arches were of varying hardness, indicated by the numerals; 10 indicates hard and 3 soft on a progressive scale.

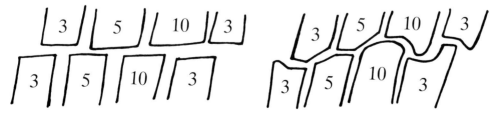

examples relate to the lee-side of a brickworks (Blakemore, Bosworth and Green, 1948), and of two aluminium smelters at Fort William, Scotland (Boddie, 1949; Sinclair, 1949), and Cornwall Island, Ontario, Canada (Krook and Maylin, 1979: Crissman, Maylin and Krook, 1980). Riet-Correa, Oliveira, Méndez and Schild (1986) reported an example in Brazilian cattle grazing close to factories grinding rock phosphate for fertiliser. As in several other instances, the factories were also close to large human urban populations.

Robinette, Jones, Rogers and Gashwiler (1957) referred to excessive wear in mule or black-tailed deer (*Odocoileus hemionus*) in the mountains of Utah, USA, close to a steel plant which emitted fluoride into the atmosphere. Chemical analysis of premolars from a young deer showed that the teeth contained 1500 parts/10^6 of fluorine which is a high figure. Typical severe dental fluorosis was described by Newman and Murphy (1979) in mule deer from an area adjacent to an aluminium plant in Washington State, USA. The bones contained over 10 times more fluoride than those of deer from another area. M2 were the most worn teeth instead of M1 (Chapter 22, p. 489). Dental fluorosis is also met with in white-tailed deer (*Odocoileus virginianus*) under similar circumstances (Kay, Tourangeau and Gordon, 1975; Suttie *et al.*, 1987).

Karstad (1967) described a double fracture of the mandible in a white-tailed deer with signs of fluorosis (Chapter 18, p. 384). The toxic elements As, Pb and Cd, as well as F, have been found in the teeth of game animals, such as roe deer, in the vicinity of Al-smelting works (Mañkovská, 1980).

The excessive tooth wear in some herds of park-maintained deer in a certain part of the German Rhineland described by Ueckermann and Scholz (1973) is probably another example of fluorosis. The excessive wear affected the premolars and M3, but not M1 and M2, of high proportions there with some consequent unevenness of the occlusal-wear plane. There was a factory in the vicinity that was known to emit polluting smoke and a much higher proportion of animals (50%) were affected in herds living 5 km from the factory than in herds living 10 km away (5%).

This story bears the hallmarks of endemic fluorosis of industrial origin although it seems that the ultimate conclusive evidence was lacking. The premolars and third molars would be largely formed after the calves were weaned and grazing on the supposedly contaminated grass. The first and second molars could largely escape the effect because their formation begins before weaning; only a certain amount of fluoride passes through the placenta and into the maternal milk (see this Chapter, p. 453). Kierdorf and Kierdorf (1986) have since shown that the bone in these animals has a very high fluoride content (3058–5528 parts/10^6).

Shupe, Olson, Peterson and Low (1984) made one of the most comprehensive studies of the prevalence of enamel fluorosis in relation to the fluoride content of the drinking water and vegetation among wild ungulates, predominantly deer (*Odocoileus* spp.), American elk (*Cervus canadensis*) and bison (*Bison bison*) in industrial and other parts of the United States. As well as in the vicinity of industrial areas, the prevalence of enamel fluorosis and fluoride in vegetation was high in areas where there are mineral-rich geothermal waters that tend to contain fluoride. Such waters may have a patchy distribution but it has to be borne in mind that wild ungulates tend to travel distances to select mineral-rich water (Fraser and Reardon, 1980). Shupe *et al.* referred to having seen dental fluorosis in moose (*Alces alces*), pronghorn (*Antilocapra americana*) and hares (*Lepus* spp.).

CHAPTER 21

Caries of the Teeth

Dental caries (*caries* – from the Latin, decay), which simply means decay or rotting of the teeth, is a form of progressive destruction of enamel, dentine and cementum initiated by microbial activity at the tooth surface. Loss of tooth substance is characteristically preceded by a softening of these tissues, brought about by partial dissolution of mineral (Silverstone, Johnson, Hardie and Williams, 1981). Thus, dental caries can be distinguished from other destructive processes that affect the crowns of teeth, dealt with in Chapter 20, such as attrition and abrasion due to mechanical wear and erosion due to chemical causes, in which as far as is known bacteria play no part.

Dental caries has been the subject of intensive study, epidemiologically and in the laboratory, because it is so common in man, particularly man living under the highly-organized social and dietary conditions that have been characteristic over the past 150 years of what is called western civilization. Most of what is known about dental caries derives either from the direct study of the human state or from experiments on animals in which various aspects of the human condition are simulated. The study of the natural disease in animals is limited to description and recording the prevalence. Clearly, however, much of the experimental work is relevant to the natural disease in animals, particularly to that in the same species.

It is appropriate here only to summarize current views of the process that gives rise to dental caries in man. It seems reasonable to extrapolate this to animals, though with the same caution that should be exercised when extrapolating to people observations made on disease in animals.

Current views centre around what is known as the chemico-parasitic theory first postulated by W. D. Miller (1890). The first stage is that a film formed of saliva and food debris and containing a sample of the bacterial flora of the mouth is left on the tooth surface; particularly on certain protected parts of the tooth crowns where food stagnation occurs. The stagnation areas are, in particular, inter-cuspal fissures and at the contact surfaces between adjacent teeth (approximal surfaces). Within a matter of minutes, this film begins to become more and more firmly attached to the tooth. This is brought about by a number of factors. Firstly, carbohydrates such as sucrose are fermented by the micro-organisms, producing acids which lower the pH; the mucin derived from saliva becomes more viscous. Secondly, this fermentation synthesizes large amounts of extracellular, complex saccharide-polymers from sugars. These polymers are sticky, gelatinous substances which help the bacteria to adhere to the tooth surface and to each other. Sucrose seems to be more cariogenic than other sugars, partly because it is more readily converted into a particularly large volume of polymer. A complex flora of bacteria grows up in the micro-environment of the film and includes many filamentous forms which form a felt-like mass, further contributing to the toughness and adhesion of the film which is now, in this fully-developed state, termed dental plaque.

Plaque provides a micro-environment to a large extent separate and different from the environment in the mouth at large. Soluble components of the diet, especially those of low molecular weight, such as the sugars glucose and sucrose, readily diffuse into the plaque and provide a substrate from which more acid is formed. The acid tends to remain localized against the tooth surface although it does of course gradually diffuse out and is neutralized by saliva which flows over the surface of the plaque.

A great deal of evidence from both natural caries and that induced experimentally in a variety of species shows that several different kinds of micro-organisms found both in the mouth at large and within dental plaque produce acid and therefore are likely to play a part in the caries process. However, a few species which not only produce acid but also large quantities of gelatinous polysaccharide have been shown to be outstandingly capable of initiating the caries process. Of these *Streptococcus mutans* is the most notable. Furthermore, it is established that there are particular strains of these organisms which are more cariogenic than others and that caries is transmissible in the sense that, if these particular strains are introduced into the mouths of colonies of animals which hitherto showed little or no caries, the caries incidence goes up markedly.

Dent (1979) was able to collect plaque from the tooth surfaces of 22 mammalian species in the London Zoo, mostly single examples. He found that the microbiological flora grown from the plaque samples was related to whether the species was herbivorous, carnivorous or omnivorous. Within these groups, the flora was broadly similar. The findings of a later study (Dent and Marsh, 1981) of an additional nine species were similar.

It seems that dental plaque forms on all surfaces of teeth of all animals, apart from the actual masticatory surfaces, but for caries to develop the following circumstances must be present: (1) the diet must contain some fermentable carbohydrate. (2) The right type of micro-organisms must be present in the mouth; these will be present if previous dietary habit is conducive for their growth and probably especially if highly cariogenic strains are present in the oral environment – for example in the mouths of other members of the same animal species. (3) It is possible that the enamel in some members of the species could be more resistant to caries than others and that well-mineralized enamel is more resistant. This does not seem to be true as a generalization except in the sense that some teeth may have deeper inter-cuspal fissures than others or that the enamel surface could be so pitted in enamel hypoplasia (see Chapter 20) that the pits provide stagnation areas. However, it has been established that if enamel has a certain minimum content of fluoride, as in areas where there is fluoride in the soil or water supply (see Chapter 20, p. 450), the teeth are notably resistant to dental caries. It is possible that the incorporation of other elements into enamel, e.g. molydenum, manganese and vanadium, may confer some caries-resistance (Büttner, 1969; Derise and Ritchey, 1974; Jenkins, 1978, pp. 250, 453).

Enamel contains a small amount of protein distributed through its structure but otherwise consists largely of crystals of hydroxyapatite which in caries are dissolved away; the enamel protein then disintegrates mechanically and needs little consideration here. However, the process of initial attack may be very slow and there is often then a stage when the enamel becomes stained brown before there is any evident loss of mineral. The brown staining is almost certainly due to some change of bacterial origin in the enamel protein. The enamel at the margins of a lesion of caries where the enamel is actually disintegrating is often similarly stained brown. There is evidence that, during this initial phase of enamel caries, the process of destruction may not only cease, or cease for a time, but enamel that has become partially demineralized may undergo a process of remineralization that is by no means fully understood. In other words, lesions have some capacity to heal. Such healing can occur, for instance, where caries has started at the contact surfaces between two adjacent teeth and one of the teeth is lost. The lesion on the tooth that remains often proceeds no further and remains indefinitely as a brown-stained area.

This account of dental caries derives largely from Silverstone, Johnson, Hardie and Williams (1981) which we recommend for further discussion of the complex biochemical happenings within dental plaque that may be concerned in caries and in periodontal disease (Chapter 24).

The type of caries so far referred to is primarily caries of enamel, the dentine is attacked secondarily.

When caries reaches dentine, it seems to advance more easily because it spreads out in the dentine, in particular at the enamel-dentine junction, so undermining the enamel. Dentine differs markedly from enamel in containing a much higher proportion of organic matter and in containing parts of living cells, the odontoblast processes, within its tubular system. Hence, there are many differences from caries in enamel. The initial attack in dentine is, as in enamel, one of acid demineralization but this leaves a mass of dentine protein, softened though still with the essential structure of dentine. This softened dentine becomes colonized by a complex flora of micro-organisms quite different from that of the advancing demineralization zone. It is a flora that produces enzymes which proteolyse the dentine matrix and convert it to breakdown products which nourish the flora. Especially in teeth that are of poor structure, it is typical of caries in human dentine that it tends to spread laterally, following poorly-mineralized growth lines. It is of interest that, in one of the few illustrations of caries in the dentine of an anthropoid ape (Joest, 1926), such lateral spread, forming a cleft, is clearly seen, as is thick bacterial plaque with a dominant structure of filamentous micro-organisms.

There is naturally a reaction in the living dentine, or dentine-pulp complex, being attacked by these processes. The odontoblast processes in the affected tubules react by producing a zone of hypermineralization which may either be situated immediately beneath the advancing caries or at some distance from it, usually at the pulp surface. The pulp itself reacts by producing reactionary dentine which amounts to a retreat of the pulp from the advancing lesion thus delaying the moment when micro-organisms reach the pulp and it is laid open, or exposed, by the process. A further series of events is then set in motion (see Chapter 23).

Once caries produces a cavity involving dentine, food stagnation tends to be continuous and cessation of the process is unlikely. Nevertheless, remineralization of dentine, like that already referred to in enamel, can occur and, especially if, by the breaking away of undermined enamel, the dentine lesion becomes exposed to mastication and gross food stagnation ceases, the process of caries may become totally arrested. The partly destroyed crown of the tooth then presents deeply-stained areas of dentine which are either quite hard or are of leather-like texture.

It is inappropriate to consider all the complexities of the processes referred to. It must suffice to point out that, in dentine caries, there are several zones. The juxta-position of two zones harbouring a quite different bacterial flora each associated with a biochemical micro-environment of quite different character is particularly thought-provoking. The deeper advancing demineralizing zone has a low pH and must be largely nourished by what fermentable carbohydrate reaches it through the more superficial zone of proteolysis which tends to have a high pH.

A different point of attack of caries, found both in man (Schamschula, Keyes and Hornabrook, 1972) and animals, is the root surface (Fig. 21.1). For the root to be exposed, some other factor has to be present, the recession of the gingiva that is a feature of advancing age or of periodontal disease (see Chapter 24). Thus, whereas caries of the crowns of teeth is most prevalent in the young, caries of the root surface is more prevalent in the latter half of life. As in enamel caries, attack appears to be by the products of bacteria harboured in plaque on the root surface. Plaque also accumulates on the gingival margin, namely in the angle between the margin and the tooth surface. The flora in that plaque is complex and substantially different from that in plaque on the crown surfaces of teeth. It could be that the flora and biochemical events in root-surface and gingival plaque are different even though they are in continuity. The features mentioned and some others suggest that root-surface caries and enamel caries ought not to be grouped together as slightly different examples of the same thing.

Experimental evidence in the hamster and rat (Hix and O'Leary, 1976) suggests that certain strains of *Streptococcus mutans* and species of the filamentous bacterium, *Actinomyces*, which are common in gingival plaque, play an important part in inducing both root-surface caries and periodontal disease.

Another type of dental caries arises as a complication of tooth wear. Where contiguous teeth are in contact, as in the post-canine dentitions of most mammals, occlusal wear is associated with considerable wear of the approximal surfaces because in mastication every tooth has some degree of individual movement; consequently, contiguous teeth rub against one another. When the occlusal enamel is lost, the approximal enamel, thinned and weakened by wear, breaks away (Fig. 21.1). Food stagnation between the teeth is thereby facilitated and the dentine, exposed partly by wear and partly by fracture, is attacked by caries.

It is often a combination of root-surface caries with caries associated with wear that leads to the gross tooth destruction in aged primates and it is very often impossible to decide which was the initial cause.

Reference has been made to the role that tooth structure, in particular the so-called quality of mineralization of the enamel, plays in offering resistance to the carious process. The idea that the perfection or otherwise of teeth, of enamel in particular, is the main factor on which caries-resistance depends arose early. These ideas gained currency in the years after the discovery of vitamins about 60 years ago; it was found, for example, that a deficiency of vitamin D in dogs, rabbits and guinea pigs, produced enamel that was not as fully mineralized as normal.

Fig. 21.1. Diagrammatic representation of caries associated with wear and gingival recession.
A: the worn occlusal surfaces, the shading being exposed dentine; wear has flattened the approximal surfaces and thinned the approximal enamel which, now weakened, is liable to fracture leaving exposed dentine which is commonly attacked by caries.
B: buccal view; the alveolar crests have receded and caries is depicted on the exposed root surfaces starting either interdentally or buccally. From Miles (1969).

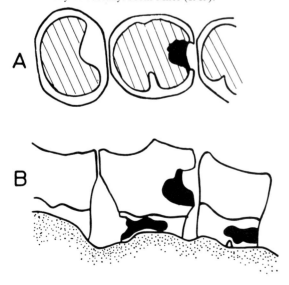

Attempts were also being made at about the same time to induce dental caries in laboratory-maintained animals and it was found in the dog (Mellanby, 1930, p. 55 *et seq.*) and rhesus monkey (Howe, 1924) that caries could only be induced, and then only sporadically, by feeding sticky sugar mixtures if the diets were also deficient in vitamins and minerals.

Nevertheless, with the discovery of dental plaque and the microbiological microcosm it contains, there were other workers who maintained that it is the chemico-physical nature of the food debris left on the tooth surface that is the main factor on which caries initiation depends.

At the time of the first edition of this book (1936), these controversies were at their peak and Sir Frank Colyer was a staunch and influential supporter of the view that the importance of tooth structure had been much exaggerated. He was concerned to emphasize that the rarity of dental caries in animals in the wild state, even though their tooth structure may show imperfections, argued against the idea that perfection of tooth structure in children, which would presumably be brought about by highly nutritious diet and control of systemic disease, would be a reliable safeguard against dental caries. Certainly Williams (1897) (Figs. 21.2 A, B, 21.3) and Walkhoff (1913) showed that what was regarded as defective enamel structure was common in the teeth of anthropoid apes in the wild as well as in contemporary man.

At the present time, the quality of tooth structure is no longer regarded as the crucial issue. The weight of evidence shows that the details of the micro-environment of dental plaque is of greater importance; in particular, the substrate within plaque and whether it supplies a continuous enough food supply for microorganisms that produce acid.

Hence, it no longer seems appropriate to reproduce the arguments Colyer presented in full though we have included the figures (Figs. 21.2–21.6) with their captions but little modified. We strongly recommend Colyer's original text to those interested in the history of these matters, and also his later work (Colyer, 1947*b*) in which he reported on 280 ground sections of teeth of 60 species from the wild. Degrees of what was interpreted as poorly-mineralized enamel were present in nearly all, with the exception of the macaques, *Ateles* and some lagomorphs; the enamel of the mustelids seemed to be better mineralized than that of the canids and felids.

It is now accepted that the so-called quality of mineralization of enamel cannot be reliably assessed by optical microscopy; much more sophisticated methods are required, for example, micro-hardness tests and chemical or X-radiation micro-analysis. The importance of the composition of enamel and of its structure in caries resistance cannot be ignored; the marked effect on caries resistance of small quantities of fluoride incorporated in the substance of enamel has already been referred to (this Chapter, p. 456).

One piece of experimental work deserves mention here. Shaw and Sognnaes (1955) fed a sugar-rich diet to a group of ten 2-year-old and three 1-year-old rhesus monkeys that had been imported from India. After periods ranging from $2\frac{1}{2}$ to $5\frac{1}{2}$ years, all the animals developed dental caries. However, it was found that the caries was confined to the teeth or parts of teeth that had developed in captivity. It is difficult to avoid the conclusion that in some way the structure or composition of the caries-susceptible teeth was different from that of the resistant teeth.

Schuman and Sognnaes (1956) found that developmental defects of the enamel surface were common in the small number of anthropoid apes from the wild they examined, in contrast to the rhesus monkey where such defects were notably absent (see Chapter 20).

Fig. 21.2. *Gorilla gorilla* (gorilla). Wild.
A: Ground section through the occlusal surface of a molar. The surface is wrinkled with numerous sulci; there are corresponding undulations in the pattern of growth lines in the enamel and the numerous dark patches of enamel suggest imperfect mineralization. From Williams (1897). ×c. 20.
B: A higher-power view of one of the sulci. From Williams (1897). ×c. 100.

A

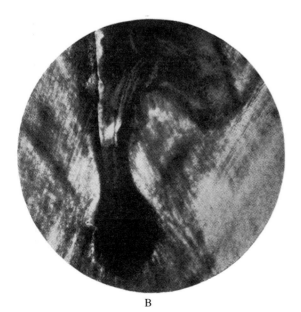

B

Fig. 21.3. *Pan troglodytes* (chimpanzee). Ground section of a premolar, showing stellate cracks in the enamel which could be the result of imperfect mineralization. From Williams (1897). ×c. 75.

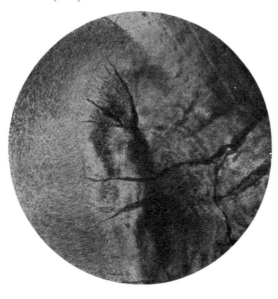

Fig. 21.4. *Gorilla gorilla* (gorilla). Ground section through a cusp of a molar. The surface is more smooth and normal than in Fig. 21.2A, but the dark areas of enamel and distinct growth lines are suggestive of periods of disturbed mineralization, probably malnutrition, during formation. ×c. 14.

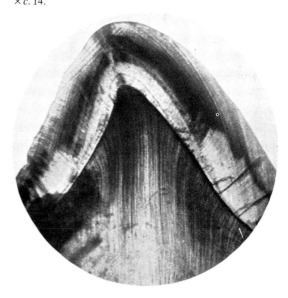

Fig. 21.5. *Pithecia* sp. (saki monkey). Ground section of a molar. The inter-cusp fissure extends into the dentine; the walls of the fissure are of patchy optical density. Both these features are known to occur in nutritional deficiencies.

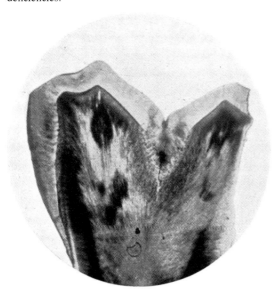

Fig. 21.6. Ground section of a tooth from a felid from the wild. There are patches of optical difference in the enamel and prominent growth lines in the dentine. Both features are known to occur in nutritional deficiencies. ×c. 40.

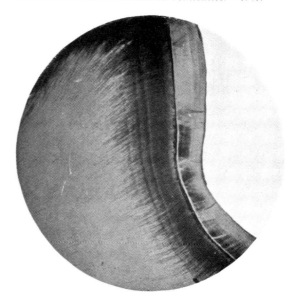

Experimental dental caries in laboratory-maintained animals

Many animal species have been used for experiments aimed at throwing light on the cause and control of dental caries in man. Guinea pigs (Howe, 1920a, b) and rabbits (Mellanby and Killick, 1926), being so readily available in laboratories, were used for some of the first experiments. Although some carious lesions on the exposed dentine on the occlusal surfaces of the molars of rabbits were induced (Mellanby, 1930, p. 60), it was quickly realized that, because the teeth are of continuous growth and subject to so much wear that the whole of the crowns of the molars is replaced about every 40 days, these species are unsuitable for caries research.

The laboratory white rat (*Rattus norvegicus*), also readily available in laboratories, began to be used at about the same time and was much more suitable because its cheek teeth are of limited growth and their occlusal surfaces are more like those of man than those of guinea pig and rabbit, because they have cusps and fissures; however, the cusps are much more like irregular bucco-lingual ridges and are separated by a bucco-lingual fissure. It was found that, although caries could be induced by appropriate diets, the cavities produced were often associated with fracture of the enamel and therefore not sufficiently similar to the common human lesion. The laboratory rat has no enamel on the tips of the molar cusps when the teeth first erupt and this was thought to be responsible for what was called fracture-caries; later it was found that the tooth wear and enamel fracture were due to the coarse abrasive character of the cariogenic diets then in use and could be avoided if finely-ground moist diets were used. The rat is now again much used in caries research (see later).

The colloquial term rat has just been used, in accord with common practice, to refer to species of the genus *Rattus*. Many other rodent species are also known colloquially as rats though usually preceded by some qualifying word.

The Syrian hamster (*Mesocricetus auratus*) (Keyes and Dale, 1944; Orland, 1946; Keyes, 1968) began to be developed as a laboratory animal and to be used, instead of the rat, for caries experiments because its cusps are covered completely by enamel when first erupted. An advantage of the hamster is that its jaws open very widely, by nearly 180°. This facilitates the examination of its mouth while restrained in a special head clamp, and the topical application of substances to the tooth crowns (Johansen, 1953). Unlike the rat, the hamster has cheek pouches and the associated habit of storing food in them. The pouches appear to be no handicap for caries experimentation. Gross caries can be induced in hamsters within 5–7 weeks if the cariogenic regime is started at weaning.

A variety of other rodent species were used for caries research at this time in the expectation that one would be found with molar crown morphology sufficiently similar to man's which might be particularly susceptible to dental caries or periodontal disease or both; for example, the rice rat (*Oryzomys palustris*) which is slightly smaller than *Rattus norvegicus*; it has enamel on the cusp tips (Gupta and Shaw, 1956a). The rice rat seem to have been favoured for research into root-surface caries (Rosen, Doff, App and Rotilie, 1981).

The cotton rat (*Sigmodon hispidus hispidus*) has enjoyed a period of popularity for caries experiments and seems to be particularly susceptible to enamel-fissure caries (Shaw, Schwiegert, Elvehjem and Phillips, 1944). It has more shallow transverse fissures between the molar cusps than *Rattus norvegicus* and the cusp tips are covered with enamel although this is speedily worn away. One reason for selecting the cotton rat is that it is weaned as early as 14 days compared to 20 days in *Rattus norvegicus*; this enables studies of the local effects of diet on the teeth to be started earlier (J. H. Shaw, Shaffer and Soldan, 1950).

The white-tailed rat (*Mystromys albicaudatus*) of South Africa, a relative of the Syrian hamster that has no cheek pouches (Ockerse, 1953; Larson and Fitzgerald, 1968), has been used for caries experimentation and so has the Mongolian gerbil (*Meriones unguiculatus*) (Gupta and Shaw, 1960; Fitzgerald and Fitzgerald, 1965; Hiatt, Gartner and Provenza, 1974), though not extensively. The crowns of the molars of the gerbil are unusually high and differ considerably in morphology from those of rats and hamsters; they have deep lingual and buccal longitudinal grooves like the molars of guinea pigs and rabbits. However, the molars of gerbils are not of continuous growth. The gerbil is better suited for periodontal research (see Chapter 24, p. 527). Both the white-tailed rat and the gerbil have litter sizes of about 3–4, that is much smaller than the strains of rat now most widely used which bear about 12 per litter.

Early attempts to use the laboratory mouse (*Mus musculus*) for caries research were not very successful. However, research into the part the immune system may play in susceptibility to dental caries has brought the wide range of closely-inbred strains of mice into the field and both smooth-surface and fissure caries can be induced (Kemp, Huis in't Veid, Havenaar and Backer Dirks, 1981). The dentition of the mouse is similar to that of the rat, though the teeth are much smaller so that sectioning and staining are obligatory for the detection of carious lesions.

A strain of mice known as the black fat mouse (PBB) is said to develop rampant caries in response to the appropriate diet (Navia, 1977). At the present time, the laboratory rat and mouse are being widely used again for caries research, e.g. for testing the efficacy of proposed anti-caries agents, because soft small-particle diets have been devized that do not lead to enamel fracture and can produce caries of the enamel on the approximal surfaces and on other smooth surfaces of the crowns of the teeth; this is known as smooth-surface caries.

Certain particular strains of the laboratory albino rat seem to be favoured in dental research, e.g. Wistar, Osborne-Mendel, Sprague-Dawley, NIH black rats, black-hooded rats (so-called because the dorsal surface of the head and neck is black). The reason for selecting these strains is rarely clear; sometimes it is simply because they happen to be readily available, and sometimes it is because genetic factors may play a part in caries susceptibility and there was therefore a need to choose the strain used for other work it was desirable to compare (Chai, Hunt, Hoppert and Rosen, 1968). The Osborne-Mendel (Grenby and Owen, 1980) and Sprague-Dawley strains are widely used because they appear to be unusually caries-susceptible. The experimental caries referred to is that of the cheek teeth; however, Stephan and Harris (1955) gave examples of the continuously-erupting incisors being affected.

Techniques are available that enable litters of rodents, mainly rats, mice and hamsters, to be born and maintained in a completely germ-free environment. The effect can then be determined of the introduction into their environment of single pure strains of micro-organisms (gnotobiosis). This technique has been valuable for determining the cariogenicity of various organisms or groups of organisms.

The primates that can be maintained in laboratories, for instance, the rhesus monkey (*Macaca mulatta*), have naturally been greatly favoured because of their resemblance to man, though they have been less used than the rodents because of the cost and difficulties of maintaining large colonies. Difficulties that developed in obtaining the rhesus monkey from the wild led to the introduction of the cynomolgus monkey, also known as the crab-eating and Java macaque (*Macaca fascicularis*, formerly *M. irus*) which is easier to maintain because smaller than the rhesus. Its dentition is similar and much used for dental research of all kinds. Both rhesus and cynomolgus monkeys breed quite well in captivity if given the right environment and it is now quite common for laboratories to breed their own colonies of macaques.

The diminishing stocks of macaques in the wild and restrictions on their importation have led to the use on an exploratory basis of the New World monkeys, marmosets (*Callithrix jacchus*), tamarins (*Sanguinus*), and squirrel monkeys (*Saimiri sciureus*), brown capuchin monkey (*Cebus apella* formerly *C. fatuellus*) (Johansen and Keyes, 1955), all of which are smaller than the macaques.

Lavelle, Shellis and Poole (1977) have provided some information on species-differences in the structure of the enamel of primates, including a review of the literature. Their own observations, which include scanning electron microscopy, are limited to 40 teeth from 16 species; as a basis for generalization, much larger numbers are needed.

An early review by Johansen and Keyes (1955) is still a useful source of information about the dental characteristics of species used in caries research. Later accounts are available in Navia (1977) and Tanzer (1981).

Spontaneous dental caries in animals

Discrimination has to be exercised in epidemiological studies of dental caries in man; not every stain or blemish of the tooth surface can be regarded as caries. Certain rigorous criteria are laid down; for example, a stained inter-cuspal fissure or pit on the occlusal surface of a tooth is only scored as caries if a sharp tapering probe stuck into it is hard to withdraw; this means that either caries has attacked the sides of the fissure and made them rough, or caries has passed beyond the fissure and into the dentine where it tends to spread out and undermine the enamel. Lesions of smooth approximal surfaces are scored as caries only if they can be detected with a fine probe, or if the surfaces are rough enough to fray a silk thread passed between the contact surfaces, or if destruction of enamel is demonstrable in radiographs. Epidemiologists follow the rule that, when in doubt, caries should not be recorded as present. Similar rigorous criteria have also to be applied to experimental caries in animals; for example, in the rat and hamster, caries is usually scored on the basis of ground slices through the tooth rows while still in the jaw, cut in the most appropriate plane. In most studies of the natural disease in animals, the number of animals or skulls available is too small for criteria of similar rigour to be applied; they are mentioned here in order to establish a perspective.

Order **PRIMATES**

Apes and Gibbons

In 1774 wild-ape skulls examined by Colyer (Table 21.1), there were only three instances of caries starting in occlusal enamel pits or fissures; all the remainder were approximal and associated with attrition leading either to fracture caries or to root-surface caries. In the gorilla, chimpanzee and gibbon, the incisors were most commonly affected; in the orang-utan, the premolars (Table 21.2). In every instance, the caries was associated either with considerable attrition leading to flaking-off of some enamel, so creating an approximal space capable of retaining food, or to exposure of the root surface by food packing between the teeth. The prevalence of caries in the incisor teeth (Fig. 21.7) is associated with more advanced attrition in that region (Colyer, 1947b).

The observations of Schultz (1939, 1941, 1944, 1950, 1956) accord with these figures in respect of relative frequence in the various species (Table 21.1. Summarized also in Table 23.1), though he stresses the importance of taking account of the ages of the animals because, being so closely related to tooth wear which in turn is related to age, the prevalence of caries is higher in aged animals and may be virtually absent from the young. (The opposite would be true of primary caries of enamel, e.g. of occlusal fissures.) Thus, in comparing numbers of animals with and without caries, only those of similar age can be compared. Schultz found that 2.4% of skulls of adult gorillas, mostly from the wild, were affected by caries and 3% of senile ones, compared with 0.7% for Colyer's group.

In chimpanzees, Schultz (1956) found that 30.6% of skulls of old animals were affected by caries, which is more than six times higher than Colyer's figure. One old male chimpanzee had lost completely about half his teeth, presumably from attrition combined with caries (Schultz, 1940).

Schuman and Sognnaes (1956) found that 6% of 78 chimpanzee skulls from the wild showed some dental caries; the age of the animals and type of lesion were not specified.

In 97 adult orang-utan skulls, Schultz (1941) recorded some caries in 4.1%; among the aged ones, 13.0% were affected, figures which are also much higher than Colyer's. Selenka (1898) found that 5.2% of the skulls of orang-utan were affected by dental caries; however, they were adult animals and therefore his figures cannot, any more than Schultz's, be compared directly with Colyer's.

Schultz's data in Table 21.1 relate to wild-shot gibbons, mainly *Hylobates lar* with a small proportion of two other species (see Chapter 3, p.20). The prevalence of dental caries in the senile ones (cranial sutures obliterated) was 8.7%, whereas in the adults only 0.9% were affected.

Fig. 21.7. *Pan troglodytes* (chimpanzee). ♂. Wild. RCS Odonto. Mus., G 139.8.
A: Caries of the upper incisors. ×c. 0.8.
B: A longitudinal ground section through a carious lesion of one of the incisors.

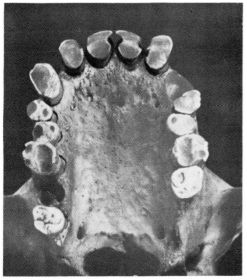

A

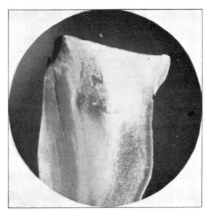

B

Table 21.1 The number of skulls showing caries and the number of carious teeth among skulls of the various genera of ANTHROPOID APES from the wild

Genus	No. of specimens (without mandibles)	Colyer			Schultz (1956)				Selenka (1898)	
		No. showing caries	No. of carious teeth	% showing caries	No. of adults	% showing caries	No. of seniles	% showing caries	No. of adults	% showing caries
Gorilla–gorilla	689 (129)	5	13	0.7	186	2.7	107	2.8	—	—
Pan–chimpanzee	465 (93)	21	76	4.5	110	12.7	62	30.6	—	—
Pongo–orang	255 (38)	6	15	2.4	97	4.1	46	13.0	194	5.2
Hylobates–gibbon	277 (5)	4	15	1.4	335	0.9	104	8.7	—	—
Symphalangus–siamang	88	0	0	0					—	—
Total	**1774 (265 without mandibles)**	**36**	**119**	**2.0**	**728**	**3.6**	**319**	**11.6**		

Table 21.2 ANTHROPOID APES: the position of carious cavities

	Gorilla		Chimpanzee		Orang		Gibbon	
	C.	Ap.	C.	Ap.	C.	Ap.	C.	Ap.
Maxilla								
Incisors	—	7	—	49	—	—	—	9
Premolars	—	—	—	3	—	11	—	—
Molars	—	—	—	1	—	1	—	2
Mandible								
Incisors	—	6	—	14	—	—	—	4
Premolars	—	—	—	7	2	—	—	—
Molars	—	—	1	1	—	1	—	—

The types of teeth affected and the carious lesions divided into two categories: crown (C) which would be caries starting in occlusal fissures or pits and also includes a minority starting in pits on the lingual or buccal surfaces of the teeth; approximal (Ap).

All such figures need to be received with some caution because tooth wear tends to be used as a criterion of age and because much caries is associated with tooth wear. Unless other criteria of age are also used, such as the obliteration of cranial sutures (Schultz, 1950), the argument about prevalence of caries and age tends to be circular.

Schultz (1935) pointed out also that specimens in museums may not be truly representative of the field populations they derive from. For instance, field collectors have a preference for normal specimens and may discard old animals with imperfect dentitions. His own collections, on which his studies are mainly based, were not subject to this caveat; every skull was collected regardless of age and condition (Schultz, 1935, p. 567).

Unfortunately, Schultz did not in his various studies specify the sites of caries although the association with wear to which he drew attention suggests that the majority of the lesions he scored in his aged groups were approximal, as in Colyer's. Indeed, several of Schultz's (1956) excellent drawings of the jaws of apes confirm this. One of a young adult female wild-shot orang-utan shows what is almost certainly primary enamel caries of four teeth, not secondary to tooth wear.

In parts of West Africa, with the spread of agriculture at the expense of forest, the chimpanzee can only survive by adapting to living close to agricultural settlements by raiding the plantations and scavenging among the village refuse heaps for food. Jones and Cave (1960) described severe deterioration of the dentition in a group of nine skulls from such areas. This deterioration was mainly due to periodontal disease and there was no evidence that the diet led to dental caries. However, in one aged female with gross periodontal disease, the surviving molars showed root-surface caries.

On the other hand, Colyer found five examples of caries (11%) in 45 chimpanzees from the Batouri district of Cameroun, an area which abounds in human habitations and abandoned banana plantations (this Chapter, p. 469). This is substantially higher than the overall prevalence (4.5%) of caries in chimpanzee skulls from the wild (Table 21.1). There seems to be no reason to doubt Colyer's view that primary caries of enamel in anthropoid apes in the wild is rare.

Comparable figures for anthropoid apes in captivity are not available. Jones and Cave (1960) described the dentition of a male chimpanzee, Jimmy, which was at least 46 years old and had spent 23 years in the London Zoo. Advanced generalized periodontal disease was present with less tooth wear than occurs in the wild and there was no caries. A female gorilla which spent about seven years in the London Zoo had only one tooth affected by caries at death; it was primary enamel caries of an occlusal fissure in a mandibular first molar (Colyer, 1947b). Glick, Swart and Woolf (1979) referred to caries of several abscessed teeth in a zoo-maintained gorilla (Chapter 23, p. 509). Schultz (1935) mentioned having seen caries of deciduous teeth in 4 captive chimpanzees. Caries of deciduous teeth of apes appears to be even more rare than caries of the permanent teeth in young animals.

Old World monkeys

In respect of dental caries in the Old World monkeys, data are available for many taxa in much more adequate numbers so that the prevalence in the wild and captive states can be compared.

Table 21.3 makes a comparison between the prevalences of caries in the wild and captive states in several genera of Cercopithecinae. However, it must be pointed out that, for a fully meaningful comparison, not only should the numbers in the groups being compared be more equal but, groups of similar age should be compared. The length of time in captivity should be similar and the types of caries scored should be specified. However, even with these reservations, it seems that the prevalence of caries is distinctly higher in captivity. Perhaps it is also worth noting that the figures suggest that, in the wild, caries is more common in *Cercopithecus* than in the other genera. This is not reflected in captivity, where the highest percentages are for *Erythrocebus* and *Mandrillus*, an observation which is weakened by the small numbers of animals in those groups.

More reliable differences emerge when the data are analysed in terms of type of tooth affected and, by inference, of type of caries (Table 21.4), and are also analysed into crown caries, which would be mostly primary caries of enamel, and approximal caries (Table 21.5) which would be mostly secondary to tooth wear; that is, would be either caries of dentine exposed by fracture of enamel (Fig. 21.8) or root-surface caries secondary to gingival recession. Nevertheless, examples of probable primary approximal caries not associated with marked tooth wear do occasionally occur in the wild (Fig. 21.9).

In all groups in the wild (Table 21.4), it is mainly the incisors that are affected (Table 21.6) that is caries often secondary to tooth wear (Figs. 21.10, 21.11). In captive animals in all the genera, except *Cercocebus* where the numbers are too small to be significant, the caries is mainly occlusal fissure caries of molars and

Table 21.3 Caries prevalence in various genera of CERCOPITHECINAE compared with the COLOBINAE, and in the captive state compared with the wild

Genus	Colyer From the wild state				Schultz (1956) From the wild state						Colyer From the captive state			
	No. of specimens	No. showing caries	No. of carious teeth	% showing caries	No. of adults	% showing caries	No. of seniles	% showing caries			No. of specimens	No. showing caries	No. of carious teeth	% showing caries
Cercopithecinae														
Cercopithecus	1533	63	223	4.1	84	2.4	28	21.4			338	26	67	7.7
Erythrocebus	42	0	0	0	—	—	—	—			54	10	23	18.5
Cercocebus	277	4	15	1.4	—	—	—	—			38	1	4	2.6
Macaca	588	10	34	1.7	107	1.9	78	12.8			305	18	50	5.9
Papio	288	1	2	0.3	43	0	31	9.7			153	15	50	9.8
Mandrillus	36	0	0	0	—	—	—	—			27	5	29	18.5
Theropithecus	21	0	0	0	—	—	—	—			5	0	0	0
Total	**2785**	**78**	**274**	**2.8**	**234**	**1.7**	**137**	**13.9**			**920**	**75**	**223**	**8.1**
Colobinae														
Colobus	—	—	—	—	51	0	8	0			—	—	—	—
Nasalis*	—	—	—	—	46	0	21	0			—	—	—	—
Presbytis*	—	—	—	—	180	0.6	92	1.1			48	0	0	0
Total	**2322****	**2**	**3**	**0.1**	**277**	**0.4**	**121**	**0.8**			**48**	**0**	**0**	**0**

* These are Asian colobines.
** About half of Colyer's sample consisted of Colobus.

premolars. Although the data in the Tables do not bring this out, primary caries of the approximal enamel was also common in young captive animals.

One example of caries was seen in the deciduous teeth of an animal from the wild.

The high prevalence of approximal caries in the wild state (Figs. 21.10, 21.11), in contrast to the captive state where crown caries is predominant (Figs. 21.12), accords with the view that in captivity the increase in caries is due to the prevalence of caries of a different kind, namely primary caries of enamel.

The generalization was made above that caries is more common in *Cercopithecus* than in the other Old World monkeys. However, the unwisdom of making rigid generalizations is illustrated by Table 21.7 which shows the wide range of percentage of specimens showing caries among the various species in the *Cercopithecus* genus in the wild. The magnitude of the difference between 1.4% for *Cercopithecus mona* and *Cerco. pogonias* combined and zero per cent for some other species does suggest that these species are more susceptible to caries than the others.

However, prevalence may also vary from district to district; as Table 21.8 shows, prevalence in three *Cercopithecus* species was several times higher in skulls from the Batouri district of Cameroun than among the skulls from other districts.

In *Cercopithecus cephus*, the caries affected incisors only; in seven specimens, the upper and in one the lower incisors. The amount of attrition was no greater than in the skulls from other districts. In the three skulls of *Cerco. nictitans nictitans* from Batouri which showed caries, the teeth affected were as follows: (a) mesial aspects of I^1; (b) mesial aspects of I^1 and the approximal surfaces of I_1 and I_2; (c) the occlusal surface of M_1.

Only the difference in respect of *Cerco. pogonias grayi* reaches statistical significance. In the Batouri skulls, tooth wear was a marked feature with some examples of cup-like wear of the exposed dentine on the occlusal surfaces of the post-canine teeth. In one instance (Fig. 21.13), there was carious destruction of the labial enamel of the upper incisors. Caries starting in the enamel is rare in the wild (Fig. 21.9).

Fig. 21.8. *Cercopithecus pogonias grayi* (crowned guenon). Three stages in the association between tooth wear and the onset of caries in the wild.
A: ♀. Wild. RCS Odonto. Mus., A 75.921. The I^1 are worn, producing a large central area of dentine surrounded by a rim of enamel which is still intact. × c. 0.8.
B: ♀. Wild. RCS Odonto. Mus., G 140.5537. The rim of enamel has broken mesially, creating a space between the teeth where food can lodge. × c. 0.8.
C: ♀. Wild. RCS Odonto. Mus., G 140.5544. Cavities between the I^1 are larger, cupped out and carious. × c. 0.8.

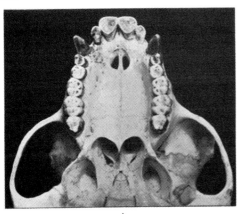

A

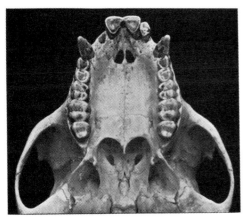

B

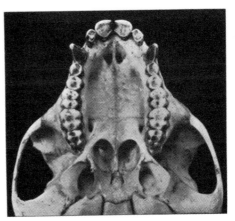

C

Table 21.4 CERCOPITHECINAE: the position of carious cavities in the genera that show caries

Genus	From the wild state								Total destruction of the crown	From the captive state								Total destruction of the crown
	Incisors		Canines		Premolars		Molars			Incisors		Canines		Premolars		Molars		
	C.	Ap.	C.	Ap.	C.	Ap.	C.	Ap.		C.	Ap.	C.	Ap.	C.	Ap.	C.	Ap.	
Cercopithecus	—	185	—	—	2	12	7	5	5 premolars, 7 molars	2	—	—	2	4	1	49	8	1 molar
Erythrocebus	—	—	—	—	—	—	—	—	—	—	—	—	—	—	—	14	9	—
Cercocebus	—	6	—	—	1	2	3	—	2 premolars, 1 molar	—	—	—	—	—	3	—	1	—
Macaca	—	10	—	—	—	8	—	15	1 molar	2	—	—	—	—	4	35	7	2 molars, 3 incisors,
Papio	—	—	—	—	—	—	2	—								44	2	1 molar
Mandrillus	—	—	—	—	—	—	—	—	—	—	—	—	—	5	1	16	1	1 premolar, 5 molars
Total	—	**201**	—	—	**3**	**22**	**12**	**20**	**16**	**4**	—	—	**2**	**9**	**9**	**158**	**28**	**13**
% of tooth		73.3			0		11.7		15.0		3.1		0.9		8.5		87.5	

C = crown; Ap = approximal.

Table 21.5 Site of caries in the CERCOPITHECINAE summarized from Table 21.4

	Wild state		Captive state	
	No.	%	No.	%
Approximal cavities	243	94.2	39	18.6
Crown cavities	15	5.8	171	81.4

Table 21.6 Analysis in terms of tooth type affected by carious lesions in 20 skulls of *CERCOPITHECUS POGONIAS GRAYI* in Table 21.7

Upper first incisors	35
Lower first incisors	22
Lower second incisors	13
Upper premolars	10
Lower premolars	2
Upper molars	4
Lower molars	5
Total no. of teeth affected	**91**

It seems possible that there was some exceptional factor in the environment of the Batouri district monkeys. The only clue to what this factor might have been is the excessive wear present in *Cerco. p. grayi*. Much of the dental caries might have been secondary to this wear. According to F. G. Merfield, who collected the Batouri specimens, the monkeys obtain much of their food from abandoned banana plantations; perhaps they may also have consumed abrasive items of food and even some fermentable and cariogenic carbohydrate obtained from refuse heaps, like the chimpanzees mentioned on page 465.

Schultz (1956) provided data on the prevalence of caries in his collection of wild-shot Old World monkeys which, although incorporated with Colyer's data in Table 21.3 (summarized also in Table 23.1), cannot be compared in detail with them because Schultz confined his analysis to adult animals. There are, as has already been mentioned, good reasons for so doing. Naturally, as caries is a cumulative condition that does not heal, lesions are bound to accumulate with age until a point is reached when teeth are lost and it is impossible, of course, to be sure whether they were lost through caries or not. However, there is no doubt from Schultz's work as well as Colyer's that the real cause of the increase in caries prevalence between young and old adults is tooth wear to which the caries is largely secondary.

Unfortunately, Schultz did not separate crown (most occlusal fissure) caries from approximal caries and his data record not the number of teeth affected but the number of skulls which show one or more carious teeth.

Lampel (1963) found that 15% of about 200 adult skulls of three genera of Cercopithecinae from the wild were affected by caries and that there was a slight increase with the amount of tooth wear. The type of caries was not specified but about twice as many of the incisors were affected as post-canine teeth. This accords with Colyer's findings.

Caries in the incisors in Old World monkeys in the wild state usually starts in dentine exposed by wear, often combined with the breaking away of the approximal enamel (Fig. 21.8). In the post-canine teeth, wear

Table 21.7 Prevalence of caries among various species of *CERCOPITHECUS* in the wild

Species	No. of specimens	No. showing caries	% showing caries
Cerco. aethiops	249	4*	1.6
C. mona + pogonias**	222	32	14.4
C. l'hoesti	15	1	6.7
C. neglectus	345	1	1.3
C. diana	8	0	0.0
C. nictitans	483	18	3.7
C. cephus	126	8	6.3
C. talapoin	15	0	0.0
Total	**1463**	**64**	**4.4**

* 3 were from the same district, suggesting that there was some regional factor.
** *Cercopithecus pogonias* includes *Cerco. pogonias grayi* which Colyer listed under *Cerco. mona* (as *Cerco. mona grayi*).

Fig. 21.9. *Cercopithecus pogonias grayi* (crowned guenon). ♂. Wild; from Batouri District. RCS Odonto. Mus., G 140.5541. Caries of the approximal surfaces of the premolars not associated with advanced attrition. Probably primary caries of enamel. ×c. 0.66.

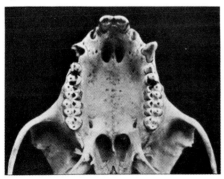

Fig. 21.10. *Cercopithecus denti* (Dent's guenon). Wild. BMNH 1907.1.2.1. Caries of the I^1. ×c. 1.3.

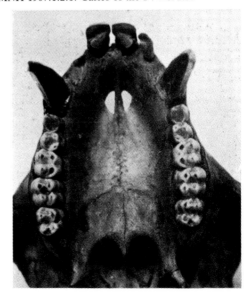

may lead to cupping out of the exposed occlusal dentine and it seems that food may be retained in the cup-like cavities because caries sometimes begins in them (Fig. 21.11), undermining the surrounding enamel as it progresses. Alternatively, by the time the teeth are worn, it is common for there to be gingival recession, usually brought about by gingivitis; food debris remains in contact with the exposed root surfaces

Table 21.8 Caries prevalence in skulls of three species of *CERCOPITHECUS* from the Batouri district compared with skulls of the same species from other places

Species	From Batouri			From other places		
	No. of specimens	No. showing caries	% showing caries	No. of specimens	No. showing caries	% showing caries
Cerco. cephus cephus	88	8	9.1	28	0	0.0
C. nictitans nictitans	125	3	2.4	70	1	1.4
C. pogonias grayi	76	20	26.3	17	2	11.8

Fig. 21.11. *Cercopithecus neglectus* (de Brazza's monkey). ♂. Wild. BMNH 1906.11.1.11. Marked attrition of the teeth. Caries has commenced in several of the cupped-out surfaces and has destroyed the crowns of the left M_1 and M_2.

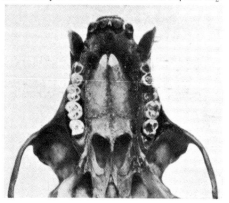

Fig. 21.12. *Cercopithecus* sp. (guenon). ♂. Captive. RCS Odonto. Mus., G 140.32. Extensive caries confined to the right side. The right upper canine is injured and, perhaps being painful, mastication was confined to the other side so that non-use of the right side led to food stagnation and consequential caries.

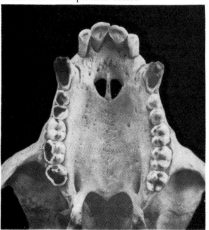
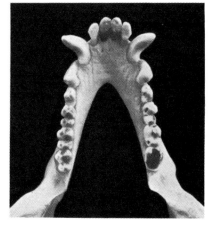

approximally, and caries commences there, gradually undermining the enamel as it advances (Fig. 21.1).

In captive Old World monkeys, the caries is substantially different; it commonly attacks the crowns of the molars and premolars before the occlusal enamel has been penetrated by wear and may start in the occlusal fissures. Figure 21.12 shows extensive caries commencing in the occlusal fissures of the post-canine teeth on the right side of a young captive adult cercopitheque. This caries is similar to the rapidly destructive caries seen in young children given biscuits or sweets or other fermentable carbohydrate snacks at bedtime. The limitation of caries to the right side was probably due to an injury that the upper canine on that side had suffered. This meant that the animal masticated on the opposite side and food stagnated around the teeth on the right. Sometimes special local conditions determine the onset of caries. Figure 10.71 shows the mandible of a captive vervet monkey in which both M_3 are rotated and only partially erupted so that there are food catchment areas under the convex distal surfaces of the second molars. Caries has affected the enamel of both these surfaces.

Although there are site-differences between dental caries in the wild and in captive animals, the microscopic anatomy of the process is the same (Fig. 21.14).

The data in Table 21.3 show that dental caries in macaques in the wild is less common than in the cercopitheques, a finding confirmed by Schultz (1935) who found no caries at all in 82 young rhesus monkeys

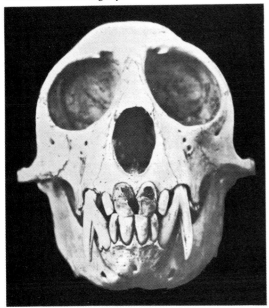

Fig. 21.13. *Cercopithecus pogonias grayi* (crowned guenon). ♂. Wild; from Batouri District. RCS Odonto. Mus., G 140.555. Caries of the labial enamel of the upper incisors. Reduced slightly.

Fig. 21.14. *Cercocebus* sp. (mangabey).
Left: Wild. Ground section through a molar with caries which almost certainly started in dentine exposed by wear of the occlusal surface. The enamel which survived on the right shows the level the wear had reached.
Right: Captive. Ground section through a carious lesion of a molar. Because the enamel over the cusp shows little wear, the caries must have started in originally intact occlusal enamel.

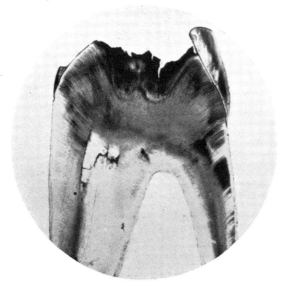

from the wild, but 13.7% of 73 old ones had an average of 2.0 carious teeth; 1.3% of 233 captive young ones had an average of 1.3 teeth affected and 23.1% of old ones had an average of 2.3 carious teeth.

Knap (1941) briefly reported finding some caries in 4% of 500 skulls of macaques from the wild in Indonesia. The species, ages of the animals and the type of caries were not given.

In 300 skulls of sub-adult rhesus monkeys which had been in captivity probably for only about a year, Colyer (1921) found 6 skulls with some caries. There were 10 carious teeth, four on the occlusal surfaces, and therefore almost certainly primary caries of enamel, and six on approximal surfaces which, because there was little post-canine attrition, would have been enamel caries also. Although 6 affected among 300 is a small proportion, these findings are a little out of accord with the general rule, i.e. virtual complete absence of caries of enamel. It seems possible that not all of these monkeys experienced the same environment immediately after capture, but they were all young and at the susceptible age for caries; the diet of a few may have included fermentable carbohydrates.

Anderson and Arnim (1936) examined the mouths of 76 young unanaesthetized rhesus monkeys; their mouths were propped open with a bar of wood. The animals had lived in captivity for unspecified periods. Three animals had carious lesions diagnosed by the same criteria and under conditions similar to those used in human epidemiological studies. The lesions were either in occlusal fissures or on approximal surfaces and appeared to be primary enamel caries unassociated with tooth wear. Some of the affected teeth were extracted; sections of them showed the microscopical appearances typical of human dental caries.

A meticulous record of dental caries in the rhesus monkey was published by Shaw, Elvehjem and Phillips (1945). They noticed dental caries in the skulls of 138 young animals which had been maintained for periods of up to three years on experimental diets, not for the study of dental caries, but to explore a relationship between diet and poliomyelitis. Many areas of stained enamel could only be diagnosed as uncertain caries even with the aid of a sharp probe; the only way to be certain was to grind these areas progressively and examine for loss of enamel and in particular for undermining of the enamel by dentine involvement. Radiography might well have been a satisfactory alternative to demonstrate these same changes. The number of carious lesions increased in an orderly fashion with the length of time the animal had been on the basic diet which had a high sucrose content. In the 2–3-year group, there was an average of 6.8 lesions per animal.

Dental caries appears to be rare in the baboon in the wild (Table 21.3). Bramblett (1969) made a systematic study of tooth wear in the skulls of yellow baboons (*Papio cynocephalus*), mostly from the wild, that enables age to be assessed with greatly improved accuracy and that would be helpful in future studies of the prevalence of dental caries. There was a high rate of attrition, especially in males, with ultimate disorganization of occlusion. C. A. Bramblett (personal communication 1982) found that caries mainly affected the incisors but was not a major factor in the destruction of the dentitions. There was no significant difference in caries prevalence between the sexes. This is one of the few studies in which sex differences were looked for.

James and Barnicot (1949) wrote a detailed account of the dentitions of 16 aged baboons that had lived in the London Zoo for periods estimated to be 6–10 years. There was considerable tooth wear and periodontal disease which was the presumed cause of the loss of some of the teeth. Thirteen of the skulls showed some dental caries, a total of 54 teeth being affected. The lesions were mostly in the post-canine dentition and nearly all had commenced on the root surfaces between the teeth. Incisors were not affected. Nine of the canines were affected whereas Colyer found none affected either in his wild state or captive groups. However, nearly all the lesions in canines had commenced in pulp cavities exposed by tooth fracture or wear that had led to death of the pulp. James and Barnicot examined sections of the carious teeth and found the changes were similar to those of caries in man except that there appeared to be less tendency to spread beneath the enamel. *Bacillus acidophilus* (*Lactobacillus*) was isolated from the carious dentine.

The dentitions of baboons in captivity commonly show considerable wear with the formation of cup-like depressions in the exposed dentine of the occlusal surfaces. Caries may start in these dentine surfaces as in the example in Figure 21.15.

In captive animals, caries can commence in occlusal fissures long before much tooth wear has occurred. Figure 19.2 shows an example in a young baboon. In the M^2, primary enamel caries is commencing in the fissures between cusps. In the M^1, which have been erupted for a longer time, the caries has reached dentine and is undermining the enamel.

Ashley-Montague (1935) described the skull of a male baboon that died in New York Zoo after an unknown period of captivity. The animal had suffered from severe osteomalacia attributed to dietary deficiency and inadequate exposure to sunlight. The tooth arches were splayed, especially in the incisor region, as if under the pressure of the occlusal stresses. The crowns of the two upper canines were carious and so were several molars. At least one molar was affected by root-surface caries, the other lesions were not described in sufficient detail to decide whether they were primary caries of enamel or not.

The study summarized in Table 21.3 emphasizes that dental caries is rare in colobus monkeys in the wild. Later work confirms this. In 67 *Colobus badius preussi* Colyer (1940) found no caries and in another, of 1485 *Colobus badius* and *Colobus guereza*, he found only two skulls in which there was any caries, in both instances starting in the cementum of the root surface (Colyer 1943b). Caries was totally absent from a group of 302 wild-shot colobids examined by Schultz (1958).

Little information is available about dental caries in captive colobids, probably because they are so difficult to rear and maintain for any length of time in menageries. However, Table 21.3 shows that Colyer did find 48 putatively-captive *Colobus* skulls in various museums. There was no caries in any of them.

New World monkeys

Caries in general is less common in New World monkeys in the wild state (Table 21.9) than in Old World monkeys (Table 21.3). In both groups, caries is usually secondary to attrition which produces food catchment areas and areas of dentine exposed by wear or fracture (Figs. 21.16–21.18), or it is root-surface

Fig. 21.15. *Papio hamadryas* (hamadryas baboon). Captive. RCS Odonto. Mus., G140.62. Marked attrition of the teeth; caries has started in the cupped-out exposed dentine of the occlusal surfaces. ×c. 0.5.

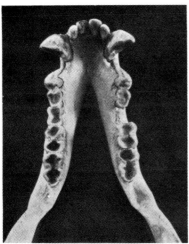

Fig. 21.16. *Cebus apella* (brown capuchin). ♀. Wild. BMNH 1903.7.7.12. Caries of both I^1.

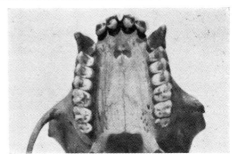

Fig. 21.17. *Cebus albifrons* (brown pale-fronted capuchin). ♂. Wild. RCS Odonto. Mus., G140.64. A carious cavity in the left M^1. Cupped out wear of the right M^1 has led to the mesial enamel breaking away but there is no caries. ×c. 1.5.

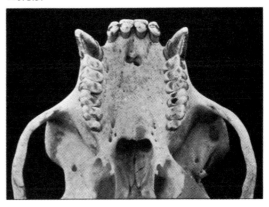

Fig. 21.18. *Cebus apella* (brown capuchin). ♂. Wild. BMNH 1903.7.7.5. Caries of the upper molars.

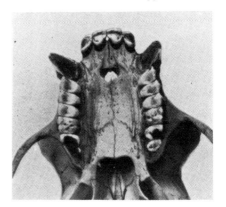

caries (Fig. 21.19). Only in three instances, in *Cebus*, did the caries primarily attack the enamel; in one, the caries started in pits on the buccal aspects of the M_1 and in another on the occlusal surfaces of the M^1; in the third, there was extensive caries of the enamel of the buccal aspects of the canine and post-canine teeth on the left side, spreading to the occlusal surfaces (Fig. 21.20).

The genera *Aotus* ($n = 117$), *Cacajao* ($n = 23$) and *Brachyteles* ($n = 25$) were free from caries both in the wild and captive states. The incidence of caries in the wild state in the other genera is given in Table 21.9. The number of skulls from the captive state was too small for inclusion. New World monkeys are difficult to keep in captivity and tend not to survive for long; hence the small numbers of skulls from the captive state available in museums. Those available showed no greater prevalence in caries but this is perhaps because it is likely that they had been but a short time in captivity before death.

In the cebid genera (Table 21.10), there were slightly more occlusal cavities than approximal caries and the molars were more affected than the incisors. These features are probably due to the greater wear of the molars and the tendency to develop cupped-out attritional cavities in the occlusal surfaces (Colyer, 1947b).

Table 21.11, in which the prevalence of dental caries in various species of *Cebus* is analysed, suggests that there are some species-differences.

Table 21.9 incorporates some data for New World monkeys from Schultz (1956). Schultz (1960) in addition recorded the prevalence of caries in another wild-shot group of 911 Central American monkeys, *Alouatta*, *Cebus* and *Ateles*. (These data are also summarized in Table 23.1.) He separated them into four age categories: juvenile, young adults, middle-aged adults and aged adults. The adult groups were separated mainly on the basis of the degree of tooth wear, together with some features such as closure of cranial sutures. In the aged group, the dentitions tended to be disorganized by abscesses secondary to tooth wear, caries or to periodontal disease and to tooth loss from those causes.

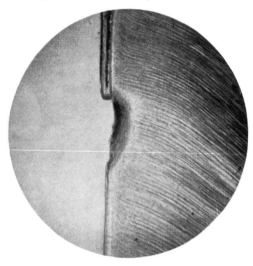

Fig. 21.19. *Lagothrix* (woolly monkey). Wild. Ground section through an incisor with caries starting at the cervical margin on the root surface just below the crown.

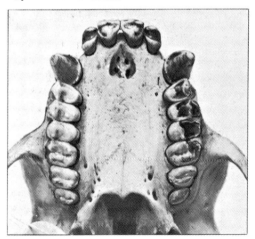
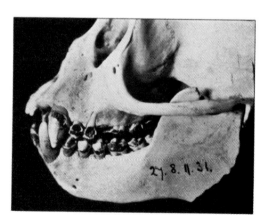

Fig. 21.20. *Cebus apella* (brown capuchin). ♀. Wild. BMNH 1927.8.11.31. Extensive caries of the left upper canine, premolars and molars.

Family and genus	Colyer			Schultz (1956)			Schultz (1960)		
	No. of specimens	No. showing caries	No. of carious teeth	% of specimens with carious teeth	No. of adults	% showing caries	No. of seniles	% showing caries	No. of specimens	No. showing caries	% showing caries	No. of specimens	No. showing caries	% showing caries	No. of seniles	% showing caries

Family and genus	No. of specimens	No. showing caries	No. of carious teeth	% of specimens with carious teeth	No. of adults	% showing caries	No. of seniles	% showing caries	No. of specimens	No. showing caries	% showing caries	No. of seniles	% showing caries
Cebidae													
Cebus	538	24	43	4.5	70	11.4	20	25.0	110	5.4	—	8	50.0
Lagothrix	82	1	3	1.2	—	—	—	—	—	—	—	—	—
Ateles	186	1	2	0.5	93	3.2	63	14.3	182	7.7	—	21	29.0
Alouatta	781	2	2	0.3	132	0	77	1.3	342	0.3	—	36	3.0
Pithecia	148	2	2	1.4	—	—	—	—	—	—	—	—	—
Saimiri	100	4	13	4.0	44	9.1	10	40.0	—	—	—	—	—
Callicebus	122	1	1	0.8	—	—	—	—	—	—	—	—	—
Total Cebidae	**1957**	**35**	**66**	**1.8**	**339**	**4.4**	**170**	**11.2**	**634**	**3.3**	—	**65**	**16.9**
Callitrichidae	211	1	4	0.5	71*	7.0	15	13.3	—	—	—	—	—

*A single callitrichid genus, *Leontocebus* (= *Leontopithecus*).

Table 21.10 Tooth types affected by caries in various taxa of NEW WORLD MONKEYS

Family and genus	Incisors		Canines		Premolars		Molars		Total destruction of the crown
	C.	Ap.	C.	Ap.	C.	Ap.	C.	Ap.	
Cebidae									
Cebus	—	9	1	—	4	1	13	10	5 molars
Lagothrix	1	—	—	—	—	—	1	—	1 molars
Ateles	2	—	—	—	—	—	2	—	—
Alouatta	—	—	—	—	—	—	1	—	1 molar
Pithecia	—	—	—	—	—	—	1	—	1 premolar
Saimiri	—	—	2	—	1	—	3	2	—
Callicebus	—	—	—	—	—	—	1	—	4 molars
Callitrichidae	—	4	—	—	—	—	—	—	—
% of lesions of both types in all taxa combined	22.85		4.3		10.0		62.85		

C = crown; Ap = approximal.

Alouatta, although experiencing nearly as much tooth wear and tooth loss as the others, showed little caries. In the middle-aged groups of *Cebus* and *Ateles*, 50% and 29% respectively of the skulls showed caries. In the aged groups, the prevalence had increased to 75% and 67%. Because Schultz's data, very properly, relate to adults only and take account of age, his percentages are naturally much higher than Colyer's (Table 21.11). Schultz confirmed a number of Colyer's statements; that caries is most common in *Cebus*, and in all three genera most common in the incisors and molars. Schultz did not distinguish types and precise sites of caries, whether initiated in enamel or approximally, although there seems no doubt that most of what he scored as caries was secondary to attrition. In an earlier paper, Schultz (1925) reproduced some good photographs of caries in *Alouatta* and *Cebus*.

Smith, Genoways and Jones (1977) made a careful examination of the dentitions of a total of 285 monkeys belonging to the same three genera, collected in the forests of Nicaragua in 1962–1968. These workers were cognisant of the difficulty of distinguishing the initial lesion of caries and mentioned that it is particularly difficult in *Alouatta* and *Ateles* because the crowns of the teeth are frequently coated with darkly-stained deposits of salivary calculus. They described and illustrated particular instances of caries in a way that is very useful but the percentage prevalences take no account of age and only enable it to be said that they confirm that caries is least common in *Alouatta* and most common in *Cebus*. They confirmed that caries in these monkeys commences in food lodgement areas approximately associated with tooth wear. However, Smith *et al.* also referred to caries commencing within the complex folding of the occlusal surfaces of the molars; this type of caries does not, from their illustrations, appear to be necessarily associated with tooth wear and appears to be fissure caries of the enamel. This needs to be looked into, especially in view of the increasing use of New World monkeys as experimental animals in caries research.

For the same reason, any study of caries in marmosets in the wild is of interest. Hershkovitz (1970), on the skulls of 904 wild-caught marmosets and tamarins of 14 species representing 3 genera and without any age differentiation, recorded simply the numbers of teeth carious, abscessed or missing. In *Saguinus fuscicollis*, there was a noteworthy incidence of caries in the upper incisors and in the canines in both jaws; in *S. oedipus*, there was a remarkably high incidence of caries in the upper canines, over half of a total of 32 animals being affected. Herschkovitz made no special comment on the lack of accord between this susceptibility of the canines and the observations of Colyer and other authors quoted here who agree that the canines are notably unaffected by caries in both the New World and Old World monkeys. When caries does occur in canine teeth, it is usually secondary to fracture of the tooth. The two species of *Saguinus* mentioned are among the so-called long-tusked species and therefore the canines could be particularly susceptible to injury. However, according to Hill (1957, p. 137), the canines are not really any longer than in the short-tusked tamarins but the incisors are shorter, making the canines more prominent and apparently longer.

It has already been mentioned (p. 474) that few data are available on captive New World monkeys. Ammons, Schectman and Page (1972) found 28 carious teeth among 17 tamarins, *Saguinus oedipus*, that had been in captivity for up to 12 months. There was one instance of a lesion on the labial surface of an upper canine tooth.

Lemurs

The skulls of 650 lemurs from the wild state were examined and two instances of caries were detected. In both cases, the approximal surfaces were attacked; in one, the caries had started in the dentine following exposure from attrition; in the other, the dentine had become exposed by fracture. Two examples of caries were found in eighty-five specimens of captive lemurs. In one, a *Lemur coronatus*, seven teeth were affected, the crowns and the approximal surfaces having been attacked.

Order **CARNIVORA**

Dental caries in the carnivores in the wild state must be exceedingly rare and few instances have been met with

Table 21.11 The prevalence of dental caries in various species of *CEBUS* from the wild

Species	No. of specimens	No. showing caries	% showing caries
Cebus apella*	194	14	7.2
C. albifrons	12	2	16.7
C. capucinus	120	3	2.5
C. nigrivittatus	16	0	0
Total	**342**	**19**	**5.6**

* In a group of 11 skulls from a certain region of Brazil (Chapada, Matto Grosso), 6 (54.5%) were affected by caries, suggesting the possibility of a special regional factor.

in the literature. Colyer examined 7635 skulls of carnivores in various museums and found only one instance, in a marbled cat (*Felis marmorata* BMNH 1855.12.24.254); it was a captive animal. The left P^4 was abnormal in shape and caries had attacked its occlusal surface (Fig. 4.5). The domestic cat appears to be occasionally affected, with lesions starting in occlusal pits, but it seems to be generally less so than in the dog.

Viverrids

The teeth of palm civets often show a remarkable degree of attrition and this is at times followed by fracture of the teeth and an appearance strongly suggestive of caries. But an examination of the exposed dentine invariably shows that it is impenetrable to pressure with a sharp-pointed instrument. In two specimens, however, there was definite cavity formation; one of these is shown in Figure 21.21. However, there were no signs of the undermining of enamel which is characteristic of advanced caries.

Mustelids

In contrast to 13 young adult badgers (*Meles meles*), killed by road traffic while on their nocturnal activities, which showed no dental caries, Andrews and Murray (1974) found caries in 2 out of 6 older animals found dead in the wild from other causes. The two were estimated to be quite aged, about 12–14 years, and each had a single lesion of the occlusal enamel of a cheek tooth. They also described a male badger aged about 10 years which had lived in captivity since infancy. Its diet had been a mixed one and had included dog biscuit which of course contains some fermentable carbohydrate. The cause of the caries in the two wild badgers can only be speculated upon. Badgers are omnivorous and their diet includes small mammals, worms, insects, roots, fruit and they are said to be fond of honey. In recent years, badgers have been known to explore areas of human habitation and to scavenge human refuse. In this way, some may consume fermentable carbohydrate foods.

Bjotvedt and Turner (1976) raised the question whether some deep pits on the worn occlusal surfaces of the post-canine teeth in sea otters (*Enhydra lutris*) were the result of caries. However, they noted that the pits showed no evidence of lateral undermining of the enamel and the dentine surfaces of the pits were not caries but were in some way the result of tooth wear (Chapter 22, p. 496).

Procyonids

An example of caries was seen in an olingo (*Bassaricyon*) in which there was considerable destruction of three molars. The chief interest in the family Procyonidae was the condition of the teeth in some of the specimens of coati (*Nasua*) from Chapada in Brazil, the same district from which there was a high prevalence of caries in *Cebus apella* already referred to (Table 21.11) were obtained. Several of the specimens showed cup-like attrition and the dentine was rough and deeply stained; in one, there was definite cavity formation.

Ursids

The only caries which Hall (1940) found among the skulls of 3761 North American carnivores of various kinds from the wild was in bears. Among 195 American black bears (*Ursus* (=*Euarctos*) *americanus*), there were five instances and among 165 grizzly bears (*U. horribilis*=*arctos*) there were three. Occlusal surfaces of molars were affected, 2–4 in each animal. Only two of the affected bears were young adults; the others, judging mainly from the considerable tooth wear that was present, were much older. In the two young animals, the caries may have commenced in enamel but, in all the others, it was dentine exposed by wear that was attacked. It seems possible that these bears had experienced much more wear than usual because in one instance wear alone had exposed the pulp-cavity. Berries and fruits form an important part

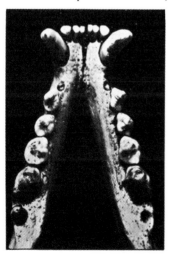

Fig. 21.21. *Paguma larvata grayi* (Himalayan palm civet). ♂. Wild. BMNH 1914.7.10.243. Mandible showing cavities in the distal portions of both M_1.

of the diet of bears in the summer and the juice of berries in particular has a low pH which could erode enamel. The question arises whether some factor of that sort played a part in producing excessive destruction of the teeth of these bears.

Hall suggested that honey (of which bears are notoriously fond; they will climb trees to obtain it) might have provided the fermentable carbohydrate, without which it is thought that dental caries whatever the species of animal does not occur, might have been responsible. This suggestion is strengthened by his finding no dental caries in 16 skulls of polar bears, which are more strictly carnivorous than the black and brown bears (Hall, 1945).

An example of a doubtful carious lesion in a brown bear (*Ursus arctos*) was seen in Lund University (Sweden). The occlusal surfaces were cupped out by wear and in one depression the exposed dentine surface was deeply stained though hard.

Caries is said to occur in captive bears but precise information does not seem to have been collected. The reputed fondness of zoo bears for buns, or rather the public fondness for feeding such unsuitable food to bears, makes this an interesting point.

Canids

Considering that so many dogs are fed meals that differ but little from those of their owners and, as a result, periodontal disease is common (Chapter 24), it is surprising that dental caries is not common also. Colyer in the first edition of this book recorded that he found only 4 instances (about 0.5%) in 762 skulls of dogs in various museums (Fig. 21.22). W.D. Miller (1890, p. 200) found 18 instances (6%) among 295 skulls, mostly of bull dogs and what he called lap dogs. At the Royal Veterinary College Hospital in London, only 3 instances of dental caries in dogs were recorded in the whole of the 5 years up to 1932 (Hare, 1934). A study by Bodingbauer (1947) comes closer to being an epidemiological study of a cross-section of a living dog population. He routinely examined for caries the 2113 dogs he saw in his veterinary practice over a period of 15 years and found that 5.8% had one or more carious teeth. Furthermore, he analysed his findings for sex differences, which were slight, and for age differences. The prevalence rose until about 3–4 years of age, then remained more or less stable or fell a little, with another rise in prevalence at about 6 years (8 years in the bitches) (Bodingbauer's fig. 4). These results are difficult to interpret because caries is a progressive condition and, unless carious teeth were, for example, being completely destroyed, exfoliated or extracted at a greater rate than new individuals were being affected, an exponential curve would be expected. The two peaks seen in the females, and of which there is a suggestion in the males, are also apparent in human epidemiological studies of dental caries; caries is most active in childhood and adolescence, then in adult life it is less common; however, in late adult life the figure rises again; this second rise is because teeth then tend to be affected by caries of a different sort, namely root-surface caries. At necropsy Hamp *et al.* (1984) found a slight peak in the incidence of caries at 3–5 years in 162 dogs ranging in age from 7 months to 14 years, with high levels only in animals over the age of 9 years. They distinguished primary enamel caries from root-surface caries but did not say whether the caries in old animals was mainly of root-surface type. As primary enamel caries is not a serious problem in the dog, studies that might throw light on this are not likely to be undertaken.

There seems to be general agreement that caries is much more common in the upper cheek teeth than the lower. In a group of 373 dogs, Kuiper and van der Gaag (1982) found 29 carious lesions among 12 dogs. All were in cheek teeth and 72% were in upper molars.

Quite out of accord with all the other studies quoted is one by Meyer and Suter (1976) who found that 35% of 200 dogs brought to their clinic had one or more carious teeth. However, they mentioned that they counted as caries every stained part or fissure, whereas Bodingbauer was more discriminating. Meyer and Suter pointed out that there are advantages in expressing the data as the percentage of teeth carious of the total number of teeth in the population studied; for instance, there were 84 000 teeth in their population of 200 dogs; 170, that is 2.02%, of the teeth were carious. This method, as an additional way of expressing data, has attractions but does not escape the necessity that Meyer and Suter's paper illustrates, of adopting precise and rigorous criteria for scoring caries. In human caries epidemiology, the practice is

Fig. 21.22. *Canis familiaris* (domestic dog).
Left: Alsatian. Formerly RCS Odonto. Mus., G 139.31. Carious left M_1.
Right: RCS Odonto. Mus., G 139.32. Carious M^1.

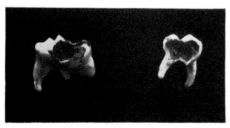

recommended of 'when in doubt do not score as caries'.

Svihla (1957) reported that six out of 22 jaws of Hawaiian dogs found at prehistoric cave sites had at least one molar tooth affected by caries. These dogs, the so-called poi dog, which have smooth round crania devoid of sagittal crests, used to be kept as a source of food and therefore were fattened on diets of starch-rich root vegetables, such as taro, poi and sweet potato, and sugar-rich fruits such as bread-fruit and bananas. During tribal wars, the women and children retreated to caves accompanied by pigs and dogs as a supply of food.

Allo (1971) mentioned finding dental caries affecting a food stagnation area of a malposed lower carnassial tooth (M_1) on a prehistoric Maori dog (Chapter 24, p. 550), where the carnassial was impacted against an adjacent M_2. Two molars in another mandible were also affected by caries.

The virus of canine distemper commonly affects the developing teeth and produces areas of enamel hypoplasia (Chapter 20, p. 445). The affected surface of the enamel may be soft and may become brown and then disintegrate after the teeth erupt. These lesions then resemble caries (Fig. 20.15) but can be distinguished by their shallowness and by the fact that many teeth are affected symmetrically.

According to W. D. Miller (1890), the occlusal surface of M^1 was the most common site to be affected by caries. Schneck (1967) agreed and so did Bodingbauer (1947); the P^4 was the next most commonly affected. Wright (1939) said M^1 and M^2 and nearly always bilaterally, M_2 and M_3 being next most affected.

Bodingbauer (1947) said he had seen caries approximally and cervically but usually starting in occlusal pits. Histologically, the lesions show the features characteristic of caries in man (Bodingbauer, 1950; Meyer, Suter and Triadan, 1980).

It might be supposed that short-jawed breeds with crowded arches would be more susceptible to caries than the long-jawed breeds in which all the adjacent teeth in the arch may be spaced. However, Bodingbauer (1947) found the exact reverse; commonest of all to be affected was the fox terrier which has quite a long jaw.

The rarity of dental caries in the dog seems highly likely to be due to the steeply-sloping surfaces of the largely-slicing teeth which present few nooks and crevices in which food debris can lodge long enough to do any harm. When caries does occur, it is said to be usually in the minute crevice that exists between the cusps of the sectorial upper teeth. E. Becker (1970, fig. 254) showed an example of caries on the mesial slope of the sectorial M_1 of a five-year-old Alsatian. It seems likely that caries commenced in a developmental defect.

Because of the shape of its teeth, the dog is an unsuitable animal for use in experiments on dental caries. Nevertheless, Gardner, Darke and Keary (1962) did try. They kept dogs on a human-type highly cariogenic diet for many months and even cut shallow cavities in the teeth to provide stagnation areas. No caries resulted. However, they then subjected the extracted teeth of the dogs to some simple test-tube experiments similar to those that W. D. Miller pioneered on human teeth well over a hundred years previously. They incubated the dogs' teeth in a mixture of human saliva and bread with a litle sugar added and found that a softening process very much like dental caries affected most of the enamel surfaces. We can deduce from this that the reason the dog appears to be resistant to dental caries is not because its tooth tissues are intrinsically resistant; it is noteworthy that the enamel of dog has a higher content of organic matter than that of man, the apes and pig (Bremer, 1939) and it is more permeable to dyes than human enamel (Fish, 1933).

We have to look for explanations other than resistance of the enamel to the carious process. The overall morphology of the teeth may be one, and another is the possibility that saliva in the dog contains antibacterial substances that prevent the growth of the bacteria on its teeth, on the activity of which dental caries depends. Indeed, it is known that canine saliva does have marked generalized antibacterial properties (Armstrong and Jenkins, 1953). It must also be borne in mind that the carnivores have very little amylase in their saliva (Young and Van Lennep, 1978). Thus starch, even cooked starch, left as debris on the teeth, may not be so readily broken down as in man into soluble sugars which can be then turned into acid by the bacterial flora on the tooth surface.

The organisms which are thought to be concerned in dental caries are not in general capable of breaking down starch, but have to wait, as it were, for starch to be broken down to sugars.

Dental caries does not appear to have been described in Canidae in the wild or for that matter when kept in zoos.

Order SIRENIA

The manatee is a marine or estuarine mammal which lives on sea- or water-weed which it masticates with high-crowned cheek teeth. Their broad occlusal surfaces become much worn. These circumstances seem to be far from favourable for dental caries to occur. Nevertheless, W. D. Miller (1893, 1894) described lesions that he regarded as dental caries on several molars of two museum skulls of the African manatee (*Trichechus senegalensis*). He did not say precisely which parts of the teeth were affected but it was probably approximally because he mentioned food retention between the teeth. He prepared stained sections of some bits of the affected dentine and one of his illustrations does show changes suggestive of authentic caries of dentine. Miller pointed out that the herbivorous manatee are said to sleep after meals with the mouth open and snout out of the water; plant debris, which might contain carbohydrate fermentable by bacteria into acids, might stay on the tooth surface or between the teeth and might under those conditions ferment and produce a local concentration of acid on the tooth surface high enough to initiate caries. There is not necessarily any large difference between the teeth of marine mammals being bathed in water and those of land mammals being constantly bathed in saliva. In both cases, the dental plaque formed on the tooth surface provides a micro-environment in which there is only a slow interchange of soluble substances between that environment and the larger environment beyond.

Order **ARTIODACTYLA**

Suids

Three examples of putative dental caries were noted in the following museum specimens of pigs from the wild:
  (i) *Sus scrofa* (European wild boar). BMNH 1908.3.8.12. There were deep cup-like wear surfaces of dentine on the occlusal surfaces of both M_2; the dentine surface was deeply stained and rough but not softened. There was some fracture of enamel at the margins of the lesional surfaces. One cavity was more convincing as caries because the enamel was slightly undermined.
  (ii) *Sus scrofa vittatus* (Indonesian wild boar). BMNH 1909.4.1.511. Four maxillary molars which were much worn and showed food packing were carious.
  (iii) *Sus scrofa* (European wild boar). RCS Osteo. Series, 1768. There were cup-like wear surfaces, fracture of weakened enamel and approximal caries.

Dental caries occurs occasionally in domestic pigs, usually in old sows (Anthony and Lewis, 1961). The lesions are not well-described but they seem mostly to be between the teeth and associated with food packing; they may be of the root-surface variety. An instance of typical occlusal caries of both M^1 in a pig was figured by Harvey and Penny (1985). Andrews (1973) described caries of all four M1, apparently primary enamel caries starting in occlusal pits, in a boar which was one of a group maintained on a sucrose-rich diet for a study of cardiovascular disease. The affected M1, the first cheek teeth to erupt, at about 5 months of age, had been erupted and therefore exposed to the effect of this diet for only seven months; this suggests that the pig is susceptible to caries, a point of interest because the so-called miniature pig, still quite a large animal, bred from a number of small forms, e.g. Vietnamese pig (Otto and Schumacher, 1978), has been used for a number of purposes in dental research (Weaver, Sorenson and Jump, 1962; Jump and Weaver, 1966) but not systematically for caries research.

Bovids

Caries sometimes occurs in young sheep, the disease starting around the necks of the incisor teeth as in the two incisors shown in Figure 21.23. There is no doubt as to the nature of the disease, the margin of the enamel showing definite signs of demineralization. This animal had been reared on a diet which included so-called artificial cake.

Although, according to Jubb and Kennedy (1970), dental caries is common in sheep, as well as cattle, detailed records in the literature are scarce.

Garlick (1954*b*) found that 343 (19%) out of 1803 adult slaughter-house cattle had some carious teeth

Fig. 21.23. *Ovis aries* (sheep). Two incisors showing caries of the root surface at the cervix. Left: labial view; right: distal view.

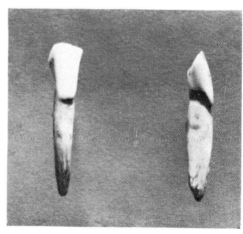

and remarked that the condition is common in cattle all over the world. In some cases, dentine on the wear tables of the cheek teeth was affected, leading to deep hollowing out of the crown and eventually to death of the pulp and root abscess; in others, there was caries of enamel which appeared to start in developmentally-defective enamel flecks (see Chapter 20, p. 450). This type of caries penetrated the thickness of the enamel without tending to spread laterally.

Cervids

Cowan (1946) found what he thought was dental caries in a single tooth among the skulls of 750 mule deer (*Odocoileus hemionus columbianus*). The tooth was an M_1 which had been fractured, exposing the enamel surface of one of the infoldings which was deeply stained and irregular. There was some possible undermining of enamel by caries of dentine. Cowan mentioned that food mass was commonly impacted in the worn occlusal lakes in these deer.

Order **PERISSODACTYLA**

Equids

Caries is extremely rare in the wild and perhaps non-existent. In the skull of a quagga (*Equus quagga boehmi* BMNH 1894.1.2.1), there were cavities suggestive of caries in the occlusal surfaces of both M^1 and the right M^3.

Dental caries is common in the cheek teeth of the domesticated horse. However, the caries is very different from that already described, mainly because the morphology of the teeth is different and very specialized (C. E. Harvey, 1985).

The enamel on the occlusal surfaces of the cheek teeth is deeply infolded (Fig. 21.24) to produce funnel-like cavities (infundibulae) lined with enamel; these infundibulae become filled with thick cementum because, not only the roots, but also the crowns of the teeth are totally enclosed in a covering of cementum. When the occlusal surfaces of the cheek teeth are worn, the enamel-lined hollows present as so-called lakes of cementum bordered by enamel. At the centres of the cementum lakes, there is always a canal which extends for more or less the whole length of the tooth

Fig. 21.24. *Equus caballus* (horse). RCS Odonto. Mus., A 231.366. On the right is a longitudinal section through an upper molar in which can be seen the complex system of columns of dentine, enamel and coronal cementum produced by deep infolding of the enamel from the occlusal surface. On the left at A is the occlusal surface of another upper molar showing early caries. Below, at B, C, D and E, are a series of transverse sectional surfaces through this tooth at levels towards the root. Arrows indicate the canals that persist in the cementum that fills the infundibulae. The unmarked cavities are sections through the complex pulp cavity.

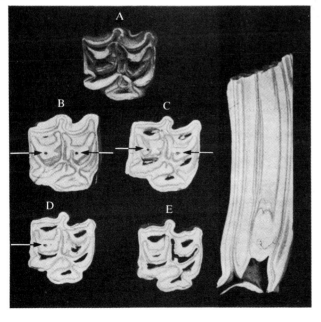

(Fig. 21.25). It is in these canals that caries starts. The surrounding cementum becomes softened and discoloured brown and is gradually destroyed by a bacterial process. In due course, the process reaches the enamel which is attacked, as is the dentine (Figs. 21.25, 24.43). Usually the process spreads laterally and remains fairly superficial but sometimes penetrates deeply into the tooth, weakening it so much that it splits longitudinally.

Some writers have used the term cementum necrosis instead of caries for this process of destruction of the cementum lakes. However, as the cementum over enamel on the crowns of teeth can hardly be living in the sense of containing living cells, necrosis does not seem to be a suitable term for this diseased state. There seems to be no objection to calling it dental caries as long as it is acknowledged that different kinds of caries exist.

Fig. 21.25. *Equus caballus* (horse). RCS Odonto. Mus., G 137.6. A semi-diagrammatic view, oblique to the occlusal wear-surface, of the surface of a longitudinal section through a carious upper cheek tooth. The sectioned surface is shaded. Caries starting on the occlusal surface in the cementum filling an infundibulum has penetrated along the infundibulum and reached the pulp cavity, p. Impacted fodder, f, is still *in situ* in both the infundibulum and pulp cavity. Another infundibulum at i is brown-stained and enlarged by caries of the cementum.

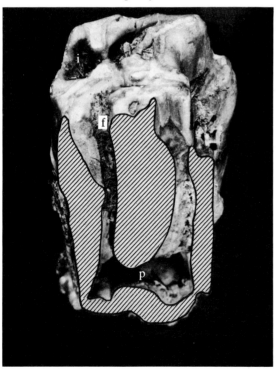

In a study of dental caries in 365 otherwise healthy slaughterhouse horses, K. Honma, Yamakawa, Yamuchi and Hosoya (1962) found a close relationship with advancing age (Fig. 21.26). The prevalence in older animals was probably substantially higher than this figure indicates because 6.7% of the animals over 11 years of age had some teeth missing, probably due to caries; no reference was made to periodontal disease in the animals. Most of the caries was in the cheek teeth; the exact sites were not specified but it can be assumed that the caries was of the usual occlusal variety.

Colyer (1906) found dental caries in 66 skulls (13.5%) of horses out of 484, mostly from animals over the age of 12 years. In some, only one tooth was

Fig. 21.26. *Equus caballus* (horse). The relationship between age and the percentage incidence of caries in slaughter-house horses in Japan. Data from K. Homna *et al.* (1962). The figure at the top of each column is the total number of horses examined in that age group.

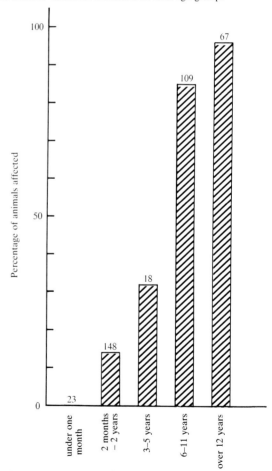

affected, most commonly an M1, and in others as many as 10 teeth were carious. Caries was more common in the upper jaw than the lower; M1 was the most common tooth to be affected, perhaps simply because, being the first cheek tooth to erupt, it is in the mouth for the longest time (Hofmeyr, 1960). The caries most commonly affected the occlusal surfaces, starting in the cementum lakes, but caries sometimes started approximally (Fig. 24.44). Colyer referred to a horse of only about 4 years of age in which the approximal surfaces of many teeth as well as the wear tables were carious. Colyer (1906) and Merillat (1906) independently described the infundibulum or central canal defect in the cementum lakes as the main predisposing cause of dental caries in the horse and suggested that fermentable carbohydrate becomes driven into these defects. Colyer mentioned that these canals are rarely seen in skulls not affected by caries.

According to Shuttleworth (1948), cheek teeth affected by occlusal caries commonly split longitudinally under the stress of biting and, until an incident of that kind, the dental caries frequently remains undiscovered. Such longitudinal splitting of the teeth can also occur in the absence of caries (E. Becker, 1970, his fig. 261; and Chapter 19, p. 435).

The incisors in the horse can be affected by caries. Lee and Stolfus (1968) described a seven-year-old stallion with caries of the labial surfaces of both upper and lower incisors, possibly caused by molasses added to his feed. Incidentally, the caries was treated by cutting away the caries, leaving undercut cavities which were filled with a silicate cement.

A useful account of dental caries in the horse was given by E. Becker (1970, p. 252) who referred to instances of the repair of carious lesions with cast-bronze inserts.

Amin and Kassem (1987) figured an example of bilateral caries in a donkey which probably began on the occlusal surfaces but had destroyed the approximal halves of adjacent upper cheek teeth.

Order **HYRACOIDEA**

The cheek teeth of hyrax (*Procavia capensis*) in captivity seem to be particularly susceptible to caries (Fig. 21.27) perhaps because the deep sulci on the occlusal surfaces are liable to retain food. The hyrax is a strict herbivore but, in captivity, it seems possible that it has access to some fermentable carbohydrate, for instance cooked starch or sugars. Caries does not occur in hyrax in the wild; Colyer found none in 308 specimens.

Order **RODENTIA**

Colyer found no caries in the 3800 skulls of a variety of rodents from the wild he examined in various museums. In captive animals, caries occurs occasionally. In those species where the cheek teeth are of limited growth, caries may occur on the approximal surfaces or in the pits or fissures between cusps on the

Fig. 21.27. *Procavia capensis* (large-toothed rock hyrax). Captive. RCS Odonto. Mus., G 138.61. Caries of the upper cheek teeth. *c*. Natural size

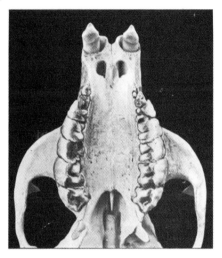

Fig. 21.28. *Dasyprocta* sp. (agouti). Captive. RCS Odonto. Mus., G 138.311. Caries of the occlusal surfaces of several cheek teeth. The crown of the left M_2 has been almost completely destroyed. × *c*. 0.56.

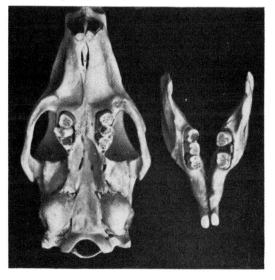

484 Section 4: Other Disorders of Teeth and Jaws

occlusal surfaces (Figs. 21.28, 21.29). Where the cheek teeth are of continuous growth, the rate of wear of the occlusal surfaces is normally much greater and the occlusal surface is worn away too quickly for caries to occur.

Schitoskey (1971) found among 180 skulls of feral nutria, *Myocaster coypus*, from the southern States of America, 20 with dental caries, in one instance affecting a lower incisor. No further details were given except that one-half of one molar was destroyed by the caries.

Zschokke and Saxer (1933) described a nutria in which several cheek teeth were carious. One carious tooth had been split by biting on food impacted within it and an abscess had resulted. Scheuring and Scheuring (1982), without referring to the published examples of dental caries in nutria just mentioned, reported some degree of caries in 67% of 250 farm-bred nutria of various ages.

Order **LAGOMORPHA**

In hutch-kept domestic rabbits, caries is said to be common though evidence for this does not seem to have been collected, apart from some provided by L. Guanziroli (personal communication to J. F. Colyer c. 1920) who found six instances of dental caries among his eleven hutch-kept rabbits. In addition to fresh vegetable foods, the animals were given a plentiful supply of so-called middlings prepared from cooked cereals which therefore contained fermentable carbohydrate.

Order **CHIROPTERA**

The majority of the common bats are mainly insectivorous but some are omnivorous and consume leaves and fruit. Bats seem to be among the species that are least likely to develop dental caries or for that matter to be studied in that respect. However, Phillips and Jones (1970) have shown that the Central American spear-nosed bat, *Phyllostomas hastatus*, is commonly affected. They found that 40% of the skulls of 52 specimens from a wide geographic range from Venezuela to Panama and Trinidad had readily-detectable caries. The condition was statistically significantly more common in the males. The lesions appeared to start in developmental enamel pits on the buccal surfaces of the upper molars or in the depressions between the occlusal ridge-like cusps. On reaching the dentine, the destructive process spread out, undermining enamel which then broke away leaving more than half the crown destroyed. Stained sections of some of the teeth showed changes in the dentine like those of caries in other mammals.

The sites where caries was starting were evidently where food stagnation would be likely; indeed, debris was found still *in situ* in some specimens, in particular between the buccal aspects of M^3 and a flap of mucosa which is normally present in this species. The nature of the debris was not commented on.

An unusual feature in this species was the presence of stained, often intersecting, cracks in the enamel surface. The cracks could often be felt with a sharp probe and sections showed that they often extended slightly into the dentine. These cracks resembled the crazing of enamel described in a number of species (Chapter 22, p. 505). They did not seem to be related to tooth wear which was not a predominant feature in these animals.

Dental caries is certainly not a common feature in other species of bat. Phillips (1971) found only one instance, an adult male *Mormoops megalophylla*, in 1508 specimens of three bat families (Emballonuridae, Noctilionidae, Chilonycteridae). That there is probably something unusual about the environment or diet of *P. hastatus*, combined perhaps with some details of its tooth morphology, which makes it susceptible, is suggested by the total absence of caries in a large number of skulls of the smaller related species, *Phyl-*

Fig. 21.29. *Marmota marmota* (alpine marmot). Captive. RCS Odonto. Mus., G 138.43. The cheek teeth are of limited growth. There are several carious cavities which commenced in the grooves between the cusps on the occlusal surfaces.

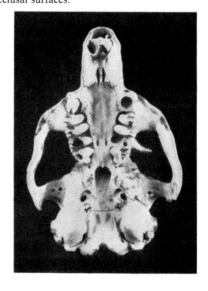

lostomas discolor, collected from the same region. In this species, the buccal enamel pits are absent from the upper molars and the mucosal flap is smaller. Enamel cracks were also less common.

The only known dietary habit of *P. hastatus* that could be relevant is that the species is particularly omnivorous and is known to feed on small vertebrates as well as on fruit and insects. In captivity, animals have been observed to crush the bones of mice and it may be that the enamel cracks, which would facilitate carious attacks, are induced by masticatory stress of that kind.

However, the omnivorous diet mentioned is not one likely to induce dental caries as it is commonly understood. The weight of evidence suggests strongly that some type of fermentable carbohydrate has to be present on the tooth surface, but it could be that other forms of progressive tooth destruction are active. Theoretically both enamel and dentine could be destroyed by a proteolytic process. Tooth enamel contains organic matter and, if that were destroyed, there is little doubt that the enamel mineral structures would tend to disintegrate. Proteolytic destruction of the organic matrix of dentine does occur in dentine caries but only after the mineral has been removed by an acid process, or at any rate a low-pH process. At one time, commencing with the work of Preiswerk (1906), many researchers were attracted by the suggestion that the initial carious attack of both enamel and dentine is a proteolytic process, i.e. the organic matter is destroyed first and then the mineral. However, this is a view that no longer has adherents because, in spite of much search, no-one has discovered any micro-organisms in caries which are capable of destroying the organic matter of intact enamel or dentine. The mineral has to be removed first.

Order **INSECTIVORA**

Hall (1940) examined the skulls of 2256 insectivores of various kinds and found no caries.

Order **MARSUPIALIA**

Among 3336 museum specimens of pouched mammals, Colyer found one instance of certain dental caries, on the incisive edge of an upper incisor in a captive wombat, *Vombatus ursinus* (RCS Odonto Mus., G141.3). There were two doubtful instances in other specimens. According to Farris (1950), dental caries is common in opossums bred for laboratory research. He gave no further details.

Fossil vertebrates

Moodie (1923) reviewed the highly anecdotal literature on dental caries in fossil vertebrates, including in a Cretaceous reptile, and the incisor of a Pliocene camel. Most of the instances are unconvincing and not described by people with wide experience of dental caries who would be able to distinguish it from other destructive processes such as tooth resorption and even post-mortem destruction by moulds (Moodie, 1923, p.225). One of the most fully documented instances in a lower molar of a Pleistocene, *Mastodon americanus*, from Ohio, USA, was described by Hermann (1908). Moodie gave a free translation of most of the paper. There were three cavities within sharp irregular margins opening up into larger cavities within the tooth where at least two of the cavities were connected. One of the cavities was on the mesial margin of the occlusal surface; the other two were in valleys or fissures between the cusps of the occlusal surface. A drawing of a section across the tooth shows how the destructive process, which appeared to have started on the crown surface, had spread out in the dentine below, in a manner that is characteristic of caries. If the tooth or section are still available, important light could be thrown on whether these lesions were produced during life. A zone of hypermineralization or of some other change in the dentine adjacent to the lesions would then be expected and could be demonstrated by microradiography or polarizing-light microscopy.

Leidy (1886) published a short note on what purported to be caries of an upper molar tooth from a *Mastodon floridanus*.

CHAPTER 22

Tooth Destruction from Causes Other than Caries

Introduction

This title covers a number of conditions, attrition, abrasion, erosion and resorption which, apart from dental caries, cause loss of tooth substance.

Wear of the occlusal surfaces of the teeth and of surfaces between teeth where adjacent teeth are in contact and move slightly against each other are normal accompaniments of mastication and examples are referred to here and there throughout this work. Attrition is the term used for such wear associated with tooth-to-tooth contact, but it must be pointed out that during mastication the sides of teeth not actually in tooth-to-tooth contact become smooth and polished. Attrition can under certain circumstances be excessive.

Abrasion is a term usually employed for wear of the teeth other than by food during mastication; hence surfaces other than the occlusal and contact surfaces may be affected. However, the term abrasion is sometimes used in a slightly different sense. When teeth are used, like the homodont dentitions of marine mammals, for the prehension of prey preparatory to swallowing it whole, rather than for mastication, the surfaces of the teeth become polished and worn; so also do the high-pointed cusps of the insectivores, used for grasping insects and piercing their hard cuticle before they are swallowed whole. It is often useful (Butler, 1981) to employ the term abrasion for this type of wear that occurs, it is said, without the teeth coming into actual contact.

The term erosion is by convention reserved for loss of tooth substance by what appear to be chemical causes. As the cause of loss of tooth substances cannot always be identified, it is impossible always to use the terms abrasion and erosion, and even attrition, with precision; the term wasting of dental tissues used by some older writers (W. D. Miller, 1907a,b) has advantages.

Another cause of loss of tooth substance is resorption by the animal's own tissues. Crazing, fine fracture lines in the enamel, will also be referred to here (p. 505) although it does not, as far as is known, lead to destruction of tooth substance.

Irregular and excessive attrition

The high-crowned, hypsodont teeth of horses and bovids erupt continuously (Chapter 16) and are continuously worn away (Ainamo and Talari, 1976). However, if any tooth is lost or becomes excessively worn, the opposing tooth may not wear away and the portion of its crown above the gingiva may become abnormally high. E. Becker (1970, Fig. 105) showed an example in an ox in which M^2 and M^3 were higher than the adjacent teeth because loss of the opposing teeth had reduced the amount of wear on them. In the horse shown in Figure 22.1, most of the teeth have worn normally, but the left P_4 has worn away almost completely and the crown of the opposing P^4 has hardly worn at all. The specimen has been prepared to show the portions of the upper teeth within the alveoli and it can be seen that the overall height of the P^4 from the apices of the roots to the occlusal surface is much greater than that of the other cheek teeth.

Generalized irregular wear due to diminished

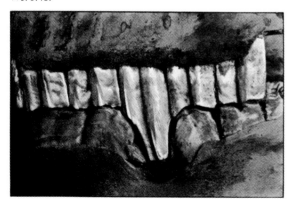

Fig. 22.1. *Equus caballus* (horse). RCS Odonto. Mus., G 40.1. Portions of the left maxilla and mandible showing extreme wear of P_4 and corresponding projection of the opposing P^4 beyond the occlusal plane of the others. The axial apex-to-occlusal-surface length of P^4 is greater than that of the adjacent teeth and it projects beyond them because it is less worn, not because it is over-erupted. × c. 0.46.

attrition is common in elderly horses and several extreme examples were figured by Kitt (1892) and Joest (1926). Another gross example in a skull of a 26-year-old ex-racehorse was described by Friel (1938); the horse had suffered for many years from periods of digestive disturbance but nevertheless had sired a record number of race winners. Diminished wear can also be caused by poor occlusion, as in the asymmetrical skull shown in Figure 13.76B. An extreme example of asymmetry was figured by Joest (1926, fig. 81).

In the horse, the lower tooth arch is much narrower than the upper; associated with this difference between the arches and with the predominantly side-to-side movement of the jaws, the wear surfaces of the lower cheek teeth slope facing buccally and the upper wear surfaces face lingually (Fig. 22.2A). Sometimes this wear pattern becomes greatly exaggerated and the slope increases, often asymmetrically, and the lingual edge of the lower wear table and the buccal edge of the upper one become jagged and sharp (Fig. 22.2B); the occlusal ends of the vertical enamel ridges can become hook-shaped (Harvey, 1985). This is known as shear mouth or shear bite and can be a source of great discomfort to the horse, producing ulceration of tongue and cheek; it is dealt with by filing down the sharp edges of the teeth (Catcott and Smithcors, 1972). Striking examples were figured by Joest (1926, fig. 121) and Kitt (1921, fig. 804).

There are many possible causes for this condition, such as lack of harmony in shape or size or relationship of the tooth arches, and other types of malocclusion of developmental origin or acquired during growth. It seems often to arise from some interference of masticatory function by painful teeth; for example, in the horse shown in Figure 22.3, there is shear bite on the right side where there are two suppurating and presumably tender molars which are likely to have encouraged mastication on the left side where the plane of the wear tables shows the normal obliquity. On the right side, the slope is increased and wear is irregular. Presumably as a result of the asymmetrical mastication, the wear table of the incisors has become oblique also. Instead of meeting edge to edge, the lower incisors are behind the upper ones on the right side and in front of them on the left, a condition known as cross bite. Several examples from the literature of shear bite in horses were reviewed by Müller (1935, p. 462). A comprehensive account of tooth wear in horses is given by Becker (1970, pp. 129–149). Examples in both horses and donkeys (*Equus asinus*) were mentioned by Amin and Kassem (1987).

The horse is also affected by a condition in which the occlusal surfaces of the cheek teeth, especially those of the upper jaw, are worn into cup-like depressions. Joest (1926, fig. 113) called the condition 'excavatio praecox' and figured an important early stage in a nine-year old horse. M1 is the most common site. Figure 22.4 shows a section through such a cup-shaped area which suggests that there has been a loss of the central enamel-cementum lake. Once affected in this way, wear can become so rapid that the tooth becomes worn below the level of its fellows and even to the level of the gum. In an examination of 484 London working horses, Colyer (1906) found five instances of this condition; in two, the right P^3 was affected and, in another, the left P^3 was level with the gum; in the fourth specimen, both right and left P^3 were in a similar condition, the remaining teeth showing marked cup-like attrition; in the fifth specimen (Fig. 22.1), the left P_4 was worn level with the gum, the opposing P^4 projecting considerably beyond the level of its neighbours. A similar example was described by M'Intosh (1929). Joest (1926, fig. 123) described the cupped-out dense polished wear tables of the cheek teeth in a 23-year-old horse.

The cupping-out of the occlusal surfaces and the irregular distribution of the wear suggest some inequality in the density of the tooth structure, and is a question worthy of investigation. That the maxillary cheek teeth of horses should be so much more liable to destruction than the mandibular teeth may, in some measure, be accounted for by the general arrangement of the tissues in the teeth (Fig. 1.8) and, in some measure, to the fact that the maxilla is the fixed point or mortar, the mandibular tooth acting as a pestle. In the centre of the maxillary tooth, there are two zones of cementum and the constant pounding of the mandibular tooth on these would, if the cementum were of feeble structure, cause excessive attrition.

Localized excessive wear similar to that in horses described above is seen in many other herbivorous animals, captive, wild and domestic. An example in the skull of a captive sub-adult Himalayan tahr (*Hemitragus jemlahicus*) is shown in Figure 22.5. The mesial half of each M_1 and the distal two-thirds of each M^1 show a marked degree of attrition and the opposing teeth are above the level of their neighbours. A similar condition of the dentition in a goat (*Capra hircus*) was described by J. Murie (1870). A good example of localized wear in an animal in the wild is an oryx skull (*Oryx leucoryx latipes*, BMNH 1934.8.4.30).

Fig. 22.2. *Equus caballus* (horse). A: Diagram of a transverse section through the jaws showing that the lower arch is narrower than the upper one so that, in mastication, the mandible is moved to one side to bring the cheek-tooth wear tables fully together and the teeth of the opposite side are out of occlusion. In the next part of the masticatory cycle, the jaw moves to full functional occlusion on the opposite side. The wear tables are not horizontal but slightly inclined.
The jaws of sheep, oxen and deer are similar to this in transverse section but usually proportionately wider.
B: The horse dentition in one-sided shear-bite. On the left side, the inclination of the wear tables is much increased so that the lingual margin of the lower teeth and the buccal margin of the upper ones have become sharp fragile edges which break and produce a saw-edge which may ulcerate the tongue and cheek of the horse.

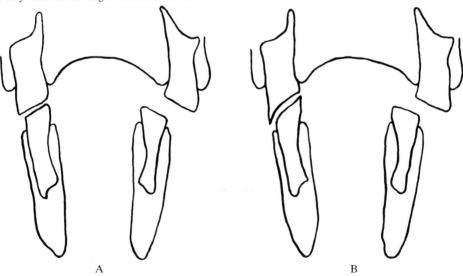

Cowan (1946) found some examples of irregular cheek-tooth wear in mule or black-tailed deer (*Odocoileus hemionus*) in the wild, mostly associated with fractures of teeth or loss of teeth. According to Newman and Yu (1976), the greatest wear in this species occurs on M1 with slightly less on the adjacent P4 and M2.

More generalized irregular wear of the teeth, producing dentitions with a grossly uneven wear table, is a feature of endemic enamel fluorosis in herbivores (see Chapter 20).

Zukowsky (1963–4) described a captive, aged (8-pointed) sika stag (*Cervus nippon*) which had to be killed because it became ill-nourished. The occlusal

Fig. 22.3. *Equus caballus* (horse). RCS Odonto. Mus., G 40.61. Showing shear-bite, or sloping wear table, affecting the whole dentition.
A: Anterior view. Instead of meeting edge to edge, the rows of incisors cross each other; the lower incisors are behind the upper ones on the right side of the specimen and in front of the upper ones on the left.
B: Palatal view. The cheek-tooth wear surfaces on the right side show the exaggerated slope and sharp buccal margins characteristic of shear-bite.
C: The mandible. On the right side where the shear-bite is present, the occlusal surface is irregular and there are two teeth showing localized periodontal disease which was the probable initiating cause of the shear-bite.

A

B

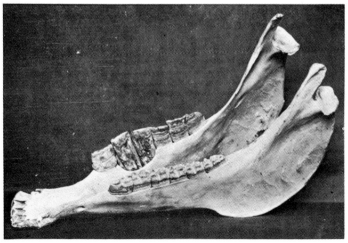
C

490 Section 4: Other Disorders of Teeth and Jaws

surfaces of the cheek teeth were irregularly worn in a mesio-distal curve so that, whereas those in the middle of the lower row and those at the ends of the upper row were worn down to the roots, those in the middle of the upper row and ends of the lower row were elongated. This curve was an exaggeration of the slightly curved or spherical nature of the occlusal plane in a herbivore. Occlusal planes are rarely quite flat.

Wear tends to be greatest at the middle of the cheek-tooth rows in the mule deer *Odocoileus hemionus* (Linsdale and Tomich, 1953) and is sometimes less in the upper jaw, especially in advanced age, so that a

Fig. 22.4. *Equus caballus* (horse). Section through a molar showing irregular wear of the tooth tissue (Colyer, 1914).

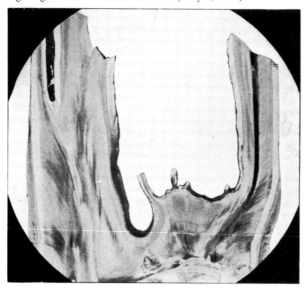

Fig. 22.5. *Hemitragus jemlahicus* (Himalayan tahr). RCS Odonto. Mus., G 39.41. Showing irregular wear of the teeth.

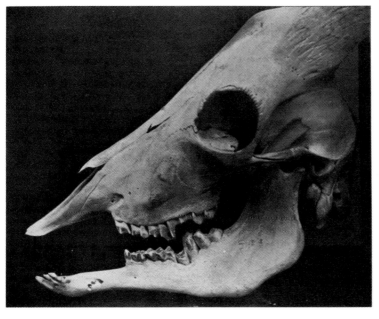

condition like that just described in a sika stag may arise with M_1 eventually becoming so worn that its roots become separated. Gingival disease or abscess formation is a common sequel. In advanced age, as in most artiodactyls, the teeth are worn nearly to the gum and their function is so greatly diminished that it is evident that this largely determines the maximum life span of the animal. W.P. Taylor (1956, p. 550) estimated that this is about 9 years for the mule deer (*Odocoileus hemionus columbianus*) in the wild but quoted known ages of 15, 16 and 22 years for this species in captivity.

K. Shaw (1981) described a similar condition in a female American elk (*Cervus canadensis*) in the wild in which mesial and distal slopes of wear converged at a point at P^3 and M^1 which fitted into a corresponding depression on M_1. Although the animal was found dead from an unknown cause, the paunch was full of forage so the occlusal condition had not been a severe handicap.

Peterson, Scheidler and Stephens (1982) referred to an unusual type of incisor tooth wear they found in Alaskan moose (*Alces alces*) that was absent from moose living in Michigan 12° further south. The sides of the teeth were worn so that they became a spaced row of irregularly cylindrical pegs. These workers suggested that winter grazing by the moose on the foliage of the cranberry bush, which is the dominant Alaskan ground cover and is perhaps contaminated with wind-deposited glacial silt, may be responsible for the incisor wear. Young and Marty (1986), without reference to this, described as a special feature of moose in another geographic area what may have been an earlier stage of the same condition.

Foley and Atkinson (1984) described a type of uneven and excessive tooth wear that affected over 50% of Defassa waterbuck (*Kobus defassa*) of all ages in the Nakura National Park, Kenya. The excessive wear predominantly affected P4 and the molars; in the upper jaw, mostly the lingual part of the occlusal surfaces and, in the lower jaw, mostly the buccal part so that the wear surfaces were sloped as in Figure 22.1B. In addition, it was common for a part of a tooth, either upper or lower, to be only slightly worn whereas the corresponding surface of the tooth in the opposite jaw was excessively worn, producing the sort of disordered occlusion seen in Figure 22.5. These abnormal patterns of wear were always symmetrical. Foley and Atkinson found no differences between the diet of these animals and that of the populations of waterbuck in other parts of Africa where this condition does not occur. They concluded that the abnormal tooth wear may have a genetic basis (founder effect: see Chapter 1, p. 10), bearing in mind that the affected population has in recent years increased greatly in numbers because of the destruction by hunters of their predators and many other competing ungulate species. Furthermore, the population has been isolated by fencing and in consequence has tended to be inbred. This would have created conditions favourable for an increase in the population of genetic abnormalities controlled by recessive genes (genetic drift, Chapter 1, p. 10), one of which in some as yet undefined way could be responsible for the irregular wear. As we have already pointed out, this type of wear could be due to a regular pattern of differences in hardness, that is mineralization, between individual teeth in the dentitions.

In sheep, excessive wear of the incisor teeth creates a serious economic problem for farmers. As in bovids in general, there are no upper incisors; the three lower incisors and the incisiform canine occlude with a dental pad, composed of fibrous tissue covered with keratinized epithelium, which replaces upper incisors. For convenience, the common practice will be adopted here of referring to the eight incisiform teeth as incisors. The incisor apparatus in the sheep will be described more fully in Chapter 24. In certain regions of New Zealand and, to a lesser extent, of Australia, the wearing down to the gum of the incisors by the age of about 4 or 5 years has been described, that is before the end of the 6 or 7 lambing seasons that can reasonably be expected of a ewe. When their incisors are worn to the gum or nearly so, ewes are said to be gummy. According to Barnicoat (1957), although this problem was known long before, it has become aggravated during the last 60 years or so. Solution of the problem is complicated by the fact that not all members of particular flocks are affected.

The condition does not appear to be a problem on Scottish sheep-farms, although Duckworth *et al.* (1962) mentioned that the deciduous incisors of penned sheep sometimes become badly worn by chewing on the metal rails of the pen. In Western Ireland, however, rapid wear of the incisors, like that in New Zealand, does present a problem (Nolan and Black, 1970).

A great many factors which might have localized distribution have been investigated. Only a limited review of the subject is appropriate here. In the first place, the tooth tissue of certain flocks could be softer or more susceptible to wear for genetic reasons, or be so because of some environmental deficiency during their development. On the other hand, Erasmuson (1985) found no difference between the wear resistance of enamel and dentine, measured on a machine, from high and low-wear farms. Sheep consume a considerable amount of soil during grazing and the abrasiveness of the soils would be variable (Healy, 1968). The amount of soil consumed would be more when grazing on short grass, for example on overstocked land, and would be less when the grass was long and lush and the rainfall ample. Top-dressing the land with gritty material fertilizers, such as ground bone, slag or limestone, which may be slow to dissolve, especially in areas of sparse rainfall (Barnicoat, 1957), would increase the rate of tooth wear. The type of grass in the pastures varies; some grasses are tougher and more abrasive than others. There are also chemical differences in the pasture juices (Cutress and Healy, 1965); some might be more acid than others or contain mineral-complexing substances (chelators) (Barnicoat and Hall, 1960); both would introduce an element of chemical erosion into the wear process.

Mitchum and Bruère (1984) have pointed out that the microbial part of the biomass of soil, principally bacteria, actinomycetes and fungi, increases greatly from the droppings of sheep and tends to be higher if the stocking rate is high. They postulated that soil micro-organisms are the principal sources of organic acids and chelating agents which, by coming into contact with the incisors, particularly in grazing, contribute to the rapid loss of tooth substance. There is evidence that excessive incisor-tooth wear is related to the density of the sheep population per hecare of land (Nolan and Black, 1970).

Baker, Jones and Wardrop (1959) showed that silica in the soil is incorporated into the substance of grasses in the form of solid mineral particles (opal-phytoliths) which are harder than the tissues of sheep's teeth. They identified opal-phytoliths in the faeces of sheep on Australian pastures but what proportion of them was derived from the grasses, from the soil or from dust on the grasses was impossible to determine although ways of doing so are now available (Armitage, 1975). To emphasize that such mineral particles could be related to tooth wear, Baker *et al.* mentioned that a sheep might consume as much as 22 lb of them annually.

Nolan and Black (1970) made several measurements of the heights of the crowns of the incisors of sheep from transverse file marks made on the labial surfaces and so measured the rate of wear at intervals through several seasons. They found that both rate of tooth wear and faecal silica content rose and fell together at different periods of the year, both being highest in ewes during the period between lambing and weaning, when of course grazing would be most intense. These findings provide some support for the view that the consumption of silica particles, ultimately derived from the soil, plays some part in producing excessive wear of the incisors.

The severity of the wear of the incisor teeth described by Bruère, West, Orr and O'Callaghan (1979) in 200 sheep on a number of farms in North Island, New Zealand, was quite remarkable. The deciduous teeth became worn down to the gum before it was time for them to be shed, the permanent incisors erupted prematurely, sometimes beside the worn remains of their predecessors, and then underwent rapid wear in their turn so that by about 4 years they were worn to the gum. It was inferred that, at an earlier stage, periodontal disease must have been present because in some cases some of the incisors were lost; such cases were categorized as broken mouth (Chapter 24, p. 559) associated with gummy mouth. There was evidence that these animals suffered from some systemic disturbance of mineralization with osteoporosis (Orr, O'Callaghan, West and Bruère, 1979). Another feature was that a number of unerupted teeth were associated with dentigerous cysts (see Chapter 25, p. 579). Furthermore, there was some evidence that a deficiency of copper was the underlying cause or at least a contributing factor; following administration of copper to the ewes and application of copper sulphate to the grazing land, the new season's lambs showed less tooth wear. Transfer of ewes to other farms produced similar improvement. It was not stated whether cheek-tooth wear was affected. Severe wear affected the deciduous incisors of yearling cattle as well as sheep on these farms.

In a group of New Zealand sheep in which the deciduous incisors were severely worn by about 50 weeks of age, Thurley (1984) noted that there was no excessive wear of the cheek teeth.

The exact manner in which the incisors meet the upper dental pad affects the amount and type of wear of the teeth. As is discussed in Chapter 24 (p. 556), there is some doubt about what is to be regarded as the normal relation of the teeth to the pad but, when the teeth occlude with the flat inferior surface of the pad, the wear surfaces of the incisors are also flat and transverse to the long axis of the crown; in this relationship, wear is likely to be maximal, especially if the incisors rub on the pad during the mandibular movements associated with chewing on the cheek teeth, as during rumination. When the incisors meet the anterior rounded edge of the pad, the wear surface tends to be oblique and its anterior edge is chisel-shaped and sharp. The wear may extend well down on the lingual surface of the tooth. If the teeth pass in front of the pad, which must be regarded as malocclusion, the incisive edges become unevenly worn with wear also on the labial surfaces because grass is gripped between the upper and lower dental pads and dragged across the incisors (Chapter 13, p. 295). This relationship of the teeth in front of the pad is known as an undershot jaw in Britain, North America and Australia; the opposite condition, where the mandible is short and the incisors too far back, is overshot. It is important to note that, in New Zealand, these two terms are used in the opposite senses.

Variations of the types of incisor wear in sheep were illustrated by Barnicoat (1957), including examples where the wear surfaces of the first incisors were oblique and faced each other to form a notch.

Chapman (1933) studied the developing pattern of wear in plaster casts he prepared every 3–4 weeks from a group of 45 lambs over a period of about two years. He found that approximal wear proceeded at a fast rate so that, after two years, the approximal enamel was worn through; the worn surfaces were not always flat but sometimes the convex mesial surface of one tooth would fit into a longitudinal concavity worn into the distal surface of the adjacent tooth.

Rudge (1970) found irregular excessive wear of the cheek teeth in a group of feral goats (*Capra hircus*) on an isolated island off the coast of New Zealand. In some animals, estimated to be as old as 10 years, some molars were worn down to the gum and others were worn to beyond the furcations between roots. The excessive wear was associated with deep gingival pockets and destruction of the bony alveolar crests between affected teeth. The grazing on the island appeared to be of poor quality and dusty with wind-blown volcanic soil. It seems likely that there were no natural predators and thus the goats survived despite their dental handicap.

Smith, Genoways and Jones (1977) found tooth attrition particularly marked in howler monkeys (*Alouatta palliata*) living on a volcanic island where the tree foliage was frequently covered in volcanic dust. They also described examples of shear bite in two howler monkeys (*A. palliata*) and a capuchin (*Cebus capucinus*) from the wild; in two instances there was a presumptively painful tooth on the opposite side. It must be borne in mind, however, that normally in non-human primates with large canines that limit side-to-side jaw movement, wear of the teeth predominantly affects the lingual parts of the cheek teeth (Chapters 21, 23). This becomes shear bite if it is exaggerated unilaterally.

Tooth wear is not a notable feature of the carnivore dentition. However, Rausch (1961) found that, among grizzly bear (*Ursus horribilis*) in the wild, the incisors were sometimes worn flat by the fifth year and, in animals reliably estimated to be over six years of age, the dentition as a whole was severely worn with many broken and some missing teeth.

The cheek teeth in wombats are of continuous growth and grow in lateral curves which in the upper jaw are concave buccally and in the lower jaw are concave lingually. Beier (1982) found five skulls of common wombat (*Vombatus ursinus*) and one of a southern hairy-nosed wombat (*Lasiorhinus latifrons*) in which wear was abnormal, so that the buccal edges of the wear tables of the upper teeth and the lingual edges of the lower ones were of almost razor sharpness. In several instances, the wear tables presented a step-like pattern of upper mesially-directed and lower distally-directed surfaces. There must have been a lack of coordination between growth of the teeth and growth of bone because the growing ends of several upper teeth projected through openings in the bone of the jaw.

W. D. Miller (1907b) described smooth facets on the sides of the crowns of the conical teeth of a pilot whale (*Globicephala*) which are puzzling when seen in isolated teeth but in fact are facets worn by contact with the antagonistic teeth of the opposite jaw. The teeth are spaced along the jaws by nearly the width of the teeth and the teeth of the opposite jaws interdigitate; hence wear is produced on the sides as well as to some extent at the tips of the cones.

C. S. Tomes (1873) described in some detail an abscessed tooth in a grampus or killer whale (*Orcinus orca*) found stranded in the River Severn. Wear had produced a large opening into the pulp which had in consequence died (see this Chapter, p. 501; Fig. 22.14). This is unusual because in cetaceans in general the tooth pulp becomes replaced by dentine seemingly independently of a reaction to wear and, as Tomes pointed out, because the homodont teeth interdigitate, wear mainly affects the surfaces between the teeth rather than the tips of the crowns. In this grampus, the jaw relationship had been altered by an asymmetry of the jaws so that some teeth, including the abscessed one, met tip to tip. Colyer (1938) from his examination of 20 grampus skulls, inclined to the view that wear of the tops of the teeth as well as the sides between the teeth is more common than Tomes supposed. It is noteworthy that, in a stranded Risso's dolphin (*Grampus griseus*) with abscessed teeth (Fraser, 1953), the teeth were much worn. In this species, there are normally no teeth in the upper jaw.

There is evidence that tooth loss from wear, and

494 *Section 4: Other Disorders of Teeth and Jaws*

fracture of the teeth and the abscesses that tend to occur as sequelae, are common in the bottle-nosed dolphin (*Tursiops truncatus*). However, as de Smet (1977) pointed out, the specimens tend to be of animals found beached, which, whether dead or alive, would be unlikely to be typical of the normal population and might consist largely of aged animals or ones sick from systemic disorders of some kind. It could be that they became enfeebled and beached because their dentitions were no longer adequate to catch and hold the slippery fish on which they prey. De Smet found several examples of badly worn and broken teeth in animals that had lived in zoos or dolphinaria for some years and were not aged. In the wild state also, there is sometimes considerable wear in animals that could not be of advanced age.

De Smet (1972) described wear and tooth loss in two stranded white-beaked dolphins (*Lagenorhynchus albirostris*). Both animals had multiple exostoses of many vertebrae which may have affected their mobility in the water.

Robineau (1981) described four adult common dolphins (*Delphinus delphis*) in which the tooth rows were very incomplete. Because there was evidence of tooth sockets filled with new bone, it seemed that there had been loss of teeth rather than failure of development. The cause can only be speculated upon; two of the dolphins, both female, showed evidence of systemic bone disease possibly of endocrine origin. Only one showed much tooth wear. Further examples of tooth wear and tooth loss in the Cetacea are given in Chapter 23.

Abrasion and erosion

The loss of tooth tissue in the form of V-shaped slots at the necks of human teeth or saucer-like cavities on the labial or buccal surfaces of the crowns, first described by John Hunter (1778), at one time excited much controversy, some holding the view that they were due to some acid erosive agent perhaps secreted by the gum; indeed some puzzling examples are still to be seen, although, since W. D. Miller (1907*a*) simulated the condition by prolonged tooth brushing on a mechanical model, it is accepted that unwise brushing techniques are the most common cause in man. Both V-shaped grooving and saucer-like wasting are sometimes met with in animals, though rarely in the wild state.

These historical examples emphasize why these two topics, abrasion and erosion, are dealt with jointly here. Although it is possible to commence the section with examples where the tooth loss is certainly due to mechanical abrasion, we almost at once have to present many instances where it is uncertain whether the loss of tooth substance is due to abrasion or to some chemical action, or to a combination of both. Towards the end of the section, instances are presented where the cause of tooth destruction is more certainly chemical in nature, perhaps through the agency of parasitic organisms.

Abnormal wear of the incisor teeth in the horse is sometimes due to various habits, such as rubbing the teeth against the edge of a manger (crib-biting) which tends to remove the enamel from the labial aspects of its lower incisors (Bahn, 1980). A good example affecting the labial aspects of both the upper and lower incisors of a horse was depicted by Joest (1926, fig. 125). Alternatively, the horse may nibble at its manger or at iron bars, thus wearing away the cutting edges (Kitt, 1892). Colyer (1906) referred to an I^1 in which a distinct notch had been worn (RCS Odonto. Mus., G 40.2). Wear of the canines in horses or of the tooth surfaces adjacent to the diastema sometimes occurs from rubbing of the bit.

Penny and Mullen (1975) found that about one-third of abattoir specimens of pig showed wear, often severe, in the region of P4 and M1 which they considered may be due to chewing on the iron bars of their pens or something similar. On the other hand, Samuel and Woodall (1988) found a few examples of excessive wear of M1 in feral pigs, even to the extent that the pulp was exposed.

Dogs may show excessive tooth wear due mainly to retrieving or playing with balls, stones or sticks. The incisors may become worn when the dog nibbles its own skin in chronic pruritus (Hofmeyr, 1960).

In the skull of a wallaby shown in Figure 22.6, there is a wedge-shaped notch on the distal aspects of I^1. I^2 and I^3 show more wear than in the normal, and it is possible that the wasting of I^1 was due to food impregnated with an abrasive material having been drawn across the tooth. Many specimens of macropods show rapid wear of the cheek teeth and, when the premolar erupts, the teeth become separated, as in Figure 22.6.

According to Michel (1904), wedge-shaped abrasion is seen in 10–15% of old domestic cattle, mostly on the distal surfaces of the canine teeth. Examples were figured by W. D. Miller (1907*b*) who described the condition in the skull of a Niata ox from Serra di Rio Grande de Sol in Brazil. Niata cattle (p. 292–294, Chapter 13) have stunted upper jaws so that the lower incisors do not occlude properly with the incisive pad. It is said that this makes it impossible for them to feed on any other than long grass. Miller suggested that the wedge abrasion often found between the teeth is due to long grass, perhaps associated with abrasive soil, being drawn between the teeth in grazing. An ox incisor (RCS Odonto. Mus., G 45.3) has the same condition.

Miller's paper is a fund of anecdotal information about tooth abrasion or tooth wasting in various species, not all of which can be quoted here. He mentions that the condition commonly affects the incisors or canines of domestic pigs; apart from grubbing with their teeth into the soil, pigs tend to rub their teeth against fences and trees. Examples in a peccary (*Tayassu* sp.) and a tapir (*Tapirus* sp.) were also referred to.

Magitot (1877, fig. 13, pl. 16) figured an example of erosion of the occlusal one-third of the labial enamel of the I_1 of a two and a half year-old cow. Both Magitot and Bland-Sutton (1884*a*), in using the term erosion, included the possibility of wear being responsible. However, from the evidence available, it is impossible to be sure that this was not an example of defective formation of enamel.

Colyer (1921) found 19 instances of excessive wear of the incisor teeth among 300 skulls of young rhesus monkeys (*Macaca mulatta*) which died after a period in captivity, probably not much more than one year (see Chapter 10, pp. 206–208). The wear had been very rapid because many pulp cavities were laid open with resulting abscess formation. It seems possible that stressful crowded conditions of captivity had induced some gnawing habit, perhaps cage-biting.

Some specimens in the collection of cercopithecids in the Congo collection of what is now the Koninklijk Museum voor Midden-Afrika have a curious kind of wear of the mandibular incisors. In the earlier stages, the wear is greater on the lingual than on the labial aspects, the cutting edges of the teeth being rounded off. With an increased loss of tissue, the tooth assumes a cone-like shape, as seen in Figure 22.7. The way the teeth are worn suggests that fibrous food is drawn from the inside of the mouth in an outward and downward direction against the incisor teeth, the outward pull causing wear of the lingual surfaces and the downward pull causing wear of the labial surfaces. When the contact points between the teeth have been worn away, it may be that the fibres pass between the teeth and so produce the rounded appearance. Stripping the fleshy part of a leaf from a central rib seems to be a possible cause. This type of wear was common in Whiteside's guenon (*Cercopithecus ascanius whitesidei*) and Wolf's monkey (*C. wolfi*) and was seen in a few specimens of Katangan guenon (*C. ascanius katangae*) and Schmidt's guenon (*C. ascanius*

Fig. 22.7. *Cercopithecus ascanius whitesidei* (Whiteside's guenon). KMMA 6920. Showing unusual wear of the mandibular incisors.

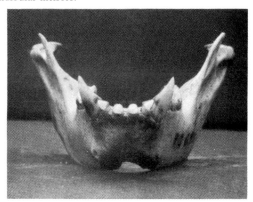

Fig. 22.6. *Macropus agilis* (agile wallaby). RCS Odonto. Mus., G 45.2. Wedge-shaped defect of distal surface of I^1. × *c*. 0.6.

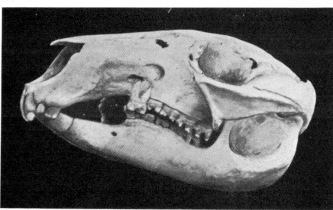

496 Section 4: Other Disorders of Teeth and Jaws

schmidti), but not in Stuhlmann's guenon (*C. mitis stuhlmanni*), De Brazza's monkey (*C. neglectus*) or black-cheeked white-nosed monkey (*C. ascanius ascanius*). The large majority of the specimens of *C. ascanius whitesidei* and *C. wolfi* came from the same district in which many of the specimens of *C. neglectus* were collected. Colyer (1943*b*) described a similar type of wear on the mandibular incisors in two *Colobus* monkeys and, in another (P-C. Mus., M243), excessive wear limited to the lingual aspects of the third molars. In general, the teeth of *Colobus* do not show a great deal of wear, probably because they tend to be strictly arboreal and leaf-eating and therefore their diet is free from the contamination with dusty or gritty soil. However, such occlusal wear of the cheek teeth as does occur is more marked on the lingual cusps of maxillary teeth and buccal cusps of the mandibular ones. The cercopithecids, on the other hand, are much more ground-living and omnivorous.

The sea otter (*Enhydra lutris*), which lives mainly on sea urchins, molluscs and crabs, which involve crunching hard exoskeleton, has a dentition like that of the badger and bears, of broad-crowned (bunodont) omnivorous type. The crowns of the teeth become much worn (Fisher, 1941). Bjotvedt and Turner (1976), in four out of a group of five jaws, found alveolar abscesses of the canines, some secondary to excessive wear and two secondary to fracture of the tooth crowns with exposure of the pulp.

The crowns of the cheek teeth of the sea lions are simple, slightly recurved cones tipped with enamel which is thick at the cervix so that the crowns appear to be rather bulbous. The tips of the crowns in all but the youngest animals are worn flat. Quite commonly there is some loss of substance of the root close to the cervix; the area of loss may encircle the root. J. Murie (1870) first drew attention to this condition by describing an instance (Fig. 22.8) in which the loss of substance was particularly severe. All the roots, including those of the canines and incisors, were encircled by deep wide grooves at the cervices, the loss of substance being particularly great on the distal aspects of the canines. The affected surfaces were blackened and highly polished. The tips of the crowns showed the usual amount of wear. There are a number of instances of some degree of similar loss of tooth substance among several species of sea lion in the collections of the British Museum (Natural History): the southern sea

Fig. 22.8. *Otaria byronia* (southern sea lion). A specimen showing a destructive process affecting the cervical regions of all the teeth (Murie, 1870).

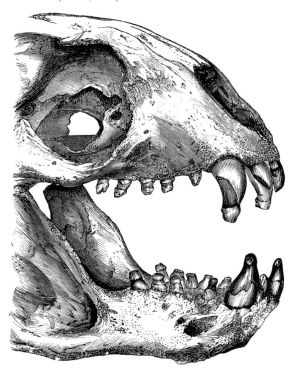

lion (*Otaria byronia*) from the Falklands, the New Zealand sea lion, (*Phocarctos hookeri*) (Fig. 22.9) and the northern sea lion (*Eumetopius jubatus*) (Fig. 22.10). The affected surfaces were stained and highly polished and, with a lens, Colyer noted scratches running in various directions. These appearances are suggestive of wear by contact with some abrasive substance. According to Lydekker (1893–4, vol. 2, p. 109), *Otaria byronia* lives in localities where freshwater rivulets enter the sea and it spends many hours in shallow water feeding on fish, crustaceans, echinoderms and shell-fish. If shell-fish formed a significant part of the diet, the occlusal attrition would be accounted for (W.D. Miller 1907b). Lecomte, a seaman quoted by Lydekker, said that it was usual to find pebbles in the stomachs of the animals. It seems possible therefore that sand and pebbles taken into the mouth while feeding in shallow waters are responsible for the curious abrasion of the teeth whether or not the animals have the habit of shaking shingle about in their mouths (Underwood, 1914). This is a matter that deserves to be looked at more closely and, bearing in mind that there is evidence that mysterious erosions of the roots of teeth in man can be due to sometimes obscure chemical causes, such as fruit eating and food regurgitation, sometimes in combination with mechanical causes such as tooth brushing (Lewis and Smith, 1973; Eccles, 1979), the possibility of some partly chemical cause in sea lions should be considered.

Fig. 22.9. *Phocarctos hookeri* (New Zealand sealion). BMNH 1897.10.10.5. The roots of the canines are highly polished, stained and slightly worn. The necks of the cheek teeth show encircling smooth-surfaced grooves. All the teeth show occlusal attrition.

Figure 22.11 shows one of several Weddell seals (*Leptonychotes weddelli*) in the British Museum (Natural History) with wear of the cutting edges and labial surfaces of the upper incisors and canines; the roots of the teeth were unaffected and the condition is quite different from that described in the sea lion.

Fig. 22.10. *Eumetopius jubatus* (northern sealion). ♂. BMNH 1908.3.11.4. The root of the upper canine in particular is highly polished and stained and the necks of the cheek teeth show encircling shallow grooves. There is a crescentic notch on the distal aspect of the third maxillary cheek tooth and the occlusal surface of the fourth tooth in particular is much worn.

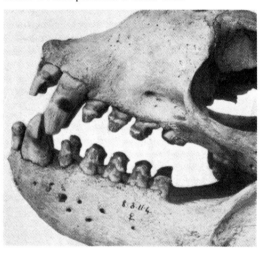

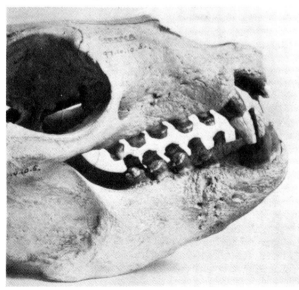

During the winter months, Weddell seals live almost entirely in the water beneath the ice, it being slightly warmer in the water. In order to breathe, they must be continuously active keeping open their breathing holes in the ice. It is well authenticated that their teeth play a crucial role in this and also in re-opening the holes that may have partly frozen over when the seals need to return to the water after they have emerged on to the ice. Wilson (1907) and Lindsey (1937) have given vivid eyewitness accounts of the way the seals use their teeth for this purpose. The seal fixes its canines and incisors in the solid ice with the mouth open at an angle of about 150° and swings the entire head from side to side until a double groove is cut and the ice can be broken off. Wilson thought the seal used the lower teeth as a more or less fixed point while the upper ones swing to and fro; he said the action was like that of a centre bit, the lower teeth being the fixed centre and upper teeth the cutting arc. He said that he would sometimes hear, while walking on the ice, the scrunching noise of seals' teeth opening up new ice to form a blow hole. Underwood (1914) mentioned having seen this activity convincingly in a cine-film, *Ninety Degrees South*, (now in the custody of the British Film Institute) made by H.G. Ponting on the 1911 Scott Expedition.

A specimen described and figured by Wilson (1907) is similar to that in Figure 22.11 though wear is more advanced; the pulps of both I^2 are exposed, the teeth are abscessed and one lower canine is broken and jagged. The cheek teeth are unaffected by the wear process. Stirling (1969) made a study of the skulls of 171 Weddell seals, predominantly to assess their ages from the growth lines in the dentine. The upper canines seemed to be affected by wear first, then I^2 and then the lower canines; wear of I^1 and the lower incisors was uncommon.

This ice-sawing use of the dentition in Weddell seals constitutes one of the most notable instances of the dependence of a species on its teeth for survival. Unless there is some elaborate group activity whereby some seals maintain a breathing hole while others range as far as they dare to search for food, a seal that lost the efficiency of its dentition could hardly survive for long. As Stirling pointed out, any seal which could not maintain an air hole would drown or, caught on the surface with a partly frozen-over hole, would die of starvation or exposure.

The incisors and canines in the upper jaw of the Weddell seal are much more inclined forwards and splayed out than in other seals (Fig. 22.12), apparently as an adaptation that has evolved in response to the need to maintain breathing holes through the ice.

W. D. Miller (1907b) described the skull (BMNH 1887.8.1.5) of a monk seal (*Monachus tropicalis*) with smooth saucer-shaped cavities on the distal aspects of the well-developed upper canines close to the gum. He noted scratches on the surface of the cavities and suggested that they are produced by tearing at seaweed charged with sand.

Fabian (1950) described what can be translated as 'boreholes' in the cementum on the sides of the teeth of sperm whales. Some were slit-like cavities and some appeared to join up beneath the surface with other adjacent ones. Keil (1959) described destructive lesions of the teeth of sperm whales (*Physeter catodon*) that had some features in common with those noted by Fabian. There are many marine invertebrates that are capable of boring into coral, shells or limestones and it is possible that some such minute marine organisms

Fig. 22.11. *Leptonychotes weddelli* (Weddell's seal). BMNH 1908.2.20.58. Wear of the cutting edges and labial surfaces of the canines and lateral incisors, which are longer and larger than the medial ones.

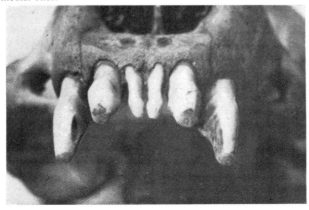

can attack the teeth of whales. Peyer (1968) referred to the invasion of the tooth substance of the elasmobranch (*Raia clavata*) by a boring marine micro-organism.

Alternatively, the lesions in the teeth of sperm whales may be due to resorption by the tissues that surround the teeth when they are partially erupted (see later under Resorption). Although the teeth of the sperm whale are large, not a great deal more than the tips of the teeth are exposed even when they have erupted.

Barnacles quite frequently colonize the surface of the teeth of sperm whales (Chapter 18) and possibly those of other odontocetes also. The mode of attachment is by some substance secreted by the barnacle. Walker (1972) has shown that, in barnacles of the genus *Balanus*, the cement substance is a protein associated with phenols and with an enzyme, polyphenol oxidase, and after secretion undergoes a tanning process which toughens it. As the cause of the destructive processes just referred to is entirely a matter of speculation, it is worth raising the possibility that the cement substances of barnacles might by prolonged contact with the tooth surface have some erosive effect on it.

Ness (1966) described a curious destructive process that had affected many teeth of four specimens of the Amazon porpoise (*Inia geoffrensis*) in the collection of the British Museum (Natural History). This porpoise is found only in the waters of the Amazon and Orinoco and predominantly in the shallow waters of their estuaries (Watson, 1981). The long narrow snout bears about 25–35 teeth on each quadrant of the jaws, probably depending on the age of the animal, new teeth being added to the posterior ends of the rows as age advances. The crowns of the anterior teeth are conical (Fig. 22.13) and the most posterior ones are flattened and more molariform with gradual transitional forms in between. The animals are piscivorous although some plant matter has been found in their

Fig. 22.12. Above: *Lobodon carcinophagus* (crabeater seal). Below: *Leptonychotes weddelli* (Weddell's seal). Right lateral and dorsal views of the upper jaw to show how the canines and incisors are more mesially inclined and splayed out in Weddell's seal than in other seals. The Weddell seal is the only seal to use its teeth to maintain breathing holes in the ice and this arrangement appears to be an adaptation for this purpose (from Stirling, 1969).

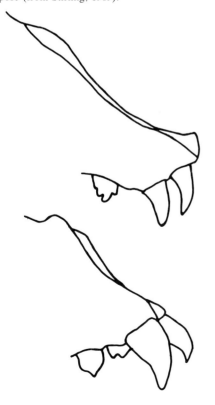
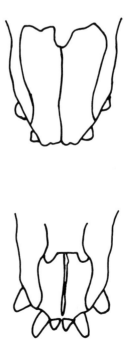

stomachs. It is said that, as the morphology of their dentitions suggests, these porpoises are unusual in that they bite their prey into pieces before swallowing it.

The enamel surface is uneven or wrinkled, and varies in colour in different specimens from straw-colour and reddish-brown to brownish-black. The colour seems to be a stain or deposit acquired after the teeth erupt because, where the enamel is worn, the surface is creamy-white; Ness found a transverse linear deposit like dental calculus at the level of the presumptive gingival margin of many specimens (Fig. 22.13 A). The tooth-destructive lesions that affected the four skulls consisted mainly of shallow cavities in the enamel with sharp margins; some extended into the dentine in such a way as to undermine the enamel. The dentine floors of the cavities were rough and brown-stained. In one specimen, the lesions were slightly different and consisted of shallow cavities extending across the labial aspects of the cervices of the teeth so that the root surface was involved, with undermining of the cervical enamel margin.

Each skull had at least one tooth of which the crown was totally destroyed, presumably by extension of the lesions already described (Fig. 22.13B). The surfaces of the remaining roots, at about the level of the cervix, presented a dish-shaped cavity, rough-surfaced and brown with an irregular rim of enamel. The pulp cavities were not exposed so that there must have been formation of reactionary dentine in response to the advancing destructive process. This observation alone seems to discount the possibility that these lesions were the result of some sort of destructive process that affected the teeth after death, for example during storage in the museum. Invasion of the tooth substance by micro-organisms that live in soil is met with following prolonged burial (Falin, 1961; Poole and Tratman, 1978). The nature of the organisms is far from being understood but they can burrow into dentine, producing fine channels which become confluent and probably in time could totally destroy tooth substance. Falin suggested that a fungus is responsible.

Ness (1966) tentatively called these lesions in the Amazon porpoise dental caries and indeed the appearances, especially with undermining of enamel,

Fig. 22.13. *Inia geoffrensis* (Amazon porpoise). BMNH 1937.5.26.1. Probably an old animal.
A: Teeth situated towards the anterior end of the left mandibular tooth row. There are smooth-surfaced grooves accentuating the cervices and, below that level, there are rough-surfaced brown-stained caries-like excavations. Two teeth have, below that level, a rim of dental calculus which would have been just above the gum margin.
B: Teeth from about the same position on the right side. The crowns of three teeth have been completely destroyed, leaving deeply-excavated irregular surfaces surrounded by a rim which would have projected just above the gum. Drawn from Ness (1966).

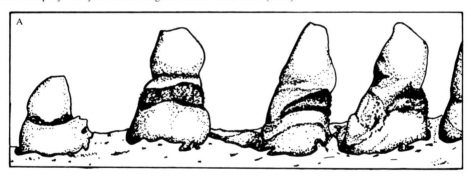

resemble the lesions of dental caries as met with in terrestrial mammals much more closely, for example, than the lesions referred to so far in this Chapter. Unfortunately, Ness was only able to make a restricted examination and presumably therefore was unable to say whether the dentine that formed the floors of these cavities was softened or not. This is a point of crucial importance because it is characteristic of what is meant by the term dental caries that the loss of tooth substance is preceded by softening of the tissue (Chapter 21, p. 457). Use of the term dental caries also today implies that the destruction was brought about by parasitic organisms living in a substrate on the tooth surface; furthermore the weight of evidence incriminating fermentable carbohydrate in dental caries is such that use of that term tends to imply the presence of carbohydrate. As Ness pointed out, there is nothing in what is known of the diet of these dolphins that is likely to supply fermentable carbohydrate. However, it is conceivable that destructive substances, acid or chelating agents, could be formed out of a predominantly protein-containing substrate on the tooth surface by a microflora quite different from that described in Chapter 21 in discussing dental caries; even some sort of macro-parasite could be envisaged. We have already referred to parasitic barnacles or some more minute marine parasites being possibly responsible for sperm-whale dental lesions.

Pilleri and Gihr (1969b) illustrated lesions in the crowns of several teeth near the anterior ends of the mandibular rows in a freshly-caught Amazon porpoise which seem closely to resemble the cervical or root-surface lesions that Ness found in one of his museum specimens. The crowns of many teeth were nearly totally destroyed and presented the dish-shaped surfaces described by Ness.

The lesions Ness described are worthy of closer examination; the value of the contribution made by Pilleri and Gihr lies mainly in their illustrations. Other specimens of Amazon porpoise housed in museums need to be examined and it is to be hoped that some can be sectioned and even that freshly-caught specimens could be obtained suitable for micro-biological examination.

There is of course, a semantic question here, such as what is meant by the term dental caries, which it is not profitable to discuss further. We have chosen to discuss these lesions, which we think it is best for the time being to categorize as caries-like, here rather than in Chapter 21 on Dental Caries, largely because it is convenient to contemplate them together with other destructive lesions the nature of which is still obscure. An account of dental caries in another marine mammal, the manatee, is included in Chapter 21 (p. 480) but in that case there was some supportive histological evidence.

Loss of tooth tissue from resorption

Tooth resorption, which occurs normally in the shedding of deciduous teeth, is brought about by cells which morphologically resemble the osteoclasts which resorb bone. Hence, the term osteoclast is used for cells which resorb hard tissue whether it be bone, cementum, dentine or enamel. Tooth resorption can occur in a number of pathological states; it is often stimulated by local inflammatory conditions, including chronic abscess, and teeth that have remained unerupted in the jaws for long periods often show areas of resorption on their surfaces, including sometimes the enamel of the crown. However, the cause is often obscure.

Most areas of resorption seem to start on the surface although it is not uncommon for it to burrow deeply into the tooth. There is another, less common, form of resorption that starts on the surface of the pulp and burrows outwards into the crown or the root.

The abscessed tooth of a killer whale (*Orcinus orca*) described by C. S. Tomes (1873) that has already been referred to (p. 493) showed a great deal of resorption of the root (Fig. 22.14).

Yablokov, Bel'kovich and Borisov (1974) reproduced a photograph of a killer whale with what they called carious injury of a tooth among the upper row. The lesion is close to the gum margin and may well have extended beneath it. In the absence of a detailed description, our inclination is to think that this lesion was yet another example of resorption, perhaps similar to the lesion in killer whale described by Tomes (Fig. 22.14).

The tooth of a sperm whale in Figure 22.15 shows a large area of resorption limited to one side of the root. The process causing the loss of tissue was probably associated with pressure from the adjacent tooth. According to M. A. C. Hinton (personal communication to Colyer, c. 1930), the teeth are set loosely in their sockets and can be moved easily; thus, a tooth may be brought into close contact with its neighbour and it is possible that when this occurs foreign bodies become lodged between the teeth and so start pathological processes in the periodontal tissues leading to resorption of the hard tissues. However, Boschma (1950b), in describing three similar examples of resorption of the lingual surfaces of sperm whale teeth, although agreeing that, as in all Cetacea, the teeth are slightly mobile in their sockets, claimed that they are nevertheless firmly implanted and are unlikely to be moved laterally into contact with one another. He pointed out that all three of his resorbed teeth had facets of wear at or near their tips. He believed that the wear was due to contact with maxillary teeth which produced undue stress on the teeth or possibly injured the gum margin, so inducing gingival inflammation and tooth resorption. Unerupted rudimentary teeth are common in the upper jaw of the sperm whale (Chapter 5, p. 101) and are known sometimes to erupt (Boschma, 1938*b*).

Under certain conditions, the nature of which is obscure, resorption may be of a burrowing type. In the skull of a mongoose-lemur in Figure 22.16A, the margin of the alveolar process covering the cheek teeth is thickened and there has been slight destruction of the alveolar process in the regions of the mandibular incisors and the maxillary canines. Both the maxillary canines have openings on the labial surfaces communicating with the pulp cavities; the left canine has an opening on the lingual surface of the crown also. However, both teeth (Fig. 22.16B) show even larger openings on the root surfaces. A section through one of the teeth (Fig. 22.16C) shows that the surface openings are part of a complex of cavities produced by burrowing resorption which produced connections between the surfaces of the pulp cavities and roots as well as with the external surfaces of the crowns. Where the resorptive process produces only connections between the pulp cavity and the crown surface, it is evident that the process was derived from the cells of the pulp but, where there are openings on the root surface also, as here, it is impossible to say whether the resorptive process initially derived from the pulp or the periodontal ligament.

In the I^1 of a guenon monkey shown in Figure 22.17, the destruction of the tissues is more advanced than in the specimen just described. The tooth had been forced into the nasal process by an injury; it is but a hollow shell, much of the dentine having been removed by resorption, as is shown by the radiograph (Fig. 22.17C). As in the previous example, there are connections between the pulp, root and crown surfaces but here, because both the crown and root of the displaced tooth were in relation to living tissues, the possibility cannot be discounted that the resorptive process derived from the cells of the tissues that invested the crown.

A sperm whale tooth in Nantucket Museum, illustrated by Pouchet and Beauregard (1889) as an example of caries, shows a lesion similar to Figure 22.15 and is almost certainly an example of resorption.

Lesions similar to those just described appear to be common in the teeth of the sperm whale but have been described as dental caries. According to Berzin (1972), the most common site of presumed origin is beneath the gum margin. From the illustrations in the literature he refers to, and from three unpublished photographs kindly sent to us by Dr A. V. Yablokov of Moscow (1984), we believe them all to be examples of focal resorption sometimes initiated by the periodontal tissues and sometimes by the tooth pulp.

Sclater (1871) described a pair of tusks of a female Indian elephant (*Elephas maximus*) from the wild which had the appearance, near where they had emerged from the gum, of having been deeply excavated. Directly after the animal had been shot, a deposit of what resembled blowfly eggs was noticed

Fig. 22.14. *Orcinus orca* (killer whale). Zool. Coll., Oxford Univ. Mus., 14465. A tooth showing resorption of the root (C. S. Tomes 1873). ×c. 0.85.

Fig. 22.15. *Physeter catodon* (sperm whale). RCS Odonto. Mus., G116.1. A tooth showing a large area of resorption. ×c. 0.42.

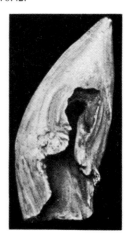

Fig. 22.16. *Lemur mongoz* (mongoose-lemur). RCS Odonto. Mus., G 171.2.
A: There are large openings on the labial surfaces of the maxillary canines. ×c. 1.5.
B: The maxillary canines removed. (a) and (b) Labial and lingual views respectively of the left tooth. (c) and (d) Labial and lingual views respectively of the right tooth. In addition to the openings on the surfaces of the crowns, there are, on the root surfaces, large openings with cupped-out surfaces typical of resorption.
C: Longitudinal ground section through the right canine showing evidence of burrowing resorption.

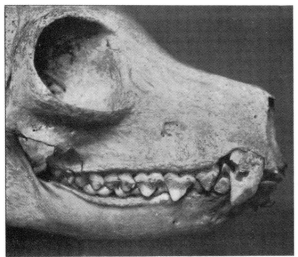
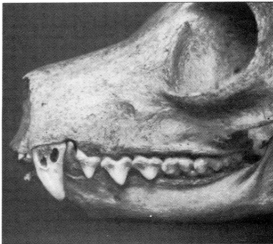

A

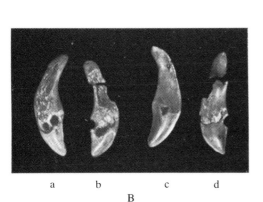

B

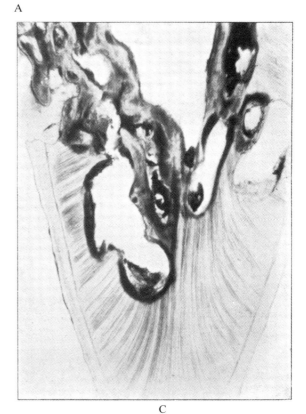

C

around the tusk and the adjacent gum margin, that is in the vicinity of the excavated areas. Although no maggots or pupae were seen, Sclater suggested that the lesions on the tusks might be due to erosion by pupae. According to C. S. Tomes (1876, p. 321), such excavations are common on the tusks of females. He mentioned that there was in the Museum of the Royal College of Surgeons such a tusk with pupae still attached. Nevertheless, Tomes proposed that the cause of the excavations might be resorption by chronically inflamed tissues of the elephant. This specimen is no longer extant.

Some years later, before the first edition of this book (1936), the Chief Conservator of Forests, Chepauk, Madras (personal communication) reported that he had examined the condition of the tusks of 30 captive elephants (Table 22.1). In his view, the egg-like bodies referred to by Sclater were those of the bott fly: 'They have nothing to do with the erosion and nowadays are not to be seen on our elephants because the tusks of the males and tusks of the females are treated daily with a mixture of dikamali (the gum resin of *Gardenia lucida*), neem-oil, garlic and camphor, which appears to be highly distasteful to flies whilst causing no inconvenience to the elephants'. According to the Inspector of Livestock attached to the Department, lesions may be caused as follows: 'Bits of half-chewed wood and even small stones from mud, which elephants often pick up, frequently stick in the pocket formed by the tight-fitting lip round the tusk. These bits of stone may set up irritation and decay may follow, as in the case of human beings'. It is, however, more probable that the damage causes an inflammatory reaction in the gum which leads to resorption of the tooth tissue.

This view is supported by a tusk (Fig. 22.18) described by Bland Sutton (1884a) which has an appearance that closely resembles lesions described by Sclater. The tusk is a small one (33 cm long) of a female Indian elephant about 25 years old which died of pulmonary tuberculosis in the Zoological Gardens, London (Garrod, 1875). At necropsy, the tip of the tusk was noticed projecting only about 2.5 cm beyond the gum. When removed, there was an area extending about 8 cm rootwards of the exposed tusk surface which was deeply and irregularly excavated. This surface (Fig. 22.18) resembles, on a giant scale, the surfaces we are familiar with of teeth undergoing resorption by osteoclasts. Furthermore, in this instance, because the lesional surface was within the bone and inaccessible to larvae or other parasites, the question of erosion of the tusk through such an agency does not arise.

Schneck and Osborn (1976) described areas of progressive subgingival tooth resorption, more colloquially categorized as neck lesions, on the cheek teeth of 6 cats of middle or advanced age. A deeply burrowing type of resorption seemed to begin immediately beneath the gingival margin; this situation renders them liable to infection, when their original character may be obscured and they are only recognizable by the resorption-bay contours of the cavities. Schneck and Osborn gave some examples from the literature which they claimed were misdiagnosed as

Table 22.1 The condition of the tusks of 30 CAPTIVE ELEPHANTS *(Elephas maximus)* in Madras

Condition	No. of specimens
Both tusks broken or partly broken	9
One tusk broken with the other affected	7
'Erosion' present on both side	5
'Erosion' present on one side	6
No 'erosion'	2
One tusk broken with no 'erosion' of the other tusk	1

Fig. 22.17. *Cercopithecus* sp. (guenon monkey). A: Labial view of left I^1 misplaced by injury. B: Distal view. C: Radiograph.
There is extensive burrowing resorption involving the pulp, producing openings on the surfaces of both crown and root.

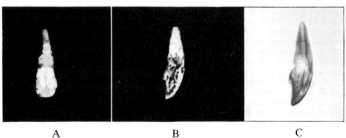

dental caries, e.g. where cervical lesions have led to the breaking off of crowns of the teeth (Builder, 1955). Some lesions described by Seawright, English and Gartner (1970) in cats suffering from vitamin A excess may have been of this type even though categorized as caries. Schlup (1982) reported a continuous increase in the prevalence of so-called neck lesions with age in a group of 200 cats. About 50% of animals over the age of 10 years had such lesions. Sections of the lesions (Schlup and Stich, 1982) showed the features of resorption; caries could be ruled out.

Similar subgingival resorptions were described by Thurley (1985) on the roots of deciduous teeth of three of a high-wear flock of sheep. Some lesions were healed by the deposition of cementum. In one tooth, a resorbing lesion had involved the enamel and penetrated to the pulp which had undergone necrosis, probably because food debris had been thrust into it.

Joest (1926, fig. 127) figured an example of extensive resorption of the buccal surface of a molar of a young horse. One margin of the area of resorption would have been subgingivally situated but whether the resorption spread from there is impossible to say. Joest (fig. 134) also figured a canine from an old horse which showed areas of resorption over the whole of both crown and root. It is reasonable to deduce that the tooth had remained unerupted.

Crazing

Crazing is a term used by Hershkovitz (1970) to describe a pattern of fine brown fracture lines found in the dental enamel of many mammals, including people over the age of 40 years; it is usually associated with some occlusal wear. It appears to be restricted to the incisor-canine complex (Smith, Genoways and Jones, 1977) and seems likely to be due to mechanical stress. In museum specimens, crazing can be simulated by the cracking of enamel that is apt to occur in dry atmospheres and which can be distinguished by the absence of the usually brown pigmentation that is characteristic of crazing that occurred during life.

Non-mammalian vertebrates

There is little in the non-mammalian literature that is relevant to this Chapter; indeed, because in most non-mammals teeth are continuously replaced (polyphyodonty), in general teeth are replaced before there is time for excessive wear or for such to be detected. However, a little evidence suggests that in advanced age the process of tooth replacement slows down and may fail completely, so that the few animals that reach really advanced age may become more or less toothless.

The almost toothless skull of an unusually large, and therefore probably aged, monitor lizard (*Varanus*) found in the wild was described by Bellairs and Miles (1961*a,b*). An aged alligator in the London Zoo is said to have become practically toothless though there is the possibility that disease played a part (Miles, 1961). Poole (1961) adduced some evidence that tooth replacement in the Nile crocodile (*Crocodylus niloticus*) may slow down with advancing age.

Fig. 22.18. *Elephas maximus* (Indian elephant). ♀. RCS Odonto. Mus., G 133.1. A tusk showing loss of the tissue from resorption. The implanted growing end is to the left so that the resorption occurred near where the tusk emerges from the gum (Bland Sutton, 1884*a*, p. 135).

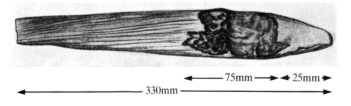

CHAPTER 23

Dento-alveolar Abscess

Introduction

Abscesses of teeth naturally involve the tissues of the alveolus in which the root of the tooth is implanted and are therefore properly termed dento-alveolar abscesses. The portal of entry of the micro-organisms that are the exciting cause is usually the tooth pulp that has been exposed by some process such as dental caries, wear or fracture of the tooth crown. The alternative route by which infection reaches the root and its surrounding tissues is the gingival margin as, for example, a complication of periodontal disease.

Examples of osteomyelitis will be included in this account because this condition often begins as an abscess. A dento-alveolar abscess is, in general, a localization of an infection; whereas osteomyelitis is a more severe suppurative infection which tends to spread within the bone. The essential difference is that in osteomyelitis there is death of bone on a macroscopic scale, producing a sequestrum of dead bone.

A condition known as lumpy jaw will also be dealt with because it can be regarded as a complex abscess which usually involves the teeth and indeed the teeth probably in most cases provide the portal of entry for the causative organisms.

Statistics showing the prevalence of tooth abscesses and other dental disorders, in both captive and wild animals, are rare, but some idea of the prevalence in captive animals is provided in the report by Kazimiroff (1938) on examples of dental disease encountered during one year at the New York Zoological Park. There were dento-alveolar abscesses in a wolf and a kangaroo, both being complications of fractures of incisors, and in a patas monkey as a complication of a fracture of a canine tooth. There were abscesses complicating generalized periodontal disease in an axis deer (*Cervus axis*) and a woodchuck (*Marmota monax*).

According to Bland Sutton (1884a), dento-alveolar abscesses in animals, at least in captivity, commonly lead to complications that can be fatal, whereas in man such complications are extremely rare. He described several instances where at autopsy there was septic pneumonia, apparently as the result of the inhalation of septic discharge into the mouth from dento-alveolar abscesses. Several instances occurred in deer, and others were in a chimpanzee, a rhesus monkey, a ratel or honey-badger (*Mellivora capensis*), and a kangaroo. In the same paper, Bland Sutton gave a colourful account of two attempts to lance the abscessed stump of a tusk in Jumbo the famous African elephant at the London Zoo.

Under captive conditions, the results of injury to the teeth tend to be severe owing to the greater liability to infection. An example of the appalling state seen at times in some of these animals is shown in the skull and mandible of a polar bear (Figs. 23.1A, B) where the pulp cavities of the incisors and canines have been exposed from wear against the bars of its den. The incisors show advanced attrition and there is suppuration with abscess openings (sinuses) in the bone over the ends of the roots. The tip of the right maxillary canine has been worn sufficiently to expose the pulp, and infection, with subsequent suppuration, has followed. The bone covering the outer aspect of the root shows signs of periostitis, and the end of the root projects into the nasal fossa (Fig. 23.1C). The left maxillary canine is in a parlous condition, having been reduced to a hollow tube. On the nasal aspect of the socket, there is small sinus and the outer wall shows a slight degree of inflammatory reaction. The severity of the reaction around these teeth is of interest, being least around the most injured tooth. This difference is probably connected with ease of drainage, the open cavity in the left tooth permitting efficient drainage, whereas the small opening into the right canine would easily become blocked and impede drainage.

In the mandible, the suppuration around the canines has been most active at the root apices. An abscess has formed around the terminal portion of the tooth and encroached more upon the outer than the inner aspect of the socket, due no doubt to the outer wall being less resistant. The abscess, as it increased, seems to have gradually spread, eventually discharging both at the gingival margin and through a large opening at the lower border of the jaw.

Harvey (1985) gave an account of alveolar abscesses and the conditions that give rise to them in captive wild animals.

Order **PRIMATES**

Family **Pongidae**

Dento-alveolar abscesses are common in the wild in all the apes that survive long enough for the dentitions to wear out, that is to show the advanced degrees of tooth wear associated with exposure of the pulp and often with caries. Such animals tend to be categorized as old because the state of the dentition is the main criterion used to assess age. Schultz (1935, 1940, 1944, 1956), who has provided the most comprehensive statistics (Table 23.1), found alveolar abscesses in 39% of aged gibbons, 61% of orang-utan (Fig. 23.2), 60% in chimpanzees and 60% of gorillas (Fig. 23.3), and compared these figures with 13% of aged colobus.

Schultz (1939) mentioned one aged male gorilla that had as many as 18 abscessed teeth. Randall (1943/1944) found what were simply described as alveolar abscesses in the skulls of about 27% of 250 fully-adult gorillas, that is with at least some wear of all the teeth. The condition was commoner in males but the cause was not discussed.

The data of Schultz, presented in more detail in Table 23.1, compared the prevalence of alveolar abscess and loss of teeth during life in the adult and senile categories. There was a marked increase in both with age. An unknown proportion of the teeth lost during life would have been abscessed. Schultz did not specify the sites of the abscesses and naturally was unable to assign a cause to each one. Undoubtedly there would have been many different causes, some due to direct trauma, many secondary to exposure of the pulp either by caries or tooth wear but, in Chapter 24, we argue that some, perhaps in some species as many as half, are likely to have been secondary to periodontal disease.

In general, the prevalence of both abscesses and ante-mortem tooth loss in the apes is considerably higher than in the monkeys, perhaps because the apes are larger and use their teeth with greater force compared with the size of their teeth.

As mentioned in Chapter 19, injuries commonly affect the canine teeth of pongids in the wild and lead to suppurative infection which may spread to other teeth (Wegner, 1962). The exposed pulps of these long-rooted canines provide a portal of entry for infec-

Fig. 23.1. *Thalarctos maritimus* (polar bear). Captive. RCS Odonto. Mus., G115.1. A, right side. B, left side. There has been suppuration in connection with the maxillary and mandibular canines on both sides. Both maxillary canines have suffered injuries which exposed the pulps, on the left side with a wide opening; perhaps because pus could thus drain through the tooth, the suppuration is less severe than on the right where the opening into the pulp cavity was small and could be easily blocked.
C: Medial aspect of the right maxilla showing the necrotic end of the canine through an opening in the bone produced by the suppuration. The pus thus drained into the nasal cavity (Colyer, 1915). ×c. 0.75.

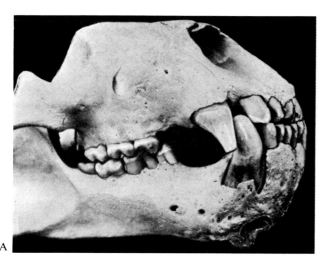

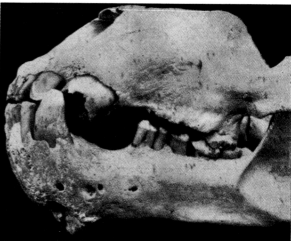

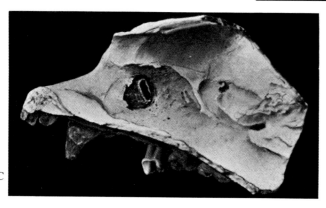

tion to reach particularly deeply into the jaw. In the mandible, the pus may not penetrate easily and may set up a suppurative condition around the main blood vessel supplying the bone in the mandibular canal and is therefore liable to set up a spreading infection accompanied by necrosis of bone, namely osteomyelitis.

In the orang-utan, suppuration around the maxillary canine tends to emerge on the facial aspect of the bone just behind the prominence produced by the huge root of the canine and then, as in the example shown in Figure 23.4, to pass down towards the mouth in the premolar region, forming a gutter-like groove in the bone.

Glick, Swart and Woolf (1979) described a fatal cranial osteomyelitis and subdural abscess in a captive

Table 23.1 Percentage frequencies of adult and senile PRIMATES from the wild with at least one carious or abscessed tooth, or with ante-mortem tooth loss (Schultz, 1956); with subsequent data from a group of *Colobus* (Schultz, 1958) from Liberia (marked*), and groups of *Cebus, Ateles* and *Alouatta* (Schultz, 1960) from Central America (marked**)–these four groups were from specified localities

	Age	No. of specimens	Caries	Abscess	Ante-mortem loss
Family Pongidae					
Gorilla	adult	186	2.7	16.1	4.3
	senile	107	2.8	59.7	20.5
Pan	adult	110	12.7	20.0	8.2
	senile	62	30.6	59.6	29.0
Pongo	adult	97	4.1	13.4	0
	senile	46	13.0	60.8	15.2
Hylobates	adult	335	0.9	7.5	3.9
	senile	104	8.7	38.5	22.1
Family Cercopithecidae					
Cercopithecus	adult	84	2.4	0	1.2
	senile	28	21.4	35.7	10.7
Macaca	adult	107	1.9	4.7	2.8
	senile	78	12.8	28.2	10.3
Papio	adult	43	0	4.6	7.0
	senile	31	9.7	25.8	19.3
Colobus	adult	51	0	3.9	0
	senile	8	0	12.5	12.5
*Colobus**	adult	274	0	6.6	1.8
	senile	28	0	28.6	14.3
Nasalis	adult	46	0	2.2	0
	senile	21	0	4.8	9.5
Presbytis	adult	180	0.6	5.6	2.2
	senile	92	1.1	19.6	3.3
Family Cebidae					
Cebus	adult	70	11.4	5.7	5.7
	senile	20	25.0	15.0	10.0
*Cebus***	adult	110	5.4	11.8	4.5
	senile	8	50.0	75.0	37.0
Ateles	adult	93	3.2	3.2	0
	senile	63	14.3	28.6	7.9
*Ateles***	adult	182	7.7	19.2	3.3
	senile	21	29.0	67.0	29.0
Alouatta	adult	132	0	3.0	0.8
	senile	77	1.3	22.1	11.7
*Alouatta***	adult	342	0.3	18.1	5.3
	senile	36	3.0	75.0	17.0

male lowland gorilla that arose as an apparent complication of an abscessed maxillary canine though a direct connection between the lesions could not be established at necropsy.

Super-family **Cercopithecoidea**

The data of Schultz in respect of various cercopithecines are given in Table 23.1. The prevalence of alveolar abscess among aged animals is high though less so in old colobines (with the exception of the *Colobus* from Liberia), probably because they are rarely affected by dental caries even when the dentition is worn.

Fig. 23.2. *Pongo pygmaeus* (orang-utan). ♂. Wild. Schultz Coll. AS 1159. Mandible showing extensive abscesses perforating the alveoli of C, P_2, M_1 and M_2.
From: Schultz, A. H. (1941). By kind permission of the A. H. Schultz Foundation.

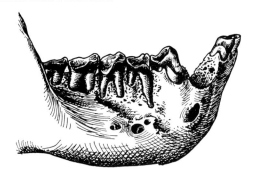

As already mentioned in respect of alveolar abscesses in pongids, there is reason to believe that a high proportion of the abscesses in Cercopithecoidea were secondary to periodontal disease. It is therefore appropriate to discuss the findings of Schultz more fully in Chapter 24 (pp. 537–539).

In a study of over 200 skulls of three species of *Cercopithecus*, Lampel (1963) found traces of a total of 14 dental abscesses in 6 adults. He did not describe the abscesses in detail nor did he indicate whether they were associated with pulp exposures from tooth wear; they appear not to have been secondary to dental caries.

Colyer (1921) referred to dental-alveolar abscesses complicating excessive wear of the incisors in young rhesus monkeys in captivity (see Chapter 22, p. 495).

Schultz (1935) compared the skulls of 290 captive macaques (*Macaca*) with 252 from the wild. Although caries was more prevalent in the captive, the prevalence of alveolar abscess was only about half that in the wild. This is discussed further in Chapter 24 (p. 539), but is difficult to account for unless periodontal disease, or perhaps tooth wear, was much higher in the wild.

Inflammatory changes were induced by Genvert, Miller and Burn (1940) around the apices of the roots of 69 teeth in 15 rhesus monkeys (*Macaca mulatta*) by exposing the coronal pulps to the mouth and so allowing them to become infected. Some were inoculated with known bacteria. The pulps died and, after

Fig. 23.3. *Gorilla gorilla* (gorilla). ♂. Wild. Schultz Coll. There is an extensive suppuration involving the left maxillary and frontal air sinuses, with multiple fistulae, probably secondary to abscesses on the canine and molars. There is loss of the molar teeth from suppuration, an abscess on the worn canine with a large perforation over its root and others below the border and on the floor of the orbit and in the palate. There has been some destruction, with repair, of the upper border of the orbit.
From: Schultz, A. H. (1956). By kind permission of the A. H. Schultz Foundation.

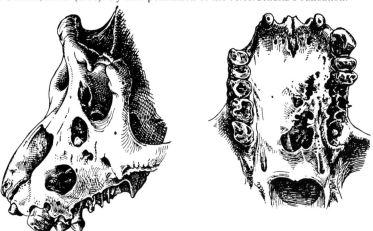

various periods of time, lesions developed which radiography and histology showed were mostly chronic granulomas. A few showed chronic suppuration with the formation of discharging sinuses. Radicular cysts developed on five teeth. An osteomyelitis of the jaw developed in one animal with spreading infection in the subcutaneous tissues of the head.

Bramblett (1969) estimated that the dentition of the male baboon (*Papio*) could last for up to about 20 years. His data suggested a lower figure for females, but this may be because the female life span tended to be shorter because of their smaller size which made them easier prey to predators such as leopards.

Although there were no abscesses with discharging sinuses as a complication of the caries that affected the majority of sixteen baboon skulls from the London Zoo (James and Barnicot, 1949), radiography revealed many less florid inflammations and abscesses at the root apices.

In his study of the dentitions of *Colobus*, Colyer (1943b) illustrated a specimen which has a cyst over the root of an upper canine tooth. Such radicular cysts usually originate from chronic infection (see Fig. 25.10).

Super-family **Ceboidea**

Schultz (1960) (Table 23.1) found a particularly high proportion of dental abscesses (67–75%) among the 65 senile *Alouatta*, *Cebus* and *Ateles* from defined areas of

Fig. 23.4. *Pongo pygmaeus* (orang-utan). ♂. BMNH 3.6.2. Suppuration in connection with the maxillary canine has produced an opening behind the prominence produced by the canine root. Pus passing towards the mouth in the region of the premolars has produced a gutter-like groove.

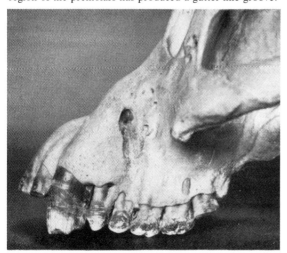

Central America compared with the corresponding species from a wider area of territory in Southern America (Schultz, 1956), and which were likely to have come from a wider variety of habitats.

Smith, Genoways and Jones (1977), in a study of the same taxa, also found alveolar abscesses to be extremely common especially in the mantled howler (*Alouatta palliata*); their material was not divided into age groups. The abscesses were often secondary to dental caries (see Chapter 21) but sometimes presumably due to localized gingival incidents such as food impaction. A common sequence of events seemed to be as follows: wear occurs between the canine and first premolar in the lower jaw by contact with the long upper canine; there is consequent gingival irritation which leads to a gingival abscess with localized bone destruction (see Chapter 24). Associated wear of the posterior slope of the upper canine sometimes exposes the pulp and leads to periapical abscess and sinus formation.

Examples of dento-alveolar abscesses in *Alouatta* and *Cebus* were illustrated by Schultz (1925).

In the maxilla in monkeys, the canine root is often covered by a thin layer of bone which the pus readily penetrates, especially on the facial aspect. The pus may then discharge into the mouth, but, as the apex of the canine lies not far below the orbit, it may penetrate the bone and form an abscess below the orbit beneath the skin (Fig. 23.5). According to Sedgwick and Cooper (1972), the long sharp canine teeth of squirrel monkeys (*Saimiri sciureus*) are sometimes clipped or filed to reduce the hazard to their handlers; the pulp thus may become exposed and infected, and an abscess pointing beneath the eye is a common sequel. Olfert (1974) noted five examples, some bilateral, in a colony of 34 males maintained for drug research.

Order **CARNIVORA**

As Wright (1939) pointed out, abscesses on the upper carnassial tooth (M^1) in the dog, usually due to either periodontal disease or traumatic damage to the tooth, may point on the face beneath the eye and, if untreated, can produce persistent suppurating openings in the skin. The maxillary air sinus may also be involved. The radiographic and microscopic anatomy of chronic abscesses on the root apices in the dog was studied by Bodingbauer (1955a).

Abscesses of teeth commonly arise from injuries and accidents to teeth apart from fractures. Hall (1940), in recording several dento-alveolar abscesses in various species of carnivores, referred to the wedging of a stick crosswise in a wolf (*Canis lupus*) between the upper carnassial teeth, both of which were abscessed.

A typical example of extensive osteomyelitis of the mandible of a young adult badger (*Meles meles*) was described by Paget (1972). There was a fractured lower premolar which must have become abscessed and been the starting point of a suppuration that had spread to the body of the mandible on the opposite side. There was a core of dead bone, the sequestrum, enclosed by a shell of new bone, the involucrum, formed by the periosteum which had been lifted from the dead bone by spreading pus formation. The animal had died, presumably of septicaemia and starvation, close to its sett.

An abscess on a P^3 in a captive spotted hyaena, *Crocuta crocuta*, said to be 41 years of age, was described by Henderson, Borthwick and Camburn (1984).

Hall (1940) described the skull of a hooded skunk (*Mephitis macroura*) in which an abscess involving the mandibular molars, probably arising from trauma, had destroyed part of the ascending ramus with considerable proliferation of reactive bone.

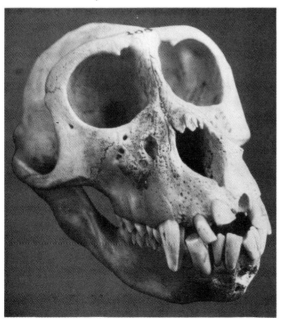

Fig. 23.5. *Macaca fascicularis* (crab-eating macaque). ♂. Formerly RCS Osteo. Series, 138. Severe suppuration in connection with the maxillary canine has made an exit on the facial aspect of the bone close to the apex of the tooth at too high a level for it to be likely that the pus drained into the mouth. It probably produced a subcutaneous abscess below the eye.

Order **PINNIPEDIA**

Abscesses occur as a result of the excessive tooth wear in the Weddell seal (*Leptonychotes weddelli*) associated with the use of the teeth in this species to maintain breathing holes through the ice (Chapter 22, p. 498). The upper teeth in particular are used for this purpose; consequently the upper jaw is the common site for abscesses which produce perforations of the bone in the palate, outer surface of the jaw or into the nasal passages (Stirling, 1969).

Abscesses associated with tooth fracture may be common in the leopard-seal (*Hydrurga leptonyx*) (p. 435, Chapter 19).

Order **CETACEA**

The reports on whales stranded on British coasts (Harmer, 1927; Fraser, 1934, 1946, 1953) record many abscesses in the jaws of a variety of species. One in a white-beaked dolphin (*Lagenorhynchus albirostris*) had produced a suppuration which spread along the lower border of the mandible producing multiple openings (Harmer, 1927). Abscesses and tooth loss are common in the bottle-nosed dolphin (*Tursiops truncatus*) also (Chapter 22, p. 494). Pilleri and Gihr (1969a) described an example from the Mediterranean in which there were many empty partially healed tooth sockets. The remaining teeth were much worn with widely exposed pulp cavities. Many teeth were abscessed and on one side of the mandible an abscess had spread into the vertical ramus where there were multiple sinuses indicative of osteomyelitis.

C. S. Tomes (1873) described in some detail an abscessed tooth in a grampus or killer whale (*Orcinus orca*). Wear had produced a large opening into the pulp which had in consequence died. This is unusual because in cetaceans in general the tooth pulp becomes replaced by dentine seemingly independently of any reaction to wear and also, as Tomes pointed out, because the homodont teeth interdigitate, wear mainly affects the approximal surfaces of the teeth rather than the tips of the crowns. In this grampus, the jaw relationship was asymmetrical so that some teeth, including the abscessed one, met tip to tip. Colyer (1938) also described a dento-alveolar abscess in the lower jaw of a grampus and, from his examination of 20 grampus skulls, inclined to the view that wear of the tops of the teeth as well as the sides between the teeth is more common than Tomes had supposed.

It is noteworthy that, in a stranded Risso's dolphin

(*Grampus griseus*) with abscessed teeth described by Fraser (1953), the teeth were much worn. In this species, there are normally no teeth in the upper jaw.

UNGULATES

In animals with hypsodont teeth set in deep alveoli, such as bovids and equids, the pus, especially in the mandible, does not easily emerge from the alveolar bone but burrows into the body of the mandible and may form a deep-seated abscess with a periosteal reaction on the surface. Sometimes the pus may travel for some distance in the bone, for instance into the coronoid process. In the lower jaw of a horse in Figure 23.6, a mass of new bone has been formed around a sinus opening on the lateral side of the bone. The sinus leads to a smooth-walled cavity in the bone around the end of the root of the third premolar which has a sponge-like mass of firmly attached cementum projecting from it. This cementum was probably formed as a reaction to the suppurative infection responsible for the sinus and the proliferation of periosteal bone.

Suppuration in connection with brachydont teeth, such as those of pigs, which are not set in particularly deep alveoli, is usually able to burrow through the alveolar bone or to emerge around the teeth. Deep-seated suppuration is more uncommon but does sometimes occur with the pus eventually emerging, for example, close to the lower border of the mandible (Fig. 23.7). According to D.J. Anthony and Lewis (1961), dento-alveolar abscesses, especially of lower incisor teeth, are commonly noted in pigs at slaughter. Jubb and Kennedy (1970) mentioned osteomyelitis of the jaws as a complication of clipping the tusks of piglets.

Fig. 23.6. *Equus caballus* (horse). RCS Odonto. Mus., G 102.4.
Above: An abscess in connection with P_3 has given rise to a deep-seated suppuration in the mandible which has produced, as a subperiosteal reaction, a mass of spongy bone on the outer surface of the bone in relation to a suppurative sinus.
Below: The lateral plate of bone has been removed, revealing an irregular mass of cementum on the end of the root of P_3. This mass resembles a radicular odontome (Chapter 25, p. 594).
× c. 0.3.

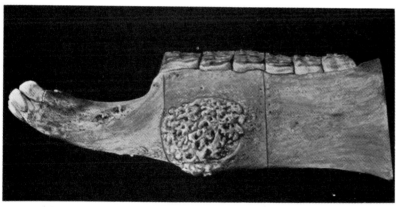

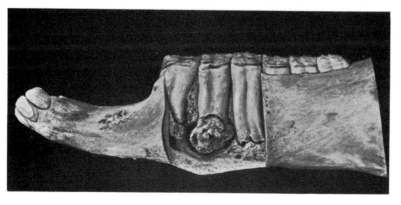

Hooijer (1941) described the sequelae of a probably dento-alveolar abscess in a hippopotamus (*Hippopotamus amphibius*) in terms of the consequential tooth loss and change of function of the surviving teeth.

Garlick (1954b) listed the dental disorders found in 7480 cattle slaughtered for beef in the United States and recorded 10 examples of abscessed permanent teeth. The sample excluded animals that were too old to be of value for milking or breeding.

There seem to be no similar data for sheep. Elsewhere, we have given accounts of tooth loss in sheep from excessive tooth wear (Chapter 22) and periodontal disease (Chapter 24) both of which are likely to be at least occasionally complicated by dento-alveolar abscesses.

Bodingbauer (1954) described an osteomyelitis of the upper jaw in a sheep. The sequence was thought to be as follows: loss of the most posterior molar from the upper jaw on one side led to irregular wear and over-eruption of the corresponding mandibular tooth, the sharp occlusal surface of which impinged on the upper gum. The gum became ulcerated with a spreading infection in the bone, leading to bone destruction and the formation of a fistula between the mouth and the nasal cavity, and finally to a fatal haemorrhage from the ulcerated palatine artery.

Among 156 mule deer (*Odocoileus hemionus columbianus*) from the wild, Cowan (1946) found eight dento-alveolar abscesses. All but two were in the mandible. One was connected with a broken M_1 and had invaded much of the body of the mandible, producing several sinuses. The other abscesses were less extensive and appeared to arise from injury to teeth, or as complications of tooth wear associated with forage-packing between teeth. This animal had severe deforming arthritis of the hind limbs.

An instance of osteomyelitis of the vertical ramus of the mandible of a duiker arising from spread of suppuration from periodontal disease of the cheek teeth is referred to in Chapter 24 (p. 551).

Paine (1909) described an example of a type of tooth abscess that was common in a cereal-growing part of South Africa where wheat chaff was often part of the feed of young horses. Much trouble resulted from the wheat awns penetrating between the deciduous molars, injuring the gum and some becoming firmly fixed there. The teeth frequently became abscessed and at this stage the effective treatment was to punch out the offending deciduous molars. However, sometimes when the abscess was on maxillary teeth, a purulent swelling appeared under the skin high up on the side of the face. In the young horse, the growing ends of the upper permanent molars are close to the facial surface of the bone and produce prominences on the surface as described in Chapter 13.

Arnone (1920) described a mandible of a horse (*Equus caballus*) from Quaternary deposits of the Val d'Arno in which on the lower margin, beneath a premolar presumed to have been lost in life by some inflammatory process, there were two spiky projections of periosteal new bone which he judged to be a

Fig. 23.7. *Sus scrofa* (domestic pig). ♂. RCS Hunterian Mus., P414. An abscess in connection with the fourth premolar, lost as a result of the inflammatory process, has given rise to a deep-seated suppuration in the mandible, which has produced an opening surrounded by reactive subperiosteal bone on the outer aspect.

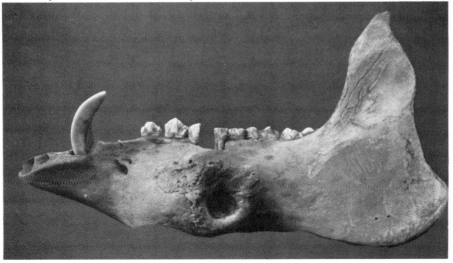

response to a suppurative process rather than neoplastic.

LUMPY JAW

A number of species, mostly herbivorous but including man, are affected by a condition for which the most convenient collective term is the simple descriptive one, lumpy jaw, which takes account of the fact that the causative micro-organism is not the same in all species. The condition has been most fully studied in cattle and therefore will be most fully dealt with here, with briefer references under other taxa.

In cattle, infection of the bone of the jaws with the filamentous, so-called higher, bacterium, *Actinomyces bovis*, is firmly established as the cause of lumpy jaw. The organisms produce a granulomatous reaction which destroys bone progressively; tissue is broken down with the formation of pus in which typical colonies of the actinomycetes are often evident to the naked eye as yellow granules. The purulent granulation tissue makes its way to the bone surface where the formation of new periosteal bone and indurated thickening of the overlying soft tissues produces the characteristic lumpiness of the swellings (Fig. 23.8) (C.E. Harvey, 1985). Other typical examples were figured by Franke (1921, p.162), Tharp, Amstutz and Helwig (1980) and E. Becker (1970, fig. 290). There is a tendency for the formation of multiple sinuses through the skin or into the mouth, usually involving the teeth. With the formation of sinuses, secondary infection with a mixed microbial flora occurs and may obscure the identification of the original causative organism. It may be that lesions do not follow a single invasion; repeated invasions which lead to sensitization of the tissues may be a factor (Tharp, *et al.*, 1980).

Actinomycosis may affect other regions and organs, for instance the gut or lung, and it is probable that the jaws are the sites of predilection because the organisms gain access to the tissues via injuries to the oral mucosa, in particular to the gingiva. It is said that

Fig. 23.8. *Bos taurus* (domestic ox). RCS Odonto. Mus., G 51.72. A mandible with typical lumpy jaw involving the cheek teeth on the left side; reported on (*c*.1930) as 'streptothrix infection'. There has been much bone destruction leaving many thickened trabeculae; the lesion is grossly expanded by periosteal new-bone formation and there are several deep openings which represent discharging sinuses. There is a fracture through the lesion and the fragments have been held apart to display the lesion. Only one tooth, M_1, appears to have been lost. On the opposite side, there is a smaller expansile lesion of the same kind related to M_1 which has been explored by removing a little lingual bone. ×*c*. 0.8.

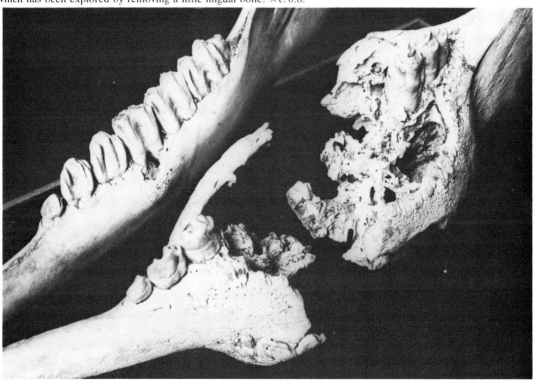

the sharp awns of grasses infected with the actinomyces are the main cause of these breaches of the mucosal surface. Many early lesions are probably indistinguishable from dento-alveolar abscesses; gross lesions with much proliferation of new bone are difficult to distinguish in museum specimens from neoplasms (Tharp, *et al.*, 1980; their fig. 24).

Uncertainty about the causative organism of lumpy jaw is contributed to by various taxonomic revisions that have taken place and by the fact that secondary infection tends to obscure the primary causative organism; furthermore, the *Actinomyces* are difficult to grow *in vitro*.

Of 15,867 dairy herd bulls being considered for breeding, 341 (2.15%) were discarded because of lumpy jaw (R. H. Becker *et al.*, 1964). The prevalence was seven times greater among those of the Guernsey breed than among the five other breeds represented, suggesting the possibility of genetic differences in susceptibility.

Sheep also can develop lumpy jaw and the cause appears to be infection with the *Actinomyces bovis*. According to A. Murie (1944), lumpy jaw, assumed to be due to infection with *A. bovis* although bacteriological investigation has not been carried out, was common among the wild dall sheep (*Ovis dalli*) in the Mount McKinley National Park in Alaska. It seems from his data that 48 (21.7%) out of 221 skulls of animals that died in the region during 1937–1941 were affected by actinomycosis of the jaws, most commonly the mandible. Included were a few doubtful examples of what Murie described as necrotic stomatitis which affects the oral mucosa and only occasionally involves the bone but is also caused by *A. bovis* (see Chapter 24, pp. 562, 563). Lumpy jaw in dall sheep was comprehensively discussed by Hoefs and Cowan (1979) who found 14 mandibles in a sample of 20 to be affected (see also Glaze, Hoefs and Bunch, 1982).

Blair (1907) described three examples of typical severe lumpy jaw with multiple sinuses in the mandibular cheek-tooth region among skulls of bighorn sheep (*Ovis canadensis*) in the wild from a region of British Columbia, Canada, at that time very remote from contact with domestic animals. Blair mentioned having seen the condition in grizzly bear.

Gross hyperplasia of the cementum of the cheek teeth involved in the lesions was a feature of actinomycosis of the jaws of a female oribi (*Ourebia orebi*), a small African antelope, which died after four years in captivity (Miles, 1951).

Lesions, mostly in the mandible, suggestive of lumpy jaw have been described in cervids in the wild, for instance in caribou (*Rangifer tarandus*; Miller, Cawley, Choquette and Broughton, 1975; Doerr and Dieterich, 1979), American elk (*Cervus canadensis*; O. J. Murie, 1930, 1951), and in American moose (*Alces alces*; Ritcey and Edwards, 1958); also in a deer in captivity (Kober, 1965). One of the fullest records is in European reindeer (*Rangifer tarandus*) introduced into the sub-antarctic island of South Georgia during the period 1911–1925. Leader-Williams (1980) found what he called simply 'mandibular lesions' in 49 (12.3%) out of 400 of wild-shot adult reindeer from two herds on the island; about one-third of the mandibles were bilaterally affected. The features of the lesions, revealed by radiography and sectioning some of the worst-affected jaws, were bone destruction, centred in most specimens around the premolars or molars (most commonly M_1), by a granulomatous tissue. There was a periosteal reaction which produced indurated thickening of the body of the mandible. In many, one or more of the cheek teeth was recorded as damaged, which we interpret to mean fractured, so that many of these lesions could have been abscesses associated with the ingress of non-specific suppurative organisms, either via a pulp exposed by wear or fracture or via gingiva injured by food packing. However, in the six jaws with particularly large lesions that were sectioned, eosinophilic masses resembling actinomycotic colonies were found in the granulomatous tissue; thus, although culture of the organism with full identification was not possible, the evidence strongly suggests that so-called lumpy jaw does occur in reindeer in the wild. Leader-Williams has provided a valuable review of the earlier literature.

Lumpy jaw occurred in a dromedary and a llama in the Berlin Zoo (Unger, 1976).

Tritschler and Romack (1965) described what can be called lumpy jaw in a horse from which a species of *Nocardia* was identified; this is a filamentous bacterium related to the *Actinomyces* which produces mostly soft-tissue lesions in a wide variety of species. There was a complex suppuration in relation to several cheek teeth in the right mandible which extended widely in, and thickened, the body of the bone and produced several discharging sinuses in the skin. Wear of the affected teeth had exposed the pulps and fodder was found not only packed in the pulp cavities but had been driven through into the abscess cavity beyond.

H. Fox (1923, p. 568) mentioned that lumpy jaw had been observed in two American tapirs (*Tapirus* sp.) and two Malayan tapirs (*T. indicus*) in the Philadelphia Zoo and that the condition had been diagnosed in tapirs in other zoological collections so that there is a possibility that this species is particularly susceptible.

Order **PROBOSCIDEA**

Bricknell (1987) figured an example of an abscess on the root of the molar of an fossil proboscidean, *Archidiskodon meridonalis*.

Orders **RODENTIA** and **LAGOMORPHA**

The growing ends of the incisors of both upper and lower jaws in rodents and lagomorphs always extend back far enough to involve the molar teeth, although to different extents in the various species.

In the marmots (*Marmota* spp. in the family Sciuridae), the mandibular incisor reaches slightly beyond the last molar and passes under this tooth towards the outer aspect of the bone (Fig. 23.9). There is a thin layer of bone between the growing end of the incisor and the socket of the last molar, so that pus around the end of the incisor has little resistance to overcome in breaking through the bone behind the last molar. The specimen shown in Figure 23.10 is also the mandible of a marmot; the left incisor was fractured and death of the pulp followed with suppuration around the growing end. Two abscess sinuses were formed, one along the inner aspect of the lower border about level with the third molar, and the other on the outer and posterior aspect of the last molar. On the outer aspect of the bone, there was a globular swelling, the surface of which was rough. A section through the globular swelling disclosed that it was solid and the result of a sclerosing osteitis.

The root of the lower incisor of the beaver extends to behind the last cheek tooth, the growing end being surrounded by dense bone; the molar teeth have longer roots than those of the marmot and extend down on the outer side of the incisor (Fig. 23.11). When suppuration occurs in connexion with the incisor, the pus seems unable to penetrate the bone and instead tracks forward along the side of the tooth, causing necrosis of the tooth and extensive suppuration in the bone (Fig. 23.12).

The tooth-bearing part of a mandible of a beaver (*Castor fiber canadensis*, BMNH 1856.12.10.689) is grossly thickened and distorted by the deposition of bone on all its surfaces. The surfaces are irregular and deeply ridged and pitted. Shallow empty tooth sockets are present and the teeth may have been exfoliated shortly before death or lost since. The specimen has no relevant history but it seems likely that a chronic low-grade infection associated with the teeth was responsible for the changes.

Bauchau and Le Boulengé-Nguyen (1985) described the mandible of a field vole (*Microtus agrestis*) among those recovered from owl pellets which was

Fig. 23.9. *Marmota marmota* (alpine marmot). Mandible; bone has been removed to show the relation of the incisor to the cheek teeth.

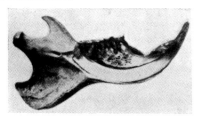

Fig. 23.10. *Marmota marmota* (alpine marmot). RCS Odonto. Mus., G110.5. The left incisor has been fractured and suppuration has followed around the growing portion. (A) Prominence due to sclerosing periosteitis. (B) Sinus discharging posterior to the last molar (Colyer, 1915). ×c. 0.86.

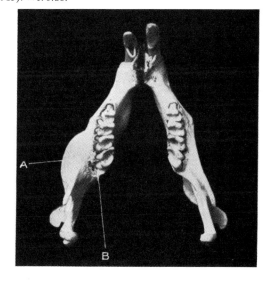

Fig. 23.11. *Castor canadensis* (American beaver). Mandible; bone has been removed to show the relation of the incisor to the cheek teeth. ×c. 0.5.

much thickened, almost certainly by a chronic suppurative process mainly along the course of the root of the broken left incisor. Because of the multiple sinus openings on the surface of the bone, they suggested that the infection was probably actinomycosis.

An example of an abscess on the roots of the mandibular cheek teeth of a rabbit, causing overgrowth of the incisors, is shown in Figure 17.8.

Order **MARSUPIALIA**

Fracture of the teeth, which is a common hazard in animals in both the wild and captive states, and commonly leads to dento-alveolar abscess, is dealt with in Chapter 19. Certain teeth in certain species are more susceptible than others, including the delicate, long, procumbent lower incisors possessed by many marsupials. The roots of these teeth are of continuous growth, but are shorter and do not have the marked curve that is characteristic of rodent incisors. In kangaroos and wallabies, the basal ends of the lower incisors are near the mental foramen and, when the incisor is abscessed, the pus often comes to the surface through that foramen (Fig. 23.13). In the much smaller potoroo (Fig. 23.14), however, the root of the incisor extends back to the medial side of the first tooth of the molar series where it produces a slight elevation of thin bone on the inner surface of the mandible. If the incisor becomes abscessed, the pus forms an opening there and may furthermore secondarily involve the molar, leading to its loss, as in Figure 23.14.

Watts and McLean (1956) isolated a variety of organisms from single localized abscesses in the maxillae of kangaroos and wallabies; most of the abscesses involved an M^1 and most of the organisms cultured were assignable either to the genera *Fusobacterium* or *Bacteroides*.

Archer (1975) mentioned the mandible of a kultarr, *Sminthopsis spenceri* (Queensland Museum, J23103), in which there were abscesses on M_1 and M_4.

Lumpy jaw is one of the most common diseases of macropods in captivity, especially in the larger ones (Horton and Samuel, 1978; C. E. Harvey, 1985, p. 301). The condition is also met with in the wild. Schürer (1980) studied 144 skulls of kangaroos in German and Australian museums and concluded that the red kangaroo (*Macropus rufus*) in captivity is particularly liable to suffer from lumpy jaw which involves the teeth and leads to their loss; for example, 15 out of 36 skulls were affected. In his view, the disorder commences as a focal periodontitis.

Griner (1983) took a similar view in analysing the marsupial necropsy records of the San Diego Zoo and Wild Animal Park over the 14 years up to 1978 in grouping together acute periodontitis and lumpy jaw without distinction. She found that 66 (21%) of the macropods were affected and made the interesting observation that, whereas about half the southern grey kangaroos (probably *Macropus giganteus* or *fuliginosus*), dama wallabies (*Macropus eugenii*) and Parma wallabies (*Macropus parma*) were affected, hardly any of the New Guinea scrub wallabies (prob-

Fig. 23.12. *Castor fiber canadensis* (American beaver). Captive. RCS Odonto. Mus., G 108.1. Mandible showing necrosis of the incisors, which have been removed and displayed at the sides of the jaw, and extensive suppuration of the surrounding bone (Colyer, 1915). ×c. 0.5.

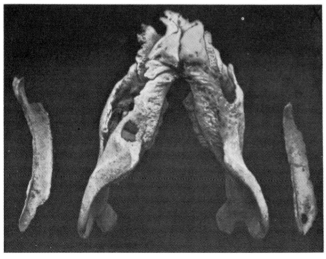

ably *Thylogale brunii* or *T. stigmatica*) and Tasmanian Bennett wallabies (*M. rufogriseus*) were; however, it may be significant that these two species were kept on open parkland.

One of the earliest accounts is by H. Fox (1923) based mainly upon his experience of many examples at the Philadelphia Zoological Gardens. The condition in marsupials resembles lumpy jaw in the bovids but often runs a more acute and rapid course and appears to start as gingival ulceration, associated often with ulceration of the lips and tongue mucosa, and with nasal discharge, rather than to start, as in bovids, as a suppuration deep in the jaws. However, in macropods the bone of the jaws does become involved and necrosis of the bone is a feature.

The causative organism was identified by Fox as a filamentous one, *Nocardia macropodidarum*. In spite of many revisions that have occurred in the systematic taxonomy of this type of organism and the possibility that it corresponds with *Sphaerophorus necrophorus* (Arundel, Barker and Beveridge, 1977), lumpy jaw in macropods is still commonly categorized as a feature of nocardiosis (Barker, Calaby and Sharman, 1963).

The evidence for the gingiva, damaged by sharp food particles such as grass awns (H. Fox, 1923, p. 572), being the portal of entry is stronger than in the bovids. Hence this condition has also to be considered in the context of periodontal disease in Chapter 24.

Scott (1925) described how 15 out of 19 recently-imported Bennett wallabies (*Macropus rufogriseus*) died at the London Zoo within about three months, 14 with multiple lesions from which a streptothrix, which Scott placed in the Order Nocardia, was isolated; in six, the primary site was the jaw, in others it was the tongue and pharynx.

Somewhat similar was an outbreak among wallabies and kangaroos in Brisbane Botanical Gardens (Tucker and Millar, 1953) in which eight animals died in a period of two months, due apparently to an organism which could not be specified more precisely than that it belonged to the genus *Nocardia*. The particular susceptibility of marsupials to whatever was the cause was emphasized by the fact that the affected animals were penned with deer that remained unaffected.

A. R. Tomlinson and Gooding (1954) described outbreaks of lumpy jaw among kangaroos in the wild in Western Australia, manifested by jaw abscesses with much external swelling, swollen gangrenous tongue, watery discharge from the eyes, multiple swellings in

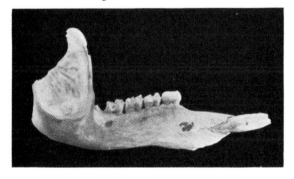

Fig. 23.13. *Wallabia bicolor* (swamp wallaby). RCS Odonto. Mus., G 110.2. Mandible affected with suppuration in connection with the incisor; the pus has found an exit through the mental foramen. ×c. 0.66.

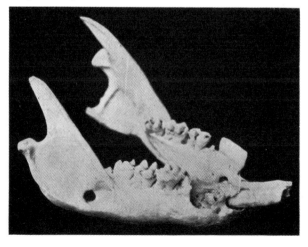

Fig. 23.14. *Potorous tridactylus* (potoroo). RCS Odonto. Mus., G 110.4. The incisors have been fractured and the pus has emerged on the inner side of the bone (Colyer, 1915). The foramen posteriorly is for the mandibular nerve. ×c. 0.66.

various parts of the body and general emaciation. The bacteriology of the lesions tenatively incriminated *Fusiformis necrophorus* which causes foot-rot in cattle. According to Arundel, Barker and Beveridge (1977), this organism, now called *Sphaerophorus necrophorus*, is the same one that Beveridge (1934) established as the cause of the group of necrotic lesions covered by the term necrobacillosis (Finnie, 1976). Furthermore, it seems to be the same as the *Fusobacterium necrophorum* Samuel (1983) isolated, sometimes in pure culture, from 81% of 50 lumpy-jaw lesions in macropods, mostly in captivity. Samuel appears to be of the opinion that the *Nocardia macropodidarum* is a different bacterium which could have been responsible for at least some of the other 29% of the lesions he studied. (See also Miller, Beighton and Butler, 1980).

Tomlinson and Gooding suggested that the infection originated in the soil and was carried into the bodies of the kangaroos by punctures of the skin and mucosa by the prickly vegetation common to the region. It is of interest that consideration was given to spreading this infection deliberately among the kangaroos to control their numbers as an alternative to poisoning their water-supplies with arsenite of soda.

The endemic character of lumpy jaw in Australian marsupials seems to be of ancient lineage. Horton and Samuel (1978) found evidence of it in six Pleistocene fossil macropod mandibles from a site estimated to be 26000 years old.

Other examples of dento-alveolar abscesses due to injuries to teeth (Chapter 19), dental caries (Chapter 21), tooth wear (Chapter 22) and periodontal disease (Chapter 24) are to be found in the chapters indicated.

CHAPTER 24

Periodontal Disease

Introduction

The tissues that invest the roots of the teeth and provide their support, the alveolar process, periodontal ligament, cementum and gum, are known collectively as the periodontium (around the teeth). The term parodontium (beside the tooth) has a history at least as long and has much otherwise to recommend it; however, periodontium and its derivatives has the wider usage. Diseases of the periodontium are referred to as periodontal diseases. Nearly all are inflammatory in origin and commence as an inflammation of the gum margin, namely gingivitis. Most are chronic and persistent, and therefore in due course may involve the deeper tissues leading to loss of bone, beginning with the crests of the alveoli. There is deepening of the gingival crevices.

Some degree of chronic gingivitis is said to be almost universal in man and fully-fledged chronic periodontal disease is as common a cause of tooth loss as dental caries. Hence periodontal disease in man has been the subject of intensive study and most of our ideas of the cause and mechanism of periodontal disease in animals derive from research directed to an understanding of the human state. The following is a brief account of the mechanisms that seem to apply in general.

Bacterial plaque has already been referred to in Chapter 21 as forming on the tooth surface. However, plaque also extends to the margin of the gingiva (Fig. 24.1). This gingival plaque is formed in the same way as coronal plaque; that is, even if we assume a completely clean surface to begin with, following a meal, a layer of saliva containing food substance remains as an adsorbed layer on the surfaces of the tooth and gingiva. This layer may become detached after a short time by the normal movement of lips and tongue but, if the foods include fermentable carbohydrate, there will be among the microbial flora organisms or their enzymes which are able to convert the carbohydrate to the sticky high-polymers. The resulting plaque will now be much more difficult to dislodge and will become a more or less permanent feature of the gingival crevice and one in which a particular flora will arise. Gram-negative organisms and filamentous forms tend to predominate, the products of which are in various degrees irritant to the marginal gingiva so that a chronic gingivitis is set up. There is evidence in man that the flora of gingival plaque is significantly different from that of the crown; e.g. organisms that produce toxic substances that are particularly harmful to cells and capable of killing them are common in gingival plaque; such organisms are *Capnocytophaga* and *Actinobacillus actinomycetemcomitans*.

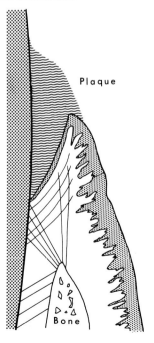

Fig. 24.1. Diagrammatic representation of the relation of plaque to the tooth surface, gingival margin and crevice. The main orientations of the fibre system in the gingiva are indicated. On the tooth side, the fibres are embedded in the cementum, which is not shown, of the root surface. The crevice is represented here as deeper and wider than it would be in health when it is an angle or shallow groove. If a gingival pocket were present, the crevice would be much deeper with a more evident difference between the enclosed environment of sub-gingival plaque in the pocket and the more exposed supra-gingival plaque on the tooth surface.

A number of protozoa parasitize the mouth in the dog, macaque and man (Kofoid, Hinshaw and Johnstone, 1929), in particular an *Entamoeba*, which has been categorized as *E. gingivalis*, and a *Trichomonas*, categorized as *T. buccalis*. However, their taxonomic status, in particular in what way they differ from protozoa found in other parts of the digestive tract, is in doubt. At one time, they were thought to play an important pathogenic role, for example in periodontal disease; the early experimental work was reviewed by Kofoid *et al.*, (1929). They are now regarded as harmless saprophytes and receive little attention.

Swelling of the gingival margin is a feature of gingivitis and this alone increases the depth of the gingival crevice which is a particular area of stagnation. The lining epithelium may be partially destroyed so that micro-ulcers occur, and serum and inflammatory cells, in other words pus, exude into the crevice. Such a state of affairs can exist it seems for a long time, perhaps several years, without the condition being noticeably progressive. In due course, however, the condition may spread beyond the gingival margin and the bone at the crest of the tooth sockets is resorbed, most commonly interdentally first, and the gingival crevice becomes deeper. Gingival or periodontal pockets appear. Various degrees of atrophy of the gingival margin occur so that, instead of the pockets deepening, the gingival margin in places recedes. Periodontitis can now be regarded as fully-fledged and, usually slowly, more of the periodontium is destroyed until the teeth become loose. If the process continues unhindered, in due course the teeth are exfoliated.

The host response that has only briefly been referred to in the above account is highly important and variable and much of the advance of knowledge of human periodontal disease in recent years consists in elucidation of the complex humoral and tissue immune responses that occur and differences in individuals probably account for the different rates of progress and forms of the disease met with. There are some forms in which nearly total absence of response makes it seem as if the initial breakdown of the periodontium occurs from within; local inflammatory signs may be slight and secondary.

Once gingival pockets have arisen, the environment and microbiological flora within the pockets subgingivally tend to be different in subtle ways from those supragingivally. Bacterial plaque is formed on the tooth surface in both situations but there are certain differences; for instance, subgingival bacterial plaque incorporates crevicular fluid (tissue exudate). Mineralization of either subgingival or supragingival plaque, or both, sometimes occurs producing a hard material, dental calculus or tartar, adherent to the tooth. Both subgingival and supragingival calculus aggravate the inflammatory process; in particular, the rough surface of subgingival calculus rubs against the inner aspect of the pocket and helps to maintain the chronic inflammation and may from time to time induce acute suppurative episodes.

There is no limit to the continuous or intermittent accretion of supragingival calculus and sometimes, if there is gross stagnation of food debris, masses of calculus of considerable size may build up on the tooth surface and, if teeth are not in occlusion, may encroach on their occlusal surfaces.

In man, the evidence is overwhelming that organisms present in microbial plaque or in the gingival crevice or pocket, or substances derived from them, constitute the primary extrinsic exciting agents in periodontal disease, although the host response plays an important part in determining the course of the disease (Page and Schroeder, 1982, p. 17). Some bacteria found commonly in periodontal pockets can elaborate and release a potent leukotoxin, that is a toxin capable of destroying the body's leukocytes (defence cells) but also capable of damaging or destroying other tissue cells. There is some evidence that certain bacteria or groups of bacteria may be associated with certain types of periodontitis though whether this is because the environment conduces a particular flora or because the bacteria are producing a specific disease type is uncertain. High blood titres of antibodies to specific organisms isolated from periodontal disease may become lower following successful treatment for the disease.

The importance of the immune system in determining the course of periodontal disease, and perhaps its onset, is high-lighted where there is an inherited immunodeficiency. There are several such disorders in man; in one, the Chédiak-Higashi syndrome, neutrophil and monocyte functions are compromised (Page and Schroeder, 1982, p. 261) and one of the features is rapidly progressive and generalized periodontal disease which develops at puberty and leads to destruction of the periodontium and tooth exfoliation. A similar disease occurs in ranch-bred mink and mice (Lavine, Page and Padgett, 1976), cattle (Renshaw, Davis, Fudenberg and Padgett, 1974), cats (Kramer *et al.*, 1975), and dogs (Renshaw *et al.*, 1975), that is in animals living in the protected environments associated with domestication. If such harmful genes arose, as mutations, in the wild, they would be quickly eliminated.

Gingivitis, and its complication periodontal disease as already described, arises from irritation of the gingival margin. The irritant is mainly chemical in the sense that the various components or products of dental plaque are chemical whether they are bacterial toxins or the product of the action of bacterial enzymes on substrates within the plaque. However, there is always a traumatic element; for instance the inflammatory changes within the gingiva render it less resistant to mechanical damage by tough or rough items of food during mastication. Sometimes the mechanical traumatic element predominates, as when fibrous food becomes wedged between teeth, although the bacterial element still operates, as is evident if a micro-abscess develops.

When wear occurs, as explained in Chapter 21 in connection with dental caries, there is often loss of the contour of the contact surfaces between the teeth, followed often by breaking away of enamel leading to food stagnation between the teeth; this leads sometimes to caries, but always to damage, chemical and mechanical, followed by inflammation and recession of the interdental papillae. This type of periodontal disease secondary to tooth wear, although essentially of the same character as what has already been described, differs in a number of respects: in particular, unless superimposed on a more generalized periodontal disease that began earlier, it tends to occur later in life. It also tends to be localized or irregular in its distribution and severity, for example periodontal abscesses are liable to occur, leading to localized deepening pockets with associated bone loss and, of course, to loss of single or groups of teeth. Periodontal disease secondary to tooth wear is undoubtedly the commonest type met with in animals, certainly in animals in the wild and as recognized in skulls where the presence of a gingivitis in a young animal can only be guessed at from loss of the alveolar crestal bone, assuming that such loss can be distinguished from damage during skull preparation.

A consideration of periodontal disease illustrates how it is taken for granted that the integrity of tissues depends upon their being separated from the external environment by an intact epithelial covering. This may be true for mammals but seems not to apply with the same strictness to reptiles. In lizards, as in most reptiles, the teeth are not socketted but are ankylosed to the margins of the jaw. The teeth are usually continuously replaced but in the large herbivorous agamid lizard, *Uromastix aegypticus*, the most anterior teeth, which are incisiform, are not replaced although the more posterior molariform teeth are. The incisiform teeth gradually wear away and, when this process is nearly complete, the bone of the jaw where the teeth are attached becomes bared and eventually provides a masticatory surface, as is shown by distinct wear facets (Throckmorton, 1979). This surface bone seems to be avascular and it is likely that its lacunae become filled with mineralized tissue. Presumably, at the junction between this masticatory bone and the margin of adjacent epithelialized soft tissue, the problem of preservation of integrity is the same as at the junction between gingiva and tooth.

There is a variety of gingivitis, known as necrotizing ulcerative gingivitis, that is characterized by an intense localized inflammation associated with necrosis of the gingival margin which leads to gross ulceration. The interdental papillae are usually first affected by the ulceration but the condition quickly spreads around the teeth and the whole mouth may become affected. The necrosis of tissue produces a marked halitosis. There is often destruction of the bony alveolar crests but pocketing and progressive destruction are unusual unless there are recurrent attacks and secondary infection. A bacterial flora in which motile spirochaete and fusiform bacteria predominate is usually found on the lesions but the exciting organisms have not been reliably identified. The condition is transmissible but only to animals whose resistance has been lowered by, for example, some systemic disease or malnutrition.

Necrotizing ulcerative gingivitis has been most fully studied in man (Page and Schroeder, 1982, p. 228) but also occurs in cats (Reed, 1975, p. 156) and dogs (p. 550). It has been described in the chimpanzee (p. 534) and a condition met with in sheep (p. 562) is probably of this nature also.

It is in the nature of this book that most of the examples are skulls and the state of the soft parts has to be deduced from details of the bone of the alveolar crests and from what deposits there may be on the teeth. Unless specimens are exceptionally well and continuously cared for, deposits on the teeth tend to be lost; indeed they can be lost by over-enthusiastic cleaning during the initial preparation of the skull.

As age advances, some recession of the actual margin of the alveolar crest has to be regarded as normal and therefore has to be distinguished from recession of the crest associated with periodontal disease. Recession as an age change tends to affect the whole circumference of the cortex of the alveolar crest. Recession due to periodontal disease is likely to be irregular, usually more marked interdentally, producing a cupping-out. In periodontal disease, the surface of the bone may not have a smooth intact cortex and areas may be irregular and rough and of cancellous structure. Sometimes the response to the disease may be tissue-proliferative and the alveolar crests may be thickened and lipped.

On teeth in skulls, the two types of calculus can usually be distinguished, mainly by their profiles; for example supragingival calculus is yellowish-white and tends to jut out with a surface facing rootwards that is at about right angles to the tooth surface and a more oblique surface facing coronally; it is situated predominantly buccally and lingually. Subgingival calculus is less bulky and tends to occur as a complete ring around the tooth; it is often black from the inclusion of altered blood.

Dental calculus is always situated at some distance from the alveolar crests, usually a minimum of two or three millimetres, like a tidemark; especially if, in its profile, the shelf-like surface corresponding with the gum margin can be recognized, the calculus is a reliable indication of the location of the gum margin in life. Hence, the distance from calculus to alveolar crest in any situation can be an indicator of the depth of pockets. Of course, it is always possible that in life there had been a period of gingival recession after the deposition of a ring of calculus.

In adult animals of many species, the sides of the crowns of the teeth may to a variable extent be covered with an extremely adherent black deposit which can only be removed by diligent scraping with a sharp instrument. Sometimes, only the worn parts of the occlusal surfaces are free of this pigmentation. Such deposits are found not only in ungulates where enamel of the crowns is covered with a thin layer of cementum, but also in primates and insectivores in which there is no coronal cementum. Harvey and Penny (1985) showed extensive black deposits in a pig, in which there is no coronal cementum. Nothing seems to be known of the nature of these pigmented deposits; possibly they consist of a thin layer of mineralized plaque in which pigment has been incorporated. Bacteria, for example some species of *Bacteroides*, which produce a black pigment which is a form of haematin, have been described in dental plaque in man and some experimental animals (Slots and Genco, 1984).

The defects in the outer alveolar plate, known as dehiscences and fenestrations, exposing the roots of the teeth, that have been described in *Gorilla* (p. 532) and *Colobus* (p. 540), introduce a further difficulty (Fig. 24.2). To the non-expert, they may be indistinguishable from the changes of the localized forms of periodontal disease. Their cause is unclear but there is no doubt that they occur, and in species additional to those mentioned, and are not the product of local disease; some are perhaps connected with the skeletal atrophy, or osteoporosis, that is a feature of advanced age.

Some information on periodontal disease in captive wild animals is available in C. E. Harvey (1985).

Experimental periodontal disease in laboratory-maintained animals

A large range of laboratory-bred species has been used in the study of periodontal disease in the search for species in which the disease induced resembles that in man in all or most details. Page and Schroeder (1982, p. 272) made a comprehensive analysis of this literature and concluded that periodontitis in animals is never identical in all particulars with that in man. In some cases, the differences are in the gross clinical manifestations of the disease but more often they are more subtle at a microscopic-structural, cytological or biochemical level; e.g. in man and many animals there is a wide variety of inflammatory cell types in the gingival tissues, in rodents it tends to consist of neutrophils only and in marmosets the variety of infiltrating inflammatory cells is simpler. Nevertheless, Page and Schroeder pointed out that, providing these differences are recognized, the various experimental animals can continue to throw light on certain selected aspects of the disease process.

Some of this experimental work is directly relevant to naturally-occurring disease processes both in the wild and in captivity. For instance, Levy et al. (1976) compared the characteristics of the periodontal disease induced in three primate species, *Callithrix jacchus*, *Cebus apella* and *Macaca mulatta*, by the intragingival injection of a substance which challenged their immune-response systems. Important species differences were noted in gross, as well as histological characteristics, which would be likely to exist in the natural disease also; for instance, although bone destruction was a dominant feature of the lesions in all three species, there was much more reactive bone formation in *Macaca mulatta*.

It would be well beyond the scope of this book to attempt a comprehensive review of all the experimental work that has been done on laboratory-maintained animals to elucidate the causes of periodontal disease. It would be inappropriate, however, to ignore this work; much of it has relevance to the natural manifestations of the disease. Our aim is to give a general account with some indication of the historical aspects together with signpost references which are likely to provide avenues into the main body of literature on the various aspects indicated.

The earliest studies of experimentally-produced periodontal disease appear to be those of McCollum, Simmonds and Kinney who in 1922 produced a variety of dental disorders in rats, including what were described as gingival crest lesions, by maintaining groups of them on 57 different diets deficient in mineral, protein or vitamin A. Better descriptions of periodontal disease induced in puppies by diets deficient in mineral and vitamins were given by Jones and Simonton (1928). Mellanby (1930, p. 25) produced periodontal disease in dogs by maintaining them on high-cereal diets deficient in vitamins A and D. These experiments were part of the research into the properties of the newly-discovered fat-soluble vitamins. The early experimental work, especially on the dog, on the relationship between periodontal disease and the other vitamins was reviewed by Becks, Wainwright and Morgan (1943).

In scurvy, due to ascorbic acid deficiency, the predominant symptoms are gingival bleeding and loosening of the teeth, producing in effect an acute form of periodontal disease, as was demonstrated in the guinea pig by Howe (1920b), Boyle, Bessey and Wolbach (1937) and Glickman (1948). Ascorbic acid, which is essential for collagen synthesis, is synthesized within the bodies of most mammals, except primates and guinea pigs, so that only those animals and man can suffer from scurvy.

King (1944) and King and Glover (1945) used the

Fig. 24.2. Composite diagram shows increasing osseous involvement with increasing age, young (left) to old (right). Buccal vertical bone defects (dehiscence to right) formed by union of fenestration (to left) with the margin of increasing horizontal bone loss. From Kakehashi, S., Baer, P. N. and White, C. L. (1963).

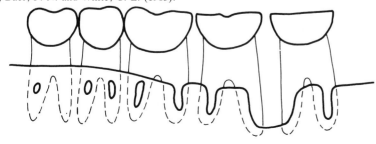

ferret and later the hamster (King and Gimson, 1948; King, 1949) for production of periodontal disease and for the study of tartar formation.

Rather surprisingly, a certain amount of plaque, which may even mineralize, may form on the teeth of rats (Rovin, Costich and Gordon, 1966; Navia, 1977), mice (Baer, Newton and White, 1964) and dogs (Listgarten and Heneghan, 1973) kept under strictly germ-free conditions; furthermore, a mild inflammatory-cell infiltration may be present in the gingiva. The plaque appears to be salivary in origin, the rheological properties of the mucin being sensitive to pH and other changes. Taubman, Buckelow, Ebersole and Smith (1981) showed that the inflammatory changes in the gingiva are the response to antigens in the diet, bedding material etc.

Periodontal disease seems not to occur in the golden or Syrian hamster (*Mesocricetus auratus*) living in its natural habitat but can readily be induced experimentally by providing sugar-rich diets. The disease can be induced more speedily than in the rat and there is the advantage that the head and teeth of the hamster are slightly larger than those of the rat and the mouth can be opened very widely for inspection during life (reviewed by Keyes, 1968). As in rats and mice, a mass of plaque forms on the tooth surface; the plaque grows into the gingival crevice and seems to enlarge it aggressively. Within about 100 days, the crests of the bony alveoli are resorbed and there are broad, widely-open pockets all round the cheek teeth filled with bacterial plaque which is often partially mineralized. The inner aspects of the pockets may become ulcerated but inflammatory change in the tissue appears to be slight with only sparse inflammatory cells. If penicillin or other antibiotics are added to the sugar-rich diet, a procedure which of course diminishes bacterial growth in the mouth, these events do not occur; there is no formation of masses of plaque.

The laboratory rat (*Rattus norvegicus*) has been much used (Thilander, 1961; Heijl, Wennström, Lindhe and Socransky, 1980) but has the disadvantage that it seems to be naturally resistant to the disease process and develops periodontal disease only under very artificial conditions such as a sucrose-rich diet and a special oral bacterial flora. The oral mucosa of the rat, including the gingiva, is particularly highly keratinized and this may account for its resistance to periodontal disease. Frandsen, Becks, Nelson and Evans (1953) produced gross degeneration of the periodontium in rats with dietary protein deprivation.

Figure 24.3 derives from the early experiment of W. G. Skillen and shows injury to the gingiva of a rat maintained on a diet of whole-wheat flour, powdered milk, maize corn, oats and beans.

The laboratory mouse (*Mus musculus*) has also been used (Baer, Newton and White, 1964) and several strains explored for susceptibility to induced periodontal disease uncomplicated by the impaction of hair or bedding material between the teeth which is a problem common to the rodents in general (Messer, 1980). The first stage of periodontal disease in the mouse and rat is the accumulation of masses of supragingival plaque which extend into the gingival crevice. At first apparently by simple physical displacement of the soft tissues by the bulky plaque, but later by actual tissue destruction, crater-like pockets are produced after about 12 months. The cellular reaction in the tissues is mild and substantially different from that found in the dog and in primates; for example abscess formation is rare even in advanced periodontal disease.

Some inbred strains of mice and hamsters bred for research appear to be more resistant to periodontal disease than others (Page and Schroeder, 1982, p. 112). However, a genetic basis for these differences is difficult to establish; for instance, Baer and Lieberman (1960) found that, although the predilection for periodontal disease was different among six strains of mice, this seemed to be dependent on hair impaction in the gingival crevices related to the grooming habit within each strain and, in one strain, to a tendency for its hair to be shed.

Fig. 24.3. Section through the tooth of a rat, showing a foreign body (awn of an oat) forced into the periodontal tissues (W. G. Skillen).

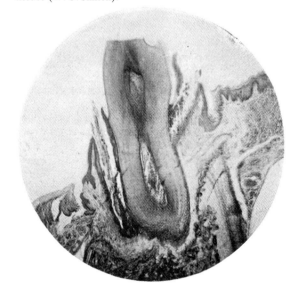

The nature of the organisms that flourish in the plaque and play a crucial part in the process has been identified; furthermore it is in the hamster that the transmissibility of the disease has been demonstrated; periodontal changes can be induced in weanling hamsters by infecting their mouths with flora from the periodontal lesions of their mothers (Keyes and Jordan, 1964). In a similar way, the infectious agent can be transmitted from a periodontal-disease-prone strain to a resistant one. Root-surface caries tends to be associated with the periodontal lesions (see Chapter 21).

To avoid destruction of the crowns of the teeth by caries, which tends to occur as well as periodontal disease on the high-sucrose diet, periodontal experiments are started after the animals have passed through the caries-prone period; that is they are started at about 70–80 days after birth (Navia, 1977, p. 315).

Hamsters kept on a conventional pelleted diet develop mild periodontal lesions; impaction of hair or bedding material between the teeth and in the gingival crevices may be a feature. However, when the diet is rich in fermentable carbohydrate of any kind, including the starch of cereals, a quite different and more rapidly-progressive type of periodontal destruction develops, but none of such florid type as is associated with sucrose (Keyes, 1968, p. 277).

Studies have been made of periodontal disease in laboratory-maintained grey or Chinese hamsters (*Cricetulus migratorius*), in particular in a strain afflicted with a spontaneous hereditary diabetes (Cohen, Shklar and Yerganian, 1963) which develops a severe cellular gingival inflammatory response with abscess formation which is not usually found in the Syrian hamster.

In the rat and hamster, and perhaps in mammals generally, some recession of the gum and alveolar crests occurs simply as a result of the passage of time or of physiological function and to some extent at least is correlated with reduction in height of the crowns of the teeth by attrition; in other words there are age changes. These are hard to distinguish from disease but have been demonstrated in rats, mice and hamsters by excluding infection with antibiotics (hamsters, Rushton, 1951, 1955) or by raising animals in totally germ-free environments (e.g. Baer, Newton and White, 1964, for mice; Amstad-Jossi and Schroeder, 1978, for rats). The situation in the rat is reviewed with particular thoroughness by Page and Schroeder, 1982, pp. 71–78.

Tonna (1972) selected the BNL (Brookhaven National Laboratory) mouse, an inbred Swiss albino strain, for study of age changes in the periodontium because it is particularly short-lived, with a mean life span of less than 12 months, and because it is highly resistant to periodontal disease.

Moskow, Rennert and Wasserman (1973) used the gerbil to explore the calculus- and periodontal disease-inducing properties of a number of diets and found that the animals became very ill, with death of many, with high protein diets that had been found to be nutritionally adequate for rats and mice (see Chapter 21).

Ockerse (1956) noted periodontal disease as well as caries in some white-tailed rats (*Mystromys albicaudatus*) he maintained on high-carbohydrate diets intended to induce caries (see Chapter 21).

Gupta and Shaw (1956a, b) found that periodontal disease was common in the marsh rice rat (*Oryzomys palustris*) maintained on a pelleted diet. The rice rat is slightly smaller than laboratory strains of brown rat (*Rattus norvegicus*). The first signs of a mild gingivitis occurred at about 16 days after birth and, by 10–15 weeks, typical periodontal disease with dental calculus and deep pockets full of food debris and hairs had developed with eventual exfoliation (Leonard, 1979). Lesions of this severity take about 15–65 weeks to develop in the hamster under similar conditions. A further advantage of the rice rat for experiments on periodontal disease is that it appears to be resistant to dental caries; hence, the periodontal lesions can be studied without the complications associated with carious destruction of the crowns of the teeth (Navia, 1977, p. 319; Gotcher and Jee, 1981). Strains of rice rat much less susceptible, and in which the disease progresses much less slowly than just described, have now been distinguished (Mulvihill, Susis, Shaw and Goldhaber, 1967). Their low susceptibility seems to be inherent because it is only slightly increased by oral inoculation of bacterial flora from susceptible strains.

Small animals, such as the rodents mentioned, may be advantageous when the study is histological but, for the examination of grosser changes during life, the advantages of larger animals are evident. Naturally, non-human primates have been greatly used and would no doubt be more used but for the cost of maintenance and increasing difficulty in obtaining them from the wild. A large amount of experimental work on periodontal disease has been done on beagle dogs chosen because they are of moderate size and short-haired, which favour economic maintenance (Andersen, 1970). They are also docile, which enables them to share cages and also to be trained to have their mouths examined, cleaned and treated while restrained in simple harness and without anaesthesia

(Saxe, Greene, Bohannan and Vermillion, 1967). Experiments (Burwasser and Hill, 1939; Egelberg, 1965) show that a soft, minced diet favours the rapid accumulation of supragingival plaque and calculus and, according to Rosenberg, Rehfeld and Emmering (1966), gingivitis in dogs can be prevented by including in the diet raw ox-tails or bone with some meat and tendon attached.

Non-human primates have been used since about 1940 as animal models for the production of periodontal disease following the report by Day, Langston and Shukers (1935) of ulcerative gingivitis in rhesus monkeys (*Macaca mulatta*) maintained on diets deficient in folic acid, and the experimental production of scurvy in that species by Tomlinson (1939, 1942). Ramfjord (1951) described gingivitis and periodontal disease induced experimentally in *M. mulatta* by the acute febrile state associated with poliomyelitis.

Several macaques besides the rhesus monkey have been employed in research into periodontal disease; for instance, crab-eating macaques (*M. fascicularis*) (Hopps and Johnson, 1976), Japanese macaques (*M. fuscata*) (Krygier, Genco, Mashino and Hausmann, 1973) and pig-tailed macaques (*M. nemestrina*) (Ive, Shapiro and Ivey, 1980). The green monkey (*Cercopithecus sabaeus*, formerly *C. aethiops sabaeus*) was used by Østergaard and Löe (1975) for the experimental production of the gingival changes of ascorbic-acid deficiency.

Avery and Simpson (1973) studied the beneficial effect of twice-weekly tooth-cleaning on gingivitis in captive baboons (*Papio anubis*).

The much smaller New World monkeys began to be used for the same purpose. Periodontal disease was induced by protein-deficient diets in capuchin monkeys (*Cebus* spp.) by J.H. Shaw (1949) and in spider monkeys (*Ateles geoffroyi*) by Goldman (1947, 1954). These early works on primates have been thoroughly reviewed by Cohen and Goldman (1960).

Captive marmosets or tamarins of various species, but predominantly cotton-top tamarins (*Saguinus oedipus*) (Dreizen, Levy and Bernick, 1972), saddleback tamarins (*Saguinus fuscicollis*), and common marmosets (*Callithrix jacchus*), as well as brown capuchins (*Cebus apella*) (both by Levy, Robertson, Dreizen, Mackler and Bernick, 1976), have been used a great deal for experiments on periodontal disease (Levy, Dreizen and Bernick, 1972); Deasy, Grota and Kennedy (1972) studied the effect of the sex hormones on gingivitis in squirrel monkeys (*Saimiri sciureus*) some of which were ovariectomized.

In recent years, such experiments have been on animals bred in captivity but this also applies to many other primates mentioned here, including the macaques.

In nearly all the species mentioned, gingivitis or periodontal disease was induced by diets containing a large proportion of the fermentable carbohydrate, for instance sucrose, that is so conducive to the formation of plaque. It is important that in other respects the diet is nutritious, for example in terms of vitamins and minerals. Furthermore, in analysing the mechanism that produces the periodontal disease, account should be taken of the fact that Saffar, Sagroun, de Tessieres and Makris (1981) have shown that the high-sucrose diet (Keyes 2000) most favoured for producing periodontal disease and caries in hamsters (Navia, 1977, p. 315) can cause a generalized osteoporosis.

Other experimental work has shown that, apart from some general nutritional states, such as vitamin deficiencies, already referred to, periodontal disease can be induced or greatly adversely affected by stress induced in rats by low temperature (Shklar, 1966) and by hypophysectomy (Shklar, Chauncey and Shapiro, 1967) and in mice by adrenal cortical steroids (Glickman, Stone and Chawla, 1953) and by deficiency of the vitamin pantothenic acid (B.M. Levy, 1947).

Occlusal stress, which influences the loosening of teeth that is a feature of periodontal disease, has been investigated experimentally in rats (Carranza, Simes, May and Cabrini, 1971), dogs (Nyman, Lindhe and Ericsson, 1978), marmosets (Sallum *et al.*, 1976), squirrel monkey (Polson, Meitner and Zander, 1976) and macaques (Stones, 1938; J.P. Moss and Picton, 1982). The experimental work on occlusal stress was reviewed by Navia (1977, p. 333).

The production of gingivitis which develops into periodontal disease is facilitated by placing ligatures of silk, wire or in the form of rubber bands around the necks of the teeth; e.g. the work of Weiner, Demarco and Bissada (1979) on rats, Kiel, Kornman and Robertson (1983) on *Macaca fascicularis* and of Heijl, Rifkin and Zander (1976) on squirrel monkeys (*Saimiri sciureus*).

The drug diphenylhydantoin (epanutin, dilantin), used for the control of human epilepsy, has the side-effect of producing gingival hyperplasia. Ferrets (King, 1952) were used for the early studies of this side-effect and later, for some reason, the cat was chosen by Ishikawa and Glickman (1961) and by later workers (Nuki and Cooper, 1972). Perhaps because they accept confinement with less docility than dogs, cats have otherwise been but little used for research into dental disease. However, Richardson (1965) studied the effect of ablation of salivary glands, antibiotics and toothbrushing on the formation of supragingival calculus in the cat.

Fuller reviews of this experimental literature are available in Navia (1977), Dreizen and Levy (1981) and Page and Schroeder (1982). A review of diseases of the mouth and teeth in laboratory primates by Ruch (1959) contains information not easily obtained elsewhere; for instance, the story is told of the dental treatment at the end of last century of the famous chimpanzee, Consul, of the Folies Bergères.

Spontaneous periodontal disease in animals
Order **PRIMATES**

In the skulls of adult primates from the wild, it is quite common to find slight generalized, horizontal recession of the alveolar crests which may be an age change (Fig. 24.2) or the result of a mild gingivitis. Localized recession of the bone, especially interdentally and in a patchy way, often without a cortex to the bony surface, is also common, especially in older animals in association with tooth wear, and is almost certainly indicative of periodontal disease. If a tidemark of calculus is present, the diagnosis is confirmed but unfortunately such calculus tends to become detached after a time and may even have been cleaned off when the skull was prepared.

As far as can be inferred from skulls in museums, the supporting tissues of most primates in the wild remain healthy until, as wear of the teeth becomes a marked feature, interdental recession of the alveolar crests becomes evident and loss of the cortical covering, indicative of gingival inflammation, becomes common.

On the better evidence from captive animals, the generalization can be made that periodontal disease is a great deal more common in captivity than in the wild.

Family **Pongidae**

Schultz, whose studies of the anthropoid apes are so noteworthy, for instance in respect of dental caries, unfortunately did not attempt the full recording of the bony changes of periodontal disease. He did, however, record dento-alveolar abscesses and was aware that these were broadly due either to the entry of infection via the pulp cavity, exposed by caries, wear or tooth fracture, or to entry of infection via the gingiva. In his records, he did not distinguish between these and his Tables refer simply to abscesses and teeth lost during life (closed sockets), as well as caries, in three age categories, juvenile, adult and senile.

It would be expected that abscess and tooth loss secondary to caries would increase with age at about the same rate as caries. Instead (Schultz, 1935, 1950, 1956), the increase between adult and senile groups in percentage of skulls with at least one dento-alveolar abscess is several-fold higher than for caries. For the gorilla (Schultz, 1956), the percentage of skulls with abscesses increased from 16.1 to 59.7 and, with caries, the increase was only from 2.7 to 2.8; tooth loss increased from 4.3 to 20.5 (Fig. 24.4; Table 23.1). The pattern was similar in respect of the other apes. Randall (1943) found unspecified alveolar abscesses in about 27% of skulls of 250 gorillas (see Chapter 23, p. 507).

These differences in incremental rates are explained if we assume that at least half the abscesses and tooth losses were due to periodontal disease which would be expected to be associated with the caries because most of it was secondary to tooth wear (Chapter 23, p. 507). Extrapolating from what will be referred to below, it seems likely that the majority of great apes of any great age in the wild are affected by gingivitis or periodontal disease associated with tooth wear of their cheek teeth.

However, as Schultz (1935) pointed out, there is a potential fallacy in basing any argument on museum collections of skulls because only in exceptional circumstances are they representative cross-sections of populations of the area in which they were collected. Hunters tended not to collect old or diseased animals or the young. Often they only collected one adult male and one adult female from any one region.

Localized areas of destruction of alveolar crest bone were met with in about 7% of specimens of orang and chimpanzee studied by Colyer; they were rare in the gibbon but common in the gorilla. In the chimpanzee, the condition was usually limited to one or two positions and the destruction was indicative of but shallow periodontal pockets. Three specimens of orang showed a fairly extensive loss of bone; in one from Sintang, there were eight pockets in the maxilla and, in two from Barita River, there were seven and four pockets respectively.

Gorilla

The chief interest of the disease in gorillas is the greater prevalence of these localized areas of periodontal disease in the eastern (mountain) gorilla (*Gorilla*

gorilla berengei) of Central Africa than in the western (lowland) gorilla (*G. g. gorilla*) of the tropical forests of West Africa. Among 652 skulls of the western gorilla, 43 (6.6%) had evidence of localized periodontal disease, mostly one or two pockets only (Figs. 24.5 and 24.6) but in one skull there were seven. Although there were only 36 skulls of the eastern gorilla available for examination, the proportion affected (44.4%) was considerably higher. Gyldenstolpe (1928) reported an even higher proportion (70%) of periodontal disease among 10 specimens of *G. g. berengei* in the Natural History Museum in Stockholm.

The areas of alveolar bone destruction in *G. g. berengei* varied a great deal in depth and were not associated with marked wear of the teeth. One young adult with but little wear of the occlusal surfaces of the cheek teeth showed as many as 10 very localized areas of interdental alveolar crest destruction (Fig. 24.7).

The interest in the high prevalence of periodontal disease in the eastern gorilla is enhanced by the finding that two groups of specimens of eastern gorillas, categorized as 19 *G. graueri* and 6 *G. rex-pygmaeorum* which, like the 10 *G. g. berengei*, were free of the disorder. The species status of *G. graueri* and *G. rex-pygmaeorum* is no longer accepted and most authorities would now regard them as *G. g. berengei* from particular localities.

According to Gyldenstolpe (1928), the staple diet of *Gorilla g. berengei* is bamboo shoots, succulent herbs and berries. The diet of the groups categorized as *G. graueri* (von Lorenz, 1917, translated by E. Schwarz, personal communication to J. F. Colyer, *c*. 1935)

Fig. 24.4. Graphic presentation of data from Schultz illustrating the relationship between dento-alveolar abscess, caries and closed sockets in primates from the wild. (A) gorilla (Schultz, 1950, p. 241), and (B) macaque (Schultz, 1935, p. 564).

In old animals, the prevalence of dento-alveolar abscess is much higher than that of caries. Even if all the closed sockets represent teeth lost from caries, the number of abscesses is disproportionately high. This would be explained if some of the abscesses were complications of periodontal disease.

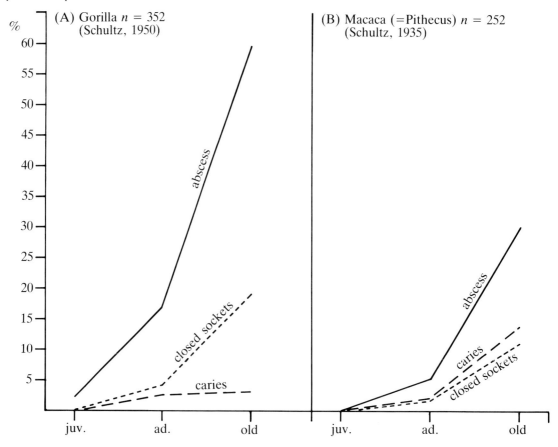

and *G. rex-pygmaeorum* is similar but without bamboo which is not available in their habitats. The diet of the western gorilla also does not include bamboo. It lives on bananas and plantains, gathered only when they are ripe, forest fruits and the succulent roots of elephant grass (P. H. G. Powell-Cotton, personal communication to J. F. Colyer) and, incidentally, is particularly fond of sweet potatoes and sugar-cane filched from cultivated plantations (Jenks, 1911).

The inference might be drawn tentatively that

Fig. 24.5. *Gorilla g. gorilla* (gorilla). ♀. Wild. RCS Odonto. Mus., G 165.1. Bone destruction between the left M_2 and M_3.

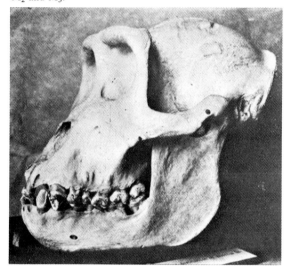

Fig. 24.6. *Gorilla g. gorilla* (gorilla). ♀. Probably wild. RCS Odonto. Mus., G 165.2. Periodontal disease has led to the loss of left M_1 and M_2.

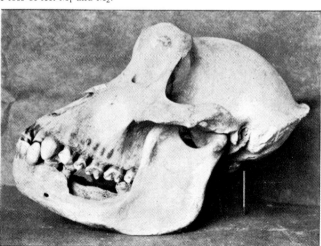

bamboo shoots contain fibrous matter which tends to become wedged between the teeth or, in some other way, is damaging to the gingiva.

The questions that observations of this kind raise are hardly likely ever to be answered because the mountain gorilla, like so many species mentioned in this book, is threatened with extinction. Even posing the question as an important one seems to invite the risk of a return to the conditions of less than 100 years ago when hunters earned their living largely by shooting animals to satisfy the needs of museums and scientific and private collectors.

Kakehashi, Baer and White (1963) recorded in considerable detail the evidence of periodontal disease in 292 western gorilla skulls in two American Museums (not the same as Schultz, 1935) dividing the material into four age groups according to the state of the dentitions, including tooth wear. Among the male skulls, the periodontal disease scores were as follows: young 9.1%, young adult 66.6%, adult 92.5%, old adult 96.9%. The scores for the females were similar though on the whole a little lower. Most of the old adults, and to a less extent the adults, showed advanced wear of the occlusal surfaces of the post-canine teeth, with total loss of cusps, frequent pulp exposure and considerable approximal wear.

These findings support the above analysis (p. 529) of the Schultz material. However, what is more important is that Kakehashi *et al.* recorded 6 details on which the diagnosis of periodontal disease was made.
1. Measurements were made of the distance of the buccal alveolar crests from the cervices of the teeth,

and expressed as horizontal bone loss. 2. The depth of approximal craters from the cervix was measured. 3. The depth measured of vertical extensions of the bone loss on the buccal aspects (dehiscences). 4. Whether the furcations between the roots of multirooted teeth were involved. 5. The presence and size of windows in the buccal alveolar plates exposing areas of root (fenestrations).

This was a pioneer system which could be improved upon; for instance, as pointed out by Page and Schroeder (1982), fenestrations are common in many species and are a feature of development rather than disease, although they may be an indication that the alveolar process lacks robustness and may be susceptible to the rapid progress of periodontal disease.

Too few gorillas are kept in captivity for much data on periodontal disease to be forthcoming.

Chimpanzee

At the present time, with the growth of the human population at the expense of forest land, some of the chimpanzees that survive do so in close relation to man, his plantations and refuse heaps. By raiding these food sources, chimpanzees and some other primates have a plentiful supply of food that is more easily masticated than what they would gather in the natural forest. This easier way of life may have extended their life span into the period when, mainly through tooth wear with consequent tooth loss, their dentitions have become so inefficient that they would no longer be able to keep themselves alive. However, it also seems likely that the new-style softer diet, probably including fermentable carbohydrate and infected material from refuse heaps, induces periodontal disease. This statement derives in part from the observations that Jones

Fig. 24.7. *Gorilla g. berengei* (eastern gorilla).
Above: formerly Rothschild Museum.
Below: ♂. NRM A62 1346. Both show destruction of the alveolar crest bone especially between the teeth.

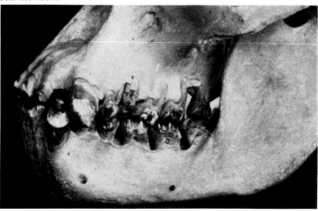

A

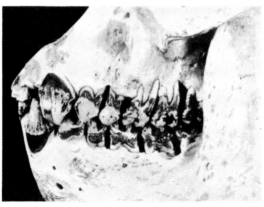

B

and Cave (1960) made on 12 skulls of adult chimpanzees shot in Sierra Leone (RCS Odonto Mus., A 69.93–A 69.935). All showed evidence of periodontal disease in the form of loss of alveolar crests in a number of places; one old adult, and in particular two which could be assessed as very old because of obliteration of the cranial sutures, showed loss of more than half the dentition, presumably from periodontal disease which severely affected the extremely worn remaining teeth. There was, incidentally, no dental caries.

The only other data on chimpanzees in the wild appear to be those of Schultz (1956) derived from the high prevalence (59.6%) of dento-alveolar abscesses in the old specimens in relation to the prevalence of caries (30.6%) (Table 23.1). Unlike in the gorilla material referred to above, the number of skulls in which teeth had been lost in life, 29.0%, closely matched the number of skulls affected by caries so that it cannot be argued that many teeth had been lost from periodontal disease.

Figure 24.8 shows the dentition of a chimpanzee which has many of the features Kakehashi et al. described in their old adult group of gorillas. Wear has destroyed the contours of the occlusal sufaces and exposed much dentine, the alveolar crests have receded and in places are thickened by the formation of reactive bone, the left lower cheek teeth had been exfoliated but only shortly before death because shallow sockets have not yet filled in. This skull came from the Batouri district (Chapter 3, p. 22). It seems likely that it was a very old animal which survived in spite of being partially toothless because it lived in an area where there were many banana plantations and animals had access to the village refuse heaps feeding on which could have hastened the progress of the periodontal disease initially associated with tooth wear. In the more arduous and competitive environment away from human habitation, this animal would not have survived.

Figure 24.9 shows the dentition of a chimpanzee from the wild which, from the amount of wear of the post-canine teeth, is fully adult but not of advanced age. There is a disproportionate amount of wear of the incisors. Judging from the localized alveolar crest destruction, there must have been deep periodontal pockets in relation to the incisors and to the first premolars.

Figure 24.10 shows the skull of an adult chimpanzee with much tooth wear and therefore of advanced age, and with much destruction of the alveolar crests from periodontal disease extending well beyond the root furcations. Several teeth have been lost and the post-canine dentition is nearing the stage of total loss that is seen in the specimen in Figure 24.11.

In captivity, advanced periodontal disease may develop in quite young chimpanzees (Fig. 24.12). Arnold and Baram (1973) found moderately heavy deposits of plaque and calculus on the labial surfaces of the incisors of four captive living chimpanzees aged 6–10 years. All had mild generalized marginal gingivitis but radiographs showed hardly any bony involvement. Two old animals, a male aged 39 and a female aged 44 years, showed advanced tooth wear and generalized

Fig. 24.8. *Pan troglodytes* (chimpanzee) Wild, Batouri District, Cameroun. P-C. Mus., M.O. Coll. (Incorrectly referred to as gorilla in Colyer, 1963). Wear of the occlusal surfaces has destroyed the cusp contours and exposed areas of dentine. Teeth have been lost, either from exposure of the pulps by wear and consequent abscesses or from periodontal disease secondary to the wear. The remaining teeth have become spaced, so facilitating gross food-wedging and aggravating the alveolar destruction.

534 Section 4: Other Disorders of Teeth and Jaws

periodontal disease with plaque and calculus on all the teeth, loss of alveolar bone exposing up to nearly half of the length of the root; in fact all the manifestations of an advanced periodontal disease of chronic rather than acute character. Each had lost several teeth, presumably from periodontal disease because there was no caries.

The findings of Page, Simpson and Ammons (1975) in respect of eight living captive chimpanzees aged 27–49 years were similar except that several teeth were carious. Five juvenile animals, aged four to eight years, had mild gingivitis with dental plaque and only shallow gingival pocket-formation and no bone loss as determined by radiography.

The infectious nature of necrotizing ulcerative gingivitis, at least in animals rendered susceptible, for example by poor hygienic conditions or poor diet, is illustrated by an outbreak of the disease in a free-ranging island colony of 42 chimpanzees described by Van Riper et al. (1967). A very sick animal was

Fig. 24.9. *Pan troglodytes* (chimpanzee). ♀. Wild. P-C. Mus. M574. Extensive destruction of the alveolar crest of the left I^1, of both I_2 and the right P3.
Below: Occlusal surfaces showing disproportionate wear of the incisors. Judging from the moderate amount of wear of the post-canine teeth, this is not an aged animal.

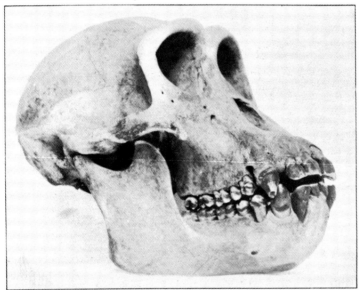

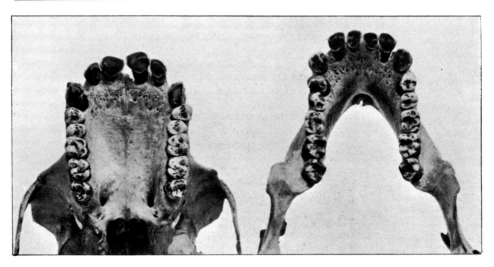

discovered with extensive gingival ulceration from which the typical microbial flora, with motile fusiform and spiral forms, was isolated. A little later more than half the colony was found to have similar ulcerative gingivitis though in a less-severe form.

Some necrotic gingival lesions that simultaneously affected two chimpanzees and a gorilla in Berlin Zoo (von Hoffman, 1876) were probably those of necrotizing ulcerative gingivitis and the same diagnosis seems likely in the case of a zoo chimpanzee described by Grzimek (1953) as suffering from foul breath and oral lesions which were not benefited by ascorbic acid.

Orang-utan and gibbon

Schwanitz, Kreft and Fleischer-Peters (1983) described a 13-year-old captive orang-utan whose primary oral disorder seemed to be gross generalized gingival hyperplasia of genetic origin because her mother and two female sibs were similarly affected. Furthermore, there was a morphological abnormality of chromosome 23 which the mother showed as well. Eruption of the permanent dentition was retarded and some teeth appeared to be absent but whether this means that they remained unerupted or were totally absent was not made clear.

Fig. 24.10. *Pan troglodytes* (chimpanzee). ♂. Wild. P-C. Mus. M4475. Several teeth have been lost during life. There is extensive destruction of the alveoli of others.

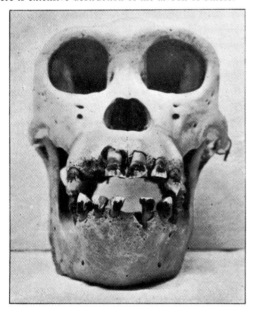
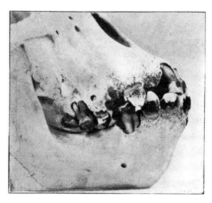
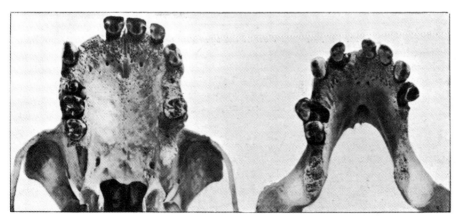

The only evidence of typical periodontal disease in orangs and gibbons is what can be deduced from the records of Schultz (1956) of dento-alveolar abscesses (Table 23.1). Among the skulls of 97 orangs of both sexes in the adult group, he found dental caries in 4.1%, abscesses in 13.4% and no teeth lost during life. In 46 in the old group, there was caries in 13.0%, abscesses in 60.8% and lost teeth in 15.2%. There was thus what seems to be an increase with age in abscesses and lost teeth which is disproportionately high compared with the increase in caries. It seems likely that at least half of the abscesses and lost teeth were from complications of periodontal disease.

Schultz (1941) illustrated two examples of periodontal disease and one of periodontal abscesses secondary to tooth wear in orangs. Two show root-surface caries. Among a collection of 144 skulls of orang-utan from West Borneo, Röhrer-Ertl, Frey and Schmidhuber-Schneider (1985) found some which, although of uncertain age, had lived long enough for the teeth to be much worn, some with exposed pulps. There was localised periodontal disease and tooth loss secondary to the wear.

Similarly, among 439 gibbons of both sexes (Table 23.1) in the adult group, there was caries in 0.9%, abscesses in 7.5% and teeth lost 3.9%, compared with,

Fig. 24.11. *Pan troglodytes* (chimpanzee). ♀. Wild. P-C. Mus. M 165. Total loss of the teeth, almost certainly from periodontal disease.

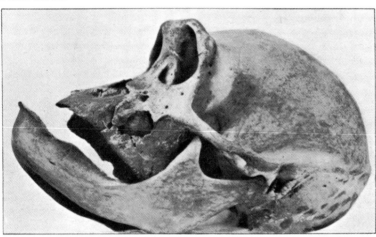

Fig. 24.12. *Pan troglodytes* (chimpanzee). ♂. Juvenile. Kept as a pet in Cameroun. RCS Odonto. Mus., G 166.14 (renumbered). Extensive destruction of alveolar bone with exposure of the roots of the teeth. The left half of the specimen still has the gingiva *in situ*.

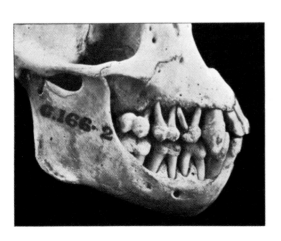
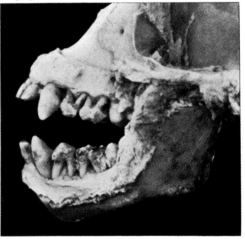

in the old group, caries 8.7%, abscesses in 38.5% and teeth lost in 22.1%. The same inference may be drawn that periodontal disease accounted for a proportion of the abscesses and teeth lost.

Super-family Cercopithecoidea

Table 24.1 shows the percentage affected by localized areas of destruction of alveolar crest bone, namely periodontal disease, in a large number of skulls of various genera of Old and New World monkeys examined by Colyer in several museum collections. It suggests that among the cercopithecoids periodontal disease is rather high, reaching 8.5% in the leaf monkey (*Presbytis*) in the wild. Examples of *Presbytis* are shown in Figures 24.13 and 24.14.

Again, as for the other categories of primates already mentioned, Schultz (1935, 1956, p.981) recorded higher percentages of alveolar abscess in his groups of old *Cercopithecus* and *Presbytis* than would be predicted from the increase of caries with advancing age (Table 23.1). For 112 specimens of *Cercopithecus*, caries increased from 2.4% in the adult group to 21.4% in the old animals and alveolar abscess increased from nil to 35.7%. In 172 *Presbytis*, caries increased from 0.6 to 1.1% and alveolar abscesses from 5.6 to 19.6%. The inference may be drawn tentatively that a proportion of the abscesses are complications of periodontal disease, but it is not possible to deduce how prevalent the disease was.

Lampel (1963) described as a conspicuous feature of the upper jaw of *Cercopithecus* the presence of root exposures unassociated with alveolar abscess or other inflammatory change. Most of these were window

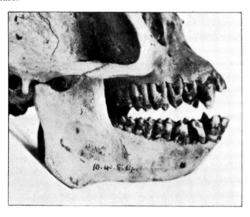

Fig. 24.13. *Presbytis frontata* (white-fronted leaf monkey). ♀. Wild. BMNH 1910.4.5.14. Considerable destruction of the alveolar process around the upper premolars and molars.

Table 24.1 The incidence of localized areas of alveolar crest destruction in various genera of OLD and NEW WORLD MONKEYS from the wild

Genus	No. of specimens	No. with localized lesions	% with localized lesions
Cercopithecoidea			
Cercopithecus	1473	22	1.5
Macaca	588	34	5.8
Papio	288	0	0.0
Colobus	1205	49	4.1
Presbytis	985	84	8.5
Ceboidea			
Cebus	538	6	1.1
Ateles	186	4	2.2
Alouatta	781	20	2.6
Callithrix	164	20	12.2

Fig. 24.14. *Presbytis cristata* (silvered leaf monkey). ♀. Wild. BMNH 1909.4.1.6. Extensive destruction of the crestal bone between the molars.

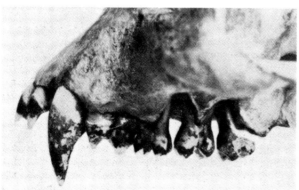

538 Section 4: Other Disorders of Teeth and Jaws

defects in the outer alveolar plate but a few extended to the alveolar crest. They were seldom seen in the mandible. These root exposures were undoubtedly the same as the fenestrations and dehiscences noted in the gorilla and which are also common in *Colobus* (p. 540).

The specimen of *Cercopithecus* in Figure 24.15 is extremely unusual in showing total loss of the dentition with apparent healing of the sockets. It is not recorded whether it was from the wild or captive. However, it is inconceivable that an animal in such a toothless state could survive in the wild; hence it must be assumed that it had been kept in captivity, perhaps for a prolonged period. Periodontal disease seems to be the most likely cause of the tooth loss, especially having regard to the severe periodontal disease in the two captive cercopithecids in Figures 24.16 and 24.17.

Although the macaques are popular for experimental oral biology in the laboratory, nothing appears to be recorded directly of their periodontal status in the wild. Among 252 skulls of macaques from the wild, Schultz (1935) found no alveolar abscesses in the young group but 5.2% of the adult group were affected and 30.1% of the old; compared to nil, 2.1% and 13.7% for caries (Fig. 24.4). Later data reported by Schultz (1956) (Table 23.1) are similar. The inference may be drawn, as was done from his records

Fig. 24.15. *Cercopithecus wolfi wolfi* (Wolf's monkey). KMMA 2744. Total loss of the teeth, presumably from periodontal disease.

Fig. 24.16. *Cercopithecus pygerythrus pygerythrus* (savannah monkey). ♀. Captive. RCS Odonto. Mus., G167.2. Gross destruction of the alveolar process; much calculus on the teeth. Bland Sutton (1885c) was of the opinion that inhalation of pus from these gingival lesions was the ultimate cause of the septic pneumonia that caused the death of this animal in the London Zoo.

Fig. 24.17. *Cercopithecus aethiops cynosuros* (Malbrook monkey). ♂. Captive. RCS Odonto. Mus., G167.15. Advanced periodontal disease. Captive for three years and four months only.

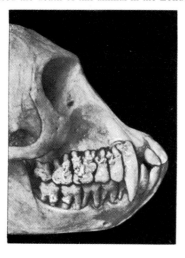

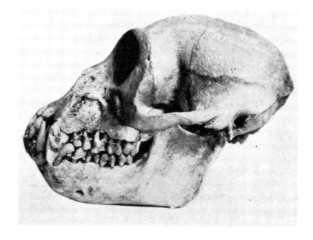

for other primates, that there is disproportion between the prevalences of caries and abscess because many of the abscesses were complications of periodontal disease.

Among 290 skulls of captive macaques, Schultz (1935) found that prevalence of caries was higher in the old animals than in the adult but, surprisingly, the prevalence of alveolar abscesses was less than half that in the wild (see also Chapter 23, p.510). It seems unlikely that periodontal disease was less prevalent in the captive so that the observation, which cannot be accounted for, weakens the argument made here that the disproportionate greater prevalence of abscesses over caries arises because many of them were periodontal abscesses.

Colyer (1921) found that supragingival calculus was present on the labial aspects of the lower incisors and there was resorption of the surface of the bone in this vicinity in the majority of 251 skulls of young rhesus monkeys with some of their deciduous teeth still functional (Fig. 24.18). These animals were wild-caught for research purposes, and had been brought to London during 1918 where they died, probably of an epidemic infection, within about 12 months. As Colyer pointed out, supragingival calculus and gingivitis confined to the front of the mouth in children is indicative of mouth-breathing and the inference may be drawn that the gingivitis in these young monkeys was brought about by mouth-breathing induced by infection of the upper respiratory tract associated with the conditions of their captivity.

Auskaps and Shaw (1957) found periodontal disease in only one crab-eating monkey (*M. fascicularis*) among 14 young and three adults which had been laboratory maintained for experiments on poliomyelitis, although some dental caries was present in all but five. The affected adults had supragingival calculus on the labial and lingual surfaces of all the teeth, the gum was swollen and there was loss of the alveolar crest bone extending to the furcations of the cheek teeth. All the others had healthy gingiva and dental plaque was found in only four of them.

The susceptibility of the rhesus monkey (*M. mulatta*) to scurvy if the diet is not rich in fresh fruit and vegetables was shown by Howitt (1931) who described the scurvy, including the characteristic swollen bleeding gingiva and loosening of the teeth, as well as subperiosteal haemorrhages and swollen joints, that affected a shipment of 39 wild-caught monkeys from India to the United States. The diet for about 70 days consisted largely of rice and dried beans. Some animals died but the remainder recovered rapidly once fruit and raw cabbage were added to the diet. Cohen and Goldman (1960) described, with histological detail, a young macaque that had suffered in exactly the same way during transportation by sea. Necrotizing ulcerative gingivitis had supervened on the gingival changes of scurvy.

Bramblett (1969 and personal communication) found much destruction of alveolar bone from periodontal disease, mostly associated with teeth damaged by trauma or by caries secondary to tooth wear, in the

Fig. 24.18. *Macaca mulatta* (rhesus monkey). Juvenile. Captive. Three specimens in the RCS Odonto. Mus., A 84.2 Series.
A: The normal condition of the labial bone covering the roots of the incisors.
B: Early stage of destruction of the alveolar crests from periodontal disease and some widespread pitting of the bone surface indicative of deep-seated infection.
C: More advanced stage of destruction and grosser pitting.

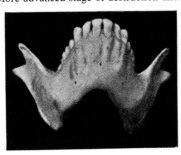

A

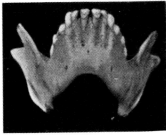 

B   C

older animals among the skulls of 60 yellow baboons (*Papio cynocephalus*), mostly from the wild.

James and Barnicot (1949) found that localized periodontal disease secondary to tooth wear and caries were part of the general deterioration of the older animals among 16 baboons (*P. hamadryas*) aged 6–12 years that died in the London Zoo.

On the other hand, in a similar group of 13 captive olive baboons (*P. anubis*), which were slightly older, and three over 12 years of age (described as middle-aged because the life span of the baboon is 26–28 years), examined in the same way, Avery and Simpson (1973) and Simpson and Avery (1974) found much more severe and generalized periodontal disease. Gingival biopsies all showed various degrees of inflammation and tissue hyperplasia. In the three older animals, there were deep pockets extending to more than half their root length. There was gross plaque and both supragingival and subgingival calculus.

Hodosh, Povar and Shklar (1971) examined, under anaesthesia, 40 captive olive baboons and geladas (*Papio anubis* and *Theropithecus gelada*) in unspecified proportions, apparently in good health. They were estimated to be aged seven to 10 years and had presumably been in captivity for at least several years. Gingivitis, sometimes with gingival hyperplasia, was present in 33, mostly localized to the incisor region. In three animals, there was periodontal disease of moderate severity, without loosening of the teeth but with gingival pockets from which pus exuded.

In 1344 skulls of three species of *Colobus*, Colyer (1943*b*) found localized areas of destruction of alveolar crest bone, indicative of periodontal disease, in 46 (3.4%), that is less commonly than in many monkeys in the wild, but in several instances the bone destruction was severe (Fig. 24.19).

As usual in his records, Schultz (1958) in his study of a group of 309 colobus monkeys – western black and white colobus (*Colobus polykomos*), red colobus (*C. badius badius*) and olive colobus (*C. verus*) – did not score periodontal disease directly. However, he did record alveolar abscesses (Table 23.1) and, as like Colyer (1943*b*) he found no caries in this group, it can be assumed that the abscesses were due to periodontal disease, apart from a few due to exposure of tooth pulps by wear or by fracture of the teeth. He found alveolar abscesses in 8.9% of the adults and 28.6% of the old animals. Four (6.8%) of the old animals had lost teeth during life; some of these would have been lost from periodontal disease. This seems to suggest a much greater prevalence than Colyer found; however, it must be pointed out that he did not make a distinction between adults and aged; indeed the over one thousand skulls he referred to may have included young ones in which periodontal disease is rare.

Colyer (1943*b*) drew attention to the existence in *Colobus* of defects in the alveolar bone over the outer aspects of the roots of the teeth unassociated with any sign of inflammatory change. The defects may take the form of windows through which the roots are exposed or defects extending from the gingival margin for as much as one-third of the length of the root. He men-

Fig. 24.19. *Colobus badius tephrosceles* (Uganda red colobus). ♂. Wild. BMNH 1930.8.1.3. Irregular destruction of bone between the teeth. Food is still present in the area of bone destruction between M_1 and M_2.

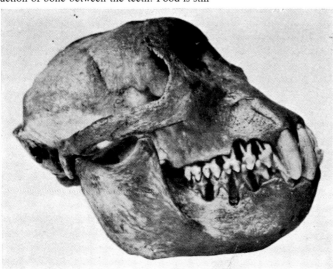

tioned that similar defects are found in the gorilla and it is quite certain that they are what Kakehashi *et al.* (1963) described as fenestrations and dehiscences (Fig. 24.2).

Schultz (1958) found similar defects in 76% of his adult *Colobus* and thought they developed in advancing age because they were not present in the juveniles. This seems quite possible because these defects could be a manifestation of the atrophy which is a feature of old age. In its extreme form, the skeletal changes are known as senile osteoporosis. It is well established, at any rate for man, that these age changes can begin long before really advanced age is reached. The other possibility is that some dietary deficiency is responsible, for example a deficiency of one or more of the factors upon which bone maintenance depends such as deficiency of protein, vitamin D or of mineral. However, judging from the effects of experimental protein deficiency (p. 528), these are more likely to affect the alveolar crests widely and to render the periodontium susceptible to periodontal disease.

Kuhn (1963) described larger defects of interdental bone in the lower molar regions of three mandibles of red colobus (*Colobus badius badius*). The defects in no way resembled the dehiscences described above although they appeared not to be associated with inflammatory change, their bony surfaces being dense and smooth. The most severe example was bilateral and was associated with the absence or loss of two molars on each side. Kuhn regarded them as defects of formation of the alveolar process, that is as developmental anomalies; slow-growing, space-occupying benign tumours, such as fibromas, seem to be at least equally likely.

Super-family **Ceboidea**

Adult New World monkeys in the wild seem, when their teeth become worn, to be no more free from periodontal disease than Old World monkeys (Table 24.1); the ultimate cause is the wear which renders the interdental gingiva susceptible to damage which admits infection, so setting up chronic gingivitis with sequential alveolar crest destruction in the particularly affected areas.

Smith, Genoways and Jones (1977) found many examples among 285 skulls of three species collected in Nicaragua; namely *Alouatta palliata*, the mantled howler monkey which is predominantly folivorous, *Ateles geoffroyi*, the black-handed spider monkey, which is predominantly frugivorous and *Cebus capucinus*, the white-throated capuchin monkey, which is predominantly frugivorous but, rather more than the other two, will also take insects. Most of the lesions were in the molar region but, in the *Alouatta* in particular, some were associated with worn canine teeth, some being so worn that the pulps were exposed; these, as well as many other lesions, were associated with openings in the buccal bone, indicative of abscesses. It was then difficult to decide whether the lesions were of periodontal or pulpal origin. Perhaps for this reason, Smith *et al.* only gave the percentages of animals affected by pulp or periodontal disorders, 27.6% of *Alouatta*, 8% of *Ateles* and 16.7% of *Cebus*. The high percentage in *Alouatta* may reflect the high proportion of canine abscesses related to wear. Unfortunately these workers did not relate their findings to either age or degrees of generalized wear of the teeth.

In all three taxa, but particularly in *Alouatta*, the bone over the buccal roots of the upper molars was thin with areas where, in the absence of any inflammatory change, the roots were exposed (dehiscences, p. 524). Smith *et al.* attributed these bone defects to resorption in response to mechanical stresses on the teeth.

Based on about 900 skulls of the same three species, Schultz (1960) gave the prevalence of alveolar abscesses in *Alouatta* as 26% in adults and 75% in old animals, for *Ateles* 28% in adults and 67% in old and for *Cebus* 18% in adult and 75% in old animals (Table 23.1). These increases with age were substantially greater than for caries so that it may be argued, as has been done several times here, that probably many of the abscesses were periodontal.

Further information about periodontal disease in the howler monkey (*Alouatta*) in the wild was provided by Hall, Grupe and Claycombe (1967) who examined 106 formalin-preserved heads and 65 skulls of these species from Northern Argentina, mostly adults with various degrees of tooth wear. Sixteen of the adult heads showed localized areas of gingival pocketing, in many of which food or hair was still wedged. Many of the skulls showed bone destruction indicative of the same disorder, and a total of 152 fenestrations exposing the roots of premolars and molars; two of the fenestrations extended to the alveolar crest and are therefore categorized as dehiscences.

Marmosets and tamarins are a diverse group of genera within the Family Callithricidae. They are basically frugivorous and herbivorous but, especially in times of drought, consume insects, rather more freely than other New World monkeys.

Information about periodontal disease among them in the wild is scarce apart from the data given in Table 24.1; Hershkovitz (1970) examined the skulls of 904 wild-caught animals, consisting of three genera from Central and South America, *Cebuella*, *Callithrix* and *Saguinus*. Periodontal disease was found among the adults but unfortunately it is impossible to derive the prevalence from the data because abscesses and periodontal disease appear to be grouped together as alveolar infections.

More information is available about periodontal disease among marmosets and tamarins in captivity. Friedman, Levy and Ennever (1972) made what can be categorized as an epidemiological study of 176 marmosets and tamarins living as a captive colony in Texas. All had been there for at least six months. Thirty-four were saddle-back tamarins (*Saguinus fuscicollis*), 108 were cotton-top tamarins (*Saguinus oedipus*) and 34 were common marmosets (*Callithrix jacchus*). Males and females were about equal in each group. Careful examination of the mouths was made under anaesthesia, scoring gingivitis on a 0–4 scale and dental calculus on a 0–3 scale. Gingivitis in some degree was present in 90% and deposits of calculus in 100%.

Assessed in terms of severity, gingivitis was most severe in *C. jacchus* and least in *S. fuscicollis*; severity was closely correlated with the assessments of abundance of calculus. Friedman *et al.* said nothing about the diet or general conditions of these animals but, knowing that marmosets are unlikely to suffer so much from gingivitis in the wild and that gingivitis depends greatly on the physical and chemical character of the diet, it seems likely that the diet of these animals diverged significantly from their diet in the wild.

J. H. Shaw and Auskaps (1954) examined the jaws of 34 common marmosets (*Callithrix* [=*Hapale*] *jacchus*) which died after unspecified periods of captivity where their diet consisted of standard cariogenic laboratory pellets, a grain mixture and fresh fruits. It is interesting that it was several weeks before the marmosets chose to consume the pellets but thereafter they consumed them freely. It is also worth mentioning that no trace of caries was found. However, 10 of the animals had heavy deposits of dental calculus and the majority showed evidence of periodontal disease ranging from mild inflammation to deep localized pockets with bone destruction, in several instances leading to loss of teeth. Several of the animals had generalized osteoporosis which affected most of the skeleton.

Ammons, Schectman and Page (1972) examined 125 living *Saguinus oedipus* under anaesthesia and scored a variety of dental disorders including caries, fractures, tooth wear and gingivitis. Seventeen were later killed and examined radiographically and histologically, as were 40 other heads. All were wild-caught and had been in captivity in a zoo for periods of 3–12 months. Dividing the mouth with a full complement of 36 teeth into 62 labial and interdental gingival units, they found that 18.2% of the total of 7440 gingival units in the living 125 animals were chronically inflamed, in most cases without severe gingival pocketing. Gingivitis was most common in the front of the mouth.

In the skulls, only 0.5% of 1287 corresponding unit areas showed the bony changes of periodontal disease. Ammons *et al.* concluded that the prevalence of progressive periodontal disease in marmosets in captivity is exceptionally low compared with other primates. They argued that the lowness of their scores for gingivitis compared with those of others for captive marmosets represents a truly better state of health of the animals they examined; better because they were housed in a high-humidity atmosphere and their diet was highly nutritious with plenty of roughage and was enriched with vitamins, in particular with D_3 for which it had become known that marmosets have a particularly high requirement. However, they also believed that the scores for bony changes of periodontal diseases of previous workers had tended to be high because they had not been aware of the bony defects, fenestrations and dehiscences, and counted them as evidence of periodontal disease.

Sub-order **Prosimii**

The bushbabies, *Galago senegalensis* and the larger *Galago crassicaudatus*, are sometimes laboratory-maintained for experimental research. According to Grant, Chase and Bernick (1973), the animals accumulate plaque and calculus on their teeth and consequent gingivitis develops which, after about 12 months, begins to become periodontal disease, that is gingival pockets and resorption of alveolar crest bone develop. These observations are based mainly on the histological study of five specimens aged 5 and 15 months and $2\frac{1}{2}$, 6 and 7 years. All but the 7-year-old were captive-born. A noteworthy feature was the amount of deposition and resorption of cementum succeeded by further deposition.

Some of the teeth in the two oldest specimens, which were near the end of their life-span, showed ankylosis of the cementum to bone, the first stage of which appeared to be deposition of mineralized particles in the periodontal ligament.

In lemurs in the wild state, where the diet is composed of leaves, fruit, insects, birds and birds' eggs, periodontal disease is uncommon although Colyer found six instances of localized crestal bone destruction among 650 skulls of assorted Lemuridae. However, in captivity the disease is common, perhaps because, in addition to fresh fruits and vegetables, bread and cooked potatoes have been known to be included in the diet.

Order **CARNIVORA**

Family **Felidae**

As far as is known, the cats, large and small, in the wild state on their strictly flesh diet are free of periodontal disease. Little information is available about their periodontal status in captivity but it seems likely that if the food given them is fresh meat, especially on the bone, they remain disease-free. Zoos are usually careful to provide their large felids with plenty of meat and bone because it was established long ago that without them their condition deteriorates and they will not breed.

In some experiments on eight tigers (*Panthera tigris*) in a Detroit zoo, in which the animals were regularly anaesthetized for oral examination and a regime of tooth cleaning, Haberstroh *et al.* (1984) demonstrated the improvement in gingival health produced by adding bones to the diet and even more by prophylactic tooth cleaning.

Cook and Stoller (1986) recorded under general anaesthesia, using criteria of assessment used in human dentistry, the periodontal status of ten captive snow leopards (*Panthera uncia*) of various ages up to 22 years. All the animals had chronic gingivitis but severe periodontal disease was only present in the two animals over 18 years of age.

Catcott (1975) and other veterinary textbooks agree that periodontal disease associated with dental calculus is common in the domestic cat (Fig. 24.20) and it seems that the manifestations and changes are similar to those in the dog. Nevertheless systematic studies are few.

Schlup (1982) found that 57.5% of 200 cats brought to a veterinary clinic for other complaints had deposits of dental calculus associated with periodontal disease in the form of gingivitis, gingival pockets or gingival recession. Thoroughbred cats of exotic breeds were more affected than what is referred to as the European short-haired cats. Reichart, Dürr, Triadan and Vickendey (1984) made a careful study of the histology of natural periodontal disease in 15 domestic cats aged one to eight years.

Quite large masses of dental calculus can build up on the crowns of the teeth. Richardson (1965) showed that a considerable amount of dental calculus could be deposited on the sides of the crowns of the carnassial teeth in an 18-week period.

It is said that domestic cats, if fed on flesh which they have to rend and cut with their teeth, remain free from the disease, but if the meat is chopped up they are apt to develop trouble. This suggests that it is the physical and not so much the chemical character of the

Fig. 24.20. *Felis catus* (domestic cat).
Above: RCS Odonto. Mus., G 147.5.
Below: RCS Odonto. Mus., G 147.51. Both are moist specimens with the gingiva *in situ*.
Above with normal gingiva. Below shows the changes of gingivitis; the gingival margin is swollen, lessening the amount of the tooth crowns exposed to view.

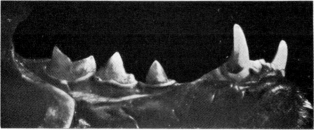

food which is responsible for the development of the disease in the cat.

Studer and Stapley (1973) found that a group of 8 kittens developed gingivitis and deposits of dental calculus when fed on moist commercial canned cat food, whereas a similar group fed dry commercial cat food did not; furthermore the gingivitis and soft calculus deposits disappeared after a period on the dry type of diet, presumably because the friction of the dry food removed the deposits and was beneficial to the gingiva. However, in evaluating experiments of this kind account must be taken of the fact that there is evidence (Seawright and Hrdlicka, 1974) that excess of vitamin A, for instance in raw liver, produces hyperaemia and swelling of the gingiva which can proceed to destruction of alveolar crest bone. The mechanism for this is unclear but liver contains little calcium so that a diet that consists largely of liver can lead to calcium deficiency.

In felids and in carnivores in general, the disease tends to spread in the maxilla more rapidly and extensively than in the mandible owing to the less-dense character of the bone of the maxilla. Burrows, Miller and Harvey (1985) and Grove (1985) have given good accounts of the periodontal diseases in domestic cats and dogs.

Families **Procyonidae** and **Ailuropodidae**

The Procyonidae and Ailuropodidae in the wild state seem to be entirely free from periodontal disease. In the captive state, the disease has been seen in racoons (Fig. 24.21), coatis (Fig. 24.22) and the red panda (*Ailurus fulgens*). The diet of racoons (*Procyon*) and coatis (*Nasua*) in the wild state is almost entirely carnivorous, except that nuts and fruit are sometimes eaten. In captivity, these animals are fed on raw meat and fat, which are not cut up, fruit, nuts, carrots and, in addition, they obtain a fair supply of bread and biscuits from visitors. The chief food of the red panda in the native state is composed of various fruits, acorns, the young shoots of bamboo roots etc. In the Zoo, its diet is fruit and meat cut up into small cubes together with a little bread and milk.

Family **Viverridae**

The Viverridae frequently acquire periodontal disease, particularly the civet (Fig. 24.23), the mongoose and the meerkat (Fig. 24.24). The food of the mongoose (*Herpestes* spp.) in nature is varied. They live upon rats, mice, snakes and lizards, buds and

Fig. 24.22. *Nasua nasua* (southern coati). Captive. RCS Odonto. Mus., G 149.2. Considerable thickening of the margin of the alveolar process.

Fig. 24.21. *Procyon cancrivorus* (crab-eating racoon). Captive. RCS Odonto. Mus., G 148.1. An early stage of bone destruction due to periodontal disease.

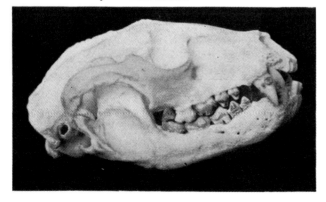

insects but at times eat fruit. The Royal Natural History (Lydekker, 1893–1894, p. 470) states the following foods were found in the stomach of a mongoose in India: a quail, a small wasp's nest, a lizard, a number of insects and part of a custard-apple. Meerkats (*Suricata suricatta*) feed chiefly on succulent bulbs which they scratch up with the long claws on their fore feet. In the captive state, mongooses and meerkats are given minced and small chunks of raw or cooked meat and bananas. Mincing and cooking the meat renders it more liable to cling around the teeth.

The propensity of meerkats to develop periodontal disease in the captive state is perhaps increased by the unusual arrangement of post-canine teeth in the mandibular arch which is seen in Fig. 24.24. The teeth are slightly obliquely set and overlap a little. This seems likely to encourage the lodgement of certain types of food they are likely to receive in captivity, although in the wild state they are immune to the disease.

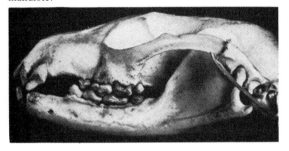

Fig. 24.23. *Paguma larvata* (masked palm civet). Captive. RCS Odonto. Mus., G 149.1. Destruction of the alveolar crests in the maxilla is more advanced than in the mandible.

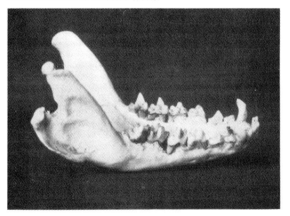

Fig. 24.24. *Suricata* sp. (meerkat). Captive. RCS Odonto. Mus., G 151.41. Mandible showing advanced irregular destruction of the alveolar crests.

Family **Ursidae**

A very old spectacled bear (Fig. 24.25) which had been in captivity at the London Zoo for nearly 14 years had severe periodontal disease associated with tooth loss and a large mass of dental calculus on the buccal surface of the first maxillary molar on the right side. The teeth on the left side show only an early stage of the disease. The feature of interest of this specimen is that the pulp chamber of the upper right canine had been exposed by attrition or fracture and the pulp had died; it is more than possible that the injury to the canine determined the severity of the disease on the right side by creating a tender area, so interfering with the functional activity of the teeth on that side of the mouth.

Rausch (1961) found that badly worn and fractured teeth with secondarily-associated periodontal disease (Chapter 22, p. 493) were a feature of advanced

Fig. 24.25. *Tremarctos ornatus* (spectacled bear). ♀. Captive. RCS Odonto. Mus., G 143.1. An advanced stage of periodontal disease and a large mass of dental calculus on the buccal surfaces of the upper cheek teeth.

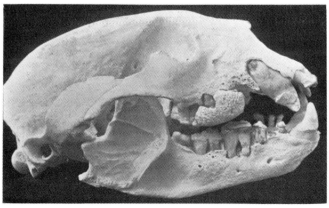

age in the black bear (*Ursus americanus*) in the wild, and even more so in the more herbivorous brown bear (*Ursus arctos*). Incisors and canines were worn down to the gum or lost and, in the post-canine teeth, the pulps were often exposed by wear.

Family **Mustelidae**

Dental disease in the mink (*Mustela vison*) in the wild has not been recorded but, among minks raised on ranches commercially for their fur, abrasion of the teeth and tooth fractures are common, probably due to cage-chewing (Lavine, Page and Padgett, 1976). Mild gingivitis, associated with dental plaque and small amounts of calculus, is also common by the age of 4 years and later there is often moderately even loss of alveolar crestal bone which appears to be partly an age change. The interest in periodontal disease in the mink derives from the occurrence among ranch-bred colonies of an autosomal recessive genetic trait in which the immune system is defective. The clinical condition that results, which includes widespread vascular changes and a susceptibility to develop multiple chronic recurrent infections, closely resemble the Chédiak-Higashi syndrome in man (see p. 523). At about puberty, the affected mink develop a severe generalized periodontitis with gross deposition of dental calculus, diffuse inflammation of the gingiva and severe irregular loss of alveolar bone (Lavine *et al.*, 1976). These changes speedily lead to tooth loss within two or three years. The periodontal destruction appears to be in part due to infection with a virus (Aleutian disease: Eklund *et al.*, 1968).

Family **Canidae**

Canids in the wild state are entirely free from the generalized form of periodontal disease which is at times found in captive animals. The natural diet of these animals is fresh meat which they have to rend with their teeth. In the captive state in zoos, these animals are now usually fed on natural raw meat and cooked meat; previously the diet included greens, carrots and oatmeal. Mincing and cooking the meat alters its physical character and renders it more liable to cling about the teeth. However, as emphasized by Krook (1976), because meat is low in Ca and high in P, and liver even more so, a diet almost wholly of meat off the bone tends to produce hyperparathyroidism which induces a generalized osteoporosis (A. D. J. Watson, 1981). Wild carnivores compensate for the P excess by consuming raw bone, so raising their Ca intake.

In view of the foregoing, it is surprising that, according to Vosburgh, Barbiers, Sikarskie and Ullrey (1982), canids in American zoos are most commonly given a soft meat-based diet, which seems to imply that the meat is not on the bone. It is also surprising that Vosburgh *et al.* found that more plaque accumulated on the teeth of two wolves (*Canis lupus*) fed on such a diet than accumulated on the teeth of a pair of wolves fed a hard dry, extruded dog food for the same period of four months.

Some destruction of the alveolar bone occurs in wild canids but it is evidently uncommon as, out of 1157 specimens examined by Colyer, only 24 instances (2.1%) were detected. It is of interest to note that, in the wild state, the tendency to pockets seems to occur more frequently in certain varieties. For instance, in 25 specimens of the Tibetan fox (*Vulpes ferrilata*) five showed gingival pocketing.

Hounds are prone to develop periodontal disease. The following notes on the diet of hounds were given to Colyer by the master of a well-known pack: 'The expense of feeding sixty or more couples of hounds is great, and the diet depends upon the prices of the various ingredients. At the present time the best hard Scotch oatmeal and molassine hound meal guaranteed to contain 25 per cent of flesh are boiled together in a cauldron and, when cold and set, this is cut up with a spade and served to the hounds. Care is taken that no bones are served, as the hounds will fight to the death over a bone. With the exception of a few stallion hounds who may have separate kennels, the bulk of the pack are only used for hunting during the prime of life; when they have passed from four to six years they are too slow, and by that time their teeth have loosened and in many cases fallen out.' Here we have a diet entirely different from the natural food of the animals: as it is a diet which is likely to cling around the teeth, it is not surprising to learn that pathological conditions develop which lead to their early loss.

It seems to be generally agreed that gingivitis and periodontal disease are common in domestic dogs (Harvey, 1985). The consensus from the early descriptions is that the first stage is a marginal gingivitis associated with, or secondary to, the accretion of supragingival calculus that in many dogs leads to recession of the gum rather than to pocket formation, conditions that are interpreted as showing high resistance to the disease process (Figs. 24.26, 24.27). In some animals, especially in advanced age, pocket formation occurs with more rapid progression to tooth loss (Figs. 24.28, 24.29). The disease tends to progress in the maxilla more rapidly and extensively than in the mandible where the bone is more dense (Fig. 24.30). Massive concretions of calculus that commonly occur on the crowns of the teeth of domestic dogs were illustrated by E. Becker (1970, pp. 275–279). Coignoul and Cheville (1984) described the laminar structure of these concretions and identified many of the types of bacteria that become incorporated into them.

Wolves in captivity are liable to develop periodontal disease similar to that in the domestic dog (Fig. 24.31).

The nearest to what might be called epidemiological studies of large unselected dog populations is an account by Talbot (1899, 1913). Over a period of 20 years, he examined the mouths of dogs from several sources. Unfortunately the numbers of animals are not specified but he gave the following percentages of animals affected by what seems to have been periodontal disease of at least moderate severity, with supragingival calculus and pocketing: 25% of dogs exhibited at shows and aged one to four years; 25% of stray dogs from the Chicago streets, estimated to be up to four years of age. Of stray dogs estimated to be aged

Fig. 24.26. *Canis familiaris* (domestic dog). RCS Odonto. Mus., G 145.5. Periodontal disease is more advanced in the maxilla than the mandible. In the mandible, the alveolar crests are thickened.

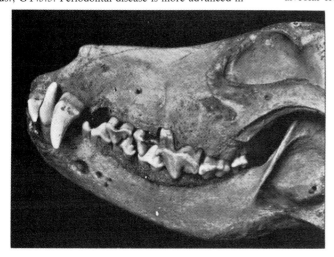

Fig. 24.27. *Canis familiaris* (domestic dog). RCS Odonto. Mus., G 145.9. The gingiva and part of the rhinarium are *in situ*. There is advanced periodontal disease. The gum is thickened and slightly nodular.

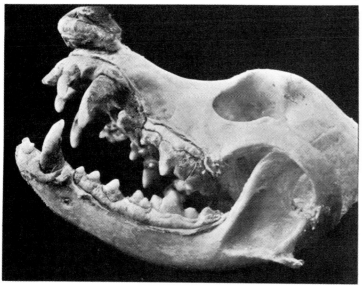

12 years or over, 95% were affected and much more severely. Finally, all of 27 pet dogs placed in hospital for treatment, mostly for severe chronic illnesses, had gross periodontal disease; many frankly suppurating teeth could be removed with the fingers.

Gad (1968) made a careful study under anaesthesia of 62 mongrel dogs ranging in age from three months to 12 years at an animal clinic. None suffered from any evident systemic disease. He recorded the periodontal status in terms of a periodontal index which expresses the severity of the periodontal disease including pocketting, a debris index which assesses the soft deposits present on the teeth and a calculus index which assesses the calculus deposits. These indices were adapted from studies in man. Healthy gingiva was found in the two youngest dogs; all the rest showed much soft deposit and calculus was present in 50. There was a statistically significant increase in periodontal disease severity with age with exfoliation of teeth or near exfoliation in the oldest animals.

Hamp and Lindberg (1977) found severe periodontal disease with many localized deep, so-called vertical, pockets at autopsy in seven out of 12 dogs aged three to 14 years of various breeds which suffered from a variety of severe disorders, malignant tumours, myocardial necrosis, meningitis and uraemia. Such a group is, however, far from representative of the canine population at large.

Among 162 adult dogs of a variety of breeds and ranging in estimated age from one to 12 years, examined at autopsy, Hamp, et al. (1975, 1984) found that 64% showed periodontal disease with bone loss. Of the 98 animals over the age of five years, 80% had severe degrees of the condition. It was not recorded to what extent this was a selected group of animals; that is, whether they were sick animals or, for instance, a more random group of stray animals from the streets. Periodontal disease was most common among the 26 poodles and was rare in the 16 long-jawed German Shepherd dogs (Alsatians). In a similar study, Smith, Zontine and Willits (1985) examined 24 dogs ranging in age from two to 18 years and of a variety of breeds. All the dogs had disease ranging from mild gingivitis to severe periodontal disease, the degree of which roughly correlated with age. Either dental plaque or calculus was present in all.

According to Wright (1939), the prevalence of periodontal disease is greatest in pekinese, pomeranians and spaniels and comparatively low in the bulldog and mastiff. This distribution does not suggest a regular relationship with short-jawed breeds that have crowded and even irregular tooth arches.

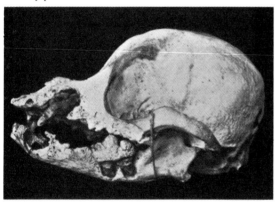

Fig. 24.28. *Canis familiaris* (domestic dog). RCS Odonto. Mus., G 145.8. Advanced periodontal disease with much destruction of bone. The remaining alveolar bone is coarsely pitted.

Fig. 24.29. *Canis familiaris* (domestic dog). RCS Odonto. Mus., G 145.6. Advanced periodontal disease with much destruction of bone. Much calculus on the crowns of the teeth.

Studies of colonies of beagles show a high prevalence of gingivitis, which is one of the reasons for the animals' popularity for experimental periodontal disease (p. 527). In a colony of 40 beagles aged one to eight years, Hull, Soames and Davies (1974) found that all showed typical gingivitis with generalized redness, swelling and bleeding of the gingival margins and most of the animals showed alveolar bone loss and periodontal disease in the form of pocketing involving particularly the buccal aspects of the cheek teeth and extending to the root furcations at an early stage. Two six-year-old dogs had lost teeth as a result of periodontal disease.

Among the findings of Sorensen, Löe and Ramfjord (1980) in respect of 74 beagles, one to 12 years of age, the solid masses of supragingival calculus on the buccal surfaces of the cheek teeth were commented on; subgingival calculus was less common and was usually an extension of the supragingival calculus into any periodontal pockets that were present.

Page and Schroeder (1981) found a much lower prevalence of periodontal disease in a similar colony of 2000 beagles maintained as a research stock. Although gingivitis was present in all, only 15 showed gingival pocketing and loss of the bone of the alveolar crests as determined by radiography. All the animals were adult but further information about age was not given. Among the 15 animals, the range of severity was considerable; in several only a few teeth were affected.

The difference in prevalence between these three different groups of beagles may in part be accounted for by differences in criteria for distinguishing gingivitis from periodontal disease and are likely also to be affected by differences in diet and in particular the carbohydrate content thereof. All three groups were maintained on commercial dog food which

Fig. 24.30. *Canis familiaris* (domestic dog). RCS Odonto. Mus., G 145.7. Irregular distribution of periodontal disease, an upper molar being but slightly affected.

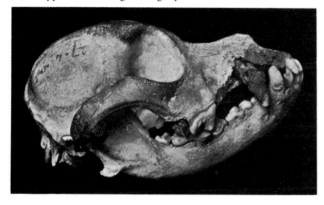

Fig. 24.31. *Canis lupus* (wolf). Captive. RCS Odonto. Mus., G 142.2. Severe periodontal disease with irregular bone destruction. Several teeth have been lost and the coarse pitting of the bone surface suggests widespread infection.

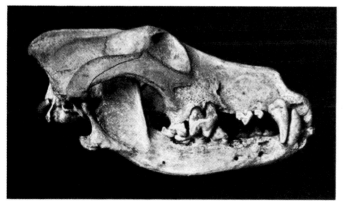

included dog meal and 'pelleted dog chow' which appears to consist mainly of meat protein and fat, cereal, barley malt and molasses. It seems that none received bones or fresh meat. It is an interesting sidelight on what even experimental biologists consider to be a suitable diet for a domesticated carnivore that one experimentalist referred to the dog chow as the 'normal diet'.

Even if fresh meat is included in the diet of dogs, account must be taken of the fact that they develop bleeding and swelling of the gum, as well as systemic toxic effects, if they receive excess vitamin A (Rodahl, 1952, cf. cats, p. 544). As already mentioned (p. 546), a diet mainly of meat or liver is liable to be grossly imbalanced in respect of calcium and phosphorus, namely calcium deficiency with phosphorus excess, and so tends to produce generalized osteoporosis. Henrikson (1968) showed that the osteoporosis produced in this way in dogs occurred early in the bone of the periodontium and thereby led to the more rapid transition of chronic marginal gingivitis to periodontal disease with resorption of the crestal bone. However, unlike in periodontal disease arising predominantly from food stagnation, the interdental crests were not affected.

A study by Allo (1971) of the jaws of dogs from prehistoric Maori sites in New Zealand suggested that periodontal disease associated with tooth wear was common in the South Island populations and almost non-existent in those on the North Island, which could be presumed to be of similar age structure; that is, were not discordant groups of young dogs only. Allo concluded that the diet of the two populations, and probably that of the associated Maoris, was different. However, it does not seem possible to infer more than that the diet of the North Island dogs contained enough abrasive material to wear the teeth.

Necrotizing ulcerative gingivitis occurs in the dog. Mikx and Van Campen (1982), by inoculating the gingiva of unaffected dogs, showed that the condition is transmissible. However, successful transmission was achieved only if the host dogs were pretreated with a gluco-corticosteroid which diminished their immune response; this accords with the view that acute ulcerative gingivitis only occurs in dogs that are not otherwise in good health.

An unusual type of organism that appears to be capable of producing acute gingival ulceration is the *Actinomyces* or organisms of similar character. Ginsberg and Little (1948) isolated what they categorized as *Actinomyces asteroides* as the cause of an acute severe ulceration of the gingiva around a mandibular canine in a puppy, associated with ulceration of the lips. This organism now seems to be re-classified as *Nocardia asteroides* (Jones and Hunt, 1983, p. 643).

Order CETACEA

Necrotic stomatitis with multiple gingival abscesses occurs in dolphins (*Tursiops truncatus*) in captivity (Colgrove *et al.*, 1975) but usually as part of more widespread infections.

Order ARTIODACTYLA

Periodontal disease, often of a severe character, is frequently seen in artiodactyls kept in captivity. Among 1237 skulls of bovids and 247 of cervids from the wild, Colyer found no instances of the generalized destruction of the tooth sockets that is common in the captive equivalents. Among the bovids, there were only nine instances of localized destructive lesions and among the cervids only one. Irregularities in position of the teeth were common, but usually the bone in the region of the irregular teeth was perfectly normal. It would therefore seem that the food obtained by the artiodactyls in the wild state is not liable to cause injury to the gingiva.

Family Suidae

In a study of 137 feral pigs, that is a population of animals living in a wild state but derived from domesticated stock, in California, Barrett (1978) found that all the animals over 4 years of age had periodontal disease severe enough to produce abscesses (Fig. 24.32). The first stage appeared to be wear and the impaction of bits of twig and acorn husks between the teeth and into the gingival crevices. Multiple periodontal abscesses leading to tooth loss were common, particularly affecting the first and second molars, and sometimes spreading beyond the jaws into the bones of the skull. Barrett concluded that deterioration of the dentition and the associated infections were the most common cause of mortality among the adult animals.

Samuel and Woodall (1988) found less severe periodontal disease in 107 feral pigs in Australia. Theirs is an important study because a comparison was made with 50 domestic pigs and because the heads were examined both before and after skeletalization. The disease was limited to the cheek teeth and was present in a similar moderately-severe form in 24% of the feral and 23% of the domestic. However, the groups were not age-matched; the proportion of fully adult animals (over 22 months of age) was much higher in the feral group. Nevertheless, if comparison is restricted to the 13 fully-adult feral animals and eight fully-adult domestic, the prevalence of the severe grades of the disease was about 37% in both groups. The prevalence of gingivitis without the involvement of the bone was much higher in the domestic pigs.

Gingivitis in at least a mild form appears to be common in the domestic pig; Spouge (1984) was unable to find any specimens free of chronic inflammatory change histologically in the marginal gingiva among commercially-slaughtered pigs.

Young laboratory-maintained miniature pigs (Chapter 21) accumulate much dental calculus, sometimes almost covering the crowns of the teeth, but notably associated with only mild gingivitis. Weaver (1964) showed that this calculus is composed mainly of calcite, a different complex calcium phosphate from the hydroxyapatite that is the principal constituent of dental calculus in man and other species. Weaver pointed out that calcite may induce less tissue reaction than hydroxyapatite or may be associated with a less-pathogenic bacterial flora. Nevertheless, these animals by the age of 10–12 years do develop gross forms of periodontal disease with tooth loss (M. E. Weaver, personal communication).

Straw, bits of wood and other foreign bodies are apt to become wedged between the teeth of domestic pigs and to cause localized periodontal tissue destruction (Harvey and Penny, 1985).

Families **Bovidae** and **Cervidae**

Periodontal disease is common in bovids and deer kept in captivity, and specimens vary in the way the disease progresses. In many of the herbivores, the bone of the maxilla is dense and this seems to restrict the spread of the disease. The skull of a captive thamin deer (*Cervus eldii*) shown in Figure 24.33 is a good example of the localization of the disease. Although there is a great deal of destruction of alveolar bone in the vicinity of the molars, it is more or less limited to the interdental regions. Furthermore, the bony surfaces where the bone has been lost are rounded and compact, not pitted or rarefied. This type of bone destruction is seen where the lesion is due to food packing without any marked added infection.

On the other hand, instances occur where the infection enters the bone and spreads diffusely, as in the duiker shown in Figure 24.34. The periodontal disease affecting the cheek teeth on the right is associated with a suppurative process which has expanded the tooth-bearing part of the lower jaw and has spread to the vertical ramus, destroying the coronoid process. The part of the process that remains is necrotic and eroded and surrounded by an involucrum of new periosteal bone (Chapter 23, p. 512). The specimen also shows some periosteal thickening of the bone at the posterior root of the zygomatic arch so evidently the suppurative process around the coronoid process was spreading beyond the limits of the lower jaw.

The great height of the crowns of the teeth in herbivores, with much of the crowns below the level of the gum, would tend to provide a pathway for infection to pass along the sides of the teeth to reach deep into the bone. This could explain how gingival infection in the herbivores can extend within the jaws even though the bone of the alveolar process is unusually dense.

An example of severe periodontal disease in the region of the upper molar teeth in a gazelle from the wild is seen in Figure 24.35.

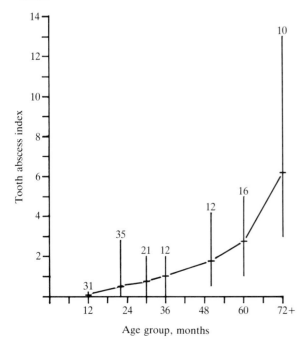

Fig. 24.32. The relationship of the tooth-abscess index to age group for 137 feral pig skulls. Horizontal bars indicate the mean and vertical bars indicate the range for each age group. Sample size is given above each range line. The index was based on the number of teeth abscessed in both upper and lower jaws. Each tooth involved was given a value of one if still present and a value of two if missing as a result of an abscess. The total gave the index value for the skull.
From: Barrett, R. H. (1978). By kind permission of the editor.

552 Section 4: Other Disorders of Teeth and Jaws

Sheep and goats
INCISIVE APPARATUS

The incisive apparatus of sheep, upon which they depend for grazing, is essential for their well-being. Two conditions affect the apparatus and eventually destroy its function and therefore present a serious problem to farmers. One is excessive incisor tooth wear already dealt with in Chapter 22. The other is periodontal disease which loosens the incisors and leads to exfoliation of them one by one, a condition known as broken mouth. This is common enough to be a serious problem to farmers (Gunn, 1970). It has been much investigated and a considerable amount of literature has grown up (Spence and Aitchison, 1985).

When ewes are judged to be past their prime as producers of lambs, they are culled; that is sold for slaughter. Normally, ewes can be expected to produce lambs for five or six seasons; to have to cull prematurely, say after four seasons, is uneconomic. Herrtage, Saunders and Terlecki (1974) found defective dentitions were one of about 6 common factors that in various combinations contributed to the decision to cull. The findings of Richardson et al. (1979) in respect of 481 culled ewes were similar; only two had healthy dentitions.

Thurley (1987) found histological evidence of acute gingivitis related to the deciduous incisors in the majority of certain groups of lambs by about 20 weeks of age. By about 35 weeks, a more chronic type of inflammation appeared to be established.

The periodontium in the incisor region of the sheep and other bovids is highly specialized; the factors concerned are so special that it cannot be taken for granted that the so-called periodontal disease that affects the incisors can be properly considered as merely a modification of the type of periodontal disease so far considered.

It is necessary to describe the incisor apparatus at some length because in recent years some features have been brought to light which are not yet particularly well-known.

In the sheep, as in bovids in general, there are no upper incisors; the three lower incisors and the incisiform canine occlude with a dental pad, composed of fibrous tissue covered with keratinized epithelium, which replaces upper incisors. For convenience, the common practice will be adopted here of referring to the eight incisiform teeth as incisors. There is an additional feature that appears to have escaped notice because it is an entirely soft-tissue feature totally absent from skulls; behind the lower incisors, the gingiva is raised and widened to form a pad with a flattened surface at about the level of the tips of the incisors (Fig. 24.36). This dental pad resembles the upper one but it is not so wide antero-posteriorly.

Cutress (1972) gave the first full description of the lower dental pad but at least two earlier writers came close to identifying it; for instance, it can be seen in one of the illustrations in Box (1935) and Hitchin (1948) referred to sheep in which the upper dental pad impinged on the gum behind the incisors. Although only described in sheep, this lower dental pad is also present in the ox and may well be present in other ruminants.

When they first erupt, the incisors are completely covered with enamel, but, with use, the incisive edge becomes worn to form a flat facet with dentine exposed

Fig. 24.33. *Cervus eldii* (thamin deer). Captive. RCS Odonto. Mus., G 156.2. Advanced periodontal disease with destruction of bone mainly interdentally.

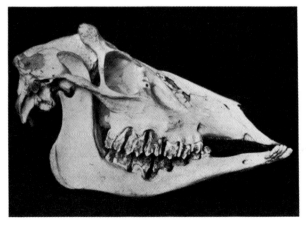

at the centre. Because the enamel on the lingual surface is thinner than on the labial side and therefore less resistant to wear, the wear surface slopes a little, making the labial edge sharper than otherwise would be the case.

The fully functional state as described by Northey, Hawley and Suckling (1975) is shown diagrammatically in Figure 24.37. It seems that, in grazing, grass is held between the upper pad and the incisors and lower pad combined; the sheep then makes a nodding movement of the head (Markham and Stewart, 1962), usually downwards; the grass seems to be torn by this action rather than cut through by the teeth. As can be judged from Figure 24.37B, this would seem to exert a tilting force on the forwardly-inclined incisors, all the more so if, as Cutress (1972) pointed out, the cheek teeth, during grazing with the incisors, are out of occlusion because the cheek teeth only come into occlusion on one side at a time, when the mandible is swung considerably laterally.

Northey et al. concluded that, when the incisor is pressed against the pad under the circumstances just described (Fig. 24.37 A), the direction of the resultant force, the line BC through what they call the centre of anchorage at X, is much more nearly through the axis of the tooth than would be supposed: in other words, the tilting thrust is less than supposed. They further argued that, when the incisors become much worn and the relationship with the upper pad becomes more like that depicted in Figure 24.37B, the resultant force is no longer through the anchorage point X and there is less tilting force. The load on the teeth is diminished

Fig. 24.34. *Sylvicapra grimmia* (Grimm's duiker). Captive. RCS Odonto. Mus., G 154.1. Destruction of alveolar crest bone is localized to M_1 and P_4 but is associated with a suppurative process that has expanded the jaw and spread to the coronoid process which is necrotic and surrounded by an involucrum of new bone.

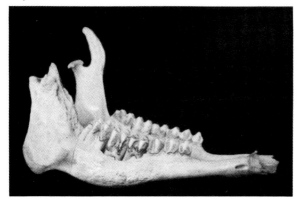

Fig. 24.35. *Gazella subgutturosa* (goitered gazelle). ♂. Wild. BMNH 1914.3.23.1.
A: Palatal view showing a loss of bone around right M^2 and loss of M^1.
B: Right lateral view.

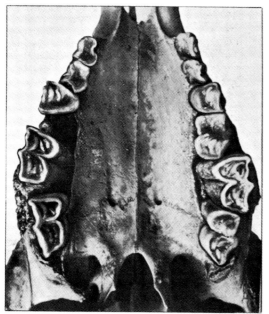

A

B

further by the associated greater contact between the upper and lower pads.

A number of factors can disturb the incisor mechanism and bring about a relationship more like that depicted in Figure 24.37 C in which there is pad-to-pad contact but the incisors pass in front of the upper pad which thus occludes with the lingual surfaces of the teeth instead of their incisive edges. In undershot jaw, the relationship may be as in Figure 24.37 C from the beginning; this could be because the upper jaw is disproportionately short or the lower jaw disproportionately long. The relationship in the Figure could arise because the periodontium gives way in the face of the tilting forces to which the teeth are subjected so that they tilt or drift forwards. This would be more likely to happen if wear of the incisors does not occur or there is some recession of the alveolar crest, as an age change or as a response to gingivitis. This would increase the height of the crowns and increase the leverage as analysed in Figure 24.37 C. Once such a state of affairs had arisen, it is likely to progress; with more loss of alveolar bone, especially labially, the leverage would increase and it is easy to envisage eventual rapid loosening, deterioration and finally exfoliation of the teeth.

Spence (1978), who studied this matter mainly by cutting thick sections through the incisor region and examining the arrangement of the collagen fibres of the periodontal ligament which passes from tooth to bone, did not fully accept the analysis of Northey *et al.* as depicted in Figure 24.37. The retrusion of the mandible to achieve adequate contact between incisors and the upper dental pad needs to be verified.

According to Spence, the direction of force exerted on the tooth is G in Figure 24.38A, which tends to rotate the tooth forward around a fulcrum X and is resisted predominantly by the periodontal ligament fibres depicted by BB. Spence believes that the collagen fibres of the ligament (Fig. 24.38 B) are well able, in a state of health, with the assistance of the buffering action of cushion-like vascular loose connective tissue and adipose tissue at the apex of the root at C in Figure 24.38A, to withstand these tilting forces.

The arrangement of the collagen bundles, as described by Spence, in the periodontal ligament and their unusual length in the cervical region (Fig. 24.38A and B) must allow considerable lateral movement of the teeth about the fulcrum X. The fibre bundles are much shorter below the level of this fulcrum. Mobility of the teeth seems to be an essential part of the incisor mechanism in the sheep and in some way, perhaps because movement absorbs the force, enables them to

Fig. 24.36. *Ovis aries* (sheep). Incisor wear and occlusion.
A: A paramedian section through the incisor region of a sheep. The incisor teeth occlude with the rostral margin of the upper dental pad which, for about half its width, is in contact with the lower dental pad, a shelf-like extension of the lingual gingiva.
B: View from above of the lower dental pad in a 5-year-old sheep; the lingual margin is indicated by one set of arrows and the contoured gingival margin by arrows at M.
From: Cuttress, T. W. (1972). By kind permission of the author.

A

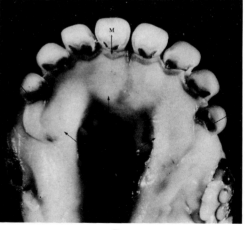

B

Fig. 24.37. *Ovis aries* (sheep). Incisor wear and occlusion.
A: Lateral view of the skull of a young sheep with the incisor region depicted as a paramedian section. The sheep is able to grip the upper dental pad with the tips of its incisors. The relative heights of the crown and root parts of the teeth are indicated. The bite force BC, analysed in the inset, exerts but little leverage on the teeth.
B: As a result of wear of the incisors, contact occurs between the upper and lower dental pads in biting, which reduces the load on the teeth.
C: The exposed crowns of the incisors are higher in proportion to the root because there has not been compensatory wear. In order to achieve pad-to-pad contact, the lower jaw is brought forward and the leverage on the teeth is great. Irreparable damage is done to the tooth anchorage system and there is loss of alveolar crest bone.
  A similar state of affairs would exist if the lower jaw were developmentally undershot.
From: Northey, R.D., Hawley, J.G. & Suckling, G.W. (1975).

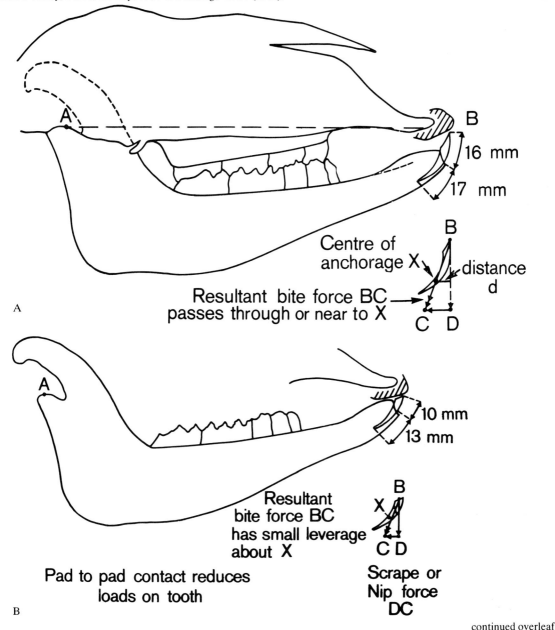

continued overleaf

withstand the stresses exerted on the teeth in grazing. The probable importance of wear of the incisors, with consequent shortening the teeth as levers, in maintaining the integrity of the incisor mechanism has already been referred to. In accord with this hypothesis, Hatt, Lyle-Stewart and Cresswell (1968) simulated natural wear of the incisor teeth by grinding the incisive edges twice yearly in flocks of sheep that were particularly susceptible to develop broken mouth. Over a period of three years, although they considered that the incisor or dental-pad relationship had improved, no convincing evidence of prevention of broken mouth was forthcoming. Nevertheless, this procedure in the extreme form of grinding the incisors of three-year ewes down to the gingival margin is said to have become widespread, especially in Australia (Spence, Hooper and Austin, 1986). Steps are being taken to ban the procedure on humanitarian grounds (Denholm and Vizard, 1986).

We have already referred to the uncertainty that exists about what is to be regarded as the normal relationship of the incisors to the upper dental pad. On the whole, the evidence seems to favour occlusion with the pad at or near its anterior edge (Morris, Whitley, Orr and Laws, 1985). However, in all of a sample of 12 ewes of the Scotch black-faced sheep, a mountain breed, Hitchin (1948) found that the incisors met the middle of the pad; he called this Type 1 occlusion. All these ewes were free of periodontal disease and the wear surface formed only a slight bevel; on the other hand, only 5.1% of 57 ewes of the lowland Cheviot breed had this type of incisor relationship; in the rest, the incisors occluded with the anterior edge of the upper dental pad and the wear surface was very oblique so that there was a sharp chisel-like bevel (Type 2). Various degrees of gingivitis and abnormal loosening of the incisors were present in nearly all these Cheviots and some teeth had been lost. The ewes in both groups were of similar ages and were from farms in the same valley. Hitchin found examples among other flocks of sheep of what he called Type 3 occlusion in which the incisors passed in front of the upper dental pad which did not really occlude with the incisors at all but closed instead on what he called the gum behind the incisors. This Type 3 must surely be regarded as an abnormality or malocclusion; it corresponds to what farmers call undershot jaw (termed overshot in New Zealand; Chapters 13 and 22; p. 295 and 492). The converse, when the incisors are so far back (or the lower jaw is so short) that they pass behind the upper pad or only occlude with its posterior part is known as overshot jaw, or in various districts as sow-mouthed, rat-mouthed or parrot-mouthed, terms which are also applied to cattle. Animals with this type of occlusion would only be able to crop long grass.

In skulls, the presence of wear surfaces extending over most of the lingual surfaces of the incisors would appear to be an indication of the existence of a Type 2 or 3 relationship with the dental pad.

There is evidence that these types of occlusion are heritable (Hitchin 1957).

It would be going too far to suggest that these different types of occlusion have been bred for artificially or have arisen by natural selection to suit types of farmland. However, whereas sheep with Types 1 and 2 occlusion would be able to crop short grass and also anything else they were offered, e.g. whole turnips as a winter supplement, those with Type 3 occlusion would almost certainly be unable to crop grass very short although they might be able to deal with turnip; thus they are likely to survive adequately on valley farms and especially Southern English farms where lush grass is likely to be available but, transferred to a hill farm and left largely to fend for themselves on the sparse

Figure 24.37 continued

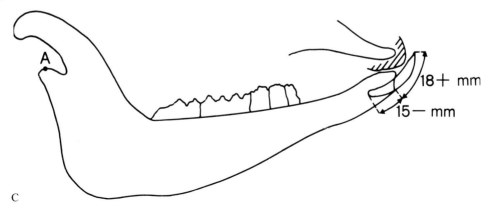

grass of a hill-side, such a dental handicap would tend to reveal itself (J. Duckworth et al., 1962).

It seems probable that the type of incisor occlusion can change during the lifetime of the animal and this seems more likely to happen if wear of the teeth does not occur. Suckling (1975) made serial photographs of the incisor relationship of a single sheep from the age of $3\frac{1}{2}$ to $5\frac{1}{2}$ years. At first, the teeth were in Type 1 relationship but evidently moved gradually forward because by $4\frac{1}{2}$ years they were occluding with the anterior margin of the dental pad and at the same time the height of the exposed part of the crown beyond the labial gingival margin had increased and there was a marginal gingivitis. At $5\frac{1}{4}$ years, the incisors were well clear of the dental pad and had increased further in height because there had been very little wear and gingival change had continued to expose more tooth. By $5\frac{1}{2}$ years, one of the incisors had been exfoliated. This certainly illustrates the remarkable speed with which the deteriorative process can occur and the inference may be drawn that absence of wear played a part. Details of the diet were not given.

Although attractive, the foregoing analyses leave out of account what may be happening to the incisors during the grinding that takes place between the cheek teeth, not only in the preparation of the bolus for swallowing but particularly during rumination. Markham and Stewart (1962) pointed out that a sheep spends eight to nine hours a day in this activity; this may involve as many as 39 000 grinding movements which, they said, include fore and aft movements of the jaw as well as scraping of the incisors across the pad. These workers regarded the resulting jiggling of the incisors during rumination as an important factor in the production of broken mouth.

Speculation about the mode of use of the incisive apparatus in sheep is increased by consideration of the finding by I. H. Simpson and Croft (1973) that 5–6-year ewes whose incisors had been clipped down to the gum seemed to suffer no gross disadvantage compared with a control group of their sound-mouthed sisters in terms of body-weight maintenance, fleece weight and lamb production.

It will be evident that the jaw mechanism in sheep is of considerable complexity and that there is room for further investigation. A further complicating feature which is but little referred to is that in the sheep the two halves of the mandible are not united but are held together by fibrous tissue which permits at least a little movement at the symphysis. In bovids and equids, as in the rodents, there is functional as well as spatial separation between the anterior grazing teeth and the posterior grinding ones; while the teeth in one of these two compartments are functioning, those in the other are not. Both must be taken account of in interpreting morphology. Murphy (1959) described anterior and posterior articular facets on both the cranial and mandibular components of the mandibular joint in the sheep which would correspond with two different grazing and grinding positions of the mandible.

Cutress and Ludwig (1969), from a review of the literature and their own observations, concluded that there is an acute form of periodontal disease, which is best termed necrotizing ulcerative gingivitis, which affects both the incisors and molar dentitions, and will be dealt with separately (p. 562); they concluded that the changes in the incisor periodontium that resulted in broken mouth were probably not primarily inflammatory but a degenerative process due to some sort of systemic metabolic disturbance. A periodontal disease of this kind, for which the term periodontosis has been employed, is well-authenticated in man.

Subsequently, Cutress (1976) studied the histology of the lesions and, noting in particular the presence of plaque and its mineralized form dental calculus, concluded that, as in the majority of forms of periodontal disease in man and other animals, the condition is plaque-initiated. In the sheep, the microbial flora involved, which has not been fully investigated, appears to be of low-grade pathogenicity, the host response is unusual in many of its details and not particularly effective in combatting the bacterial attack (Cutress and Schroeder, 1982).

Spence, Aitchison, Sykes and Atkinson (1980) and Atkinson, Spence, Aitchison and Sykes (1982) agree with this. They found that sheep, both from farms where broken mouth does not occur and farms where broken mouth appears to be endemic, develop a chronic gingivitis in relation to the permanent incisors almost as soon as they erupt, and that this is sometimes preceded by an acute phase. The chronic gingivitis is associated with subgingival plaque and shows little progression except on broken-mouth farms were many ewes develop severe invasive periodontitis with large amounts of plaque and destruction of the collagen bundles in the periodontal ligament that are so important for supporting the incisors against grazing forces; there is consequent loosening and the condition progresses until the incisors begin to be lost, one at a time. This analysis leaves unexplained why the gingivitis progresses to tooth loss only in certain flocks. It leaves a role, of course, for increased trauma induced by certain relationships between incisors and the tooth pads or by certain types of feed, such as whole turnips, to play a part in hastening loosening and loss of the teeth. Page and Schroeder (1982) have made valuable

558 Section 4: Other Disorders of Teeth and Jaws

additions to the analysis of the condition by comparing with a particular type of periodontal disease, juvenile periodontitis, in man. They believe that some of the features of the incisor periodontal disease in sheep derive from the pathogenic characteristics of the associated microbial flora which so far has not been much studied.

Page and Schroeder (p. 234) pointed out that, as a result of the shallowness of the incisor sockets in proportion to the length of the roots and the distance of the alveolar bone from the lingual aspect of the coronal half of the root (Fig. 24.38A), periodontal pockets involving one-half to two-thirds of the root length can form without resorption of alveolar bone.

In the literature on excessive incisor wear (gummy mouth, Chapter 22), comments on the gingival condition of the affected animals are few. Perhaps this can be taken as evidence that the teeth affected by wear remain free of periodontal disease; it could support the view that moderate degrees of wear are beneficial to the mechanism of the periodontium. Hardly any papers discuss both broken and gummy mouth; in fact the two conditions appear to have attracted different sets of workers. There is even the possibility of some

Fig. 24.38. *Ovis aries* (sheep).
A: Diagrammatic representation of forces applied to an incisor of a sheep. MJ, mandibular joint region; T, force applied by contraction of muscles around MJ; R, reactant force to T; G, force applied in grazing; A, component force of R down long axis of tooth; B, component force of R at right angles to tooth axis; X, tooth fulcrum; BB, force balancing the applied force B; C, highly-vascular loose adipose connective tissue which acts as a cushion.
B: Diagrammatic representation of the principal collagen fibre groups in the cervical part of the periodontal ligament as seen in a transverse section. Various fibre groups are numbered. Some fibres, 1, extend from the bone, pass between the teeth and in front of them to form, with fibres 9, a sort of horizontal sling. 2 indicates long ligament fibres that pass from the bone to the lingual surfaces of the teeth. 8 indicates fibres that link the teeth to each other.
From: Spence, J. A. (1978). By kind permission of the editor.

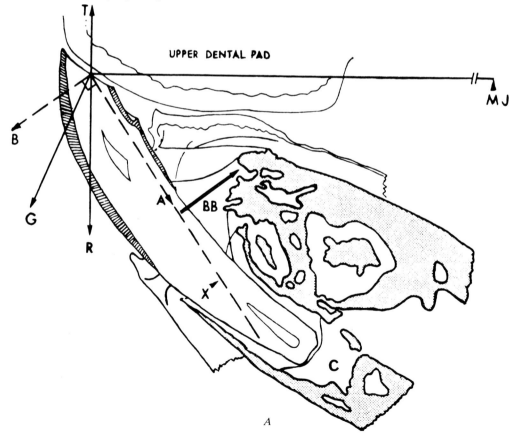

confusion in the literature; a paper by C. Richardson *et al.* (1979) places 'incisors loose, missing or reduced to stumps' in the category of broken mouth. An exception is the work of Orr *et al.* (1979) who described a flock of sheep which suffered from remarkably rapid incisor wear which was concurrent with periodontal disease and tooth loss to a not fully specified extent. However, these animals appeared to be suffering from a widespread systemic osteoporosis. A later investigation by Orr, Christiansen and Kissling (1986) into a possible relation between incisor wear and periodontal disease highlighted the complexity of the problem.

In seeking the ultimate cause of broken mouth, consideration needs to be given to evidence that it is a disorder that has arisen only in recent times. F. T. Harvey, an experienced veterinary surgeon with a farming practice, in Cornwall, writing in 1920 mainly about periodontal disease in the horse, mentioned that the disease commonly affected the incisors of the ox but he had been unable to find any in the sheep. This statement seems bound to have embraced broken mouth in sheep. As far as the scientific literature is concerned, references to the condition do not appear until about 1943. According to Hitchin and Walker-Love (1959), in 1921 only about 5% of five-year-old ewes in a Scottish flock had broken mouth but, in 1950, 37% of the 5-year ewes on the same farm were affected. All this suggests that broken mouth in sheep has only become a serious widespread problem in the last few decades. One factor in this may have been more intense breeding programmes in which, in order to secure some commercial advantage such as better wool or twin-lambing, lowland breeds bearing genes for under-shot jaw which may have been able to cope with a lush environment had been unintentionally introduced into hill flocks where the less-favourable type of occlusion had proved unequal to the environment.

A second possibility, suggested by the many

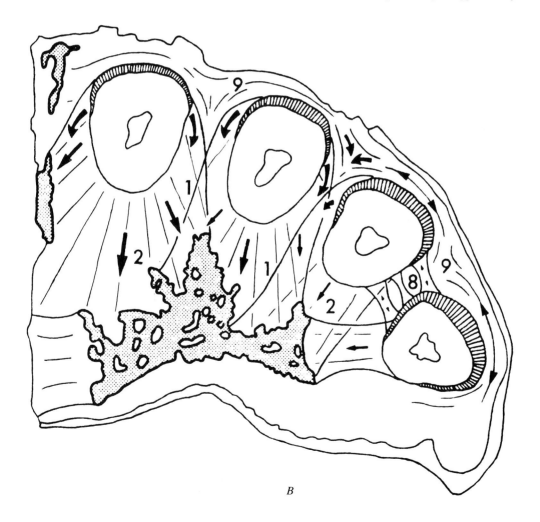

B

reports of dental plaque in association with the periodontal disease of the incisors, is that supplements in the form of artificial foods, sheep cubes for example, and silage plays a part. The use of these seems to have been one of the changes in farming practice that has occurred in the past 40 years or so. The composition of sheep cubes no doubt varies, but basically they consist of some form of protein and a large proportion of cereal. One account (D. N. Jones, 1958) referred to starch-equivalent 66% and digestible crude protein 14%, together with vitamins and minerals. It thus seems highly likely that some fermentable carbohydrate is present, as there would be in silage, in the preparation of which molasses is commonly added. It seems unlikely that the sheep would use their incisors significantly in taking the cubes but the ingredients in the cubes would be part of the regurgitated bolus in rumination and, during rumination, the carbohydrate, if any, might have more prolonged contact with the incisors and lead to the formation of plaque on their crowns.

POST-CANINE REGION

In much of the literature on incisor-tooth loss in sheep, there is no mention of the cheek teeth. It is now gradually emerging that the cheek teeth are often affected by a type of periodontal disease that more closely resembles that of other species than that which leads to incisor-tooth loss (Cutress, 1976). I. G. Shaw (1981) pointed out that cheek-tooth disease tends to go unrecognized and neglected because it is difficult to examine that part of the dentition in the living animal. From his description of food packing between the teeth, pus exuding from the gum, abscess formation and occasional thickening of the tooth-bearing body of the mandible, it seems that, when periodontal disease affects the cheek teeth, it is much more in the nature of a suppurative inflammation than when the incisors are affected. According to Shaw, the effects on the well-being of sheep of deterioration of the cheek-tooth dentition from periodontal disease is much more severe than from broken mouth; affected ewes may deteriorate rapidly during pregnancy and lactation because they are unable to masticate the extra food required.

The foregoing refers mainly to lowland flocks, but Aitchison and Spence (1984) have shown that periodontal disease can be nearly as prevalent in hill-farm flocks. They found that about 60% of 478 abattoir specimens of predominantly hill-farm culled ewes, of mean age seven to eight years, showed cheek-tooth periodontal disease. They also established that incisor- and cheek-tooth disease tended to occur in association.

The main difference between lowland and hill sheep is probably that animals on low-ground farms remain economically productive for longer despite tooth loss because grazing is easier (Spence, Sykes, Atkinson and Aitchison, 1985).

Andrews (1981) described the effect of removal of diseased teeth from five adult sheep which were in poor general condition. Most suffered from various degrees of broken mouth and many of the cheek teeth were loose. Removal of the loosest teeth was followed by speedy improvement in general condition, which enabled the ewes to produce further lambs and so avoid being culled.

Although black deposits are common on most parts of the crowns of the teeth of mature sheep, apart from the occlusal surfaces (see above p. 524), dental calculus does not seem to be a predominant feature of periodontal disease. However, the severe periodontal disease J. R. Baker and Britt (1984) found among the sheep on the Scottish island of North Ronaldsay, Orkneys, was associated with thick deposits of yellowish-brown calculus on cheek teeth, even on the occlusal surfaces of the lower first cheek tooth; however, that tooth is not normally in occlusion. The periodontal disease in these animals may be related to the unusual nature of their diet which consists largely of seaweed they gather at the shore line. This is inevitably associated with taking sand into the mouth which could damage the gingiva.

Porter, Scott and Manktelow (1970) described periodontal disease affecting all parts of the dentition in sheep kept on pastures that had been dressed with superphosphate. The changes, which included food impaction between the molar teeth, were greater in the sheep kept on pasture that had received a heavier dressing. Incidentally, these workers reported some improvement in the periodontal disease following selenium administration.

Rabkin (1946) examined the mouths of 36 freshly-slaughtered sheep ranging from six months to an estimated 14 years. No morbid oral changes were found in the younger animals but, in those estimated to be over six years of age, some cheek teeth were lost, many others were loose with traces of grass impacted between them and associated with deep gingival pockets and alveolar crest destruction. In some instances, there was irregular excessive wear of the occlusal surfaces. Radiographs of transverse slices through the mandible illustrated these changes. In the oldest animals, a form of osteoporosis was present with deposition of fat in the enlarged marrow spaces.

Similar changes were observed in three goats estimated to be aged between five and seven years.

Periodontal disease in feral goats associated with excessive tooth wear is referred to in Chapter 22 (p. 493).

Vigal and Machordom (1987) recorded that about 90% of Spanish ibex (*Capra pyrenaica*) over the age of ten years from the wild from two National Game Reserves in Spain showed the bony changes of periodontal disease. However, a difference between the two groups that could not be explained was that, whereas the changes were common in one group by the age of three years, the other group seemed not to develop the disease until six years of age.

There is some evidence that periodontal disease affected the post-canine dentition of the domesticated sheep of Iron Age times (Brothwell, 1981).

Cattle

In spite of a comment by F. T. Harvey (1920) to the effect that periodontal disease commonly affects the incisors of cattle (*Bos taurus*), the literature contains no evidence that the disease is widespread. Garlick (1954b) made no mention of periodontal disease in his list of the dental disorders encountered in 7480 abattoir cattle in the USA.

In many regions of Brazil and neighbouring Bolivia, a particularly severe form of periodontal disease known as cara inchada (swollen face) is endemic among the young zebu (*Bos indicus*) or crossbred cattle (Dobereiner, Tokarnia and Rose, 1976; Blobel *et al.*, 1987). Only the deciduous molar teeth are affected and the disease is confined to lowland regions where the pasture is predominantly guinea grass (*Panicum maximum*). If affected animals are transferred to higher country, where guinea grass does not grow, unless they are in extremis the disease regresses. On the other hand, adult animals transferred from the disease-free high ground to the lowlands remains free of the disease.

The condition appears to begin as a mild gingivitis which begins about as soon as the deciduous teeth erupt and is common to both highland and lowland cattle. However, in approximately 25% of the lowland calves, gingival pockets develop associated with destruction of the alveolar crest bone beneath. Grain stalks become wedged between the teeth and a much more diffuse inflammation develops with some new formation of bone on the surface of the jaws. This inflammatory thickening gives rise to the name swollen face. In the later stages, the animals become very debilitated and many die. The incisors do not exhibit the characteristic features of the disease although sometimes they do seem to be affected by a more ordinary inflammatory process and maggots (*Cochliomyia hominivorax*) may infest the gingival pockets.

The cause is not understood. There is the possibility that the guinea grass contains irritant properties or harbours a particular pathogenic microbial flora to which by adult life the cattle have acquired immunity. However, Camargo, Veija and Conrad (1982) reported evidence that the forage in the affected highland areas is deficient in copper, cobalt, zinc and phosphorus, and that the provision of salt licks containing these elements, in particular copper, can diminish the prevalence of cara inchada.

Other bovids

Pekelharing (1974) examined the dried mandibles of 274 feral chamois (*Rupicapra rupicapra*) estimated to be over three years of age. They were descended from seven animals from the Austrian Alps which had been freed in the New Zealand mountains 60 years previously. Twenty per cent, with equal sex prevalence, showed the gross bony changes of periodontal disease, including tooth loss and the consequent occlusal dysharmonies. He pointed out that this was an underestimate of the true prevalence because, as the soft tissues were missing, the initial phases of the disease before bone involvement were not recorded. G. Niethammer (1971) had noted that the skulls of the males, but not the females, in the same feral population were larger than in the present European stock and also that there was a higher rate of tooth wear in the New Zealand animals. He also noted a number of examples of absence of cheek teeth but Pekelharing suggested that this was tooth loss from periodontal disease rather than congenital absence.

Localized periodontal disease with fibrous plant material wedged between some cheek teeth was found by Gainer (1982) in two skulls of Nyasa wildebest (*Connochaetes taurinus johnstoni*) from the wild. As these appear to be the only examples among over 1000 skulls, they probably do not signify more than sequelae of the occasional chance of food wedging between the teeth.

A single instance of gingival hyperplasia in the incisor region of a greater kudu (*Tragelaphus strepsiceros*) was described by Gewalt and Pauling (1971).

Deer

Kratochvil (1984) found changes indicative of localized periodontal disease in a collection of 1140 skulls of roe deer (*Capreolus capreolus*) killed in Bohemia in 1972–1977. Animals under one year of age were affected and there was an increase in prevalence with age. Thirty-six per cent of skulls estimated to be over 12 years of age were severely affected. Some teeth were lost from the disease with consequent irregular wear and over-eruption of remaining teeth.

Feldhamer (1982) found evidence of periodontal disease in two out of 191 skulls of the Japanese sika deer (*Cervus nippon*) that had been introduced into Maryland State, USA. The changes in both were gross so that the prevalence of lesser degrees of periodontal disease could have been greater.

The final disintegrative stages of wear with secondary periodontal disease were described by Schuh and Niebauer (1982) in four mandibles of park-maintained European elk (*Alces alces*) estimated to be aged nine, 10, 12 and 16 years. R. O. Peterson, Scheidler and Stephens (1982) found that 35% of wolf-killed American moose (*Alces alces*) estimated to be over 10 years of age had moderate or advanced periodontal disease. However, the value of these data is slight because the diagnostic criteria were not given.

Localized cheek-tooth periodontal lesions, mostly associated with tooth wear and food impaction between the teeth, were found by Henrichsen and Dieterich (1984) in about 22% of skulls of free-living mature musk-oxen (*Ovibos moschatus*) compared with more than 70% of skulls of captive animals. In interpreting localized periodontal disease and bone loss, it should be borne in mind that impacted accumulations of food in the mandibular buccal vestibule have been described in free-living deer (Couvillion, Nettles, Rawlings and Joyner, 1986). It seems likely that some of these accumulations commence as food impactions between the teeth but in any case they produce damaging ulceration of the oral mucosa.

Acute ulcerative gingivitis and necrotic stomatitis

Adult sheep in various parts of New Zealand are affected by an acute form of gingivitis which can, by ulceration and necrosis, involve the bone and periodontium generally, with chronic abscess formation and loss of teeth. The cheek teeth are affected predominantly but the incisors may also be affected. The condition was first described by Salisbury, Armstrong and Gray (1953) who isolated spirochaetes and fusiform bacilli from the lesions, which supports the view that the condition is essentially the same as the acute necrotizing ulcerative gingivitis met with, for instance, in dogs (p. 524) and man. The enzootic parodontal disease described by Hart and Mackinnon (1958) and Mackinnon (1959) in sheep seems likely to have been the same condition.

Spais, Lazaridis, Papasteriadis and Leontides (1969) described the outbreak of a similar acute disorder that affected about 25% of a flock of 115 sheep in Greece. Seven died. At autopsy, there was some evidence of generalized loss of mineral from the skeleton and the authors attributed the disorder to over-acidified apple pulp with which the animals were fed.

It may be that the periodontal disorder Anderson and Bulgin (1984) described as associated with the death of about 25% of a flock of 300 ewes in the United States was of the same character. The sheep became anorexic and emaciated with a terminal diarrhoea. At autopsy, all showed severely destructive periodontal disease of the cheek dentition with plant fibres, in particular the awns of cheatgrass (*Bromus tectorum*), packed within ulcerated gingival pockets. The inference is drawn that pastures should be managed so that this particular coarse grass is eliminated. McSweeney and Ladds (1988) demonstrated how damaging the seeds of spear grass (*Heteropogon contortus*) are to the gingiva and oral mucosa in general in sheep and goats but less so in cattle.

There is some evidence that the acute form of periodontal disease occurs in sheep on selenium-deficient land and the administration of selenium to flocks appears to diminish its prevalence (Cutress and Ludwig, 1969). For a fuller discussion of this condition, Page and Schroeder (1982) is recommended. It must be pointed out that selenium in doses likely to occur on land contaminated by industrial processes is toxic to grazing animals (Harr and Muth, 1972).

Reference was made in Chapter 23 (p. 576) to necrotic stomatitis in sheep due to infection of the mucosa, including the gingiva, with *Actinomyces bovis*. The relation of this to what is now called actinobacillosis, due to infection with *Actinobacillus lignieresi* of the soft tissues of the face and mouth (woody tongue when the tongue is affected) in sheep and cattle is by no means clear from the literature. As in actinomycosis, the tissues and the pus contain discrete rosette-shaped colonies of the filamentous organism (Shahan and Davis, 1942). It seems sufficient here to indicate that helpful descriptions of the condition are given by Tharp, Amstutz and Helwig (1980), Jones and Hunt (1983, p. 639), and Blood, Radostits and Henderson (1983, p. 670).

Necrotic stomatitis is the term that occurs in the literature (e.g. A. Murie, 1944, pp. 117, 160) for what seems to begin as a necrotic ulceration of the gingiva which may spread to the mucosa elsewhere in the mouth and sometimes spreads into the bone, leading to loss of teeth. It occurs predominantly in ungulates in the wild, both young and old. Murie mentioned examples in Dall sheep (*Ovis dalli*), deer (*Odocoileus*), American moose (*Alces alces*), American elk (*Cervus canadensis*), and caribou (*Rangifer tarandus*) (however, see the section on marsupials, p. 570). Although Murie concluded that the examples he saw in wild Dall sheep were caused by *Actinomyces bovis* (Chapter 23, p. 572), other instances have been regarded as examples of necrobacillosis, which implies that an organism variously called *Bacillus necrophorus* or *Sphaerophorus necrophorus* is responsible. There is also *Actinobacillus lignieresi*, already referred to as the cause of woody tongue in sheep, which appears sometimes to produce lesions of other parts of the oral mucosa. It is to be noted that an organism belonging to this genus, *Actinobacillus actinomycetecomitans*, is one of the organisms thought to be mainly responsible for the tissue destruction in the periodontal disease in man (Chapter 24, p. 522).

There seems to be insufficient knowledge available to separate these disorders on a rational basis, apart from the difficulties created by revisions that have occurred in the taxonomic classification of the various organisms concerned, based on advances of knowledge of their structure and behaviour. On the whole, it seems likely that there is a variety of different organisms, though mainly of filamentous habit, which can produce invasive necrosis of tissues, and also that the clinical disorder produced depends on the initial portal of entry of the organisms.

It is worth adding a single example to those already mentioned. Cass (1947) described the condition in a young white-tailed deer (*Odocoileus virginianus*), shot in a national park in Wyoming, as typical of the condition observed in many American elk (*Cervus canadensis*) by O.J. Murie (1930, 1951). The animal was in poor condition and there was a large impacted food mass in the right buccal vestibule in relation to the upper molars which were displaced and loose. Fodder was packed between them, the gum was receded and ulcerated and the tongue was swollen and protruded. No bacteriological examination was possible but, by comparison with records of similar examples in the literature of which he provided a valuable review, Cass placed the condition in the category of necrobacillosis, a name based on the work of Beveridge (1934) whose thorough study strongly implicated an organism that he called *Bacillus necrophorus*, which was then renamed *Actinomyces necrophorus*, the name employed by Cass, and which has also been categorized as *Sphaerophorus necrophorus* and more recently (Samuel, 1983) as *Fusobacterium necrophorum* (see Chapter 23, p. 520).

Order **PERISSODACTYLA**

In the days before the First World War when there were some 250 000 working horses in London (Rolph, 1980), periodontal disease was described as the scourge of the horse, more probably of the manger-fed urban cab horses rather than of the farm animals which would have a chance to enjoy some meadow grazing.

The following account is based mainly on the examination of 484 skulls of horses of various ages that had been worked in London (Colyer, 1906). About 77.5% of the horses were over 12 years of age, 13.8% were between six and 12 years, 7% were between three and six years and 1.6% were under the age of three years. Approximately one-third of the total presented some degree of periodontal disease, ranging from slight destruction of the alveolar crests to extremely gross changes. Most of the skulls appear to have come from knackers' yards so they were not a true epidemiological sample of the working population of London horses, as of course the age structure of the group illustrates. Various selective processes would have been at work, some indeed may have reached the knacker's yard for the very reason that they had dental disease which would have affected their value as working animals, if only because they would be inefficient feeders and therefore unable to maintain the demanding work-output expected of cab-horses. Nevertheless, it is clear that periodontal disease was a common disorder.

Not all of this material was skeletalized, many were examined while the soft tissue was *in situ*. The sequence of events in the course of the disease appeared to be as follows: first there is loss of the tag of gum which normally neatly fills the space between adjacent teeth. If the tag becomes inflamed, whether the initial injury is mechanical, as from the spiky ends of machine-cut dry chaff, or chemical irritation by bacteria growing in soft food-debris left on the gum surface, the architecture of the gum between the teeth tends to be destroyed and a triangular gap is made between the teeth in which chaff or hay can become wedged (Fig. 24.39). Perhaps because it tends to swell as it becomes sodden, this wedged fodder, which may stay for several days before it finally rots or disintegrates, may create a gap between the teeth and so make future wedging of fodder more likely to occur. Thus, the process tends to become continuous or chronic and a suppurative inflammation tends to spread in the periodontal tissues between the teeth (Figs. 24.40, 24.41, 24.42), and to extend to the buccal and lingual aspects. Bone is resorbed and finally the suppuration may become widespread and gross, the teeth may fall out or, worse still, the whole of the horse's dentition becomes disorganized with uneven wear. The horse reaches a sorry state of bad health, partly due to being unable to feed and partly due to infection.

Sometimes caries would be present as well to add to the process of painful destruction of the dentition (Fig. 24.43). For example, the animal of which the skull is depicted in Figure 24.44, when seen alive by Colyer, had pus pouring from the mouth; a further discharge was flowing from the nostrils and, on opening the mouth, the sulcus between the cheeks and teeth was seen to be clogged with food. The animal was killed, and the head prepared partly as a moist and partly as a dry specimen. The margin of the gum is thickened and detached from the bone. Between the second and third molars there is a large space communicating with the maxillary air sinus, the lining membrane of the cavity being thickened and inflamed. The left half of the skull and the mandible show that the alveolar process has been extensively resorbed, the greatest loss of tissue being in the region of the last two maxillary molars. The space between these teeth communicates with the air sinus. The buccal aspects of the posterior teeth show signs of caries, which is no doubt due to the lodgment of food in the sulcus. It would have been difficult to believe that such an extreme condition could have arisen from so slight injury to the mucoperiosteum had we not been able to follow the disease in the horse throughout the progressive stages shown in the illustrations.

The teeth in advanced cases often show resorption of the roots and infection of the pulp tissues (Fig.

Fig. 24.39. *Equus caballus* (horse). RCS Odonto. Mus., G 157.53. Fodder is still *in situ* between the teeth. Wedging of fodder in this way has led to destruction of the interdental gum and bone, so creating spaces for further accumulations of fodder.

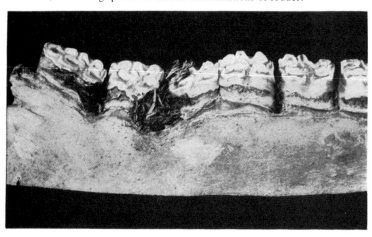

Fig. 24.40. *Equus caballus* (horse). RCS Odonto. Mus., G 157.3. Upper left tooth arch. An early stage of bone destruction in periodontal disease.

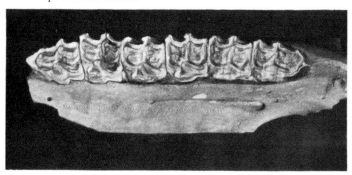

24.45). When there is a combination of periodontal disease and caries, the cheek teeth are particularly liable to split axially.

To enable them to be worked hard, London horses were fed intensively and, so that they could masticate and digest the largest quantity of food in the shortest time, it was usually given in the form of chaff, namely hay and straw cut into short lengths. It is probable that this contributed to the onset of periodontal disease. Horses in their natural habitat gather stalks of grass but these would be long and pliable and unlikely to damage the oral mucosa; chaff on the other hand, perhaps containing fragments of thorn or hedgewood which, in grazing, a horse would avoid or, if it took into its mouth, would detect and discharge with its sensitive mobile lips, would be liable to damage the mucosa and be driven between the teeth.

However, even grazed horses and those working in rural environments may suffer gingival damage. H. Kirby (personal communication to J. F. Colyer c. 1930) expressed the view that the periodontal

Fig. 24.41. *Equus caballus* (horse). RCS Odonto. Mus., G 157.4. Bone destruction has advanced sufficiently to loosen the attachment of M^1.

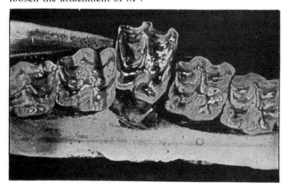

Fig. 24.43. *Equus caballus* (horse). RCS Odonto. Mus., G 157.01. The muco-periosteum of the palate is in place. The teeth are carious. The first stage of periodontal disease is shown in the slight destruction of the muco-periosteum between the right M^2 and M^3 (pointer).

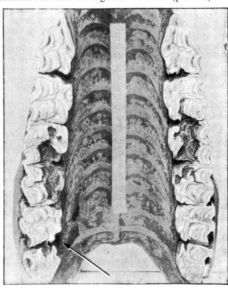

Fig. 24.42. *Equus caballus* (horse). RCS Odonto. Mus., G 157.6. Suppuration in the mandible below a site of fodder wedging.

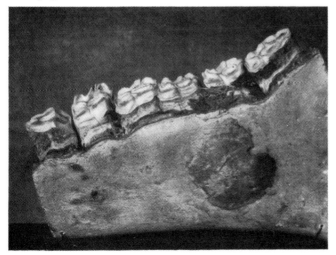

disease he found under those circumstances was 'caused largely by the awns of worthless grasses in hay of inferior quality, e.g. brome grass, false brome, barley grass, couch grass, false oat grass etc. In India, I have seen many cases caused by spear grass, and in South Africa by *stik-grass* [Boer name]. In America, needle grass or porcupine grass causes the same condition'.

Table 24.2 The incidence of periodontal disease in a sample of 645 HORSES in Germany (from Voss, 1937)

Age (years)...	Up to 4	5–6	7–8	9–10	11–12	13–15	16–18	19 & more	Unknown age
No. of skulls	32	98	134	119	85	96	34	12	37
No. with periodontal disease	10	26	31	39	29	47	29	11	1
% with periodontal disease	19.2	26.3	23.1	32.9	34.1	48.9	53.6	91.7	2.8

Fig. 24.44. *Equus caballus* (horse). RCS Odonto. Mus., G 157.94.
Above: Right maxilla with mucosa *in situ*. Advanced periodontal disease associated with dental caries. A probe leads into a suppurating sinus that passes deeply into the jaw.
Below: Left lateral view of the other half of the same specimen. All soft tissue has been removed. Periodontal disease has destroyed the alveolar crests, especially of the upper molars; the gap between M^2 and M^3 suggests food wedging. Note the irregular wear tables.

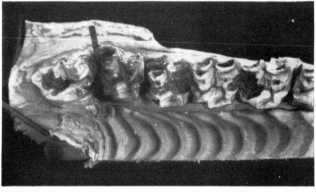

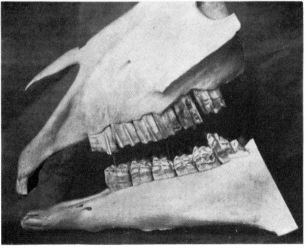

According to Thomson (1905) horses, running wild in countries where there are prolonged droughts, die more from periodontal disease than from actual starvation. They dig up roots and, when chewing these, grit and sand get between the gum and the teeth.

Voss (1937) pointed out that the columnar hypsodont crowns of the teeth in the horse taper a little towards the roots so that, as wear reduces their height, the occlusal wear tables become smaller and slight spacing occurs between the teeth which increases the likelihood of fibrous foods such as grass becoming impacted between the teeth.

Voss examined the mouths of 645 horses of estimated age 4–19 years and over. As Table 24.2 shows, 19.2% of those up to 4 years of age had some gingivitis and the prevalence rose progressively with age until, in the group aged 19 years and over, 91.7% showed gingivitis or peridontal disease. The processes he described are similar to those noted by Colyer.

Voss figured examples from necropsies, some with grass or hay still wedged in gingival pockets and, in particular, between teeth, and some with deposits of tartar on the sides of the crowns of the teeth and abscesses over the sides of the roots (gumboils). He also figured radiographs showing the bony involvement, vertical splitting of worn and sometimes carious teeth and, in very old horses, involvement of the furcations between the teeth when the entire crown had been worn away and only short roots remain.

Fig. 24.45. *Equus caballus* (horse). Two cheek teeth removed from advanced periodontal disease showing resorption of cementum. From a photograph lent to Colyer by W. L. Little.

The far-reaching effects of periodontal disease in the horse have been described by F. T. Harvey (1920) and Pillers (1933). Harvey drew attention to the frequency of the disease in colts, and quoted instances of severe local and systemic diseases in which there was reason to believe that the infection gained an entrance through the alveolar region. Pillers considered that chronic toxaemia is a common termination of the disease in middle life of cart horses and that many cases of arthritis and synovitis are secondary to oral sepsis. He stated that he had seen cases of recurrent colic recover by attention to the teeth and cases of brachial neuritis clear up when a suppurating tooth had been removed.

Food packing between the teeth and various other complications of periodontal disease or dental caries, such as loose or abscessed teeth, must produce discomfort and difficulty in mastication which are not easily detected. However, grinding the teeth together may be a sign of mild dental discomfort; if the discomfort is much more severe, the horse may not complete the act of mastication but may allow partially-masticated food to drop from its mouth; this is known as cudding (Little, 1913). A horse that does this will in due course become malnourished.

Groves (1966) found that the skulls of 12 equids of three species (two onagers, *E. hemionus onager* and *E. h. khur*, and the wild ass *E. africanus africanus*) that had lived in captivity since youth were smaller than 13 wild-shot specimens. Their cheek teeth showed irregular wear, some of which was due to the loss of the opposing teeth through periodontal disease. In some instances, over-eruption of molars prevented the incisors from being brought together. Groves attributed the failure of the skulls to grow to normal size to the defective dentition which must have affected the animals' nutrition.

Amin and Kassem (1987) examined 20 skulls of donkeys and ten of horses ranging from five to 20 years of age and found that deposits of dental calculus, often ring-like, were common on the buccal aspects of the teeth. The calculus was composed of calcium carbonates and phosphates; no magnesium salts were detected.

Order **RODENTIA**

Because the brown rat (*Rattus norvegicus*) in the wild lives so commonly within man's habitations and to some extent shares his diet or lives on his garbage, it would be interesting to know whether periodontal disease is common. Nothing seems to be recorded, however. Periodontal disease seems to be common in captivity; for example among animals maintained and bred in laboratories for various experimental purposes, providing they are kept long enough for the disease to develop fully, that is into their second year of life which is not by any means always the case. One of the fullest descriptions, including the histology, is that of Gottlieb (1922) although he provided no details of the diet or ages of the 200 animals he examined and no information about incidence. Bössmann, Deerberg, Preuss and Rehm (1981) found that gross periodontal disease with interdental food and hair impaction was universal at autopsy in a colony of rats maintained, as a study of senility, until natural death at 27–40 months of age. The crowns of many molars were destroyed by wear but these workers made the interesting observation that the drinking water was acidified with hydrochloric acid, 'a common hygienic measure in laboratory and breeding colonies'. Such a practice could, of course, either destroy the tooth crowns or enhance the effect of wear (see Chapter 22).

An exceptionally severe form of periodontal disease that affected rats maintained for a year on a powdered commercial diet was described by Robinson (1985). The diet was prepared by a new process which inadvertently reduced some ingredients to long, pointed vegetable fibres. These fibres were found in the granulomatous periodontal lesions and also within similar lesions deep in the maxilla and turbinate bones, apparently having to spread from the periodontal lesions, often resulting in oro-nasal fistulae.

A high prevalence of loss of alveolar crest bone, that is periodontal disease, was found by Sheppe (1966) in the skulls of 43 house mice (*Mus musculus*) gathered from a heavily infested farm food-store. They were divided into young (probably 1–3 months old) and adult, partly on the basis of tooth wear. Destruction of alveolar bone in various degrees up to very severe, where the roots of the teeth were socketted at the bottom of a trough of bone, and often affecting one or more quadrant more severely than the others, was present in 83% of the 26 adults and 37% of the 19 young. A clue to the particularly high prevalence of periodontal disease was provided by the finding of tufts of hairs wedged so firmly into the bony pockets that they had survived preparation and cleaning of the skulls. The hair would have been derived from grooming and could indicate some disorder of hair growth, perhaps a genetic defect because it is likely that the whole population in the food-store arose from a small number of animals, even a single pair, which bred rapidly in the presence of the ample food supply. Similarly, some other kind of genetic defect of the immune system could have increased the susceptibility to periodontal disease.

Quite severe destruction of the crests of the cheek tooth alveoli was found by Sheppe (1965) in nine out of 25 skulls of adult deer mice (*Peromyscus oreas*) collected within an area of about a quarter of a square mile in British Columbia. In some, there was merely flattening of the interdental crests but, in others, nearly the whole of the roots of the teeth were exposed. The condition was also present in two skulls collected from the same general area 15 years previously but was absent in several hundred skulls of this species in the Museum of Zoology, University of British Columbia. Furthermore, 26 skulls of the related species which has similar food habits, *Peromyscus maniculatus austerus*, collected from the same small area at the same time as the affected group were not affected. Until the matter is more fully studied, some transmissible infection seems to be the most likely cause although the possibility of age change has to be considered.

Shklar and Person (1975) described what must be accepted as mild periodontal disease, limited to the interdental gingiva, in all 20 adult wild-caught little pocket mice, a North American species (*Perognathus longimembris*). There was recession of the interdental gingiva with, in some instances, the formation of pockets within which there were micro-ulcers associated with inflammatory cells in the connective tissue. The pocket mouse is small, weighing up to about 11 g, and gained fame by being included in the Apollo 17 space mission for the study of the effect of cosmic radiation on tissues.

Shalla (1972) examined the skulls of 100 lemmings (*Lemmus trimucronatus*) recovered from the regurgitated gastric pellets of the snowy owl found on the Alaskan tundra. Twenty-one showed bilateral cheek-tooth alveolar crest resorption, mostly little more than blunting of the crests. In the absence of a fuller description, there seems no reason to suppose anything other than an age change. There were several fractures of tooth cusps associated with some local bone loss indicative of infection.

The Mongolian gerbil (*Meriones ungiculatus*) is a burrowing root- and seed-eating rodent, slightly larger than the house mouse, which since about 1954 has been bred in laboratories for various kinds of research. It has a potential life span of five years which is much longer than that of most of the small rodents; for instance the life span of the laboratory rat and hamster is two to three years and the mouse about two years. It is not recorded whether periodontal disease ever affects the gerbil in its natural habitat but the disease occurs commonly spontaneously in laboratory-maintained animals. Gupta and Shaw (1960) found only a mild gingivitis in animals less than one year old, mild in the sense that it was much less than in rice rats and hamsters maintained on similar standard laboratory diets for a similar period. However, Moscow, Wasserman and Rennert (1968), in a study of 61 animals ranging in age up to five years, found that, with advancing age, all stages of the disease developed, with gross accumulations of dental plaque, some mineralized, approximal impaction of hair and wood-shaving bedding material and ultimately spontaneous exfoliation of the teeth.

Order **INSECTIVORA**

Periodontal disease was found in 15 out of 27 American least shrews (*Cryptotis parva*), minute insectivores weighing up to about six g each, maintained as a laboratory colony for eight months in order to study their reproductive physiology (Patton, Hooper, Mock and Doyle, 1971). The changes ranged from gingival inflammation discovered histologically to gross periodontal destruction with much dental calculus and with some tooth loss. Infection in a few instances had spread to form subcutaneous abscesses. Culture revealed a mixed type of bacterial infection. The diet seems to have been ground fresh meat and some canned dog food.

Hedgehogs (*Erinaceus europaeus*), especially when kept in captivity, frequently develop periodontal disease in a severe form. Their natural food is varied, including insects, worms, snakes' eggs and, to a certain extent, roots and fruit. In the captive state, they are given minced raw meat with milk as well as bread and milk.

Zuhrt (1958) found gross calculus on the crowns of the cheek teeth and sometimes of the incisors as well in 8.4% of 310 skulls of mature hedgehogs; mostly in the Berlin Zoological Museum and presumably mostly from the wild. Poduschka and Poduschka (1986) confirmed this and extended their study to hedgehogs of five genera, *Erinaceus*, *Aethechinus*, *Atelerix*, *Hemiechinus* and *Paraechinus*, bred and kept in captivity. The skulls of animals killed on the roads were also examined. Gross calculus on all surfaces of the teeth, sometimes leaving only the tips of the pointed cusps exposed, was common but more so in captive animals. Supragingival and the more blackish subgingival forms of calculus could be distinguished. Serial examination of captive animals showed that gross amounts of calculus could accumulate in a few months. These findings are surprising; it is known that hedgehogs living on the fringes of towns tend to make regular night raids on human garbage containers. Whether this adversely affects their dentitions would be worthy of investigation.

The golden moles (*Chrysochloridae*) of South Africa are a family represented by about 17 species; they have a life style similar to that of the European mole. The dentition presents some unusual features. The teeth do not erupt until the animal is fully grown and they do so almost simultaneously. Broom (1907) drew attention to what he regarded as the early loss of the dentition in two species living in the Stellenbosch area. This is best described in his own words [*Chrysochloris hottentota* is more usually known as *Amblysomus hottentotus* (Corbet and Hill, 1986)].

> Both *Chrysochloris hottentota* and *asiatica* are found in the same gardens, but they apparently keep certain regions to themselves, *asiatica* preferring the drier and sandy soil and *hottentota* the richer garden soil. When the animal becomes aged which may be when it is a year or two old, and at a time when it is in full sexual vigour, all the teeth begin to become loose and fall out, the molars as well as the anterior teeth, so that the animal becomes practically toothless. I have seen one specimen with every tooth of the first set shed and not one of the second set yet through the gum. Many other specimens have only three or four functional teeth. After being for some time in this practically toothless state a few teeth of the second set come through the gum.

The toothless animals appear to remain in good health and their stomachs are full of worms broken into segments, as in the fully-toothed animals. The toothless state is rarely seen in animals from other districts. Among 11 specimens from Stellenbosch, of which 10 are in the Odontological Museum of the Royal College of Surgeons, 7 showed some destruction of alveolar bone (Fig. 24.46); all 89 specimens from other districts were unaffected in this way. Stellenbosch is a highly cultivated arable area in which worms are likely to propagate freely. According to W. Floyd (personal communication to J. F. Colyer c. 1930), the golden mole makes for newly-manured ground.

The reason for the inclusion here of this account of the loss of the dentition in the golden mole is that the tooth loss appears to be due to a disorder of the periodontium; there appears to be no suggestion that the teeth become worn. We wonder whether the explanation is that the loss of teeth is an ageing phenomenon and is not more widespread because animals in most areas are unable to burrow and survive

570 Section 4: Other disorders of Teeth and Jaws

once they reach the age when their teeth begin to deteriorate. In the soft, cultivated soil of the Stellenbosch, rich in worms, the aged animals may be able to burrow and acquire food more easily and thus survive into the stage when their dentitions are totally lost or nearly so.

Order **CHIROPTERA**

Phillips (1971) has described in unusual detail evidence of periodontal disease he found in Davis' long-tongued bat *Glossophaga alticola*. Among 2400 specimens of several genera, he found typical periodontal disease only in this species. Some were skulls and some were specimens stored in alcohol. Evidence of periodontal disease in the form of recession of the alveolar crest bone was common in old animals where the dentition was worn. Phillips pointed out that the first stage of the disease process, simple gingivitis before the bone was involved, could not be detected, even in the alcohol-preserved specimens, because the colour changes and swelling, upon which the diagnosis depends, were lost. No calculus was found on the teeth, nor evidence of food lodgement between the teeth or in gingival pockets.

The diet of the glossophagine bats consists mainly of nectar and the pulp and juice of soft ripe fruit.

Dehiscences and fenestrations were found over the roots of upper teeth of all the adult long-nosed bats, *Leptonycteris*, examined and were also common in other genera of the glossophagine bats.

Fig. 24.46. *Amblysomus hottentotus* (Hottentot mole). Wild. RCS Odonto. Mus., G160.2. From Stellenbosch; showing loss of the teeth from periodontal disease. Enlarged.

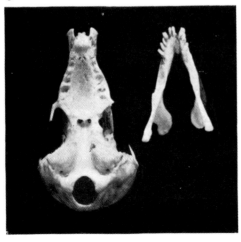

Phillips described for the first time a condition which, as far as the literature goes, is unique to bats, namely the invasion of the gingiva by the protonymphs of mites (genus *Radfordiella*). The mites appear to penetrate the gingival crevice on the lingual aspects of the upper post-canine teeth. The margins of the long lesions are smooth but the surfaces are irregular and pitted. In severe infestations with the protonymphs, the maxillary air sinus becomes exposed to the mouth. The normal maxillary sinus in bats is almost filled with a loose connective tissue; this is destroyed in mite infestation. Histological sections show the protonymphs *in situ* in the tissues and that a feature of the lesions is resorption of the roots of the teeth.

What may be called mite-induced periodontal disease was found in several genera of bats, *Leptonycteris*, *Anoura* and *Monophyllus*. Although at least two species of the mite *Radfordiella* were identified, the exact taxonomy of the mites has not been established, nor whether they correspond with species that infest the skin of bats. These are known to be highly host-specific; that is, different species of bat, even though living contiguously, seem only to be host to particular species of mite.

It is clear that, for reasons unknown, periodontal disease is common in bats generally and deserves study, preferably of specimens with soft tissues *in situ*. Vierhaus (1980–81) found that about one-fifth of 255 skulls of 20 Central European species of bat showed marked destruction of the cheek tooth alveolar crests, which could reasonably be interpreted as periodontal disease, leading to some loss of teeth. *Myotis daubentonii* seemed to be particularly susceptible, one third of 60 adult skulls being affected.

Order **MARSUPIALIA**

Amongst marsupials, captive kangaroos and wallabies often become victims of periodontal disease. An early stage in a wallaby is shown in Figure 24.47; in the maxilla, the septa between the molars have been destroyed, most severely in relation to the first molar; in the mandible on the other hand, no changes are evident.

A more advanced stage is shown in the skull of a swamp wallaby in Figure 24.48. The disease here is more widespread and the mandibular teeth are also affected. There is a slight deposit of dental calculus. A further stage in the progress of the disease is shown in a rock wallaby in Figure 24.49. In the maxilla, the bone of the alveolar crests has been destroyed extensively in the neighbourhood of the third molar and, in places, the cortical bone is thinned and pitted with tiny perforations. In the mandible, the disease is not so advanced and there is marked thickening of the bone below the first molar. The final stage is shown in another rock wallaby (Fig. 24.50); all the molars have been lost with the exception of one in the maxilla and one in the mandible, the bone around both being extensively destroyed.

W. A. Miller (1977) described the skull of a fully-adult bandicoot (*Isoodon macrourus*) from which nearly all the teeth had exfoliated, leaving only a few fragments which were revealed by radiography. The

Fig. 24.47. *Macropus rufogriseus* (Bennett's wallaby). Captive. RCS Odonto. Mus., G 161.1. Moderate destruction of the alveolar crests, more marked in the maxilla than in the mandible.

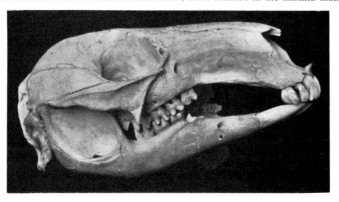

Fig. 24.48. *Wallabia bicolor* (swamp wallaby). Captive. RCS Odonto. Mus., G 161.2. More advanced periodontal disease, affecting both upper and lower jaws.

Fig. 24.49. *Petrogale* sp. (rock wallaby). Captive. RCS Odonto. Mus., G 161.3. Severe periodontal disease with uneven destruction of alveolar crests. The remaining alveolar bone is pitted.

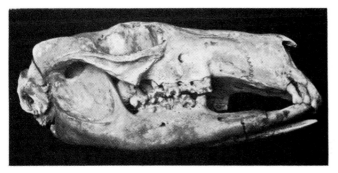

provenance of the animal was unknown but it seems likely to have been captive, or even kept as a pet, because a nearly-edentulous marsupial would not be likely to survive in the wild for long. Periodontal disease is the most likely cause of the tooth loss.

Lyne and Mort (1981) referred to changes in the cheek-tooth alveolar bone, with tooth loss, in laboratory-maintained bandicoots of the *Isoodon* genus; the changes, from the description and illustrations, seem to be severe periodontal disease. The crowns of the upper cheek teeth of one (AGL 1213) appear to be completely invested with dental calculus. Lesser degrees of alveolar destruction were common among 102 captive animals. Similar tooth loss was common in the bandicoot *Perameles* when caged for longer than two years (Lyne, 1982). *Isoodon* in the wild is omnivorous but with a preference for insects; *Perameles* is more strictly insectivorous. The diet of these captive animals consisted mainly of such items as minced beef, eggs and meal worms.

Examples of severe destruction of alveolar bone in marsupials, such as the two wallabies in Figures 24.51 and 24.52, are of special interest because wallabies and kangaroos, particularly under captive conditions, are susceptible to infection with the *Sphaerophorus necrophorus* or whatever other bacteria are responsible for so-called lumpy jaw (see Chapter 23). In a skull, that would be very difficult to distinguish from bone destruction due to severe localized periodontal disease. Indeed, W. A. Miller

Fig. 24.50. *Petrogale* sp. (rock wallaby). Captive. RCS Odonto. Mus., G 161.4. Nearly all the cheek teeth have been lost from periodontal disease.

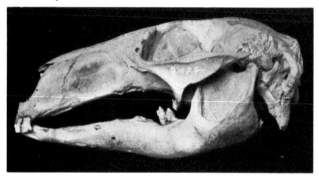

Fig. 24.51. *Petrogale penicillata assimilis* (brush-tailed wallaby). ♂. Wild. BMNH 1926.3.11.4. Much destruction of the alveolar crestal bone of the upper cheek teeth, even more marked on the palatal aspect. The cortical bone over the roots of the teeth is deeply pitted. In the mandible, destruction is not so advanced.

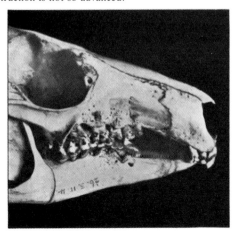

and Beighton (1979) offered the opinion, based only on the Figures in the first edition of this book, that the specimens in Figures 24.51 and 24.52 are examples of lumpy jaw rather than periodontal disease.

The idea that animals living in their natural habitat are unlikely to suffer from periodontal disease receives a severe knock when the macropods are considered. Many texts mention that periodontal disease is common in macropods in the wild as well as in the captive state but without presenting evidence. However, Miller and Beighton (1979), by studying 119 skulls of the eastern grey kangaroo (*Macropus giganteus*) from the wild located in Australian museums, established that they suffered not only from localized alveolar-crest destruction associated with food packing between the teeth but also from gross depositions of calculus on the crowns of the teeth. In many specimens, the remains of fibrous food were still *in situ* between the teeth. It seems that deposition of calculus begins while the teeth are erupting because it was found on the occlusal surfaces of the erupting teeth; however, as soon as the teeth come into occlusion the calculus is worn away from the occlusal surfaces and thereafter is found on the lingual and buccal surfaces only. In the macropods, the cheek teeth move into the functional arch from behind (molar progression); then in due course as they move forwards they reach a point when they are out of occlusion with those of the opposing arch. When they reach this non-functional state, the sides of the teeth are still covered with calculus. There is less calculus on the incisors. The calculus tends to be dark in colour and even black; Miller and Beighton produced some evidence that the incorporation of manganese is responsible for the pigmentation of the calculus.

Further study of the dentitions of macropods in the wild is desirable to establish whether dental calculus, food packing between the teeth and the consequent periodontal disease are a wide-spread phenomena and not confined to certain groups, perhaps those living close to human habitation, where there might be access to items of diet especially conducive to dental plaque formation.

It seems highly likely that there is an association between periodontal disease and so-called lumpy jaw in macropods, the periodontal lesions providing portals of entry for the exciting organism of lumpy jaw (Wallach, 1971).

Fig. 24.52. *Petrogale* sp. (rock wallaby). ♀. Wild. BMNH 1908.8.8.08. A specimen similar to that in Figure 24.51 but the destructive process is more advanced so that the right M^1 has been lost and the root of the right M^2 is almost denuded of bone. In the mandible, the disease is less marked and limited to M_1 and M_2.

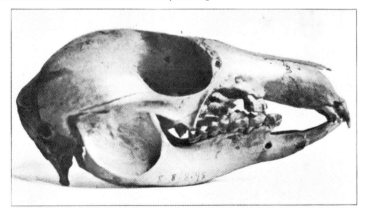

CHAPTER 25

Odontomes

Introduction

The term 'Odontome' was first used by Paul Broca in a paper read before the Academy of Sciences of Paris in 1867 in which, under the title 'Recherches sur un nouveau groupe de Tumeurs désigné sous le nom d'Odontomes', he described some 'of the hard excrescences which hypertrophy and abnormal growth of the tissues produce'. In his *Traité des Tumeurs*, Broca (1869) published a more complete account of such growths, classifying them under four headings:

(1) *Odontomes embryoplastiques*, or those arising during the early stages of development before the membrana eboris is formed.
(2) *Odontomes odontoplastiques*, or those arising after the formation of the membrana eboris and the formative elements of the tooth.
(3) *Odontomes coronaires*, or those arising during the formation of the crown of the tooth.
(4) *Odontomes radiculaires*, or those arising subsequently to the completion of the crown and therefore only causing a deformity of the root.

The suffix *ome* or *oma*, although most widely used to connote neoplasia, as in carcinoma, sarcoma and lipoma, has never been stringently restricted to that meaning. Quite early on, it was also used to connote a swelling, as in granuloma, or what can now be referred to as tumour-like; the word tumour, from simply connoting a swelling, has gradually come to connote neoplasia.

A considerable advance in our knowledge of odontomes was made by J. Bland Sutton (1888*a*). He said:

> In the most extended sense an odontome may be defined as a tumour composed of dental tissues

(enamel, dentine and cementum), in varying proportions and different degrees of development, arising from tooth germs, or from teeth still in the process of growth. It is customary to restrict the term, at least from a clinical point of view, to those hard tumours, found in the jaws, composed of fully developed enamel, dentine, cementum or varieties of these tissues. Nevertheless, the careful observations made by several workers during the past few years render a re-consideration of the matter imperative: for the term should apply not only to solid tumours, but to certain cystic forms as well. This extension of the term is of some importance, for we can include several curious aberrations of tooth-development.

The classification he suggested was constructed upon the nature of the particular cells of the tooth germ from which the tumour arose. It was as follows:
(A) Aberrations of the enamel organ.
   (1) Epithelial odontome.
(B) Aberrations of the follicle.
   (1) Follicular (dentigerous) cyst.
   (2) Fibrous odontome.
   (3) Cementomata.
   (4) Compound follicular odontome.
(C) Aberrations of the dentine papilla.
   (1) Radicular odontome.
(D) Aberrations of the whole tooth germ.
   (1) Composite odontome.

A *Report on Odontomes* by Gabell, James and Payne (1914), prepared under the aegis of the British Dental Association, extended the scope of the term odontome to include certain gross malformations of the teeth which seem to be examples of abnormal development. These authors, while following in principle Bland Sutton's classification, considered that certain alterations were necessary; alterations in a large measure due to 'the recognition of the wider function of the enamel organ and the discovery of the epithelial lining in certain cysts'. The classification they put forward was as follows:
(A) Epithelial odontomes. Where the abnormal development takes place in the dental epithelium alone.
(B) Composite odontomes. Where the abnormal development takes place primarily in the dental epithelium, and secondarily in the dental papilla, and may occur in the follicle also.
(C) Connective tissue odontomes. Where the abnormal development takes place in the dental tissues of mesoblastic origin alone.

The *Report on Odontomes* is still of considerable value because it described, illustrated and gave the locations of many museum specimens of odontomes in the Dental and Medical Schools in Great Britain; although most of the specimens are from man, other species are well represented.

Sprawson (1937) put forward arguments in favour of extending the term odontome to include all abnormalities of excessive growth derived from the dental formative organs; these included supernumerary teeth, connate teeth (strictly only where one of the elements was supernumerary) and even extra cusps and extra roots. This view had much influence during the following decade or two and was valuable in bringing out clearly the existence of a progressive series of conditions starting with slight disorders of development such as extra cusps and extra parts of teeth and continuing with supernumerary teeth, conditions where no suggestion of neoplasia arises; the series continues with disorders of development which may consist of little more than an erupted cluster of supernumerary teeth (compound composite odontomes) and ones (complex composite odontomes) which consist of a jumbled mass of tooth tissue exhibiting none of the continuous growth characteristics of neoplasia. The series continues, but as a less regular progression, with the true neoplasms, usually arising from one only of the dental formative elements, that is from the enamel organ or the dental papilla, such as ameloblastomas, odontogenic fibromas.

Hamartomas

Considered in this way, it became clear that most of the calcified odontomes fit very well the term hamartoma (Gk. *hamartion*, a bodily defect, to go wrong), introduced by Albrecht (1904) to describe tumour-like masses that are composed of a mixture of tissues normal to the part in which they are occurring and that are due to an anomaly in tissue development. Although these lesions resemble neoplasms in many ways, such as in clinical presentation and microscopic appearance, they are not true neoplasms; their growth potential is limited and growth ceases when the host has reached full maturity (Lucas and Pindborg, 1967).

The inclusion of the suffix *oma* in the term hamartoma, coined specifically to exclude the connotation of neoplasm, and the use of tumour-like in its definition illustrate the point made earlier about the changing meaning of these terms.

Thus, at the present time, the term odontome is usually employed only for hamartomatous or tumour-like formations; true neoplasms of the tooth formative tissue are categorized as odontogenic tumours.

We are concerned in this chapter with odontomes in that sense. Sprawson's classification of odontomes with the inclusion of such things as supernumerary teeth and extra roots is no longer popular and we have not followed it here.

Odontogenic tumours will not be dealt with systematically though there will be a few incidental references to them. For accounts of the odontogenic tumours, we direct readers to Gorlin, Meskin and Brodey (1963) and Gorlin (1970) and, for other tumours that may occur in the mouth or jaws, to McClure et al. (1978).

Enamel pearls

These are small, usually hemispherical, nodules of enamel formed on the surfaces of the roots of teeth, presumably by remnants of the epithelial root sheath (debris of Mallassez, 1885) which persisted in the periodontal ligament and happened to differentiate into ameloblasts. Such nodules are well documented in man and are often obscured by being overgrown by cementum; there seem to be no records of their prevalence in other species. Hooijer and Eulderink (1975) figured an isolated P_2 of a wild sheep (*Ovis ammon*) with about 60 pearls, up to about 2.0 mm in diameter, distributed over the greater part of the root surface. Such pearls must be regarded as trifling examples of aberrant tooth development.

Radicular cysts

The dental or radicular cyst is an epithelium-lined cyst, usually at the end of the root of a tooth, arising in granulation tissue in response to infection that has gained access to the periodontal ligament via a pulp cavity exposed by dental caries or by fracture of the tooth. The epithelium derives from the remnants of the epithelial root sheath that persist in the periodontal ligament. Such cysts are common in man, because of the prevalence of dental caries with consequent death of the pulp, but rare in animals. Because of the origin of the epithelium from the enamel organ, dental cysts were classified as odontomes, e.g. by Sprawson and most earlier authorities. However, it is not usual to do so at the present time because the stimulus for their production appears to be inflammatory and they cannot be regarded as anomalies of development. They will, therefore, not be considered systematically here and in any case, as already mentioned, there are few recorded instances in animals, although radicular cysts have been produced experimentally in rhesus monkeys (Chapter 23, p. 510).

Dentigerous cysts

A dentigerous cyst is an epithelium-lined cyst which forms around the crown of a tooth so that the crown appears to have erupted into it. Although some such cysts may arise from the envelopment of the crown of a successional tooth by a dental cyst on the root of its deciduous predecessor (Sprawson, 1927), most of them probably arise as a developmental overgrowth of the enamel organ and are thus usually still included in the category of odontomes.

Invagination

Colyer (1926b) and Rushton (1937) brought out more clearly than previous workers that a disorder common to many malformed teeth and odontomes is an enamel-lined invagination into the substance of the tooth (Fig. 25.1). Degrees of this disorder range from shallow pits in the crown to deeply-dilated pits, the morphology of the crown otherwise being normal, and to dilated invaginations which extend deeply enough to involve the root and distort the shape of the tooth. In the most severe form, the dilated invagination is the main feature of a more or less globular mass of dental tissue in the so-called dilated odontome.

Teratomatous odontomes

Pitts (1933) pointed out that odontomes indistinguishable from those found in the jaws sometimes occur in other sites; he termed those in the jaws as gnathic odontomes and those elsewhere as teratomatous odontomes using the term teratoma to imply that a mass of tissue represents the abortive formation of a twin of the host individual, what Bland Sutton (1922, p. 529) referred to as 'memorials of a lost individual', ovarian jaws being the typical example. This view of teratomas and their origin is not, however, shared by Willis (1962), a respected senior student of these matters in both man and other vertebrates.

Examples

DENTIGEROUS CYSTS

Such cysts have been recorded in various animal species and appear to involve particularly commonly the mandibular incisors and canines in sheep. The cyst may involve a tooth of the normal series, which of course remains unerupted, or can involve a supernumerary tooth. They can be a feature of composite odontomes. Usually the whole crown of the tooth projects into the cyst cavity but sometimes some of the root also, and occasionally the whole tooth lies loose within the cavity.

Dentigerous and dental cysts tend to increase slowly and steadily in size. The driving force for this increase in size is the continuous proliferation which is an inherent feature of the basal layer of the stratified squamous epithelium that lines these cysts and is associated with the continuous desquamation of dead cells on the surface of the lining. These free cells undergo autolysis into substances of lower molecular weight than those of the blood and fluids of tissues in general. Thus the osmotic pressure exerted by the cyst fluid is higher than that of the adjacent tissues. The wall of the cyst comes to act as a semipermeable membrane through which water and crystalloids from the tissues pass to the cyst fluid in response to the difference in osmotic pressure. The consequent high hydrostatic pressure within the cyst leads to increase in its size (Toller, 1948, 1967).

The general characteristics of dentigerous cysts are shown in the specimen in Figure 25.2 which is the incisor part of the mandible of a sheep, which has been expanded by a pair of more or less symmetrical cysts on either side. Each cyst contains an incisor tooth firmly attached to its medial wall but not implanted into the bone. The cysts appeared to have become secondarily infected because they contained pus in which *Staphylococcus pyogenes albus* and a bacillus were identified.

Fig. 25.1. Diagram based upon radiographs and sections through a range of human specimens arranged to show degrees of enamel-lined invaginations. (1) A small invagination; in the crown of a tooth of more or less normal morphology; (2) a similar but more extensive invagination; (3) an invagination which extends into and beyond the end of the root; (4) a gross dilatation typical of the dilated odontome; (5) A dilatation extending from an opening in a supernumerary element of the crown; (6) A dilatation extending from an invagination of the crown of a molar tooth; (7) a dilatation extending from an opening in the root area; (8) a similar but much more gross dilatation with the formation of a dilated mass attached to a molar tooth of more or less normal morphology. Based on Rushton (1937).

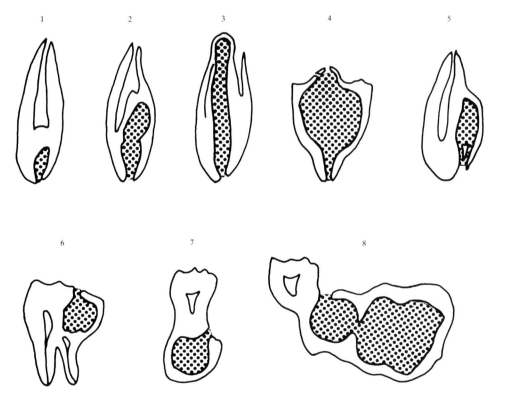

578 Section 4: Other Disorders of Teeth and Jaws

The *Report on Odontomes* (Gabell *et al.*, 1914) described bilateral dentigerous cysts in the incisor region of the mandible of a pig (formerly RCS Pathological Series, 2197) and two examples of dentigerous cysts in the mandibular incisor region of sheep, one being the bilateral example shown in Figure 25.2. Bland Sutton (1922, p. 239) said that he had seen them in sheep and pigs: 'In sheep they are common and generally affect the incisors and as a rule are bilateral'. He gave no examples.

The portion of a mandible in Figure 25.3 is also from a sheep, about three years of age. The anterior part of the left side is distended by a cyst about the size of a large walnut; the cyst is strictly confined to that side and has not encroached upon the symphysis. I_1 and I_2 are in place but the latter is in an irregular position. Two deciduous teeth remain but they have been displaced by the growth; one of these is situated between I_1 and I_2 and this tooth looks as if the pulp chamber had been exposed during life. I_3 is embedded in the cyst wall, but there is no trace of the canine.

There are two other dentigerous cysts in the incisor region of sheep in the Royal College of Surgeons Odontological Museum (G 55.2 and G 55.3); G 55.2 was described by C. S. Tomes (1872) and is a particularly-complete varnished specimen with the teeth and cyst lining still *in situ*. There is also the mandible of a bushbuck (*Tragelaphus scriptus*) (G 56.1) in which the entire right cheek-tooth row, except for P_2, has been replaced by a cyst.

Dolamore (1914) described a dentigerous cyst in a sheep (Fig. 25.4) which may have arisen as a dental cyst

Fig. 25.2. *Ovis aries* (sheep). Formerly RCS Special Pathology 4187.1. Portion of the mandible expanded on each side by a cyst. × c. 0.67.

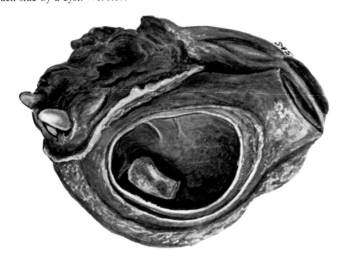

Fig. 25.3. *Ovis aries* (sheep). 3 years old. RCS Odonto. Mus., G 55.1. Portion of the mandible expanded by bilateral dentigerous cysts.

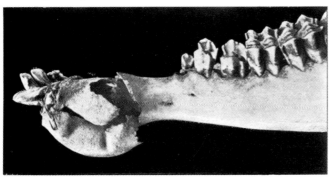

on a deciduous predecessor as postulated by Sprawson (1937). The specimen consists of the anterior part of the mandible bearing the deciduous incisors. Beneath the incisors on the left is a large cyst which has expanded the bone and involves the crown of a permanent incisor. The i_1 on that side is absent and it seems possible that the cyst was originally a dental cyst that arose on the root of the tooth, later lost as a result of the disease process, and then enveloped the crown of the successor tooth.

In view of these multiple examples of dentigerous cysts in sheep, it is of great interest that Orr, O'Callaghan, West and Bruère (1979) found that dentigerous cysts involving permanent incisors were a feature of five out of 23 ewes in New Zealand that were affected by a syndrome consisting of an association between excessive deciduous incisor tooth wear and focal dystrophic changes in the bone of the anterior part of the mandible, which were probably of systemic origin (Chapter 22, p. 492). In one of the ewes, there was a small radicular cyst on a permanent incisor. According to Blood, Radostits and Henderson (1983, p. 147), examples of this syndrome, jaw cysts associated with tooth wear, have been described in Russian literature.

It would be tempting to think that Orr *et al.* were dealing with a heritable condition or susceptibility that arose as a mutation which would be likely to spread among an inbred flock, but Dyson and Spence (1979) described similar cysts, though unassociated apparently with a dystrophic bone condition, in 12 out of a Scottish flock of 750 ewes. The crown of a permanent incisor projected into the cyst lumen and often the deciduous incisor was still present with its root in the wall of the cyst.

Rieck (1986) described 17 instances of a hitherto unknown congenital polycystic condition of the jaws that affected three different breeds of cattle in Germany.

E. Becker (1970, p. 181) described in detail a dentigerous cyst involving several lower incisors in a female camel.

Bland Sutton (1888a) gave an account of a tree porcupine (*Coendu prehensilis*) in which there was a large dentigerous cyst in the mandible extending from the incisor region to behind the cheek teeth. The crowns of two teeth projected into the cyst and the inferior alveolar nerve was traceable in its lateral wall.

A dentigerous cyst in the maxilla of a horse and containing 104 teeth of various sizes connected by soft tissue was described by Treman (1907).

Bilateral dentigerous cysts involving the M_2 were described in a young baboon (*Papio hamadryas*) by Manning (1972).

Dentigerous cysts are rare in animals from the wild state; one example was in the skull of a howler monkey (*Alouatta belzebul*, RCS Odonto. Mus., G 55.51) in which there was a large cyst in connection with the right M^3. The tooth was misplaced horizontally with the root facing distally and only a small part of the crown projected into the cyst cavity. Immediately distal to M^2, there was an opening into the cyst and another in the zygomatic surface of the bone. The facial surface bulged outwards and there had been some destruction of the floor of the orbit as the result of suppuration. The zygomatic process of the maxilla had been lost and there were indications that the zygomatic arch had been fractured.

Fig. 25.4. *Ovis aries* (sheep). Juvenile. RCS Odonto. Mus., G 55.4. Radiograph of the mandible distended by a cyst in the incisor region.

COMPOSITE ODONTOMES

Complex composite odontomes are irregular masses composed of all three dental tissues, enamel, dentine and cementum, and arise from the uncoordinated growth of tooth primordia. The overall regulation of morphogenetic activity that normally leads to the formation of a tooth of particular shape is evidently defective. However, as a rule, the tissues of which the odontomes are composed are of more or less normal structure so that both the differentiation of the formative cell types and actual formation of mineralized tissues seem to proceed in a normal orderly fashion.

Compound composite odontomes consist of multiple tooth elements of various sizes and degrees of morphological differentiation, ranging from seed-like irregular masses to differentiated toothlets with recognizable roots and with crowns which may be simple cones or multituberculate forms, sometimes with invaginations. Evidently they arise from overgrowth of what can be called tooth primordial material with not quite complete absence of normal co-ordinated growth. Neither complex nor compound composite odontomes exhibit the continuous growth that is characteristic of neoplasia, although they do show the continuous formation of dentine and cementum that is characteristic of normal dentine and cementum.

An example of a complex composite odontome is shown in Figure 25.5. The mass, which measures 60 mm in the longest diameter, occupied the position of the right M_2 in an African buffalo. The upper surface has the appearance of a crater of which a portion of the rim has been lost. The top of the wall (A) has been subjected to considerable wear. The worn surface provides a cross-section of the structure of the mass and shows the appearance of a number of small denticles fused together by a less-dense cementum-like tissue. Towards the base, the mass resembles branching coral. The odontome is cleft by a fissure which extends almost to its base, suggesting the possibility that it arose from two elements or there was a process of dichotomy.

Another example from a buffalo of this type of odontome, which is associated with a cyst, is shown in

Fig. 25.5. *Synceros caffer aequinoctialis* (Abyssinian buffalo). RCS Odonto. Mus., G 57.2. A composite odontome from the position of the right M_2. (A) indicates an area of wear by mastication.

Figure 25.6. The cyst is situated in the hard palate on the right side and has almost filled the maxillary air sinus, reducing it to a slit-like space between the thin bony wall of the cyst and the expanded floor of the nasal fossa. The inner surface of the bony cyst wall is elaborately ridged with a lace-like pattern, suggesting that new bone has been laid down on that surface; in other words, there was a period when the cyst was no longer expanding and was even diminishing in size. This is in accord with the observation that the inferior surface of the odontome (Fig. 25.6B) contained within the cyst shows a large area of wear and thus that part of the odontome must have been exposed in the mouth. This means, of course, that the cyst was open to the mouth which would lead to its partial regression.

The cyst cavity is about 190 mm wide at its greatest diameter and about 76 mm deep. The three premolars have erupted and M^2 and M^3 are buried in the lateral wall of the cyst, M^2 has been displaced and is lying horizontally. M^1 is missing but must have been erupted because the two buccal sockets are present. However, they have been partly destroyed by the cyst. Thus the evidence suggests that the full complement of teeth was present and that the odontome was therefore wholly an additional element (Colyer, 1905).

The odontome contained within the cyst consisted of one large main mass with a greatest diameter of 11.4 cm and four additional smaller pieces; other pieces may have been lost in the preparation of the specimen. The odontome is irregular in shape with both spiky and rounded projections. The medial surface shown in Figure 25.6B is deeply concave; several of the projections from this surface are tooth-like and are covered with enamel. The structure revealed by a ground section through one of the smaller pieces of odontome (Fig. 25.6C) is suggestive of closely-packed minute denticles each with a small central pulp cavity.

An odontome in a captive Tasmanian devil (Fig. 25.7) is situated in the right maxilla and is associated with a cyst which has expanded the bone in three directions as shell-like expansions or loculi which appear to have been perforated in several places. There are expansions into the jugal bone, into the jugal fossa and into the orbit. The main part of the cyst appears to be quite open laterally though it is impossible to be certain whether or not some thin bone has been lost in the preparation of the specimen. Each main loculus has subsidiary loculi in its wall so the appearance is of a polycystic condition. The mineralized odontome at the centre of the lesion is still firmly embedded in bone and presents an irregular nodular surface, part of which appears to be enamel.

The erupted M^3 above which the odontome is

Ch.25. Odontomes   581

Fig. 25.6. *Synceros caffer caffer* (Cape buffalo). RCS Odonto. Mus., G 57.1.
A: The left maxilla is distended by a cyst. A composite odontome was contained in the cyst.
(a) and (a) indicate two sockets for the roots of M^1 partly destroyed by the cyst; (b) unerupted M^2 (c) an extension of the cyst into the pterygoid fossa; M^3 is buried in its wall; (d) the position occupied by a small satellite piece of odontome. × c. 0.33.
B: The composite odontome contained in the cyst. The pointer A indicates a wide surface showing wear from contact with the tongue. The main surface is deeply concave and presents several tooth-like projections covered with enamel.
C: A ground section through one of the smaller fragments of odontome found within the cyst shown in A. The appearance is that of closely-packed minute denticles cut in cross-section. Each denticle has a small central pulp cavity.

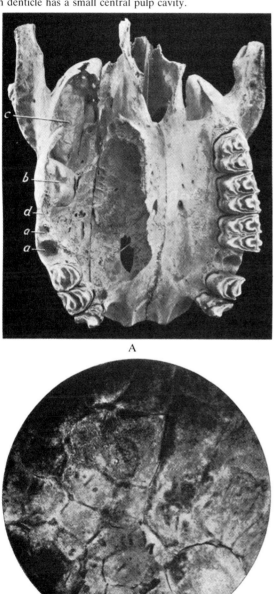

A

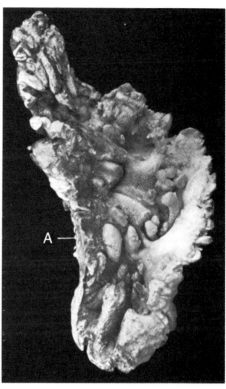

B

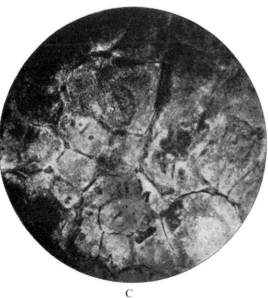

C

situated has been broken in life and most of the crown has been lost. The small M^4 has erupted buccally to it. The M^3 appears to have been shattered by trauma before it was completely formed which invites the question whether the trauma had anything to do with the formation of the odontome.

A composite odontome from a sheep skull found in open country in New Zealand is shown in Figure 25.8. It is attached to an otherwise normally-developed mandibular right third molar. The odontome covers the posterior, medial and part of the anterior surfaces of the tooth, the larger part of the mass being posterior. The specimen has suffered from exposure to the weather, but it is evident that the external, and part of the anterior, surfaces were covered by a thin layer of cementum. The crown of the tooth and part of the odontome have been worn away by exposure to mastication. Elsewhere, the surface of the odontome is covered with cementum which has a granular surface. Ground sections of the specimen (Figs. 25.8B, C) show that the odontome is composed of a complex mixture of enamel, dentine and cementum with formations that resemble denticles of various sizes as well as loops of enamel the convexities of which face the external surface of the mass. The concavities of the loops are covered with dentine which in turn is covered with cementum.

It seems that the morphology of the third molar has not been affected by the growth of the odontome around it. The section (Fig. 25.8B) shows a distinct line of demarcation between the odontome and the cementum covering the root.

The portion of the mandible of an ox in Figure 25.9 has two odontomes embedded in the bone below the roots of the I_2. The i_1 are in position and appear to be normal. The left odontome consists of an irregular rounded mass with some projecting spikes. It overlaps the incisal edge of, and is firmly attached to, a tooth of normal permanent incisor shape (Fig. 25.9B) apart from where the odontome is attached. The mass on the right side is more globular and is not associated with a tooth (Fig. 25.9C). On one surface of the mass, there is a deep pit from the margins of which four spikes of enamel project. The odontomes have not been sectioned but they appear to be composed of enamel and dentine with a covering of cementum.

An odontome from an ox, morphologically closely similar to that depicted in Figure 25.9C was figured by Kitt (1892).

Two complex and composite odontomes in cattle were described by van der Gaag and Gruys (1972). One was bilateral, each partially enveloping the crown of one of the I_1 and emerging with it into the mouth. The other was between an M^2 and M^3, attached to both by what may have been the deposition of surface cementum.

Only a few odontomes have been described in primates apart from man. Figure 25.10 shows one in a colobus monkey. There is a large cyst over the root of the right maxillary canine, the margins of the cavity are slightly everted and, deep down in the cavity, there are a number of small masses of tooth tissue. The canine is normal in shape but the root has been bared by the cyst.

The skull of a Chacma baboon (*Papio ursinus*) (RCS Odonto. Mus., G 57.31) has a small irregularly-

Fig. 25.7. *Sarcophilis harrisii* (Tasmanian devil). Captive. RCS Odonto. Mus., G 57.4. A: An odontome embedded in the zygomatic process of the right maxilla. B: Palatal view.

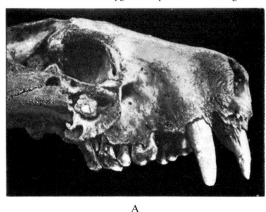

A

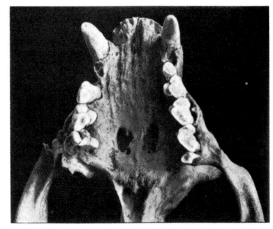

B

shaped composite odontome in the bone just above a P³.

Figure 25.11 shows the skull of a young captive baboon with a complex composite odontome in the right maxilla. The mass (Fig. 25.11B) is composed basically of three elements. One (a) resembles the malformed crown of an I² with a small tubercle projecting from its enamel-covered labial surface; what can be regarded as the cervical part of the crown is fused to the rest of the mass, presenting a lower zone of pitted

Fig. 25.8. *Ovis aries* (sheep). RCS Odonto. Mus., G 60.1. A: A composite odontome attached to an M₃.
B: A section through the odontome, showing its complex architecture with enamel loops with convexities facing the external surface, dentine and cementum with small denticles in cross-section. To left is a root of the molar in longitudinal section. There is a distinct line of demarcation between the root and the odontome.
C: A ground section of part of the odontome. An enamel-lined cavity is filled with cementum.

A

B

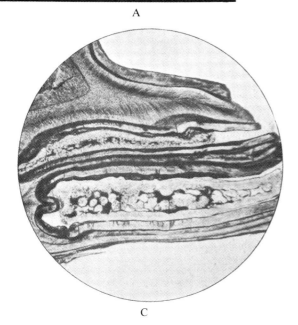

C

584 Section 4: Other Disorders of Teeth and Jaws

enamel and an upper zone marked by irregular transverse ridges. The part (b) consists of a fused cluster of three enamel-covered denticles between one pair of which there is a tongue-shaped fold. The third part (c) is an irregularly lobulated mass projecting from the cluster of denticles just mentioned. On the labial surface of the odontome, there is an opening leading deep into the interior of the mass. A radiograph shows that there is a complex pulp cavity with extensions into at least three of the elements of the mass.

In the specimen just described, there were indications of the formation of several denticles but they remained fused to the main mass; in the skull of a subadult dog shown in Figure 25.12, the odontome situated between P^1 and the upper canine consists of three separate elements. It is therefore to be categorized as a compound composite odontome. There is a small irregular mass of dental tissues and two minute spindle-shaped denticles. The odontome has displaced the canine slightly. The formation of its root is not complete and has not yet erupted whereas all three of the

Fig. 25.9. *Bos taurus* (ox). Museum of Comparative Pathology, University of Liverpool.
A: Ventral surface of portion of a mandible with two odontomes embedded in the bone. Windows have been cut to expose the odontomes.
B: The odontome from the left side of the mandible. Left, ventral surface. Middle, medial surface. Right, lingual surface.
C: The odontome from the right side of the mandible. Left, ventral surface; (en) indicates enamel. Right, lingual surface.

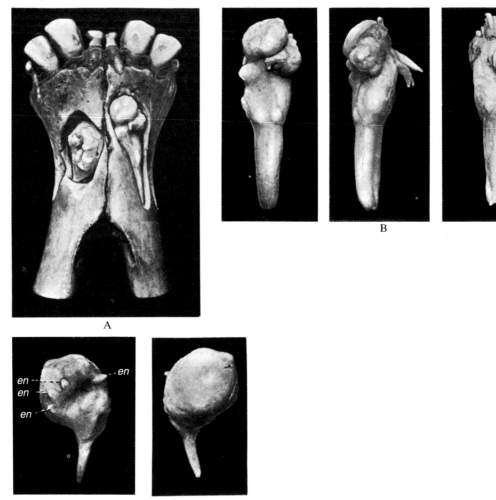

other canines are erupted. The position this odontome occupies is a common situation for supernumerary teeth to occur in the dog (see Chapter 4, p. 86).

Figure 25.13 shows a remarkable jumble of distorted toothlets in the mandible of a hippopotamus which seems to fall into the category of compound odontomes. The canine is directed forwards horizontally instead of vertically. The toothlets are partly embedded in the bone lateral to the canine and premolar roots and partly exposed on the surface. The toothlets comprise: (A) a conical toothlet, (B) a curved tooth shaped a little like a miniature canine. The end of the root is uppermost and has the funnel-shape indicative of continuing formation; (C) is a mass that looks like a coiled serpent and appears to consist of two denticles, each a little like (B), joined together; (D) is a small U-shaped denticle.

Compound composite odontomes may be associated with cysts, as in the bilateral examples in a 12-year-old thar (Himalayan goat) described by Bland Sutton (1888*b*). Each maxilla contained a cystic odontome which had expanded into the maxillary air sinus, reducing it to a slit (Fig. 25.14). The walls of the cysts were thick and fibrous and contained innumerable small nodules of dental hard tissue. Other similar nodules were free within the lumina (Fig. 25.14B).

Figure 25.15 shows a portion of the mandible of a horse in which the positions of P_3 and P_4 are occupied by a large mass of dental tissues which has expanded the bone. A part of the upper surface of the mass has been worn down by mastication showing that part of it had emerged into the mouth, so admitting infection and accounting for the evidence of suppurative inflammation in the surrounding bone, which is pierced in several places by sinuses. P_2 lies unerupted in the bone in front of the odontome and is covered by a thick layer of cementum which in one place forms a projection overlapping the odontome. The sectioned surface of the mass contains in the substance of its lower part two teeth which are probably P_3 and P_4.

The mass can be removed from the jaw and, in the

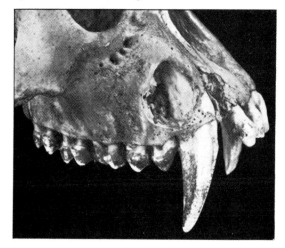

Fig. 25.10. *Colobus guereza gallarum* (Abysinnian guereza). ♂. ZMB no number. Upper jaw with a large cyst over the root of the right canine.

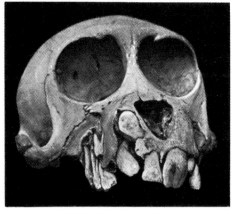
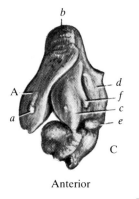
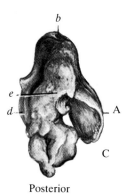

Fig. 25.11. *Papio* sp. (baboon). RCS Odonto. Mus., G 57.32.
A: A composite odontome in the right maxilla has been exposed by removal of the bone over I^2.
B: The two surfaces of the odontome. It consists basically of three elements. One element (A) resembles the crown of an incisor with a tubercle (a) projecting from its labial surface, and is fused at (b) to the other elements. (c), (d) and (e) indicate three fused denticles which comprise the second element. (f) is a tongue-like fold between two of the denticles. The third element is the irregular mass (C) below. × *c.* 2.0.

bony bed in which it rests, there are several irregular masses of mineralized tissue which can be picked out (Fig. 25.15C). Sections of these smaller masses show that they are composed of cementum containing lacunae, the remains of encapsulated cells, which are typical of the horse. Bland Sutton (1888a), who first described the specimen, categorized the masses as a cementome, i.e. an odontome composed of cementum only. However, dentine and possibly enamel are identifiable with the substance of the large mass so it is now categorized as a compound composite odontome. Additional comments were made by Colyer (1952).

A number of composite odontomes similar to the masses described here are depicted in E. Becker (1970).

The difficulty in making a distinction between composite odontomes, abnormally-shaped teeth and supernumerary teeth, with the inference that they represent different degrees of the same type of disorder, is illustrated by 79 skulls of crab-eating fox (*Cerdocyon thous*) in the British Museum of Natural History which are described in Chapter 4 (p. 94). There were two instances of mandibular disto-molars, one of a connate normal molar with a supernumerary tooth, and one (Fig. 4.57) where there was a cluster of denticles in place of the M_3. One denticle had the morphology of a miniature M_3, another was a single-rooted denticle and the third a tooth with three large cusps on a single root.

Two long-nosed mongoose skulls (*Crossarchus obscurus*) have been mentioned in Chapter 4 (p. 69). In one (Fig. 4.10), M^1 and M^2 on both sides have extra cusps on the buccal aspects of the crowns. In the other (Fig. 4.11), the right M^2 is larger than the contralateral tooth and has a distinct fissure crossing the occlusal surface and passing on to the lingual surface where it extends to the end of the root. The fissure thus partially divides the tooth into a larger mesial and a smaller

Fig. 25.12. *Canis familiaris* (domestic dog). RCS Odonto. Mus., G57.3. Lingual to the upper canine, there is a small compound odontome composed of three elements.

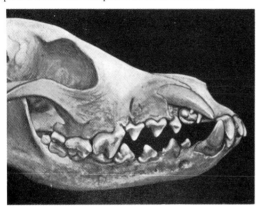

Fig. 25.13. *Hippopotamus amphibius* (hippopotamus). RCS Odonto. Mus., G77.5. Left side of the mandible with four irregular-shaped masses of tooth tissue. (A) Cone-shaped denticle; (B) denticle shaped like a canine; (C) spiral-shaped denticle; (D) denticle of U-shape. × c. 0.25.

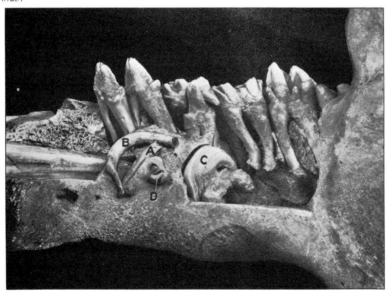

distal portion. We would categorize this as connation but it could alternatively be interpreted as an example of partial dichotomy of a tooth primordium, that is a stage towards the formation of two teeth. As an only partially-regulated overgrowth of dental formative tissues, it can be categorized as an odontome.

In the banded mongoose of Figure 25.16, the right M^2 has an extra cusp on the buccal aspect of the crown; the left M^2 is over-large and of very complex shape and is composed of several elements. From the occlusal surface (above), it seems to consist of a more or less normal molar which is rotated and has additional cuspal elements on its anatomical mesial surface. Seen from its positional mesial surface (below), however, it is evident that the fissures seen on the occlusal surface extend deeply for the whole length of the roots, partially dividing the tooth into a cluster of single-rooted, single-cusped toothlets. There has been a considerable degree of regulation of the overgrowth in the direction of formation of separate supernumerary teeth; the amount of regulation can easily be imagined. Nevertheless, comparing this specimen with the others described here, it is sensible to place it among the odontomes.

Encapsulated complex masses of enamel and dentine with a radial architecture and still undergoing active formation were described in the mandibles of two rats by Bullock and Curtis (1930) who categorized the lesions as neoplasms. However, the tissues, in addition to being partially organized into toothlets, seemed to be of near-normal structure which is unlikely in neoplasia. It is possible that the lesions were hamartomatous overgrowths of tooth tissue, that is complex composite odontomes in the sense used here.

Baer and Kilham (1964) described encapsulated masses of dentine and cementum with a fused-nodule structure in the jaws of young hamsters 12–18 months after experimental infection with rat virus. Some lesions were in relation to the roots of the mandibular molars; others were related to the growing ends of the upper incisors. The lesions appeared to contain no enamel and all that can be said is that they were odontome-like. The virus had other multiple effects; growth was stunted and various secondary infections

Fig. 25.14. *Hemitragus jemlahicus* (Himalayan thar). Formerly RCS Special Pathology 4182.1.
A: Part of the right maxilla has been removed to display a large polycystic mass in which are embedded numerous masses of tooth tissue, some with the form of denticles.
B: Some of the masses removed from the cyst walls or from their lumina. From Bland Sutton (1888b).

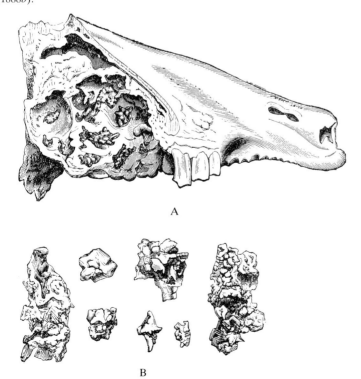

Fig. 25.15. *Equus caballus* (horse). RCS Odonto. Mus., G 59.2.
A: Portion of the mandible with a large mass of hard tissue in the position of P_3 and P_4.
B: The mandible with mass removed to show the irregular character of the wall of the cavity.
C: Six nodular pieces of hard tissue removed from the wall of the cavity. From Bland Sutton (1888a).

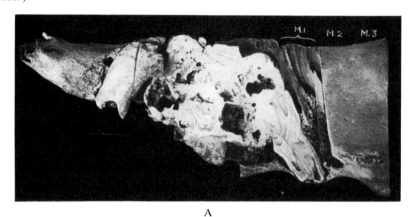

A

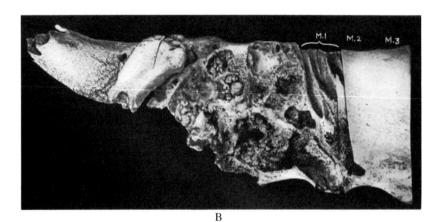

B

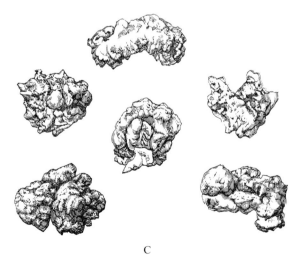

C

Ch.25. Odontomes 589

occurred. None of the control animals developed odontome-like lesions.

Dubielzig, Beck, Wilson and Ribble (1986) described localized areas of dentine dysplasia, possibly containing some enamel, in the region of the root bifurcations of P1 and M1 on both sides of the mandibles of two unrelated dogs. These lesions can be placed loosely under the general heading of odontomes because they represent disorganized growth of dental tissues and were probably developmental in origin. Both dogs were in terminal renal failure and one also had hyperparathyroidism secondary to the renal disease, but it seems unlikely that the dentine dysplasia and renal disease were causally connected.

DILATED COMPOSITE ODONTOMES

As already mentioned, many tooth abnormalities show enamel-lined invaginations which tend to be dilated and are therefore categorized as dilated composite odontomes. In 1942, Colyer described an odontome in a sub-adult male gorilla from the wild (RCS Odonto. Mus., G 61.2) that shows the basic features of this type of odontome (Fig. 25.17). It consists of an M_3 with a normal crown and a mesial root that has not yet grown to its full length. The distal root is expanded into an ovoid mass, 15×10 mm, with a rounded surface and an opening on the buccal side surrounded by irregular nodular enamel. A section through the specimen shows that the opening leads into an irregular space of

Fig. 25.16. *Mungos mungo* (banded mongoose). BMNH 1882.5.26.1.
A: View of palate. The right M^2 has an extra cusp on the buccal aspect; the left M^2 is abnormal in shape.
B: Drawings of the left M^2. Above, occlusal surface; positional lingual and buccal surfaces are indicated. Below, mesial surface; the lettering enables toothlets that appear in both views to be identified. $\times c. 3.0$.

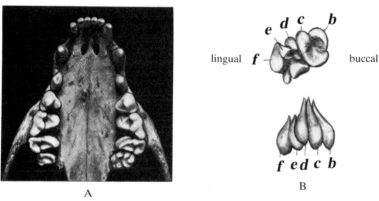

Fig. 25.17. *Gorilla gorilla* (gorilla). ♂ RCS Odonto. Mus., G 61.2.
Left: Mesio-buccal view of right M_3. The distal root is expanded into an ovoid mass.
Right: A mesio-distal section showing that the mass has the structure of a dilated composite odontome (Colyer, 1942b).

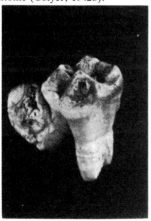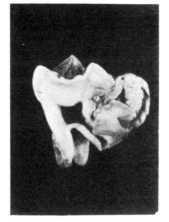

angular outline, lined with enamel and surrounded by a thick zone of dentine. The specimen thus has the structure of a dilated composite odontome. Its interest lies in the fact that it may be inferred that the development and formation of the tooth had proceeded normally until about one-half of the crown had been formed. Then, in a strictly localized area, the tooth-formative process must have become disorganized with general overgrowth and with growth of enamel-formative tissue into the interior of the overgrown dentine-formative tissue. In the way that is characteristic of the vast majority of mineralized odontomes, growth had ceased at about the time when the development and growth of the parent tooth was complete.

A much-worn upper molar of a horse (RCS Odonto. Mus., G 66.31), like the odontome in a gorilla just described, has a dilated expansion on its palatal surface, the rest of the morphology of the tooth being normal. This molar resembles one more fully recorded by Heinze and Sajonski (1964).

A deep dilated invagination in the crown of a maxillary molar in a hamster was described by Sharawy and Lobene (1968). The globular dilatation, which was filled with food debris and micro-organisms, communicated with the occlusal surface by a narrow opening. As is often the case, the dentine between the invagination and the pulp was thin and, presumably in consequence, bacterial toxins had penetrated and brought about death of the pulp.

Figure 25.18 shows an example in a molar of a Barbary sheep. Deep invaginations lined with enamel and filled with cementum are a normal feature of the crowns of the teeth of ruminants and equids; in other words, a process of invagination normally takes place in the enamel organ that is rather like that envisaged to be responsible for the anomalous invaginations of odontomes. This molar has an additional element on its lingual aspect which contains an expanded globular invagination open to the occlusal surface. The cavity is lined by what appears to be cementum but there is one area (f in Fig. 25.18C) lined with enamel. On the root portion of the additional element, there is a mass of cementum containing another cavity which is connected by a narrow canal with an opening in a depression (h in Fig. 25.18C) on the wall of the coronal cavity.

The odontome shown in Figure 25.19 is another example of disordered development with the principal feature an invagination of the enamel organ. It is a P^4 of a Cape buffalo which is of nearly normal morphology apart from a hemispherical prominence on the rootward half of the mesial aspect of the crown. There is a similar smaller prominence on the distal aspect. As a preliminary to interpretation of sectioned surfaces cut through this specimen, it is necessary to refer to the arrangement of the tissues in the normal ruminant premolar (Fig. 25.20). The arrangement is similar to that already described for the sheep's molar but there is only one enamel-lined infolding, crescent-shaped in cross-section, in the crown instead of two (Fig.

Fig. 25.18. *Ammotragus lervia* (Barbary sheep). RCS Odonto. Mus., G 60.3.
A: An odontome in connection with the left M^1.
B: Diagrammatic view of the occlusal surface of the specimen. Buc., buccal surface; Dist., distal surface; Ling., lingual surface; Mes., mesial surface; (c) normal cementum-filled occlusal infoldings of enamel; (d) dentine; (e) enamel; (q) reactionary dentine laid down by the pulp in response to wear and now exposed by the wear; (b) empty cavity in the expanded odontomatous part of the tooth. The arrow indicates the plane of the oblique longitudinal section seen in C.
C: Drawings showing the tissues identified in the surfaces of the odontome divided as indicated by the arrow in B. Left, the buccal surface. Right, the lingual surface; (a) cementum-like tissue lining the expanded cavity; (f) an area of enamel lining; (d) dentine; (e) enamel; (c) cementum; (p) pulp cavity; (g) a cavity in the mass of cementum at the base of the tooth which opens by a narrow canal into a depression (h) in the wall of the cavity in the crown.

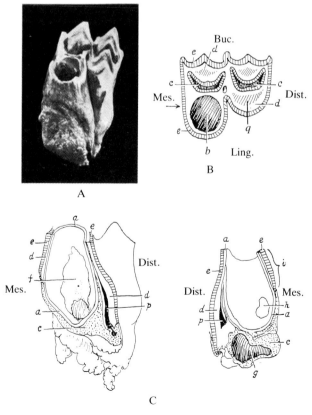

25.20 A); the coronal infolding normally has an extension towards each root. The cementum usually nearly fills the invagination leaving only a central canal, the infundibulum but, sometimes, as in Figure 25.20 B, it simply lines the enamel leaving a moderately-large cavity open to the mouth.

The sectioned surfaces of the odontomatous premolar (Fig. 25.19 B) cut approximately in the same mesio-distal plane as the normal tooth (Fig. 25.20 B) show a complex arrangement of tissues with four additional layers of enamel compared with the normal tooth. These additional layers can be interpreted as the lining of two channels, one on the buccal side of a central cavity, which is lined with thicker enamel, and one on its palatal side. The channels have an outer thin lining of enamel, which are the four additional layers in the section, and an inner lining of cementum. At the level of the prominences seen on the outside of the specimen (Fig. 25.19 A), these channels open into expanded cavities which communicate with one another. Here there is no lining of enamel between the cementum and the dentine. It seems possible to interpret the specimen as follows: If we regard the outer incomplete layers of enamel as representing the normal cementum-lined enamel-infolding of a ruminant premolar and call it the primary invagination, the innermost and more obvious enamel-lined cavity is a secondary invagination, derived from the enamel organ, that extends into the cementum that normally lines, or more or less fills, the primary infolding of enamel.

A specimen which may belong to the same category of disorder as the two foregoing is shown in Figure 25.21. It was categorized as a cementome by Woods (1907) who first described it and that view was endorsed by Gabell *et al.* (1914) and by Colyer in the first edition of this book. We offer an alternative interpretation for consideration. The specimen is without history. It is an upper molar of a horse. On the buccal aspect of the root, there are two irregularly-rounded

Fig. 25.19. *Synceros caffer caffer* (Cape buffalo). RCS Odonto. Mus., G 61.1.
A: An odontome in connection with a right P^4.
B: Drawings (compare with Fig. 25.20) showing the interpretation of the structure and tissues identified and divided longitudinally in the mesio-distal plane. (e) indicates the outer enamel of the crown; (c) its outer covering of cementum; (d) dentine; (e_1) the enamel lining the infolding of a normal tooth, here grossly dilated; (c_1) the cementum-infilling into which the abnormal secondary invagination has occurred lined with thicker enamel at (e_2) and filled with cementum at (c_3).

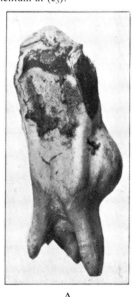
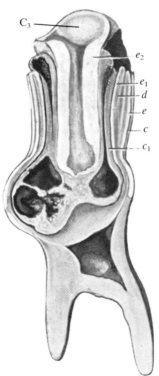
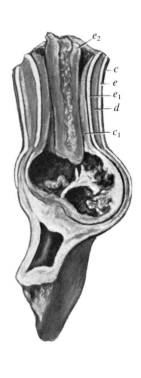

592 Section 4: Other Disorders of Teeth and Jaws

projections; the lower one is about the size of a pea and the one nearer the end of the root is about the size of a walnut. The smaller projection has two openings of irregular shape leading into a cavity; the larger one has a single larger opening into a cavity within. The worn occlusal surface depicted in Figure 25.21 C presents the usual features of the horse-molar wear-table, including two cementum-filled enamel-infoldings but there is a slit-like opening to the buccal side of these, where there ought to be dentine.

It does not seem to be established whether the occlusal slit communicates with the cavities in the projections on the root but, if it does, this specimen could be a dilated odontome. This would be confirmed if there are at least traces of enamel lining the cavities. However, it is always possible that an aberrant invagination of enamel organ produced the deformity of the tooth but did not actually proceed to form enamel. A section through part of one of the projections shows only that the wall of the cavity is composed of cementum-like tissue with vascular channels (Fig. 25.21 B).

A similar type of disorder of development seems sometimes to involve the infundibulum of the horse incisor, leading to a dilated deformity of the crown. An example was figured by Kitt (1892).

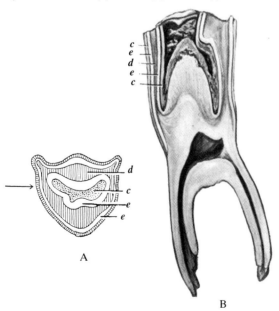

Fig. 25.20. *Synceros caffer* (Cape buffalo). Sections through a normal maxillary premolar for comparison with Figure 25.19.
A: Transverse section through the crown. Buccal surface above, lingual surface below.
B: Longitudinal section in the mesio-distal plane indicated by the arrow in A. (e) enamel; (d) dentine; (c) cementum.

Figure 25.22 shows another variety of dilated odontome, in an oryx. On the palatal aspect of the right maxillary premolars, there is an approximately-ovoid odontome which has two large openings with irregular margins. Part of this surface has been worn by mastication. The surfaces against P^2 and P^3 present grooves congruent with the teeth and there is a tongue-like projection of the odontome which passes between two teeth. The openings lead into a single large cavity. A section through the wall (Fig. 25.22 B) shows only layers of cementum without enamel or dentine. Specimens of this kind cannot be thoroughly explored unless multiple sections are made but, by analogy with other dilated odontomes, it is unlikely to be composed of cementum only. The congruity with the adjacent teeth and the tongue-like extension suggests that the odontome developed after those teeth had been formed. On the other hand, the congruity could be produced by the later deposition of cementum on the surface of the odontome as can occur on a normal tooth.

Richard Owen, in his *History of British Fossil Mammals and Birds* (1846), described as abnormal a tooth in a fossil horse (Fig. 25.23) which had features somewhat similar to those referred to above:

> The tooth was a fossil molar of a large sized horse, *Equus fossilis*, from the tertiary formations near Cromer. The tooth was from the lower jaw, and presented a swelling near the base of the implanted portion. To ascertain the cause of this enlargement a transverse section was made, when a spherical cavity was exposed large enough to contain a pistol-ball. The inner surface was smooth; the parieties of this cavity, composed of dentine and enamel of the normal structure, were from one to two lines thick, and were entire and imperforate. The water percolating the stratum in which this tooth had lain had found access to the cavity through the porous texture of its walls, and had deposited on its interior a thin ferruginous crust, but the cavity had evidently been the result of some inflammatory and ulcerative process in the original formative pulp of the tooth.

The better understanding that we now have of the way in which teeth develop and are formed does not allow us to accept Owen's suggestion that the cause is inflammatory in the sense in which we use the word today. We conceal our present ignorance of the exact details of the processes involved by saying that these disorders are due to the absence of the normal regulative processes that normally order the behaviour of cells during development.

As a rare phenomenon, true neoplasms, especially malignant ones, may arise in connection with odontomes; for example Brodey and Morris (1960) described in a young dog a compound composite odontome, consisting of a large number of small denticles of various sizes; another cluster of larger denticles, each consisting of a small root surmounted by a conical or globular crown, erupted in the palate. Four years later, a rapidly-growing undifferentiated carcinoma arose among the unerupted denticles and the associated soft tissues.

Fig. 25.21. *Equus caballus* (horse). Museum of Comparative Pathology, University of Liverpool.
A: Buccal view of a left M^2. The occlusal surface is below. There are irregularly-rounded projections on this surface. Openings, (a), (b) and (c), lead to cavities in the tooth.
B: Ground section of part of the wall of the cavity at (a) in A, showing the cavity surface on the right and that the wall is lined with cementum.
C: Drawing of the occlusal surface of the specimen on the left. (Mes.) indicates the mesial surface. For comparison, the occlusal surface of a normal horse molar in the same orientation is on the right. (e) enamel; (d) dentine; (c) cementum. (Woods, 1907.)

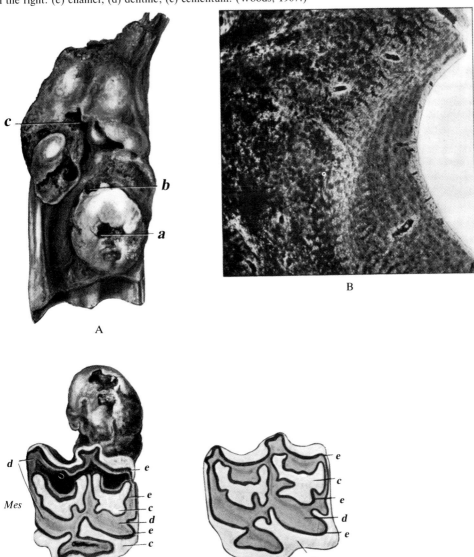

Section 4: Other Disorders of Teeth and Jaws

RADICULAR ODONTOMES

The term radicular odontome was used by Broca (1869) to include odontomes which arose subsequent to the completion of the crown. It is still a useful descriptive morphological term and some radicular odontomes, because they involve either cementum alone or dentine and cementum, may fall into the connective tissue or mesodermal category of later classifications. However, what Bland Sutton (1888a) regarded as the commonest variety of this type of odontome, those on the roots of teeth of continuous growth in rodents, is strictly a composite odontome because, in these teeth, enamel formation still proceeds on at least one side of the root; hence abnormal growth involves enamel, dentine and cementum.

The most notable example is one described by Bland Sutton (1885a,b, 1888a) (Fig. 25.24A) in a marmot (*Marmota* sp., there is doubt whether Bland Sutton referred to the species correctly). The abnormal growth of tissue on the basal ends of the upper incisors had produced resorption of the bone of the palate so that the masses project beyond the surface and were originally mistaken for supernumerary teeth (Bland Sutton, 1884a). Unfortunately the specimen has been damaged and one of the upper incisors lost so that it is only possible to depict three incisors in Figure 25.24B. The upper incisor is shorter than normal and the radicular half has a crumpled appearance with a nodular expansion at what would have been the growing extremity. However, growth would have ceased at the time, perhaps shortly before death, when the ends of the teeth ulcerated through the soft tissue of the

Fig. 25.22. *Oryx beisa* (Beisa oryx). RCS Odonto. Mus., G 60.2.
A: A dilated composite odontome situated on the lingual aspect of two premolars with which its surface is congruent, even to the extent of a process projecting between the two teeth. Its surface is worn by mastication and there are two irregular openings into a single central dilated cavity.
B: A ground section through part of the wall of the odontome. (a) outer layer of cementum containing many lacunae arranged in rows; (b) a layer of more trabecular cementum with spaces that had contained vascular tissue; (c) inner layer of cementum similar to that of (a). No enamel was detected.

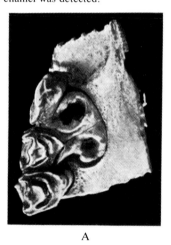

A

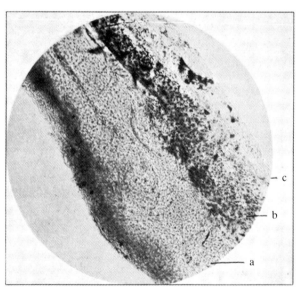

B

Fig. 25.23. *Equus* sp. (fossil horse). An abnormal-shaped lower molar (Owen, 1846; Bland Sutton, 1885b).

palate. The radicular parts of the lower incisors are also transversely ridged and the enamel at the growing ends curls over towards the pulp cavity. The cheek teeth, which are of limited growth, are normal

Fig. 25.24. *Marmota* sp. (marmot).
A: Palatal view showing the bases of the incisors projecting through the bone of the hard palate. (Bland Sutton, 1884*a*, 1885*a*, *b*, 1888*a*).
B: Three of the incisors removed from the specimen. The left upper incisor above is shorter than normal and has a nodular expansion at its basal end. The two lower incisors are below.

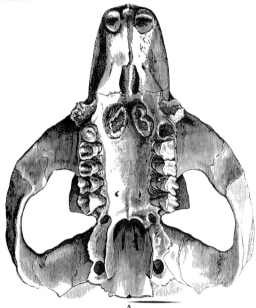

A

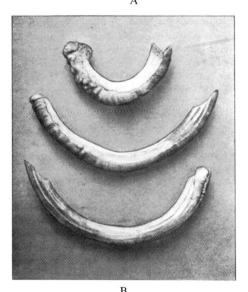

B

although there is some evidence of a destructive process at the alveolar margins of the bone.

Whether this specimen can properly be regarded as evidence of disordered growth of the kind we met in odontomes, or is the product of some interruption in the eruption of the incisors leading to apparent overgrowth of the basal ends, is uncertain. Bland Sutton (1893) did not include the specimen in his later general account of odontomes.

Nevertheless, there are many specimens that do appear to be examples of odontomatous overgrowth. Bland Sutton (1888*a*) illustrated three; one was in a young prairie marmot (or black-tailed prairie dog, *Cynomys ludovicianus*), with masses of hard tissue connected with the growing ends of the incisors on both sides of the upper and lower jaws. Figure 25.25 is a section through one of the masses which Bland Sutton interpreted as a disorderly type of dentine. The second example was a mass on the end of the root of an incisor in an adult North American porcupine (*Erethizon dorsatum*). The third was a similar mass on the end of an incisor of an agouti (*Dasyprocta* sp.) which microscopy showed was composed mainly of cementum with some patches of dentine (see also Bland Sutton, 1885*a*). All three examples were associated with suppuration which appeared to be sequential to the eruption of part of the odontome into the mouth. Bland Sutton (1922) illustrated a radicular odontome on the basal extremity of the continuously-growing mandibular canine of a boar.

An example similar to those just referred to is shown in another prairie dog in Figure 25.26. There is a nodular growth on the basal ends of both mandibular incisors. On the left, the mass has grown out on the lateral aspect of the coronoid process and, on the right, it has penetrated the medial aspect. The cheek teeth are affected by caries and there is some resorption of the bone of the alveolar margins.

Masses of tooth tissue showing various degrees of disorganized growth, and therefore simulating radicular odontomes, are common in injured teeth. The examples of carnivore canines (e.g. Fig. 19.9) in Chapter 19 show this particularly well, as do the many examples of injuries to teeth of continuous growth such as the tusks of elephant (e.g. Fig. 19.27).

Irregular localized masses of cementum produced on the surface of roots as an inflammatory reaction can also simulate radicular odontomes. Figure 23.6 shows an example in a horse where, projecting from the root of a mandibular premolar involved in an abscess, there is a sponge-like mass of firmly attached cementum which was probably formed as a reaction to the suppurative infection.

596 Section 4: Other Disorders of Teeth and Jaws

CEMENTOMAS

Radicular odontomes apparently formed of cementum only and attached to roots may be called cementomas. It is often difficult to distinguish normal cementum from bone except when the tissue is an integral part of a normal tooth. When the cementum is of abnormal structure because it was formed in response to an inflammatory process or as a result of anomalous development or perhaps as a low-grade neoplastic process, it is often impossible to say with certainty whether what may be called the parent tissue is cementum or bone, especially bearing in mind that ankylosis, that is disappearance of the periodontal ligament and fusion between the cementum and bone of the jaw, can occur. Ankylosis could occur as a transient phenomenon.

Large so-called cementomas are particularly common in the horse. Figure 25.27 shows an example from an old cart horse. It is a mass measuring 14 × 10 cm in the maxilla over the sockets of the second and third molar teeth. Those teeth are no longer present and evidently had not been in continuity with the mass. The

Fig. 25.25. *Cynomys ludovicianus* (black-tailed prairie dog). Juvenile. A section through an odontome on the basal end of a mandibular incisor (Bland Sutton, 1885a, b, 1888a). The mass appears to be composed of a disordered type of dentine and has involved the roots of molars seen in transverse section on the left.

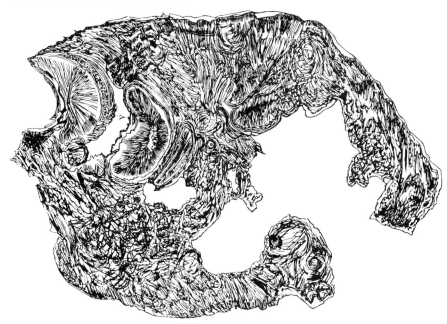

Fig. 25.26. *Cynomys ludovicianus* (black-tailed prairie dog). RCS Odonto. Mus., G 62.2. Left, the two halves of the mandible. There is a mass of hard tissue at the basal end of each incisor. Right, the incisors removed from the jaw.

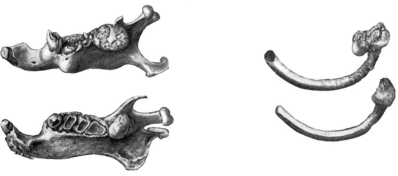

mass must have filled most of the maxillary air sinus. Its surface is rounded and nodular. A ground section of part of the mass (Fig. 25.27B) showed a dense structure resembling cementum but with many channels which must have carried blood vessels. This specimen is one of four in the Museum of Comparative Pathology of the University of Liverpool described by Woods (1907).

Figure 25.28 is a similar example from the same Museum although in this horse the rounded mass of the odontome, apparently composed of cementum only, is attached to the buccal side of the tooth, an upper third premolar.

The odontome in Figure 25.29 arose in connection with the mandibular molar of a horse; it measures about 10.5 cm × 6.0 cm. It was originally described by C. S. Tomes (1872) who mentioned that it weighed 10 oz (283 g). The tooth is partly surrounded by the mass and on the surface there is the impression of another tooth which the mass must have displaced. A ground surface of the odontome (Fig. 25.29A) and two fracture surfaces in the opposite plane (Figs. 25.29B and C) show that it has a compact outer shell from which stalactite-like ribs project inwards to a more compact centre, leaving multiple spaces of various sizes between the ribs. Although the greater part of the mass seems to be cementum, in some areas, it is dense enough to suggest dentine, as pointed out by Tomes, so this may be a composite odontome and not a cementoma.

Figure 25.30 shows yet another enormous odontome from a horse. The mass weighs 70 oz (nearly 2 kg) and is composed of cementum in which the lacunae show richly-branching canaliculi and many spaces which had contained vascular tissue (Bland Sutton, 1891).

Joest (1926, fig. 143) figured a cementoma, small by comparison with those just described, attached to the side of a P³ of a horse.

Fig. 25.27. *Equus caballus* (horse). Museum of Comparative Pathology, University of Liverpool.
A: Part of the left maxilla viewed from the palato-medial aspect. There is a large composite odontome which has encroached on the maxillary air sinus.
B: Ground section of part of the odontome. The cementum is very vascular, the lacunae numerous but small and fairly regular (Woods, 1907).

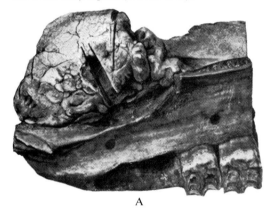

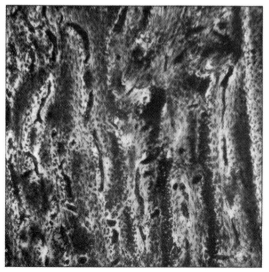

Fig. 25.28. *Equus caballus* (horse). Museum of Comparative Pathology, University of Liverpool. Oblique palatal view of the upper jaw. There is an odontome involving the buccal aspect of the left P³. The right P² (pointer) is misplaced with occlusal surface facing buccally. Palatal mucosa and bone have been removed to display the root.

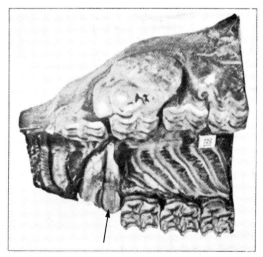

598 Section 4: Other Disorders of Teeth and Jaws

An odontome in the position of the maxillary canine in a horse, but encroaching on the incisors, is shown in Figure 25.31. Only part of the outer portion of the mass has been sectioned and that seemed to be composed of cementum-like tissue. A similar odontome was described by Magitot (1877, his plate X, fig. 7 and plate XVIII, fig. 8).

Cementomas in animals other than the horse are rare. Woods (1907) described one that affected the lower incisor of an ox. There was no history. The tooth seemed to be of normal morphology but most of the crown was enclosed in a mass of cementum which a ground section (Fig. 25.32) shows to be up to 1.4 cm thick. It seems by no means certain that this is an example of aberrant development and therefore an odontome. It seems possible that it was a tooth that had remained buried in the jaw for many years. Under those circumstances, the coronal cementum might slowly continue to be formed; however, in that case the cementum would probably show a system of growth lines concentric to the tooth. Instead, the cementum (Fig. 25.32) appears to be trabecular with marrow spaces. The mass was continuous with the smoother-surfaced cementum over the root, which argues against the mass being bone ankylosed to the crown of the tooth.

Fig. 25.29. *Equus caballus* (horse). RCS Odonto. Mus., G 59.1.
A: An odontome in connection with a mandibular molar. The odontome was fractured into two pieces in the plane indicated by the arrow. (a) indicates a hollow where the mass fitted against the root of an adjacent tooth.
B: The fractured surface of the left portion.
C: A section of the right portion cut in a horizontal plane.

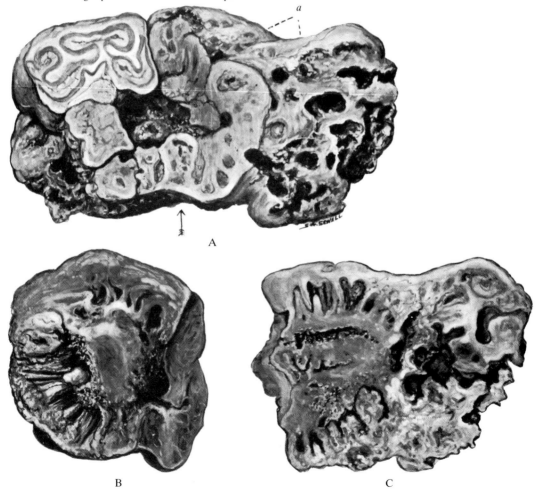

Ch.25. Odontomes 599

Figure 25.33 shows a mandibular premolar of an hippopotamus with a small cementoma at the bifurcation between the two roots.

ECTOPIC TEETH

Teeth sometimes occur in places remote from their normal situation in the jaws (see Chapter 9, p. 153). Apart from the vertical ramus of the mandible and high up in the posterior part of the maxilla, where they probably arise from a simple over-extension backwards of the dental lamina, and in ovarian cysts, one of the commonest ectopic situations is the region of the ear. Most of the examples in the literature are in the horse and sheep. In the horse, the usual situation is behind the inner ear in the mastoid region, hence the term mastoid or tympanic teeth. The teeth are usually misshapen and may consist of either clusters of toothlets or of unorganized masses of enamel, dentine and

Fig. 25.30. *Equus caballus* (horse). Museum of the Royal Veterinary College. A large odontome weighing 70 oz., at one time thought to be a fossilized brain (Bland Sutton, 1891).

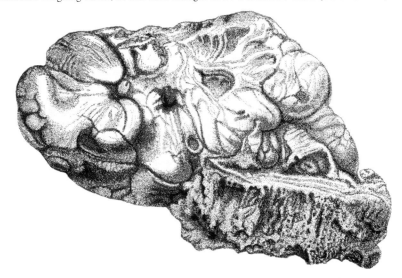

Fig. 25.31. *Equus caballus* (horse). Museum of the Royal Veterinary College. An odontome in the position of the right maxillary canine.

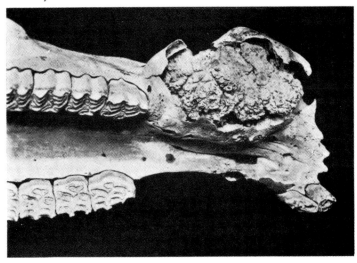

Fig. 25.32. *Bos taurus* (ox). Transverse section through the crown of an incisor surrounded by an overgrowth of cementum which has a trabecular structure enclosing vascular tissue (Woods, 1907).

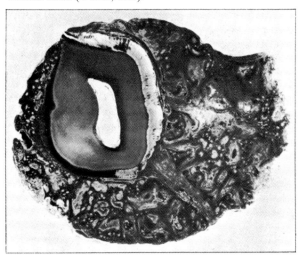

Fig. 25.33. *Hippopotamus amphibius* (hippopotamus). RCS Odonto. Mus., G 58.1.
Left: A mandibular premolar with a mass of cementum attached to the bifurcation between the roots. ×c. 0.5
Right: A surface of the sectioned tooth. ×c. 0.5.

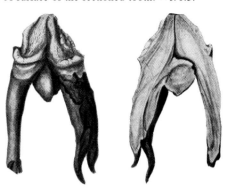

Fig. 25.34. *Equus caballus* (horse). The mastoid region of the skull with a cluster of fused teeth (Bland Sutton, 1904).

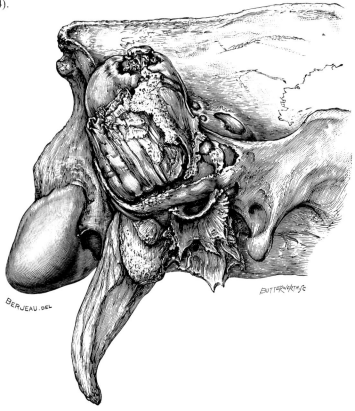

cementum; in other words, they have the structure of a complex composite odontome. The example shown in Figure 25.34 appears to be composed of a fused cluster of at least four teeth (Bland Sutton, 1904); the sectioned example in Figure 25.35 consisted of 2 masses, together weighing 219 g, with the structure of complex composite odontomas.

There are two examples (G 66.2, G 66.3) of mastoid compound odontomes in horses in the RCS Odontological Museum.

Magitot (1877) described two instances and his figures show how the mastoid odontomes or teeth may produce considerable elevations on the intracranial surface of the bone. Two examples in the horse were figured by Kitt (1892). An important review of the early literature on ectopic teeth in the horse is available in Lanzilotti-Buonsanti and Generali (1873) who gave details of 34 instances of teeth, mostly malformed in the temporal region or ear, many associated with cysts or abscesses.

In sheep, the location of the ectopic teeth tends to be different, and they appear to be parts of branchial fistulae opening just beneath the external ear (Kitt, 1892). One or more teeth, often identifiable as incisors, occur in the cutaneous tissues around the fistula. Often the teeth erupt through the skin. According to Bland Sutton (1922, p. 527) who reviewed the literature, the teeth may be part of a miniature mandible and there may be a rudimentary tongue and lip.

Figure 25.36 shows the mandible of a lamb with two additional rows of teeth on the lateral aspect of the left jaw. Close examination shows that the upper row has the morphology of teeth of the right side and the lower row of teeth of the left side. This is in fact a teratoma consisting of what can be called a partially-duplicated mandible (Colyer, 1952). A similar example was described by Grajcarek and Willer (1985).

Double mouth deformity is an entity encountered in adults of many species of fish (reviewed by Nomura, 1978) and has occurred as an epidemic in trout (*Salmo trutta*) hatcheries (Swan, 1968). Typically there is an extra mouth, more or less complete with jaws and teeth, on the ventral aspect of the head. In a dace (*Leuciscus haskonensis*) described by Y. Honma, Chiba and Yoshie (1981), an extra mouth opening was present but without the bony jaws.

Gupta (1978) described a guinea pig in which there was a pair of additional incisors associated with a piece of bone situated in the mandibular symphysis. This could be regarded as the teratomatous formation of part of a mandible.

A teratomatous cyst in a calf, containing two caniniform teeth and situated in the symphysis between the two halves of the mandible, was described by J. R. Baker (1981).

FIBROUS ODONTOMES

Bland Sutton (1885*a,b*, 1888*a*) described as fibrous odontomes a number of localized expansile lesions that appeared to be composed of fibrous tissue, not all of which incorporated teeth in their substance. The designation fibrous odontomes was followed by later writers, as will be recounted here, although there is now considerable doubt whether many of them are correctly so-designated; indeed, the category fibrous odontome is not recognized by most contemporary

Fig. 25.35. *Equus caballus* (horse). Two masses of dental tissue from the temporal fossa of a mare (Bland Sutton, 1904). The larger mass (A) weighed 175 g and shows a sectioned surface. The mass (C) weighed 44 g. (B) is a sectioned surface of it. The sectioned surfaces show a disorderly structure of infolded-enamel sheet, dentine and cementum. Surface (B) suggests slightly the cementum-filled infoldings of enamel characteristic of horse cheek teeth.

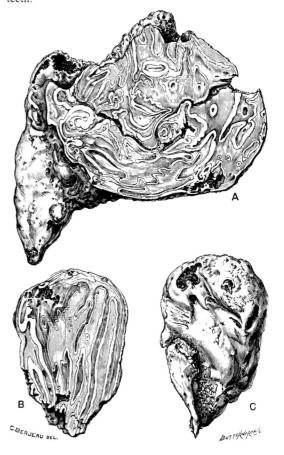

602 Section 4: Other Disorders of Teeth and Jaws

authorities and Sprawson (1937) excluded this category from his classification.

Some lesions that have been described as fibrous odontomes appear to be dentigerous cysts with unusually thick fibrous walls.

Figures 25.37 and 25.38 show two lesions described by Bland Sutton (1888a) as fibrous odontomes. One (Fig. 25.37) is in a dasyure (marsupial cat) in which there were bilateral, rounded encapsulated masses the size of walnuts in the maxillae. Microscopy showed that they were composed of lamellae of dense fibrous tissue enclosing a central softer portion into which the roots of a molar projected. Bland Sutton wrote:

In many parts of this dense capsule, tracts of bony tissue were found. Near the centre of the mass the fibrous layers became attached to the neck of the tooth and seemed to blend with the periodontal membrane. The central softer part of the tumour contained numerous giant cells and embryonic connective tissue; in some ... sections a direct continuity of this tissue with the pulp of the tooth was clearly made out.

The skeleton of this animal was everywhere softened by rickets, the lower jaw being so flexible that it could be twisted as easily as gutta-percha.

In the second case, a goat, there were four swellings, one in each maxilla and two in the mandible, the latter being situated near the angles of the bone. The bone of the skull was so soft that it could be cut easily

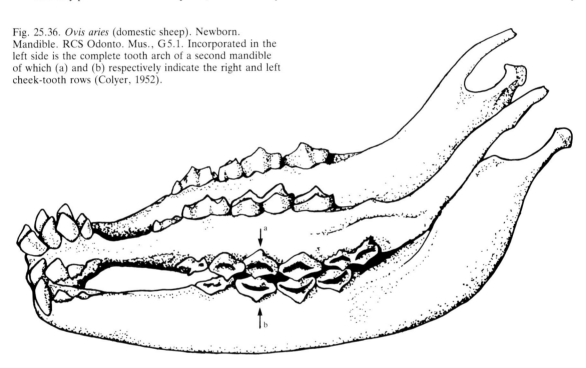

Fig. 25.36. *Ovis aries* (domestic sheep). Newborn. Mandible. RCS Odonto. Mus., G 5.1. Incorporated in the left side is the complete tooth arch of a second mandible of which (a) and (b) respectively indicate the right and left cheek-tooth rows (Colyer, 1952).

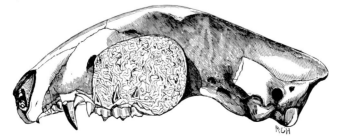

Fig. 25.37. *Dasyurus* sp. (marsupial cat). Part of a skull showing a fibrous odontome in section (Bland Sutton, 1888a).

with a knife. A section was made through the right maxilla (Fig. 25.38). Three teeth were involved in the swelling which had pushed up the floor of the air sinus and had expanded the other walls of the bone and formed a rounded projection externally. In the mandible, each swelling was composed of a spongy mass with a small internal cavity, filled with a pulpy-looking material. There were no traces of teeth in the swelling (see also Bland Sutton 1885a,c).

Bland Sutton later (1889) reported other similar lesions, also referred to as fibrous odontomes, in two bears from the London Zoo which had to be killed because of severe rickets. 'They were the most rickety mammals I have ever examined. The skull bones ... were greatly thickened, yet so soft that they were readily cleft by means of a stout knife. The follicles of the molar teeth in each upper jaw were greatly enlarged and projected as rounded fibrous tumours into the orbits and zygomatic fossae.' One of the enlarged tooth follicles is shown in Figure 25.39. Bland Sutton compared the lesions with examples of human rickets which at that time was a common condition in children. Gabell *et al.* (1914) appear to have accepted Bland Sutton's view that these lesions could be called fibrous odontomes and were associated with rickets.

Before the discovery of the antirachitic factor, i.e. the vitamin D complex, in 1917, the term rickets cannot have had the precision of meaning it has today. It is possible that the animals described by Bland Sutton were suffering from true rickets with lesions that happened to occur in the jaws and to involve unerupted teeth (see Chapter 16, pp. 333–334). The jaw lesions were not odontomes in the sense that we use the term today.

However, Bland Sutton (1888a) also put forward the suggestion, in respect of some of the circumscribed lesions which contained spicules of cementum or bone within the fibrous stroma, that, had the animals lived, each lesion might have become a mass of cementum or bone; in other words, that they were cementomas or osteomas in process of being formed; if the mass was connected with teeth, it was more likely to be a cementoma. This comes close to the concept behind the terms cementifying fibroma and ossifying fibroma as used at the present time.

There is in the RCS Odontological Museum the half mandible of a young goat (G 65.2) which has been divided transversely to display a large ovoid lesion which has expanded the vertical ramus and extends forward to involve the last molar which has exfoliated. The lesion appears to be strictly circumscribed and is encapsulated by a thin cortex of bone, except medially where it has broken through the bone. Within, there is a mostly dense homogeneous mass of hard tissue containing a few globular and irregular spaces. Most of its periphery is separated from the investing cortex by a narrow space bridged by trabeculae of the lesional tissue.

At the anterior end of the jaw and replacing the incisors, the bone is expanded by an irregular cyst-like cavity, widely open at the symphysis. Although empty now, this cavity could have contained a tissue mass like the other lesion. The specimen has no history but it resembles the specimens described and illustrated by Spencer (1889). In the absence of more information or histological evidence, it is impossible to do more than speculate that the larger more complete lesion could be a cementum-forming or bone-forming neoplasm; however, the appearances are also consistent with those of an osteodystrophy (Chapter 16, p. 334), such as fibrous dysplasia, and the existence of two similar lesions makes this alternative more likely.

Spencer (1889) gave an account of a disorder in a herd of goats which tended to run an acute course of a

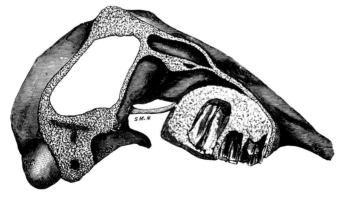

Fig. 25.38. *Capra hircus* (goat). Skull with a fibrous odontome connected with the teeth (Bland Sutton, 1888a).

few months and either ended in death or in such debilitation that the animals had to be killed. The predominant features were bilateral swellings of the lower jaw, usually between the angle and the molar teeth, which increased in size until the mouth could not be closed and the animal could no longer feed or masticate. The swellings when cut were firm and pink; spicules of bone produced a grating sensation when the cut surface was scraped with a knife. The upper jaws and long bones were sometimes affected also. Enlargement of the epiphyseal plates, characteristics of rickets, was absent. The affected animals were kept for milking in South London on a diet of hay and corn, supplemented by any garbage that they could scavenge as they wandered about the streets, conditions conducive to nutritional deficiency. Spencer drew attention to the similarity between this disorder in goats and the condition known as big-head or bran-disease in horses which occurs when they are fed exclusively on bran and is now considered to be due to a high phosphorus–low calcium diet (Jones and Hunt, 1983, p. 1171). In a group of horses described by Walthall and McKenzie (1976) with lesions of this kind principally in the jaws, there was evidence to suggest that the cause was the high oxalate content of certain pasture grasses. The multiple cysts of the jaws of horses described by Espersen (1962) appear to be another type of osteodystrophic lesion.

Knaggs (1923) re-interpreted the condition in goats described by Spencer (1889) as an example of osteitis fibrosa. Figure 25.40 shows one of the lesions in a goat that Knaggs interpreted in this way. Osteitis fibrosa was a term that was then beginning to be employed as part of the recognition of a group of bone disorders now known as the osteodystrophies (Andrews, 1985b). Hunter and Turnbull (1931–32) showed that the diffuse variety of osteitis fibrosa in which there are lesions in several bones is due to hyperparathyroidism. The terminology at the present time is monostotic fibrous dysplasia for the solitary lesions of similar histology which have a different ill-understood aetiology unassociated with hyperparathyroidism. It is beyond the scope of this book to elaborate further or speculate on the nature of these non-odontomatous conditions. We have only dealt with the matter because the so-called fibrous odontomes were dealt with by Colyer in the first edition and references to them occur in the literature.

Halliwell and Hahn (1980) described seven instances of fibro-osseous lesions in the tooth-bearing regions of the jaws of the greater kudu (*Tragelaphus strepsiceros*) which bore some resemblance to the lesions just described. Most were in the mandible, some bilateral and in several there was renal disease so that the authors suggest that the lesions may be manifestations of hyperparathyroidism secondary to

Fig. 25.39. *Ursus* sp. (bear). RCS Odonto. Mus., G 48.1. A: An upper molar partially enclosed in what Bland Sutton (1889) described as a greatly thickened fibrous follicle. The animal suffered from severe rickets.

B: A demineralized section of the tissue associated with the tooth, showing a matrix of fibrous tissue in which short trabeculae of bone have been formed. Active bone formation is proceeding on some surfaces.

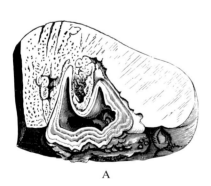

A

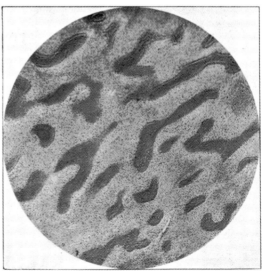

B

renal failure. When the jaw lesions are part of widespread similar focal changes in the skeleton and parathyroid hyperplasia, as well as advanced renal disease are demonstrated at necropsy, as in a female wolf described by Feeback, Jensen and Kosanke (1986), the sequence of events is not in doubt.

PSEUDO-ODONTOMES

Schour, Bhaskar, Greep and Weinman (1949) described odontome-like structures (pseudo-odontomes) that formed in the ends of the roots of the incisors in the ia mutant rat in which there is a defect in the process of osteoclastic bone resorption. Because the eruption of teeth depends on bone resorption as well as bone deposition, the incisor teeth fail to erupt (hence the name of the mutant, incisor-absent or ia). The teeth continue to grow but within the confines of a bony crypt which does not increase in size. The consequent compression leads to gross distortion of the growing dental tissues which produce a disorganized and elaborately folded mass of tissue closely resembling in structure a complex composite odontome.

Similar distortions of forming teeth, though of much less degree, occur in the grey-lethal mouse, a strain described by Grüneberg (1936). Sokoloff and Zipkin (1967) described pseudo-odontomatous masses on the roots of the incisors of an inbred strain of mouse (STR/IN) which formed if the animals were kept on a soft diet. The bone of this strain of mice is more dense than in normal mice and it seems possible that the mechanism of formation of these odontomatous masses is somewhat similar to those in the ia rats. Because of the soft diet, the teeth are not shortened in proportion to the growth at the basal end of the tooth where in consequence the tissues become folded and compressed (see Chapter 16, p. 333).

Burn, Orten and Smith (1941) found that growth of the incisors is grossly disrupted in rats maintained for up to 12 months on vitamin A-deficient diets and, towards the end of that period, odontome-like masses may develop in the vicinity of the basal ends of the incisors. There is no doubt that these lesions have many of the features of complex composite odontomes but, because the animals were given vitamin A supplements occasionally if their general condition became sufficiently poor, it is possible that the lesions represent a phase of repair. The formative tissues were so disorganized and suppressed during the deficiency phase that, if formation did re-commence, the formation of a tissue mass, even one growing from several focal centres, could be envisaged. It would be interesting to know whether a prolonged period of vitamin A

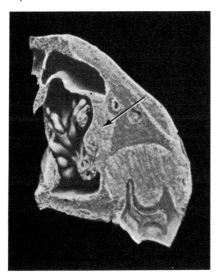

Fig. 25.40. *Capra hircus* (goat). RCS Formerly Series of Special Pathology 715. A transversely-sectioned upper jaw showing lesions described by Knaggs (1923) as osteitis fibrosa. There is a mass of lesional tissue embracing the roots of a molar tooth, and another (pointer) adjacent to the nasal fossa on the left. Further sections through the jaw may have shown the lesions to be confluent.

Fig. 25.41. *Gadus morrhua* (cod). RCS Odonto. Mus., G 65.1. Growing from the left upper jaw, a flat spheroidal mass covered with large numbers of teeth pointing in various directions but mainly radially away from the centre of the mass. A piece of the mass has at some time been removed for histological examination.

supplementation would lead to a return of normal regulation of formation of the incisors or simply to the continuation of formation of the odontome-like mass.

ODONTOMES IN NON-MAMMALS

No examples of odontomes in Reptilia or Amphibia were encountered in a search of recent literature; this may be because interest in non-mammals has not been as intense as in mammals.

According to Budd and Roberts (1978), neoplasia of tooth-germ origin, as well as odontomes, occur occasionally in fish.

Williamson (1918) described as a dentigerous cyst a globular cystic mass 3 cm in diameter protruding from the mouth of a cod (*Gadus morrhua*) and attached to the right premaxilla. The wall of the cyst was fibrous tissue with an outer covering of skin. However, within the fibrous wall of the greater part of the cyst there was a thin bony capsule which appears to have been continuous with the bone of the jaw. The cavity of the cyst was lined with a thin vascular membrane and contained about 1700 minute teeth, most of which were loose in the lumen but some were attached to the bony capsule. Several clusters of teeth protruded through holes in the bony capsule into the subdermis of the overlying skin. The specimen has the features of a dentigerous cyst though the nature of the bony capsule is uncertain.

The specimen in Figure 25.41, also from a cod, has a mass covered with large numbers of teeth attached to the tooth-bearing surface of the upper jaw. The teeth seem to be of normal morphology. Whereas the teeth of the jaw proper are directed towards the throat, many of the teeth covering the mass are pointing in various directions, suggesting that tooth development was not fully ordered (Chapter 1). Most of the teeth on the mass, however, are pointing radially away from its centre, in accord with centrifugal growth of the mass from the centre. The nature of the tissue that forms the mass itself is not recorded although a piece was removed for histological examination. The mass could be categorized as either a hamartoma or as a compound composite odontome.

References

The italic numerals indicate the pages where the items are referred to.

Abler, W.B. & Scanlon, P.F. (1975). A fourth pair of mandibular molars in a white-tailed deer. *Journal of Wildlife Diseases*, **11**, 76–78. *113*

Ainamo, J. (1970). Morphogenetic and functional characteristics of coronal cementum in bovine molars. *Scandinavian Journal of Dental Research*, **78**, 378–386. *331*

Ainamo, J. & Talari, A. (1976). Eruptive movements of teeth in human adults. In *The Eruption and Occlusion of Teeth*, edit. D.F.G. Poole & M.V. Stack. Butterworths, London, pp. 97–107. *331, 348, 487*

Aitchison, G.U. & Spence, J.A. (1984). Dental disease in hill sheep: an abattoir survey. *Journal of Comparative Pathology*, **94**, 285–300. *560*

Aitchison, J. (1963). Changing incisor dentition of bulldogs. *The Veterinary Record*, **75**, 153–154. *86*

Aitchison, J. (1964). Incisor dentitions of short-muzzled dogs. *The Veterinary Record*, **76**, 165–169. *86*

Albrecht, E. (1904). Uber Hamartome. *Verhandlungen der Deutschen Pathologischen Gesellschaft*, **7**, 153–157. *575*

Alexander, M.M. & Dozier, H.L. (1949). An extreme case of malocclusion in the musk rat. *The American Naturalist*, **42**, 252–254. *360*

Ali, S.M. (1977). Appearance of tusks in elephants. *Newsletter of the Zoological Survey of India*, **3**, 62–64. *345*

Allen, G.M. (1916). Bats of the genus *Corynorhinus*. *Bulletin of the Museum of Comparative Zoology Harvard*, **60**, 333–356. *135*

Allen, G.M. (1920). Dogs of the American Aborigines. *Bulletin of the Museum of Comparative Zoology Harvard*, **63**, 431–517. *264, 266*

Allen, J.A. (1922–25). Primates collected by the American Museum Congo Expedition. *Bulletin of the American Museum of Natural History*, **47**, 4, 283–499. *61, 182, 183*

Allo, J. (1971). The dentition of the Maori dog of New Zealand. *Record of the Auckland Institute and Museum*, **8**, 29–45. *87, 479, 550*

Amin, A.E. & Kassem, M.M. (1987). Topographical anatomical studies on the teeth with special reference to some surgical affections in donkey and horse. *Assiat Veterinary Medical Journal*, **18**, 212–218. *483, 487, 567*

Ammons, W. F., Schectman, L. R. & Page, R. C. (1972). Host tissue response in chronic periodontal disease. I. The normal periodontium and clinical manifestations of dental and periodontal disease in the marmoset. *Journal of Periodontal Research*, **7**, 131–143. *476, 542*

Amstad-Jossi, M. & Schroeder, H. E. (1978). Age-related alterations of periodontal structures around the cemento-enamel junction and of the gingival connective tissue composition in germ-free rats. *Journal of Periodontal Research*, **13**, 76–90. *527*

Andersen, A. C. (Editor) (1970). *The Beagle as an Experimental Dog*. Iowa State University Press, Ames, Iowa, U.S.A. *527*

Anderson, B. C. & Bulgin, M. S. (1984). Starvation associated with dental disease in range ewes. *Journal of the American Veterinary Medical Association*, **184**, 737–738. *562*

Anderson, B. G. & Arnim, S. S. (1936). The incidence of dental caries in seventy-six monkeys. *Yale Journal of Biology and Medicine*, **9**, 443–444. *472*

Anderson, B. G., Seipel, C. M. & Van Wagenen, G. (1953). Malocclusion in androgen-treated monkeys. *American Journal of Orthodontics*, **39**, 187–192. *334*

Anderson, S. & Jones, J. K. (1967). *Recent Mammals of the World. A Synopsis of Families*. Ronald Press, New York. *xii, 131*

Andre, W. (1784). A description of the teeth of the *Anarrhichas lupus* L. and of those of *Chaeton nigricans* L. to which is added, an attempt to prove that the teeth of cartilaginous fishes are perpetually renewed. *Philosophical Transcriptions of the Royal Society*, **74**, 279–289. *7*

Andresen, V. (1931). Bio-mechanische Orthodontie für den Allgemein-Zahnarzt und für den Schulzahnarzt. Normbegriff and Optimumsbegriff. *Fortschritte der Orthodontik*, **1**, 276–286. *292*

Andrews, A. H. (1973). Dental caries in an experimental domestic pig. *The Veterinary Record*, **94**, 257–259. *480*

Andrews, A. H. (1981). Clinical signs and treatment of aged sheep with loose mandibular or maxillary cheek teeth. *The Veterinary Record*, **108**, 331–333. *560*

Andrews, A. H. (1982). The use of dentition to age young cattle. In *Ageing and Sexing Animal Bones from Archaeological Sites*, edit. B. Wilson, C. Grigson & S. Payne. British Archaeological Reports, British Series, 109, pp. 141–153. *342, 345*

Andrews, A. H. (1985*a*). Anatomy of the oral cavity, eruption and developmental abnormalities in ruminants. In *Veterinary Dentistry*, Chapter 16, edit. C. E. Harvey & W. B. Saunders. Philadelphia, pp. 235–257. *294, 342*

Andrews, A. H. (1985*b*). Osteodystrophia fibrosa in goats. *The Veterinary Annual*, **25**, 226–230. *604*

Andrews, A. H. & Murray, R. R. (1974). Dental caries in the European badger (*Meles meles* L). *The Veterinary Record*, **95**, 163–165. *477*

Andrews, A. H. & Noddle, B. A. (1975). Absence of premolar teeth from ruminant mandibles found at archaeological sites. *Journal of Archeological Science*, **2**, 137–144. *115*

Anneroth, G. & Ericsson, S. G. (1967). An experimental histological study of monkey teeth without antagonist. *Odontologisk Revy*, **4**, 345–359. *332*

Annett, H. E. (1939). Note on a new recessive lethal in cattle. *Journal of Genetics*, **37**, 301–302. *294*

Anon (1963). Editorial comment. *The Veterinary Record*, **75**, 668. *87*

Anthony, D. J. & Lewis, E. F. (1961). *Diseases of the Pig*, 5th edn. Baillière, Tindall & Cox, London. *480, 513*

Anthony, R. (1933). Recherches sur les Incisives supérieures des Eléphantidae actuels et fossiles. (Éléphants et Mastodontes). *Archives du Muséum National d'Histoire Naturelle*, sixth series, **10**, 62–124. *126, 127, 128*

Anthony, R. & Prouteaux, M. (1929). Un crâne d'éléphant d'Afrique (*Loxodonta africana* Blum.) à quatre incisives supérieures. *Archives du Muséum National d'Histoire Naturelle*, sixth series, **4**, 15–34. *126*

Archer, M. (1975). Abnormal dental development and its significance in dasyurids and other marsupials. *Memoirs of the Queensland Museum*, **17**(2), 251–265. *141, 142, 146, 148, 327, 328, 329, 518*

Armitage, P. L. (1975). The extraction and identification of opal phytoliths from the teeth of ungulates. *Journal of Archaeological Science*, **2**, 187–197. *492*

Armstrong, P. A. & Jenkins, G. N. (1953). Studies of the antibacterial substances in dog saliva. *Journal of Dental Research*, **32**, 733 (abstract). *479*

Arnall, L. (1961). Some aspects of dental development in the dog. – II. Eruption and extrusion. *Journal of Small Animal Practice*, **1**, 259–267. *341*

Arnbjerg, J. (1986). Schmelz- und Wurzelhypoplasien nach Staupe. *Kleintierpraxis*, **31**, 323–326. *447*

Arnold, L. & Baram, P. (1973). Periodontal disease in chimpanzees. *Journal of Periodontology*, **44**, 437–442. *533*

Arnone, L. (1920). Studio ne fossili sugli effetti della carie dei denti. *Annali di Odontologia*, **5**, 25–27. *514*

Arundel, J. H., Barker, I. K. & Beveridge, I. (1977). Diseases in marsupials. In *The Biology of Marsupials*, chapter 8, edit. B. Stonehouse & D. Gilmore. Macmillan, London, pp. 415–519. *519, 520*

Arvy, L. (1977). Contribution to the knowledge of morphological anomalies in cetacean teeth. *Investigations on Cetacea*, **8**, 245–254. *101, 102*

Asahi, M. & Mori, M. (1980). Abnormalities in the dentition of the racoon dog, *Nyctereutes procyonides*. *Zoological Magazine*, **89**, 61–64. *84*

Ash, C. (1954). *Whaler's Eye*. Allen & Unwin, London. *391*

Ashley-Montague, M. F. (1935). A rare case of osteomalacia with severe caries of the maxillary canine teeth in a captive baboon. *International Journal of Orthodontia and Dentistry for Children*, **21**, 672–681. *472*

Ashton, E. H., Flinn, R. M., Griffiths, R. K. & Moore, W. J. (1979). The results of geographic isolation on the teeth and skull of the green monkey (*Cercopithecus aethiops sabaeus*) in St. Kitts – a multivariate retrospect. *Journal of Zoology, London*, **188**, 533–555. *195, 196*

Ashton, E. H. & Zuckerman, S. (1950*a*). The influence of geographic isolation on the skull of the green monkey (*Cercopithecus aethiops sabaeus*). I. A comparison between the teeth of the St. Kitts and African green monkey. *Proceedings of the Royal Society*, B, **137**, 212–238. *18, 35, 195, 197*

Ashton, E. H. & Zuckerman, S. (1950*b*). Some quantitative dental characteristics of the chimpanzee, gorilla and the orang-utan. *Philosophical Transactions of the Royal Society*, **234**B, 471–484. *18, 197*

Ashton, E. H. & Zuckerman, S. (1951*a*). The influence of geographic isolation on the skull of the green monkey

(*Cercopithecus aethiops sabaeus*). II. The cranial dimensions of the St. Kitts and African green monkey. *Proceedings of the Royal Society*, B, **138**, 205–213. 195, 196

Ashton, E. H. & Zuckerman, S. (1951*b*). The influence of geographic isolation on the skull of the green monkey (*Cercopithecus aethiops sabaeus*). III. The developmental stability of the skulls and teeth of the St. Kitts and African green monkey. *Proceedings of the Royal Society*, B, **138**, 213–218. 35, 195, 196

Ashton, E. H. & Zuckerman, S. (1951*c*). The influence of geographic isolation on the skull of the green monkey (*Cercopithecus aethiops sabaeus*). IV. The degree and speed of dental differentiation in the St. Kitts green monkey. *Proceedings of the Royal Society*, B, **138**, 354–374. 18, 195, 196

Ashton, E. H. & Zuckerman, S. (1952). Overall dental dimensions of hominoids. *Nature, London*, **169**, 571–572. 18

Atkinson, P. J., Spence, J. A., Aitchison, G. & Sykes, A. R. (1982). Mandibular bone in ageing sheep. *Journal of Comparative Pathology*, **92**, 51–67. 557

Aulerich, R. J. & Swindler, D. R. (1968). The dentition of the mink (*Mustela vison*). *Journal of Mammalogy*, **49**, 488–494. 340, 341

Auskaps, A. M. & Shaw, J. H. (1957). Studies on the dentition of the cynomolgus monkey. *Journal of Dental Research*, **36**, 432–436. 539

Avery, B. E. & Simpson, D. M. (1973). The baboon as a model system for the study of periodontal disease: clinical and light microscopic observations. *Journal of Periodontology*, **44**, 675–686. 528, 540

Bader, R. S. (1965). Fluctuation asymmetry in the dentition of the house mouse. *Growth*, **29**, 291–300. 10

Baer, P. N. & Kilham, L. (1964). Rat virus and periodontal disease. IV. The aged hamster. *Oral Surgery*, **18**, 803–811. 587

Baer, P. N. & Lieberman, J. E. (1960). Periodontal disease in six strains of inbred mice. *Journal of Dental Research*, **39**, 215–225. 526

Baer, P. N., Newton, W. L. & White, C. L. (1964). Studies on periodontal disease in the mouse. VI. The older germ-free mouse and its conventional control. *Journal of Periodontology*, **35**, 388–396. 526, 527

Bahn, P. G. (1980). Crib-biting: tethered horses in the Palaeolithic? *World Archeology*, **12**, 212–217. 494

Baker, G., Jones, L. H. P. & Wardrop, I. D. (1959). Cause of wear in sheep's teeth. *Nature, London*, **1**, 1583–1584. 492

Baker, J. R. (1981). Congenital intermandibular cyst in a calf. *The Veterinary Record*, **109**, 424–426. 601

Baker, J. R. & Britt, D. P. (1984). Dental calculus and periodontal disease in sheep. *The Veterinary Record*, **115**, 411–412. 560

Baldi, B. (1948). Anomalia dentaria in un bovino. *La Nuova Veterinaria*, **24**, 19–21. 115

Barker, S., Calaby, J. H. & Sharman, G. B. (1963). Diseases of Australian laboratory marsupials. *The Veterinary Bulletin*, **33**, 539–544. 519

Barnicoat, C. R. (1957). Wear in sheep's teeth. *New Zealand Journal of Science and Technology*, Section A, **38**, 583–632. 491, 492, 493

Barnicoat, C. R. & Hall, D. M. (1960). Attrition of incisors of grazing sheep. *Nature, London*, **185**, 179. 492

Barrett, R. H. (1978). The feral hog at Dye Creek Ranch, California. *Hilgardia*, **46**, 283–355. 550, 551

Barrett-Hamilton, G. E. H. & Hinton, M. A. C. (1913). On a collection of mammals from the Inner Hebrides. *Proceedings of the Zoological Society of London*, **1913**, 821–839. 139

Barrette, C. (1986). Mechanical analysis of the malformed, yet functional, mandibular joints of a wild timber wolf, *Canis lupus*. *Archives of Oral Biology*, **31**, 351–356. 380

Barrington, E. P. & Meyer, J. (1966). Recovery of the rat dental organ from experimental lathyrism. *Journal of Periodontology*, **37**, 453–467. 333

Barrow, M. V., Simpson, C. F. & Miller, E. J. (1974). Lathyrism: a review. *Quarterly Review of Biology*, **49**, 101–128. 332

Bate, C. S. (1856–7). On absorption. *Transactions of the Odontological Society of London*, **1**, 21–30. 312

Bateman, J. A. (1970). Supernumerary incisors in mustelids. *Mammalian Review*, **1**, 81–86. 71, 72

Bateson, W. (1892). On numerical variation in teeth with a discussion of the conception of homology. *Proceedings of the Zoological Society of London*, **1892**, 102–115. 99, 100

Bateson, W. (1894). *Materials for the Study of Variation*. Macmillan, London. 1, 2, 6, 7, 8, 20, 22, 23, 31, 49, 50, 53, 54, 55, 57, 83, 84, 86, 89, 94, 95, 97, 98, 99, 100, 118, 143, 145, 146, 147, 148

Batt, R. A. L. (1978). Abnormal dentition and decrease in body weight in the genetically obese mouse (genotype, *ob/ob*). *International Journal of Obesity*, **2**, 457–462. 333

Bauchau, V. & Le Boulengé-Nguyen, P. (1985). Une curieuse anomalie du crâne chez *Microtus agrestis* (L., 1766), *Lutra*, **28**, 22–25. 517

Bauer, W. (1925). Die Veränderungen der Zähne und Kiefer bei experimenteller Hunderachitis (mit vergleichenden Untersuchungen being kindlicher Rachitis). *Zeitschrift für Stomatologie*, **23**, 407–420. 447

Beale, T. (1839). *The Natural History of the Sperm Whale*. John van Voorst, London. 388

Beaver, T. D., Feldhamer, G. A. & Chapman, J. A. (1981). Dental and cranial anomalies in the river otter (Carnivora: Mustelidae). *Brimleyana*, **7**, 101–109. 74

Becker, E. (1970). Zähne. In *Handbuch der Speziellen Pathologischen Anatomie der Haustiere*, 3rd edn., vol. 5, *Alimentary system*, edit. J. Dobberstein, G. Pallaske & H. Stünz. Paul Parey, Berlin, pp. 83–313. 309, 352, 372, 388, 432, 435, 445, 479, 483, 487, 515, 547, 579, 586

Becker, R. B. & Arnold, P. T. D. (1949). 'Bulldog head' cattle. Prognathism in grade Jersey strain. *Journal of Heredity*, **40**, 282–286. 294

Becker, R. H., Wilcox, C. J., Simpson, C. F., Gilmore, L. O. & Fechheimer, N. S. (1964). Genetic aspects of actinomycosis and actinobacillosis in cattle. *Florida Agricultural Experiment Station Technical Bulletin* No. 670; and *Ohio Agricultural Research Bulletin* No. 938. 516

Becks, H., Wainwright, W. W. & Morgan, A. F. (1943). Comparative study of oral changes in dogs due to deficiencies of pantothenic acid, nicotinic acid, and unknowns of the B vitamin complex. *American Journal of Orthodontics and Oral Surgery*, **29**, 183–207. 525

Beecher, R. M. & Corruccini, R. S. (1981). Effects of dietary consistency on craniofacial and occlusal development in the rat. *The Angle Orthodontist*, **51**, 61–69. 156

Beecher, R. M., Corruccini, R. S. & Freeman, M. (1983). Craniofacial correlates of dietary consistency in a non-human primate. *Journal of Craniofacial Genetics and Developmental Biology*, **3**, 193–202. 157, 229

Beemer, A. M. & Kuttin, E. A. (1969). An anomaly of teeth causing starvation and paralysis in chinchillas. *Refuah Veterinarith*, **26**, 175–177. 362

Beier, K. (1982). Zahnanomalien bei Wombats (*Lasiorhinus latifrons* Owen, 1845 und *Vombatus ursinus* Shaw, 1800). *Zoologischer Anzeiger*, **209**, 111–119. 493

Bellairs, A. d'A. (1981). Congenital and developmental diseases. In *Diseases of the Reptilia*, chapter 14, vol. 2, edit. J. E. Cooper & O. F. Jackson. Academic Press, London and New York, pp. 469–485. 158

Bellairs, A. d'A. & Miles, A. E. W. (1961a). Apparent failure of tooth replacement in monitor lizards. *British Journal of Herpetology*, **2**, 186–189. 505

Bellairs, A. d'A. & Miles, A. E. W. (1961b). Apparent failure of tooth replacement in monitor lizards. Addendum. *British Journal of Herpetology*, **3**, 14–15. 505

Bennejeant, C. (1936). *Anomalies et Variations Dentaires chez les Primates*. Paul Vallier, Clermont-Ferrand. 19, 22, 23, 24, 26, 28, 37, 40, 44, 47, 49, 50, 55, 56, 59, 60, 350

Bennett, A. G. (1931). *Whaling in the Antarctic*. W. Blackwood & Sons, London. 389

Bennett, I. C. & Law, D. B. (1965). Incorporation of tetracycline in developing dog enamel and dentin. *Journal of Dental Research*, **44**, 788–793. 438

Bennett, S. (1885). Casual communications. *Transactions of the Odontological Society of Great Britain*, **17**, 82. 23

Bennett, S. (1886). On some recent additions to the museum of the Odontological Society. *Transactions of the Odontological Society of Great Britain*, new series, **18**, 193–194. 23

Benson, S. B. (1943). Occurrence of upper canines in mountain sheep, *Ovis canadensis*. *American Midland Naturalist*, **30**, 786–789. 114

Berdjis, C. C., Greenberg, L. D., Rinehart, J. F. & Fitzgerald, G. (1960). Oral and dental lesions in vitamin B_6, deficient rhesus monkeys. *British Journal of Experimental Pathology*, **41**, 198–205. 442

Berkovitz, B. K. B. (1968a). Some stages in the early development of the post-incisor development of *Trichosaurus vulpecula* (Phalangeroidea: Marsupialia). *Journal of Zoology, London*, **154**, 403–414. 13

Berkovitz, B. K. B. (1968b). Supernumerary deciduous incisors and the order of eruption of the incisor teeth in the albino ferret. *Journal of Zoology, London*, **155**, 445–449. 71, 339

Berkovitz, B. K. B. (1968c). The development of incisor teeth of *Setonix brachyuris* (Macropodidae: marsupialia) with special reference to the prelacteal teeth. *Archives of Oral Biology*, **13**, 170–190. 13

Berkovitz, B. K. B. (1972). Tooth development in *Protemnodon eugenii*. *Journal of Dental Research*, **51**, 1467–1473. 13

Berkovitz, B. K. B. (1976). Theories of tooth eruption. In *The Eruption and Occlusion of Teeth*, edit. D. F. G. Poole & M. V. Stack. Butterworths, London, pp. 193–204. 356, 364

Berkovitz, B. K. B. & Musgrave, J. H. (1971). A rare dental anomaly in an adult male orang-utan (*Pongo pygmaeus*): bilateral supernumerary maxillary premolar. *Journal of Zoology, London*, **164**, 266–268. 29

Berkovitz, B. K. B. & Thomson, P. (1973). Observations on the aetiology of supernumerary upper incisors in the albino ferret. *Archives of Oral Biology*, **18**, 457–463. 71

Berzin, A. A. (1972). *The Sperm Whale*. Israel Program for Scientific Translations, Jerusalem. U.S. Dept. of Commerce, Washington. 504

Beveridge, W. I. B. (1934). A study of twelve strains of *Bacillus necrophorus*, with observations on the oxygen intolerance of the organism. *Journal of Pathology and Bacteriology*, **38**, 467–490. 520, 563

Bhaskar, S. N., Schour, I., MacDowell, E. C. & Weinmann, J. P. (1951). The skull and the dentition of screw tail mice. *Anatomical Record*, **110**, 199–299. 333

Biddulph, C. H. (1937). A mugger (*Crocodilus porosus*) with a broken lower jaw. *Journal of the Bombay Natural History Society*, **39**, 421–442. 393

Binns, W. H. & Escobar, A. (1967). Defects in permanent teeth following pulp exposure of primary teeth. *Journal of Dentistry for Children*, **34**, 4–14. 439

Björk, A. (1950). Some biological aspects of prognathism and occlusion of the teeth. *Acta Odontologica Scandinavica*, **9**, 1–40. 175

Bjotvedt, G. & Turner, C. G. (1976). Tooth trauma and feeding behaviour in prehistoric Aleutian sea otters. *Veterinary Medicine and Small Animal Clinician*, **71**, 831–883. 477, 496

Blair, W. R. (1907). Actinomycosis in the black mountain sheep. *Eleventh Annual Report of the New York Zoological Society*, pp. 137–141. 516

Blakemore, F., Bosworth, T. J. & Green, H. H. (1948). Industrial fluorosis of farm animals in England, attributable to the manufacture of bricks, the calcining of ironstone, and to enamelling processes. *Journal of Comparative Pathology*, **58**, 267–291. 454

Bland Sutton, J. (1884a). Comparative dental pathology (1). *Transactions of the Odontological Society of Great Britain*, new series, **16**, 88–145.
356, 366, 495, 504, 505, 506, 594, 595

Bland Sutton, J. (1884b). Bone disease in monkeys. *Transactions of the Pathological Society of London*, **35**, 468–472. 333

Bland Sutton, J. (1884c). Bone disease in wild animals. *Transactions of the Pathological Society of London*, **35**, 472–476. 333

Bland Sutton, J. (1885a). Tumours in animals. *Journal of Anatomy and Physiology, Normal and Pathological*, **19**, 415–475. 594, 595, 596, 601, 603

Bland Sutton, J. (1885b). Comparative dental pathology. *Transactions of the Odontological Society of Great Britain*, new series, **17**, 42–73. 356, 366, 594, 596, 601

Bland Sutton, J. (1885c). Injuries and diseases of the jaws in animals. *Transactions of the Odontological Society of Great Britain*, new series, **17**, 158–216.
361, 372, 384, 538, 603

Bland Sutton, J. (1888a). Odontomes. *Transactions of the Odontological Society of Great Britain*, new series, **20**, 32–85.
435, 574, 579, 586, 588, 594, 595, 596, 601, 602, 603

Bland Sutton, J. (1888b). On a remarkable case of odontomes in a thar, *Capra jemlaica* (Himalayan goat). *Transac-

tions of the Odontological Society of Great Britain, new series, **20**, 185–192. *585, 587*
Bland Sutton, J. (1889). On the relation of rickets to some forms of odontomes. *Transactions of the Odontological Society of Great Britain*, new series, **21**, 138–144. *603, 604*
Bland Sutton, J. (1891). A casual communication. *Transactions of the Odontological Society of Great Britain*, new series, **23**, 215–216. *597, 599*
Bland Sutton, J. (1893). *Tumours Innocent and Malignant*. Cassell, London. *595*
Bland Sutton, J. (1904). On teeth in the temporal bone of horses; with a note on the cervical teeth of sheep. *Transactions of the Odontological Society of Great Britain*, new series, **36**, 10–21. *600, 601*
Bland Sutton, J. (1910). The diseases of elephants' tusks in relation to billiard balls. *The Lancet*, **2**, 1534–1537. *426*
Bland Sutton, J. (1922). *Tumours Innocent and Malignant*, 7th edn. Cassell, London. *576, 578, 595, 601*
Blobel, H., Döbereiner, J., Valadao Rosa, I., Gomes Ferreira Lima, F. & dos Santos Dutra, I. (1987). Bakteriologische Untersuchungen an der "Cara inchada", einer periodontalen Erkrankung bei Rindern in Brasilien. *Tierärztliche Umschau*, **42**, 152–157. *561*
Blood, D.C., Radostits, O.M. & Henderson, J.A. (1983). *Veterinary Medicine*, 6th edn. Baillière Tindall, London. *562, 579*
Blüntschli, H. (1913). Die fossilen Affens Patagoniens und der Ursprung der platyrrhinen Affen. *Verhandlungen der Anatomischen Gesellschaft, Jena*, **27**, 33–66. *55*
Boddie, G.F. (1949). Effects of fluorine compounds in the Fort William area. In *Industrial Fluorosis, a study of the hazard to man and animals near Fort William, Scotland.* Medical Research Council Memorandum No. 22. His Majesty's Stationery Office, London. *452, 453, 454*
Bodingbauer, J. (1946). Die Staupe-Schmelzhypoplasien (Staupegebiss des Hundes) als Fehlerquelle in der experimentellen Kariesforschung. *Zeitschrift für Stomatologie*, **43**, 293–309. *445, 448*
Bodingbauer, J. (1947). Vergleichende Betrachtunen über das Vorkommen der Karies beim Menschen und beim Hunde. *Zeitschrift für Stomatologie*, **44**, 333–349. *478, 479*
Bodingbauer, J. (1949a). Die Staupe-Schmelzhypoplasien (Staupegebiss) des Hundes. *Schweizer Archiv für Tierheilkunde*, **91**, 84–116. *445, 447, 448*
Bodingbauer, J. (1949b). Domestikationsanomalien: Dentitio difficilis (*Homo*). – Retention des letzten Molaren (*Sus scrofa domesticus*). *Zeitschrift für Stomatologie*, **46**, 106–119. *348*
Bodingbauer, J. (1950). Histopathologische und bakteriologische Studien über die Zahnbein-Karies des Hundes. *Wiener Tierärztliche Monatsschrift*, **37**, 323–329. *479*
Bodingbauer, J. (1954). Durch Zahnausfall und anschliessende Usurbildung bedingte Osteomyelitis beim Schaf. *Wiener Tierärztliche Monatsschrift*, **41**, 496–504. *514*
Bodingbauer, J. (1955a). Die apicale Paradentitis des Hundes. *Zentralblatt für Veterinärmedizin*, **2**, 368–382. *511*
Bodingbauer, J. (1955b). Schmelzhypoplasien bei Tieren. *Wiener Tierärztliche Monatsschrift*, **42**, 83–100. *445*
Bodingbauer, J. (1963). Oligodontia and polydontia in prehistoric dogs. *The Veterinary Record*, **75**, 668–672. *86, 87*
Bodingbauer, J. & Hager, G. (1959). Zur Frage der Aetiologie der Zahnunterzahl (Oligodontie) des Hundes. I. Teil. *Wiener Tierärztliche Monatsschrift*, **46**, 214–321. *86, 87*
Bolk, L. (1916). Problems of human dentition. *American Journal of Anatomy*, **19**, 91–148. *20, 22*
Borg, K. (1967). Om förekomsten av hörn tänder i överkaken hos radjur. *Zoologisk Revy*, **29**, 96–94. *112*
Bortone, S.A. (1972). Pugheadedness in the pirate perch, *Aphredoderus sayanus* (Pisces: Aphredoderidae). With implications on feeding. *Chesapeake Science*, **13**, 231–233. *159*
Boschma, H. (1938a). On the teeth and some other particulars of the sperm whale (*Physeter macrocephalus* L.). *Temminckia, Leiden*, **3**, 151–278. *101*
Boschma, H. (1938b). Double teeth in the sperm whale (*Physeter macrocephalus* L.). *Zoologische Mededelingen*, **20**, 211–221. *102, 502*
Boschma, H. (1950a). Maxillary teeth in specimens of *Hyperoodon rostratus* (Müller) and *Mesoplodon grayi* (von Haast) stranded on the Dutch coasts. *Koniklijke Nederlandse Akademie van Wetenschappen. Series C. Biological and Medical Sciences, Proceedings*, **53**, 775–786. *101*
Boschma, H. (1950b). Absorption of tooth tissue in the sperm whale. *Koniklijke Nederlandse Akademie van Wetenschappen. Series C. Biological and Medical Sciences, Proceedings*, **53**, 289–293. *502*
Boschma, H. (1951). Rows of small teeth in ziphioid whales. *Zoologische Mededelingen*, **31**, 140–148. *101*
Bössmann, K.L., Deerberg, F., Preuss, V. & Rehm, S. (1981). Dental and periodontal alterations in ageing Han:wist rats. *Zeitschrift für Versuchstierkunde*, **23**, 305–311. *568*
Bourdelle, M.E. (1904). Anomalie de dents molaires chez un cheval. *Revue vétérinaire*, **29**, 546–555. *122*
Box, H.K. (1935). Experimental traumatogenic occlusion in sheep. *Oral Health*, **25**, 9–15. *552*
Boyle, P.E., Bessey, O. & Wolbach, S.B. (1937). Experimental production of the diffuse alveolar bone atrophy type of periodontal disease by diets deficient in ascorbic acid (vitamin C). *Journal of the American Dental Association*, **24**, 1768–1777. *525*
Bradley, O.C. (1902). On two cases of dental anomaly. *Journal of Anatomy and Physiology*, **36**, 356–367. *84, 86*
Bradley, O.C. (1907). Dental anomalies and their significance. *Proceedings of the 25th General Meeting, National Veterinary Association 1907 at Great Yarmouth*, pp. 1–19. *88, 121*
Braithwaite, R.W. (1979). Natural selection in *Rattus* molars. *Journal of Zoology, London*, **189**, 545–548. *131, 132*
Bramblett, C.A. (1967). Pathology in the Darajani baboon. *American Journal of Physical Anthropology*, **26**, 331–340. *371*
Bramblett, C.A. (1969). Non-metric skeletal age changes in the Darajani baboon. *American Journal of Physical Anthropology*, **30**, 161–172. *334, 336, 337, 472, 511, 539*
Brandt, G.W. (1941). Achondroplasia in calves. *Journal of Heredity*, **32**, 183–186. *294*
Brash, J.C., McKeag, H.T.A. & Scott, J.H. (1956). *The Aetiology of Irregularity and Malocclusion of the Teeth*, 2nd edn. Dental Board of the United Kingdom, London. *156, 158*
Brazaitis, P. (1981). Maxillary regeneration in a marsh croco-

dile, *Crocodylus palustris*. *Journal of Herpetology*, **15**, 360–362. 393

Bremer, G. (1939). Untersuchungsmethodik und vergleichende chemische Analysen an Zähnen von Homo, Affe, Hund und Schwein. *Uppsala Läkareförenings Förhandlingar*, **44**, 219–245. 479

Breuer, R. (1932). Ein seltener Fall von Selbsthilfe der Natur. *Zeitschrift für Stomatologie*, **30**, 334–343. 379, 386

Breuer, R. (1943). Zwei Fälle, bemerkenswerter anatomischer Befunde an Schädeln kleinerer Säugetiere. *Zoologischer Anzeiger*, **144**, 34–37. 91

Bricknell, I. (1987). Palaeopathology of Pleistocene proboscideans in Britain. *Modern Geology*, **2**, 295–309. 427, 516

Briedermann, L. (1965). Die Altersbestimmung erlegten Schwarzwildes. *Arbeitsgemeinschaft für Jagd und Wildforschung, Berlin*, No. 22. 343, 345

Broca, M. P. (1867). Recherches sur un nouveau group de tumeurs désigné sous le nom d'odontomes. *Comptes Rendus Hebdomadaires des Séances de l'Académie des Sciences*, **65**, 1117–1121. 574

Broca, P. (1869). *Traité des Tumeurs*, Volume 2, Chapter 10. Des odontomes en général, pp. 275–330. Chapter 11. Des principales variétés d'odontomes, pp. 331–374. Paris. An English translation of these two chapters appeared as a series of well-indexed parts in *Monthly Review of Dental Surgery* (1873–1876), vol. 2 between pp. 18–452, vol. 3 between pp. 26–471, and vol. 4 between pp. 25–133. 574, 594

Brockie, R. E. (1964). Dental abnormalities in European and New Zealand hedgehogs. *Nature, London*, 4939, 1355–1356. 138

Brodey, S. & Morris, A. L. (1960). Undifferentiated carcinoma in the maxilla of a dog. *Journal of the American Veterinary Medicine Association*, **137**, 553–559. 593

Brokx, P. A. (1972). The superior canines of *Odocoileus* and other deer. *Journal of Mammalogy*, **53**, 359–366. 111

Broom, R. (1907). A contribution to the knowledge of the Cape golden moles. *Transactions of the South African Philosophical Society*, **18**, 283–311. 569

Brothwell, D. (1981). Disease as an environmental parameter. In *The Environment of Man: the Iron Age to the Anglo-Saxon Period*, edit. M. Jones & G. Dimbleby. British Archaeological Reports, British Series, 87, pp. 231–247. 561

Brown, G. (1902). *Dentition as Indicative of the Age of the Animals of the Farm*. Royal Agricultural Society of England. 342, 343

Brown, W. A. B., Christofferson, P. V., Massler, M. & Weiss, M. B. (1960). Postnatal tooth development in cattle. *American Journal of Veterinary Research*, **21**, 7–34. 451

Bruère, A. N., West, D. M., Orr, M. B. & O'Callaghan, M. W. (1979). A syndrome of dental abnormalities of sheep: 1. Clinical aspects on a commercial sheep farm in the Wairarapa. *New Zealand Veterinary Journal*, **27**, 152–158. 492

Buchalczyk, T. (1961). Einseitige Gebissanomalie bei *Neomys fodiens* Pennant 1771. *Acta Theriologica*, **4**, 277. 139

Buchalczyk, T., Dynowski, J, & Szteyn, S. (1981). Variations in number of teeth and asymmetry of the skull in the wolf. *Acta Theriologica*, **26**, 23–30. 85

Budd, J. & Roberts, R. J. (1978). Neoplasia in teleosts. In *Fish Pathology*, chapter 5, edit. R. J. Roberts. Baillière Tindall, London, p. 109. 606

Builder, P. L. (1955). Opening paper. *The Veterinary Record*, **67**, 386–392. 505

Bull, G. & Payne, S. (1982). Tooth eruption and epiphysial fusion in pigs and wild boar. In *Ageing and Sexing Animal Bones from Archaeological Sites*, edit. B. Wilson, C. Grigson & S. Payne. British Archaeological Reports, British Series, **109**, pp. 55–71. 342, 343

Bullen, F. T. (1898). *The Cruise of the 'Cachalot' Round the World after Sperm Whales*. Smith, Elder & Co., London. 389

Bullock, D. & Rackham, J. (1982). Epiphysial fusion and tooth eruption of feral goats from Moffatdale, Dumfries and Galloway, Scotland. In *Ageing and Sexing Animal Bones from Archaeological Sites*, edit. B. Wilson, C. Grigson & S. Payne. British Archaeological Reports British Series, **109**, pp. 73–80. 345

Bullock, F. D. & Curtis, M. R. (1930). Spontaneous tumours of the rat. *Journal of Cancer Research*, **14**, 1–115. 587

Bunch, T. D., Hoefs, M. & Glaze, R. L. (1984). Upper canines in Dall's sheep (*Ovis dalli dalli*). *Journal of Wildlife Diseases*, **20**, 158–161. 114

Burn, C. G., Orten, A. H. & Smith, A. H. (1941). Changes in the structure of the developing tooth in rats maintained on a diet deficient in vitamin A. *Yale Journal of Biology and Medicine*, **13**, 817–830. 605

Burrows, C. F., Miller, W. H. & Harvey, C. E. (1985). Oral medicine. In *Veterinary Dentistry*, Chapter 4, edit. C. E. Harvey. W. B. Saunders, Philadelphia, pp. 34–58. 64, 88, 115, 544

Burwasser, P. & Hill, T. J. (1939). The effects of hard and soft diets on the gingival tissues of dogs. *Journal of Dental Research*, **18**, 389–393. 528

Busch, F. (1890a). Ueber Verletzungen, Abscesse und Dentikel am Stosszahn des Elephanten. *Deutsche Monatsschrift Zahnheilkunde*, **8**, 62–65. 405

Busch, F. (1890b). Ueber Verletzungen, Abscesse und Dentikel am Stosszahn des Elephanten. *Verhandlungen der Deutschen Odontologischen Gesellschaft*, **1**, 42–84. 405, 421

Busch, F. (1890c). Zur Physiologie und Pathologie der Zähne des Elefanten. *Verhandlungen der Deutschen Odontologischen Gesellschaft*, **1**, 246–315. 405

Busch, F. (1891). Weiteres über die Zähne der Hufthiere. *Verhandlungen der Deutschen Odontologischen Gesellschaft*, **2**, 196–232. 363

Busch, F. (1892). Ueber die Bezahnung der schwimmenden Säugethiere (Cetaceen Sirenen). *Verhandlungen der Deutschen Odontologischen Gesellschaft*, **3**, 41–79. 391

Butler, P. M. (1939). 1. Studies of the mammalian dentition. Differentiation of the post-canine dentition. *Proceedings of the Zoological Society of London*, **109B**, 1–36. 2, 3

Butler, P. M. (1963). Tooth morphology and primate evolution. In *Dental Anthropology*, edit. D. R. Brothwell. Pergamon, London, pp. 1–13. 2

Butler, P. M. (1967). Dental merism and tooth development. *Journal of Dental Research*, **46**, 845–850. 2

Butler, P. M. (1978a). Molar cusp nomenclature and homology. In *Development, Function and Evolution of Teeth*, chapter 26, edit. P. M. Butler & K. A. Joysey. Academic Press, London, pp. 439–453. 15

Butler, P. M. (1978b). The ontogeny of mammalian heterodonty. *Journal de Biologie Buccale*, **6**, 217–227. *2*

Butler, P. M. (1981). Dentition in function. In *Dental Anatomy and Embryology*, edit. J. W. Osborn. Blackwell Scientific Publications, Oxford, pp. 329–352. *14, 15, 486*

Butler, P. M. (1982). Some problems of the ontogeny of tooth patterns. In *Teeth: Form, Function and Evolution*, edit. B. Kurtén. Columbia University Press, New York, pp. 44–50. *2, 15*

Büttner, W. (1969). Trace elements and dental caries in experiments on animals. *Caries Research*, **3**, 1–13. *456*

Byrd, K. E. (1978). Radiographic analysis of dental development and eruption sequences in the pygmy marmoset, *Cebuella pygmaea*. *Journal of Dental Research*, **48**, 384. *339*

Byrd, K. E. (1981). Sequences of dental ontogeny and callitrichid taxonomy. *Primates*, **22**(1), 103–118. *339*

Cadéac, C. (1910). *Pathologie Chirurgicale de l'Appareil Digestif. Bouche, Pharynx, Oesophage, Estomac, Intestin*. Paris. *115, 121, 122, 276, 308, 309, 310, 311*

Caldwell, D. K. (1964). Tusk twinning in the Pacific walrus. *Journal of Mammalogy*, **45**, 490–491. *99*

Camargo, W. V. de A., Veiga, J. S. & Conrad, J. H. (1982). Cu, Mo, S, Zn periodontitis in Brazilian cattle. In *Trace Element Metabolism in Man and Animals. Proceedings of the 4th International Symposium on T.E.M. in Man and Animals*, edit. J. M. Cawthorne, J. McC. Howell & C. L. White. Perth. *561*

Carranza, F. A., Simes, R. J., Mayo, J. & Cabrini, R. L. (1971). Histometric evaluation of periodontal bone loss in rats. I. The effect of marginal irritation, systemic irradiation and trauma from occlusion. *Journal of Periodontal Research*, **6**, 65–72. *528*

Carroll, F. D., Gregory, F. D. & Rollins, W. C. (1951). Thyrotropic-hormone deficiency in homozygous dwarf beef cattle. *Journal of Animal Science*, **10**, 917–921. *294*

Casinos, A., Fitella, S. & Grau, E. (1977). Notes on cetaceans of the Iberian coasts: II. A specimen of sperm whale, *Physeter macrocephalus* Linné, 1758, with partial amputation of the lower jaw. *Saugetierkundliche Mitteilungen*, **25**, 238–240. *389*

Cass, J. S. (1947). Buccal food impaction in white-tailed deer and *Actinomyces necrophorus* in big game. *Journal of Wild Life Management*, **11**, 91–94. *563*

Catcott, E. J. (1968). *Canine Medicine*. American Veterinary Publications, Illinois. *350, 372, 438, 445*

Catcott, E. J. (Editor) (1975). *Feline Medicine and Surgery*. Second edition. American Veterinary Publications, Santa Barbara, California. *543*

Catcott, E. J. & Smithcors, J. F. (Editors). (1972). *Equine Medicine and Surgery*, 2nd edn. American Veterinary Publications, Illinois. *487*

Cave, A. J. E. & Blackwood, H. J. J. (1965). Traumatic deformity of hippopotamus tusks. *Journal of Zoology, London*, **146**, 447–450. *404*

Cave, A. J. E. & Bonner, W. N. (1987). Facial dystrophy in a leopard seal (*Hydrurga leptonyx*). *British Antarctic Survey Bulletin*, no. 75, pp. 67–71. *371*

Chai, C. K. (1970). Effect of inbreeding in rabbits: skeletal variations and malformations. *Journal of Heredity*, **61**, 3–8. *319*

Chai, C. K. & Degenhardt, K.-H. (1962). Developmental abnormalities in inbred rabbits. *Journal of Heredity*, **53**, 174–182. *319*

Chai, C. K., Hunt, H. R., Hoppert, C. A. & Rosen, S. (1968). Hereditary basis of caries resistance in rats. *Journal of Dental Research*, **47**, 127–137. *462*

Chapman, A. (1933). A comparative study of growth and dentition. With special reference to the interproximal attrition. *Proceedings of the 8th Australian Dental Congress*, Adelaide, pp. 396–411. *493*

Chapman, D. I. & Chapman, N. G. (1969). The incidence of congenital abnormalities in the mandibular dentition of fallow deer (*Dama dama* L.). *Research in Veterinary Science*, **10**, 485–487. *111*

Chapman, D. I. & Chapman, N. G. (1973). Maxillary canine teeth in fallow deer, *Dama dama*. *Journal of Zoology, London*, **170**, 143–162. *111*

Chase, J. E. & Cooper, R. W. (1969). *Saguinus nigricollis* – physical growth and dental eruption in a small population of captive-born individuals. *American Journal of Physical Anthropology*, **30**, 111–116. *338*

Chaturvedi, Y. (1966). Malocclusion in rodent incisors. *Proceedings of the Zoological Society, Calcutta*, **19**, 87–103. *361*

Cheselden, W. (1733). *Osteographia or the Anatomy of the Bones*. London. *365, 367*

Chiasson, R. B. (1955). Dental abnormalities of the Alaskan fur seal. *Journal of Mammalogy*, **36**, 562–564. *98*

Child, G. (1965). Additional records of abnormal dentition in Impala. *Mammalia*, **29**, 622–623. *115*

Child, G. (1969). The incidence of abnormal tooth formulae in two impala populations. *Mammalia*, **33**, 541–543. *115*

Child, G., Sowls, L. & Mitchell, B. L. (1965). Variations in the dentition, ageing criteria and growth patterns in wart hog. *Arnoldia*, **38**(1), 1–23. *107*

Choate, J. R. (1968). Dental abnormalities in the short-tailed shrew, *Blarina brevicauda*. *Journal of Mammalogy*, **49**, 251–258. *139*

Christeller, E. (1923). Die Formen der Ostitis fibrosa und der verwandten Knochenerkrankungen der Säugetiere, zugleich ein Beitrag zur Frage der 'Rachitis' der Affen. *Ergebnisse der allgemeinen Pathologie und pathologischen Anatomie des Menschen und Tiere*, **21**(2), 1–184. *333*

Christy, C. (1924). *Big Game and Pygmies*, p. 83. Macmillan, London. *426*

Clarke, R. (1956). Sperm whales of the Azores. *Discovery Report*, **28**, 239–298. *391*

Clarke, R. (1966). The stalked barnacle *Conchoderma*, ectoparasitic on whales. *Norsk Hvalfangst-Tidende*, **55**, 153–168. *391*

Clements, E. M. B. & Zuckerman, S. (1953). The order of eruption of the permanent teeth in the Hominoidea. *American Journal of Physical Anthropology*, **11** (new series), 313–337. *335*

Clutton-Brock, J. (1981). *Domesticated Animals from Early Times*. Heinemann & British Museum (Natural History), London. *11, 85, 264*

Cohen, D. W. & Goldman, H. M. (1960). Oral disease in primates. *Annals of the New York Academy of Sciences*, **85**, 889–909. *528, 539*

Cohen, M. M., Shklar, G. & Yerganian, G. (1963). Pulpal and periodontal disease in a strain of Chinese hamsters with hereditary diabetes mellitus. *Oral Surgery*, **16**, 104–112. *527*

Coignoul, F. & Cheville, N. (1984). Animal model of human disease. Calcified microbial plaque. Dental calculus of dogs. *American Journal of Pathology*, **117**, 499–501. 547

Colgrove, G.S., Sawa, T.R., Brown, J.T., McDowell, P.F. & Nachtigall, P.E. (1975). Necrotic stomatitis in a dolphin. *Journal of Wildlife Diseases*, **11**, 460–464. 550

Colyer, J.F. (1905). On some recent additions to the Museum. *Transactions of the Odontological Society of Great Britain*, New Series, **37**, 11–14. 350, 580

Colyer, J.F. (1906). Variations and diseases of the teeth of horses. *Transactions of the Odontological Society of Great Britain*, New Series, **38**, 42–74.
118, 309, 482, 483, 488, 494, 563, 564

Colyer, J.F. (1907). Some specimens recently presented to the Museum of the Society. *Transactions of the Odontological Society of Great Britain*, New Series, **39**, 154–163. 325, 388, 444

Colyer, J.F. (1913). Some recent additions to the Odontological collection of the Royal College of Surgeons Museum. *Proceedings of the Royal Society of Medicine*, **6**, Odontological Section, 140–152. 373, 395, 398, 447

Colyer, J.F. (1914). Abnormalities in the position of the teeth in animals. *The Dental Record*, **34**, 1–28. 310, 312, 490

Colyer, J.F. (1915). Injuries of the jaws and teeth in animals. *The Dental Record*, **35**, 61–92, 157–168.
23, 396, 397, 398, 399, 405, 408, 421, 422, 423, 426, 427, 429, 433, 434, 435, 508, 517, 518, 519

Colyer, J.F. (1919). Variations in position of the teeth in New World monkeys. *Proceedings of the Royal Society of Medicine*, **12**, Section of Odontology, 39–54.
53, 215, 216, 217, 218, 219, 220, 221, 222, 223, 224, 227, 228, 230, 231, 232, 233

Colyer, J.F. (1921). A note on the dental condition of three hundred skulls of *Macacus rhesus*. *British Dental Journal*, **42**, 481–487.
51, 206, 371, 442, 472, 495, 510, 539

Colyer, J.F. (1926a). The pathology of the teeth of elephants. *The Dental Record*, **46**, 1–15, 73–80. 129

Colyer, J.F. (1926b). Abnormally-shaped teeth from the region of the premaxilla. *Proceedings of the Royal Society of Medicine*, **19**, Section of Odontology, 39–57. 576

Colyer, J.F. (1928). Variations and abnormalities in the position of the teeth in *Erythrocebus*, the patas monkey. *The Dental Record*, **49**, 9–11. 47, 48, 197, 198, 199, 200

Colyer, J.F. (1931). *Abnormal Conditions of the Teeth of Animals in the Relationship to Similar Conditions in Man*. Dental Board of the United Kingdom, London.
xv, 23

Colyer, J.F. (1936). *Variations and Diseases of the Teeth of Animals*. John Bale, Sons & Danielsson, London.
x, xi, 2, 20, 34, 62, 73, 83, 85, 96, 159, 292, 402, 432, 441, 444, 458, 478, 504, 591, 604

Colyer, J.F. (1938). Dento-alveolar abscess in a grampus (*Orca gladiator* Bonn). *The Scottish Naturalist*, **230**, 53–55. 493, 512

Colyer, J.F. (1940). Variations of the teeth of Preuss's colobus. *Proceedings of the Royal Society of Medicine*, **33**, 757–768. 35, 38, 40, 473

Colyer, J.F. (1942a). Unusual deformity in a lion. *The Dental Record*, **62**, 246–248. 400

Colyer, J.F. (1942b). An odontome in a gorilla. *The Dental Record*, **62**, 249. 589

Colyer, J.F. (1943a). The history of the Odontological Museum. *British Dental Journal*, **75**, 1–7, 31–34. x

Colyer, J.F. (1943b). Variations of the teeth of the colobus monkey. *The Dental Record*, **63**, 109–115, 134–139, 159–164, 183–188, 206–216, 232–237.
34, 35, 36, 37, 38, 40, 41, 159, 178, 179, 180, 181, 182, 183, 372, 374, 402, 403, 444, 473, 496, 511, 540

Colyer, J.F. (1944). Multiple tusks in an elephant. *The Dental Record*, **64**, 63–65. 129

Colyer, J.F. (1945a). Comminuted fracture of the mandible in a bay duiker. *The Dental Record*, **65**, 54. 382

Colyer, J.F. (1945b). Variations in number of teeth of animals. *The Dental Record*, **65**, 121–125. 46, 73, 74

Colyer, J.F. (1947a). Positional variations of the teeth in animals. *British Dental Journal*, **82**, 179–184.
159, 163, 180, 181, 182, 184, 185, 186, 218

Colyer, J.F. (1947b) Dental disease in animals. *British Dental Journal*, **82**, 1–10, 31–35. 441, 458, 465, 474

Colyer, J.F. (1948). Variations of the teeth of the green monkey in St. Kitts. *Proceedings of the Royal Society of Medicine*, **41**, 845–848. 35, 45, 47, 195, 197

Colyer, J.F. (1951a). A rare displacement of a canine in a tiger. *Proceedings of the Royal Society of Medicine*, **44**, 58–59. 240, 241

Colyer, J.F. (1951b). Asymmetry of the dental arch in a toque monkey (*Macaca sinica*). *Proceedings of the Royal Society of Medicine*, **44**, 60. 205

Colyer, J.F. (1952). Three specimens in the Odontological Section of the Royal College of Surgeons' museum. *Proceedings of the Royal Society of Medicine*, **45**, 753–756. 126, 129, 586, 601, 602

Colyer, J.F. (undated) Notes on the dental conditions of a collection of monkeys from Uganda. (Unpublished typescript on file, Odontological Museum, Royal College of Surgeons, London). 183

Colyer, J.F. & Miles, A.E.W. (1957). Injury to and rate of growth of an elephant tusk. *Journal of Mammalogy*, **38**, 243–247. 406, 407

Colyer, J.F. & Sprawson, E. (1931). *Dental Surgery and Pathology*, 6th edn. Longmans Green & Co., London. 333

Combe, C. (1801). Account of an elephant's tusk, in which the iron head of a spear was found imbedded. *Philosophical Transactions of the Royal Society of London*, **91**, Part I, 165–168. 421, 426

Compagno, L.J.V. (1967). Tooth pattern reversal in three species of sharks. *Copeia*, No. 1, 242–244. 6

Conaway, C. & Landry, S.O. (1958). Rudimentary mandibular teeth of *Scalopus aquaticus*. *Journal of Mammalogy*, **39**, 58–64. 353

Cook, R.A. & Stoller, N.H. (1986). Periodontal status in snow leopards. *Journal of the American Veterinary Medical Association*, **189**, 1082–1083. 543

Cooper, J.E., Arnold, L. & Henderson, G.M. (1982). A developmental abnormality in the Round Island skink *Leiolopisma telfairii*. *Dodo, Journal of the Jersey Wildlife Preservation Trust*, **19**, 78–81. 159

Cope, Z. (1959). *The Royal College of Surgeons of England. A History*. Anthony Blond, London. 419

Corbet, G.B. (1975). Examples of short- and long-term changes of dental patter: in Scottish voles (Rodentia: Microtinae). *Mammal Review*, **5**, 7–21. 133

Corbet, G.B. (1986). Temporal and spatial variation of dental pattern in the voles, *Microtus arvalis*, of the Orkney Islands. *Journal of Zoology, London*, **208**, 395–402. 132

Corbet, G.B. & Hill, J.E. (1986). *A World List of Mam-*

malian Species, 2nd edn. London: British Museum (Natural History) and Comstock. *xii, 135, 569*

Corruccini, R. S. & Beecher, R. M. (1982). Occlusal variation related to soft diet in a nonhuman primate. *Science*, **281**, 74–76. *157*

Corruccini, R.S. & Beecher, R.M. (1984). Occlusofacial morphological integration lowered in baboons raised on soft diet. *Journal of Craniofacial Genetics and Developmental Biology*, **4**, 135–142. *157*

Cotton, W. R. & Gaines, J. F. (1974). Unerupted dentition secondary to congenital osteopetrosis in the Osborne-Mendel rat. *Proceedings of the Society for Experimental Biology and Medicine*, **146**, 554–561. *333*

Couvillion, C. E., Nettles, V. F., Rawlings, C. A. & Joyner, R. L. (1986). Elaeophorosis in white-tailed deer: pathology of the natural disease and its relation to oral food impactions. *Journal of Wildlife Diseases*, **22**, 214–223. *562*

Cowan, I.M. (1946). Parasites, diseases, injuries and anomalies of the Columbian black-tailed deer, *Odocoileus hemionus columbianus* (Richardson) in British Columbia. *Canadian Journal of Research*, **24**, 71–103. *481, 488, 514*

Crew, F. A. E. (1924). The bull-dog calf: a contribution to the study of achondroplasia. *Proceedings of the Royal Society of Medicine*, **17**, (Section of Comparative Medicine), 39–58. *294*

Crissman, J.W., Maylin, G.A. & Krook, L. (1980). New York State and U.S. Federal fluoride pollution standards do not protect cattle health. *Cornell Veterinarian*, **70**, 183–192. *451, 453, 454*

Cunningham, G. (1889). Implantation of teeth. *Journal of the British Dental Association*, **10**, 740–761. *422*

Cunningham, G. (1911). Mr. George Cunningham and children's teeth. *The Dental Surgeon, London*, **7**, 510–511. *422*

Cutress, T.W. (1972). The incisive apparatus of the sheep. *Research in Veterinary Science*, **13**, 74–76. *552, 553, 554*

Cutress, T. W. (1976). Histopathology of periodontal disease in sheep. *Journal of Periodontology*, **47**, 643–650. *557, 560*

Cutress, T. W. & Healy, W. B. (1965). Wear of sheep's teeth. II. Effects of pasture juices on dentine. *New Zealand Journal of Agricultural Research*, **8**, 753–762. *492*

Cutress, T. W. & Ludwig, T. G. (1969). Periodontal disease in sheep. 1. Review of the literature. *Journal of Periodontology*, **40**, 529–534. *557, 562*

Cutress, T. W. & Schroeder, H. E. (1982). Histopathology of periodontitis ('broken mouth') in sheep: a further consideration. *Research in Veterinary Science*, **33**, 64–69. *557*

Danforth, C. H. (1958). The occurrence and genetic behavior of duplicate lower incisors in the mouse. *Genetics*, **43**, 140–148. *131*

Darling, A. I. & Levers, B. G. H. (1973). Submerged human deciduous molars and ankylosis. *Archives of Oral Biology*, **18**, 1021–1040. *332*

Darwin, C. (1859). *On the Origin of Species*, 1st edn. John Murray, London. *7*

Darwin, C. (1868). *The Variation of Animals and Plants under Domestication*, 1st edn. John Murray, London. *7, 134, 283, 292, 294*

Dawson, C.E. (1961–64). A bibliography of anomalies of fishes. *Gulf Research Reports*, **1**, 308–398. *159*

Dawson, C.E. (1965–69). A bibliography of anomalies of fishes. Supplement 1. *Gulf Research Reports*, **2**, 169–171. *159*

Dawson, C. E. (1970–71). A bibliography of anomalies of fishes. Supplement 2. *Gulf Research Reports*, **3**, 215–239. *159*

Dawson, C.E. & Heal, E. (1976). A bibliography of anomalies of fishes. Supplement 3. *Gulf Research Reports*, **5**, 35–41. *159*

Dawson, M. (1956). Extra teeth in a hare. *Journal of Mammalogy*, **37**, 453–454. *134*

Day, P.L., Langston, W.C. & Shukers, C.F. (1935). Leukopenia and anemia in the monkey resulting from vitamin deficiency. *Journal of Nutrition*, **9**, 637–644. *528*

Dean, H.T. (1934). Classification of mottled enamel diagnosis. *Journal of the American Dental Association*, **21**, 1421–1426. *452*

Dearden, L.C. (1954). Extra premolars in the river otter. *Journal of Mammalogy*, **35**, 125–126. *74*

Deasy, M.J., Grota, L.J. & Kennedy, J.E. (1972). The effect of estrogen, progesterone and cortisol on gingival inflammation. *Journal of Periodontal Research*, **7**, 111–124. *528*

De Beaufort, F. (1964). Denture du chevreuil *Capreolus capreolus*: incisives et canines supernuméraires. *Mammalia*, **28**, 519–520. *112*

De Calesta, D. S., Zemlicka, D. & Cooper, L. D. (1980). Supernumerary incisors in a black-tailed deer. *The Murrelet*, **61**, 103–104. *111*

Degerbøl, M. (1929/1930). Om 3 tilfaede af dobbelte Stødtaender hos Hvaltrossen (*Odobaenus rosmarus* L.). *Videnskabelige Meddelelser fra Dansk naturhistorick Forening l København*, **88**, 287–292. *98, 99*

Degerbøl, M. (1961). On a find of a Preboreal domestic dog (*Canis familiaris* L.) from Star Carr, Yorkshire. *Proceedings of the Prehistoric Society*, **27**, 35–54. *11*

Dekeyser, P. L. & Derivot, J. (1956). Sur la présence de canines supérieures chez les Bovidés. *Bulletin de l'Institut Française d'Afrique Nord*, **18A**, 1272–1281. *114*

De la Torre, L. & Dysart, M.P. (1966). Fusion of incisor enamel organs in a Guatamalan opossum. *Journal of Mammalogy*, **47**, 709–710. *149*

della Serra, O. (1951). Variacoes do articulado dos dentes incisivos nos macacos do genero *Alouatta* Lac. 1799. *Papéis Avulsos do Departmento de Zoologia, São Paulo*, **10**, 139–146. *224, 226*

della Serra, O. (1955*a*). O articulado dos dentes labias nos simios da familia *Cebidae* Swainson, 1835 (Primates – Mammalia). *Papéis Avulsos do Departmento de Zoologia, São Paulo*, **12**, 165–188. *213, 219, 220, 221, 228*

della Serra, O. (1955*b*). Osteomielite após fratura bilateral do corpo da mandibula do porco do mato (*Tayassu peccari* Fischer), com cura espontanea. *Papéis Avulsos do Departmento de Zoologia, São Paulo*, **12**, 303–312. *388*

Denholm, L.J. & Vizard, A.L. (1986). Trimming the incisor teeth of sheep: another view. *The Veterinary Record*, **119**, 182–184. *556*

Deniz, E. & Payne, S. (1982). Eruption and wear in the mandibular dentition as a guide to ageing Turkish Angora goats. In *Ageing and Sexing Animal Bones from Archaeological Sites*, edit. B. Wilson, C. Grigson & S. Payne. British Archaeological Reports, British Series, 109, pp. 155–214. *345*

Dent, V. E. (1979). The bacteriology of dental plaque from a variety of zoo-maintained mammalian species. *Archives of Oral Biology*, **24**, 277–282. 456

Dent, V. E. & Marsh, P. D. (1981). Evidence for a basic plaque microbial community on the tooth surface in animals. *Archives of Oral Biology*, **26**, 171–179. 456

Derise, N. L. & Ritchey, S. J. (1974). Mineral composition of normal human enamel and dentin and the relation of composition to dental caries: microminerals. *Journal of Dental Research*, **53**, 853–858. 456

de Smet, W. M. A. (1972). Sur deux dauphins à bec blanc, *Lagenorhynchus albirostris* (Gray), de la côte belge. *Bulletin de l'Institut Royale des Sciences Naturelles de Belgique*, **48**, 2–17. 494

de Smet, W. M. A. (1977). The fate of old bottle-nosed dolphins, *Tursiops truncatus*, in nature as revealed by the condition of their skeletons. *Aquatic Mammals*, **5**, 78–86. 494

Dippenaar, N. J. (1978). Dental abnormalities in *Crocidura mariquensis* (A. Smith, 1844) (Mammalia: Soricidae). *Annals of the Transvaal Museum*, **31**, 165–168. 139

Dobereiner, J., Tokarnia, C. H. & Rose, I. V. (1976). Cara inchada, a periodontal disease of cattle in Brazil. *Proceedings of the 9th International Congress on Diseases of Cattle*. Paris, pp. 1133–1139. 561

Döcke, F. (1959). Zahn und Bissanomalien beim Wild- und Farmfuchs. *Archiv für Tierzucht*, **2**, 130–161. 84

Doerr, J. G. & Dieterich, R. A. (1979). Mandibular lesions in the Western Arctic caribou herd of Alaska. *Journal of Wildlife Diseases*, **15**, 309–318. 387, 516

Dolamore, W. H. (1914). On dentigerous cysts. *Proceedings of the Royal Society of Medicine*, Odontological Section. **7**, 75–78. 578

Dolgov, V. A. & Rossolimo, O. L. (1964). Dental abnormalities in *Canis lupus* (Linnaeus 1758). *Acta Theriologica*, **8**, 237–244. 83, 85

Domning, D. P. & Hayek, L.-A. C. (1984). Horizontal tooth replacement in the Amazonian manatee (*Trichechus inunguis*). *Mammalia*, **48**, 105–127. 345

Donald, H. P. & Wiener, G. (1954). Observations on mandibular prognathism. *The Veterinary Record*, **66**, 479–482. 294

Dönitz (1868). Uber einen monströsen Fuchsschädel. *Sitzungsberichte der Gesellschaft Naturforschender Freunde zu Berlin*, 1868, 21–22. 263

Dönitz (1872). Ueber Gebiss-Abnormalitäten bei *Cervus axis* und *Canis mesomelas*. *Sitzungsberichte der Gesellschaft Naturforschenden Freunde zu Berlin*, 1872, 54. 89, 113

Dorst, J. & Dandelot, P. (1972). *A Field Guide to the Larger Mammals of Africa*, 2nd edn. Collins, London. xii

Douma-Petridou, E. (1984). Dental inclinations and skull deformations on specimens of the family Mustelidae from Greece. *Biologia Gallo-Hellenica*, **11**, 129–132. 74

Dozier, H. L. (1943). Occurrence of ringworm disease and lumpy jaw in the muskrat in Maryland. *Journal of the American Veterinary Medical Association*, **102**, 451–453. 361

Dreizen, S. & Levy, B. M. (1981). *Handbook of Experimental Stomatology*. C.R.C. Press, Florida. 529

Dreizen, S., Levy, B. M. & Bernick, S. (1972). Studies on the biology of the periodontium of marmosets. XII. The effect of an experimentally produced malasorption syndrome on the marmoset periodontium. *Journal of Periodontal Research*, **7**, 251–254. 528

Dubielzig, R. R. (1979). Effect of canine distemper virus on the ameroblastic layer of the developing tooth. *Veterinary Pathology*, **16**, 268–270. 445

Dubielzig, R. R. (1980). Lesions of canine distemper virus in the developing tooth. *Laboratory Investigation*, **42**, 113 (abstract). 445

Dubielzig, R. R., Beck, K. A., Wilson, J. W. & Ribble, G. A. (1986). Dental dysplasia in two young uremic dogs. *Veterinary Pathology*, **23**, 333–335. 589

Dubielzig, R. R., Higgins, R. J. & Krakowka, S. (1981). Lesions of the enamel organ of developing dog teeth following experimental inoculation of gnotobiotic puppies with canine distemper virus. *Veterinary Pathology*, **18**, 684–689. 446

Dubielzig, R. R., Wilson, J. W., Beck, K. A. & Robbins, T. (1986). Dental dysplasia and epitheliogenesis imperfecta in a foal. *Veterinary Pathology*, **23**, 325–327. 450

Duckworth, J., Benzie, D., Cresswell, E., Hill, R. & Boyne, A. W. (1962). Studies of the dentition of sheep. III. A study of the effects of vitamin D and phosphorus deficiencies in the young animal on the productivity, dentition and skeleton of Scottish blackface ewes. *Research in Veterinary Science*, **3**, 408–415. 491, 557

Duckworth, W. L. H. (1907). A note on the dentition of some New Guinea skulls. *Transactions of the Odontological Society of Great Britain*, New Series, **39**, 58–68. 27, 175

Duckworth, W. L. H. (1912). On the natural repair of fractures, as seen in the skeletons of anthropoid apes. *Journal of Anatomy, London*, **46**, 81–85. 371

Duerst, J. U. (1930). Vergleichende Untersuchungsmethoden am Skelett bei Saugern. *Abderhaldens Handbuch der Biologischen Arbeitsmethoden*, **7**, 125–539. 159

Dupas, L. (1903). Duplicité dentaire par soudure d'une mitoyenne avec une autre incisive surnuméraire. *Receuil de Médecine Vétérinaire*, **10**, série 8, 695–696. 120

Dupas, L. (1904). Avortement d'une mitoyenne inférieure ayant causé la transposition du coin de remplacement voisin et empêché la chute du coin de lait correspondant. *Bulletin et Mémoires de la Société Centrale de Médecine Vétérinaire*, 58, 123–124; annexe to *Receuil de Médecine Vétérinaire*, 11, série 8. 122

Dyer, D. L. (1981). An analysis of bony changes following trauma to a black bear mandible. *Journal of Wildlife Diseases*, **17**, 97–100. 380

Dyson, D. A. & Spence, J. A. (1979). A cystic jaw lesion in sheep. *The Veterinary Record*, **105**, 467–468. 579

Eaglen, R. H. (1985). Behavioral correlates of tooth eruption in Madagascan lemurs. *American Journal of Physical Anthropology*, **66**, 307–315. 339

Eccles, J. D. (1979). Dental erosion of non-industrial origin. A clinical survey and classification. *Journal of Prosthetic Dentistry*, **42**, 649–653. 497

Edmund, A. G. & Hoffstetter, R. (1970). *Essonodontherium gervaisi* es un sinonimo de *Megatherium americanum* Cuvier (Xenarthra, Mammalia). *Ameghiniana*, 7(4), 317–328. 134

Edwards, M. B. (1980). Effects of N-methylnitrosourea on oral and dental tissues in the inbred Lewis rat. *Archives of Oral Biology*, **25**, 59–65. 7

Egelberg, J. (1965). Local effect of diet on plaque formation and development of gingivitis in dogs. I. Effect of hard and soft diets. *Odontologisk Revy*, **16**, 31–41. 528

Eklund, C. M., Hadlow, W. J., Kennedy, R. C., Boyle, C. C. & Jackson, T. A. (1968). Aleutian disease of mink:

properties of the etiologic agent and the host responses. *Journal of Infectious Disease*, **118**, 510–526.   *546*

Elkan, E. (1981). Pathology and histopathological techniques. In *Diseases of the Reptilia*, chapter 3, vol. 1, edit. J. E. Cooper & O. F. Jackson. Academic Press, London and New York, pp. 75–91.   *393*

Elzay, R. P. & Hughes, R. D. (1969). Anodontia in a cat. *Journal of the American Veterinary Medical Association*, **154**, 667–670.   *63*

Emele-Nwaubani, J. C. & Ihemelandu, E. C. (1984). Anodontia of the incisor and canine teeth in a cryptorchid West African dwarf goat. *Tropical Veterinarian*, **2**, 172–174.   *117, 297*

Emmerson, M. A. & Hazel, L. N. (1956). Radiographic demonstration of dwarf gene-carrier beef animals. *Journal of the American Veterinary Medicine Association*, **128**, 381–390.   *294*

Epstein, H. (1971). *The Origin of the Domestic Animals of Africa*, vol. 2. Africana Publishing Corporation, London.   *296*

Erasmuson, A. F. (1985). Abrasion of ovine teeth from high-wear and low-wear farms. *New Zealand Journal of Agricultural Research*, **28**, 225–331.   *491*

Erdheim, J. (1906). Tetania parathyreopriva. *Mitteilungen aus den Grenzgebieten der Medizin und Chirurgie*, **16**, 632–744.   *438*

Eschler, W. & Lüps, P. (1984). Überzählige Eckzahn bei einem Steinbok (*Capra ibex*). *Zeitschrift für Jagdwissenschaften*, **30**, 201–208.   *117*

Eskelsen, A. M. & Fichter, E. (1985). Duplication of molars in a pronghorn. *Tebiwa*, **22**, 54–56.   *117*

Espersen, G. (1962). *Multiple Kæbecyster Hos Hesten*. C. Fr. Mortensen, Copenhagen.   *604*

Evans, S. E. (1983). Mandibular fracture and inferred behavior in a fossil reptile. *Copeia*, No. 3, 845–847.   *393*

Every, R. G. (1970). Sharpness of teeth in man and other primates. *Postilla*, **143**, 1–30.   *357*

Every, R. G. (1975). Significance of tooth sharpness for mammalian, especially Primate, evolution. *Contributions to Primatology*, **5**, 293–325.   *357*

Every, R. G. & Kühne, W. G. (1971). Bimodal wear of mammalian teeth. In *Early Mammals*, edit. D. M. Kermack & K. A. Kermack. *Journal of the Linnean Society, Zoology*, **50**, Supplement, 23–26.   *357*

Ewer, R. F. (1973). *The Carnivores*. Cornell University Press, Ithaca, New York.   *62*

Fabian, H. (1933). Merkmale und Grenzen in Domestikationsfrage am Gebiss. *Deutsche Zahnheilkunde*, **84**, 1–99.   *71, 84, 264*

Fabian, H. (1950). 'Bohr' höhlen an Pottwalzähnen. *Zoologischer Anzeiger*, **145**, 147–162.   *498*

Falin, L. I. (1961). Histological and histochemical studies of human teeth of the bronze and stone ages. *Archives of Oral Biology*, **5**, 5–13.   *501*

Farbiszewska, I. & Makarzec, B. (1960). Gebissvariabilität bei *Microtus arvalis*. *Acta Theriologica*, **3**, 302–307.   *132*

Farris, E. J. (1950). *The Care and Breeding of Laboratory Animals*. Wiley, New York.   *485*

Feeback, D. L., Jensen, J. M. & Kosanke, S. D. (1986). Renal hyperostotic osteodystrophy associated with unilateral renal aplasia in a captive maned wolf. *Journal of Wildlife Diseases*, **22**, 595–600.   *605*

Feldhamer, A. G. (1982). Cranial and dental abnormalities in sika deer. *Acta Theriologica*, **27**, 350–353.   *111, 562*

Feldhammer, G. A. & Chapman, J. A. (1980). Mandibular dental anomalies in white-tailed deer. *Brimleyana*, **4**, 161–163.   *112*

Fine, M. A. (1964). An abnormal P_2, in *Canis* cf. *C. latrans* from the Hagerman fauna of Idaho. *Journal of Mammalogy*, **45**, 483–485.   *88*

Finnie, E. P. (1976). Necrobacillosis in kangaroos. In *Wildlife Diseases*, edit. L. A. Page. Plenum Press, London and New York, pp. 511–518.   *520*

Finsch, O. (1887). Abnorme Eberhauer, Pretiosen im Schmuck der Südsee-Völker. *Mitteilungen der Anthropologischen Gesellschaft, Wien*, **17**, 153–159. *364*

Fischer, M. (1867). Note sure une déformation pathologique de la machoire inférieur du cachalot. *Journal de l'Anatomie et de la Physiologie (Paris)*, **4**, 382–388. *391*

Fish, E. W. (1933). *An Experimental Investigation of Enamel Dentine and the Dental Pulp*. Bale, Sons & Danielsson, London.   *479*

Fish, E. W. & Harris, L. J. (1934). The effects of vitamin C deficiency on tooth structure in guinea-pigs. *Philosophical Transactions of the Royal Society of London*, Series B, **223**, 489–510.   *362*

Fisher, E. M. (1941). Notes on the teeth of the sea otter. *Journal of Mammalogy*, **22**, 428–433.   *496*

Fitzgerald, D. B. & Fitzgerald, R. J. (1965). Induction of dental caries in gerbils. *Archives of Oral Biology*, **11**, 139–140.   *461*

Fleischer, G. (1967). Beitrag zur Kenntnis der innerartlichen Ausformung und zwischenartlicher Unterschiede von Gebiss und Zahnen einiger Arten der Gattung *Canis*. *Zeitschrift für Säugetierkunde*, **32**, 150–159.   *88*

Flower, W. H. (1869). Exhibition of the skull of a hyrax with abnormal dentition obtained by Mr. Blanford. *Proceedings of the Zoological Society of London*, 1869, 603–604.   *124*

Flux, J. E. C. (1980). High incidence of missing posterior upper molars in hares (*Lepus europaeus*) in New Zealand. *New Zealand Journal of Zoology*, **7**, 257–259. *134*

Foley, R. A. & Atkinson, S. (1984). A dental abnormality among a population of Defassa waterbuck (*Kobus defassa* Ruppell 1835). *African Journal of Ecology*, **22**, 289–294.   *491*

Fordyce, R. E. (1982). Dental anomaly in a fossil squalodont dolphin from New Zealand, and the evolution of polydonty in whales. *New Zealand Journal of Zoology*, **9**, 419–426.   *102*

Foster, J. B. (1964). Evolution of mammals on islands. *Nature, London*, **202**, 234–235.   *10*

Fowle, C. D. & Passmore, R. C. (1948). A supernumerary incisor in the white-tailed deer. *Journal of Mammalogy*, **29**, 301.   *111*

Fox, H. (1923). *Disease in Captive Wild Mammals and Birds*. J. P. Lippincott Co., Philadelphia.   *516, 519*

Fox, M. W. (1963). Developmental abnormalities of the canine skull. *Canadian Journal of Comparative Medicine and Veterinary Science*, **27**, 219–222.   *266, 271*

Fox, R. R. & Crary, D. D. (1971). Mandibular prognathism in the rabbit: genetic studies. *Journal of Heredity*, **62**, 23–27.   *319*

Frandsen, A. M., Becks, H., Nelson, M. M. & Evans, H. M. (1953). The effects of various levels of dietary protein on the periodontal tissues of young rats. *The Journal of Periodontology*, **24**, 135–142.   *526*

Franke, G. (1921). Über Wachstum und Verbildungen des

Kiefers und der Nasenscheidewand auf Grund vergleichender Kiefer-Messungen und experimenteller Untersuchungen über Knochenwachstum. *Zeitschrift für Laryngologie, Rhinologie und ihre Grenzgebiete*, **10**, 1–205. 515

Franklin, M.C. (1950). *The Influence of Diet on Dental Development in the Sheep*. Bulletin No. 252. Commonwealth Scientific and Industrial Research Organization, Australia, Melbourne. 295, 451, 452, 453

Fraser, D. & Reardon, E. (1980). Attraction of wild ungulates to mineral rich springs in central Canada. *Holarctic Ecology*, **3**, 36–39. 454

Fraser, F.C. (1934). Report on Cetacea stranded on the British coasts between 1927 and 1932. *British Museum (Natural History) Publication*, 11. 102, 512

Fraser, F.C. (1946). Report on Cetacea stranded on the British coasts between 1933 and 1937. *British Museum (Natural History) Publication*, 12. 102, 103, 385, 512

Fraser, F.C. (1953). Report on Cetacea stranded on the British coasts between 1938 and 1947. *British Museum (Natural History) Publication*, 13.
101, 385, 493, 512, 513

Free, S.L., Bergstrom, A.S. & Tanck, J.E. (1972). Mandibular and dental anomalies of white-tailed deer. *New York Fish and Game Journal*, **19**, 32–46. 112, 113

Freedman, L. (1957). The fossil Cercopithecoidea of South Africa. *Annals of the Transvaal Museum*, **23**, 121–262. 49

Freedman, L. (1962). Quantitative features of deciduous dentition of *Papio ursinus*. *South African Journal of Science*, **58**, 229–236. 336

French, T.W. (1984). Dental anomalies in three species of shrews from Indiana. *Proceedings of the Indiana Academy of Science*, **94**, 635–640. 139

Friant, M. (1935). Description et interpretation de la dentition d'une jeune Indris. *Comptes Rendues de l'Association les Anatomistes*, **30**, 205–213. 60

Friedlowsky, A. (1869). Über Missbildungen von Saugethierzähnen. *Sitzungsberichte der Kaiserlichen Akademia der Wissenschaften, Mathematisch-Naturwissenhaftliche Classe*, **59**, 333–350. 126, 361, 363, 366

Friedman, L.A., Levy, B.M. & Ennever, J. (1972). Epidemiology of gingivitis and calculus in a marmoset colony. *Journal of Dental Research*, **51**, 803–806. 542

Friel, S. (1938). The malocclusion of a famous horse. *The Dental Record*, **58**, 655–666. 487

Frisch, J.E. (1963). Dental variability in a population of gibbons *Hylobates lar*. In *Dental Anthropology*, edit. D.R. Brothwell. Pergamon Press, Oxford, pp. 15–28. 32

Frisch, J.E. (1965). Trends in the evolution of the Hominoid dentition. *Bibliotheca Primatologica*, **3**, 1–130. 20, 22

Frum, W. (1946). Abnormality in dentition of *Myotis lucifugus*. *Journal of Mammalogy*, **27**, 176. 135

Fullagar, P.J., Rogers, T.H. & Mansfield, D. (1960). Supernumerary teeth in the badger. *Proceedings of the Zoological Society, London*, **133**, 494. 73

Fuller, W.A. (1954). The first premolar and canine tooth in bison. *Journal of Mammalogy*, **35**, 454–456. 115

Fulmer, B.A. & Ridenhour, R.L. (1967). Jaw injury and condition of king salmon. *California Fish and Game*, **53**, 282–285. 393

Gabell, D.P., James, W.W. & Payne, J.L. (1914). *The Report on Odontomes*. British Dental Association, London. 575, 578, 591, 603

Gad, T. (1968). Periodontal disease in dogs. I. Clinical investigation. *Journal of Periodontal Research*, **3**, 268–272. 548

Gainer, R.S. (1982). Dental anomalies in Nyasa wildebeest. *Journal of Mammalogy, London*, **63**, 526–527. 561

Gainer, R.S. (1986). Tertiary dentine in the root of an elephant tusk. *Journal of Zoo Animal Medicine*, **17**, 69–72. 348, 405

Gardner, A.F., Darke, B.H. & Keary, G.T. (1962). Dental caries in domesticated dogs. *Journal of the American Veterinary Medical Association*, **140**, 433–435. 479

Gardner, A.F., Dasler, W. & Weinmann, J.P. (1958). Masticatory apparatus of albino rats in experimental lathyrism. *Journal of Dental Research*, **37**, 492–515. 333

Garlick, N.L. (1954a). The teeth of the ox in clinical diagnosis. II. Gross anatomy and physiology. *American Journal of Veterinary Research*, **15**, 385–394. 115

Garlick, N.L. (1954b). The teeth of the ox in clinical diagnosis. III. Developmental anomalies and general pathology. *American Journal of Veterinary Research*, **15**, 500–508. 428, 432, 450, 480, 514, 561

Garlick, N.L. (1955). The teeth of the ox in clinical diagnosis. IV. Dental fluorosis. *American Journal of Veterinary Research*, **16**, 38–44. 451

Garrod, A.H. (1875). Report on the Indian elephant which died in the Gardens on July 7th, 1875. *Proceedings of the Zoological Society of London*, 1875, 542–543. 504

Gaskin, D.E. (1972). *Whales, Dolphins and Seals*. Heinemann, London. 391

Gasson, P. (1980). An abnormality in the dentition of a fox (*Vulpes vulpes*) from south Essex. *Journal of Zoology, London*, **192**, 562–564. 90

Gatineau, M. (1956). Variabilité du nombre des alveoles radiculaires du maxillaire supérieure chez le mulot (*Apodeumus sylvaticus*, L.) et chez la souris blanc (*Mus musculus*, L., albinos). *Mammalia*, **20**, 427–438. 133

Gaunt, W.A. & Miles, A.E.W. (1967). Fundamental aspects of tooth morphogenesis. In *Structural and Chemical Organization of Teeth*, chapter 4, vol. 1, edit. A.E.W. Miles. Academic Press, London, pp. 151–197. 2, 3

Gehr, F. (1954). A woodchuck who needed tooth-straightening. *Natural History, American Museum of Natural History, New York*, **63**, 310–311. 359, 360

Geiger, G. (1976). Zahnanomalien beim Hauskaninchen. *Saugetierkundliche Mitteilungen*, **24**, 110–112. 134

Geiger, G. (1980). Polyodontie der Zangen bei einem Reh. *Acta Theriologica*, **25**, 265–275. 114

Genvert, H., Miller, H. & Burn, C.G. (1940). Experimental production of apical lesions of teeth in monkeys, and their relation to systemic disease. *Yale Journal of Biology and Medicine*, **13**, 649–661. 510

Geoffroy Saint-Hilaire, I. (1838). *Anomalies du Système Dentaire*. Paris. 56

Getty, R. (1975). *Sisson and Grossman's the Anatomy of Domestic Animals*, 5th edn. W.B. Saunders, London. 343, 344

Gewalt, W. & Pauling, H. (1971). Zahnfleischwucherungen beim Grossen Kudu (*Tragelaphus strepsiceros* Pall.) *Zoologische Garten, Leipzig*, **40**, 298–301. 561

Gidley, J.W. (1901). Tooth characters and revision of the North American species of the genus *Equus*. *Bulletin of the American Museum of Natural History*, **14**, 91–141. 157

Gilmore, L.O. (1949). The inheritance of functional causes

of reproductive inefficiency: a review. *Journal of Dairy Science*, **32**, 71–91. 294

Gingerich, P. D. & Schoeninger, M. J. (1979). Patterns of tooth size variability in Primates. *American Journal of Physical Anthropology*, **51**, 457–466. 18

Gingerich, P. D. & Winkler, D. A. (1979). Patterns of variation and correlation in the dentition of the red fox, *Vulpes vulpes*. *Journal of Mammalogy*, **60**, 691–704. 92

Ginsberg, A. & Little, A. C. W. (1948). Actinomycosis in dogs. *Journal of Pathology and Bacteriology*, **60**, 563–571. 550

Glass, B. P. (1953). Variation in the lower incisors of the Mexican freetail bat *Tadarida mexicana* (Saussure). *Proceedings of the Oklahoma Academy of Science*, **34**, 73–74. 135

Glass, G. E. & Todd, N. B. (1977). Quasi-continuous variation of the second upper premolar in *Felis bengalensis* Kerr 1792 and its significance for some fossil lynxes. *Zeitschrift für Säugetierkunde*, **42**, 36–44. 63

Glasstone, S. (1952). The development of halved tooth germs. A study in experimental embryology. *Journal of Anatomy, London*, **86**, 12–15. 5, 8

Glaze, R. I., Hoefs, M. & Bunch, T. D. (1982). Aberrations of the tooth arcade and mandible in Dall's sheep from southwestern Yukon. *Journal of Wildlife Diseases*, **18**, 305–309. 516

Glick, C., Swart, G. & Woolf, A. (1979). Dental caries, periodontal abscesses, and extensive cranial osteitis in a captive lowland gorilla (*Gorilla gorilla gorilla*). *Journal of Zoo Animal Medicine*, **10**, 94–97. 465, 509

Glickman, I. (1948). Acute vitamin C deficiency and periodontal disease. I. The periodontal tissues of the guinea pig in acute vitamin C deficiency. *Journal of Dental Research*, **27**, 9–23. 525

Glickman, I., Stone, I. C. & Chawla, T. N. (1953). The effect of the systemic administration of cortisone upon the periodontium of white mice. *Journal of Periodontology*, **24**, 161–166. 528

Goethe, J. W. (1823). *Zur Naturwissenschaft*, vol. 2, part 1. *Zur Morphologie*, p. 7. Stuttgart and Tübingen. 421

Goethe, J. W. (1949). *Gedenkausgabe der Werke Brief und Gespräche*, vol. 11, p. 670. Commemorative reprint. Tag-und Jahrshafte, Zürich. 421

Goldman, H. M. (1947). Periodontosis in the spider monkey – a preliminary report. *Journal of Periodontology*, **18**, 34–40. 528

Goldman, H. M. (1954). The effects of dietary protein deprivation and of age on the periodontal tissues of the rat and spider monkey. *Journal of Periodontology*, **25**, 87–96. 528

Gorlin, R. J. (1970). Odontogenic tumors. In *Thoma's Oral Pathology*, chapter 11, 6th edn., edit. R. J. Gorlin & H. M. Goldman. C. V. Mosby Co., St Louis, pp. 481–515. 576

Gorlin, R. J., Meskin, L. H. & Brodey, R. (1963). Odontogenic tumors in man and animals: pathologic classification and clinical behaviour – a review. *Annals New York Academy of Science*, **108**, 723–771. 576

Gotcher, J. E. & Jee, W. S. S. (1981). The progress of the periodontal syndrome in the rice rat. 1. Morphometric and autoradiographic studies. *Journal of Periodontal Research*, **16**, 275–291. 527

Gottlieb, B. (1922). Die Paradental pyorrhoe der Rattenmolaren. *Vierteljahrsschrift für Zahnheilkunde*, **38**, 273–291. 568

Goubaux, A. (1854). Des aberrations dentaire chez les animaux domestiques. *Recueil de Médecine Vétérinaire*, **1**, Series 4, 70–80. 121

Graf, M., Grundbacher, B., Gschwendtner, J. & Lüps, P. (1976). Grössen- und Lagevariation des zweiten Prämolaren bei der Hauskatze *Felis silvestris* f. *catus*. *Revue Suisse de Zoologie*, **83**, 952–956. 64

Grajcarek, K. & Willer, S. (1985). Embryonale Duplikation des Unterkieferfortsatzes beim Schaf. *Monatshefte für Veterinärmedizin*, **40**, 284. 601

Grant, D. A., Chase, J. & Bernick, S. (1973). Biology of the periodontium in primates of the galago species. I. The normal periodontium in young animals. II. Inflammatory periodontal disease. III. Lability of cementum. IV. Changes in ageing. V. Ankylosis: types and sequential event. *Journal of Periodontology*, **44**, 540–550. 542

Grant, H. T. (1956). Underdeveloped mandible in a herd of dairy shorthorn cattle. *Journal of Heredity*, **47**, 165–170. 294

Greer, K. R. & Yeager, L. E. (1967). Sex and age indicators from upper canine teeth of elk (wapiti). *Journal of Wildlife Management*, **31**, 408–417. 111, 112

Gregory, P. W., Roubicek, C. B., Carroll, F. C., Stratton, P. O. & Hilston, N. W. (1953). Inheritance of bovine dwarfism and the detection of heterozygotes. *Hilgardia*, **22**, 407–450. 294

Gregory, K. E., Koch, R. M. & Swiger, L. (1962). Malocclusion – a hereditary defect in cattle. *Journal of Heredity*, **53**, 168–170. 294

Grenby, T. H. & Owen, D. (1980). A gnotobiotic study to distinguish between heredity and the oral microflora as transmitters of dental caries activity in laboratory rats. *Caries Research*, **14**, 434–440. 462

Grew, N. (1681). *Catalogue and Description of Natural and Artificial Rareties Belonging to the Royal Society and Preserved at Gresham College*. London. 419

Grigson, C. (1974). The craniology and relationship of four species of *Bos*. I. Basic craniology: *Bos taurus* L. and its absolute size. *Journal of Archaeological Science*, **1**, 353–379. 157

Grigson, C. (1982). Sex and age determination of some bones and teeth of domestic cattle. In *Ageing and Sexing Animal Bones from Archaeological Sites*, edit. B. Wilson, C. Grigson & S. Payne. British Archaeological Reports, British Series, **109**, 7–23. 342, 344

Grigson, C. (forthcoming). Dental anomalies in an isolated herd of feral cattle at Chillingham, Northumberland, England. 115

Griner, L. A. (1983). *Pathology of Zoo Animals*. Zoological Society of San Diego. 518

Grinstead, B. G. (1971). Effects of pugheadedness on growth and survival of striped bass, *Morone saxatilis* (Walbaum) introduced into Canton reservoir, Oklahoma. *Proceedings of the Oklahoma Academy of Science*, **51**, 8–12. 159

Grove, T. K. (1985). Periodontal disease. In *Veterinary Dentistry*, chapter 5, edit. C. E. Harvey. W. B. Saunders, Philadelphia, pp. 59–78. 544

Groves, C. (1981). *Ancestors for the Pigs. Taxonomy and Phylogeny of the genus Sus*. Canberra: Technical Bulletin of the Department of Prehistory, Research School of Pacific Studies, Australian National University. 283

Groves, C. P. (1966). Skull-changes due to captivity in certain *Equidae*. *Zeitschrift für Säugetierkunde*, **31**, 44–46.
567

Grummt, W. (1961). Zur Morphologie der Kleinäugigen Wühlmaus *Pitymys subterraneus* de Selys-Longchamps. *Zoologische Anzeiger*, **166**, 26–32.
132

Grüneberg, H. (1936). The relations of endogenous and exogenous factors in bone and tooth development. The teeth of the grey-lethal mouse. *Journal of Anatomy*, **71**, 236–244.
333, 605

Grüneberg, H. (1951). The genetics of a tooth defect in the mouse. *Proceedings of the Royal Society of London, Series B*, **138**, 437–451.
2

Grüneberg, H. (1963). *The Pathology of Development; A study of Inherited Skeletal Disorders in Animals*. Blackwell Scientific Publications, Oxford.
131, 132, 333

Grüneberg, H. (1966). The molars of the tabby mouse, and a test of the 'single-active X-chromosome' hypothesis. *Journal of Embryology and Experimental Morphology*, **15**, 223–244.
3

Grüneberg, H. & Lea, A. J. (1940). An inherited jaw anomaly in long-haired dachshunds. *Journal of Genetics*, **39**, 285–296.
270

Grzimek, B. (1953). Mundfaule bei Schimpansen. *Zoologische Garten, Leipzig*, **20**, 197.
535

Gudger, E. W. (1937). Abnormal dentition in sharks, Selachii. *Bulletin of the American Museum of Natural History*, **37**, 249–280.
7

Guilday, J. E. (1961). Abnormal lower third molar in *Odocoileus*. *Journal of Mammalogy*, **42**, 552–553.
113

Guilday, J. E. (1962). Supernumerary molars of *Otocyon*. *Journal of Mammalogy*, **43**, 455–462.
84

Guilday, J. E. (1982). Dental variation in *Microtus xanthognathus*, *M. chrotorrhinus*, and *M. pennsylvanicus* (Rodentia: Mammalia). *Annals of the Carnegie Museum Natural History, Pittsburgh*, **51**, 211–230.
132

Gunn, R. G. (1970). A note on the effect of broken mouth on the performance of Scottish blackface hill ewes. *Animal Production*, **12**, 517–520.
552

Günther, F. (1957). Erbliche Zahnanomalien beim Kaninchen. *Berliner und Münchener Tierärzliche Wochenschrift*, **70**, 154–155.
361

Gupta, B. N. (1978). Duplication of lower incisors in a guinea pig. *Veterinary Pathology*, **15**, 683–684.
131, 601

Gupta, O. M. & Shaw, J. H. (1956a). Periodontal disease in the rice rat. I. Anatomic and histopathologic findings. *Oral Surgery, Oral Medicine and Oral Pathology*, **9**, 592–603.
461, 527

Gupta, O. M. & Shaw, J. H. (1956b). Periodontal disease in the rice rat. II. Methods for the evaluation of the extent of periodontal disease. *Oral Surgery, Oral Medicine and Oral Pathology*, **9**, 727–735.
527

Gupta, O. P. & Shaw, J. H. (1960). Dental anatomy and characteristics of periodontal lesions in the Mongolian gerbil. *Journal of Dental Research*, **39**, 1014–1022.
461, 569

Gyldenstolpe, N. (1928). Zoological results of the Swedish expedition to central Africa 1921. Vertebrata. 5. Mammals from the Birunga volcanoes, north of Lake Kivu. *Arkiv för Zoologi*, **20**, 1–76.
24, 167, 530

Habermehl, K. H. (1957). Über das Gebiss des Hausschweines (*Sus scrofa dom.* L.) mit besonderer Berücksichtigung der Backenzahnwurzeln. *Zentralblatt für Veterinärmedizin*, **4**(8), 794–810.
107

Habermehl, K. H. (1975). *Die Altersbestimmung bei Haus- und Labortieren*. Paul Parey, Berlin & Hamburg.
339, 342, 343, 344, 345

Haberstroh, D. V. M., Ullrey, D. E., Sikarski, J. G., Richter, N. A., Colmery, B. H. & Myers, T. D. (1984). Diet and oral health in captive Amur tigers (*Panthera tigris altaica*). *Journal of Zoo Animal Medicine*, **15**, 142–146.
543

Hadjimarkos, D. (1970). Effect of selenium in the hamster. In *Trace Element Metabolism in Animals*, edit. C. F. Mills. E. & S. Livingston, Edinburgh, pp. 215–218.
451

Haft, J. S. (1963). Malformation of molars in *Microtus breweri* (Baird). *Journal of Mammalogy*, **44**, 270–272.
132

Hall, B. K. & Hanken, J. (1985). Repair of fractured lower jaws in the spotted salamander: do amphibians form secondary cartilage? *Journal of Experimental Zoology*, **233**, 359–368.
393

Hall, E. R. (1928). Records of supernumerary teeth in bears. *University of California Publication in Zoology*, **30**(11), 243–250.
77

Hall, E. R. (1940). Supernumerary and missing teeth in wild mammals of the orders Insectivora and Carnivora, with some notes on disease. *Journal of Dental Research*, **19**, 103–143.
63, 67, 71, 74, 83, 84, 88, 92, 139, 477, 485, 511, 512

Hall, E. R. (1945). Dental caries in wild bears. *Transactions of the Kansas Academy of Science*, **48**, 79–84.
478

Hall, E. R. & Dalqvest, W. W. (1950). *Pipistrellus cinnamomeus* Miller 1902 referred to genus *Myotis*. *University of Kansas Publications of the Museum of Natural History*, **1**(25), 581–590.
135

Hall, P. M. (1985). Brachycephalic growth and dental anomalies in the New Guinea crocodile (*Crocodylus novaeguineae*). *Journal of Herpetology*, **19**, 300–303.
158

Hall, R. D., Beattie, R. J. & Wyckoff, G. H. (1979). Weight gains and sequence of dental eruptions in infant owl monkeys (*Aotus trivirgatus*). In *Nursery Care of Non-human Primates*, edit. G. C. Ruppenthal & D. J. Reese. Plenum Press, London & New York, pp. 321–328.
338

Hall, W. B., Grupe, H. E. & Claycombe, C. K. (1967). The periodontium and periodontal pathology in the howler monkey. *Archives of Oral Biology*, **12**, 359–365.
541

Halliwell, W. H. & Hahn, F. F. (1980). Fibro-osseus lesions in the mandible and maxilla of Greater Kudus. In *Comparative Pathology in Zoo Animals*, edit. R. J. Montali & G. Migaki. Plenum Press, New York, pp. 573–578.
604

Hamerstrom, F. N. & Camburn, F. L. (1950). Weight relationships in the George Reserve deer herd. *Journal of Mammalogy*, **31**, 5–17.
333

Hamilton, J. E. (1939). The leopard seal *Hydrurga leptonyx* (de Blainville). *Discovery Reports*, **18**, 239–264.
435

Hamp, S.-E. & Lindberg, R. (1977). Histopathology of spontaneous periodontitis in dogs. *Journal of Periodontal Research*, **12**, 46–54.
548

Hamp, S.-E., Olsson, S.-E., Farsö-Madsen, K., Viklands, P. & Fornell, J. (1984). A macroscopic and radiologic investigation of dental diseases of the dog. *Veterinary Radiology*, **25**, 86–92.
478, 548

Hamp, S.-E., Viklands, P., Farsö-Madsen, K. & Fornell, J. (1975). Prevalence of periodontal disease in dogs. I.

Clinical and roentgenographical observations. *Journal of Dental Research*, **54**, Special Issue A, Abstract L.19. *548*

Hancox, M. (1988). Dental anomalies in the Eurasian badger. *Journal of Zoology, London*, **216**, 606–608. *73*

Handley, C. O. (1959). A revision of American bats of the Genera *Enderma* and *Pleocotus*. *Proceedings of the United States National Museum*, **110**, article 3417, 95–246. *135*

Hansen, R. M. (1956). Extra incisors in the rodent *Dicrostonyx groenlandicus*. *Journal of Mammalogy*, **37**, 549–550. *131*

Hard, G. C. & Atkinson, F. F. V. (1967a). 'Slobbers' in laboratory guinea-pigs as a form of chronic fluorosis. *Journal of Pathology and Bacteriology*, **94**, 95–101. *362*

Hard, G. C. & Atkinson, F. F. V. (1967b). The aetiology of 'slobbers' (chronic fluorosis) in the guinea-pig. *Journal of Pathology and Bacteriology*, **94**, 103–112. *362*

Hare, T. (1934). Caries of the teeth of animals. *The Veterinary Record*, **14**, 681–682. *478*

Harmer, S. F. (1927). Report on Cetacea stranded on the British coasts 1913–1926. *British Museum (Natural History) Publication*, 10. *512*

Harr, J. R. & Muth, O. H. (1972). Selenium poisoning in domestic animals and its relationship to man. *Clinical Toxicology*, **5**, 175–186. *562*

Harris, A. H. & Fleharty, E. D. (1962). Extra tooth in the long-tailed vole. *Journal of Mammalogy*, **43**, 267–268. *131*

Harrison, D. L. (1982). An unusual dental anomaly in the Arabian hyrax, *Procavia capensis jayakari* (Hyracoidea: Procaviidae). *Mammalia*, **46**, 395–396. *363*

Harrison, D. L. & Bates, P. J. J. (1985). An unusual dental anomaly in an African hedgehog (*Erinaceus albiventris* Wagner, 1941) (Insectivora: Erinaceidae). *Mammalia*, **49**, 432–434. *95, 319*

Harrison, R. J. & King, J. & E. (1965). *Marine Mammals*. Hutchinson University Library, London. *95*

Hart, K. E. & Mackinnon, M. M. (1958). Enzootic paradontal disease of adult sheep in the Bulls-Santoft area. *New Zealand Veterinary Journal*, **6**, 118–123. *562*

Harvey, C. E. (1985). *Veterinary Dentistry*. W. B. Saunders & Co., Philadelphia. *481, 487, 507, 515, 518, 525, 546*

Harvey, C. E. & Penny, R. H. C. (1985). Oral and dental disease in pigs. In *Veterinary Dentistry*, chapter 18, edit. C. E. Harvey. W. B. Saunders, Philadelphia. *283, 480, 524 551*

Harvey, F. T. (1920). Some points in the natural history of alveolar or periodontal disease in the horse, ox and sheep. *The Veterinary Record*, **32**, 457–463. *559, 561, 567*

Hatt, S. D., Lyle-Stewart, W. & Cresswell, E. (1968). Periodontal disease in sheep. *The Dental Practitioner*, **19**, 123–127. *556*

Hauck, E. (1942). Untersuchungen über die Form und Abänderungsbreite des Hundegebisses. III. Die Zahnzahl. Der Zahnbogen. Die Zahngrosse. Der Zahnstein. V. Die Zahnstellung. *Wiener Tierärztliche Monatsschrift*, **29**, 468–469 and 534–539. *86, 263, 275*

Healy, W. B. (1968). The influence of soil type on ingestion of soil by grazing animals. *Transactions of the 9th International Congress of Soil Science*, **3**, 437–445. *451, 492*

Heath, G. (1976). Upper canine teeth in fallow deer. *Deer*, **3**, 485. *111*

Hedrick, P. W. (1983). *Genetics of Populations*. Science Books International, Boston. *10*

Heezen, B. C. (1957). Whales entangled in deep sea cables. *Deep-Sea Research*, **4**, 105–115. *390*

Heijl, L., Rifkin, B. R. & Zander, H. A. (1976). Conversion of chronic gingivitis to periodontitis in squirrel monkeys. *Journal of Periodontology*, **46**, 710–716. *528*

Heijl, L., Wennström, J., Lindhe, J. & Socransky, S. S. (1980). Periodontal disease in gnotobiotic rats. *Journal of Periodontal Research*, **15**, 405–419. *526*

Heinze, W. & Sajonski, H. (1964). Odontoma simplex partiale in der Schmelzeinstülpung eines Pferdemolaren. *Zentralblatt für Allgemeine Pathologie und Pathologische Anatomie*, **105**, 479–486. *590*

Heizer, E. E. & Hervey, M. C. (1937). Impacted molars – a new lethal in cattle. *Journal of Heredity*, **28**, 123–128. *294*

Hellwing, S. & Ghizela, G. (1963). Small mammals from the outskirts of Jassy. *Travaux du Muséum d'Histoire Naturelle 'Grigore Antipa'*, **4**, 505–519. *133*

Helminen, M. (1958). Upper canine teeth in a Finnish white-tailed deer. *Archivum Societatis Zoologicae Botanicae Fennicae Vanamo*, **13**, 146–148. *111*

Hemmer, H. (1964). Strukturveränderungen am Schädel eines Leoparden (*Panthera pardus*) infolge Unterkieferbruchs. *Säugetierkundliche Mitteilungen*, **12**, 52–55. *379*

Hemmer, H. (1966). Mitteilung zum Backzahnwechsel der Hauskatze (*Felis catus* L.). *Der Zoologische Garten*, **32**, 323–327. *339*

Henderson, A. M. & Greene, D. L. (1975). Dental field theory: an application to primate evolution. *Journal of Dental Research*, **54**, 344–350. *2*

Henderson, G. M., Borthwick, R. & Camburn, M. A. (1984). Maxillary dento-alveolar abscess in a spotted hyaena. *Journal of Zoo Animal Medicine*, **15**, 99–108. *512*

Henrichsen, P. (1981). Dental anomalies in musk oxen from Greenland. *Videnskabilige Meddelelser fra Dansk Naturhistorisk Forening i Kjøbenhavn*, **143**, 13–123. *114*

Henrichsen, P. (1982). Population analysis of musk oxen, *Ouibos moschatus* (Zimmermann 1780), based on occurrence of dental anomalies. *Säugertierkundliche Mitteilungen*, **30**, 260–280. *450*

Henrichsen, P. & Dieterich, R. A. (1984). Dental lesions in muskoxen from natural, introduced and captive populations. *Biological Papers, University of Alaska*. Special Report No. 4. *562*

Henrikson, P.-Å. (1968). Periodontal disease and calcium deficiency. An experimental study in the dog. *Acta Odontologica scandinavica*, **26**, Supplement 50, 1–132. *550*

Henry, C. B. (1935). Observations on the eruption of the molar teeth. *The Dental Record*, **55**, 697–707. *108, 283*

Hensel, R. (1879). Über Homologieen und Variaten in der Zahnformeln einiger Saugetiere. *Morphologisches Jahrbuch*, **5**, 529–561. *84*

Heran, I. (1970). Anomalies in the position of the lower teeth in the fisher, *Martes pennanti* Erxl., (Mammalia; Mustelidae). *Canadian Journal of Zoology*, **48**, 1465. *251*

Heran, I. (1971). Some notes on dentition in Mustelidae. *Vestnik Ceskoslovenske Spolecnosti Zoologicke*, **35**, 199–204. *73*

Hermann, R. (1908). Caries bei Mastodon. *Anatomischer Anzeiger*, **33**, 305–313. *485*

Herold, W. (1955a). Studien an Inselpopulationen der Waldmaus *Apodemus sylvaticus* L. *Wissenschaftliche Zeitschrift der Humboldt-Universitat, Berlin*, **5**, 143–149. *133*

Herold, W. (1955b). Zahnverschmelzung bei einer Gelbhalsmaus (*Apodemus flavicollis* Melch.) *Zeitschrift für Säugetierkunde*, **20**, 184–186. *133*

Herold, W. (1956). Uberzahlige Schneidezähne bei einem Luchs, *Lynx lynx* (Linné, 1758). *Säugertierkundlichen Mitteilungen, Stuttgart*, **4**, 81–82. *66*

Herold, W. (1960). Über die Wurzeln der Oberkiefer-Molaren bei *Rattus norvegicus* (Berkenhout) und *Rattus rattus* L. *Zeitschrift für Säugetierkunde*, **25**, 15–23. *133*

Herold, W. (1964). Über zwei Zahnwurzel-Varianten des M^3 bei der Waldmaus. *Zeitschrift für Säugetierkunde*, **29**, 251–253. *133*

Herold, W. & Zimmerman, K. (1960). Molaren-Abbau bei der Hausmaus (*Mus musculus* L.) *Zeitschrift für Säugetierkunde*, **25**, 81. *131*

Herrick, E. H. (1951). Another cat anomaly. *Transactions of the Kansas Academy of Science*, **54**, 546–547. *64*

Herrtage, M. E., Saunders, R. W. & Terlecki, S. (1974). Physical examination of cull ewes at points of slaughter. *The Veterinary Record*, **94**, 257–260. *552*

Hershkovitz, P. (1970). Dental and periodontal diseases and abnormalities in wild-caught marmosets (Primates, Callithridae). *American Journal of Physical Anthropology*, **32**, 377–394. *58, 388, 476, 505, 541*

Hiatt, J. L., Gartner, L. P. & Provenza, D. V. (1974). Molar development in the Mongolian gerbil (*Meriones unguiculatus*). *American Journal of Anatomy*, **141**, 1–22. *461*

Hickman, G. C. (1979). An inverted tooth in a white rhinoceros. *The Lammergeyer*, No. 27, 46–47. *387*

Hill, W. C. O. (1953a). Dental variations in the genus *Mandrillus* Ritgen. *Mammalia*, **17**, 208–217. *49, 428*

Hill, W. C. O. (1953b). *Primates, Comparative Anatomy and Taxonomy. I. Strepsirhini*. Edinburgh University Press. *61*

Hill, W. C. O. (1957). *Primates, Comparative Anatomy and Taxonomy. III. Pithecoidea, Platyrrhini (Families Hapalidae and Callimiconidae)*. Edinburgh University Press. *476*

Hill, W. C. O. (1970). *Primates, Comparative Anatomy and Taxonomy. VIII. Cynopithecinae*. Edinburgh University Press. *48*

Hillman, D. (1979). Displaced abomasums and black teeth. *California Veterinarian*, **33**, 31–33. *450*

Hilzheimer, M. (1906). Variationen des Canidengebisses mit besonderer Berücksichtigung des Haushundes. *Zeitschrift für Morphologie und Anthropologie*, **9**, 1–40. *84*

Hilzheimer, M. (1908). Einige Zahnanomalien wilder Tiere. *Anatomischer Anzeiger*, **32**, 442–445. *29, 68, 134*

Hitchin, A. D. (1948). Relationship of occlusion and parodontal disease in the sheep. A preliminary report on an investigation. *The Dental Record*, **68**, 251–254. *295, 552, 556*

Hitchin, A. D. (1957). Occlusion in sheep – some breeding experiments. *The Dental Practitioner*, **7**, 172–177. *295, 556*

Hitchin, A. D. & Morris, I. (1966). Geminated odontome – connation of the incisors in the dog – its etiology and ontogeny. *Journal of Dental Research*, **45**, 575–583. *8, 9*

Hitchin, A. D. & Walker-Love, J. (1959). Broken mouth in hill sheep. *Agriculture*, **66**, 5–8. *559*

Hix, J. O. & O'Leary, T. J. (1976). The relationship between cemental caries, oral hygiene status and fermentable carbohydrate intake. *Journal of Periodontology*, **47**, 398–404. *457*

Hodosh, M., Povar, M. & Shklar, G. (1971). Periodontitis in the baboon (*Papio anubis*). *Journal of Periodontology*, **42**, 594–596. *540*

Hoefs, M. (1974). Abnormal dentition in Dall sheep (*Ovis dalli dalli* Nelson). *Canadian Field Naturalist*, **88**, 227–229. *117, 516*

Hoefs, M. & Cowan, I. M. (1979). Ecological investigation of a population of Dall sheep (*Ovis dalli dalli* Nelson). *Syesis*, **12**, Supplement 1, 1–81. *516*

Hoenig, J. M. & Walsh, A. H. (1983). Skeletal lesions and deformities in large sharks. *Journal of Wildlife Diseases*, **19**, 27–33. *159*

Hofmeyr, C. F. B. (1960). Comparative dental pathology (with particular reference to caries and paradontal disease in the horse and the dog. *Journal of the South African Veterinary Medical Association*, **31**, 471–480. *482, 494*

Holloway, P. J., Shaw, J. H. & Sweeney, E. A. (1961). Effects of various sucrose:casein ratios in purified diets on the teeth and supporting structures of rats. *Archives of Oral Biology*, **3**, 185–200. *10*

Honma, K., Yamakawa, M., Yamauchi, S. & Hosoya, S. (1962). Statistical study of the occurrence of dental caries of domestic animals. I. Horse. *Japanese Journal of Veterinary Research*, **10**, 31–36. *482*

Honma, Y., Chiba, A. & Yoshie, S. (1981). An aberrant dace, *Leuciscus hakonensis* from the Uono river, Niigata. *Japanese Journal of Ichthyology*, **28**, 340–342. *601*

Honma, Y. & Ikeda, I. (1971). A pug-headed specimen of black porgy, *Acanthopagrus schlegeli*, from the rivermouth of Asa-kawa, Shikoku. *Japanese Journal of Ichthyology*, **18**, 36–38. *159*

Honma, Y. & Yoshie, S. (1978). Studies on Japanese chars of the genus *Salvelinus* – IX. A pug-headed specimen of char, *Salvelinus leucomaenis*, from Umezu-gawa river, Sado Island. *Annual Report of the Sado Marine Biological Station, Niigata University*, No. 8, 1–5. *159*

Hooijer, D. A. (1941). Note on a diseased dental condition in *Hippopotamus amphibius* L. *Proceedings of the Section of Sciences. Nederlandse Akademie van Wetenschappen*, **44**, 1147–1150. *514*

Hooijer, D. A. (1948). Prehistoric teeth of man and the orang-utan from central Sumatra, with notes on the fossil orang-utan from Java and southern China. *Zoologische-Mededeelingen Museum, Leiden*, **29**, 175–301. *30*

Hooijer, D. A. (1952). Notes on the dentition of the golden monkey *Rhinopithecus*. *Journal of Mammalogy*, **33**, 258–260. *35, 38, 188*

Hooijer, D. A. (1961). Dental anomaly in *Tapirus terrestris*, L. *Bijdragen tot de Dierkunde*, **31**, 63–64. *123*

Hooijer, D. A. & Eulderink, F. (1975). Numerous radicular enamel pearls in a premolar of a wild sheep. *Oral Surgery, Oral Medicine, Oral Pathology*, **39**, 637–640. *576*

Hooper, E. (1946). Extra teeth in a shrew. *Journal of Mammalogy*, **27**, 394. *139*

Hooper, E. T. (1956). Extra teeth in the pygmy mouse, *Raiomys musculus*. *Journal of Mammalogy*, **36**, 298–299. *131*

Hooper, E. T. (1957). Supernumerary teeth in *Peromyscus truei*. *Journal of Mammalogy*, **38**, 522. *131*

Hopps, R. M. & Johnson, N. W. (1976). Cell dynamics of experimental gingivitis in macaques. Cell proliferation within the inflammatory infiltrate. *Journal of Periodontal Research*, **11**, 210–217. 528

Horton, D. R. & Samuel, J. (1978). Palaeopathology of a fossil macropod population. *Australian Journal of Zoology*, **26**, 279–292. 346, 518, 520

Howe, P. R. (1920a). Dental caries. *Dental Cosmos*, **62**, 921–929. 461

Howe, P. R. (1920b). The effect of scorbutic diets upon the teeth, *Dental Cosmos*, **62**, 586–590. 461, 525

Howe, P. R. (1924). Studies on dental disorders following experimental feeding with monkeys. *Journal of the American Dental Association*, **11**, 1149–1160. 458

Howell, A. B. (1925). Asymmetry in the skulls of mammals. *Proceedings of the United States National Museum*, **67**, article 27, 1–18. 167, 372

Howitt, B. F. (1931). Spontaneous scurvy in monkeys. *Archives of Pathology*, **11**, 574–583. 539

Hrdlička, A. (1907). Anatomical observations on a collection of orang skulls from Western Borneo. *Proceedings of the United States National Museum*, **31**, 539–568. 29

Huang, C. M., Mi, M. P. & Vogt, D. W. (1981). Mandibular prognathism in the rabbit: discrimination between single-locus and multifactorial models of inheritance. *Journal of Heredity*, **72**, 296–298. 319

Hübner, O. (1930). Überzählige Zähne bei Anthropomorphen. *Zeitschrift für Stomatologie*, **28**, 397–408. 29, 30

Huchzermeyer, F. W. (1986). Osteomalacia in young captive crocodiles (*Crocodylus niloticus*). *Journal of the South African Veterinary Association*. **57**, 167–168. 158

Hull, P. S., Soames, J. V. & Davies, R. M. (1974). Periodontal disease in a beagle dog colony. *Journal of Comparative Pathology*, **84**, 143–150. 549

Humphreys, E. R., Robins, M. W. & Stones, V. A. (1985). Age-related and ^{224}Ra-induced abnormalities in the teeth of male mice. *Archives of Oral Biology*, **30**, 55–64. 7

Humphreys, H. F. (1926). Particulars relating to the broken tusk of wild elephant. *British Dental Journal*, **47**, 1400–1402. 405, 406

Hunter, D. & Turnbull, H. M. (1931–32). Hyperparathyroidism: generalized osteitis fibrosa. *British Journal of Surgery*, **19**, 203–284. 604

Hunter, J. (1778). *A Practical Treatise on the Diseases of the Teeth. Section II. The Decay of the Teeth, by Denudation*. London. 494

Hunter, J. (1787). Observations on the structure and oeconomy of whales. *Philosophical Transactions of the Royal Society of London*, **77**, 371–450. 101

Hurme, V. O. (1960). Estimation of monkey age by dental formula. *Annals of the New York Academy of Science*, **85**, 795–799. 336

Hurme, V. O. & Van Wagenen, G. (1953). Basic data on the emergence of deciduous teeth in the monkey (*Mulatta mulatta*). *Proceedings of the American Philosophical Society*, **97**, 291–315. 336

Hurme, V. O. & Van Wagenen, G. (1961). Basic data on the emergence of permanent teeth in the rhesus monkey (*Macaca mulatta*). *Proceedings of the American Philosophical Society*, **105**, 105–140. 336

Hutterer, R. (1977). Zahnreduktion bei der Zwergspitzmaus, *Sorex minutus*. *Säugetierkundliche Mitteilungen*, **25**, 158–160. 139

Huxley, T. H. (1880). On the cranial and dental characters of the Canidae. *Proceedings of the Zoological Society of London*, 1880, 238–288. 94

Hylander, W. L. & Kay, R. F. (1975). Maxillary premolar reduction in the golden monkey (*Rhinopithecus roxellanae*). *Journal of Dental Research*, **54**, 1242. 35, 38

Ingles, J. M., Bates, P. J. J. & Harrison, D. L. (1981). Dental anomalies in Indian and African gerbils (Rodentia: Gerbillinae). *Mammalia*, **45**, 266–268. 133, 353

Inke, G. & Ernst, T. (1965–66). Überzählige Eckzähne und Distomolaren im Milchgebiss der Altweltaffen. *Morphologisches Jahrbuch*, **108**, 199–211. 31, 48, 50

Ishikawa, J. & Glickman, I. (1961). Gingival response to the systemic administration of sodium diphenyl hydantoin (dilantin) in cats. *Journal of Periodontology*, **332**, 149–158. 528

Ive, J. C., Shapiro, P. A. & Ivey, J. L. (1980). Age-related changes in the periodontium of pigtail monkeys. *Journal of Periodontal Research*, **15**, 420–428. 528

Jackson, H. H. T. (1928). A taxonomic review of the American long-tailed shrews (genera *Sorex* and *Microsorex*). *North American Fauna (Washington)*, **51**, 1–238. 139

Jackson, J. (1975). Mandibular dental abnormalities in roe deer (*Capreolus capreolus*) from the New Forest. *Journal of Zoology, London*, **177**, 491–493. 112

Jackson, J. (1976). Mandibular and maxillary dental abnormalities in fallow deer *Dama dama* from the New Forest. *Journal of Zoology, London*, **180**, 518–523. 111

James, W. W. (1960). *The Jaws and Teeth of Primates*. Pitman Medical Publishing Co. Ltd., London. 19

James, W. W. & Barnicot, N. A. (1949). Dental lesions in baboons from the Society's Gardens. *Proceedings of the Zoological Society of London*, **119**, 743–753. 432, 472, 511, 540

Jayo, M., Leipold, H. W., Dennis, S. M. & Eldridge, F. E. (1987). Brachygnathia superior and degenerative joint disease: a new lethal syndrome in Angus calves. *Veterinary Pathology*, **24**, 148–155. 294

Jenkins, G. N. (1978). *The Physiology and Biochemistry of the Mouth*, 4th edn. Blackwell's Scientific Publications, Oxford. 456

Jenks, A. E. (1911). Bulu knowledge of the gorilla and chimpanzee. *American Anthropologist*, **13** (new series), 56–64. 531

Jenks, J. A., Leslie, D. M. & Gibbs, H. C. (1986). Anomalies of the skull of a white-tailed deer fawn from Maine. *Journal of Wildlife Diseases*, **22**, 286–289. 288

Jöchle, W. (1958). Ringförmige Eberhauer von Hausschweinen im Bereich der Südsee und auf Neuguinea. *Berliner und Münchener Tierärztliche Wochenschrift*, **71**, 94–95. 364

Joest, E. (1926). *Handbuch der Speziellen Pathologischen Anatomie der Haustiere*, 2nd edn., vol. 1
115, 121, 122, 292, 294, 309, 334, 346, 352, 457, 487, 488, 494, 505, 597

Johansen, E. (1953). A system for recording oral conditions and dental lesions in the live Syrian hamster. *Journal of Dental Research*, **32**, 578–584. 461

Johansen, E. & Keyes, P. H. (1955). Production and evaluation of experimental animal caries: a review. In *Advances in Experimental Caries Research*, edit. R. F. Sognnaes. American Association for the Advancement of Science, Washington, D.C., pp. 1–46. 462

Johnson, D. H. (1952). The occurrence and significance of

extra molar teeth in rodents. *Journal of Mammalogy*, **33**, 70–72. *131*

Johnston, G. W., Dreizen, S. & Levy, B. M. (1970). Dental development in cotton ear marmoset (*Callithrix jacchus*). *American Journal of Physical Anthropology*, **33**, 41–48. *338, 339*

Jones, D. N. (1958). A survey of management systems and the incidence of 'broken mouths' in Blackface sheep in north-east of Scotland. *Hill Farming Research Organization, Edinburgh. First Reports, 1954–58*, pp. 57–71. *560*

Jones, F. W. (1923). *The Mammals of South Australia. Part 1. The Monotremes and the Carnivorous Marsupials.* Government Printer, Adelaide. *140*

Jones, G. S. (1978). *Microtus longicaudus* with grooved incisors. *The Murrelet*, **59**, 104–105. *131*

Jones, G. S. (1979). An unusual incisor malocclusion of *Rattus exulans* from Java, Indonesia. *Malayan Nature Journal*, **33**, 123–124. *360*

Jones, J. K. (1957). A dental abnormality in the shrew, *Crocidura lasiura*. *Transactions of the Kansas Academy of Science*, **60**, 88–89. *139*

Jones, J. K. (1960). Absence of third upper premolar in *Eutamias*. *Journal of Mammalogy*, **41**, 268–269. *133*

Jones, M. R. & Simonton, F. V. (1928). Mineral metabolism in relation to alveolar atrophy in dogs. *Journal of the American Dental Association*, **15**, 881–911. *525*

Jones, T. C. & Hunt, R. D. (1983). *Veterinary Pathology*, 5th edn. Lea and Febiger, Philadelphia. *334, 550, 562, 604*

Jones, T. S. & Cave, A. J. E. (1960). Diet, longevity and dental disease in the Sierra Leone chimpanzee. *Proceedings of the Zoological Society of London*, **135**, 147–155. *440, 465, 532*

Jowsey, J. & Gershon-Cohen, J. (1964). Effect of dietary calcium levels on production and reversal of experimental osteoporosis in cats. *Proceedings of the Society for Experimental Biology and Medicine*, **116**, 437–441. *334*

Jowsey, J. & Raisz, L. G. (1968). Experimental osteoporosis and parathyroid activity. *Endocrinology*, **83**, 384–396. *334*

Jubb, K. V. F. & Kennedy, P. C. (1970). *Pathology of Domestic Animals*, 2nd edn. Academic Press, London and New York. *480, 513*

Jump, E. B. & Weaver, M. E. (1966). The miniature pig in dental research. In *Swine in Biomedical Research*, edit. L. K. Bustad, R. O. McClellan & M. P. Burns. Battelle Memorial Institute, Pacific North West Laboratory, Washington, pp. 543–557. *480*

Junge, G. C. A. (1942). A remarkable double tusk of a Sumatran elephant. *Zoologische Mededeelingen*, **23**, 107–108. *126*

Kahmann, H. (1965). Die Alveolenmuster der Oberkiefzahnreihe der Waldmaus, Hausratte und Hausmaus aus Populationen der grossen Tyrrhenischen Inseln. *Zeitschrift für Säugetierkunde*, **34**, 164–183. *133*

Kakehashi, S., Baer, P. N. & White, C. L. (1963). Comparative pathology of periodontal disease. I. Gorilla. *Oral Surgery, Oral Medicine and Oral Pathology*, **16**, 397–406. *525, 531, 533, 541*

Karstad, L. (1967). Fluorosis in deer (*Odocoileus virginianus*). *Bulletin of the Wildlife Diseases Association*, **3**, 42–46. *384, 454*

Kay, C. E., Tourangeau, P. C. & Gordon, C. C. (1975). Industrial fluorosis in wild mule and whitetail deer from Western Montana. *Fluoride*, **8**, 182–191. *454*

Kazimiroff, T. (1938). A report on the dental pathology found in animals that died in the New York Zoological Park in 1938. *Zoologica (New York)*, **24**, 297–304. *506*

Keep, M. E. (1985). Dental abnormality in a free-ranging lion. *Lammergeyer*, No. 35, 33–35. *64*

Keil, A. (1959). Eigentümliche Defecte an Pottwalzähnen. *Deutsche Zahnärztliche Zeitschrift*, **14**, 523–526. *498*

Keith, Sir A. (1931). *New Discoveries Relating to the Antiquity of Man*, Williams and Norgate, London, pp. 103 and 380. *159*

Kelly, P. A. & Fairley, J. S. (1982). An accessory cusp on the third upper molar of wood mice *Apodemus sylvaticus* from the West of Ireland. *Journal of Zoology, London*, **198**, 532–533. *132*

Kemp, E. M., Huis in't Veld, J. H. J., Havenaar, R. & Backer Dirks, O. (1981). Experimental dental caries in mice. *Proceeding 'Symposium on Animal Models in Cariology'*, edit. J. M. Tanzer. Special Supplement Microbiology Abstracts, Information Retrieval Inc., Washington, D.C., pp. 121–130. *461*

Keyes, P. H. (1968). Odontopathic infections. In *The Golden Hamster, its Biology and its Use in Medical Research*, chapter 19, edit. R. A. Hoffman, P. F. Robinson & H. Magalhaes. Iowa State Uiversity Press, Iowa. *461, 526, 527*

Keyes, P. H. & Dale, P. P. (1944). A preliminary survey of the pouches and dentition of the Syrian hamster. *Journal of Dental Research*, **23**, 427–444. *461*

Keyes, P. H. & Jordan, H. V. (1964). Periodontal lesions in the Syrian hamster. III. Findings related to an infectious and transmissible component. *Archives of Oral Biology*, **9**, 377–400. *527*

Kiel, R. A., Kornman, K. S. & Robertson, P. B. (1983). Clinical and microbiological effects of localized ligature-induced periodontitis on non-ligated sites in the cynomolgus monkey. *Journal of Periodontal Research*, **18**, 200–211. *528*

Kierdorf, U. & Kierdorf, H. (1986). Erste Untersuchungsergebnisse zum Auftreten von Dentalfluorose beim Rehwild (*Capreolus capreolus* L.) in Nordrhein-Westfalen. *Zeitschrift für Jagdwissenschaften*, **32**, 55–56. *454*

Kierdorf, H. & Kierdorf, U. (1987). Einwurzeliger vierter Prämolar mit kegelförmiger Krone im Unterkiefer eines Rehbockes (*Capreolus capreolus* L.). *Zeitschrift für Jagdwissentschaften*, **33**, 268–272. *113*

King, J. D. (1944). Experimental investigation of parodontal disease. I. The significance of the blood supply in gingival health and disease of the ferret. *British Dental Journal*, **77**, 213–221, 245–254. *525*

King, J. D. (1949). Histological observations on parodontal disease in the golden hamster (*Cricetus auratus*): calculus, food particles and other foreign bodies as aetiological factors. *Journal of Pathology and Bacteriology*, **61**, 413–425. *526*

King, J. D. (1952). Experimental production of gingival hyperplasia in ferrets given epanutin (sodium diphenylhydantoinate). *British Journal of Experimental Pathology*, **33**, 491–499. *528*

King, J. D. & Gimson, A. P. (1948). Lesions of the gum and alveolar bone, due to calculus or other debris, in the golden hamster (*Cricetus auratus*). *British Journal of Nutrition*, **2**, 111–118. *526*

King, J. D. & Glover, R. E. (1945). The relative effects of dietary constituents and other factors upon calculus formation and gingival disease in the ferret. *Journal of Pathology and Bacteriology*, **57**, 353–362. 525

Kinghorn, R. G. & Yeager, L. E. (1951). Plural canines in elk. *Journal of Mammalogy*, **32**, 112–113. 112

Kirkdahl, M. E. (1967). Some comparative aspects of the reduction of the premolars in the insectivora. *Journal of Dental Research*, **46**, 805–808. 138

Kirkpatrick, T. H. (1964). Molar progression and macropod age. *Queensland Journal of Agricultural Animal Science*, **22**, 319–328. 346

Kirkpatrick, T. H. (1965). Age estimation in the grey kangaroo, the red kangaroo, the eastern wallaroo, and the red-necked wallaby, with notes on dental abnormalities. *Queensland Journal of Agricultural and Animal Sciences*, **22**, 301–317. 346

Kitt, T. (1892). Anomalien der Zähne unserer Hausthiere. *Verhandlungen der Deutschen Odontologischen Gesellschaft*, **3**, 111–196.
115, 121, 122, 309, 311, 349, 352, 487, 494, 582, 592, 601

Kitt, T. (1921). Anomalien der Zähne. *Lehrbuch der Pathologischen Anatomie der Haustiere für Tierärzte und Studierende der Tiermedizin*, chapter 9, 5th edn, vol. 1. Stuttgart, pp. 487–542. 115, 121, 487

Klatt, B. (1921). Mendelismus, Domestikation and Kraniologie. *Archiv für Anthropologie*, neue Folge, **18**. 264

Knaggs, R. L. (1923). Osteitis fibrosa. *British Journal of Surgery*, **10**, 487–500. 604

Knap, M. (1941). Carie bij apen en die diëetproeven met het oog op caries bij den Macacus. *Nederlandsch-Indische Tandheelkundig Tijdschrift*, **9**, 103–112. English summary. 472

Knudsen, P. A. (1965). Congenital malformations of upper incisors in exencephalic mouse embryos, induced by hypervitaminosis A. II. Morphology of fused upper incisors. *Acta Odontologica Scandinavica*, **23**, 391–409. 9

Knudsen, P. A. (1966a). Congenital malformations of lower incisors and molars in exencephalic mouse embryos, induced by hypervitaminosis A. *Acta Odontologica Scandinavica*, **24**, 55–89. 9

Knudsen, P. A. (1966b). Malformations of upper incisors in mouse embryos with exencephaly, induced by trypan blue. *Acta Odontologica Scandinavica*, **24**, 647–675. 9

Kober, W. (1965). Aktinomykose beim Damwild. *Zeitschrift für Jagdwissenschaften*, **11**, 110–111. 516

Kofoid, C. A., Hinshaw, H. C. & Johnstone, H. G. (1929). Animal parasites of the mouth and their relation to dental disease. *Journal of the American Dental Association*, **16**, 1436–1437. 522

Komada, N. (1980). Replacement of dentition in ayu, *Plecoglossus altivelis*. *Japanese Journal of Ichthology*, **27**, 144. Japanese. English summary. 159

Kowalski, M. (1987). Une prémolaire supérieure supplementaire chez *Apodemus sylvaticus* (Linnaeus, 1758). *Mammalia*, **51**, 611–613. 131

Kramer, J. W., Davis, W. C., Prieur, D. J., Baxter, J. & Norsworthy, G. D. (1975). An inherited disorder of Persian cats with intracytoplasmic inclusions in neutrophils. *Journal of the American Veterinary Medical Association*, **166**, 1103–1104. 523

Krapp, F. & Lampel, G. (1973). Zahnanomalien bei Altweltaffen (Catarrhina). *Revue Suisse de Zoologie*, **80**, 83–150. 20, 163, 165, 166, 169, 175, 176, 183, 200

Kratochvil, Z. (1965). Das Vorkommen von überzähligen Schneidezähnen bei dem Luchs *Lynx lynx* (L. 1758). *Zoologické Listy*, **14**, 186–187. 63

Kratochvil, Z. (1984). Veränderungen am Gebiss des Rehwildes (*Capreolus capreolus* L.). *Folia Zoologica*, **33**, 209–222. 112, 113, 561

Kratochvil, Z. (1987). Oligodontie bei Hasenartigen, (Lagomorpha). *Zeitschrift für Säugetierkunde*, **52**, 196–197. 134

Kraus, B. S., Ames, M. D. & Clark, G. R. (1969). Effects of maternal rubella on dental crown development. *Clinical Pediatrics*, **8**, 204–215. 10

Krausman, P. R. (1978). Dental anomalies of Carmen Mountains white-tailed deer. *Journal of Mammalogy*, **59**, 863–864. 111

Kremenak, G. R. (1969). Dental eruption chronology in dogs: deciduous tooth gingival emergence. *Journal of Dental Research*, **48**, 1177–1184. 341

Kreshover, S. J. (1944). The pathogenesis of enamel hypoplasia: an experimental study. *Journal of Dental Research*, **23**, 231–238. 438

Krogman, W. M. (1930). Studies in growth changes in the skull and face of anthropoids. I. The eruption of the teeth in anthropoids and Old World apes. *American Journal of Anatomy*, **46**, 303–312. 335, 336

Krommenhoek, W. & Slob, A. (1967). Variabilität in aantal en form van de alveolen der molaren uit de bovenkaak bij de bosmuis, *Apodemus sylvaticus*, L. *Lutra*, **9**, 41–51. 133

Kronman, J. H. (1971). Tissue reaction and recovery following experimental tooth movement. *Angle Orthodontist*, **41**, 125–132. 157

Krook, L. (1976). Periodontal disease in dogs and man. *Advances in Veterinary Science and Comparative Medicine*, **20**, 171–190. 546

Krook, L. & Maylin, G. A. (1979). Industrial pollution. Chronic fluoride poisoning in Cornwall Island cattle. *The Cornell Veterinarian*, **69**, Supplement 8. 452, 453, 454

Krumrey, W. A. & Buss, I. O. (1968). Age estimation, growth, and relationships between body dimensions of the female African elephant. *Journal of Mammalogy*, **49**, 22–31. 345

Krutsch, P. H. (1953). Supernumerary molars in the jumping mouse (*Zapus princeps*). *Journal of Mammalogy*, **34**, 265–266. 131

Krygier, G., Genco, R. J., Mashimo, P. A. & Hausmann, E. (1973). Experimental gingivitis in *Macaca speciosa* monkeys: clinical bacteriologic and histologic similarities to human gingivitis. *Journal of Periodontology*, **44**, 454–463. 528

Kubota, K. & Togawa, S. (1964). Numerical variations in the dentition of some pinnipeds. *Anatomical Record*, **150**, 487–501. 96, 97, 98, 100

Kuehn, D. W., Fuller, T. K., Mech, L. D., Paul, W. J., Fritts, S. H. & Berg, W. E. (1986). Trap-related injuries to gray wolves in Minnesota. *Journal of Wildlife Management*, **50**, 90–91. 434

Kuhn, H. J. von (1963). Ein angeborener Unterkiefer-defekt bei *Procolobus badius badius* (Kerr 1792). *Folia primatologica*, **1**, 172–177. 541

Kuiper, J. D. & van der Gaag, I. (1982). Cariës, emailhypoplasie en tandverkleuringen bij de hond. *Tijdschrift Diergeneeskunde*, **107**, 457–462. 445, 478

Kükenthal, W. (1893). Die Bezahnung. In Vergleichend-anatomische und entwicklungsgeschichtliche Untersuchungen an Walthieren. *Denkschriften der Medicinisch-Naturwissenschaftlichen Gesellschaft*, Jena, **3**, 387–448. *101*

Kumar, R., Singh, G., Manohar, M. & Nigam, J. M. (1977). Bilateral fracture of mandible in camel. *Indian Veterinary Journal*, **54**, 477–578. *385*

Kunstler, J. & Chaine, J. (1906). Variations des formules dentaires chez les primates. *Comptes Rendus des Séances de la Société de Biologie*, 1906, 99–101. *50*

Kurtén, B. (1963). Return of a lost structure in the evolution of a felid dentition. *Societas Scientiarum Fennica Commentationes Biologicae*, **26**, part 4. *7, 63, 64*

Kvam, T. (1985). Supernumerary teeth in the European lynx, *Lynx lynx lynx*, and their evolutionary significance. *Journal of Zoology, London*, **206**, 17–22. *63*

Kyle, M. G., Davis, G. B. & Thompson, K. G. (1985). Renal osteodystrophy with facial hyperostosis and 'rubber jaw' in an adult dog. *New Zealand Veterinary Journal*, **33**, 118–120. *334*

Lafosse, P. E. (1772). *Cours d'Hippatique, ou Traité Complet de la Médecine des Chevaux*. Paris, p.32. *121*

Lambiris, A. J. L. (1978). Notes on an unusual injury to a leopard toad, *Buo pardalis*. *Journal of the Herpetological Association of Africa*, No. 17, 11–12. *393*

Lampel, G. (1963). Variationsstatistische und morphologische Untersuchungen am Gebiss der Cercopithecinen. *Acta Anatomica*, **49**, 5–122.
35, 46, 47, 190, 192, 337, 469, 510, 537

Landry, S. O. (1957). Factors affecting the procumbency of rodent upper incisors. *Journal of Mammalogy*, **38**, 223–234. *358*

Lantz, G. C. & Cantwell, H. D. (1984). Facial deformity in a dog. *Journal of the American Veterinary Medical Association*, **184**, 585–586. *271*

Lanzilotti-Buonsanti, N. & Generali, G. (1873). Contribuzione alla Patologia delle cosi dette cisti dentaree del cavallo. *Gazzetta Medico-Veterinaria Milano*, **3**, 274–297. A translation into English is available by G. Fleming in *Veterinarian* (1874): **47**, 692–699. *601*

Larson, R. H. & Fitzgerald, R. J. (1968). Caries development in the African white-tailed rat (*Mystromys albicaudatus*) infected with a streptococcus of human origin. *Journal of Dental Research*, **47**, 746–749. *461*

Lavelle, C. L. B. & Moore, W. J. (1973). The incidence of agenesis and polygenesis in the primate dentition. *American Journal of Physical Anthropology*, **38**, 671–680. *20, 22*

Lavelle, C. L. B., Shellis, R. P. & Poole, D. F. G. (1977). *Evolutional Changes to the Primate Skull and Dentition*, C. C. Thomas, Springfield, U.S.A., pp. 197–277. *20, 462*

Lavine, W. S., Page, R. C. & Padgett, G. A. (1976). Host response in chronic periodontal disease. V. The dental and periodontal status of mink and mice affected by Chédiak-Higashi syndrome. *Journal of Periodontology*, **47**, 621–635. *523, 546*

Lawlor, T. E. (1979). *Handbook to the Orders and Families of Living Animals*. Mad River Press, Eureka, California. *140*

Lawrence, B. (1934). Wild coyote with an undershot jaw. *Journal of Mammalogy*, **15**, 319–320. *263*

Leader-Williams, N. (1980). Dental abnormalities and mandibular swellings in South Georgia reindeer. *Journal of Comparative Pathology*, **90**, 315–330. *516*

Lechtleitner, R. R. (1958). An extra molar in *Erethizon*. *Journal of Mammalogy*, **39**, 447–448. *131*

Lee, R. C. & Stolfus, T. A. (1968). Cosmetic repair of dental caries in a horse. *Veterinary Medicine*, **63**, 1057–1059. *483*

Leggett, W. C. (1969). Pugheadedness in landlocked Atlantic salmon (*Salmo salar*). *Journal of the Fisheries Research Board, Canada*, **26**, 3091–3093. *159*

Leidy, J. (1886). Caries in the mastodon. *Proceedings of the Academy of Natural Sciences of Philadelphia*, 1886, 38. *485*

Leidy, R. A. (1985). Pugheadedness in the California roach *Hesperoleucus symmetricus* (Baird and Girard). *California Fish and Game*. 1985, **71**, 117–122. *159*

Lensink, C. L. (1954). Deformed jaw in an Alaskan brown bear (*Ursus*). *Journal of Mammalogy*, **35**, 438–439. *379*

Leonard, E. P. (1979). Animal Model: periodontitis in the rice rat (*Oryzomys palustris*). *American Journal of Pathology*, **96**, 643–646. *527*

Lesbre, F. X. (1897/1898). Contribution a l'étude de l'ossification du squelette des mammifères domestiques. *Annales de la Société d'Agriculture, Science et Industrie de Lyon*, Series 7, **5**, 1–106. *343*

Levy, B. A. (1968). Effects of experimental trauma on developing first molar teeth in rats. *Journal of Dental Research*, **47**, 323–327. *404, 439*

Levy, B. M. (1947). The effect of pantothenic acid deficiency on the periodontal structures of mice. *Journal of Dental Research*, **26**, 443. Abstract. *528*

Levy, B. M., Dreizen, S. & Bernick, S. (1972). *The Marmoset Periodontium in Health and Disease*. Monograph in Oral Science, volume 1. S. Karger, Basel. *528*

Levy, B. M., Robertson, P. B., Dreizen, S., Mackler, B. F. & Bernick, S. (1976). Adjuvant induced destructive periodontitis in nonhuman primates. A comparative study. *Journal of Periodontal Research*, **11**, 54–60. *525, 528*

Lewis, K. J. & Smith, B. G. N. (1973). The relationship of erosion and attrition in extensive tooth loss. Case reports. *British Dental Journal*, **135**, 400–404. *497*

Lindsey, A. A. (1937). Notes on the crab-eater seal. *Journal of Mammalogy*, **19**, 456–461. *498*

Linsdale, J. M. & Tomich, P. Q. (1953). *A Herd of Mule Deer*. University of California, Berkeley, p. 74. *490*

Lippens, L. (1948). Congo four-tusker. *Field*, **191**, 274. *126*

Listgarten, M. A. & Heneghan, J. B. (1973). Observations on the periodontium and acquired pellicle of adult germ-free dogs. *Journal of Periodontology*, **44**, 85–91. *526*

Little, W. L. (1913). Periodontal disease in the horse. *Journal of Comparative Pathology and Therapeutics*, **26**, 240–249. *567*

Loeffler, K., Brosi, C., Oelschläger, W. & Feyler, L. (1979). Flurose beim Hund. *Kleintier-praxis*, **24**, 167–171. *450*

Long, C. A. (1961). A dental abnormality in *Sorex palustris*. *Journal of Mammalogy*, **42**, 527. *139*

Long, C. A. & Long, C. F. (1965). Dental anomalies in North American badgers, genus *Taxidea*. *Transactions of the Kansas Academy of Science*, **68**, 145–155. *74, 249, 373*

Long, J. O. & Cooper, R. W. (1968). Physical growth and dental eruption in captive-bred squirrel monkeys *Saimiri sciureus* (Leticin, Columbia). In *The Squirrel Monkey*, edit. L. A. Rosenblum & R. W. Cooper. Academic Press, London & New York, pp. 193–205. *338*

Lönnberg, E. (1930). The homologies of the incisors of the

higher primates in the light of some 'anomalies' in the dentition of gibbons. *Arkiv för Zoologi*, **22A**, No. 6, 1–6.   *31*

Lönnberg, E. (1937). On the occurrence of upper canines in a young *Gazella granti* Brooke. *Arkiv för Zoologi*, **29B**, no. 1, 1–3.   *114*

Lönnberg, E. (1939). Remarks on some members of the genus Cebus. *Arkiv för Zoologi*, **31A**, no. 23, 1–24.   *56*

Lönnberg, E. (1940). Notes on some members of the genera *Lagothrix* and *Ateles*. *Arkiv för Zoologi*, **32A**, no. 25, 1–14.   *53, 54, 55*

Loughlin, T. R. (1982). Functional adaptation of eruption sequence of pinniped postcanine teeth. *Journal of Mammalogy*, **63**, 523–525.   *342*

Lucas, R. B. & Pindborg, J. J. (1976). Odontogenic tumour and tumour-like lesions. In *Scientific Foundations of Dentistry*, chapter 20, edit. B. Cohen & I. R. H. Kramer, Heinemann Medical Books Ltd., London, pp. 241–250.   *575*

Lumsden, A. G. S. (1978). Development of the mouse molar dentition in intra-ocular homografts. *Ph.D. thesis, University of London*, p. 148.   *5*

Lumsden, A. G. S. (1979). Pattern formation in the molar dentition of the mouse. *Journal Biologie Buccale*, **7**, 77–103.   *2, 4, 5, 6*

Lumsden, A. G. S. (1981). The dentition of mammals: the orders of placental mammals. In *Dental Anatomy Embryology*, edit. J. W. Osborn. Blackwell Scientific Publications, Oxford, pp. 399–419.   *13, 140*

Lumsden, A. G. S. (1984). Tooth morphogenesis: contributions of the cranial neural crest in mammals. *INSERM colloque*, **125**, 29–40.   *5*

Lüps, P. (1974). Dreiwurzlige Praemolaren bei Rotfüchsen. *Zeitschrift für Jagdwissenschaften*, **20**, 163–165.   *84*

Lüps, P. (1977). Gebiss- und Zahnvariationen am einer Serie von 257 Hauskatzen (*Felis silvestris* f. *catus*, L. 1758). *Zoologische Abhandlungen Staatliches Museum für Tierkunde in Dresden*, **34**, 155–165.   *62*

Lüps, P. (1980). Vergleichende Untersuchung am zweiten oberen vorbackenzahn P^2 der Hauskatze *Felis silvestris* f. *catus*. *Zeitschrift für Säugetierkunde*, **45**, 245–249.   *63*

Lusted, L. B., Pickering, D. E., Fisher, D. & Smyth, F. S. (1953). Growth and metabolism in normal and thyroid ablated infant rhesus monkeys (*Macaca mulatta*). *American Journal of Diseases of Children*, **86**, 426–435.   *334*

Lutz, W. (1988). Verbiegungen des Gesichtsschädels beim Wildschwein (*Sus scrofa scrofa* L.) als mögliche Folge einer Rhinitis atrophicans. *Zeitschrift für Jagdwissenschaften*, **34**, 125–131.   *283*

Lydekker, R. (1899). The dental formula of the marsupial and placental carnivora. *Proceedings of the Zoological Society of London*, 1899, 922–928.   *13*

Lydekker, R. (Editor). (1893–94). *The Royal Natural History*. 6 volumes. F. Warne, London.   *275, 497, 545*

Lyne, A. G. (1982). Observations on skull growth and eruption of teeth in the marsupial bandicoot *Perameles nasuta* (Marsupalia; Peramelidae). *Australian Mammalogy*, **5**, 113–126.   *572*

Lyne, A. G. & Mort, P. A. (1981). A comparison of skull morphology in the marsupial bandicoot genus *Isoodon*; its taxonomic implications and notes on a new species, *Isoodon arnhemensis*. *Australian Mammalogy*, **4**, 107–133.   *572*

Lyons, H. (1944). *The Royal Society 1660–1940. A History of its Administration under its Charters*. Cambridge University Press, Cambridge.   *419*

McClure, H. M., Chapman, W. L., Hooper, B. E., Smith, F. G. & Fletcher, O. J. (1978). The digestive system. In *Pathology of Laboratory Animals*, edit. K..Benirscke, F. M. Garner & T. C. Jones. Springer-Verlag, New York, pp. 176–317.   *576*

McCollum, E. V., Simmonds, N. & Kinney, E. M. (1922). The relation of nutrition to tooth development and tooth preservation. I. A preliminary study of gross maxillary and dental defects in 220 rats on defective and deficient diets. *John Hopkins Hospital Bulletin*, **33**, 202–215.   *525*

McDonald, H. G. (1978). A supernumerary tooth in the ground sloth, *Megalonyx* (Edentata Mammalia). *Florida Scientist*, **41**, 12–14.   *134*

M'Intosh, W. C. (1929). On abnormal teeth in certain mammals, especially in the rabbit. *Transactions of the Royal Society of Edinburgh*, **56**, pt. 2, 333–404.   *86, 121, 358, 359, 360, 361, 366, 488*

MacKinnon, J. (1981). The structure and function of the tusks of babirusa. *Mammal Review*, **11**, 37–40.   *366*

Mackinnon, M. M. (1959). A pathological study of the enzootic paradontal disease of mature sheep. *New Zealand Veterinary Journal*, **7**, 18–26.   *562*

McSweeney, C. S. & Ladds, P. W. (1988). Focal ulceration of the mouth in ruminants fed spear grass. *Australian Veterinary Journal*, **65**, 28–30.   *562*

Magitot, E. (1877). *Traité des Anomalies du Système Dentaire chez l'Homme et chez les Mammifères*. G. Masson, Paris.   *117, 121, 495, 598, 601*

Major, C. I. F. (1904). On dental peculiarities of certain mammals. *Proceedings of the Zoological Society of London*, **1**, 416–424.   *114*

Malassez, L. (1885). Sur l'existence d'amas épithéliaux autour de la racine des dents chez l'homme adulte et à l'état normal (débris épithéliaux paradentaires). *Archives de Physiologie Normale et Pathologue*, **5**, series 3, 129–148.   *576*

Maňkovská, B. (1980). The concentration of four toxic elements in the teeth of roe deer from the area of an aluminium plant. *Biológia (Bratislava)*, **35**, 819–822.   *454*

Manning, P. J. (1972). Dentigerous cysts of the mandible in a baboon. *Oral Surgery*, **33**, 428–430.   *579*

Manville, R. H. (1954). Malocclusion in the rat. *Journal of Mammalogy*, **35**, 427.   *360*

Manville, R. H. (1963). Dental anomalies in North American lynx. *Zeitschrift für Saugetierkunde*, **28**, 166–169.   *62, 63, 66*

Markham, J. H. A. & Stewart, W. L. (1962). Dental conservation in sheep. *The Veterinary Record*, **74**, 971–978.   *553, 557*

Marlborough, D. & Meadows, B. S. (1966). A 'pug-headed' perch (*Perca fluviatalis* L.) from the river Lea. *The London Naturalist*, No. 45, 98–99.   *159*

Marshall, W. H. (1952). Note on missing teeth in *Martes americana*. *Journal of Mammalogy*, **33**, 116–117.   *71*

Martini, L. (1877). A case of anomalous dentition in a monkey. *British Journal of Dental Science*, **20**, 661–662.   *175, 176*

Maser, C. & Shaver, C. (1976). Acute incisor malocclusion in a Townsend ground squirrel. *The Murrelet*, **57**, 17–18.   *358*

Matschke, G. H. (1967). Ageing European wild hogs by dentition. *Journal of Wildlife Management*, **31**, 109–113. *343*

Matthews, L. H. (1938). The sperm whale *Physeter catodon*. *Discovery Reports*, **17**, 93–168. *102, 104*

Mazak, V. (1964). Anomalies of the second lower molars in *Mustela nivalis nivalis* Linnaeus, 1766 (Mammalia; Mustelidae). *Nytt Magazine for Zoologi, Oslo*, **12**, 24–29. *72*

Maylin, G. A., Eckerlin, R. H. & Krook, L. (1987). Fluoride intoxication in dairy calves. *Cornell Veterinarian*, **77**, 84–98. *451*

Mech, L. D., Frenzel, L. D., Karns, P. D. & Kuehn, D. W. (1970). Mandibular dental anomalies in white-tailed deer from Minnesota. *Journal of Mammalogy*, **51**, 804–806. *111, 112*

Meek, A. & Gray, R. A. H. (1911). Animal remains from Corstopitum. *Archaeologia Aeliana*, 3rd series, **7**, 220–227. *115*

Meester, J. (1959). Dental abnormalities in African shrews. *Annals of the Transvaal Museum*, **23**, 411–412. *139*

Mellanby, M. (1929). *Diet and the Teeth: An Experimental Study. Part I. Dental Structure in Dogs. Medical Research Council Special Report Series, No. 140.* His Majesty's Stationery Office, London. *341, 445, 446*

Mellanby, M. (1930). *Diet and the Teeth: An Experimental Study. Part II. B. Diet and Dental Structure in Mammals other than the Dog. Medical Research Council Special Report Series, No. 153.* His Majesty's Stationery Office, London. *458, 461, 525*

Mellanby, M. (1937). Der Ernährungsfaktor in der Kariesprophylaxe. *Deutsche Zahn-, Mund- und Kieferheilkunde*, **4**, 1–28. *447*

Mellanby, M. & Killick, E. M. (1926). A preliminary study of factors influencing calcification processes in the rabbit. *Biochemical Journal*, **20**, 902–926. *461*

Mellett, J. S. (1981). Mammalian carnassial function and the 'Every effect'. *Journal of Mammalogy*, **62**, 164–166. *357*

Menveux, M. (1898). Arrière-molaires supplémentaires chez le cheval. *Bulletin et Mémoires de la Société Centrale de Médecine Vétérinaire*, **16**, new series, 487–491. Annexe to *Recueil de Médecine Vétérinaire*, 1898, **1**, series 8. *122*

Menzel, B. W. (1974). A mouthless carp from Texas. *Transactions of the American Fisheries Society*, **103**, 142–143. *393*

Merillat, L. A. (1906). *Veterinary Surgery. Vol. 1. Animal Dentistry and Diseases of the Mouth.* A. Eger, Chicago. *483*

Merriner, J. V. & Wilson, W. L. (1972). Jaw deformity (crossbite) of Atlantic menhaden, *Brevoortia tyrannus*, from Virginia. *Chesapeake Science*, **13**, 62–63. *159*

Messer, H. H. (1980). Alveolar bone loss in a strain of mice. *Journal of Periodontal Research*, **15**, 193–205. *526*

Meyer, A. B. (1896/1897). Säugethiere vom Celebes- und Phillipinen-Archipel. *Abhandlungen und Berichte des Königliches Zoologischen und Anthropologisch-Ethnographischen Museums zu Dresden*, **6**, 1–36. *364, 366*

Meyer, P. (1975). Beispiele angeborene Zahn- und Gebissanomalien beim Europäischen Reh (*Capreolus capreolus*, Linné 1758) nebst einigen Bemerkungen zu deren Genese und Terminologie. *Zeitschrift für Jagdwissenschaften*, **21**, 89–105. *111, 112, 114*

Meyer, P. (1979a). Überzähliger, missgestalteter Backenzahn sowie Schmelzperlen im Unterkiefer eines Damhirsches (*Dama dama* [L.]). *Zeitschrift für Jagdwissenschaften*, **25**, 36–38. *113*

Meyer, P. (1979b). Hochgradige Oligodontie bei einem Rehbock (*Capreolus capreolus* L.). *Zeitschrift für Jagdwissenschaften*, **25**, 179–181. *114*

Meyer, P. (1979c). Überzählige Schneidezähne im Unterkiefer eines Rehbock (*Capreolus capreolus* L.). *Zeitschrift für Jagdwissenschaften*, **25**, 242–244. *114*

Meyer, P. (1979d). Hochgradige Oligodontie kombinert mit maskierter Hyperodontie bei einem Rehbock (*Capreolus capreolus* L.). *Zeitschrift für Jagdwissenschaften*, **25**, 244–246. *114*

Meyer, P. (1985). Beidseitiger P^1 im Unterkiefer eines Rehs (*Capreolus capreolus* L., 1758). *Zeitschrift für Jagdwissenschaften*, **31**, 120–123. *112*

Meyer, R. & Suter, G. (1976). Epidemiologische und morphologische Untersuchungen am Hundegebiss. I. Mitteilung: epidemiologische Untersuchungen. *Schweizer Archiv für Tierheilkunde*, **118**, 307–317. *478*

Meyer, R., Suter, G. & Triadan, H. (1980). Epidemiologische und morphologische Untersuchungen am Hundegebiss. II. Mitteilung: morphologische Untersuchungen. *Schweizer Archiv für Tierheilkunde*, **122**, 503–517. *447, 479*

Meyer, V. H. & Becker, H. (1967). Eine erbliche Keiferanomalie beim Rind. *Deutsche Tierärztliche Wochenschrift*, **74**, 309–310. *294*

Michel, (1904). Keilformige Defekte. *Deutsche zahnärztliche Wochenschrift*, **6**, 496–499. *494*

Mijsberg, W. A. (1931). Überzählige Milchmolaren und Prämolaren im Gebiss des Siamangs und des Menschen. *Zeitschrift für Anatomie und Entwicklungsgeschichte*, **96**, 98–118. *22*

Mijsberg-Van Rooijen, J. H. N. & Mijsberg, W. A. (1931). Die Enstehung des Menschenskinnes nach Untersuchungen am Siamang. *Zeitschrift für Anatomie und Entwicklungsgeschichte*, **95**, 708–733. *336*

Mikx, F. H. M. & van Campen, G. J. (1982). Microscopical evaluation of the microflora in relation to necrotizing ulcerative gingivitis in the beagle dog. *Journal of Periodontal Research*, **17**, 576–584. *550*

Miles, A. E. W. (1951). Actinomycosis of the jaws in an antelope (*Ourebia kenyae*). *Proceedings of the Royal Society of Medicine*, **44**, 865–866. *516*

Miles, A. E. W. (1954). Malformations of the teeth. *Proceedings of the Royal Society of Medicine*, **47**, 817–826. *6, 7, 10, 12, 430*

Miles, A. E. W. (1961). Ageing in the teeth and oral tissues. In *Structural Aspects of Ageing*, chapter 21, edit. G. H. Bourne. Pitman Medical Publications, London. *331, 505*

Miles, A. E. W. (1963). Pigmented enamel. *Proceedings of the Royal Society of Medicine*, **56**, 918–920. *361*

Miles, A. E. W. (1964). The Odontological Museum. *Annals of the Royal College of Surgeons of England*, **34**, 50–58. *362*

Miles, A. E. W. (1966). Double teeth. *Proceedings of the Royal Society of Medicine*, **59**, 680. *7*

Miles, A. E. W. (1969). The dentition of the Anglo-Saxons. *Proceedings of the Royal Society of Medicine*, **62**, 1311–1315. *458*

Miles, A. E. W. (1972). The Odontological Museum of the Royal College of Surgeons of England. *British Dental Journal*, **133**, 115–119. *x, 422*

Miles, A. E. W. (1973). Comparative jaw and tooth pathology. *British Dental Journal*, **134**, 341–345. *367, 405*

Miles, A. E. W. & White, J. W. (1959). One hundred years of

the Odontological Museum. *Proceedings of the Royal Society of Medicine*, **52**, 853–858.   *x*

Miles, A. E. W. & White, J. W. (1960). Ivory. *Proceedings of the Royal Society of Medicine*, **53**, 775–780.   *426*

Milhaud, G., Zundel, E. & Crombet, M. (1980). Étude expérimentale de la fluorose caprine. I. Protocole expérimental, comportement des animaux, lésions dentaires (1). *Receuil de Médecine Vétérinaire*, **156**, 37–46.   *451*

Miller, F. L., Cawley, A. J., Choquette, L. P. E. & Broughton, E. (1975). Radiographic examination of mandibular lesions in barren-ground caribou. *Journal of Wildlife Diseases*, **11**, 465–470.   *516*

Miller, F. L. & Tessier, G. D. (1971). Dental anomalies in caribou, *Rangifer tarandus*. *Journal of Mammalogy*, **52**, 164–174.   *111, 113*

Miller, W. A. (1977). Extreme tooth loss in a specimen of *Isoodon macrourus* (Peramelidae: Marsupialia). *Australian Wildlife Research*, **4**, 229–232.   *571*

Miller, W. A. & Beighton, D. (1979). Bone abnormalities in two groups of macropod skulls: a clue to the origin of lumpy jaw. *Australian Journal of Zoology*, **27**, 681–689.   *573*

Miller, W. A., Beighton, D. & Butler, R. (1980). Histological and osteological observations on the early stages of lumpy jaw. *Comparative Pathology in Animals*. R.J. Montali & G. Migaki, New York, pp. 231–237   *520*

Miller, W. D. (1890). *The Micro-organisms of the Human Mouth*. S.S. White Manufacturing Co., Philadelphia. Reprinted in facsimile (1973) by S. Karger, Basel.   *456, 478, 479*

Miller, W. D. (1890–91). Studies on the anatomy and pathology of the tusk of the elephant. *Dental Cosmos*, **32**, 337–348, 421–526, 673–679, and **33**, 169–175, 421–440.   *405, 422, 426*

Miller, W. D. (1893). Caries of the teeth in an African manatee (*Manatus senegalensis*). *Dental Cosmos*, **35**, 519–521.   *480*

Miller, W. D. (1894). Caries der Tierzähne. *Verhandlungen der Deutschen Odontologischen Gesellschaft*, **5**, 15–28.   *480*

Miller, W. D. (1899). Some very rare cases of gunshot and spear wounds in the tusks of elephants. *Dental Cosmos*, **41**, 1239–1244.   *422, 426, 434*

Miller, W. D. (1907a). Experiments and observations on the wasting of tooth tissue variously designated as erosion, abrasion, chemical abrasion, denudation, etc. *Dental Cosmos*, **49**, 1–23, 109–124, 225–247.   *487, 494*

Miller, W. D. (1907b). Further investigations of the subject of wasting. *Dental Cosmos*, **49**, 677–689.   *487, 493, 494, 497, 498*

Mills, C. F. (Editor) (1970). *Trace Element Metabolism in Animals*. E. & S. Livingstone, Edinburgh.   *451*

Mills, J. R. E. (1978). The relationship between tooth patterns and jaw movements in the Hominoidea. In *Development, Function and Evolution of Teeth*, chapter 21, edit. P. M. Butler & K. A. Joysey. Academic Press, London & New York, pp. 341–353.   *357*

Milton, J. L., Silberman, M. S. & Hankes, G. H. (1980). Compound comminuted fractures of the rami of the mandible in a tiger (*Panthera tigris*): a case report. *Journal of Zoo Animal Medicine*, **11**, 108–112.   *379*

Mitchell, W. D. (1903). Some notes upon the dentition of the elephant and injuries thereto. *Dental Review, Chicago*, **17**, 83–110.   *405*

Mitchum, G. D. & Bruère, A. N. (1984). Solubilization of sheep's teeth: a new look at a widespread New Zealand problem. *Proceedings of the 14th Seminar (at Bulls, N.Z. 1984) Sheep and Beef Cattle Society of the New Zealand Veterinary Association*, pp. 44–56.   *492*

Mivart, St. G. (1890). Note on canine dental abnormalities. *Proceedings of the Zoological Society of London*, 1890, 376–378.   *94*

Miyao, T. (1973). Supernumeraries and missing teeth in nine species of the suborder Microchiroptera. *Journal of the Mammal Society of Japan*, **5**, 230–233.   *135*

Mohr, E. (1958). Zur Kenntnis des Hirschebers, *Babirussa babyrussa* Linné 1758. *Zoologischen Garten*, **25**, 50–69.   *366*

Møller, I. J. (1982). Fluorides and dental fluorosis. *International Dental Journal*, **32**, 135–147.   *452*

Molnar, S. & Ward, S. C. (1975). Mineral metabolism and microstructural defects in primate teeth. *American Journal of Physical Anthropology*, **43**, 3–18.   *440*

Moodie, R. L. (1923). *Paleopathology: an Introduction to the Study of Ancient Evidences of Disease*. University of Illinois Press, Urbana, Illinois.   *485*

Moore, W. J. (1965). Masticatory function and skull growth. *Journal of Zoology, London*, **146**, 123–131.   *156*

Moran, R. J. & Fairbanks, T. (1966). Vestigial first molar in Michigan elk. *Journal of Mammalogy*, **47**, 514.   *112*

Morot, C. (1885). Anomalies dentaires par diminution de nombre. Deux cas chez le cheval. – Absence de coins. *Bulletin et Mémoires de la Société Centrale de Médecine Vétérinaire*, **3**, new series, 125–130; annexe to *Receuil de Médecine Vétérinaire*, **2**, series 7.   *122*

Morot, C. (1886a). Anomalie dentaire chez le mouton: aspect caniniforme des coins de remplacement. *Bulletin et Mémoires de la Société Centrale de Médecine Vétérinaire*, **4**, new series, 319–321; annexe to *Receuil de Médecine Vétérinaire*, **3**, series 7.   *117*

Morot, C. (1886b). Augmentation du nombre des incisives chez une vache et chez un chevreau. *Bulletin et Mémoires de la Société Centrale de Médecine Vétérinaire*, **4**, new series, 321–322; annexe to *Receuil de Médecine Vétérinaire*, **3**, series 7.   *115, 117*

Morot, C. (1886c). Anomalie dentaire chez un jeune veau. – Développement parfait d'une canine à la machoire supérieure. *Bulletin et Mémoires de la Société Centrale de Médecine Vétérinaire*, **4**, new series, 680–681; annexe to *Receuil de Médecine Vétérinaire*, **3**, series 7.   *115*

Morot, C. (1887). Sur plusieurs cas d'anomalies numériques des dents incisives observés chez les ruminants domestiques (boeuf, mouton, chèvre). *Bulletin et Mémoires de la Société Centrale de Médecine Vétérinaire*, **5**, new series, 401–406; annexe to *Receuil de Médecine Vétérinaire*, **4**, series 7.   *115, 117*

Morot, C. (1888a). Absence d'une incisive persistante et situation hétérotopique d'une canine à la machoire inférieure d'un vieux cheval hongre. *Bulletin et Mémoires de la Société Centrale de Médecine Vétérinaire*, **6**, new series, 137–138; annexe to *Receuil de Médecine Vétérinaire*, **5**, series 7.   *122*

Morot, C. (1888b). Incisive supplémentaire caniniforme à la machoire supérieure d'un cheval hongre âgé d'une douzaine d'années. *Bulletin et Mémoires de la Société Centrale de Médecine Vétérinaire*, **6**, new series, 138–139; annexe to *Receuil de Médecine Vétérinaire*, **5**, series 7.   *119*

Morot, C. (1888c). Deux cas d'incisives composées et

irrégulières: dent double chez un cheval et dent triple chez un mouton. *Bulletin et Mémoires de la Société Centrale de Médecine Vétérinaire*, **6**, new series, 139–141; annexe to *Receuil de Médecine Vétérinaire*, **5**, series 7. *117, 121*

Morot, C. (1889). Anomalies dentaires diverses. *Bulletin et Mémoires de la Société Centrale de Médecine Vétérinaire*, **7**, new series, 479–486; annexe to *Receuil de Médecine Vétérinaire*, **6**, series 7. *121*

Morot, C. (1896). Inclusion complète d'une arrière-molaire supplémentaire dans le maxillaire supérieure droit d'une jument. *Bulletin et Mémoires de la Société Centrale de Médecine Vétérinaire*, **14**, new series, 791–792; annexe to *Receuil de Médecine Vétérinaire*, **3**, series 8. *122*

Morot, C. (1898a). Anomalie deforme des incisives. *Revue Vétérinaire*, **23**, 49. *115*

Morot, C. (1898b). Inclusion d'une arrière-molaire. *Revue Vétérinaire*, **23**, 49. *122*

Morot, C. (1898c). Anomalie par diminution numérique. *Revue Vétérinaire*, **23**, 49–50. *115*

Morris, D. (1965). *The Mammals, a Guide to the Living Species*. Hodder & Stoughton, London. *xii*

Morris, P.L., Whitley, B.D., Orr, M.B. & Laws, A.J. (1985). A clinical study of periodontal disease in sheep. *New Zealand Veterinary Journal*, **33**, 87–90. Erratum p. 131. *556*

Morzer Bruyns, W.F.J. (1973). On abnormal teeth in the strap-toothed whale, *Mesoplodon layardi* (Gray, 1865). *Säugetierkundliche Mitteilungen*, **21**, 75–77. *368, 369*

Moskow, B.S., Wasserman, B.H. & Rennert, M.C. (1968). Spontaneous periodontal disease in the Mongolian gerbil. *Journal of Periodontal Research*, **3**, 69–83. *569*

Moskow, B.S., Rennert, M.C. & Wasserman, B.H. (1973). Interrelationship of dietary factors and heredity in calculus formation and periodontal lesions in the gerbil. *Journal of Periodontology*, **44**, 81–84. *527*

Moss, J.P. & Picton, D.C.A. (1967). Experimental mesial drift in adult monkeys (*Macaca irus*). *Archives of Oral Biology*, **12**, 1313–1320. *157, 332*

Moss, J.P. & Picton, D.C.A. (1982). Short term changes in the mesiodistal position of the teeth following removal of approximal contacts in the monkey (*Macaca fascicularis*). *Archives of Oral Biology*, **27**, 273–278. *157, 528*

Moss, M.L. (1978). Analysis of developmental processes possibly related to human dental sexual dimorphism. In *Development, Function and Evolution of Teeth*, chapter 11, edit. P.M Butler & K.A. Joysey. Academic Press, London & New York, pp. 135–148. *9, 10*

Mróz, K., Gasiorowska, I., Wyluda, E. & Pastuszewska, B. (1979). An appraisal of hardness of the enamel incisors in rats fed with krill meal. *Bulletin de l'Academie Polonaise des Sciences. Série des Sciences Biologiques*, Cl. 2, **27**, 131–133. *361*

Müller, H. (1935). Anomalien der Zahnstellung, Kieferform und Okklusion beim Tier. Part 3. *Deutsche Zahn-, Mund- und Kieferheilkunde*, **2**, 462–467. *487*

Mulvihill, J.E., Susis, F.R., Shaw, J.H. & Goldhaber, P. (1967). Histological studies of the periodontal syndrome in rice rats and the effect of penicillin. *Archives of Oral Biology*, **12**, 733–744. *527*

Mumford, R.E. (1963). Unusual dentition in *Myotis occultus*. *Journal of Mammalogy*, **44**, 275. *135, 136*

Mummery, J.H. (1924). *The Microscopic and General Anatomy of the Teeth, Human and Comparative*, 2nd edn. Humphrey Milford & Oxford University Press, London. *xii*

Murie, A. (1944). *The Wolves of Mount McKinley*. Fauna of the National Parks of the United States. Fauna Series No. 5. *516, 562*

Murie, J. (1865). On deformity of the lower jaw in the cachalot (*Physeter macrocephalus*, Linn.). *Proceedings of the Zoological Society of London*, 1865, 390–396. *391*

Murie, J. (1867). Notes on some diseased dental conditions in animals. *Transactions of the Odontological Society of Great Britain*, **6**, 37–69. *363, 391*

Murie, J. (1870). On some abnormal and diseased dental conditions in animals. *Transactions of the Odontological Society of Great Britain*, New Series, **2**, 257–305. *426, 488, 496*

Murie, O.J. (1930). An epizootic disease of elk. *Journal of Mammalogy*, **11**, 214–220. *516, 563*

Murie, O.J. (1951). *The Elk of North America*. The Stackpole Company, Pennsylvania. (Third printing, 1966). *111, 516, 563*

Murphy, R.C. (1947). *Logbook for Grace*. Macmillan, New York, p. 118. *389*

Murphy, T. (1959). The axis of the masticatory stroke. *Australian Dental Journal*, **4**, 104–111. *557*

Murray, A. (1868). A nine-tusked elephant. *Journal of Travel and Natural History*, **1**, 265–275. *127*

Nachtsheim, H. (1936). Erbliche Zahnanomalien beim Kaninchen. *Zuchtungskunde*, **11**, 273–287. *134, 319, 360, 361*

Nakamura, K. (1968). Studies on the sperm whale with deformed lower jaw with special reference to its feeding. *Bulletin of the Kanagawa Prefectural Museum*, **1**, 13–14. *390, 391*

Nansen, F. (1925). *Hunting and Adventure in the Arctic*. J.M. Dent, London, pp. 230–260. *391*

Napier, P.H. (1976). *Catalogue of Primates in the British Museum (Natural History) and elsewhere in the British Isles. Part I: Families Callitrichidae and Cebidae*. British Museum (Natural History), London. *xii*

Napier, P.H. (1981). *Catalogue of Primates in the British Museum (Natural History) and elsewhere in the British Isles. Part II: Family Cercopithecidae, Subfamily Cercopithecinae*. British Museum (Natural History), London. *xii*

Napier, P.H. (1985). *Catalogue of Primates in the British Museum (Natural History) and elsewhere in the British Isles. Part III: Family Cercopithecidae, Subfamily Colobinae*. British Museum (Natural History), London. *xii, 180, 182*

Nass, G.G. (1977). Intra-group variations in the dental eruption sequence of *Macaca fuscata fuscata*. *Folia Primatologia*, **28**, 306–314. *337*

Nathusius, H. (1864). *Vorstudien für Geschichte und Zucht der Haustiere zunachst am Schweineschädel*. Berlin. *283*

Navia, J.M. (Editor) (1977). *Animal Models in Dental Research*. University of Alabama Press, Alabama. *462, 526, 527, 528, 529*

Neal, B.J. & Kirkpatrick, R.D. (1957). Anomalous canine tooth development in an Arizona peccary. *Journal of Mammalogy*, **38**, 420. *109*

Neeley, K.L. & Harbaugh, F.G. (1954). Effects of fluoride ingestion on a herd of dairy cattle in the Lubbock, Texas, area. *Journal of the American Veterinary Medical Association*, **124**, 344–350. *452*

Nehring, A. (1882). Über einige Canisschädel mit auffälliger

Zahn-Formel. *Sitzungsberichte der Gesellschaft Naturforschender Freunde zu Berlin*, 1882(5), 65–68. *84*

Nehring, A. (1884). Über eine grosse Wolfsähnliche Hunde-Rasse der Vorzeit (*Canis fam. decumanus*) und über ihre Abstammung. *Sitzungsberichte der Gesellschaft Naturforschender Freunde zu Berlin*, 1884(9), 153–165. *264*

Nellis, C. H. (1972). Dental anomalies in coyotes from central Alberta. *Canadian Journal of Zoology*, 50, 1259–1262. *83, 84, 264, 342, 447*

Ness, A. R. (1966). Dental caries in the platanistid whale *Inia geoffrensis*. *Journal of Comparative Pathology*, 76, 271–278. *499, 500, 501*

Neuenschwander, A. & Lüps, R. (1975). Zahnvariationen bei einer Mauswiesel-Populations, *Mustela nivalis* Linné, 1766, aus hessen. *Säugetierkundliche Mitteilungen*, 23(2), 85–93. *71*

Neuville, M. H. (1928). Recherches sur le genre 'Steno' et remarques sur quelques autres Cétacés. *Archives du Muséum National d'Histoire Naturelle*, 6e séries, 3, 69–241. *102, 104, 105*

Neuville, M. H. (1929). Note sur une anomalie dentaire du cachalot. *Bulletin du Muséum National d'Histoire Naturelle*, 2e séries, 1, 342–346. *102, 104*

Neuville, H. (1932). Recherches comparatives sur la dentition des Cétodontes. Etude de morphologie et d'éthnologie. *Annales des Sciences Naturelles. Zoologie*, 10e séries, 15, 185–361. *102, 103, 104, 435*

Neuville, H. (1936). Ectopies de gorilla et d'orang utan. *Bulletin du Muséum National d'Histoire Naturelle*, 2e séries, 8, 225–232. *165, 174*

Newman, J. R. & Murphy, J. J. (1979). Effects of industrial fluoride on black-tailed deer (preliminary report). *Fluoride*, 12, 129–135. *454*

Newman, J. R. & Yu, M.-H. (1976). Fluorosis in black-tailed deer. *Journal of Wildlife Diseases*, 12, 39–41. *489*

Niethammer, G. (1971). Die Gemsen Neuseelands. *Zeitschrift für Säugetierkunde*, 36, 228–238. *115, 561*

Niethammer, J. (1968). Eine Pfeifhase, *Ochotona rufescens* (Gray, 1842), ohne Stiffzähne. *Säugetierkundliche Mitteilungen*, 16, 160–162. *134*

Nisbet, D. I., Butler, E. J., Robertson, J. M. & Bannatyne, C. C. (1968). Osteodystrophic disease of sheep. III. Dental mal-occlusion in young sheep. *Journal of Comparative Pathology*, 78, 73–77. *296*

Nissen, H. W. & Riesen, A. H. (1945). The deciduous dentition of chimpanzee. *Growth*, 9, 265–274. *335, 336*

Nissen, H. W. & Riesen, A. H. (1964). The eruption of the permanent dentition of chimpanzee. *American Journal of Physical Anthropology*, 22, 285–294. *335, 336*

Nolan, T. & Black, W. J. M. (1970). Effect of stocking rate on tooth wear in ewes. *Irish Journal of Agricultural Research*, 9, 187–196. *491, 492*

Nomura, H. (1978). Two mouthed characin fish, *Astyanax schubarti* Britski, 1964, from Mogi Guacu river, São Paulo State, Brazil. *Revista Brasiliera de Biologia*, 38, 753–755. *601*

Nordby, J. E., Terrill, C. E., Hazel, L. N. & Stoehr, J. A. (1945). The etiology and inheritance of inequalities in the jaws of sheep. *Anatomical Record*, 92, 235–254. *295*

Northey, R. D., Hawley, J. G. & Suckling, G. W. (1975). The tooth to pad relationship in sheep – some mechanical considerations. *New Zealand Journal of Agricultural Research*, 18, 133–138. *553, 554, 555*

Nuki, K. & Cooper, S. H. (1972). The role of inflammation in the pathogenesis of gingival enlargement during the administration of diphenylhydantoin sodium in cats. *Journal of Periodontal Research*, 7, 102–110. *528*

Nyman, S., Lindhe, J. & Ericsson, I. (1978). The effect of progressive tooth mobility on destructive periodontitis in the dog. *Journal of Clinical Periodontology*, 5, 213–225. *528*

Oboussier, H. & Storkmann, W. (1975). Begissanomalien bei *Oryx dammah* CRETZSCHMAR, 1826 und *Gazella rufifrons kanuri*, SCHWARZ 1914 (Bovidae Artiodactyla). *Säugetierkundliche Mitteilungen*, 23, 16–19. *115, 117*

Ockerse, T. (1953). Experimental dental caries in the white-tailed rat in South Africa. *Journal of Dental Research*, 32, 74–77. *461*

Ockerse, T. (1956). Experimental periodontal lesions in the white-tailed rat in South Africa. *Journal of Dental Research*, 35, 9–15. *527*

Olfert, E. D. (1974). Tooth root abscess and fistula in the squirrel monkey (*Saimiri sciureus*). *Canadian Veterinary Journal*, 15, 171–172. *511*

Ooë, T. & Esaka, S. (1981). Le chevauchement des dents jugales chez les phocques: modalités et évolution post-natale. *Mammalia*, 45, 497–503. *281*

Orland, F. J. (1946). A study of the Syrian hamster, its molars and their lesions. *Journal of Dental Research*, 25, 445–453. *461*

Orr, M. B., Christiansen, K. H. & Kissling, R. C. (1986). A survey of excessively worn incisors and periodontal disease in sheep in Dunedin City, Silverpeaks, Bruce and Clutha Counties. *New Zealand Veterinary Journal*, 34, 111–115. *559*

Orr, M. B., O'Callaghan, M. W., West, D. M. & Bruère, A. N. (1979). A syndrome of dental abnormalities of sheep. II. The pathology and radiology. *New Zealand Veterinary Journal*, 27, 276–278. *492, 559, 579*

Osborn, H. F. (1907). *Evolution of Mammalian Molar Tooth to and from the Triangular Type*. MacMillan, London and New York. *15*

Osborn, J. W. (1969). Dentine hardness and incisor wear in the beaver (*Castor fiber*). *Acta Anatomica*, 72, 123–132. *357, 359, 360*

Osborn, J. W. (1978). Morphogenetic gradients: fields versus clones. In *Development, Function and Evolution*, chapter 14, edit. P. M. Butler & K. A. Joysey. Academic Press, London & New York, pp. 171–201. *2, 3, 4, 11*

Osborn, J. W., Editor (1981a). *Dental Anatomy and Embryology*. Blackwell Scientific Publications, Oxford. *xii, 2*

Osborn, J. W. (1981b). Development of dentition. Developmental controls. In *Dental Anatomy and Embryology*, edit. J. W. Osborn. Blackwell Scientific Publications, Oxford, pp. 255–260. *2, 3*

Osborn, J. W. (1984). From reptile to mammal: evolutionary considerations of the dentition with emphasis on tooth attachment. In *The Structure, Development and Evolution of Reptiles*, edit. M. W. J. Ferguson. Zoological Society of London, London, pp. 549–574. *2, 4, 155*

Østergaard, E. & Loe, H. (1975). The collagen content and gingival tissues in ascorbic acid-deficient monkeys. *Journal of Periodontal Research*, 10, 103–114. *528*

Otto, B. & Schumacher, G. H. (1978). Gebissanomalien beim Vietnamesischen Hängebauchschwein. *Zeitschrift Versuchstierkunde*, 20, 122–131. *107, 480*

Owen, R. (1840–1845). *Odontography*. Vol. 1 text, vol. 2 atlas. Hippolyte Bailliere, London. *404, 406, 421*

Owen, R. (1846). *A History of British Fossil Mammals and Birds*. Van Voorst, London. *592, 594*

Owen, R. (1853). *Descriptive Catalogue of the Osteological Series contained in the Museum of the Royal College of Surgeons. Volume II. Mammalia Placentalia*. Taylor & Francis, London. *292*

Owen, R. (1866/1868). *On the Anatomy of Vertebrates*. London. *343*

Owen-Smith, N. (1966). The incidence of tuskless elephant in Mana Pools Game Reserve. *African Wild Life*, 20, 69–73. *129*

Page, R. C. & Schroeder, H. E. (1981). Spontaneous chronic periodontitis in adult dogs. A clinical and histopathological survey. *Journal of Periodontology*, 52, 60–73. *549*

Page, R. C. & Schroeder, H. E. (1982). *Periodontitis in Man and Other Animals*. Karger, Basel. *523, 524, 525, 526, 527, 529, 532, 557, 558, 562*

Page, R. C., Simpson, D. M. & Ammons, W. F. (1975). Host tissue response in chronic inflammatory periodontal disease. IV. The periodontal and dental status of a group of aged great apes. *Journal of Periodontology*, 46, 144–155. *534*

Paget, R. J. (1972). A case of osteomyelitis in the skull of a badger (*Meles meles*). *Journal of Zoology, London*, 168, 423–424. *512*

Paine, R. (1909). Alveolar periostitis in young equines. *Journal of Comparative Pathology and Therapeutics*, 22, 152–154. *514*

Pande, P. G. (1945). Dental and osseous changes in spontaneous fluorosis in rats. *Indian Journal of Medical Research*, 33(1), 121–128. *361*

Paradiso, J. L. (1966). Notes on supernumerary and missing teeth in the coyote. *Mammalia*, 30, 120–128. *83, 84, 88*

Parmalee, P. W. & Bogan, A. E. (1977). An unusual dental anomaly in a mink. *Journal of the Tennessee Academy of Science*, 52, 115–116. *72*

Pastuszewska, B., Wylunda, E. & Buraczewski, S. (1979). Observations on the abnormal growth of incisors in laboratory rats and mice fed with krill meal. *Bulletin de l'Academie Polonaise des Sciences. Série des Sciences Biologiques. Cl. 2*, 27, 125–129. *361*

Paterson, J. S. (1967). The guinea pig or cavy (*Cavia porcellus* L.). In *The UFAW Handbook of the Care and Management of Laboratory Animals*, 3rd edn., edit. W. Lane-Petter, A. N. Worden, B. F. Hill, J. S. Paterson & H. G. Vevers. E. & S. Livingstone, Edinburgh. *362*

Patmore, A. (1984). *Your Obedient Servant. The Story of Man's Best Friend*. Hutchinson, London. *269*

Patton, N. M., Hooper, B. E., Mock, O. B. & Doyle, R. E. (1971). Periodontal disease in the least shrew. *Journal of Periodontology*, 42, 597–599. *569*

Pavlinov, I. Y. (1975). Tooth anomalies in some Canidae. *Acta Theriologica*, 20, 507–519. *84, 90, 91, 92, 93, 94*

Pekelharing, C. J. (1968). Molar duplication in red deer and wapiti. *Journal of Mammalogy*, 49, 524–526. *112*

Pekelharing, C. J. (1974). Paradontal disease as a cause of tooth loss in a population of chamois (*Rupicapra rupicapra* L.) in New Zealand. *Zeitschrift für Säugetierkunde*, 39, 250–255. *561*

Penny, R. H. C. & Mullen, P. A. (1975). Atropic rhinitis of pigs: abattoir studies. *The Veterinary Record*, 96, 518–521. *283, 494*

Penzhorn, B. L. (1984). Dental abnormalities in free-ranging Cape Mountain zebras (*Equus zebra zebra*). *Journal of Wildlife Diseases*, 20, 161–166. *435*

Perrin, W. F. (1969). The barnacle, *Conchoderma auritum*, on a porpoise (*Stenella graffmani*). *Journal of Mammalogy*, 50, 149–151. *391*

Peterson, R. L. (1955). *North American Moose*, University of Toronto Press, Toronto, p. 86. *111*

Peterson, R. L. (1968). Notes on an unusual specimen of *Scotophilus* from Vietnam. *Canadian Journal of Zoology*, 46, 1079–1081. *135*

Peterson, R. O., Scheidler, J. M. & Stephens, P. W. (1982). Selected skeletal morphology and pathology of moose from the Kenai Peninsular, Alaska, and Isle Royale, Michigan. *Canadian Journal of Zoology*, 60, 2812–2817. *491, 562*

Petit, M. (1938). Le dent de loup des équides. *Revue de Médecine Vétérinarie de Toulouse*, 90, 619–628. *121, 352*

Petter, F. & Tostain, O. (1981). Variabilité de la 3e molaire supérieure d'*Holochilus brasiliensis* (Rongeurs, Cricetidae). *Mammalia*, 45, 257–259. *132*

Peyer, B. (1968). *Comparative Odontology*. University of Chicago Press, Chicago, p. 77. *499*

Phillips, C. J. (1971). The dentition of glossophagine bats: development, morphological characteristics, variation, pathology and evolution. *Miscellaneous Publication of the University of Kansas Museum of Natural History*, 54, 138 pp. *135, 137, 484, 570*

Phillips, C. J. & Jones, J. K. (1968). Dental anomalies in N. American bats. 1. Emballonuridae, Noctilionidae and Chilonycteridae. *Transactions of the Kansas Academy of Science*, 71, 509–520. *135, 137*

Phillips, C. J. & Jones, J. K. (1970). Dental anomalies in N. American bats. II. Possible endogenous factors in dental caries in *Phyllostomus hastatus*. *University of Kansas Science Bulletin*, 48, 791–799. *484*

Piechocki, R. (1977). Zahnanomalien beim Elbebiber, *Castor fiber albicus*. *Hercynia N.F., Leipzig*, 14, 187–195. *131*

Pilleri, G. (1983). The occurrence of extra premolar teeth in *Castor canadensis*. *Investigations on Beavers*, 1, 61–63. *131*

Pilleri, G. & Gihr, M. (1969a). Über adriatische *Tursiops truncatus* (Montagu 1821) und vergleichende Untersuchungen über mediteranne und atlantische Tümmler. *Investigations on Cetacea*, 1, 66–73. *512*

Pilleri, G. & Gihr, M. (1969b). Zur Anatomie und Pathologie von *Inia geoffrensis* de Blainville 1817 (Cetacea, Susuidae) aus dem Beni, Bolivien. *Investigations on Cetacea*, 1, 94–105. *375, 501*

Pilleri, G. & Gihr, M. (1976). A symmetrical fusion in the teeth of the La Plata dolphin, *Pontoporia blainvillei*. *Investigations on Cetacea*, 7, 147–148. *104*

Pillers, A. W. N. (1933). Some observations on periodontal disease in adult cart-horses. *The Veterinary Record*, 13, 1404–1410. *567*

Pindborg, J. J. (1970). *Pathology of the Dental Hard Tissues*. Munksgaard, Copenhagen. *438*

Pitts, A. T. (1933). Some reflections on the nature of odontomes. *British Dental Journal*, 54, 217–237. *576*

Pocock, R. I. (1916a). Some dental and cranial variations in the Scotch wild cat (*Felis sylvestris*). *Annals and Magazine of Natural History, London*, 8th series, 18, 272–277. *62*

Pocock, R. I. (1916b). On the tooth change, cranial characters and classification of the snow-leopard or ounce (*Felis uncia*). *Annals and Magazine of Natural History, London*, 8th series, **18**, 306–316. *339*

Pocock, R. I. (1929). Dogs. In *Encyclopaedia Britannica*, 14th edn., vol. 7. Encyclopaedia Britannica Co. Ltd., London. *85*

Pocock, R. I. (1940). Description of a new race of puma (*Puma concolor*), with a note on an abnormal tooth in the genus. *Annals and Magazine of Natural History, London*, 11th series, **6**, 307–313. *64*

Poduschka, W. & Poduschka, C. (1986). Zahnstein, Zahnfleischerkrankungen and Zahnanomalien bei Erinaceinen (Mammalia: Insectivora). *Zeitschrift für Angewandte Zoologie*, **73**, 231–243. *139, 569*

Pohle, H. (1923). Uber der Zahnwechsel der Bären. *Zoologischer Anzeiger*, **55**, 266–277. *341*

Pohle, H. (1940). Ein Beuteldachs mit falschem Gebiss. *Zeitschrift für Säugetierkunde*, **15**, 285–288. *146*

Pollock, S. (1951). Slobbers in the rabbit. *Journal of the American Veterinary Medical Association*, **119**, 443–444. *362*

Polson, A.M., Meitner, S.W. & Zander, H.A. (1976). Trauma and progression of marginal periodontitis in squirrel monkeys. IV. Reversibility of bone loss due to trauma alone and trauma superimposed upon periodontitis. *Journal of Periodontal Research*, **11**, 290–298. *528*

Poole, D.F.G. (1961). Notes on tooth replacement in the Nile crocodile, *Crocodilus niloticus*. *Proceedings of the Zoological Society of London*, **136**, 131–140. *505*

Poole, D.F.G. & Miles, A.E.W. (1967). The history and general organization of dentitions. In *Structural and Chemical Organization of Teeth*, vol. 1, edit. A.E.W. Miles. Academic Press, New York. *358*

Poole, D.F.G. & Tratman, E.K. (1978). Post-mortem changes in human teeth from late Upper Palaeolithic/Mesolithic occupants of an English limestone cave. *Archives of Oral Biology*, **23**, 1115–1120. *501*

Porter, W.L., Scott, R.S. & Manktelow, B.W. (1970). The occurrence of paradontal disease in sheep in relation to superphosphate top dressing, stocking rate and other related factors. *New Zealand Veterinary Journal*, **18**, 21–27. *560*

Pouchet, G. & Beauregard, H. (1889). Recherches sur la cachalot. Anatomie I–V. *Nouvelles Archives du Muséum d'Histoire Naturelle, Paris*, 3rd Series, **1**, 1–96. *102, 104, 391, 504*

Pratt, K.A. & Knight, R.L. (1981). Abnormal incisor growth in a hoary marmot. *The Murrelet*, **62**, 92. *358*

Preiswerk, G. (1906). *Atlas and Textbook of Dentistry including Diseases of the Mouth*. W.B. Saunders, London, p. 210. *485*

Prior, R. (1968). *The Roedeer of Cranborne Chase*. Oxford University Press, Oxford. *112*

Pruitt, W.O. (1957). Tooth reduction in the tundra shrew. *Journal of Mammalogy*, **38**, 21. *139*

Purohit, R.K., Dudi, P.R., Chouhan, D.S. & Sharma, C.K. (1984). Amputation of anterior fragment of irreparable mandibular fracture in camel (*Camelus dromedarius*) (a report of five clinical cases). *Indian Veterinary Journal*, **61**, 989–991. *386*

Purser, A.F., Wiener, G. & West, D.M. (1982). Causes of variation in dental characters of Scottish Blackface sheep in a hill flock, and relations to ewe performance. *Journal of Agricultural Science, Cambridge*, **99**, 287–294. *295*

Pycraft, W.P. (1913). *The Courtship of Animals*. Hutchinson, London. *373*

Quay, W.B. (1953). Anomalies of the cheekteeth in the red-backed vole, *Clethrionomys gapperi* (Vigors, 1830). *Säugetierkundliche Mitteilungen*, **1**, 110–115. *217*

Rabkin, S. (1946). Comparative alveolar pathology in sheep and goats. *Journal of Dental Research*, **25**, 513–520. *560*

Ramfjord, S. (1951). Effects of acute febrile diseases on the periodontium of rhesus monkeys with reference to poliomyelitis. *Journal of Dental Research*, **30**, 615–626. *528*

Randall, F.E. (1943/1944). The skeletal and dental development and variability of the gorilla. *Human Biology*, **15**, 236–54 and 307–37, **16**, 23–76. *335, 375, 392, 507, 529*

Ranstead, J.M. (1946). The inheritance of 'parrot-beak' in New Zealand milking shorthorns. *New Zealand Journal of Science and Technology*, **28A**, 206–216. *294*

Rantanen, A.V. & Pulliainen, E. (1970). Dental conditions in wild red foxes (*Vulpes vulpes* L.). *Annali Zoologici Fennici*, **7**, 290–294. *83*

Ratcliffe, P.R. (1970). The occurrence of vestigial teeth in badger (*Meles meles*), roe deer (*Capreolus capreolus*) and fox (*Vulpes vulpes*) from the county of Argyll, Scotland. *Journal of Zoology, London*, **162**, 521–525. *73*

Ratti, P. & Habermehl, K.H. (1977). Untersuchungen zur Altersschätzung und Altersbestimmung beim Alpensteinbock (*Capra ibex ibex*) im Kanton Graubünden. *Zeitschrift für Jagdwissenschaften*, **23**, 188–213. *115, 342*

Rausch, R.L. (1961). Notes on the black bear, *Ursus americanus* Pallas, in Alaska, with particular reference to dentition and growth. *Zeitschrift für Säugetierkunde*, **26**, 77–107. *77, 341, 493, 545*

Reade, L.L. (1930). Rogue elephants with malformed tusks. *Journal of the Bombay Natural History Society*, **33**, 979–980. *128*

Reed, J.H. (1975). The digestive system. In *Feline Medicine and Surgery*, 2nd edn., chapter 8, edit. E.J. Catcott. Santa Barbara California, American Veterinary Publications, pp. 154–158. *372, 524*

Reed, O.M. (1973). *Papio cynocephalus* age determination. *American Journal of Physical Anthropology*, **38**, 309–314. *336, 337*

Reichart, P.A., Durr, U.-M., Triadan, H. & Vickendey, G. (1984). Periodontal disease in the domestic cat. A histopathologic study. *Journal of Periodontal Research*, **19**, 67–75. *543*

Reif, W.E. (1980). A mechanism for tooth pattern in sharks: the polarity switch model. *Roux's Archives of Developmental Biology*, **188**, 115–122. *5, 7*

Reiland, S. (1978). Growth and skeletal development of the pig. *Acta Radiologica*, supplement 358, 15–22. *342, 343*

Reinwaldt, E. (1958). Über und unterzählige Schneiderzähne bei Musteliden. *Säugetierkundliche Mitteilungen*, **6**(3), 97–100. *72*

Reinwaldt, E. (1961). Über Zahnanomalien und Zahnformel der Gattung *Sorex* Linné. *Arkiv för Zoologi*, **13**, 533–539. *139*

Reinwaldt, E. (1962). Uber einige Anomalien im Gebiss des Rotfuchses, *Vulpes vulpes vulpes* Linné. *Arkiv för Zoologi*, **15**(25), 371–375. *83, 84, 263*

Remane, A. (1921). Beitrage zur Morphologie des Anthropoiden-gebisses. *Archiv für Naturgeschichte*, **87**, A11, 1–179. *23, 24, 28, 31, 33*

Remane, A. (1926). Eine seltsame Gebissanomalie bei einem Stummelaffen. *Zeitschrift für Säugetierkunde*, **1**, 114–120. *39, 40, 443*
Renshaw, H. W., Chatburn, C., Bryan, G. M., Bartsch, R. C. & Davis, W. C. (1975). Canine granulocytopathy syndrome. Neutrophil dysfunction in a dog with recurrent infections. *Journal of the American Veterinary Medical Association*, **166**, 443–447. *523*
Renshaw, H. W., Davis, W. C., Fudenberg, H. H. & Padgett, G. A. (1974). Leukocyte dysfunction in the bovine homologue of the Chediak-Higashi syndrome of humans. *Infection and Immunity*, **10**, 928–937. *523*
Resnick, D. & Niwayama, G. (Editors) (1981). *Diagnosis of Bone and Joint Disorders*, vol. 2. W. B. Saunders, Philadelphia. *334*
Rice, D. W. (1963). Progress report on biological studies of the larger Cetacea in the waters off California. *Norsk Hvalfangst-Tidende*, **52**, 181–187. *391*
Rice, R. W. & Oyen, O. J. (1979). Supernumerary molars in baboons (*Papio cynocephalus anubis*). *Texas Journal of Science*, **31**, 267–270. *49*
Richardson, C., Richards, M., Terleckis, S. & Miller, W. M. (1979). Jaws of adult culled ewes. *Journal of Agricultural Science, Cambridge*, **93**, 521–529. *295, 552, 559*
Richardson, R. L. (1965). Effect of administering antibiotics, removing the major salivary glands and toothbrushing on dental calculi formation in the cat. *Archives of Oral Biology*, **10**, 245–253. *528, 543*
Rieck, G. W. (1985). Hypotrichie-Hypondontie-Syndrom beim Rind (Kurzbericht). *Deutsche tierärztliche Wochenschrift*, **92**, 301–344. *115*
Rieck, G. W. (1986). Multiple adamantinogene Zysten der Alveolarforsätze beim Rind, die *Odontodysplasia cystica congenita*, eine angeborene Erkrankung der Kiefer und des Zahnapparates. *Journal of Veterinary Medicine*, Series A, **33**, 588–599. *579*
Riehl, R. & Schmitt, P. (1985). The skull in normal and pugheaded females of the mosquitofish *Heterandria formosa* Agassiz, 1853 (Teleostei, Poeciliidae). *Gegenbaurs morphologisches Jahrbuch, Leipzig*, **131**, 261–270. *159*
Riet-Correa, F., Oliveira, J. A., Mendez, M. C. & Schild, A. L. (1986). Fluoride intoxication in cattle due to industrial pollution caused by processing rock phosphate. *Fluoride*, **19**, 61–64. *454*
Ritcey, R. W. & Edwards, R. Y. (1958). Parasites and diseases of the Wells Gray moose herd. *Journal of Mammalogy*, **39**, 139–145. *516*
Ritchie, J. (1940). Upper canine teeth in the Indian antelope (*Antilope cervicapra*). *Nature, London*, **145**, 859. *114*
Ritchie, J. & Edwards, A. J. H. (1913). On the occurrence of functional teeth in the upper jaw of the sperm whale. *Proceedings of the Royal Society of Edinburgh*, **33**, 166–168. *101*
Ritter, W. (1963). Durch Röntgenstrahlen induzierte Zahnverschmelzungen bei der Maus. II. Mitteilung: historische Befunde. *Deutsche Zahnärztliche Zeitschrift*, **18**, 1063–1068. *9*
Robel, D. (1971). Zur Variabilität der Molarenwurzeln bei des Oberkiefers bei Inselpopulationen der Waldmaus (*Apodemus sylvaticus* [L.], 1758). *Zeitschrift für Säugetierkunde*, **36**, 172–179. *133*
Robens, J. F. (1970). Teratogenic effects of hypervitaminosis A in the hamster and the guinea pig. *Toxicology and Applied Pharmacology*, **16**, 88–99. *10*
Robineau, D. (1981). Sur quelques cas d'edentation partielle chez *Delphinus delphis* et leur signification. *Aquatic Mammals*, **8**(2), 33–39. *494*
Robinette, W. L. (1958). Unusual dentition in mule deer. *Journal of Mammalogy*, **39**, 156–157. *111*
Robinette, W. L., Jones, D. A., Rogers, G. & Gashwiler, J. S. (1957). Notes on tooth development and wear for Rocky mountain mule deer. *Journal of Wildlife Management*, **21**, 134–153. *454*
Robins, M. W. & Rowlatt, C. (1971). Dental abnormalities in aged mice. *Gerontologia*, **17**, 261–272. *7, 405*
Robinson, M. (1985). Dietary related periodontitis and oral-nasal fistulation in rats. *Journal of Comparative Pathology*, **95**, 489–498. *568*
Robson, F. D. (1975). On vestigial and normal teeth in the scamperdown beaked whale, *Mesoplodon grayi*. *Tuatara*, **21**, 105–107. *101*
Rodahl, K. (1952). Hypervitaminosis A in the dog. *Federation Proceedings. Federation of American Societies for Experimental Biology*, **11**, 130. *550*
Röhrer-Ertl, O., Frey, K. W. & Schmidhuber-Schneider, I. (1985). Über den Zahnstatus innerhalb einer Population von *Pongo satyrus borneesis* Von Wurmb, 1784 aus Skalau in West-Borneo. *Zoologische Abhandlungen*, **41**, 37–60. *536*
Rolph, C. H. (1980). *London Particulars*. Oxford University Press, Oxford, p. 160. *563*
Romer, A. S. (1962). *The Vertebrate Body*. W. B. Saunders, Philadelphia and London. *106*
Rosen, S., Doff, R. S., App, G. & Rotilie, J. (1981). A topographical scoring system for evaluating root surface caries. In *Animal Models in Cariology. Proceedings of the Symposium on Animal Models in Cariology*, edit. J. M. Tanzer. *Special Supplement Microbiology Abstracts*. Information Retrieval Inc., Washington D.C., pp. 175–182. *461*
Rosenberg, H. M., Rehfeld, C. E. & Emmering, T. E. (1966). A method for the epidemiological assessment of periodontal health-disease state in beagle hound colony. *Journal of Periodontology*, **37**, 208–213. *528*
Rossman, L. E., Garber, D. A. & Harvey, C. E. (1985). Disorders of teeth. In *Veterinary Dentistry*, chapter 6, edit. C. E. Harvey. W. B. Saunders, Philadelphia, pp. 79–105. *352, 445*
Rovin, S., Costich, E. R. & Gordon, H. A. (1966). The influence of bacteria in the initiation of periodontal disease in germfree and conventional rats. *Journal of Periodontal Research*, **1**, 193–203. *526*
Ruch, T. E. (1959). *Diseases of Laboratory Primates*, W. B. Saunders, New York. *529*
Rudge, M. R. (1970). Dental and periodontal abnormalities in two populations of feral goats (*Capra hircus* L.) in New Zealand. *New Zealand Journal of Science and Technology*, **13**, 260–267. *115, 493*
Ruprecht, A. L. (1965a). A case of the bilateral traumatic regeneration of the *Ossis intermaxillares* in the red deer (*Cervus elaphus* Linnaeus, 1758). *Acta Theriologica*, **10**, 237–242. *375*
Ruprecht, A. L. (1965b). Supernumerary premolar in *Mustela putorius*. *Acta Theriologica*, **10**, 242. *72*
Ruprecht, A. L. (1965c). Anomalies of the teeth and asym-

metry of the skull in *Erinaceus europaeus* Linnaeus, 1758. *Acta Theriologica*, **10**, 234–236. *138*

Ruprecht, A. L. (1965d). An abnormal skull of the European hare, *Lepus europaeus* Pallas, 1778. *Acta Theriologica*, **10**, 237–239. *319*

Ruprecht, A. L. (1978a). Überzahlige Praemolar bei der Waldmaus, *Apodemus sylvaticus* (Linnaeus, 1758). *Säugetierkundliche Mitteilungen*, **26**, 79. *131*

Ruprecht, A. L. (1978b). Überzahlige Schneidezähne bei der Breitflügelfledermaus *Eptesicus serotinus* (Schreber, 1774). *Säugetierkundliche Mitteilungen*, **26**, 235–236. *135*

Ruprecht, A. L. (1978c). Dental variations in the common polecat in Poland. *Acta Theriologica*, **23**, 239–245. *71*

Rushton, M. A. (1937). A collection of dilated composite odontomes. *British Dental Journal*, **63**, 65–86. *576, 577*

Rushton, M. A. (1951). The epithelial downgrowth on the molar roots of golden hamsters. *British Dental Journal*, **90**, 87–93. *527*

Rushton, M. A. (1955). Dental effects of dietary aureomycin. *British Dental Journal*, **98**, 313–317. *527*

Ruysch, F. (1719). *Thesaurus Anatomicus Decimus*. *Amstelodami*. Tab. II, Figs. 7 and 8, p. 76, cited by Busch (1890a). *421*

Ryel, L. A. (1963). The occurrence of certain anomalies in Michigan white-tailed deer. *Journal of Mammalogy*, **44**, 79–98. *112*

Saffar, J. L., Sagroun, B., de Tessieres, C. & Makris, G. (1981). Osteoporotic effect of a high-carbohydrate diet (Keyes 2000) in golden hamsters. *Archives of Oral Biology*, **26**, 393–397. *528*

Sahlertz, I. (1877–1878). Om nogle anomalier i saelernes tandsaet. *Videnskabelige Meddelelser Kjobenhavn*, 1877–1878, 275–304. *96*

Sakai, T. (1982). Morphogenesis of molar cusps and tubercles in certain primates. In *Teeth: Form, Function and Evolution*, chapter 22, edit. B. Kurtén. Columbia University Press, New York, pp. 307–322. *20*

Salisbury, R. M., Armstrong, M. C. & Gray, K. G. (1953). Ulcero-membranous gingivitis in sheep. *New Zealand Veterinary Journal*, **1**, 51–52. *562*

Sallum, A. W., do Nacimento, A., Bozzo, L. & de Toledo, S. (1976). The effect of dexamethasone in traumatic changes of the periodontium of marmosets (*Callithrix jacchus*). *Journal of Periodontology*, **47**, 63–66. *528*

Samuel, J. L. (1983). Jaw disease in macropod marsupials: bacterial flora isolated from lesions and from mouths of affected animals. *Veterinary Microbiology*, **8**, 373–387. *520, 563*

Samuel, J. L. & Woodall, P. F. (1988). Periodontal disease in feral pigs (*Sus scrofa*) from Queensland, Australia. *Journal of Wildlife Diseases*, **24**, 201–206. *494, 550*

Santoné, P. (1937). Über die Folgen von umschriebenen traumatischen Verletzungen in Gewebe der Zahnlage. *Deutsche Zahn-, Mund- und Kieferheilkunde*, **4**, 323–337 and 602–614. *404, 439*

Saxe, S. R., Greene, J. C., Bohannan, H. M. & Vermillion, J. R. (1967). Oral debris, calculus and periodontal disease in the beagle dog. *Periodontics*, **5**, 217–225. *528*

Scammon, C. M. (1874). *The Marine Mammals of the North-western Coast of North America*. J. J. Carmany and Co., New York, p. 79. *390*

Schamschula, R. G., Keyes, P. H. & Hornabrook, R. W. (1972). Root surface caries in Lufa, New Guinea. 1. Clinical observations. *Journal of the American Dental Association*, **85**, 603–608. *457*

Schaurte, V. W. T. (1965). Beitrage zur Kenntnis des Gebisses und Zahnbaues der afrikanischen Nashörner. *Säugetierkundliche Mitteilungen*, **13**, 327–341. *124*

Scheuring, W. & Scheuring, R. (1982). Dental caries in nutria (*Myocastor coypus*, Mol.) from closed husbandries. *Medycyna Weterynaryjna*, **38**, 363–365 (Polish with English summary.) *484*

Schitoskey, F. (1971). Anomalies and pathological conditions in the skulls of nutria from southern Louisiana. *Mammalia*, **35**, 311–314. *131, 483*

Schlup, D. (1982). Epidemiologische und morphologische Untersuchungen am Katzengebiss. 1. Mitteilung: Epidemiologische Untersuchungen. *Kleintier-praxis*, **27**, 87–94. *435, 505, 543*

Schlup, D. & Stich, H. (1982). Epidemiologische und morphologische Untersuchungen am Katzengebiss. II. Mitteilung: morpologische Untersuchungen der 'neck lesions'. *Kleintierpraxis*, **27**, 179–188. *505*

Schneck, G. W. (1967). Caries in the dog. *Journal of the American Veterinary Medical Association*, **150**, 1142–1143. *479*

Schneck, G. W. & Osborn, J. W. (1976). Neck lesions in the teeth of cats. *The Veterinary Record*, **99**, 100. *505*

Schneider, K. M. (1959). Zum Zahndurchbruch des Löwen (*Panthera leo*) nebst Bemerkungen über das Zahnen einiger andere Grosskatzen und der Hauskatze (*Felis catus*). *Der Zoologische Garten*, **22**, 240–361. *339, 340*

Schour, I., Bhaskar, S. N., Greep, R. O. & Weinmann, J. P. (1949). Odontome-like formations in a mutant strain of rats. *American Journal of Anatomy*, **85**, 73–111. *605*

Schuh, E. & Niebauer, G. W. (1982). Zahn- und Zahnbetterkrankungen beim alternden Eich (*Alces alces*). Ein Beitrag zur vergleichenden Parodontologie. *Zeitschrift für Jagdwissenschaften*, **28**, 123–130. *562*

Schultz, A. H. (1925). Studies on the evolution of the human teeth. *The Dental Cosmos*, **67**, 1053–1063.
53, 55, 56, 57, 58, 476, 511

Schultz, A. H. (1935). The eruption and decay of the permanent teeth in primates. *American Journal of Physical Anthropology*, **19**, 489–581.
163, 164, 179, 190, 213, 217, 222, 229, 335, 336, 337, 338, 342, 465, 472, 507, 510, 529, 530, 531, 537, 538, 539

Schultz, A. H. (1939). Notes on diseases and healed fractures of wild apes and their bearing on the antiquity of pathological conditions in man. *Bulletin of the History of Medicine*, **7**, 571–582. *371, 463, 507*

Schultz, A. H. (1940). Growth and development of the chimpanzee. *Carnegie Institute of Washington, Publication 518, Contributions to Embryology*, **28**, 1–63.
335, 463, 507

Schultz, A. H. (1941). Growth and development of the orangutan. *Contributions to Embryology*, **29**, 57–110. Carnegie Institution of Washington Publication.
29, 30, 31, 336, 387, 463, 510, 536

Schultz, A. H. (1942). Growth and development of the proboscis monkey. *Bulletin of the Museum of Comparative Zoology at Harvard College*, **89**, no. 6, 279–314. *336, 338*

Schultz, A. H. (1944). Age changes and variability in gibbons. *American Journal of Physical Anthropology*, New Series, **2**, 1–129.
20, 22, 31, 32, 33, 177, 336, 387, 392, 463, 507

Schultz, A. H. (1950). Morphological observations on gorillas. In *The Anatomy of the Gorilla*, part 5, edit. W. K. Gregory. Columbia University Press, New York, pp. 227–251. *23, 164, 165, 335, 463, 465, 529, 530*

Schultz, A. H. (1956). The occurrence and frequency of pathological and teratological conditions and of twinning among non-human primates. *Primatologia*, **1**, 965–1014.
20, 28, 29, 30, 31, 53, 58, 61, 165, 167, 371, 392, 393, 463, 464, 465, 466, 469, 474, 475, 507, 509, 510, 511, 529, 533, 536, 537, 538

Schultz, A. H. (1958). Cranial and dental variability in *Colobus* monkeys. *Proceedings of the Zoological Society of London*, **130**, 70–105.
35, 37, 38, 41, 178, 180, 337, 392, 473, 509, 540, 541

Schultz, A. H. (1960). Age changes and variability in the skulls and teeth of Central American monkeys *Alouatta*, *Cebus* and *Ateles*. *Proceedings of the Zoological Society of London*, **133**, 337–390.
52, 53, 55, 57, 58, 222, 226, 227, 338, 388, 474, 475, 509, 511, 541

Schultz, A. H. (1964). A gorilla with exceptionally large teeth and supernumerary premolars. *Folia Primatologica*, **2**, 149–160. *23, 24*

Schultze, C. (1970). Developmental abnormalities of the teeth and jaws. In *Thoma's Oral Pathology*, chapter 3, 6th edn., edit. R. J. Gorlin & H. M. Goldman. C. V. Mosby Co., St. Louis, pp. 96–183. *440, 444*

Schulze, W. (1957). Fraktur eines Stosszahnes bei einem Indischen Elefantenbullen. *Der Zoologische Garten*, **23**, 13–18. *406*

Schumacher, S. (1929). Uber das 'Wandern' des Zähne bei Gemse und Reh. *Vierteljahrsscrift fur Zahnheilkunde*, **45**, 505–530. *157*

Schuman, E. L. & Sognnaes, R. F. (1956). Developmental microscopic defects on the teeth of subhuman primates. *American Journal Physical Anthropology*, **14**, 193–214.
439, 442, 459, 463

Schürer, U. (1980). Unterschiede in der Häufigkeit von Kiefererkrankungen bei Riesenkänguruhs. *Der Zoologische Garten*, **50**, 433–437. *518*

Schwanitz, G., Kreft, K. & Fleischer-Peters, A. (1983). Familiäre Zahnstellungsanomalien beim Orang-Utan. *Fortschritte der Kieferorthopädie*, **44**, 172–183. *535*

Schwann, H. (1906). A list of mammals obtained by Messrs. R. B. Woosnam and R. E. Dent in Bechuanaland. *Proceedings of the Zoological Society of London*, 1906, 101–111. *131*

Schwartz, J. H. (1982). Methodological approach to heterodonty and homology. In *Teeth: Form, Function and Evolution*, chapter 10, edit. B. Kurten. Columbia University Press, New York, pp. 123–144. *12*

Schwartz, J. H. (1983). Premaxillary–maxillary suture asymmetry in a juvenile *Gorilla*. *Folia primatolica*, **40**, 69–82. *12*

Schwartz, J. H. (1984). Supernumerary teeth in anthropoid primates and models of tooth development. *Archives of Oral Biology*, **29**, 833–842. *4, 6, 30*

Sciaky, I. & Ungar, H. (1961). Osteolathyrism in the incisors of rats. *Annals of Dentistry*, **20**, 42–50. *333*

Sciulli, P. W., Doyle, W. J., Kelley, C., Siegel, P. & Siegel, M. I. (1979). The interaction of stressors in the induction of fluctuating asymmetry in the laboratory rat. *American Journal of Physical Anthropology*, **50**, 279–284. *10*

Sclater, P. L. (1871). On the tusks of *Elephas indicus*. *Proceedings of the Zoological Society of London*, 1871, 145–146. *504*

Scott, H. H. (1925). A streptothrix disease of wallabies. *Proceedings of the Zoological Society of London*, 1925, 799–814. *519*

Searle, A. G. (1954). Genetical studies on the skeleton of the mouse. XI. The influence of diet on variation within pure lines. *Journal of Genetics*, **52**, 413–24. *131*

Seawright, A. A., English, P. B. & Gartner, R. J. W. (1970). Hypervitaminosis A of the cat. *Advances in Veterinary Science and Comparative Medicine*, **14**, 1–27. *505*

Seawright, A. A. & Hrdlicka, J. (1974). Pathogenetic factors in tooth loss in young cats on a high daily oral intake of vitamin A. *Australian Veterinary Journal*, **50**, 133–141. *544*

Sedgwick, C. J. & Cooper, R. W. (1972). Dental fistula in the squirrel monkey (*Saimiri sciureus*). *Journal of Zoo Animal Medicine*, **3**(2), 26–27. *511*

Seddon, H. R. (1945). Chronic endemic dental fluorosis in sheep. *Australian Veterinary Journal*, **21**, 2–8. *450, 451*

Selenka, E. (1898). Rassen, Schädel und Bezahnung des Orangutan. Menschenaffen (Anthropomorphae). *Studien über Entwicklung und Schädelbau*, vol. 1. C. W. Kreidels Verlag, Wiesbaden, pp. 1–91.
20, 29, 30, 336, 440, 441, 463, 464

Selenka, E. (1899). Schädel des Gorilla und Schimpanse. Menschenaffen (Anthropomorphae). *Studien uber Entwicklung und Schadelbau*, vol. II. C. W. Kreidels Verlag, Wiesbaden, pp. 93–160. *20, 23, 335, 336*

Seliger, W. G., Erickson, J. A. & Denney, R. N. (1969). Histology and anatomy of the wapiti canine. *Journal of Mammalogy*, **50**, 350–355. *111*

Selye, H. (1957). Lathyrism. *Revue Canadienne de Biologie*, **1**, 1–81. *332*

Setzer, H. W. (1957). An extra tooth in *Crocidura*. *Journal of Mammalogy*, **38**, 258–259. *139*

Severinghaus, C. W. & Cheatum, E. L. (1956). Life and times of the white-tailed deer. In *The Deer of North America*, edit. W. P. Taylor. Stackpole Co., Harrisburg, Pennsylvania, pp. 57–186. *288*

Shabestari, L., Taylor, G. N. & Angus, W. (1967). Dental eruption pattern of the beagle. *Journal of Dental Research*, **46**, 276–278. *341*

Shah, R. M. (1979). Current concepts on the mechanism of normal and abnormal secondary palate formation. In *Advances in the Study of Birth Defects, vol. 1, Teratogenic Mechanisms*, chapter 5, edit. T. V. N. Persaud, M. T. P. Press, Lancaster, pp. 69–84. *10*

Shadle, A. R., Ploss, W. R. & Marks, E. M. (1944). The extrusive growth and attrition of the incisor teeth of *Erethizon dorsatum*. *Anatomical Record*, **90**, 337–341. *356*

Shahan, M. S. & Davis, C. L. (1942). The diagnosis of actinomycosis and actinobacillosis. *American Journal of Veterinary Research*, **3**, 321–329. *562*

Shaler, N. S. (1873). Notes on the right and sperm whales. *The American Naturalist*, **7**, 1–4. *389*

Shalla, C. L. (1972). Preliminary evidence of periodontal disease in *Lemmus trimucronatus* skulls from northern Alaska. *Journal of Dental Research*, **51**, 1075–1079. *568*

Sharawy, A. M. & Lobene, R. R. (1968). Invaginated odontome in hamster's maxillary molar. *Oral Surgery*, **26**, 679–684. *590*

Shaw, I. G. (1981). Tooth loss in sheep. In *The Veterinary*

Annual, edit. G. S. G. Grunsell & F. W. G. Hill. Scientechnica, Bristol, pp. 126–133. *560*

Shaw, J. C. M. (1927). Four cases of fourth molar teeth in South African baboons. *Journal of Anatomy, London*, **62**, 79–85. *49*

Shaw, J. E., Griffiths, D. & Osterholtz, M. (1963). Relationship between body weight and occurrence of the fused molar and supernumerary molar traits in the rice rat. *Archives of Oral Biology*, **8**, 777–779. *133*

Shaw, J. H. (1949). Vitamin C deficiency in the ringtail monkey. *Federation Proceedings*, **8**, 396. *528*

Shaw, J. H. & Auskaps, A. M. (1954). Studies in the dentition of the marmoset. *Oral Surgery, Oral Medicine and Oral Pathology*, **7**, 671–677. *542*

Shaw, J. H., Elvehjem, C. A. & Phillips, P. H. (1945). Survey of the incidence of dental caries in the rhesus monkey. *Journal of Dental Research*, **24**, 129–136. *472*

Shaw, J. H., Schweigert, B. S., Elvehjem, C. A. & Phillips, P. H. (1944). Dental caries in the cotton rat. II. Production and description of the carious lesions. *Journal of Dental Research*, **23**, 417–425. *461*

Shaw, J. H., Shaffer, N. M. & Soldan, L. W. (1950). The postnatal development of the molar teeth in the cotton rat. *Journal of Dental Research*, **29**, 197–207. *461*

Shaw, J. H. & Sognnaes, R. F. (1955). Developmental factors in experimental animal caries. In *Advances in Experimental Caries Research*, edit. R. F. Sognnaes. American Association for the Advancement of Science, Washington, pp. 82–106. *459*

Shaw, K. (1981). Unusual tooth wear in elk at Glacier National Park, Montana. *Great Basin Naturalist*, **41**, 368–369. *491*

Shearer, T. R., Kolstad, D. L. & Suttie, J. W. (1978). Electron probe microanalysis of fluoritic bovine teeth. *Australian Journal of Veterinary Research*, **39**, 1393–1398. *453*

Shearer, T. R., Britton, J. L., Desart, D. J. & Suttie, J. W. (1980). Microhardness of molar teeth in cattle with fluorosis. *American Journal of Veterinary Research*, **41**, 1543–1545. *451, 453*

Sheppe, W. (1964). Supernumerary teeth in the deer mouse, *Peromyscus*. *Zeitschrift für Säugetierkunde*, **29**, 33–36. *131*

Sheppe, W. (1965). Periodontal disease in the deer mouse, *Peromyscus*. *Journal of Dental Research*, **44**, 506–508. *568*

Sheppe, W. (1966). Periodontal disease and supernumerary teeth in a population of *Mus musculus*. *Journal of Mammalogy*, **47**, 519–520. *131, 568*

Shermis, S. (1985). Canine fracture avulsion in *Smilodon californicus*. *Bulletin of the Southern California Academy of Sciences*, **84**, 86–95. *435*

Shermis, S. (1985–1986). Alveolar osteitis and other oral diseases in *Smilodon californicus*. *Ossa*, **12**, 187–196. *376*

Shklar, G. (1966). Periodontal disease in experimental animals subjected to chronic cold stress. *Journal of Periodontology*, **37**, 377–383. *528*

Shklar, G., Chauncey, H. H. & Shapiro, S. (1967). The effect of testosterone on the periodontium of normal and hypophysectomized rats. *Journal of Periodontology*, **38**, 203–210. *528*

Shklar, G. & Person, P. (1975). The pocket mouse (*Perognathus longimembris*). Unique model of periodontal disease. *Journal of Periodontology*, **46**, 723–730. *568*

Shlosberg, A., Bartana, U. & Egyed, M. N. (1980). Fluorosis in dairy cattle due to high fluoride rock phosphate supplement. *Fluoride*, **13**, 57–64. *450*

Shore, R. C., Berkovitz, B. K. B. & Moxham, B. J. (1984). Histological study, including ultrastructural quantification, of the periodontal ligament of the lathyritic rat mandibular dentition. *Archives of Oral Biology*, **29**, 263–273. *333*

Short, H. L. (1964). Notes and comments on mandibular malformations in deer. *Journal of Mammalogy*, **45**, 319–321. *288*

Shupe, J. L., Ammerman, C. B. & Peeler, H. T. (1974). *Effects of Fluorides in Animals*. Report of Subcommittee on Fluorosis, Committee on Animal Nutrition, National Research Council, National Academy of Sciences, Washington D.C., pp. 1–70. *451, 452*

Shupe, J. L., Christofferson, P. V., Olson, A. E., Allred, E. S. & Hurst, R. L. (1987). Relationship of cheek tooth abrasion to fluoride-induced permanent incisor lesions in livestock. *American Journal of Veterinary Research*, **48**, 1498–1593. *453*

Shupe, J. L. & Olson, A. E. (1980). Fluoride toxicosis. In *Bovine Medicine and Surgery*, 2nd edn., vol. 1, edit. H. E. Amstutz. American Veterinary Publications, Santa Barbara, California, pp. 475–488. *384, 450*

Shupe, J. L., Olson, A. E., Peterson, H. B. & Low, J. B. (1984). Fluoride toxicosis in wild ungulates. *Journal of the American Veterinary Medical Association*, **185**, 1295–1300. *454*

Shupe, J. L., Olson, A. E. & Sharma, R. P. (1972). Fluoride toxicity in domestic and wild animals. *Clinical Toxicity*, **5**, 195–213. *451, 453*

Shuttleworth, A. C. (1948). Dental diseases of the horse. *The Veterinary Record*, **60**, 563–567. *483*

Sicher, H. & Weinmann, J. P. (1944). Bone growth and physiologic tooth movement. *American Journal of Orthodontics*, **30**, 109–132. *157*

Silver, I. A. (1969). The ageing of domestic animals. In *Science in Archaeology*, 2nd edn., edit. D. Brothwell & E. S. Higgs. Thames & Hudson, London, pp. 283–302. *342, 343, 344*

Silverstone, L. M., Johnson, N. W., Hardie, J. M. & Williams, R. A. D. (1981). *Dental Caries: Aetiology, Pathology and Prevention*. Macmillan, London. *455, 457*

Simpson, D. M. & Avery, B. E. (1974). Histopathologic and ultrastructural features of inflamed gingiva in the baboon. *Journal of Periodontology*, **45**, 500–510. *540*

Simpson, I. H. & Croft, T. J. (1973). Tooth clipping of ewes – no advantage for prime lamb production. *Agricultural Gazette of New South Wales*, **84**, 270–272. *557*

Sinclair, K. J. (1949). The toxicity of fluorine compounds; a review of the literature. In *Industrial Fluorosis; a Study of the Hazard to Man and Animals near Fort William, Scotland. Appendix A*. Medical Research Council Memorandum, no. 22, His Majesty's Stationery Office, London, pp. 97–119. *453, 454*

Sinha, R. K., Singh, S. P. & Singh, S. B. (1980). A deformed specimen of *Catla catla* (Ham). *Journal of the Inland Fisheries Society of India*, **12**, 116–117. *159*

Skrentny, T. T. (1964). Preliminary study of the inheritance of missing teeth in the dog. *Wiener Tierärztliche Monatsschrift*, **51**, 231–244. *86*

Slaughter, B. H., Pine, R. H. & Pine, N. E. (1974). Eruption of cheek teeth in Insectivora and Carnivora. *Journal of Mammalogy*, **55**, 115–125. *339, 341, 342, 346*

Slavkin, H. G. (1974). Embryonic tooth formation; a tool for developmental biology. *Oral Sciences Review*, **4**, 7–136. 5

Slipjer, E. J. (1979). *Whales*, 2nd edn. Hutchinson, London. 101, 390

Slots, J. & Genco, R. J. (1984). Black-pigmented *Bacteroides* species, *Capnocytophaga* species, and *Actinobacillus actinomycetemcomitans* in human periodontal disease: virulence factors in colonization, survival, and tissue destruction. *Journal of Dental Research*, **63**, 412–421. 524

Smith, M. A., Bellairs, A. d'A. & Miles, A. E. W. (1953). Observations on the premaxillary dentition of snakes with special reference to the egg-tooth. *Linnean Society's Journal – Zoology*, **42**, 260–268. 9

Smith, J. D., Genoways, H. & Jones, J. K. (1977). Cranial and dental anomalies in three species of platyrrhine monkeys from Nicaragua. *Folia Primatologica*, **28**, 1–42. 52, 53, 55, 56, 57, 58, 215, 217, 222, 224, 227, 476, 493, 505, 511, 541.

Smith, H. C. (1984). Malocclusion of incisor teeth in a red squirrel, *Tamiasciurus hudsonicus*. *The Canadian Field-Naturalist*, **98**, 506–507. 358

Smith, M. M., Zontine, W. J. & Willits, N. H. (1985). A correlative study of the clinical and radiographic signs of periodontal disease in dogs. *Journal of the American Veterinary Medical Association*, **186**, 1286–1290. 548

Smith, P., Soskolne, W. A. & Ornoy, A. (1978). Morphological and histological changes in the developing dentition of aborted human foetuses with a maternal history of rubella. In *Development, Function and Evolution of Teeth*, Chapter 12 edit. P. M. Butler & K. A. Joysey. Academic Press, London, pp. 149–156. 10

Sofaer, J. A. (1969). Aspects of the tabby-crinkled-downless syndrome. I. The development of tabby teeth. *Journal of Embryology and Experimental Morphology*, **22**, 181–205. 3, 9

Sofaer, J. A. (1975). Genetic variation and tooth development. *British Medical Bulletin*, **31**, 107–110. 2, 10

Sofaer, J. A. (1976). The influence of heredity. In *Scientific Foundations of Dentistry*, edit. B. Cohen & I. R. H. Kramer. Heineman, London, pp. 1–12. 2, 10

Sofaer, J. A. (1977). Coordinated growth of successively initiated tooth germs in the mouse. *Archives of Oral Biology*, **22**, 71–72. 3

Sofaer, J. A. (1978). Morphogenetic influences and patterns of developmental stability in the mouse vertebral column. In *Development, Function and Evolution of Teeth*, Chapter 16, edit. P. M. Butler & K. A. Joysey. Academic Press, London, pp. 215–227. 10

Sofaer, J. A. (1981). Racial differences of tooth morphology. In *A Companion to Dental Studies. Dental Anatomy and Embryology*, vol. 1, book 2, edit. J. W. Osborn. Blackwell Scientific Publications, Oxford, pp. 151–154. 19

Sofaer, J. A., MacLean, C. J. & Bailit, H. L. (1972). Heredity and morphological variation in early and late developing human teeth of the same morphological class. *Archives of Oral Biology*, **17**, 811–816. 3

Sofaer, J. A. & Shaw, J. H. (1971). The genetics and development of fused and supernumerary molars in the rice rat. *Journal of Embryology and Experimental Morphology*, **26**, 99–109. 9, 133

Sokoloff, L. & Zipkin, I. (1967). Odontogenic hamartomas in an inbred strain of mouse (STR/IN). *Proceedings of the Society of Experimental Biology and Medicine*, **124**, 147–149. 605

Sondaar, P. Y. (1977). Insularity and its effect on mammal evolution. In *Major Patterns in Vertebrate Evolution*, edit. M. K. Hecht, P. C. Goody & B. M. Hecht. Plenum Publishing, New York, pp. 671–707. 10

Sorenson, W. P., Löe, H. & Ramfjord, S. P. (1980). Periodontal disease in the beagle dog. A cross sectional clinical study. *Journal of Periodontal Research*, **15**, 380–389. 549

Southwick, C. H. (1954). Canine teeth in a Wisconsin white-tailed deer fawn. *Journal of Mammalogy*, **35**, 456–457. 111

Spais, A., Lazaridis, T., Papasteriadis, A. & Leontides, S. (1969). Study on an out-break of paradontal disease in sheep. *Hellenic Veterinary Medicine*, **12**, 54–56. 562

Spaul, E. A. (1964). Deformity in the lower jaw of the sperm whale (*Physeter catodon*). *Proceedings of the Zoological Society of London*, **142**, 391–395. 390, 391

Spence, J. A. (1978). Functional morphology of the periodontal ligament in the incisor region of the sheep. *Research in Veterinary Science*, **25**, 144–151. 554, 558

Spence, J. A. & Aitchison, G. U. (1985). Early tooth loss in sheep: a review. *The Veterinary Annual*, **25**, 125–133. 552

Spence, J. A., Aitchison, G. U., Sykes, A. R. & Atkinson, P. J. (1980). Broken mouth (premature incisor loss) in sheep: the pathogenesis of periodontal disease. *Journal of Comparative Pathology*, **90**, 275–292. 557

Spence, J. A., Sykes, A. R., Atkinson, P. J. & Aitchison, G. U. (1985). Skeletal and blood biochemical characteristics of sheep during growth and breeding: a comparison of flocks with and without broken mouth. *Journal of Comparative Pathology*, **95**, 505–524. 560

Spence, J. A., Hooper, G. E. & Austin, A. R. (1986). Trimming incisor teeth of sheep. *The Veterinary Record*, **119**, 617. 556

Spencer, W. G. (1889). A disease of the bones of goats, having points of resemblance to mollities ossium, osteitis deformans, and multiple sarcoma of bone. *Transactions of the Pathological Society of London*, **40**, 449–457. 603, 604

Spinage, C. A. (1971). Two records of pathological conditions in the impala (*Aepyceros melampus*). *Journal of Zoology, London*, **164**, 269–270. 372

Spouge, J. D. (1984). A method of schematic three-dimensional reconstruction for studying the gross morphology of epithelial residues in periodontal ligament. *Archives of Oral Biology*, **29**, 253–255. 551

Sprawson, E. (1927). Further investigations of the pathology of dentigerous cysts with a new treatment based thereon. *Proceedings of the Royal Society of Medicine*, **20**, section on Odontology, 67–78. 576

Sprawson, E. (1937). Odontomes. *British Dental Journal*, **63**, 177–202. 6, 575, 576, 579, 602

Stager, K. E. (1943). Remarks on *Myotis occultus* in California. *Journal of Mammalogy*, **24**, 197–199. 135

Stannus, H. S. (1911). Diseases of elephants' tusks. *Lancet*, **1**, 617. 405

Stanton, A. M. (1949). Four tusked elephant from 'Bulletin du Corps des Lieutenants Honoraires de Chasse'. *Sudan Wild Life and Sport*, **1**, 13. 126

St Clair, L. E. & Jones, N. D. (1957). Observations on the cheek teeth of the dog. *Journal of the American Veterin-*

ary Medical Association, **130**, 275–279. *88, 266*

Steele, D. G. & Parama, W. (1979). Supernumerary teeth in moose and variations in tooth number in North American Cervidae. *Journal of Mammalogy*, **60**, 852–854. *111*

Stegeman, L. C. (1956). Tooth development and wear in *Myotis*. *Journal of Mammalogy*, **37**, 58–63. *135*

Stein, G. H. W. (1963). Anomolien des Zahnzahl und ihre geographische Variabilität bei Insectivora: I. Maulwurf, *Talpa europaea* L. *Mitteilungen der Zoologischer Museum, Berlin*, **39**, 223–244. *139*

Stephan, R. M. & Harris, M. R. (1955). Location of experimental caries on different tooth surfaces in the Norway rat. In *Advances in Experimental Caries Research*, edit. R. F. Sognnaes. American Association for the Advancement of Science, Washington, pp. 47–65. *462*

Stieve, V. H. (1941). Über erbliche Unterkieferanomalien beim Reh (*Capreolus capreolus capreolus*). *Zoologischer Anzeiger*, **133**, 1–19. *288*

Stirling, I. (1969). Tooth wear as a mortality factor in the Weddell seal, *Leptonychotes weddelli*. *Journal of Mammalogy*, **50**, 559–565. *498, 499, 512*

Stockard, C. R. (1941). Achrondroplasia of the extremities. In *The Genetic and Endocrine Basis for Differences in Form and Behavior*, edit. C. R. Stockard & collaborators. The Wistar Institute of Anatomy and Biology, Philadelphia, pp. 45–69. *268*

C. R. Stockard & collaborators (1941). *The Genetic and Endocrine Basis for Differences in Form and Behavior*. The Wistar Institute of Anatomy and Biology, Philadelphia. *266*

Stockard, C. R. & Johnson, A. L. (1941). The bulldog achondroplasic skull. In *The Genetic and Endocrine Basis for Differences in Form and Behavior*, edit. C. R. Stockard & collaborators. The Wistar Institute of Anatomy and Biology, Philadelphia, pp. 272–288. *266, 268*

Stockhaus, K. (1962). Zur Formenmannigfaltigkeit von Haushundschadeln. *Zeitschrift für Tierzücht und Züchtungsbiologie*, **77**, 223–228. *88*

Stones, H. H. (1938). An experimental investigation into the association of traumatic occlusion with parodontal disease. *Proceedings of the Royal Society of Medicine*, **31**, 479–495. *528*

Strauss, G. (1985). Gebissanomalien bei allen drei Nachkommen zweier Gehaubter Kapuziner (*Cebus apella*). *Milu, Berlin*, **6**, 218–223. *218*

Stuart, C. T. (1984). Abnormal dental development in male Hyrax, *Procavia capensis capensis* (Hyracoidea: Procaviidae). *Säugetierkundliche Mitteilungen*, **31**, 268–269. *363*

Studer, E. & Stapley, R. B. (1973). The role of dry foods in maintaining healthy teeth and gums in the cat. *Veterinary Medicine/Small Animal Clinician*, **68**, 1124–1126. *544*

Suckling, G. W. (1975). A photographic record of the progress of periodontal disease in one sheep. *New Zealand Journal of Agricultural Research*, **18**, 131–132. *295, 557*

Suckling, G. W. (1979). Mineralization of the enamel of ovine permanent central incisor teeth using microhardness and histological techniques. *Calcified Tissue International*, **28**, 121–129. *450*

Suckling, G. W. (1980). Defects of enamel in sheep resulting from trauma during tooth development. *Journal of Dental Research*, **59**, 1541–1548. *439*

Suckling, G. W., Elliott, D. C. & Thurley, D. C. (1983). The production of developmental defects of enamel in the incisor teeth of penned sheep resulting from induced parasitism. *Archives of Oral Biology*, **28**, 393–399. *450*

Suckling, G. W., Elliott, D. C. & Thurley, D. C. (1986). The macroscopic appearance and associated histological changes in the enamel organ of hypoplastic lesions of sheep incisor teeth resulting from induced parasitism. *Archives of Oral Biology*, **31**, 427–439. *450*

Suckling, G. W. & Purdell-Lewis, D. (1982). The macroscopic appearance, microhardness and microradiographic characteristics of experimentally produced fluorotic lesions in sheep enamel. *Caries Research*, **16**, 227–234. *453*

Suttie, J. S., Dickie, R., Clay, A. B., Nielsen, P., Mahan, W. E., Baumann, D. P. & Hamilton, R. J. (1987). Effects of fluoride emissions from a modern primary aluminium smelter on a local population of white-tailed deer (*Odocoileus virginianus*). *Journal of Wildlife Diseases*, **23**, 135–143. *454*

Suttie, J. W., Hamilton, R. J., Clay, A. C. Tobin, M. L. & Moore, W. G. (1985). Effect of fluoride ingestion on white-tailed deer (*Odocoileus virginianus*). *Journal of Wildlife Diseases*. **21**, 283–288. *451*

Svihla, A. (1957). Dental caries in the Hawaiian dog. *Occasional Papers of Bernice P. Bishop Museum, Honolulu*, **22**(2), 7–13. *479*

Swan, M. A. (1968). Double-mouth deformity in a trout (*Salmo trutta*) and its cause. *Journal of Zoology, London*, **156**, 449–455. *601*

Swindler, D. R. (1976). *Dentition of Living Primates*. Academic Press, London and New York. *177*

Swindler, D. R. (1978). *The Teeth of Primates*. Carolina Biological Supply Co., Burlington, N. Carolina. *18, 19, 180*

Swindler, D. R., Gaven, J. A. & Turner, W. M. (1963). Molar tooth size variability in African monkeys. *Human Biology*, **35**, 104–122. *35*

Swindler, D. R. & Sassouni, V. (1962). Open bite and thumb sucking in rhesus monkeys. *The Angle Orthodontist*, **32**, 27–37. *205*

Symons, N. B. B. (1951). Studies on the growth and form of the mandible. *Dental Record*, **71**, 41–53. *156*

Talbot, E. S. (1899). *Interstitial Gingivitis or so-called Pyorrhea Alveolaris*. S. S. White, Philadelphia. *547*

Talbot, E. S. (1913). *Interstitial Gingivitis and Pyorrhea Alveolaris*. Ransom and Randolph, Toledo, pp. 125–150. *547*

Talbot, L. M. & Talbot, M. H. (1963). The wildebeest in western Masailand, East Africa. *Wildlife Monographs*, **12**, 88 pp. *115*

Talent, L. G. (1975). Pugheadedness in the longspine combfish *Zaniolepus latipinnis*, from Monterey Bay, California. *California Fish and Game*, **61**, 160–162. *159*

Tanzer, J. M. (ed.) (1981). *Animal Models in Cariology. Proceedings of the Symposium on Animal Models in Cariology. Special Supplement Microbiology Abstracts*. Information Retrieval Inc., Washington D.C. *462*

Taubman, M. A., Buckelew, J. M., Ebersole, J. L. & Smith, D. J. (1981). Periodontal bone loss and immune response to ovalbumin in germ-free rats fed antigen-free diets with ovalbumin. *Infection and Immunity*, **32**, 145–152. *526*

Taylor, F. J. (1965). Supernumerary incisors in *Lynx rufus* Schreber. *Journal of Mammalogy*, **46**, 507. *66*

Taylor, R. M. S. (1986). Some unusual dental conditions in sheep. *Acta Theriologica*, **31**, 552–556. *117*

Taylor, W. P. (1956). *The Deer of North America. The White-tailed, Mule and Black-tailed Deer, Genus Odocoileus*. The Wildlife Management Institute, Washington, D.C. *491*

Tchernov, E. (1984). Commensal animals and human sedentism in the Middle East. In *Animals and Archaeology: 3. Early Herders and their Flocks*, edit. J. Clutton-Brock & C. Grigson. *British Archaeological Reports, Oxford, International Series*, **202**, 91–115. *131*

Tegetmeier, W. B. (1901). Singular dentition in a fox. *The Field*, No. 2511, 189. *90*

Ten Cate, A. R. (1972). Developmental aspects of the periodontium. In *Developmental Aspects of Oral Biology*, chapter 14, edit. H. C. Slavkin & L. A. Bavetta. Academic Press, New York, pp. 309–324. *5*

Ten Cate, A. R. & Mills, C. (1972). The development of the periodontium. *Anatomical Record*, **173**, 69–78. *5*

Tharp, V. L., Amstutz, H. E. & Helwig, J. H. (1980). Actinomycosis and actinobacillosis. In *Bovine Medicine and Surgery*, 2nd edn., vol. 1, edit. H. E. Amstutz. American Veterinary Publications, Santa Barbara, California, pp. 255–260. *515, 516, 562*

Thilander, H. (1961). Periodontal disease in the white rat. Experimental studies with special reference to some aetiological and pathogenic features. *Transactions of the Royal Schools of Dentistry, Stockholm and Umeå*, **6**, 1–99. *526*

Thomas, O. (1888a). *Catalogue of the Marsupialia and Monotremata in the Collection of the British Museum (Natural History)*. British Museum (Natural History), London. *142, 143, 146, 147*

Thomas, O. (1888b). On the homologies and succession of the teeth in the Dasyuridae, with an attempt to trace the history of the evolution of mammalian teeth in general. *Philosophical Transactions of the Royal Society of London*, **178**, 443–462. *146*

Thomson, G. (1905). Discussion of J. F. Colyer's paper. *Transactions of the Odontological Society of Great Britain*, New Series, **38**, 76. *567*

Thomson, J. H. (1867). Letter to Dr J. E. Gray. *Proceedings of the Zoological Society of London*, 1867, 246–247. *391*

Throckmorton, G. S. (1979). The effect of wear on the cheek teeth and associated dental tissues of the lizard *Uromastix aegypticus* (Agamidae). *Journal of Morphology*, **160**, 195–208. *524*

Thurley, D. C. (1984). The pathogenesis of excessive wear in the deciduous teeth of sheep. *New Zealand Veterinary Journal*, **32**, 25–29. *492*

Thurley, D. C. (1985). Erosion of the non-occlusal surfaces of sheep's deciduous teeth. *New Zealand Veterinary Journal*, **33**, 157–158. *505*

Thurley, D. C. (1987). Gingivitis around the deciduous teeth of sheep. *Journal of Comparative Pathology*, **97**, 375–383. *552*

Tims, H. W. M. & Henry, C. B. (1923). *A Manual of Dental Anatomy, Human and Comparative*. J. A. Churchill, London. *xii*

Todd, T. W., Wharton, R. E. & Todd, A. W. (1938). The effect of thyroid deficiency upon bodily growth and skeletal maturation in the sheep. *American Journal of Anatomy*, **63**, 37–78. *334*

Toller, P. A. (1948). Experimental investigation into factors concerning the growth of cysts of the jaws. *Proceedings of the Royal Society of Medicine*, **41**, 681–688. *577*

Toller, P. A. (1967). Origin and growth of cysts of the jaws. *Annals of the Royal College of Surgeons*, **40**, 306–336. *577*

Tomes, C. S. (1872). Description of an odontome. *Transactions of the Odontological Society of Great Britain*, New Series, **4**, 103–109. *578, 597*

Tomes, C. S. (1873). On a case of abscess of the pulp in a grampus (*Orca gladiator*). *Transactions of the Odontological Society of Great Britain*, New Series, **5**, 39–46. *493, 501, 502, 512*

Tomes, C. S. (1876). *A Manual of Dental Anatomy, Human and Comparative*. J. & A. Churchill, London. *504*

Tomes, C. S. (1877). Notes upon the condition of an elephant's molar which had been injured by a rifle-ball. *Transactions of the Odontological Society of Great Britain*, New Series, **9**, 89–95. *432, 434*

Tomes, C. S. (1898). Partial suppression of teeth in a very hairy monkey (*Colobus caudatus*). *Transactions of the Odontological Society of Great Britain*, New Series, **30**, 30–35. *38, 40, 443*

Tomes, C. S. (1904). *A Manual of Dental Anatomy*. J. & A. Churchill, London. *13*

Tomes, J. (1848). *A Course of Lectures on Dental Physiology and Surgery*. Parker, London, p. 245. *426*

Tomes, J. (1859). *A System of Dental Surgery*. J. & A. Churchill, London, p. 244. *8*

Tomlinson, A. R. & Gooding, C. G. (1954). A kangaroo disease. Investigations into 'lumpy jaw' on the Murchison, 1954. *Journal of the Department of Agriculture, Western Australia*, **3**, 715–718. *519, 520*

Tomlinson, T. H. (1939). Oral pathology in monkeys in various experimental dietary deficiencies. *Public Health Reports, United States Public Health Service*, **54**, 431–439. *528*

Tomlinson, T. H. (1942). Pathology of artificially induced scurvy in the monkey – with and without chronic calcium deficiency. *Public Health Reports, United States Public Health Service*, **57**, 987–993. *528*

Tonna, E. A. (1972). Paradontal inflammation and aging in the laboratory mouse. *Journal of Periodontology*, **43**, 403–410. *527*

Trebbau, P. & Van Bree, P. J. H. (1974). Notes concerning the freshwater dolphin *Inia geoffrensis* (de Blainville, 1817) in Venezuela. *Zeitschrift für Säugetierkunde*, **39**, 50–55. *104*

Treman, A. J. (1907). A dentigerous cyst containing 104 teeth. *American Veterinary Review*, **31**, 618–619. *579*

Tritschler, L. G. & Romack, F. E. (1965). Nocardiosis in equine mandibles associated with bilateral anomalies of the inferior dentition. *Veterinary Medicine/Small Animal Clinician*, **60**, 605–608. *516*

Tucker, R. & Millar, R. (1953). Outbreak of nocardiosis in marsupials in Brisbane Botanical Gardens. *Journal of Comparative Pathology*, **63**, 143–146. *519*

Ueckermann, E. & Scholz, H. (1973). Anomale Backenzahnabnutzung beim Rehwild (*Capreolus capreolus* L.) in Nordrhein-Westfalen. *Zeitschrift für Jagdwissenschaften*, **19**, 142–146. *454*

Underwood, A. S. (1914). Erosion of seals' teeth. *British Journal of Dental Science*, **57**, 519–520. *497, 498*

Unger, H. (1976). Kamele. In *Zootierkrankheiten von Wild-*

tieren im Zoo, Wildpark, Zirkus und in Privathand sowie ihre Therapie, edit. H.-G. Klös & E. M. Lang. Paul Parey, Berlin, pp. 187–194. *516*

Ussher-Smith, J. H., King, G., Pook, G. & Redshaw, M. (1976). Comparative physical development in six hand-reared lowland gorillas, *Gorilla gorilla gorilla*, at Jersey Zoological Park. *The Jersey Wildlife Preservation Trust Thirteenth Annual Report*, pp. 63–70. *335*

Vahlsing, H. L., Kim, S. & Feringa, E. R. (1977). Cyclophosphamide-induced abnormalities in the incisors of the rat. *Journal of Dental Research*, **56**, 809–816. *7*

Van Bree, P. J. H. & Duguy, R. (1970). Sur quelques aberrations pathologiques chez les petits Cétacés. *Der Zoologische Garten*, **39**, 11–15. *384*

Van Bree, P. J. H. & Jansen, F. X. J. (1962). Over drie abnormale hamsterschedels. *Natuurhistorisch Maandblad*, **51**, 44–46. *361*

Van Bree, P. J. H. & Sinkeldam, E. J. (1969). Anomalies in the dentition of fox, *Vulpes vulpes* (Linnaeus, 1758), from continental western Europe. *Bijdragen tot de Dierkunde*, **39**, 3–5. *83, 84, 90, 91, 262*

Van der Gaag, I. & Gruys, E. (1972). Een adamantinoom bij de kat en het rund en twee odontomen bij runderen. (An adamantinoma in the cat and the cow and two odontomes in cows.) *Tijdschrift Diergeneeskunde*, **97**, 22–37. *582*

Van Deusen, H. M. & Jones, J. K. (1967). Marsupials. In *Recent Mammals of the World*, edit. S. Anderson & J. K. Jones. Ronald Press, New York, pp. 61–86. *140*

Van Dissel-Scherft, M. C. & Vervoort, W. (1954). Development of the teeth in fetal *Balaenoptera physalus* (L.). (Cetacea, Mystacoceti). II. *Koninklijka Nederlandse Akademie van Wetenschappen. Series B. Biological and Medical Sciences, Proceedings*, **57**, 203–210. *101*

Van Eaton, J. (1947). A common-cat skull with supernumerary premolar teeth. *The American Midland Naturalist*, **38**, 504–505. *66*

Van Gelder, R. G. & McLaughlin, C. A. (1961). An unusual incisor in *Mephitis mephitis*. *Journal of Mammalogy*, **42**, 422–423. *71*

Van Laar, V. (1980). Extra praemolaren in de Bovenkaak van de bosmuis *Apodemus sylvaticus* (Linnaeus, 1758). *Lutra*, **23**, 12. *131*

Van Riper, D. C., Fineg, J., Day, P. W., Douglas, J. D. & Derwelis, S. K. (1967). Vincent's disease in the chimpanzee. *Journal of the American Veterinary Medical Association*, **151**, 905–906. *534*

Van Valen, L. (1964). Nature of the supernumerary molars of *Otocyon*. *Journal of Mammalogy*, **45**, 284–286. *84*

Van Vuren, D. (1984). Abnormal dentition in the American bison, *Bison bison*. *The Canadian Field Naturalist*, **98**, 366–367. *115*

Van Vuren, D. & Coblentz, B. E. (1988). Dental anomalies on Aldabra Atholl. *Journal of Zoology, London*, **216**, 503–506. *297*

Vaughan, T. A. (1978). *Mammalogy*, 2nd edn. Saunders College Publishing Co., Philadelphia. *140*

Velu, H. (1932). Le darmous (ou dermes); fluorose, spontanée des zones phosphatées. *Archives de l'Institut Pasteur d'Algérie*, **10**, 41–118. *451*

Verme, L. J. (1968). Possible hereditary defects in Michigan white-tailed deer. *Journal of Mammalogy*, **49**, 148. *113*

Verstraete, F. J. M. (1985). Anomalous development of the upper third premolar in a dog and a cat. *Journal of the South African Veterinary Association*, **56**, 131–134. *66, 88*

Vesmanis, I. & Vesmanis, A. (1980). Über eine interesante Zahnanomalie bei der Hausspitzmaus, *Crocidura russula* (Hermann, 1780). *Zoologische Beiträge, Berlin*, **26**, 241–244. *139*

Vierhaus, H. (1980–1981). Zum vorkommen parodontaler Erkrankungen bei mitteleuropäischen Fledermäusen. *Myotis*, **18–19**, 190–196. *570*

Vierhaus, H. (1983). Wie Vampirfledermäus (*Desmodus rotundus*) ihre Zähne schärfen. *Zeitschrift für Säugetierkunde*, **48**, 269–277. *357*

Vigal, C. R. & Machordom, A. (1987). Dental and skull anomalies in the Spanish wild goat, *Capra pyrenaica* Schinz, 1838. *Zeitschrift für Jagdwissenschaften*, **52**, 38–50. *561*

Virchow, H. (1940). Uberzahlige Wangenzahn im Unterkiefen eines Rehes. *Anatomischer Anzeiger*, **89**, 225–238. *112*

Vogel, C. (1966). Morphologische Studien am Geschichtsschädel catarrhiner Primaten. *Bibliotheca Primatologica*, **4**, pp. *1–226*. *49*

von Braunschweig, A. (1974). Komplizierter beiderseitiger Unterkieferbruch bei einem Rothirsch. *Zeitschrift für Jagdwissenschaften*, **20**, 212–214. *383*

von Braunschweig, V. D. (1980). Brachygnathie (Kurzkiefrigheit) bei einem Rehbock. *Zeitschrift für Jagdwissenschaften*, **26**, 43–47. *288*

Von den Driesch, A. (1976). *A Guide to Measurement of Animal Bones from Archaeological Sites*. Peabody Museum Bulletin, **1**, 136 pp. *159*

von Hoffman, G. (1876). Über einige Sectionsbefunde an anthropomorphen Affen. *Sitzungsberichte der Gesellschaft naturforschender Freunde zu Berlin*, pp. 139–152. *535*

Von Lorenz, L. R. (1917). Beitrag zur Kenntnis der Affen und Halbaffen von Zentralafrika. *Annalen des K. K. Naturhistorischen Hofmuseums, Vienna*, **31**, 169–241. *530*

von Metnitz, J. (1903). Osteodentine, vasodentine and abscess cavities in dentine. *Quarterly Circular, London*, Sept. 289–304. *405*

Vosburgh, K. M., Barbiers, R. B., Sikarskie, J. B. & Ullrey, D. E. (1982). A soft versus hard diet and oral health in captive timber wolves (*Canis lupus*). *Journal of Zoo Animal Medicine*, **13**, 104–107. *546*

Voss, H. J. (1937). *Die Zahnfachentzündung des Pferdes*. Ferdinand Enke, Stuttgart. *566, 567*

Walker, G. (1972). The biochemical composition of the cement of two barnacle species, *Balanus hameri* and *Balanus crenatus*. *Journal of the Marine Biological Association (U.K.)*, **52**, 429–435. *499*

Walkhoff, O. (1913). *Die Erdsalze im inhrer Bedeutung fur die Zahnkaries*. Herman Meusser, Berlin, pp. 6–7. *458*

Wallace, J. T. (1968). Analysis of dental variation in wild-caught California house mice. *The American Midland Naturalist*, **80**, 360–380. *131*

Wallace, J. T. & Bader, R. S. (1966). Dental agnesis in wild caught mice. *Journal of Mammalogy*, **47**, 733–734. *131*

Wallach, J. D. (1971). Lumpy jaw in captive kangaroos. *International Zoo Year Book*, **11**, 13. *573*

Wallis, W. D. (1934). A gorilla skull with abnormal denture. *American Naturalist*, **68**, 179–183. *393*

Walthall, J. C. & McKenzie, R. A. (1976). Osteodystrophia fibrosa in horses at pasture in Queensland: field and laboratory observations. *Australian Veterinary Journal*, **52**, 11–16. *604*

Watson, A. D. J. (1981). Nutritional osteodystrophies in dogs. *The Veterinary Annual*, **21**, 209–218. *546*

Watson, L. (1981). *Sea Guide to the Whales of the World*. Hutchinson, London. *369, 389, 435, 500*

Watts, P. S. & McLean, S. J. (1956). Bacteroides infection in kangaroos. *Journal of Comparative Pathology*, **66**, 159–166. *518*

Weaver, M. E. (1964). X-ray diffraction study of calculus of the miniature pig. *Archives of Oral Biology*, **9**, 75–81. *551*

Weaver, M. E., Sorenson, F. M. & Jump, E. B. (1962). The miniature pig as an experimental animal in dental research. *Archives of Oral Biology*, **7**, 17–24. *480*

Webb, G. J. W. & Messel, H. (1977). Abnormalities and injuries in the estuarine crocodile, *Crocodylus porosus*. *Australian Wildlife Research*, **4**, 311–319. *393*

Wedl, C. (1870). *Pathologie der Zähne*. Felix, Leipzig, p. 240. *422*

Wegner, R. N. (1910). Uberzahlige Incisiven bei Affen. *Zeitschrift für Morphologie und Anthropologie, Stuttgart*, **12**, 353–359. *23, 36, 50*

Wegner, R. N. (1962). Abszesse an den Eckzähnen bei den Anthropoiden. *Deutsche Zahn-, Mund- und Keiferheilkunde*, **37**, 433–444. *428, 507*

Weiner, G. S., Demarco, T. J. & Bissada, N. F. (1979). Long term effect of systemic tetracycline administration on the severity of induced periodontitis in the rat. *Journal of Periodontology*, **50**, 619–623. *528*

Weninger, M. (1948). Zur Reduktion der Prämolaren (drei Prämolaren in Oberkiefen eines Gorilla). *Zeitschrift für Stomatologie, Wien*, **5**, 223–231. *23*

Werdelin, L. (1987). Supernumerary teeth in *Lynx lynx* and the irreversibility of evolution. *Journal of Zoology, London*, **211**, 259–266. *63*

Whitaker, J. O. (1971). *Microtus pinetorum* with grooved incisors. *Journal of Mammalogy*, **52**, 827. *131*

Whittaker, D. K., Molleson, T., Daniel, A. T., Williams, J. T., Rose, P. & Resteghini, R. (1985). Quantitative assessment of tooth wear, alveolar-crest height and continuing eruption in a Romano-British population. *Archives of Oral Biology*, **30**, 493–501. *332*

Wiener, G. & Gardner, W. J. F. (1970). Dental occlusion in young bulls of different breeds. *Animal Production*, **12**, 7–12. *294*

Wiener, G. & Purser, A. F. (1957). The influence of four levels of feeding on the position and eruption of incisor teeth in sheep. *Journal of Agricultural Science, Cambridge*, **49**, 51–55. *296*

Whitehead, G. K. (1985). Verkürzter Unterkiefer bei Sikawild. *Zeitschrift für Jagdwissenschaften*, **31**, 247–248. *288*

Williams, J. L. (1897). A contribution to the study of pathology of enamel. *Dental Cosmos*, **39**, 169–196, 269–301, 353–374. *458, 459, 460*

Williamson, H. C. (1918). A dentigerous cyst on the upper jaw of a cod. *Journal of Pathology and Bacteriology*, **22**, 255–256. *606*

Williamson, M. (1981). *Island Populations*. Oxford University Press, Oxford, p. 48. *10*

Willis, R. A. (1962). The embryonic tumours and teratomas. In *The Borderland of Embryology and Pathology*, 2nd edn., chapter 11. Butterworth, London, pp. 423–465. *576*

Wilson, E. A. (1907). 1. Mammalia. Cctacca. *National Antarctic Expedition 1901–1904. Natural History. Volume 2. Zoology*. Trustees of the British Museum, London, pp. 1–69. *498*

Wing, E. S. (1965). Abnormal dentition in several white-tailed deer jaws. *Journal of Mammalogy*, **46**, 348–350. *112, 113*

Wintheiser, J. G., Clauser, D. A. & Tappen, N. C. (1977). Sequence of eruption of permanent teeth and epiphyseal union in three species of African monkeys. *Folio Primatologia*, **27**, 178–197. *337*

Witkop, C. J. & Wolf, R. O. (1963). Hypoplasia and intrinsic staining of enamel following tetracycline therapy. *Journal of the American Medical Association*, **185**, 1008–1011. *438*

Wolfe, J. L. & Layne, J. N. (1968). Variation in dental structures of the Florida mouse *Peromyscus floridanus*. *American Museum Novitates*, **2351**, 1–7. *132*

Wolsan, M. (1984a). An incisor with double crown in a stoat (*Mustela erminea* L.). *Zeitschrift für Säugetierkunde*, **49**, 57. *72*

Wolsan, M. (1984b). Dental abnormalities in the pine marten *Martes martes* (L.) (Carnivora, Mustelidae) from Poland. *Zoologische Anzeiger, Jena*, **213**, 119–127. *71*

Wolsan, M. (1984c). The origin of extra teeth in mammals. *Acta Theriologica*, **29**, 128–133. *6, 7*

Wolsan, M., Ruprecht, A. L. & Buchalczyk, T. (1985). Variation and asymmetry in the dentition of the pine and stone martens (*Martes martes* and *M. foina*) from Poland. *Acta Theriologica*, **30**, 79–114. *71*

Wood, H. E. (1938). Causal factors in shortening tooth series with age. *Journal of Dental Research*, **17**, 1–13. *157*

Woods, J. A. (1907). Odontomes in horse and ox. *The Dental Record*, **27**, 234–235. *591, 593, 597, 598, 600*

Wright, J. G. (1939). Some observations on dental disease in the dog. *The Veterinary Record*, **51**, 409–421. *479, 511, 548*

Yablokov, A. B., Bel'kovich, V. M. & Borisov, V. I. (1974). *Whales and Dolphins*. Translated from the Russian. Nauka, Moscow, p. 93. *501*

Yilmaz, R. S., Darling, A. I. & Levers, B. G. H. (1980). Mesial drift of human teeth assessed from ankylosed deciduous molars. *Archives of Oral Biology*, **25**, 127–131. *157*

Young, J. A. & Van Lennep, E. W. (1978). *The Morphology of Salivary Glands*. Academic Press, London & New York. *479*

Young, W. G. & Marty, T. M. (1986). Wear and microwear on the teeth of a moose (*Alces alces*) population in Manitoba, Canada. *Canadian Journal of Zoology*, **64**, 2467–2479. *491*

Zakrzewski, R. J. (1969). Dental abnormality in the genus *Castor*. *Journal of Mammalogy*, **50**, 652–653. *131*

Zejda, J. (1960). The influence of age on the formation of the upper third molar in the bank vole, *Clethrionomys glareolus* (Screber, 1780). (Mammalia, Rodentia). *Zoologigické Listy*, **9**, 159–166. *132*

Zejda, J. (1965). Zur Variabilität der Molarwurzeln des Oberkiefers von 4 *Apodemus* Arten. *Zeitschrift für Morphologie und Ökologie der Tiere*, **54**, 699–706. *133*

Zeman, W. V. & Fielder, F. G. (1969). Dental malocclusion

and overgrowth in rabbits. *Journal of the American Veterinary Medical Association*, **155**, 1115–1119.   *362*

Zietzschmann, O. (1943). *Ellenberger-Baum. Handbuch der Vergleichenden Anatomie der Haustiere.* Springer, Berlin.   *342, 344*

Zschokke, W. & Saxer, E. (1933). Ein Fall von Zahnkaries mit Fistelbildung bei einem Sumpfbiber. *Schweizer Archiv für Tierheilkunde*, **75**, 24–26.   *484*

Zuckerkandl, E. (1891). Über das epitheliale Rudiment eines vierten Mahlzahnes beim Menschen. *Sitzungsberichte der Akademie der Wissenschaften in Wien*, Abt. I, 100, pt. 3, 315–350.   *49*

Zuckerman, S. (1928). Age-changes in the chimpanzee, with special reference to the growth of brain, eruption of teeth, and estimation of age; with a note on the Taungs ape. *Proceedings of the Zoological Society of London*, 1928, pt. 1, 1–42.   *335*

Zuhrt, R. (1958). Zahnfleischerkrankung beim Igel als Todesursache. *Deutsche Zoologischen Garten*, **24**, 74–80.   *569*

Zukowsky, L. (1963–1964). Über eine Gebissanomalie bei *Cervus nippon mantchuricus* Swinhoe. *Der Zoologische Garten*, **29**, 5–8.   *489*

INDEX

abrasion, *see under* tooth wear
ABSENT TEETH, *see also* numerical variation under taxonomic headings
  and size of jaw 87–88
  developmental vs loss 18 62
  field concept 2 3
achondroplasia, *see under* developmental anomalies
Acrobates, *see* Phalangeridae
actinomycosis, *see* lumpy jaw
Acinonyx, *see* Felidae
Aepyceros, *see* Bovidae
Aepyprymnus, *see* Macropodidae
Aethecinus, *see* Insectivora
Ailurus, *see* Procyonidae
Albrecht, E. 575
Alcelaphus, *see* Bovidae
Alces, *see* Cervidae
Aleutian disease 523
Alouatta, *see* Cebidae
alveolar bone, *see also* periodontium, and dehiscences and fenestrations
  independence of jaw bones 155 Fig 9.3
  origin from tooth primordium 5 155
alveolar bulb 154–155, in manatee 155 Fig 9.3, pig 155, sheep Fig 9.2
alveolar crest, *see* periodontium
Amblonyx, *see* Mustelidae
Amblysomus, *see* Insectivora
Ameghino, F. 13
Ammodorcas, *see* Bovidae
Amphibia, *see* non-mammalian dentitions, and *Bufo pardalis*
anisognathism 287
ankylosis
  of injured tooth, in glutton 396 Fig 19.5, in jaguar 402
  of retained deciduous teeth 332
  in bush-babies 542, lemur 237
anodontia, *see* absent teeth

Anoura, *see* Chiroptera
ant-eater, banded, *see* *Myrmecobius fasciatus*
ant-eater, spiny, *see* Monotremata
Antechinomys, *see* Dasyuridae
Antechinus, *see* Dasyuridae
antelope, *see* Bovidae
Antidorcas, *see* Bovidae
Antilocapra, pronghorn, *see also* Ruminantia
  dental formula 117
  enamel fluorosis 454
  numerical variation 117
Antilope, *see* Bovidae
Aonyx, *see* Mustelidae
Aotus, *see* Cebidae
ape, Barbary, *see* Cercopithecidae
  black, *see* Cercopithecidae
apes, great, *see* Pongidae
Apodemus, *see* Rodentia
Arctictis, *see* Viverridae
Arctocephalus, *see* Otariidae
Arctogalidea, *see* Viverridae
Arctonyx, *see* Mustelidae
arthritis, in fluorotic animals 453, lion 445
ARTIODACTYLA, even-toed ungulates, *see also* Suidae, Tayassuidae, Camelidae, Ruminantia and *Hippopotamus amphibius*
  hypsodont and brachydont teeth, infection and dento-alveolar abscess in 513
  normal dentition 15 106–107
  origins 106
ass, *see* Equidae
asymmetry, *see also under* taxonomic headings
  developmental anomalies, *see under* developmental anomalies
  experimental 10

injury, as result of, in *Cercopithecus* 372, gorilla 372, sealion 372, leopard seal 372, impala 372
of position anomalies, *see under* position anomalies
tooth size, in greyhound 86
ATAVISM
  disto-molars in *Canis latrans* 83 Fig 4.53
  extra teeth 7
  horns in hornless breeds 7
  in bats 135
  in pongids 22
  M_2 in felids 63 64
  metaconid on M_1 in felids 64
  multiple tusks in elephants 128
  reactivation of an embryonic field 63
  reverse mutation 7
Atelerix, *see* Insectivora
Ateles, *see* Cebidae, and experimental animals
Atilax, *see* Viverridae
attrition, *see under* tooth wear
aye-aye, *see* Daubentoniidae

babirusa, *see* Suidae
baboons, *see* Cercopithecidae, and experimental animals
Babyrousa, *see* Suidae
badger, *see* Mustelidae
Baiomys, *see* Rodentia
Balaenoptera, *see* Mysticeti
Balantiopteryx, *see* Chiroptera
bandicoot rat, *see* Rodentia
bandicoots, *see* Peramelidae
Bandicota, *see* Rodentia
Bassaricus, *see* Procyonidae
Bassaricyon, *see* Procyonidae
Bateson, W. 2 6 7 20 22 95–96, 143 146–147
Bathygeridae, *see* Rodentia
bats, *see* Chiroptera

645

Bdeogale, see Viverridae
bears, see Ursidae
beaver, see Rodentia
Berardius, see Odontoceti
bettong, see Macropodidae
Bettongia, see Macropodidae
binturong, see Viverridae
Bison, see Bovidae
blackbuck, see Bovidae
Bland Sutton, J. 504 507 574–575, 601 603
Blarina, see Insectivora
Blastocerus, see Cervidae
boar, wild, see Suidae
bobcat, see Felidae
Bolk, L. 20 22 353
BONE DISORDERS 49 210 236–237 333–334 350 392–393 492 494 579 Fig 10.161
 fibrous osteodystrophy 334, see also under odontomes
 hyperparathryroidism, see separate entry
 osteomalacia 158 334 472–473
 osteopetrosis 236–237 333 334 346 348 Fig 10.161
 osteoporosis, see also dehiscences and fenestrations, alveolar 158 505 524–525 528 540–541 542 544 546 550 559 560
 renal rickets 334 589 604–605
 rickets 156 218 236 237 241 333–334 350 438 441 602–605 Fig 11.5
 yaws, in gorilla 392–393
bone growth, in eruptive process 333, see also jaw growth
boodie, see Macropodidae
Bos, see Bovidae and *Bos taurus* domestic cattle
Bos taurus, domestic unhumped cattle, see also Bovidae
 achondroplasia 293 294
 ancestry 299
 bulldog head 294
 canines 115
 cara inchada 561
 cementoma 598 Fig 25.32
 Chédiak-Higashi syndrome 523
 congenital polycystic jaw 579
 connation 115
 dental caries 480–481
 dento-alveolar abscess 292 514
 dwarfism, hypopituitary 294
 ectodermal dysplasia 115 481
 enamel fluorosis 450–454
 enamel hypocalcification 450
 enamel hypoplasia 450 495
 eruption sequence and chronology 342 Tabs 16.10 16.11
 experimental animals 451
 feral cattle 115
 geographical variation, founder effect 115
 inheritance 294 450 516
 injury, teeth 432

lumpy jaw 515–516 Fig 23.8
mesial drift 157
necrotic stomatitis 561
numerical variation 115 Fig 6.13
odontomes 115 582 Fig 25.9
periodontal disease 559 561
position anomalies 292–295 Figs 13.25 13.27–13.29
 overjet (parrot mouth) 294
 traumatic displacement 427
 underjet 294–295
retained deciduous teeth 115
teratoma 601
tooth wear 157 494
wedging, of object across tooth arch 372
breeds: Ayrshire 294, Chillingham 115, Dexter 294, Friesien 294, Galloway 294, Hereford 294, Jersey 294, Nganda 294, Niata 292–294 494 Fig 13.28, C. Darwin and 292– 292, R. Owen and 292, prehistoric 115, short-horn 294, Telemark 294
BOVIDAE, cattle antelopes, gazelles, etc., see also Ruminantia and separate entries for domestic animals: *Bos taurus* (unhumped cattle), *Ovis aries* (sheep), and *Capra hircus* (goats)
bone disorder, osteodystrophy 604
canines 114
captivity, effects of 551 Fig 24.34
dental caries 480–481
dental formula 114
dentigerous cyst 578
dento-alveolar abscess 513 551
enamel fluorosis 450–454
enamel hypoplasia 450
enamel pearls 576
eruption, over-eruption 348
 sequence and chronology 342 345
 geographical variation 115 450 491 561
gingival hyperplasia 561
hypsodonty, in dento-alveolar abscess 513 551
injury, mandible 372 374 379 382–383 384 Figs 18.9 18.14 18.15
lumpy jaw 519
mesial drift 157 Fig 9.5
necrotic stomatitis 561 563
normal dentition 114 297 299 301 302 303 305 591 Figs 6.17 13.34 13.35 13.44 13.50 25.20
numerical variation 114–117 450 Figs 6.15 6.18
odontomes 580 585 590–592 602–604 Figs 25.5 25.6 25.14 25.18 25.19 25.22 25.38 25.40
osteomyelitis 514 551
periodontal disease 550–552 561 562 Figs 24.34 24.35
position anomalies 292–307 Figs 13.25–13.68 Tabs 13.3 13.4 13.5

 impaction 348
retained deciduous teeth 298 352 Fig 13.41
tooth shape anomalies 116 117 540 Figs 6.16 6.17 6.18
tooth wear 157 488 491 Figs 9.5 22.5
wedging of plant material, interdentally 297 561
Aepyceros impala 115 372, *Alcelaphus* hartebeest 115 297 379 Figs 13.34 18.9, *Ammodorcas* dibatang 116, *Ammotragus* Barbary sheep 590 Fig 25.18, *Antidorcas* springbok 305, *Antilope* blackbuck 303, *Bison bison* American bison 114 115 454, *Bos indicus* cattle, domestic, humped 292 561 Fig 13.26, *Bos namadicus* Indian fossil cattle 292, *Bos primigenius* extinct wild cattle 292, *Bos taurus* domestic cattle, see separate entry, *Capra hircus* domestic goat, see separate entry, *Capra ibex* ibex 115 117, *Cephalophus* duikers 297–299 514 551 832–383 Figs 13.35 13.37–13.42 Tab 13.3, *Connochaetes* wildebeest 115 348 561, *Gazella* gazelles 114 117 157 303–305 551 Figs 6.18 9.5 13.57–13.62 24.35 Tabs 13.4 13.5, *Hippotragus* roan antelope 305, *Kobus* water buck 303 491, *Madoqua* dik-diks 114 116 117 301 302 Figs 6.17 13.50–13.53, *Nesotragus* royal antelope and sunis 114 116 301 Figs 6.16 13.48 13.49, *Oryx* oryx and gemsbok 115 305 306 384 488 592 Figs 13.6 18.14 25.22, *Ourebia* oribi 114 116 299 516 Figs 6.15 13.43, *Ovibos* musk ox 114 450 562, *Ovis ammon* wild sheep 576, *Ovis aries* domestic sheep see separate entry, *Ovis canadensis* bighorn sheep 114 516, *Ovis dalli* dall sheep 114 117 516 563, *Pelea* Vaal reedbuck 302, *Procapra* Tibetan gazelle 114 117 303, *Raphicerus* grysbok and steinbok 299–301 Figs 13.44–13.47, *Redunca* reedbuck 302–303 384 Figs 13.54–13.56 18.15, *Rupicapra rupicapra* chamois 115 157 561 in New Zealand 115 561, *Sylvicapra* duiker 297 298 551 Figs 13.36 34.34 Tab 13.3, *Syncerus* African buffalo 580 590–591 Figs 25.5 25.6 25.19 25.20, *Tragelaphus* bushbucks and sitatungas 306–307 561 578 604 Figs 13.64–13.68
brachydont teeth, definition 14
 dento-alveolar abscess and 513
 in bats 134, cervids 111, suids 107 513
Brachyteles, see Cebidae
Broca, P. 574 594

broken mouth in sheep 295 492 552, 556–560
buffalo, see Bovidae
Bufo pardalis, leopard toad, injury to head 393
bulldog head, see developmental anomalies
bunodont teeth, definition 13, in suids 107
bushbaby, see Lorisidae, and experimental animals
bushbuck, see Bovidae
bush dog, see Canidae

Cacajao, see Cebidae
Caenolestoidea, shrew-opossums numerical variation 151
calculus, see dental calculus
Callicebus, see Cebidae
Callimico, see Callitrichidae
Callithrix, see Callitrichidae, and experimental animals
CALLITRICHIDAE, marmosets and tamarins, see also Primates
  bone disorder, osteoporosis 542
  captivity, effects of 476 542
  connation 58
  dehiscences 542
  dental calculus 542
  dental caries 476 Tabs 21.9 21.10
  dento-alveolar abscess 542
  diet 541
  eruption sequence 338–339
  injury, mandible 388
  normal dentition 58 232
  numerical variation 58
    disto-molars 58
  occlusal overload 528
  periodontal disease 525 541–542 Tab 24.1
    experimental 528 542
  position anomalies 232 Figs 10.147–10.151 Tab 10.20
    underjet Fig 10.148 Tab 10.20
  Callimico Goeldi's marmoset 58 339, *Callithrix* marmosets 58 338 339 525 528 542, as experimental animals 462 525 528 542, *Cebuella* pygmy marmoset 234 339 Fig 10.150, *Leontopithecus* golden tamarin 339, *Saguinus* tamarins 58 232 233 234 338 339 476 528 542 548 Fig 10.148–10.151, as experimental animal 462 528 542
Callorhinus, see Otariidae
Caloprymnus, see Macropodidae
Caluromys, see Didelphidae
CAMELIDAE, camels and llamas
  dental caries in fossil 485
  dentigerous cyst 589
  injury, mandible 385–386
  lumpy jaw 516
  normal dentition 285
  position anomalies 285 Figs 13.9 13.10

Camelus camel and dromedary 385–386 485 589, *Lama* llama and guanaco 285 516 Figs 13.9 13.10
CANIDAE, dogs, wolves, foxes etc, see also *Canis familiaris* domestic dog, and Carnivora
  ancestral forms 83 87
  atavism 83 Fig 4.53
  captivity, effects of 264 506 546 547
  connation 88 89 90 91 94
  dental caries 479
  dental formulae 12 83 94
  dento-alveolar abscess 506 512
  diet 264
  domestication 85 264–266
  enamel hypoplasia 89
  eruption, non-eruption 347
  evolutionary change 83
  geographical variation 264 Tab 11.14
  injuries, jaws 379
  normal dentition 83 279
  numerical variation 83–94 Tab 4.3
    disto-molars 88 89 90 91 92 94
  odontome 94
  osteometrics 262 Tabs 11.13 11.14
  periodontal disease 546–550 Fig 24.31
  position anomalies 258–279 Figs 11.62–11.90 Tabs 11.11–11.16
    impaction 264
    overjet 260–263 Figs 11.61 11.62 Tab 11.12
    underjet 260–263 Fig 11.63 Tab 11.12
  retained deciduous teeth 88 91
  skull deformities 263 380
  tooth size anomalies 89 90 92 Fig 4.53 Tabs 4.9 4.10
  Alopex Arctic fox 84 92–94 263 264 Figs 4.54 4.55 Tabs 4.3 11.11, *Canis* wolves and jackals 258 Tabs 4.3 11.11, *Canis adustus* side-striped jackal 83 89 90 Figs 4.45 4.46 4.47 Tab 4.9, *Canis aureus* jackal 84 88 262 263 Tab 11.12, *Canis familiaris* domestic dog, see separate entry, *Canis latrans* coyote 83 84 88 263 264 342 447 Tab 11.12, *Canis lupus* wolf 83 85 88 263 264 267 272 380 434 506 512 546 547 Figs 11.69 24.31 Tabs 4.5 11.12, *Canis mesomelas* black-backed jackal 89 90 264 Figs 4.48 11.62 11.65, *Canis pallipes* Indian wolf 84, *Canis simensis* simian jackal 264, *Cerdocyon* Tabs 4.3 11.11, *Cerdocyon thous* crab-eating fox 83 94 264 265 266 586 Figs 4.56 4.57 11.66 11.68 Tab 4.3, *Chrysocyon* maned wolf Tab 11.11, *Cuon* Asiatic wild dog 94 263 347 Tabs 4.3 11.11 11.12, *Dusicyon* 83 263 Tabs 4.3 11.11,

Dusicyon gymnocercus (*Pseudalopex azarae*) Azara's fox 94, *Dusicyon mimax* see *Cerdocyon thous*, *Dusicyon vetulus* hoary fox 94, *Lycaon* hunting dog 263–265 Tabs 11.11 11.12 11.13 11.14 *Nyctereutes* racoon dog 83 84 94 263 264 Tabs 4.3 11.11, *Otocyon* bat-eared fox 83 84 94 263 264 Tabs 4.3, 11.11, dental formula 12 94, *Speothos* bush dog 83 94 264, Fig 11.64 Tabs 4.3 11.11, *Vulpes* foxes 263, Tabs 4.3 11.11, *Vulpes chama* Cape fox 84, *Vulpes cinereo-argenteus* grey fox 84 92, *Vulpes ferricata* Thibetan fox 546, *Vulpes macrotis* kit fox 92, *Vulpes vulpes* farm silver fox 84, *Vulpes vulpes* red fox 83 84 89–92 261–264 379 Figs 4.49–4.53 11.61, 11.63 11.67 18.11 Tabs 4.4 11.12 11.13 11.14

Canis familiaris domestic dog
  achondroplasia 266–268
  asymmetry 86 270
  congenital ectodermal defects 88
  connation 9 86 Figs 4.40 4.41
  calculus 546 547 548
  dental caries 478–479
  dental formula 83
  dento-alveolar abscess 511
  diet 445–447 479 546 549–550 Fig 20.14
  distemper 445–447 479 Figs 20.15 20.16
  domestication 85 264–266 Fig 11.69
  ectodermal dysplasia 88
  enamel hypoplasia 445–447 479 Figs 20.14–20.16
  eruption, chronology and sequence 341 342 Tabs 16.7 16.8
    deciduous dentition 341 Tabs 16.7 16.8
    non-eruption 347–348
  experimental animals 266 479 526–528
  hemifacial microsoma 271
  inheritance 266 268 270
  injuries, jaws 511
  necrotizing ulcerative gingivitis 550
  neoteny 266
  normal dentition Figs 11.69 11.90
  numerical variation 85–88 Figs 4.40–4.44 11.89 Tabs 4.3 4.5, 4.6 4.7 4.8
    in short-faced dogs 87 269 Figs 11.72 11.73
  osteometrics 264 266 270 272 274 275 Tabs 11.15 11.16
  periodontal disease 511 546–550, Figs 24.26–24.30
  periodontal index 548
  position anomalies 264–276 Figs 11.69–11.90 Tab 11.16

Canis familiaris (cont.)
    impaction 264
      in long-faced dogs 270 275 Figs 11.70 11.76 11.77 Tab 11.15
      in other dogs 270–275 Figs 11.78–11.82 11.90 Tab 11.15
      in short-faced dogs 266–270 275 Figs 11.70–11.75 Tab 11.15
      overjet 270 271 272
      underjet 274
    retained deciduous teeth 269
    skull deformities 263 266–275
    tooth shape anomalies 88 Figs 4.43 4.44
    tooth wear, associated with periodontal disease 550
    tumours 548
    uraemia 548
    breeds: Airedale Fig 11.69, Alsatian (German shepherd) 266 479 548 Fig 21.22, bassett hound 266 267, beagle 271 341 549 Tabs 16.7, 16.8, bloodhound Fig 4.44, borzoi 270 Fig 11.70 Tab 11.15, Boston terrier 268, boxer 86, bulldog 86 87 266 269 270 478 548 Figs 11.70–11.75 Tab 11.15, bull-terrier 86 447 Figs 4.40 20.16, dachshund 266 267 270, dingo 85, Tab 4.5, Doberman 86 87 Tab 4.8, early historic 86, Eskimo dog 87 Tab 4.5, foxhound 270 271, Tab 11.15, fox terrier 270 272 479 Tab 11.15, greyhound 86 87 270 Fig 4.42 Tabs 4.5 11.15, hound 546 Tab 4.5, Lakeland terrier 9, mastiff 86 270 548 Tabs 4.5, 11.15, miniature griffon 266 268–269, pariah dog 275, Pekingese 86 266 267 268 269 548, poi dog 479, Pomeranian 548, poodle Tab 4.5, prehistoric 86 87, prehistoric, in New Zealand 87 478 550, pug 86 266, retriever 270 Tab 11.15, Russian greyhound Tab 11.16, saluki 266, sealyham 271, sheepdog Tab 4.5, South American pug-like 260, spaniel 548 Tab 4.5, terrier Tab 4.5

Capra hircus domestic goat, *see also* Bovidae
  bone disorder, osteodystrophy 603–604 Fig 25.38
  diet 603
  enamel fluorosis, experimental 451
  eruption sequence and chronology 345
  experimental animals 451
  feral goats 115 297 345 493 561
  geographical variation 297
  numerical variation 115 117
  periodontal disease 297 493 560 561
  position anomalies 296–297 Fig 13.33
    underjet 296 Fig 13.32
  tooth wear 488 493 560
  wedging, of plant material, interdentally 297
  breeds: Angora 345, Nubian 296 Figs 13.32 13.33, West African dwarf 296

Capra, see Bovidae and *Capra hircus* domestic goat
Capreolus, see Cervidae
Captivity, effects of
  bone disorders, *see* separate entry
  dental caries 461 465 471–478 483 484 485 Figs 19.2 21.8 21.12 21.29 Tabs 21.3 21.4
  dento-alveolar abscess 506 518–520
  developmental anomalies, *see* separate entry
  enamel hypoplasia 38–40 439 441 442
  eruption 336 346 350 Figs 10.90 10.102 11.5 16.2–16.6 Tabs 16.1 16.2 16.4 16.5
  injuries 394–402 429–432 435 Figs 19.1–19.3 19.8–19.10 19.46 19.47 19.50 19.51 19.56
  overgrowth 366
  periodontal disease 529 535 539 540 542–547 551 567 569–572 Figs 24.12 24.18 24.21 24.22 24.24 24.33 24.34 24.49 24.50
  position anomalies 169–171 175 177 194–195 197–200 202 205–208 210–211 212 218–219 236–237 239–241 247–248 256–257 264 Figs 10.67–10.76 10.86–10.91 10.99 10.100 10.117 10.159–10.162 11.54 11.55 Tab 11.1
  retained deciduous teeth 350 Figs 10.69 19.73 10.88 10.89 10.90
  in *Babyrousa* 366, Callitraichidae 476 542, Canidae 264 506 546 547, Cebidae 218–219 346 474–475 Figs 10.117 16.5, *Cercocebus* 202 431 Fig 19.50, *Cercopithecus* 194–195 350 396 430 471 Figs 10.67–10.70 19.3 19.47 21.12, Cervidae 551 Figs 24.33 24.34, *Colobus* 38–40 350 473, coypu 435 Fig 19.56, Didelphidae 485, Equidae 567, *Erinaceus europaeus* 569, *Erythrocebus* 197–200 202 350 Figs 10.73 10.76, Felidae 239–241 477 543, *Gorilla* 465 572, *Hylobates* 177; Insectivora 569, Lemuridae 236–237 476 542–543 Figs 10.159–10.162, leopard 241 397–398 Fig 19.8, lion 241 346 398–402 434–435 Figs 11.5 19.9 19.10 19.55; *Lorisidae* 542, *Macaca* 205–208 346 350 432, 539 Figs 10.76–10.91 16.4 19.51 24.18, Macropodidae 518–520 570–571 Figs 24.49 24.50, *Mandrillus* 212 346 428 Fig 16.2, Mustelidae 477 546, *Oryctolagus* 484, *Pan* 169–171, 346 535 Figs 16.3 24.12, *Papio* 210–211 346 350 395–396 432 472–473 539 540 Figs 10.99 10.100 10.103 19.2, Peramelidae 571–572, *Pongo* 175 535, primates 529, Procyonidae 256–257 544 Figs 11.54 11.55 24.21 24.22, Rodentia 461 483 Figs 21.28 21.29, *Theropithecus* 540, Ursidae 346 394–395 429–430 478 545 Figs 16.6 19.1 19.46 24.24, Viverridae 247–248 Tab 11.1, Vombatidae 485

cara inchada 561
Carabelli, cusps of 15 19–20 24 56, in *Cebus* 56, fossil hominids 20, *Gorilla* 24, Tertiary primates 20
caracal, *see* Felidae
caries, *see* dental caries
caribou, *see* Cervidae
carnassial teeth,
  liable to damage 434
  in bears 77, canids 2 83, felids 62, 434, hyaenas 67, mustelids 71, procyonids 74, viverrids 67
CARNIVORA, *see* Canidae, Felidae, Procyonidae, Viverridae, Hyaenidae and Ursidae
  dental caries 476
  enamel hypoplasia 444–447
  injury, tooth, in trapped animals 434
  jaw shape 238
  normal dentition 62 238
  numerical variation 62 Tab 4.1
  position anomalies 238–279 Tab 11.1
Castor, *see* Rodentia
cat, domestic, *see Felis catus*
cat, ring-railed, *see* Procyonidae
cats, wild, *see* Felidae
cattle, *see* Bovidae and *Bos taurus*
Cavia, *see* Rodentia, and experimental animals
CEBIDAE, New World monkeys, *see also* Primates
  bone disorder, fibrous osteodystrophy 334
  captivity, effects of 474 475
  Carabelli cusps 56
  cusp pattern of molars 20
  dental caries 473–476 Tabs 21.9 21.10 23.1
  dental calculus 476 542
  dento-alveolar abscess 474 476 511 Tab 23.1
  diet 541
  dehiscences and fenestration 541 542
  eruption sequence 338–339
  geographical variation 231–232
  normal dentition 52
  numerical variation 52–58 Tabs 3.6, 3.7
  periodontal disease 474 541–542 Tab 24.1

CEBIDAE (cont.)
  position anomalies 213–232 Tab 10.17
  tooth shape and size anomalies 52–58
  tooth wear 493 541
Alouatta howler monkeys
  connation 58
  dehiscences and fenestration 541
  dental calculus 474
  dental caries 474 476 Tabs 21.9 21.10 23.1
  dentigerous cyst 579
  dento-alveolar abscess 511 Tab 23.1
  diet 541
  eruption sequence 338
  injury, mandible 388
    zygomatic arch 388
  normal dentition 223 Figs 10.128 10.129
  numerical variation 53 56–58 Tab 3.6
    disto-molar 57
  periodontal disease 541 Tab 24.1
  position anomalies 223–227 Figs 10.130–10.136 Tab 10.17
    impaction 227
    underjet Tabs 10.17 10.19
  retained deciduous tooth 226
  shear bite 493
  tooth size anomalies 58 Fig 3.67
  tooth wear, excessive 493
Alouatta belzebul black-and-red howler 223 579 Tab 10.19, *A. caraya* black howler 56 Figs 3.67 10.133 Tab 10.19, *A. fusca* brown howler Figs 10.129 10.132 10.134 Tab 10.19, *A. palliata* mantled howler 57 58 224 226 227 493 511 541 Fig 10.136, *A. seniculus* red howler 57 223 Figs 10.128 10.135 Tab 10.19
Aotus trivirgatus douroucouli monkey
  dental caries Tab 21.9
  eruption sequence 338 Tab 16.4
  normal dentition 232
  position anomalies 232 Fig 10.146
Ateles spider monkeys
  bone disorder, fibrous osteodystrophy 334
  connation 53 Fig 3.61
  dental calculus 474
  dental caries 474 476 Tabs 21.9 21.10 23.1
  dento-alveolar abscess 511 Tab 23.1
  diet 541
  enamel mineralization 458
  eruption sequence 338
  injury, mandible 388
    zygomatic arch 388
  numerical variation 52–55 221–222 Figs 3.61 3.62 10.123

  disto-molar 55
  osteometrics 221
  periodontal disease 474 476 541 Tab 24.1
  experimental 528
  position anomalies 221–223 Figs 10.123 10.126 10.127 Tab 10.17
    impaction 222
    underjet 221 Fig 10.127 Tab 10.17
  tooth shape and size anomalies 55
Ateles belzebuth brown, long-haired and white-whiskered spider monkeys 54–55 221 Figs 3.62 10.125, *A. fusciceps* brown-headed spider monkey 54 55, *A. geoffroyi* black-handed, Nicaraguan and Mexican spider monkeys 52 53 55 222 528 541 Figs 3.61 10.124, 10.126, as experimental animals 528, *A. paniscus* black-faced and red-faced spider monkeys 53 54 55 221 Figs 10.123 10.127
Brachyteles woolly spider monkey
  dental caries 474
  numerical variation 52 53 221 Figs 3.60 Tabs 3.6 10.122
  position anomalies 220–221 Tab 10.17
    overjet 221
    underjet 220–221 Fig 10.122 Tab 10.17
  tooth shape and size anomalies 53
Cacajao uakari monkeys
  dental caries 474
  normal dentition 227
  position anomalies 227–229 Tab 10.17
    overjet 228
    underjet 228 Tab 10.17
Callicebus titi monkeys
  connation 58
  dental caries Tabs 21.9 21.10
  geographical variation 231–232
  normal dentition 229–230 Fig 10.142
  numerical variation 58 231 Tab 3.6
  position anomalies 229–232 Figs 10.142–10.145 Tab 10.17
    underjet 231 Fig 10.143 Tab 10.17
  retained deciduous teeth 232
  tooth shape and size anomalies 58, Tab 3.6
Cebus capuchin monkeys
  bone disorders 218
  captivity, effects of 218–219 Fig 10.117
  Carabelli cusps 56
  dental caries 474 476 477 Figs 21.16–21.18 21.20 Tabs 21.9–21.11 23.1

  diet 541
  enamel hypoplasia 444 Fig 20.12
  eruption, non-eruption 346 Fig 16.5
    sequence 338
  experimental 462 528
  injury, mandible 388
    skeletal 371
    zygomatic arch 388
  normal dentition 213 217 Fig 10.106
  numerical variation 52–53 55–56 217 Fig 3.63 Tab 3.6
    disto-molar 55–56
  periodontal disease 541 Tab 24.1
  position anomalies 213–219 Figs 10.107 10.112 10.116 10.117 Tabs 10.17 10.18
    impaction 217 Figs 10.113–10.115
    overjet 213
    underjet 213 Tab 10.17
  shear bite 493
  tooth shape and size anomalies 56, Figs 3.64–3.66
  tooth wear, excessive 493
Cebus albifrons brown pale-fronted capuchin 56 215 Figs 10.108 10.109 Tabs 10.18 21.11, *C. apella* brown capuchin 55–56 217 218 477 528 Figs 3.65 3.66 10.106 10.107 10.112 10.113 10.115 10.117 20.12 21.16 21.18 21.20 Tabs 10.18 21.11, as experimental animal 462 525 528, *C. capucinus* white-throated capuchin 56 215 217 493 541 Figs 3.63 3.64 10.110 10.111 10.114 Tabs 10.18 21.11, *C. nigrivittatus* weeper capuchin Fig 10.116 Tabs 10.18 21.11
Lagothrix woolly monkeys
  bone disorder, fibrous osteodystrophy 334
  dental caries 473 474 Fig 21.19 Tabs 21.9 21.10
  numerical variation 52 53 Tab 3.6
    disto-molar 53
  periodontal disease Fig 10.119
  position anomalies 219–220 Figs 10.118–10.121 Tab 10.17
    overjet 219
    underjet 219 Fig 10.121 Tab 10.17
Pithecia saki monkeys
  dental caries Tabs 21.9 21.10
  enamel hypomineralization Fig 21.5
  eruption sequence 338
  normal dentition 227 Fig 10.137
  numerical variation 58 Tab 3.6
  position anomalies 227–229 Fig 10.138 Tab 10.17
    overjet 228
    underjet 228 Tab 10.17

CEBIDAE (cont.)
  Saimiri squirrel monkeys
    dental caries Tabs 21.9 21.10
    dento-alveolar abscess 511
    diet 157
    disto-molar 55
    eruption sequence 338 Tab 16.4
    experimental animal 462 528
    gingivitis 528
    numerical variation Tab 3.6
    occlusal overload 528
    position anomalies 229 Figs 10.139–10.141 Tab 10.17
      overjet 229
      underjet 229 Tab 10.17
Cebuella, see Callitrichidae
Cebus, see Cebidae, and experimental animals
cementomas, *see under* odontomes and taxonomic headings
cementum, coronal 102 481 524 591 592 598 Figs 1.8 5.8 21.24 25.20 25.21 25.32
  binding enamel plates in elephants 125
  in fluorosis 451 453
cementum hyperplasia
  joining teeth 9, in elephant 129, sperm whale 102 Fig 5.8
  repairing fractured teeth 405 421 505
Cephalophus, see Bovidae
Cercartetus, see Phalangeridae
Cercocebus, see Cercopithecidae
CERCOPITHECIDAE, Old World monkeys, *see also* Primates
  dental caries 465–473 Tabs 21.3–21.5 23.1
  dental formula 34
  dento-alveolar abscess 510–511 Tab 23.1
  enamel fluorosis 451 Fig 20.18
  enamel hypoplasia 439–444
  eruption sequence 336–338
  geographical variation 35 45 47 182 195–197 510 Tabs 10.14 23.1
  normal dentition 22 34 188
  numerical variation 34–51 Tabs 3.3 3.4
  periodontal disease 537–541 Tab 24.1
  position anomalies 177–213 Tabs 10.3 10.4 10.12
    underjet 177 Tab 10.3
  tooth shape and size anomalies 34–51
  Cercocebus mangabey monkeys
    captivity, effect of 202
    dental caries 467 471 Fig 21.14 Tabs 21.3 21.4
    numerical variation 48 Fig 10.77 Tab 3.3
    position anomalies 200–202 Tabs 10.3 10.15 10.77
      underjet 200 Tab 10.3

  tooth shape anomalies 48 431 Figs 3.52 10.50
Cercopithecus grivet, vervet, Malbrook, tantalus, green, etc monkeys
  asymmetry of skull 372
  bone growth defective 193
  captivity, effects of 194–195 471 538 Figs 10.67–10.71 21.12 24.16 24.17 Tab 10.12
  connation 44 45 47 Figs 1.5 3.45 3.47 3.50
  dehiscences and fenestration 537–538
  dental caries 465 Fig 10.71 Tabs 21.3 21.4 21.6–21.8
  dento-alveolar abscess 510 Tab 23.1
  enamel fluorosis 451 Fig 20.18
  eruption sequence 336
  geographical variation 35 45 47 195–197 Tab 10.14
  injury, skull 372
    tooth 430 Fig 19.47
  normal dentition 188 Figs 10.56 10.60
  numerical variation 35 44–46 Figs 3.46 3.48 Tabs 3.3 10.14
    disto-molar 45 46
  osteometrics 196–197
  periodontal disease 537–538 Fig 24.15 Tab 24.1
  position anomalies 45 188–197 Figs 10.57 10.59 10.63–10.71 Tabs 10.3 10.4 10.12–10.14
    in captivity 194–195 Figs 10.67–10.71 Tab 10.12
    impaction 194–195 Fig 10.70 10.71 Tab 10.12
  overjet 189–190 Figs 10.58–10.61
  underjet 189 193 Figs 10.66 10.67 Tab 10.12
  resorption 502 Fig 22.17
  root anomalies 35 47 Tab 10.14
  scavenging refuse, effect of 469
  tooth shape anomalies 44 47 Figs 3.47 3.49
  tooth size anomalies 44 46 195 Fig 3.48
  tooth wear 495 Fig 22.7
Cercopithecus aethiops grivet, vervet, Malbrook and tantalus monkeys 44 45 46 47 189 337 372 396 471 Figs 1.5 3.48–3.50 10.63–10.66 10.68 10.71 19.3 24.16 24.17 Tabs 10.12 10.13,
C. ascanius Whiteside's guenon etc 45 46 189 337 495 496 Figs 3.46 22.7 Tab 10.13, *C. campbelli see C. mona, C. cephus* moustached monkey 46 467 Figs 10.59 10.60 Tabs 10.12 10.13 21.7 21.8, *C. denti* Dent's monkey Fig 21.10 Tab 10.13,

C. diana diana monkey 35 46 47 192 Tab 10.12, *C. erythrotis* red-eared monkey 46 Figs 3.49 10.62, *C. lhoesti* L'Hoest's monkey Tab 10.12, *C. mitis* diadem monkey 46 47 Figs 10.57 20.18 Tab 10.13, *C. mona* Campbell's and mona monkeys 35 46 47 190 192 430 Figs 10.70 10.47 Tabs 10.12 21.7, *C. neglectus* De Brazza's monkey 188 496 Figs 10.61 21.7 21.11 Tabs 10.12 10.13 21.7, *C. nictitans* greater white-nosed monkey 35 46 47 192 467 Fig 22.7 Tabs 10.12 10.13 21.7 21.8, *C. petaurista* lesser white-nosed monkey 46 Fig 10.69, *C. pogonias* crowned guenon 44 47 467 469 Fig 21.8 Tabs 10.13 21.6–21.9 21.13, *C. sabaeus* green monkey 35 45 46 47 195 Figs 3.47 10.58 10.67 10.68 Tab 10.14, as experimental animal 528, on St Kitts, West Indies 35 45 47 195, *C. talapoin* talapoin Tab 10.12, *C. wolfi* Wolf's monkey, fire-bellied guenon 44 495 538 Figs 3.45 24.15 Tab 10.13
Colobus colobus monkeys
  bone disorder, osteoporosis 524–525 541
  captivity, effects of 473
  connation 41
  dehiscences 524 540–541
  dental caries 473 Tab 21.3 23.1
  dento-alveolar abscess 510 511 Tab 23.1
  ectodermal dysplasia 39–40
  enamel hypoplasia 38 39 40 442–444 Figs 3.32 3.33 20.11
    traumatic 402–403 Fig 19.13
  eruption sequence 337
  geographical variation 35 182 510 Tab 23.1
  gingival fibroma 54
  injury, maxilla 372 Fig 18.2
    zygomatic arch 392
  mesial drift Fig 10.45
  mesio-dens 36
  normal dentition 178 Figs 3.34 10.40 10.41
  numerical variation 34–38 40 Figs 3.29–3.33 10.45 Tabs 3.3 3.5
    disto-molar 37
  osteometrics Tabs 10.6 10.8
  periodontal disease 540–541 Tab 24.1
  position anomalies 178–184 Figs 10.43–10.48 Tabs 10.3 10.4 10.7
    overjet 181 Fig 10.43
    underjet 179–180 Fig 10.42 Tabs 10.5 10.6

CERCOPITHECIDAE (cont.)
   radicular cyst 511 582 Fig 25.10
   retained deciduous teeth 39 346 349 350 443 Figs 3.29 3.33
   tooth shape anomalies 35 38 40 Figs 3.31 3.32 3.34–3.38 Tab 3.5
   tooth size anomalies 38–41 Fig 3.33
   tooth wear 496
   Colobus angolensis Angolan colobus 179 180 Fig 10.43 Tabs 10.5 10.7, *C. badius* red colobus 35 36 37 38 40 54 177 179 182 183 337 350 403 473 540 541 Figs 10.42 24.19 Tabs 3.5 10.5 10.6 10.7 10.8, geographical variation in *C. badius preussi* 35, *C. guereza* eastern black-and-white colobus 35 36 37 38 40 41 179 180 182 183 337 402–403 442 443 473 511 Figs 3.2 3.32 3.37 3.38 10.40 10.41 10.44–10.47 19.13 20.11 25.10 Tabs 3.5 10.5 10.7, *C. kirki* Kirk's colobus 36 38 40, Tabs 3.5 10.5 10.7, *C. polykomos* western black-and-white colobus 35 36 37 38 40 179 180 337 372 444 540 Figs 3.29–3.31 3.36 10.48 18.2 Tabs 3.5 10.5 10.7, *C. satanus* black colobus 179 Tabs 10.5 10.7

Erythrocebus patas monkeys
   captivity, effects of 197
   connation 47
   dental caries 465 467 Fig 21.4 Tab 21.3 21.4
   dento-alveolar abscess 506
   injury, tooth 506–507
   numerical variation 47 48 Figs 3.51
     disto-molars 47
   position anomalies 178 197–200 Figs 10.72–10.76 Tab 10.3
     impaction 199 348 Figs 10.73–10.76
     underjet 199 Fig 10.76
   retained deciduous teeth 350 Figs 10.73 10.74 10.76
   root anomalies 47 48 Figs 10.73 10.75
   tooth shape anomalies 47–48 Fig 3.51

Macaca macques and rhesus monkeys
   bone growth, defective 205 Figs 10.86 10.87
   captivity, effects of 205–208 539 Fig 24.18
   connation 51 208
   dental calculus 539
   dental caries 458 462 472 Tabs 21.3 21.4
   dento-alveolar abscess 507 510–511 Fig 23.5 Tab 23.1
   diet 442 528
   enamel hypoplasia 440 442 Figs 10.90 20.9
   enamel mineralization 458
   eruption, endocrine infuence 334
     non-eruption 346 Figs 10.9 16.4
     sequence and chronology 336–337 Tabs 16.2 16.3
   experimental 458 462 510–511 525, 528 539
   injury, developing tooth 396 Fig 19.4
     skeletal 371–372
     tooth 430–432 Figs 19.48 19.51
   interglobular dentine 440
   mesio-dens 50
   mouth-protozoal parasites 522
   necrotizing ulcerative gingivitis, experimental 539
   normal dentition 202–203 Fig 10.78
   numerical variation 35 49 50 Fig 3.55 Tab 3.3
     disto-molars 50
   osteomyelitis 511
   periodontal disease 538–539 Tab 24.1
   position anomalies 203–208 Figs 10.79–10.91 Tabs 10.3 10.4 10.15 10.16
     open bite 205
     underjet 203 Tab 10.3
   radicular cysts 511
   retained deciduous teeth 205 350 Figs 10.88 10.90
   root anomalies 51 Figs 3.51 3.59
   scurvy 539
   tooth shape anomalies 50 Figs 3.56–3.58
   tooth wear, abrasion 495
   Macaca andamensis Andaman macque 50, *M. arctoides* bear macaque Fig 10.87, *M. assamensis* Assam macaque Tab 10.16, *M. cyclopis* Taiwan macaque Tab 10.16, *M. fascicularis* crab-eating macaque 50, 462 522 528 539 Figs 10.81 10.83 10.84, 10.89 23.5 Tab 10.16, as experimental animal 462 528, *M. fuscata* Japanese macaque 337 528 Tab 10.16, as experimental animal 528, *M. mulatta* rhesus monkey 49 50 204 206 334 336 371 396 430 458 459 472 495 507 510–511 525 528 539 Figs 3.57–3.59 10.79 10.82 10.85 10.91 19.4 19.48 24.18 Tabs 10.16 16.2 23.1, as experimental animal 451 458 462 510 525 528 539, *M. nemestrina* pig-tailed macaque 50 528 Tab 10.16, as experimental animal 528, *M. nigra* black ape 442 Figs 3.56 10.80 10.86 20.9 Tabs 10.16 10.80, *M. philippinensis* Phillipine macaque Tab 10.16, *M. radiata* bonnet macaque Tab 10.16, *M. silenus* lion-tailed macaque Tab 10.16, *M. sinica* toque monkey 50 431 Fig 19.51 Tab 10.16, *M. sylvanus* Barbary ape 49 50, Fig 3.55.

Mandrillus mandrills
   captivity, effects of 212
   connation 49 Fig 3.54
   dental caries 465 Tabs 21.3 21.4
   eruption, non-eruption 346 Fig 16.2
   injury, traumatic displacement of canine 428–429
   numerical variation 49 Tab 3.3
   position anomalies 212 Tab 10.3
     underjet Tab 10.3
   root anomaly 49
   tooth shape and size anomalies 49
   Mandrillus leucophaeus drill 49, *M. sphinx* mandrill 49 346 Figs 3.54 16.2

Nasalis langurs
   dental caries Tabs 21.3 23.1
   dento-alveolar abscess Tab 23.1
   eruption sequence 336 338
   numerical variation 35 Tab 3.3
   position anomalies Tabs 10.3 10.4

Papio baboons
   bone disorders 49 210 350 441 473
   captivity, effects of 210–211 472–473 539 540 Figs 10.99 10.100 19.2
   connation 49
   dental caries 472–473 540 Tabs 21.3 21.4 23.1
   dentigerous cyst 579
   dento-alveolar abscess 511 Tab 23.1
   diet 157
   enamel hypoplasia 210–211 441–442 Figs 10.103 20.6–20.8
   eruption, non-eruption 49 210 346, Figs 10.102 10.103
     over-eruption Fig 10.94
     sequence 337 Fig 16.1
   experimental animal 528
   gingival hyperplasia in captivity 540
   injury, developing tooth 395–396, Fig 19.2
     skeletal 371
     tooth 432
     traumatic displacement of canine 430 Fig 19.49
   interglobular dentine 440 442 Fig 20.8

CERCOPITHECIDAE (cont.)
  normal dentition 208 441 Figs 10.92 10.99 20.5
  numerical variation 48 49 Fig 3.53, Tab 3.3
    disto-molars 48 49
  odontome 582–584 Fig 25.11
  osteometrics 211
  periodontal disease 539–540 Tab 24.1
  position anomalies 208–211 Figs 10.93–10.104 Tabs 10.3 10.4
    impaction 209–210 Figs 10.91 10.98 10.103
    underjet Tab 10.3
  retained deciduous teeth 49 350 Figs 10.95 10.101–10.103
  tooth shape and size anomalies 49
  tooth wear 472 540 Fig 21.15
  Papio anubis olive baboon 48 49 528 540 Figs 10.93 10.98, *P. cynocephalus* yellow baboon 48 49 336 337 472 540 Figs 10.100 16.1, *P. hamadryas* hamadryas or Arabian baboon 48 211 472 540 579 Figs 10.92 10.96 10.104 21.5, *P. ursinus* chacma baboon 48 49 336 350 395 582–583 Figs 3.52 10.94–10.97 10.99 10.101 19.12 20.7
  Presbytis leaf monkeys and langurs
    connation 41 42 Figs 3.39 3.40
    dento-alveolar abscess Tab 23.1
    eruption sequence 337
    mesial drift 436 Fig 19.57
    numerical variation 35 41 42 43 Tab 3.3
      disto-molars 41
    osteometrics 185 Tab 10.10
    periodontal disease 537 Figs 24.13 24.14 Tab 24.1
    position anomalies 184–187 Figs 10.49–10.54 Tabs 10.3 10.9–10.11
      overjet 184 Fig 10.49 Tabs 10.9 10.10
      underjet 177 Tabs 10.9 10.10 10.11
    root anomalies 44 Fig 3.44
    tooth shape anomalies 42 43 44 Figs 3.40–3.43
  Presbytis aygula grey leaf monkey 43 44 Fig 3.42, *P. comata* Tab 10.9, *P. cristata* silvered leaf monkey 42 186 Figs 10.49–10.51 24.14 Tabs 10.9 10.11, *P. entellus* Hanuman and Ajax langurs 41 184 185 Fig 3.39 Tabs 10.9–10.11, *P. frontata* white-fronted leaf monkey Fig 24.13 Tab 10.9, *P. melalophos* banded leaf monkey, crested lutong 42 43 44 337 Figs 3.42 3.44 Tabs 10.9 10.11, *P. obscura* dusky leaf monkey 42

Fig 10.54 Tab 10.9, *P. phayrei* Phayre's leaf monkey 42 Figs 3.40 10.53 Tab 10.9, *P. pileata* capped langur 42 186 Fig 3.41 Tab 10.9, *P. potenziania* Mentawai islands leaf monkey Tab 10.9, *P. rubicunda* maroon leaf monkey 43 44 337–338, Fig 10.52 Tab 10.9, *P. thomasi* Thomas's leaf monkey 187 Tab 10.9, *P. vetulus* purple-faced langur 41 436 Fig 19.57 Tab 10.9
Procolobus verus olive colobus
  numerical variation 35 38
  periodontal disease 540
  position anomalies 183
  tooth shape anomalies 41
Pygathrix snub-nosed monkeys
  numerical variation Tab 3.3
  position anomalies 187–188 Tabs 10.3 10.4
Rhinopithecus snub-nosed monkeys
  connation 40
  eruption sequence 338
  numerical variation 35 37 38 Tab 3.3
    disto-molar 37
  position anomalies 187–188 Fig 10.55 Tab 10.3
  tooth size anomalies 40
Simias pig-tailed langur
  position anomalies 188 Tab 10.3
Theropithecus gelada
  dental caries Tab 21.3
  periodontal disease, in captivity 540
  numerical variation Tab 3.3
  position anomalies 212–213 Fig 10.105 Tab 10.3
  underjet Tab 10.3
Cercopithecus see Cercopithecidae, and experimental animals
CERVIDAE, deer, *see also* Ruminantia
  achondroplasia 288
  antlers 110
  bulldog skull deformity 288
  canines 110 111 112 117 291
  captivity, effects of 551 Fig 24.33
  dental caries 481
  dental formula 110
  dento-alveolar abscess 507 514
  diet 333
  enamel fluorosis 451 454
  enamel pearls 113
  eruption, retarded 333
  feral populations 288 561
  geographical variation 112 288 562
  injury, mandible 383–384 387 454
  lumpy jaw 516 519
  mesial drift 157
  necrotic stomatitis 563
  necrotic ulcerative gingivitis 563

normal dentition 110 290 291 Tab 13.2
numerical variation 110–114 Figs 6.10 6.12
  disto-molars 113
osteomyelitis 514
periodontal disease 550 551–552 561–562
position anomalies 288–292 Figs 13.14–13.24
  parrot mouth 288
  pseudo-arthrosis 375
septic pneumonia and dento-alveolar abscess 507
sub-fossil forms 112 113
tooth shape anomalies 113 114 Fig 6.11
tooth wear 157 454 488–491
ulceration 562
wedging, of plant material, interdentally 514 562
Alces alces European elk and American moose 111 454 491 516 562 563, *Blastocerus* marsh deer 290, *Capreolus capreolus* roe deer 111 113 114 157 288 290 291 454 561–562 Figs 13.19–13.32 Tab 13.2, *Cervus axis* spotted deer 113 507, *Cervus canadensis* wapiti (American elk) 111 112 113 454 491 516 563, *Cervus dama* fallow deer 111 113, *Cervus elaphus* red deer 112 375 383–384, introduced into New Zealand 112, *Cervus eldii* thamin deer 551 Fig 24.33, *Cervus nippon* sika deer 111, 288 489–490 562, *Elaphodus* tufted deer 110, *Hippocamelus* guemal 289 Fig 13.18, *Hydropotes* Chinese water deer 110 113 291–292 Figs 6.12 13.24, *Mazama* brocket deer 111 289, *Moschus* musk deer 110 288 Fig 13.13, *Muntiacus* muntjacs 110 113 114 289, *Odocoileus* American deer 289, *Odocoileus hemionus* mule deer 111 288 454 481 488–491 514, *Odocoileus virginianus* white-tailed deer 111 112 113 288 289 333 384 451 454 563 Fig 13.17, *Rangifer tarandus* caribou and reindeer 111 387 516 563
Cervus, see Cervidae
Cetacea, whales, *see also* Odontoceti and Mysticeti 100
Chaeropus, see Peramelidae
chamois, *see* Bovidae
Chédiak-Higashi syndrome 523, in cattle 523, cat 523, laboratory mouse 523, man 523 546, mink 523 546
cheek (post-canine) teeth
  problems of deciduous or permanent 12
  tooth succession 12 13

cheek (post-canine) teeth (cont.)
    problems in Metatheria 13
cheetah, *see* Felidae
CHEIROGALEIDAE, mouse and
    dwarf lemurs, *see also* Primates
  dental formula 60
  normal dentition 60
  numerical variation 60 Tab 3.8
  position anomalies 232–237
  Cheirogaleus major greater dwarf
    lemur 60, *Cheirogaleus trichotis*
    hairy-eared dwarf lemur 60,
    Microcebus murinus lesser mouse
    lemur 60
Cheirogaleus, see Cheirogaleidae
Cheseldon, W. Fig 17.25
chevrotain, *see* Tragulidae
chimpanzee, *see Pan troglodytes*
Chinchilla, see Rodentia
chipmunk, *see* Rodentia
Chironectes, see Didelphidae
CHIROPTERA, bats
  alveolar dehiscences 570
  ancestry 134–135
  atavism 135
  connation 136 137 138
  dental caries 484–485
  dental formulae 134 135 136 137 138
  diet 484 570
  enamel cracks 484
  normal dentition 134
  numerical variation 134–138 Tab 7.3
  periodontal disease 570
    mite induced 570
  tooth wear 570
  Anoura Geoffroy's long-nosed bats
    136 570, *Balantiopteryx* least sac-
    winged bats 137 138, *Choero-
    niscus* long-tailed bats 137,
    Desmodus vampire bat, tooth
    wear, thegosis, 357, *Eptesicus*
    serotine bats 135, *Glossophaga*
    long-tongued bat 136 137 570,
    Hylonycteris Underwood's long-
    tongued bat 137, *Leptonycteris*
    long-nosed bats 137 570,
    Lichonycteris brown long-nosed
    bat 137, *Lionycteris* little long-
    tongued bat 136, *Lonchophylla*
    Brazilian long-tongued bat 135
    136, *Monophyllus* Barbados long-
    tongued bat 137 570, *Mormoops*
    ghost-faced bats 138 484, *Myotis*
    mouse-eared and Daubenton's
    bats 135 136 570 Fig 7.3,
    Peropteryx sac-winged bats 137,
    Phyllostomus spear-nosed bats
    484, *Pipistrellus* pipistrelles 135,
    Plecotus long-eared bats 135,
    Pteronotus naked-backed and
    moustached bats 138,
    Rhynchonycteris proboscis bat
    137, *Saccopteryx* white-lined bats
    137, *Scotophilus* yellow bats 135,
    Tadarida freetail bats 135

Choeroniscus, see Chiroptera
Choeronycteris, see Chrioptera
Chrysochloris, see Insectivora
chuditch, *see* Dasyuridae
civets, *see* Viverridae
cleft palate 10 12 158
  tooth anomalies adjacent to 10 12
Clethrionomys, see Rodentia
coati, *see* Procyonidae
coatimundi, *see* Procyonidae
cobalt deficiency and cara inchada 561
Coendu, see Rodentia
colobus monkeys, *see* Cercopithecidae
Colyer, J.F. (later Sir F.)
  frontispiece vi
  preface to original edition xv–xvi
  ix–xi 2 3 4 35 62 85 96 128–129 140
    143 159 162 180 183 195 206 286
    292 402 418 432 458 495
Conepatus, see Mustelidae
CONNATION 6 7–9 Fig 1.5, *see also*
    under taxonomic headings
  breeding experiment in dog 9
  cementum hyperplasia, joining by 9
  developmental stages 9
  dichotomy concept 6 86
  dichotomy in *Rhinopithecus* 40
  fusion concept 8 9 71
  geographic isolation 133
  induced by trauma 400–401
  induced by X-irradiation 9
  inheritance 9 133
  odontome concept 575 576 586–587
    Fig 25.16
  roots in 8 47
  in *Cercopithecus* 47 Fig 1.5, dog 9
    Fig 1.5, mouse 9, rice rat 9 133,
    viper 9
Connochaetes, see Bovidae
continuous growth, teeth of
  canines 364–368
  cheek teeth 362 368–369
  incisors 130 332 355–362 364
  injury 428–435
  overgrowth 355–369
  in rodents and lagomorphs 355–362,
    other mammals 362–369
copper deficiency, and cara inchada
    561, tooth wear 492
coyote, *see* Canidae
coypu, *see* Rodentia
crib-biting abrasion, in bear 507,
    horses 494, mink 546, pigs 494,
    rhesus monkey 485, sheep
    491–493
Cricetulus, see Rodentia, and
    experimental animals
Cricetus, see Rodentia
Crocidura, see Insectivora
Crossarchus, see Viverridae
Cryptotis, see Insectovora, and
    experimental animals
Ctenodactylidae, *see* Rodentia
Ctenomys, see Rodentia
Cunningham, G. 422

cusp terminology 13–15
cyclopia, in sheep 154 Fig 9.2
Cynogale, see Viverridae
Cynomys, see Rodentia
Cystaphora, see Phocidae
cysts, *see* dentigerous cysts, and
    radicular cysts

Dactylopsilia, see Phalangeridae
Darwin, C. 7 134 283 292 294
Dasycercus, see Dasyuridae
dasyure, *see* Dasyuridae
DASYURIDAE, marsupial mice and
    cats, etc, *see also* Marsupialia
  bone disorder 602 Fig 25.37
  connation 147 148 Fig 8.9
  dental formulae 147 148
  dento-alveolar abscess 518
  field theory 147
  injury, mandible 379 Fig 18.10
  lumpy jaw 518
  normal dentition 146 327–328
  numerical variation 146–148 Figs
    8.8–8.10
    disto-molars 147 Fig 8.8
  position anomalies 327–329 Tab
    15.1
    open bite 329
  tooth shape and size anomalies 147
    148 Fig 8.8
  Antechinomys dunnart 148,
    Antechinus antechinus 148,
    Dasycercus mulgara 148,
    Dasyuroides kowari 148 328,
    Dasyurus marsupial cats
    (chuditch, satanellus and tiger-
    cat) 146 147 Figs 8.8 8.9
    18.10, odontome v. osteo-
    dystrophy 602 Fig 25.37,
    Myoictis three-striped marsupial
    mouse 148 328, *Phascogale* tuan
    and wambenger 148 328,
    Phascolosorex red-bellied and
    narrow-striped marsupial mice
    147 148 Fig 8.10, *Sarcophilus
    harrisii* Tasmanian devil 148 328,
    odontome 580–582 Fig 25.7,
    Sminthopsis dunnart 148 329 518,
    Thylacinus Tasmanian wolf 148
Dasyuroides, see Dasyuridae
Dasyurus, see Dasyuridae
Daubentonia, see Daubentoniidae
DAUBENTONIIDAE, aye-ayes, *see
    also* Primates
  normal dentition 59 363 Fig 17.10
  numerical variation 60 Tab 3.8
  overgrowth, incisors 363 Fig 17.11
  position anomalies 232–237 Tab
    10.21
  Daubentonia madagascariensis aye-
    aye 60 363 Figs 17.10 17.11 Tab
    3.8
DECIDUOUS TEETH
  connation 9 26 29 33
  dental caries in pongids 465

DECIDUOUS TEETH (cont.)
  dental formulae, Eutherian 12
    Metatherian 13
  enamel fluorosis 451 457
  eruption sequence and chronology
    335–346 Tabs 16.1 16.2 16.4–16.8
  extra canine, in orang-utan 29 Fig
    3.20
  in crypt of successor 428
  in marsupials 13 346
  injury to and malformation of
    successor 403 439 441 Figs 19.12
    19.13
  retained 17 39 49 88 91 115 121 133
    205 226 232 236–237 251 269 298
    332 346 349–353 359 397–398 443
    Figs 3.29 3.33 10.73 10.74 10.76
    10.88 10.90 10.95 10.101–10.103
    13.41 13.69 13.70 13.79
  rudimentary, in Phalangeridae 13
  shedding of 332
deer, *see* Cervidae
degu, *see* Rodentia
DEHISCENCES and fenestrations,
    alveolar
  age change as 524–525 541
  dietary deficiency and 541
  osteoporosis and 524 541
  in bats 570, *Gorilla* 524 532,
    monkeys 524 537–538 541–542,
Delphinus, see Odontoceti
Dendrohyrax, see Hyracoidea
Dendrolagus, see Macropodidae
dental bacterial plaque, *see* dental
    plaque
DENTAL CALCULUS, *see also*
    under taxonomic headings and
    under experimental animals
  bacterial flora, haematin producing
    524
  black surface deposits 524 573
  coronal (supragingival) 523 524 542
    569
    chemical composition 524 551 567
      573
    in man 524
  experimental 526–529
    in germ-free animals 526
  lamina structure 547
  loss in skull preparation 524 529
  ring indicative of gingival margin
    524
  subgingival 523 524 569
DENTAL CARIES 455–485, *see also*
    under taxonomic headings and
    under experimental animals
  age and 458 463
  arrested caries 456–457
  cariogenic micro-organisms 456 457
    462 479
  chemico-parasitic theory 456
  chemico-physical properties of food
    debris 458
  definition 455 501
  dental plaque, *see* separate entry

distinction from action of marine
    parasites 498–501 Fig 22.13
distinction from post-mortem
    change 485 501
in enamel hypoplasia pits 456
enamel protein in 456
epidemiology, criteria for 462
experimental 461 462, *see also*
    under experimental animals
fissure 461
in fossil mammals and reptiles 485
gingival recession and 457
gnotobiosis 462
in man 455 458 461 462 471 478
    479
museum collections
    unrepresentative 465
of dentine, primarily 457 458 463
    Fig 21.1
of dentine, secondarily 457 463
of enamel, primarily 457 458 463
    465
of root surface 457 458 461 463 527
    Fig 21.1
periodontal disease and 457 527
poor tooth structure and 457
    458–460
  Colyer's views 458
proteolytic process as 485
reaction to by dento-pulp complex
    457
refuse scavenging and 465 469 477
remineralization 457
resistance, fluorotic enamel and 456
    459
resistance, other elements and 456
smooth surface 461
tooth wear and 457–458 465 Fig
    21.1
dental cysts, *see* dentigerous cysts, and
    radicular cysts
DENTAL FORMULAE, *see also*
    under taxonomic headings
  Eutherian dental formula 12 135 Fig
    1.6
  full in canids 83
  Metatherian dental formula 13
  premolars in primates 12
DENTAL PLAQUE
  bacterial flora and 456 457 521–524
    527
  black surface deposits 524 560
  coronal (supragingival) 456 457 479
    521 528 529
  dental caries and 456–458
  experimental 526 528 529
  gingival 521–524 Fig 24.1
  in germ-free animals 526
  in sheep 557 559–560
  subgingival 522–524 Fig 24.1
dentigerous cysts, *see also* under
    taxonomic headings 492 576–579
    602 Figs 25.2–25.4 25.6
  in fish 606
  growth mechanism 576–577

relationship to radicular cysts
    578–579
DENTINE
  dysplasia 589
  interglobular 438 440 442 Fig 20.8
  ivory, lozenge pattern structure in
    elephants 408 425 Figs 19.21
    19.41
  reactionary, absence of 241 430 Fig
    19.36
  reparative 405, in bear 430 Fig
    19.46, elephant tusks 409 422 424
    Fig 19.21
DENTO-ALVEOLAR ABSCESS,
    see also under taxonomic
    headings, 506–517
  attrition, secondary to in old age
    507 510 511
  canines susceptible to, in Cebidae
    511, pongids 507
  in captivity 506 507
  categories 506–517
  experimentally produced in rhesus
    monkey 510
  septic pneumonia from 507
  deposits on teeth, *see also* dental
    calculus and dental plaque
  black 524 560
  haematin producing bacteria 524
  in man 524
Desmodus, see Chiroptera
development, *see* tooth development
DEVELOPMENTAL ANOMALIES
  achondroplasia, in cattle 293 294,
    deer 288, dogs 266–268, hare 319,
    human dwarfs 266
  asymmetry of 3 10, in gorilla 167
    Fig 10.14, horse 310 311 Fig
    13.76, pigs 283, rabbit 319, spider
    monkey 221–223 Figs 10.123
    10.126 10.127, woolly monkey
    220, Fig 10.121
  bilaterality of 25 41 44 45 46 49 75
    77 84 98 119
  bulldog head and similar skull
    deformities, bulldogs 86 87 266
    269 270 478 548 Figs 11.70–11.75
    Tab 11.15, bull-terriers 86 447
    Figs 4.40 10.16, coyote 263,
    crocodile 158, fish 159, fox 263,
    goats 297, hare 319, Jersey cattle
    294, Niata cattle 292–294 Fig
    13.28, pigs 282–283, rabbit 361,
    roe deer 288, snakes 158, white-
    tailed deer 288
  captivity and 264
  cleft palate 10 12 158
  cyclopia, in sheep 154 Fig 9.2
  domestication and 10 266 282–283
  double jaws, in fish 601, sheep 601
  experimental teratogenesis 10
  genetic vs environmental causes 1
    9–11
  geographic isolation and 7 10 297
  hemifacial microsoma 271

DEVELOPMENTAL ANOMALIES (cont.)
  incidence statistics, unreliability of 20
  jaw disproportion, *see also* jaw growth, and position anomalies 153 156 159
    terminology 160
    in dogs 265–279 Figs 11.70–11.82, fish 158–60, foxes 264, pigs 282–283 Tab 13.1, reptiles 158–159, sheep 601, wolves 264
  maternal rubella in man 10
  multiple character 10 20
diastemata
  in mesial drift 157
  in horses 157, New World monkeys 221 223 230, rodents 157
dibatang, *see* Bovidae
Diceros, *see* Rhinocerotidae
dichotomy concept, *see* extra teeth, and connation
Dicrostonyx, *see* Rodentia
DIDELPHIDAE, opossums, *see also* Marsupialia
  captivity, effect of 485
  connation 149 Fig 8.12
  dental caries, in captivity 485
  dental formulae 13 149
  normal dentition 149 329
  numerical variation 149–150
  position anomalies 329 Figs 15.21–15.24 Tabs 15.1 15.2
  tooth shape and size anomalies 149–150 Figs 8.13 8.14
  root anomalies 149–151
  Caluromys woolly opossum 150, *Chironectes* water opossum 151, *Didelphis* American opossums 149–150 328 329 485 Figs 8.12 8.13 15.21 15.24 Tab 15.2, *Glironia* bushy-tailed opossum 151, *Lestodelphys* Patagonian opossum 151, *Lutreolina* little water opossum 150 Tab 15.2, *Marmosa* mouse opossums 151 329 Tab 15.2, *Metachirus* four-eyed opossums 150 Fig 15.23 Tab 15.2, *Monodelphis* short-tailed opossums 150 151 Fig 8.14 Tab 15.2, *Philander* four-eyed opossum 150 Tab 15.2
Didelphis, *see* Didelphidae
DIET, *see also* vitamin deficiency, and vitamin A excess, and bone disorders
  alveolar dehiscences and 541
  bone growth and 156–157
  dental caries and, in bats 484, dogs 479 546, guinea pigs 458, rhesus monkeys 458
  distemper enamel hypoplasia and, in dogs 445–447 Fig 20.14
  enamel mineralization and, in dogs, rabbits and guinea pigs 458–459,
  see also under enamel mineralization
  eruption and, in white-tailed deer 333, golden mole *Amblysomus* 569
  jaw length and tooth position and, in foxes 264
  normal, of bandicoots 572, gerbils 568, guinea pigs 362, marmosets and tamarins 541, New World monkeys 541, raccoons 544
  overgrowth, tooth wear and, in rats and mice 361 362
  periodontal disease and, in bandicoots 572, bats 570, cats 543–544, dogs 546 549–550, guinea pigs 525, golden mole *Amblysomus* 569 Fig 24.46, gorilla 530–531, hamster 527, hedgehog 569, lemurs 540, macaques 528, raccoons 544, rats 526 568, sheep 560
  periodontal disease, experimental and 525–528
  pigment formation in rats 361
  pseudo-odontomes and, in rats and mice 605
  tooth size and, in mice 131, rats 10
dik-dik, *see* Bovidae
dilaceration, *see* injury, tooth
dilambdodont teeth, in bats 134
dingo, *see under Canis familiaris*
Diprotodontia 140
Distoechurus, *see* Phalangeridae
disto-molars, *see also* under numerical variation, and under taxonomic headings
  definition 20
  inheritance in rice rat 133
  in man 22
  in *Otocyon* bat-eared fox 84
  morphogenetic field concept 6
dogs, wild, *see* Canidae
dolphins, *see* Odontoceti
DOMESTICATION 7 10 523
  Darwin on domestication of pigs 283
  of *Canis* 10 85 264–266 Fig 11.69, pigs 283, *Vulpes vulpes* farm silver fox 84
donkey, *see* Equidae
Dorcopsis, *see* Macropodidae
Dorcopsulus, *see* Macropodidae
dormouse-possum, *see* Phalangeridae
double teeth, *see* connation and cementum hyperplasia
douroucouli, *see* Cebidae
drill, *see* Cercopithecidae
dromedary, *see* Camelidae
Dryopithecus molar fissure pattern 20
duck-billed platypus, *see* Monotremata
duiker, *see* Bovidae
dunnart, *see* Dasyuridae

echelon appearance of cheek tooth arches, in *Ateles* 222, *Cebus* 218
Fig 10.116, *Macaca* 202 204 Fig 10.83, ruminants 286 289 Figs 13.35 13.44
echidna, *see* Monotremata
Echymipera, *see* Peramelidae
ectodermal dysplasia, cat 63–64, in cattle (calf) 115, *Colobus* 39–40, dog 88, in horse (foal) 450
ectopic teeth 23–24 153 346 599–601 Figs 3.3 16.3
EDENTATA, sloths
  normal dentition 134
  Megalonyx ground sloth, fossil, extra caniniform tooth 134, *Megatherium americanum* ground sloth 134, connation 134
Eira, *see* Mustelidae
Elaphodus, *see* Cervidae
ELEPHANTIDAE, elephants, mammoths and mastodons
  alveolar tube 155
  dental formula 125
  molars
    eruption sequence and chronology 125 345 Tab 16.12
    fusion, cemental 129
    injury 432–434 Figs 19.53 19.54
    position anomalies 312–315 Figs 13.84 13.85
    shape anomalies 314–315 Fig 13.86
  normal dentition 125 312
  tusks
    abscesses in and on 426–427 507 Figs 19.43–19.45
    connation 128 Figs 6.7 6.8
    dilaceration 407 Figs 19.19 19.20
    eruption 345
    erosion 504
    foreign bodies in 420–427 434 Figs 19.31–19.42
    growth rate 406
    injury 8 128–129 405–420 Figs 19.18–19.44, in mammoth 428–429
    ivory lozenge pattern 408 425
    numerical variation 126–129 418 Figs 6.42–6.46 6.49 Tab 6.3
    pulp stones 424 426 Fig 19.40
    relation to jaws 405 Fig 19.17
    resorption 504–505 Fig 22.18
    shape anomalies 128 312
    spiral tusks 419–420 Figs 19.28–19.30
    tusklets 8 412–418 Figs 19.24–19.26
  Elephas maximus Indian elephant 128 312 314 345 406 504–505 Figs 6.44 6.46 6.47 13.86 19.17 19.26 19.27 19.39 Tab 6.3
  Loxodonta africana African elephant 126 127 312 406 407 Figs 6.42 6.43 13.84 19.18 19.19 19.23–19.25 19.28–10.30 Tab 16.12, 'Jumbo' 312–314 507

ELEPHANTIDAE (cont.)
  Mastodon fossil elephant, dental caries 485
  Tetrabelodon fossil elephant, multiple tusks 127
elephants, *see* Elephantidae
Elephas, *see* Elephantidae
elk, *see* Cervidae
emergence, tooth, *see* eruption
ENAMEL
  absence in some Odontoceti 102
  crazing 484–485 487 505
  in civets 444 Fig 20.13
  in dog 479
  in pig 479
  structure, species difference in primates 462
  thickness, influence on tooth size and sex dimorphism 10
  wrinkled, in orang utan 440 441, some odontoceti 105 Fig 5.12
ENAMEL FLUOROSIS 361 362 438 450–454
  coronal cementum 451 453
  enamel brittleness 452
  experimental 451
  F compared with Se, Mn, Mo 451
  health, effect on 452
  mottled enamel 451
  of deciduous teeth 451 457
  osteo-arcthritis 453
  periodontal disease, relation to 452
  pigmentation, loss of rat incisors 361
  placental passage of F 453 454
  poisoning, fluoride 362
  pollution and 450 451 453–545
  ranking scale 452
  skeletal changes 450 451 453
  sources of F 450 451 454
  wear, rapid irregular 452–454 Figs 20.19 20.20
  in cattle 450 453, *Cercopithecus* 451 Fig 20.18, deer 451 454, goats, experimental 451, guinea pig 362, herbivores 438 451, horse 451, man 450 451 452, pronghorn 454, rabbit 454, rodents 361 262 450, sheep 451–453, sheep, experimental 451
enamel hypocalcification 437 450
ENAMEL HYPOPLASIA, *see also* under taxonomic headings
  bone disorders and 334 350 438 441
  chronology 438 440 442
  dental caries and 456
  distemper enamel hypoplasia in dogs 445–447
  ectodermal dysplasia in foal 450
  endemic fluorosis, *see* enamel fluorosis
  genetic aspects 437 440 443 444 447 450
  geographical variation in 450
  hypoparathyroidism and 438

  in man 437 438 440 444 447
  infection and 437 438 439
  interglobular dentine and 438 440 442
  longitudinal grooving 439 441 442 445 Fig 20.14
  Mellanby's categories of 440 445 Fig 20.14
  nutrition and 437 438 441 442 445 458
  placental barrier protection 438
  retained deciduous teeth 350
  ring-type 438–442 444 447 450 Figs 10.20 10.90
  systemic origin 437–438 450
  tetracycline induced 438
  tooth absence and 38 443 447 450 Figs 3.32 20.15
  traumatic origin 402–403 437 439 Figs 19.12 19.13
  experimental 403–404 439
enamel-lined invaginations 576 Fig 25.1
  dilated composite odontomes 589–593
ENAMEL MINERALIZATION 458–459
  dental caries and 458–459
  factors affecting 458
  methods of study 458–459
  in *Ateles* 458, chimpanzee Figs 21.3, felids 458 Fig 21.6, gorilla Figs 21.2 21.4, guinea pigs 458, lagomorphs 458, man 458, mustelids 458, *Pithecia* Fig 21.5, pongids 458 459 Figs 21.2–21.4, rhesus monkey 458 459, waterbuck 491
enamel pearls and root nodules 79 113 576 Fig 4.6
  in African civet Fig 4.6, bears 79, fallow deer 113, man 576, wild sheep 576
endocrine system, *see also* hyperparathryroidism and hypoparathryroidism
  disorder, in dolphin 494
  hypopituitary dwarfism, in cattle 294
  parathyroids 334 347 438
  periodontal disease and 527 528
  thyroid 334
Enhydra, *see* Mustelidae
epithelial root sheath 576
  and root anomalies 102
Eptesicus, *see* Chiroptera
EQUIDAE, horses, asses, zebras, *see also Equus caballus* domestic horse, and Perissodactyla
  captivity, effects of 567
  dental calculus, chemical composition 567
  dental caries 481 483
  dental formula 118
  dento-alveolar abscess 514
  eruption, over-eruption 567
  injury, tooth 435

  normal dentition 15 118 481 487 Figs 1.8 21.24 22.2
  odontome 592 Fig 25.23
  periodontal disease 567
  position anomalies 307
  shear bite 487
Equus africanus wild ass 567, *E. asinus* domestic donkey 121 483 487 567, *E. hemionus* onager 481 567, *E. quagga* quagga 481, *Equu* sp fossil horse, dento-alveolar abscess 514–515, odontome 592 Fig 25.23, *E. zebra* mountain zebra 435
Equus, *see* Equidae and *Equus caballus* domestic horse
Equus caballus domestic horse, *see also* Equidae
  arrested growth of developing tooth 396–397 Fig 19.7
  asymmetry of skull 310 311 Fig 13.76
  bone disorders 334 604
  bran disease 604
  canines 121 Fig 6.32 Tab 6.1
  cementomas 596–598 Figs 25.27–25.31
  connation 119–121 Figs 6.27 6.28 6.29–6.35
  cysts, multiple 604
  dental caries 481–483 564 Figs 21.25 21.26 24.43 24.44
  dentigerous cysts 579
  dento-alveolar abscess 513 514 Fig 23.6
  ectodermal dysplasia 450
  enamel fluorosis 451
  enamel hypoplasia 448 Fig 20.17
  eruption, non-eruption 346
    over-eruption 348 567 Figs 16.10 22.1
    sequence and chronology 345
  infundibulum 481 483 Fig 21.24
  injury, jaw 372 388
    tooth 342 435 483 567
  lumpy jaw 516
  mesial drift 157 348 Fig 16.9
  numerical variation 118–122 235 447 450 Figs 6.20–6.26 6.33 13.69 20.17
    disto-molars 122
  odontomes 585–586 590–592 601 Figs 25.15 25.23 25.34 25.35
  osteometrics 309 Fig 13.76
  periodontal disease 559 563–567 Figs 24.39–24.45 Tab 24.2
  position anomalies 307–311 Figs 13.72–13.75
  impaction 348 Fig 16.9
  overjet (parrot mouth) 309 Fig 13.72
  shear bite 487 Figs 22.2 22.3
  underjet 309
  resorption, root 564 Fig 24.45
  subgingival burrowing 505

Equus caballus (cont.)
  retained deciduous teeth 121 307–398 352 Figs 13.69 13.70
  tooth wear 157 487–488 Figs 22.1–22.3
    bit wear 494
    crib-biting 494
    occlusal cupping out 487–488 Fig 22.4
    wedging, of plant material, interdentally 297 563–567 Figs 24.39 24.42 24.44
  wolf teeth 121–122 352 Tab 6.2
Erethizion, see Rodentia, and experimental animals
Erignathus, see Phocidae
Erinaceus, see Insectivora
erosion, see under tooth wear
ERUPTION 11 153–155 331–354 356
  anomalies 11, see also various entries below
    in golden mole *Amblysomus* 569
  bone disorders (fibrous osteodystrophy, osteopetrosis, osteomalacia, renal rickets and rickets) 333 334 346 Figs 16.3–16.1
  bone growth and resorption in 11 332 333
  captivity 334 336 346 350 Figs 10.90 10.102 11.5 16.2–16.6 Tabs 16.1 16.2 16.4 16.5
  chronology 335–346 Tabs 16.1–16.12
  continuous 331–332
  criteria 331 334
  distal drift in rodents 157
  enamel hypoplasia 350
  endocrine control 334 347
  emergence 154 331 334
  eruptive force 332–333
  experimental studies 332 356
  horizontal replacement, see separate entry under mesial drift and horizontal replacement
  hypsodont teeth 331 348 Figs 16.9 16.10
  impaction 160 332 348 Fig 16.9, see also under position anomalies, and under various taxonomic headings
  lathyrism 332–333
  malnutrition 333 334
  mesial drift, molar progression, see separate entries under mesial drift and horizontal replacement
  non-eruption 332 333 346–348 Figs 16.2–16.8
  over-eruption 332 348–349 355 Fig 16.10
  over-growth, see separate entry
  partial eruption 346
  post-emergence movements 154
  re-inclusion 332, in *Gorilla* 353–354 Fig 16.11
  retained deciduous teeth 332 333 349–353 Figs 16.6 16.7, see also under deciduous teeth
  sequence 335–346 Tabs 16.1–16.12
  sequence, primitive mammalian 346
  sex differences 336 338 Tabs 16.2 16.3
  tooth wear and 331 334 346
  in baboons 346 350 Fig 16.1, bears 341 346 Fig 16.6, bovids 342–345 348 352, cat 339 346 Fig 16.7 Tab 16.5, cattle 342 Tabs 16.10 16.11, cebids 338 346 Fig 16.5 Tab 16.4, cercopithecids 336–338 348 349 Fig 16.1 Tab 16.2, chimpanzee 335–336 346–347 Fig 16.3 Tab 16.1, civet 348, *Coendu* tree porcupine 348, *Cuon* Asiatic wild dog 347 Fig 16.8, deer 333, dogs 341–342 350 Tabs 16.7 16.8, *Eira tayra* 348, elephant 345 Tab 16.12, felids 339 Tab 16.5, *Gorilla* 335 346 353–354 Fig 16.11, horse 345 346 348 349 352 Figs 16.9 16.10, hyaenas 339, hylobatids 336, Indriidae 350, insectivores 346 353, lagomorphs 332 345–346 356, Lemuridae 339 350, macaques 331 346 Fig 16.4 Tabs 16.2 16.3, man 331 348, manatees 345, mandrill 346 Fig 16.2, marmosets and tamarins 338–339, marsupials 331 346 353, mustelids 339–341 Tab 16.6, orang-utan 336, pigs 342 348 Tab 16.9, pongids 335–336 Tab 16.1, *Procyon* racoon 348, rhinoceros, fossil 352, rodents 331 332 333 345–346 352 356, seals 342, sheep 345, viverrids 339
Erythrocebus see Cercopithecidae
Eumetopias, see Otariidae
Eutamias, see Rodentia
EXPERIMENTAL ANIMALS
  dental calculus 526–529 551
  dental caries 461–462 528
  dento-alveolar abscess 510–511
  enamel defects 404 439 451
  enamel fluorosis 451
  eruption 333
  environmental deprivation 10
  gingival hyperplasia 528
  injury, tooth germs 404 439
  necrotizing ulcerative gingivitis 539
  periodontal disease 525–529 542 550–551 568–569 Fig 24.3
  tooth development 7 8
  tooth growth 356
  tooth size 131
Ateles geoffroyi spider monkey 528
Bos taurus domestic cattle 451
Callithrix jacchus marmoset 462 525 528 542
Canis familiaris dog 479 526 527–528
Capra hircus domestic goat 451
Cavia porcellus domestic guinea pig 461 525
Cebus apella brown capuchin monkey 462 525 528
Cercopithecus sabaeus green monkey 528
Cricetulus migratorius grey or Chinese hamster 527
Cryptotis parva American least shrew 569
Erethizon dorsatum North American porcupine 356
Felis catus domestic cat 404 439 528 529 543–544 Fig 24.20
Galago bushbabies 542
Macaca mulatta rhesus monkey 451 458 462 510 525 528 539
Macaca fascicularis crab-eating macaque 462 528
Macaca fuscata Japanese macaque 528
Macaca nemestrina pig-tailed macaque 528
Meriones unguiculatus Mongolian gerbil 461 527 528
Mesocricetus auratus Syrian hamster 7 461 462 526 527 528
Mus musculus mouse 7 9 10 131 461–462 526 527 528
Mustela putorius furo domestic polecat and ferret 526 528
Mystromys albicaudatus white-tailed rat 461 527
Oryctolagus cuniculus rabbit 8 333 356 461
Oryzomys palustris rice rat 9 461 527
Ovis aries domestic sheep 404 439 451
Papio anubis olive baboon 528
Rattus norvegicus laboratory rat 10 333 356 404 461 462 525–528 568–569 Fig 24.3
Saguinus tamarin 462 528 542
Saimiri squirrel monkey 462 528
Sigmodon hispidus cotton rat 461
Sus scrofa miniature domestic pig 480 550–551
EXTRA TEETH, see also numerical variation under taxonomic headings, 6–7
  atavism 7 22
  cleft palate, adjacent to 10 12
  conical with several roots 24
  cytotoxic mechanisms 7
  development, asynchronous 6
  dichotomy concept 3 6–7 8–9
  disto-molars 6
  domestication 7
  ectopic 23
  haplodont 3 6 17
  in mice
    aged 7
    mutants 3 9 131 Fig 7.1
  in rice-rat 9

EXTRA TEETH (cont.)
  jaw size in dogs and 86
  morphogenetic field concept 2 3 6 Fig 1.3
  odontome concept 575 576
  retained deciduous teeth, mistaken for 17
  roots of 6
  supplemental 6 17
  trauma in elephants and 128–129
  tuberculate 6 17
  types of 6 17
    bizarre, produced by X-irradiation 9

false joint, see pseudo-arthrosis
FELIDAE, cats, lion, tiger, ocelot etc, see also Felis catus domestic cat, and Carnivora
  ancestral forms 63 64
  atavism 63 64
  bone disorders, rickets 241 Fig 11.5
  captivity, effects of 239–241 477 Fig 11.5 Tab 11.4
  dental caries, in captivity 477
  dental formula 62 238
  dento-alveolar abscess 240
  diet, effect on periodontal disease 543 544
  enamel hypomineralization 458 Fig 21.6
  enamel hypoplasia 402 444–445
  eruption, anomaly 239–240 Fig 11.2
    non-eruption 346 Fig 11.5
    sequence and chronology 339 Tab 16.5
  geographic variation 63 239
  injury, avulsion of canines 435
    developing tooth 397–402 Figs 19.8–19.12
    mandible 375–379 Fig 18.4–18.8
    maxilla 372 Fig 18.1
    tooth 434 435 Fig 19.55
    zygomatic arch 376
  jaw shape 238
  normal dentition 62 240 Fig 11.1
  numerical variation 62–67 Figs 4.1–4.5
    disto-molars 64
  osteometrics of tooth spacing 238–239 Tab 11.3
  periodontal disease 543–544
  position anomalies 238–241 Figs 11.1–11.6 Tabs 11.1–11.4
  pseudo-arthrosis, un-united fracture, in fossil sabre-tooth 376
  quasi-continuous variation 63
  root anomalies 64 66
  tooth shape and size anomalies 63–64
  Acinonyx jubatus cheetah 63 64 375 Fig 18.4 Tab 11.4, *Felis bengalensis* leopard cat 63 66, *F. caracal* caracal 63, *F. catus* domestic cat, see separate entry,

F. chaus jungle cat 66, *F. concolor* puma 63 64 67 Tab 16.5, *F. geoffroyi* Geoffroy's cat 67, *F. issidorensis* fossil cat 63 64, *F. libyca* African and Indian wild cat 63–64 66, *F. lynx* lynx 62 63 64 66, *F. marmorata* marbled cat 66 477 Fig 4.5, *F. pardalis* ocelot 67 Tab 11.4, *F. rubiginosus* rusty-spotted cat 63, *F. rufa* bobcat 62 66, *F. silvestris* European wild cat 63–64 66, *F. viverrina* fishing cat 66, *F., wiedi* Margay cat 67, *F. yagouaroundi* yagouaroundi 67, *Panthera leo* lion 64 238–241 339 372 398–402 434 435 444–445 Figs 11.1 11.5 19.9 19.10 19.55 Tabs 11.2–11.4 16.5, *Panthera onca* jaguar 402 Fig 19.12 Tab 16.5, *Panthera pardus* leopard 64 241 376–379 397 398 Figs 4.1 4.2 11.4 18.5–18.8 19.8 Tabs 11.2 11.4 16.5, *Panthera tigris* tiger 64 239–241 339 379 543 Figs 11.2–11.3 Tab 11.2 16.5, *Panthera uncia* snow leopard 339 543, *Prionailurus chinensis* Chinese leopard cat 66, *Smilodon californicus* sabre tooth tiger 376 435

Felis, see Felidae and *Felis catus* domestic cat
Felis catus domestic cat
  Chédiak-Higashi syndrome 523
  connation 66
  dental caries 477
  diet, effect on periodontal disease 543–544
  ectodermal dysplasia 63–64
  enamel defects produced by experimental trauma 404 439
  eruption sequence and chronology 339 Tab 16.5
  experimental 404 439 528 529 543–544 Fig 24.20
  feral 62 63 64
  gingival hyperplasia, experimental 528
  injury, tooth 435
    tooth germs, experimental 404 439
  necrotizing ulcerative gingivitis 524
  numerical variation 62 64–66 Figs 4.3 4.4
  periodontal disease 543–544
  position anomalies 240–241 Fig 11.6
  resorption 505
  tooth size anomalies 65
  wedging, of object across tooth arch 372
fenestrations, see dehiscences and fenestrations, alveolar
ferret, see Mustelidae, and experimental animals
ferret-badger, see Mustelidae

fibrous osteodystrophy, see bone disorders
FISH, see also non-mammalian dentitions
  change from juvenile to adult dentition 159
  contralateral morphology of teeth 6
  dentigerous cyst 606
  double mouth deformity 601
  injury, mouth parts 393
  malocclusion 159
  odontome, composite 606
  pug-head skull deformity 159
  teeth attacked by boring organisms 499
  traumatic dichotomy of tooth families 7
fisher, see Mustelidae
fluorine, see enamel fluorosis
founder effect, see under geographic variation
foxes, see Canidae

Galago, see Lorisidae, and experimental animals
Galictis, see Mustelidae
Gazella, see Bovidae
gazelle, see Bovidae
gelada, see Cercopithecidae
gemsbok, see Bovidae
genet, see Viverridae
genetic drift, see under geographic variation
genetics, see inheritance
Genetta, see Viverridae
GEOGRAPHIC VARIATION
  connation 133
  dental caries 465
  dento-alveolar abscess 510 Tab 23.1
  enamel hypoplasia 450 454
  geographic isolation, founder effect and genetic drift 7 10 32 35 45 47 63 84 111 112 114 115 131 133 134 135 138 139 142 195–197 231–232 288 297 450 491 561 562 568 Figs 3.47 10.58 10.67 10.68 Tab 10.14
  islands, Aldabra Atoll 297, Corsica 133, Greenland 114 450 562, Hebrides 139, Kerguelen 63, Middle Island, Australia 142, New Zealand 112 115 138 491 561, Sardinia 133, St Kitts 35 45 47 195–197
  numerical variation 134 135 138 139 142 Figs 3.47
  periodontal disease 529–531 561 562 568 Tab 11.14
  position anomalies 167 182 195–197 231–232 239 264 288 297 319 Figs 10.58 10.67 10.68 Tabs 10.14 11.14
  root anomalies 49 134
  tooth shape anomalies 132 134
  tooth size anomalies 10
  tooth wear 491

GEOGRAPHIC VARIATION (cont.)
in bats 135, beavers 131, bovids 115 491, canids 264 Tab 11.14, cats, feral and domestic 63, cattle, Chillingham 115, prehistoric and Roman 115, *Cercopithecus* 35, *Cercopithecus sabaeus* green monkey, on St Kitts 35 45–47 195–197 Figs 3.47 10.58 10.67 10.68 Tab 10.14, chamois 115 561, chimpanzee 22 167 465, *Colobus* 35 182 510 Tab 23.1, *Colobus badius preussi* red colobus 35, deer 111–112 288 454 491 562, foxes 84, gibbons 32, goats, feral 297, gorilla 529–531, hare 134, hedgehogs 138, ibex 115, lions 239, mice, feral 568, moles 139, musk ox 114 450 562, New World monkeys 231–232, Old World monkeys 35 45–47 182 195–197 510 Figs 3.47 10.58 10.67 10.68 Tabs 10.14 23.1, rodents 131 132 133, sheep 491, shrews 139, silver foxes 84, titi monkeys 231–232, wallabies 142, water buck 491
gerbil, *see* Rodentia, and experimental animals
Gerbillus, *see* Rodentia
gibbons, *see* Hylobates
gingiva, *see also* periodontium
 gingival fibroma 541
 gingival hyperplasia 540 561
  induced by dilantin (epanutin, diphenylhydantoin) 528
 gingivitis 522 525
  grass damage and 519
  reptilian state relevant to 523–524
  tooth wear and 523
gliders, *see* Phalangeridae
Glironia, *see* Didelphidae
Globicephala, *see* Odontoceti
Glossophaga, *see* Chiroptera
glutton, *see* Mustelidae
goat, *see Capra hircus*
Goethe, J.W. 421
Gorilla gorilla gorilla, *see also* Pongidae
 alveolar dehiscences and fenestrations 524 532
 asymmetry, arch 167 Fig 10.14
 skull 372
 captivity, effects of 465 572
 Carabelli, cusps of 24
 dental caries 463 Tabs 21.1 21.2 23.1
 dento-alveolar abscess 507 509–510 Fig 23.3 Tab 23.1
 enamel hypomineralized Figs 21.1 21.4
 enamel hypoplasia 439 440 Fig 20.1
 eruption, non-eruption 346 Fig 3.3
  retarded 23
  sequence 335

 geographical variation in periodontal disease 529–531
 injury, canines 428 432 Fig 19.52
 jaw 167 375 380–381 392–393 Fig 18.3 18.13
  zygomatic arch 392
 nectotizing ulcerative gingivitis 535
 normal dentition Fig 10.1
 numerical variation 22–24 167 Figs 3.1–3.3 3.6 16.11
  disto-molars 20 23 Fig 3.6
 odontomes 589 Fig 25.17
 osteomyelitis 509–510
 periodontal disease 529–532 Figs 24.2–24.7
  diet and 530–531
 position anomalies 162–167 Figs 10.3–10.13 Tabs 10.1 10.2
  ectopic teeth 23 Fig 3.3
  impaction 23 Tab 10.2
  underjet 163 Fig 10.2 Tab 10.1
 re-inclusion 353–354 Fig 16.11
 root anomalies 22
 tooth shape anomalies 22 24–26 432 Figs 3.4 3.5 3.7 19.52
 tooth size anomalies 24 25 Figs 3.5 3.7
 tooth wear 531
 yaws 392–393
Grampus, *see* Odontoceti
great apes, *see* Pongidae, and individual taxa
grison, *see* Mustelidae
growth, *see* jaw growth
grysbok, *see* Bovidae
guanaco, *see* Camelidae
guemal, *see* Cervidae
guinea pig, *see* Rodentia, and experimental animals
Gulo, *see* Mustelidae
gummy mouth, *see under Ovis aries* domestic sheep
gundi, *see* Rodentia
Hagmann, G. 54–55 58 76 285
Halichoerus, *see* Phocidae
hamartoma 115 575 587 606, *see also* odontomes
hamster, Chinese and common, *see* Rodentia, and experimental animals
hamster, Syrian, *see Mesocricetus auratus*, and experimental animals
haplodont teeth 7
hare, *see* Lagomorpha
hartebeest, *see* Bovidae
hedgehogs, *see* Insectivora
Helarctos, *see* Ursidae
Helogale, *see* Viverridae
Hemiechinus, *see* Insectivora
Herpestes, *see* Viverridae
Heterohyrax, *see* Hyracoidea
Hippocamelus, *see* Cervidae
Hippopotamus amphibius, hippopotamus
 dental formula 110

 dento-alveolar abscess 514
 normal dentition 110 285
 numerical variation 110 Figs 6.6 6.8
 odontome, denticles 285
 overgrowth, tusks and incisors 363–364 Figs 17.13 17.14
 position anomalies 285 Figs 6.6 13.7 13.8
 tooth shape and size anomalies 110 Figs 6.6 6.7
Hippotragus, *see* Bovidae
Histriophoca, *see* Phocidae
hog, forest, *see* Suidae
homoeosis 2 153
horse, domestic, *see Equus caballus*
horse, fossil, *see* Equidae
Hunter, J. 101 315 355 365 421 494
hyaenas, *see* Hyaenidae
HYAENIDAE, hyaenas, *see also* Carnivora
 dental formula 67
 dento-alveolar abscess 512
 eruption sequence 339
 normal dentition 67 241–242 Fig 11.7
 numerical variation 67 Tab 4.1
 position anomalies 241–242 Figs 11.8 11.9 Tab 11.1
 Crocuta crocuta spotted hyaena 67 242 339 512 Fig 11.9, *Hyaena brunnea* brown hyaena 241–242, *Hyaena hyaena* striped hyaena Fig 11.8
Hydromyinae, *see* Rodentia
Hydropotes, *see* Cervidae
Hydrurga, *see* Phocidae
Hylobates gibbons, *see also* Pongidae
 captivity, effects of 177
 connation 32 33
 dental caries 463 Tabs 21.1 21.2 23.1
 dento-alveolar abscess 507 Tab 23.1
 enamel hypoplasia 439–441 Figs 20.1 20.4
 eruption sequence 336
 geographical variation 32
 injury, mandible 387
 normal dentition 20 32
 numerical variation 22 31–32 Figs 3.23 3.24 Tab 3.1
  disto-molars 20 22 31
 periodontal disease 529 536–537
 polydactyly 20
 position anomalies 175–177 Figs 10.35–10.38 Tabs 10.1 10.2
  impaction 176 177 Tab 10.2
  overjet 175
  underjet 175
 sutural bone 20
 tooth shape anomalies 32–33 Figs 3.24 3.26
 tooth size anomalies 22 32 Fig 3.25
Hylobates agilis dark-handed gibbon 31 32, *H. concolor* black gibbon 31 439 Fig 20.2, *H. hoolock*

Hylobates gibbons (cont.)
  hoolock gibbon 32 Figs 3.23 3.25
    3.26 10.36, *H. lar* white-handed
    gibbon 20 31 32 177 441 Figs
    10.35 10.37 20.4, *H. moloch*
    grey gibbon 20 31 32 Figs 3.24
    10.38
Hylochoerus, see Suidae
Hylonycteris, see Chiroptera
Hyperoödon, see Odontoceti
hyperparathryroidism 604–605
  in vitamin D deficiency 438
  secondary to renal rickets 334 589
    604–605
hypoparathyroidism, in cat 347,
  experimental 438
HYPSODONT TEETH, definition 14
  bovids 114 331 348 487 513 551
  conducive to deep infection of bone
    513
  equids 331 348 487 513 567 Figs
    16.9 16.10
  eruption 331 348 Figs 16.9 16.10
  wear of 348 355
HYRACOIDEA, hyraxes
  dental caries 483 Fig 21.27
  dental formula 124
  normal dentition 124 125 312 Fig
    6.39
  numerical variation 124 Figs 6.37
    6.38
    disto-molar 124
  overgrowth, incisor 363 Fig 17.12
  position anomalies 312 Figs 13.80–
    13.83 Tab 13.6
  tooth shape anomalies 124 125 Figs
    6.38–6.41
  Dendrohyrax tree hyraxes 124 125
    312 Figs 6.38 6.40 6.41 13.80 Tab
    13.6, *Heterohyrax* small-toothed
    rock hyrax 312 Tab 13.6,
    Procavia large-toothed rock
    hyraxes 124 125 312 483 Figs 6.37
    6.39 12.28 13.81–13.83 Tab 13.6,
hyrax, *see* Hyracoidea
Hystricomorpha 130
Hystrix, see Rodentia

Ichneumia, see Viverridae
Ictonyx, see Mustelidae
impaction, *see under* eruption, and
  under position anomalies under
  taxonomic headings
impala, *see* Bovidae
Indri, see Indriidae
indri, *see* Indriidae
INDRIIDAE, indrisoid lemurs, *see
  also* Primates
  dental formula 60
  normal dentition 59 60 234–235 Fig
    10.152
  numerical variation 59 60 Tab 3.8
  position anomalies 232–237 Fig
    10.153 10.158 Tab 10.21
  retained deciduous teeth 350

Indri indri indri 60, *Propithecus
  diadema* diadem sifaka 234 Fig
  10.152, *Propithecus verrauxi*
  Coquerel's sifaka 60 350 Figs
  10.153 10.158
INHERITANCE 1–3
  absent teeth 87–88 131 Fig 7.1
  achondroplasia and bulldog head
    266 293 294 361
  asymmetry in tooth size 10
  atavism, *see* separate entry
  connation 9 133
  cross breeding in nineteenth century
    pigs 283
  dental caries 462
  dentigerous cysts 579
  disto-molars 133
  enamel hypoplasia 437 444 450
  eruption 333 341
  extra molars 9
  genetic drift, *see* geographic
    variation
  genetic vs environmental influence 1
    9–11 131 153 Fig 7.1
  immuno-deficiency (Chédiak-
    Higashi syndrome) 523 546
  incisor occlusion 294 295 361 556
  lumpy jaw 516
  over-breeding 86 87 269
  overgrowth 319 361
  periodontal disease 526
  reverse mutation 7
  tooth size 10 132
  in cat 523, cattle 294 450 516 523,
    chinchillas 361, dog 9 86 87–88
    266 268–270 341, fox 84, hamsters
    526, mink 523 546, mouse 131 523
    526 Fig 7.1, musk ox 450, pigs
    283, polecats and ferrets 71,
    rabbit 319 333 361, rat 333 462,
    rice rats 9 133, sheep 295 556
    579
Inia, see Odontoceti
INJURY, SKULL 371–393, *see also*
  under taxonomic headings
  cranium 371 372
  mandible 371 372 375–391
    displacement of fragments
      375–376
    fibrous union 361
  maxilla 372–375
    fibrous union 373
  in non-mammalian vertebrates 393
  zygomatic arch 372 376 388 392
INJURY, TOOTH 394–436, *see also*
  under taxonomic headings, and
  under captivity, effects of
  developing 394–404 439
    dilaceration 394 402 407 429 Figs
      19.1 19.8 19.12
    experimental 404
    longitudinal splitting of molars 435
      483 567
  teeth of continuous growth 428–435
    dilaceration 429

INSECTIVORA, hedgehogs, moles,
  shrews
  captivity, effect of 569
  connation 139 Tab 7.4
  dental calculus 569
  dental caries 485 569
  dental formulae 138
  diet 569
  eruption sequence 346
  geographical variation 138 139
  normal dentition 138
  numerical variation 138–139 353
    Tab 7.4
    disto-molars 139
  periodontal disease 569–570
  position anomalies 319
  retained deciduous teeth 353
  tooth shape anomalies 138–139
  tooth wear 486
  Aethecinus African hedgehogs 569,
    Amblysomus hottentotus
    Hottentot golden mole, diet 569,
    eruption 569, periodontal disease
    569 Fig 34.46, *Atelerix* African
    hedgehogs 569, *Blarina
    brevicauda* short-tailed shrew 139,
    Chrysochloris asiatica Cape
    golden mole, periodontal disease
    569, *Crocidura* white-toothed
    shrews 139, *Cryptotis parva*
    American least shrew,
    periodontal disease, experimental
    569, *Erinaceus albiventris* four-
    toed African hedgehog 319,
    Erinaceus europaeus European
    hedgehog 138 Tab 7.4, diet 569,
    periodontal disease, in captivity
    569, *Hemiechinus* steppe
    hedgehogs 138 569, *Neomys
    fodiens* water shrew 139,
    Paraechinus desert hedgehogs
    569, *Scalopus aquaticus* prairie
    mole 353, *Scapanus latimanus*
    broad-footed mole 139, *Setifer
    setosus* hedgehog tenrec, primitive
    mammalian eruption sequence
    346, *Sorex araneus* common
    shrew 139, *Sorex minutus* pygmy
    shrew 139, *Talpa europaea*
    European mole 139
islands, *see under* geographic variation
Isoodon, see Peramelidae
ivory, lozenge pattern structure 408
  425 Figs 19.21 19.41

jackals, *see* Canidae
jaguar, *see* Felidae
jaw disproportion 153 156 159 160 266
  269 271 Fig 11.70, *see also* jaw
  growth, and position anomalies
JAW GROWTH 155–157
  environmental influences 153 156
  in captivity 156 236–237
  diet 156 157
  genetic aspects 156

JAW GROWTH (cont.)
  jaw components 155 Fig 9.4
  malocclusion 155 156 159 223 249 271 309–310
  occlusion, development of 145–155
  rickets 156 218 237

kangaroo, see Macropodidae
kinkajou, see Procyonidae
koala, see *Phascolarctos cinereus*
Kobus, see Bovidae
kowari, see Dasyuridae

laboratory animals, see experimental animals
Lagenorhynchus, see Odontoceti
LAGOMORPHA, rabbits and hares, see also *Oryctolagus cuniculus* domestic rabbit
  achondroplasia 319
  bulldog skull deformity 319
  dental formulae 133
  enamel fluorosis 454
  eruption 346
  geographical variation 134
  jaw mechanism 360
  normal dentition 133 346
  numerical variation 134
  overgrowth, incisor 358 361 Fig 17.7 cheek tooth 362
  position anomalies 318–319 Figs 14.8–14.10
  Lepus capensis hare 134 319 358 454 Fig 17.7, geographical variation, founder effect 134, *Lepus nigricollis* Indian hare 318 Figs 14.8 14.9, *Lepus peguensis* Burmese hare 319, *Lepus timidus hibernicus* Irish hare 134 Fig 17.7, *Ochotona rufescens* Afghan pika 134, *Oryctolagus cuniculus* rabbit 134, see also separate entry *O. cuniculus* domestic rabbit
Lagorchestes, see Macropodidae
Lagostrophus, see Macropodidae
Lagothrix, see Cebidae
Lama, see Camelidae
langurs, see Cercopithecidae
lathyrism 332–333
lemming, see Rodentia
lemurs, see Lemuridae
lemurs, dwarf and mouse, see Cheirogaleidae
LEMURIDAE, lemurs, see also Primates
  ankylosis 237
  bone disorders 236–237 350 Fig 10.161
  captivity, effect of 236–237 476 542–543
  connation 59–60 Fig 3.68
  dental caries, in captivity 476
  dental formula 59
  diet 542
  enamel hypoplasia 444 Fig 10.162

  eruption sequence and chronology 339
  normal dentition 59
  numerical variation 59–60 Tab 3.8
  periodontal disease 542–543
  position anomalies 232–237 Figs 10.154 10.159–10.162 Tab 10.21
  resorption 502 Fig 22.16
  retained deciduous teeth 236–237 350
  tooth shape anomalies 60 Fig 3.68
  Hapalemur gentle lemur 235, *Lemur catta* ring-tailed lemur 60 236 339 350 Figs 10.160 10.161, *Lemur coronatus* crowned lemur 476, *Lemur macaco* black lemur 60 237 444 Figs 10.154 10.162, *Lemur mongoz* mongoose-lemur 59 236 502 Figs 10.159 22.16, *Lemur variegatus* ruffed lemur 59 Fig 3.68, *Lepilemur* weasel lemur 59 235
Leontopithecus, see Callitrichidae
leopard, see Felidae
Leptonychotes, see Phocidae
Leptonycteris, see Chiroptera
Lepus, see Lagomorpha
Lestodelphys, see Didelphidae
Lichonycteris, see Chiroptera
life spans, of baboon 511, chimpanzee 533, chimpanzee in captivity 534, galago in captivity 542, gerbil, Mongolian, laboratory 568–569, hamster, Syrian, in laboratory 568, leopard, snow, in captivity 543, mole, golden, *Amblysomus hottentotus* 569, mouse, BNL strain 527, mouse, laboratory 568, rat, laboratory 543
limited growth, teeth of 355
lion, see Felidae
Lionycteris, see Chiroptera
llama, see Camelidae
Lobodon, see Phocidae
Lonchophylla, see Chiroptera
lophodont teeth, definition 15
  bilophodont teeth in rodents 132, tapirs 126
Loris, see Lorisidae
LORISIDAE, lorises, pottos and bushbabies, see also Primates
  ankylosis 542
  captivity, effects of 542
  connation 61 Fig 3.69
  dental formula 60
  normal dentition 59–60
  numerical variation 60 61 Tab 3.8
  periodontal disease 542
  position anomalies 232–237 Figs 10.155–10.157 Tab 10.21
  Galago bushbabies 61 542 Figs 3.69 10.155, as experimental animals 542, *Loris tardigradus* slender loris 60, *Nycticebus coucang* slow loris 60 61 Fig 10.156,
  Periodicticus potto potto 60 61 Fig 10.157
Loxodonta, see Elephantidae
LUMPY JAW
  Actinomyces bovis or other causative agent 515–516 519 520
  incisor overgrowth and, in hamster and in musk rat 361
  periodontal disease and, in kangaroos and wallabies 572–573, Figs 24.51 24.52
  in bears 516, bovids 519, cattle 515–516 Fig 23.8, deer 516 519, dromedaries and llamas 516, horse 516, kangaroos and wallabies 518–520 572 573 Fig 24.51, *Macropus titan* fossil kangaroo 520, oribi 516, rodents 517–518, sheep, wild and domestic 516, *Sminthopis* dunart 518, tapirs 516
lutong, see Cercopithecidae
Lutra, see Mustelidae
Lutreolina, see Didelphidae
Lydekker, R. 13 497
Lyncodon, see Mustelidae
lynx, see Felidae

Macaca, see Cercopithecidae, and experimental animals
macaques, see Cercopithecidae
Macrogalidea, see Viverridae
MACROPODIDAE, kangaroos, wallabies, betongs and potaroos, see also Marsupialia
  captivity, effects of 518–520 570–571 Figs 24.49 24.50
  connation 141 143 Fig 8.2
  dental formula 141
  dental plaque 573
  dento-alveolar abscess 506 518 Fig 23.13
  eruption, anomaly 324 346 Fig 15.10 sequence 346
  geographical variation 142
  injury, tooth 518 Fig 23.14
  lumpy jaw 518–520 572 Fig 24.51
  mesial drift 320–321 346 462 507 Fig 15.3
  normal dentition 141 320 518 Fig 15.1
  disto-molars 142 143
  periodontal disease 570–573 Figs 24.47–24.52
  position anomalies 320–324 Figs 15.2–15.15 Tab 15.1
  rudimentary precursors 13
  septic pneumonia 507
  tooth replacement 320 346
  tooth shape anomalies 141 Figs 8.1–8.3
  tooth wear 320 494 Fig 22.6
  Aepyprymnus rufous rat kangaroo 143, *Bettongia* bettongs 141 143 325, *Caloprymnus* desert rat-

MACROPODIDAE (cont.)
  kangaroo 143, *Dendrolagus* tree
    kangaroo 142 143 324 326 Fig
    15.15, *Dorcopsis* forest wallaby
    324 Fig 15.16, *Dorcopsulus*
    Papuan forest wallaby 142 Fig
    8.3, *Lagorchestes* hare-wallabies
    142 321 324 Figs 15.4 15.8 15.9,
    Lagostrophys banded hare-
    wallaby 143 323 324 353 Fig
    15.10, *Macropus* kangaroos and
    wallabies 141 142 320 321–325 346
    494 518–519 Figs 8.1 8.2 15.1–
    15.3 15.5–15.7 15.11–15.14 22.6
    24.47, *Macropus titan* fossil
    kangaroo, eruption anomaly 346,
    lumpy jaw 520, *Onychogalea* nail-
    tailed wallaby 142 322, *Petrogale*
    rock-wallaby 142 570–573 Figs
    24.49–24.52, *Potorous* potaroo
    143 518 Fig 23.14, *Setonyx* short-
    tailed wallaby 13, *Thylogale*
    dusky wallaby, pademelon 519,
    Wallabia swamp wallaby 142 520
    Fig 23.13 24.48
Macropus, see Macropodidae
Macrotis, see Peramelidae
Madoqua, see Bovidae
Magitot, E. 121 601
malocclusion, see position anomalies
manatee, see Trichechidae
mandrill, see Cercopithecidae
mangabeys, see Cercopithecidae
manganese
  and caries resistance 456
  in dental calculus 573
  in soil compared with F 451
Marmosa, see Didelphidae
marmoset, see Callitrichidae, and
  experimental animals
marmot, see Rodentia
Marmota, see Rodentia
marsupial cats, see Dasyuridae
marsupial mice, see Dasyuridae
MARSUPIALIA (Metatheria), see
  also Dasyuridae, Peramelidae,
  Didelphidae, Caenolestoidea,
  Macropodidae, Phalangeridae,
  Vombatidae, *Myrmecobius
  fasciatus* and *Phascolarctos
  cinereus*
  basic dental formula 13 140
  connation 140
  dental caries 485
  eruption sequence 346
  numerical variation 140
  periodontal disease 570–573
  position anomalies 320–329
  tooth homology 13 140
  tooth succession 13 264 346
marten, see Mustelidae
Martes, see Mustelidae
maternal rubella in man 10
Mazama, see Cervidae
Megachiroptera, see Chiroptera

Megalonyx, see Edentata
Megatherium, see Edentata
Meles, see Mustelidae
Mellanby, M. 445 447 458 525
Mellivora, see Mustelidae
Melogale, see Mustelidae
Melursus, see Ursidae
Mephitis, see Mustelidae
Merfield, F.G. 469
Meriones, see Rodentia, and
  experimental animals
meristic series 2 153
Mesembriomys, see Rodentia
MESIAL DRIFT and
  HORIZONTAL
  REPLACEMENT
  horizontal replacement and molar
    progression 284 320–321 345 346
    462 507 473 Figs 15.2–15.4
  in elephant 345 Tab 16.12,
    macropods 320–321 346 462 507
    573 Figs 15.2–15.4, manatees
    345, wart hog 107 284
  mesial drift 107 157 288 348 436 493
    Figs 9.5 10.45 16.9 19.57
    in cattle 157, colobus monkeys
      Fig 10.45, gazelle 157 Fig 9.5,
      horse 157 348 Fig 16.9, man
      157, *Presbytis* 436 Fig 19.57,
      roe deer 157, ruminants 157
      288, sheep 157 493
mesio-dens 36 50
Mesocricetus auratus Syrian hamster,
  see also Rodentia
  bone disorder 528
  dental calculus 526
  dental caries 461 527 528
  dental plaque formation 457 527
  diet, effect of 527
  eruption, continuous 331
  experimental animal 7 461 462
    526–528
  life span 568
  numerical variation, extra incisors
    caused by carcinogens 7
  odontomes 587 590
  periodontal disease 526–528 569
  periodontium, age changes in germ-
    free state 527
Mesoplodon, see Odontoceti
Metachirus, see Didelphidae
Metatheria, see Marsuspialia
Microcebus, see Cheirogaleidae
Microchiroptera, see Chiroptera
Microtus, see Rodentia
Miller, W.D. 405 422 426 456 478 487
  493–495
mink, see Mustelidae
molarization, see morphogenetic field
  concept
mole rat, see Rodentia
moles, see Insectivora
molybdenum and caries resistance
  456, in soil compared with F 451
Monachus, see Phocidae

mongan, see Phalangeridae
mongoose, see Viverridae
monkeys, New World, see Cebidae,
  and experimental animals
monkeys, Old World, see Cercopithecidae, and experimental
  animals
Monodelphis, see Didelphidae
Monophyllus, see Chiroptera
MONOTREMATA (Prototheria),
  platypus, echidna, spiny anteater
  13
  Ornithorynchus duck-billed
    platypus, dental formula and
    replacement by horny plates 13
  Tachyglossus echidna or spiny
    anteater, specialized tongue 13
Moodie, R.L. 485
moose, see Cervidae
Mormoops, see Chiroptera
MORPHOGENETIC FIELD
  CONCEPT 2–5 Figs 1.1–1.4, see
  also tooth development
  atavism and 63
  Bateson's views in accord with
    147
  Butler's views 2
  evolutionary theory and 2
  gradients of capacity 2 6
  in dog 2 Fig 1.3
  in mouse Fig 1.4
  inhibitory influences 4 Fig 1.2
  molarization 2 4 84
  numerical variation 2 6 17
    and field instability 35 77
  odontogenic clones 3 4 Fig 1.4
  secondary field effect 3
  shape anomalies and 37
  tooth size anomalies and 2 3 17 77
MORPHOLOGICAL DEFINITIONS
  AND NOMENCLATURE 11–15
  20
  bilophodont teeth 132 126
  brachydont teeth 14
  bunodont teeth 13
  Carabelli, cusps of 15 19–20
  carnassial teeth 2 62 67 71 74 77 83
  cusp terminology 13–15
  dilambdodont teeth 134
  haplodont teeth 3
  hypsodont teeth 14
  lophodont teeth 15
  selenodont teeth 110 111 114
  squalodont teeth 101
  special terms 13
  symbols 12
  tribosphenic crowns 15 Fig 1.7
Moschus, see Cervidae
mouse, laboratory, see *Mus musculus*
mouse, wild and commensal, see
  Rodentia
mulgara, see Dasyuridae
Mungos, see Viverridae
Muntiacus, see Cervidae
muntjac, see Cervidae

Mus, see Rodentia, and *Mus musculus* laboratory mouse
Mus musculus laboratory mouse, see also Rodentia
  asymmetry, experimental 10
  Chédiak-Higashi syndrome 523
  connation 9
  cultured tooth germs 4
  dental caries, experimental 461–462
    susceptible strains 462
  dental calculus 526
  dental plaque 526
  diet, effect of 131
  enamel, hypoplasia, in tuberculosis 438
  eruption, continuous 331
  exancephaly and connation 9
  inheritance 131 Fig 7.1
  injury, tooth, fractured incisors, new tooth alongside 405
  irradiation, effects of 7 9
  numerical variation 131 Tab 7.1
    extra incisors, irradiation 7, aged mice 7
  overgrowth, incisor 361
  periodontal disease 526
  periodontium, age changes in 527
  pseudo-odontome 605
  tooth size anomalies 131
  tuberculosis 438
  wedging of hairs and bedding material, interdentally 526
  strains and mutants: black fat (PBB) 462, BNL short-lived 527, grey lethal 333 605, obese 333, STR/In 605 Tabby 3 9
musk ox, see Bovidae
museum specimens, unrepresentative of populations 64–65 465
Mustela, see Mustelidae, and experimental animals
MUSTELIDAE, badgers, martens, ferrets, weasels, otters, skunks, see also Carnivora
  ankylosis 396 Fig 19.5
  bone disorder, osteopetrosis 348
  bone growth, defective 249
  captivity, effects of 477 546
  Chédiak-Higashi syndrome 523 546
  connation 71 72 Fig 4.13
  dental calculus 526 546
  dental caries 477
  dental formulae 72 73 74
  dento-alveolar abscess 507 512
  enamel mineralization 458
  eruption, non-eruption and osteopetrosis 348
    sequence and chronology 341 Tab 16.6
  experimental 525–526
  gingival hyperplasia, experimental 528
  injury, developing tooth 396 Fig 19.5
    jaw 373
    tooth 512
  normal dentition 71 248 251 253–255 Figs 11.31 11.34 11.36 11.43 11.46 11.48 11.49
  numerical variation 71–74 Figs 4.14 4.15 Tabs 4.1 4.2
    disto-molars 73
  osteometrics 249
  osteomyelitis 512
  periodontal disease 525–526 546
  position anomalies 248–255 Figs 11.28–11.50 Tabs 11.1 11.7 11.8
    underjet 249 251 Figs 11.29 11.30
  retained deciduous teeth 251
  root anomalies 73
  septic pneumonia 507
  tooth shape and size anomalies 71–74
  tooth wear, crib-biting abrasion 546
    excessive 496
  Amblonyx oriental small-clawed otters 74 Fig 4.15, *Aonyx* clawless otters 74 255 Tab 11.7, *Arctonyx* hog badgers 74 249 Fig 11.29 Tab 11.7, *Conepatus* hog-nosed skunks 73 255 Figs 11.46 11.47 Tab 11.7, *Eira* tayras 73 251 348 Fig 11.31 Tab 11.7, *Enhydra* sea otter 74 255 477 496 Tab 11.7, *Galictis* grisons 73 Fig 4.14 Tabs 11.7 11.8, *Gulo* wolverine or glutton 396 Fig 19.5 Tab 11.7, *Ictonyx* striped polecats 74 253–254 Figs 11.39–11.42 Tab 11.7, *Lutra* otters 74 249 255 Fig 11.49 Tabs 11.7 11.8, *Lyncodon* Patagonian weasel Tab 11.7
  Martes martens 71 72 251 Tabs 4.2 11.7 11.8, *Meles meles* European badger 73 477 512 Tab 11.7, *Mellivora capensis* honey badger 73–74 251 507, Figs 11.34 11.35 Tab 11.7, *Melogale* ferret-badgers 74 Tab 11.7, *Mephitis* skunks 71 72 251 512 Figs 11.31 11.32, Tab 11.7, *Mustela* weasels, stoats and polecats 71 72 341 523 546 Tabs 11.7 11.8 16.6, *M. putorius furo* domestic polecat and ferret 71 72 249 251 339 341 525–526 528 Figs 4.13 11.30 Tab 11.8, as experimental animals 526 528, *Mydaus* stink badgers 74 Tab 11.7, *Poecilictis* Saharan striped weasels 74 251 253 Figs 11.36–11.38 Tab 11.7, *Poecilogale* white-naped weasels 74 254 Figs 11.43 11.44 Tab 11.7, *Pteroneura* giant otter 74 Tab 11.7, *Spilogale* spotted skunk 74 254 Fig 11.45 Tab 11.7, *Taxidea taxus* American badger 74 249 373 Tab 11.7, *Vormela peregusna* marbled polecat 72 Tab 11.7, extra teeth 72
Mydaus, see Mustelidae

Myocaster, see Rodentia
Myoictis, see Dasyuridae
Myomorpha 130
Myotis, see Chiroptera
Myrmecobiidae, banded ant-eaters, see *Myrmecobius fasciatus*
Myrmecobius fasciatus, banded ant-eater
  connation 148
  dental formula 140 148
  normal dentition 140 148
  numerical variation 148
  tooth shape and size anomalies 148, Fig 8.11
MYSTECETI, whale bone whales
  connation 101 Fig 5.5
  rudimentary teeth 100 101 102
  Balaenoptera musculus blue whale 101 102 Fig 5.5
  Balaenoptera physalis common rorqual 101 102 Fig 5.5
Mystromys, see Rodentia, and experimental animals

Nasalis see Cercopithecidae
Nasua, see Procyonidae
Nasuella, see Procyonidae
necrotic stomatitis 562–563, in cattle 561, deer 563, dolphin (captive) 550, sheep, dall 516 563, sheep, domestic 562
NECROTIZING ULCERATIVE GINGIVITIS 524 550 557 560 562–563
  causative organisms 550 562
  experimental, in macaque 539
  selenium 562
  wedging of plant material, interdentally 562
  in deer 563, dog 550 562, gorilla 563, man 524 562, sheep 557 562–563
Neomys, see Insectovora
neoplasms 575 580 587 592–593 606
neoteny in brachycephalic dogs 266
Nesotragus, see Bovidae
nomenclature, tooth, see morphological definitions and nomenclature
non-mammalian dentitions, see also Reptilia, fish and *Bufo pardalis*
  jaw deformities 158–159
  malocclusion 158–159
  polyphyodonty xi 505 523
  sited away from jaw margins xi
  tooth number increasing with age xi
NUMERICAL VARIATION, see absent teeth, extra teeth, and under taxonomic headings
nutria, see Rodentia
Nycticebus, see Lorisdae

OCCLUSION, normal, see also position anomalies and jaw growth co-ordination between

OCCLUSION (cont.)
  jaws important 154–156 159 191–192
  in growth 159
  in sheep 295 552–560 Figs 24.36–24.38
  in tooth succession 159
ocelot, *see* Felidae
Ochdontomys, *see* Rodentia
Odobaenus rosmarus walrus, *see also* Pinnipedia
  dental formula 98
  numerical variation, duplicated canines 98–99 Figs 5.2 5.3 5.4
Odocoileus, *see* Cervidae
ODONTOCETI, toothed whales
  ancestral forms 102
  asymmetrically curved jaws 391
  barnacles parasitizing teeth 389 391 499 501
  bone disorder and tooth loss 494
  boreholes in cementum 498–499
  caries-like lesions 499–501 Fig 22.13
  connation 102–105 Figs 5.6 5.11
    teeth joined by cementum 102 Fig 5.8
  coronal cementum 102
  dental formula 12
  dento-alveolar abscess 493 494 512
  injury, fractured teeth 435
    jaw 375 385 388–391
  mode of feeding 390–391
  necrotic stomatitis, in captivity 550
  normal dentition 368–369 388 500, Fig 22.13
    enamel-less teeth and enamel-tipped teeth 102
    wrinkled enamel 102 105 Fig 5.12
  numerical variation 100–102 Fig 5.7
  osteomyelitis 375 512
  overgrowth 368–369 Fig 17.18
  pseudo-arthrosis following fracture 384
  resorption 499 501–502 504 Figs 22.14 22.15
  root anomalies 102–103 105 Figs 5.9 5.10
  rudimentary teeth 100–101 102
  skeletal exostoses 494
  tooth shape anomalies 104 391
  tooth staining 500
  tooth wear, excessive 493–494
  Berardius bairdii beaked whale 391, *Delphinus delphis* common dolphin 102 494 Fig 5.7, *Globicephala* pilot whale 493, *Grampus griseus* Risso's dolphin 399 493, *Hyperoödon ampullatus* bottle-nosed whale 101 102 391, *Hyperoödon rostratus* bottle-nosed dolphin 391, *Inia geoffrensis* Amazon porpoise 102 104–105 375 391, caries-like lesions 499–501 Fig 22.13, *Lagenorhynchus albirostris* white-beaked dolphin 385 494 512, *Mesoplodon bidens* Sowerby's beaked whale 101, *Mesoplodon grayi* beaked whale 101, *Mesoplodon layardii* strap-toothed whale 369 Fig 17.18, *Orcinus orca* (= *O. gladiator*) killer whale 104 493 501–502 512 Fig 22.14, *Phocoena phocoena* common porpoise 102 384, Fig 5.6, *Physeter catodon* sperm whale 101 102 103 104 435 498 499 501 502 504 Figs 5.8 5.9 5.10 22.15, jaw caught in marine cable 390–391, jaw deformity 388–391, mandibular pseudo-arthrosis following fracture 384, mode of feeding 390–391, *Pontoporia blainvillei* La Plata dolphin 104 Fig 5.11, *Stenella graffmani* spotted dolphin 391, *Steno bredanensis* rough-toothed dolphin 102 105 Fig 5.12, *Tursiops truncatus* bottle-nosed dolphin 102 384 385 550 494 512 550
odontocoele 430 Fig 19.46
odontogenesis imperfecta, in dog 447
odontogenic tumours 575 576 587 603
ODONTOMES 2 574–606, *see also* under taxonomic headings
  cementomas 575 596–599 603
  classifications 574–575
  complex composite compound 575 579–589
  composite 575 579–594
  compound 94 575 579–589 593
  dilated 576 589–593 Fig 25.1
  developmental anomalies and 4 396 586 587
  enamel pearls and root nodules 79 113 576 Fig 4.6, *see also* separate entry
  fibrous 575 601–605
  hamartomas 575–576 587 606
  injury response, simulated by 396 595 Figs 19.4 19.9 19.27 23.6
  invagination 576 Fig 25.1
  malignant tumour and mastoid 593 599–601
  non-mammals 606
  osteodystrophy, simulated by 334 335 601–605
  radicular 576 594–595
  teratomatous 576 601 Fig 25.36
  unorganized development as 2 11
  virus-induced 587
odontopagy, *see* connation
oligodonty, *see* absent teeth
olingo, *see* Procyonidae
Ommatophoca, *see* Phocidae
onager, *see* Equidae
Ondatra, *see* Rodentia
Onychogalea, *see* Macropodidae
opossums, *see* Didelphidae
orang-utan, *see Pongo pygmaeus*
Orcinus, *see* Odontoceti
oribi, *see* Bovidae
Ornithorynchus, *see* Monotremata
Oryctolagus cuniculus domestic rabbit, *see also* Lagomorpha
  asymmetry of jaws 319
  bone disorder, osteopetrosis 333
  bulldog skull deformity 319 361
  cultured tooth germs 5 8
  dental caries 461 484
  dental formula 133
  dento-alveolar abscess 358 518 Fig 17.8
  diet 361 458
  enamel mineralization 458
  eruption sequence 346
  growth rate, cheek tooth 362
    incisor 356
  inheritance 319 361
  injury, mandible 361
  normal dentition 133
  numerical variation 134 Fig 7.2
  overgrowth, incisor 319 356 358 Figs 17.2 17.3 17.4
  position anomalies 318–319
    underjet 318–319 361
  pseudo-arthrosis 361
  tooth shape anomalies 134
Oryx, *see* Bovidae
Oryzomys, *see* Rodentia, and experimental animals
Osborn, H.F. 15
osteo-dentine 404 422
osteodystrophy, *see* bone disorders
osteomalacia, *see* bone disorders
OSTEOMETRICS 159–160 195–197
  alveolare 159 221
  basal length 159 272–274 295 Tabs 10.6 10.10
  cheektooth length 159 210 238 258 282 Tabs 10.8 11.3 11.6 11.10 11.13 11.14 13.1
  definitions
    alveolare, basal length, basion, basicranial axis, cheektooth length, glenoid-incisor length, gnathion, length of mandible opisthion, palatal breadth, palatal length, prosthion 159
  glenoid-incisor length 159 249 309
  gnathic index Tabs 10.6 10.10
  gnathion-gonion 196
  length of cranium 243 274 Tabs 10.7 10.8
  length of mandible 159 180 196 243 249 274 295 309 Tab 10.6
  maxilla length 193
  other dimensions 159 196 221
  palatal index Fig 13.76 Tab 10.6
  palatal length 159 210 238 258 274 283 Tabs 11.3 11.6 11.10 11.13 11.14 13.1
  palate length/cheekteeth length 210 238 247 248 264 283 Tabs 11.3 11.6 11.10 11.13 11.14 13.1

OSTEOMETRICS (cont.)
  premolar length Tab 11.16
  in *Ateles* spider monkeys 221, bears 258 Tab 11.10, *Cercopithecus* 193 195–197, *Colobus* 180 Tabs 10.6 10.8, dog, domestic 264 266 270 272 274 275 Tabs 11.15 11.16, dog, hunting 264, Tabs 11.13 11.14, fox 262 264 Tabs 11.13 11.14, hog-badgers 249, horse 309 Fig 13.76, jackal 262, lion 238–239 Tab 11.3, *Papio* 211, pigs, domestic 283 Tab 13.1, pigs, wild 282, *Presbytis* 185 Tab 10.10, sheep 295, viverrids 243 247 Tab 11.6
osteomyelitis, in badger 512, deer 514, dolphins 375 512, duiker *Sylvicapra* 514 551, gorilla 509–510, *Macaca* 511, pig 513, sheep 514
osteopetrosis, *see* bone disorders
osteoporosis, *see* bone disorders
Otaria, *see* Otariidae
OTARIIDAE, sealions and eared seals, *see also* Pinnipedia
  asymmetry of skull 372
  bilaterality of anomalous development 98
  connation 97
  dental formulae 96 97
  normal dentition 280 496 Fig 12.1
  grooved incisors 95 Fig 5.1
  numerical variation 96 97 98 Tab 5.1
    disto-molars 97
  position anomalies 281
  tooth shape anomalies 97
  tooth wear, abrasion v. erosion 496–497 Figs 22.8 22.9 22.10
  Arctocephalus southern fur seals 97 98, *Callorhinus ursinus* northern fur seal 96 97 98, *Eumetopias jubatus* northern sealion 96 97 372 497 Fig 22.10, *Otaria byronia* South American sealion 97 496–497 Figs 5.1 22.8 Tab 5.1, *Phocarctos hookeri* New Zealand sealion 497 Fig 22.9, *Zalophus californicus* Californian sealion 97 280 Fig 12.1
otter, *see* Mustelidae
otter-civet, *see* Viverridae
Ourebia, *see* Bovidae
OVERGROWTH 355–370, *see also under* teeth of continuous growth
  actinomycosis and 361
  cheek tooth overgrowth, in chinchilla 362, guinea pig 362, rabbit 362, wombat 493
  diet and 361 362 363
  experimental 361
  incisor overgrowth, in aye-aye 363 Figs 17.10 17.11, bandicoot rat 361, beaver 356 Fig 17.1, chinchilla 361, guinea pigs 356 362, hamster 361, hare 358 361 Fig 17.7, hippopotamus 363–364 Figs 17.13 17.14, hyraxes 363 Fig 17.12, marmot 358, muskrat 360 361, porcupine 356, rabbit 319 356 358 361 Figs 17.2 17.3 17.4, rat, laboratory 356 360 361, rat, black 358 Fig 17.6, rat, Polynesian 360–361, rodents and lagomorphs 355–362 358, squirrels 358
  inheritance 319 361
  malocclusion and 362
  slobbers 362
  tooth wear and 356–362 366 368
  tusk overgrowth, in pigs, domestic 364–365 Fig 17.15, pigs, wild 365–368 Fig 17.17, strap-toothed whale 368–369 Fig 17.18
Ovibos, *see* Bovidae
Ovis aries, domestic sheep, *see also* Bovidae
  bone disorders 296 492 559 560
  connation 117
  cyclopia 154 Fig 9.2
  dental calculus 557 560
  dental caries 480 Fig 21.23
  dental pads, upper and lower 552 Fig 24.36
  dental plaque 557 560
  dentigerous cyst 492 577–579 Figs 25.2–25.4
  dento-alveolar abscess 514
  enamel defects, experimental 404 439 451
  enamel fluorosis 451–453
  enamel hypoplasia 450
  eruption sequence and chronology 334 345
  experimental animal 404 439 451
  gummy mouth 491 492 552 558
  incisor function and relationship 295 491 552–557 560 Figs 24.36–24.38
  inheritance 295 556
  jaw joint movements 551
  lumpy jaw 516
  mesial drift 157 493
  necrotic ulcerative gingivitis 557 562–563
  numerical variation 117
  odontome 582 Fig 25.8
  osteometrics 295
  osteomyelitis 514
  periodontal disease, broken mouth 295 492 552 556–560
    cheek teeth 560 561
    incisors 295 492 552 556–560
  position anomalies 295–296 556 599 601 Figs 13.29 13.30
    open mouth 296
    overjet 295–296 556 Fig 13.31
    underjet 295–296 554 Figs 13.32 24.37
  resorption 505
  teratoma 601 Fig 25.36
  tooth shape anomalies 117 Fig 6.19
  tooth wear 157 491–493 552 556 558
    crib-biting abrasion 491
    wedging, of plant material, interdentally 562
  breeds: Blackface 296, Cheviot 556, Leicester 295, Merino 117 295–296, Rambouillet 295, Ronaldsay 560, Scotch black-face 556, Welsh 296
Ovis, *see* Bovidae and *Ovis aries* domestic sheep
Owen, R. 292 404 421 592

pademelon, *see* Macropodidae
Paguma, *see* Viverridae
Pan troglodytes chimpanzee, *see also* Pongidae
  captivity, effects of 169–171 465 535 Fig 24.12
  connation 26 29 Fig 3.9
  'Consul' of the Folies Bergères 529
  dental caries 463 465 Tabs 21.1 21.2 23.1
    scavenging of refuse and 465
  dento-alveolar abscess 507 Tab 23.1
  diet, effects of 465 532–533
  enamel, hypomineralized Fig 21.3
  enamel hypoplasia 439 440 Figs 10.20 20.3
  eruption, non-eruption 346 Fig 16.3
    sequence and chronology 335–336 Tab 16.1
  geographical variation 22
  injury, mandible 379 Fig 18.11
    tooth 396 Fig 19.6
  'Jimmy' of London Zoo 465
  necrotizing ulcerative gingivitis 524 534–535
  normal dentition Fig 10.11
  numerical variation 22 26–28 Figs 3.8 3.10 3.12 3.14 Tab 3.1
    disto-molars 20 22 Figs 3.8 3.11 3.12 3.28
  periodontal disease 529 532–535 Figs 24.8–24.12
    scavenging of refuse and 465 532–533 Fig 24.8
  pneumonia, septic 507
  position anomalies 167–171 Figs 10.16–10.19 10.21 Tabs 10.1 10.2
    impaction Tab 10.2
    overjet Fig 10.20
    underjet 167–168 Tab 10.1
  root anomalies 28 Figs 3.13–3.15
  tooth shape anomalies 28–29 Figs 3.13–3.18
  tooth size anomalies 28 Figs 3.14–3.18
  tooth wear Fig 24.8
panda, red, *see* Procyonidae
Panthera, *see* Felidae
Papio, *see* Cercopithecidae, and experimental animals

Paradoxurus, see Viverridae
Paraechinus, see Insectivora
para molars, definition 20
parrot mouth (syn. rat- sow-mouth, incisor overjet) 295, in cattle 294, deer 288, horse 309, sheep 295 556
peccary, see Tayassuidae
Pelea, see Bovidae
Perameles, see Peramelidae
Peroryctes, see Peramelidae
PERAMELIDAE, bandicoots, see also Marsupialia
　captivity, effect of 571–572
　dental calculus 572
　dental formula 146
　diet 572
　normal dentition 146
　periodontal disease, in captivity 572
　position anomalies 327 Fig 15.20 Tab 15.1
　Chaeropus pig-footed bandicoot 146, *Echymipera* spiny bandicoot 146 327, *Isoodon* golden bandicoot 146 571–572, *Macrotis* rabbit-bandicoot 146 327 Fig 15.20, *Perameles* long-nosed bandicoot 146 572, *Peroryctes* New Guinea bandicoots 146, *Rhynchomeles* Ceram bandicoot 146
Periodicticus, see Lorisidae
PERIODONTAL DISEASE 521–573, see also under taxonomic headings and under experimental animals
　age change and 524 527 529
　Aleutian disease 546
　alveolar dehiscences and 524–525 541
　bacterial flora and 523 527
　black surface deposits 524 560
　broken mouth in sheep 295 492 552 556–560
　experimental 525–529
　diet, deficient 525–528
　diet, fermentable carbohydrate 527–528
　diet, high-protein 527
　diet, high-sucrose 526–528
　diet, vitamin A excess 9 544
　germ-free animals 526 527
　hormonal challenge 528
　occlusal overload 528
　stress 528
　genetic immune-system disorders, see Chédiak-Higashi syndrome
　gingival recession and 524 527 529
　gingivitis 522 523 524, see also necrotizing ulcerative gingivitis, prevented by tooth cleaning 527–529 543
　gnotobiosis 527
　hereditary diabetes, in 527
　in man 521–525 557 558 562 563
　museum collections unrepresentative 465 529
　protozoal parasites and 522
　refuse scavenging and 567 569
　root surface caries and 527
　species differences 525
　subgingival ligature 528
　tooth wear, secondary to 523 540
　transmissibility 527
periodontium 521 522
　age changes 524 527 529 568 569
　development *in vitro* 5
　and sheep incisors 552–554 Figs 24.37 24.38
periodontosis 557
PERISSODACTYLA, odd-toed ungulates, see also Equidae, Rhinocerotidae, Tapiridae
　normal dentition 15 106–107
　origins 106
Perognathus, see Rodentia
Peromyscus, see Rodentia
Peropteryx, see Chiroptera
Petaurista, see Rodentia
Petaurus, see Phalangeridae
Petrogale, see Macropodidae
Phacochoerus, see Suidae
Phalanger, see Phalangeridae
PHALANGERIDAE, possums, gliders and phalangers, see also Marsupialia
　connation 143 144 145 Figs 8.4 8.5
　dental formula 145
　intermediate teeth 143 145 Tab 8.1
　normal dentition 143 325–327 Fig 15.17
　numerical variation 143–146 Figs 8.6 8.7
　position anomalies 325–327 Figs 15.18 15.19 Tab 15.1
　rudimentary deciduous precursors 13
　Acrobates, pygmy glider 145, *Cercartetus* dormouse-possum 145, *Dactylopsila* striped and large-tailed possums 144 Fig 8.4 Tab 8.1, *Distoechurus* feather-tailed possum 145, *Petaurus* fluffy gliders 144 Tab 8.1, *Phalanger* phalangers 144 145 327 Figs 8.7 15.19 Tab 8.1, *Pseudocheirus* mongan and ringtails 144 325–327 Figs 15.17 15.18 Tab 8.1, *Schoinobates volans* greater glider 143, *Trichosurus* mountain and brush-tailed possums 13 144 145 327 Figs 8.5 8.6 Tab 8.1
Phascogale, see Dasyuridae
Phascolarctos cinereus koala, see also Marsupialia
　dental formula 146
　numerical variation 146
　position anomalies 327 Tab 15.1
Phascolosorex, see Dasyuridae
Philander, see Didelphidae
Phoca, see Phocidae
Phocena, see Odontoceti
Phocarctos, see Otariidae
PHOCIDAE, earless seals, see also Pinnipedia
　age change in occlusion 281
　asymmetry of skull 372
　connation 99 100
　dental formulae 99 100
　dento-alveolar abscess 512
　injury, fractured abscessed teeth 435 512
　normal dentition 281
　numerical variation 96 99 100 Tab 5.1
　position anomalies 281 498
　tooth wear, abrasion v. erosion 498 ice-sawing abrasion of incisors 497–498 512 Figs 22.11 22.12
　Cystaphora cristata hooded seal 100, *Erignathus barbatus* bearded seal 100, *Halichoerus grypus* grey seal 99, *Histriophoca fasciata* ribbon seal 100, *Hydrurga leptonyx* leopard seal 281 372 435 512, *Leptonychotes weddelli* Weddell's seal 100 281, mode of life 498, tooth wear, ice-sawing abrasion of incisors 497–498 512 Figs 22.11 22.12, *Lobodon carcinophagus* crabeater seal Fig 22.12, *Monachus* monk seals 100 281 498, *Ommatophoca rossi* Ross seal 100, *Phoca groenlandica* common, harp, ringed seals 96 99 100 281
phosphorus deficiency and cara inchada 561
Phyllostomus, see Chiroptera
Physeter, see Odontoceti
pig, domestic, see *Sus scrofa*
pigmentation of teeth 361
pigs, wild, see Suidae
pika, see Rodentia
PINNIPEDIA, seals, sealions and walruses, see also Phocidae, Otariidae and *Odobaenus rosmarus*
　ancestry 95 342
　dental formula 95
　eruption sequence 342
　normal dentition 95
　numerical variation 95–100 Tab 5.1
　position anomalies 281–282
　tooth wear, abrasion v. erosion 496–498
Pipistrellus, see Chiroptera
Pithecia, see Cebidae
placental barrier 438 453 454
plaque, see dental plaque
Plecotus, see Chiroptera
Poecilictis, see Mustelidae
Poecilogale, see Mustelidae
polecat, see Mustelidae
Polyprotodontia 140

PONGIDAE, great apes and gibbons,
    see also Gorilla gorilla, Pan
    troglodytes, Pongo pygmaeus,
    Hylobates and Symphalangus
    syndactylus
  connation 22
  dental caries 463–465 Fig 21.7 Tabs
    21.1 21.2
    deciduous teeth 465
  dental formula 20
  dento-alveolar abscess 507–510 Tab
    23.1
  enamel, hypomineralized 458 459,
    Figs 21.3 21.4
  enamel hypoplasia 439–441
  eruption sequence and chronology
    335–336
  interglobular dentine 440
  normal dentition 20 161–162 188 217
  numerical variation 20–34
    disto-molars 20 22
  periodontal disease 529 Fig 24.4
  position anomalies 161–177 Tabs
    10.1 10.2
    impaction Tab 10.2
  root anomalies 22
  tooth shape and size anomalies
    20–34
  tooth wear 458 463
Pongo pygmaeus orang-utan, see also
    Pongidae
  captivity, effects of 175 535
  connation 30 Fig 3.21
  dental caries 463 465 536–537 Tabs
    21.1 21.2 23.1
  dento-alveolar abscess 507 509 Fig
    23.2 23.4 Tab 23.1
  enamel hypoplasia 440
  enamel, wrinkled 440 441
  eruption sequence and chronology
    336
  gingival hyperplasia 535
  injury, mandible 384 387 Fig 18.16
    skeletal 371 384
    tooth 428
    zygomatic arch Fig 18.16
  numerical variation 29–11 Figs 3.19
    3.20 Tab 3.1
    disto-molars 20 30
  periodontal disease 529 535–536
  position anomalies 171–175 Figs
    10.23–10.33 Tabs 10.1 10.2
    underjet 171
  root anomalies 31 Fig 3.22
  tooth shape and size anomalies 31
Pontoporia, see Odontoceti
porcupine, see Rodentia, and
    experimental animals
porpoises, see Odontoceti
POSITION ANOMALIES, see also
    under taxonomic headings,
    153–329
  asymmetry of tooth arches 167 183–
    184 186–187 193–194 200 205 220–
    223 262 263 270 276 Figs 10.14

    10.48 10.54 10.66 10.77 10.121
    10.123 10.126 11.66 11.77 Tab
    11.16
  captivity 169–171 175 177 194–195
    197–200 202 205–208 210–211 212
    218–219 236–237 239–241 247–248
    256–257 264 Figs 10.67–10.76
    10.86–10.91 10.99 10.100 10.117
    10.159–10.162 11.54 11.55 Tab
    11.1
  in cyclopia, in sheep 154 Fig 9.2
  domestication 10 264–266 283
  ectopic teeth 23–24 153 346 599–601
    Figs 3.3 16.3
  extra teeth and 18
  environmental aspects 1 153 182
  eruption processes and 153–155, Fig
    9.1
  genetic aspects 153 156
  geographic variation 10 167 182
    195–197 231–232 239 264 288 297
    319 Figs 10.58 10.67 10.68 Tabs
    10.14 11.14
  impaction 160 332 348 Fig 16.9
  in man 157 158
  inter-arch relationships 153 155–159
    180 190
  jaw growth and 1 11 153 155–157
    184 193 200 210
  malocclusion 157–158 160
    jaw growth and 155 156 159 223
      249 271 309–310
    in canids 261–276, cattle 294–295,
      non-mammals 158–159, sheep
      295 492–493 552–560 Figs
      24.36–24.38
  open bite, in marsupial 329
  open mouth, in sheep 296
  parrot mouth (syn. rat-mouth, sow-
      mouth, incisor overjet, overshot
      jaw) 295, in cattle 294, deer 288,
      horse 309, sheep 295 556
  shear-bite, in capuchin monkey 493,
      horse 487, howler monkey 493
  terminology
    incisor occlusion, Hitchin's classes
      295
    incisor overjet 160 260–263 271–
      272 295 309
    incisor underjet 160 260–263 272
      274 295 309
    inferior protrusion 160 262
    inferior retrusion 160 261 262
    overshot jaw 294 295
    superior protrusion 160 260
    superior retrusion 262
    undershot jaw 294 295
  transposition 153
possum, see Phalangeridae
Potamochoerus, see Suidae
potaroo, see Macropodidae
Potorous, see Macropodidae
Potos, see Procyonidae
potto, see Lorisidae
prairie dog, see Rodentia

Presbytis, see Cercopithecidae
PRIMATES, see also Pongidae,
    Cercopithecidae, Cebidae,
    Callitrichidae and Prosimii
  alveolar recession 529
  ancestral form 29
  captivity, effects of 529
  dental caries Tab 23.1
  dento-alveolar abscess Tab 23.1
  injuries, skeletal 371
  periodontal disease, in captivity 529
  root anomalies, grooved roots 19
    webbed roots 18
  tooth size quantification 18
Proboscidea, see Elephantidae
Procapra, see Bovidae
Procavia, see Hyracoidea
Procolobus, see Cercopithecidae
Procyon, see Procyonidae
PROCYONIDAE, raccoons, coatis
    and pandas, see also Carnivora
  captivity, effects of 256–257 544 Figs
    11.54 11.55 24.21 24.22
  connation 75 76 Fig 4.19
  dental caries 477
  dental formula 74
  diet 544
  normal dentition 74 255 256 Fig
    11.51
  numerical variation 74–76 Figs 4.16
    4.17 4.18
  periodontal disease in captivity 544
    Figs 24.21 24.22
  position anomalies 255–258 Figs
    11.52–11.60 Tabs 11.1 11.9
    impaction 348 Figs 11.53 11.54
      11.55
    underjet 258 Fig 11.60
  tooth shape anomalies 76 Figs 4.17
    4.18
  tooth wear, cup-like 477
  Ailurus fulgens red panda 76 544,
    Bassaricus astutus ring-tailed cat
    74–75, Bassaricyon olingos 256
    477, Nasua coatis 75 76 255–257
    477 Figs 4.17–4.19 11.51 11.56–
    11.58, Nasuella little coatimundi
    255, Potos kinkajou 76 256 257
    258 260 261 Figs 11.52 11.53
    11.59 11.60, Procyon raccoons 75
    256–257 259 544 Figs 4.16 11.54
    11.55 24.21
Proechimys, see Rodentia
pronghorn, see Antilocapra
Propithecus, see Indriidae
PROSIMII, see also Primates,
    Daubentoniidae, Indriidae,
    Lorisidae, Tarsidae,
    Cheirogaleidae and Lemuridae
  normal dentition 59 232 234 Fig
    10.152
  numerical variation 59 235 Tab 3.8
  periodontal disease 542–543
  position anomalies 232–237 Tab
    10.21

668 *Index*

protozoal parasites in mouth 522
Pseudocheirus, see Phalangeridae
pseudo-arthrosis, un-united fracture 361 375 376 384 392
   in fossil sabre-tooth 376
pseudo-odontomes 605–606
   grey-lethal mouse mutant 605
   ia-rat mutant 605
   STR/IN mouse mutant 605
   vitamin A deficient rats 605
Pteroneura, see Mustelidae
Pteronotus, see Chiroptera
pulp stones 412 424 426 Fig 19.40
puma, *see* Felidae
Pygathrix, see Cercopithecidae

quagga, *see* Equidae

rabbit, domestic, *see Oryctolagus cuniculus*
rabbit, wild, *see* Lagomorpha
raccoon, *see* Procyonidae
raccoon-dog, *see* Canidae
radicular cysts 510 576–577 Fig 25.10
   experimental 510 576
   growth of 576–577
   inflammatory origin 576
   in man 576
Rangifer, see Cervidae
Raphicerus, see Bovidae
rat, laboratory, *see Rattus norvegicus*
rat, wild, *see* Rodentia, and experimental animals
Rattus, see Rodentia and *Rattus norvegicus* laboratory mouse, and experimental animals
Rattus norvegicus, laboratory rat, *see also* Rodentia
   asymmetry, bilateral, experimental 10
   dental caries 461
      susceptible strains 462
   dental plaque formation 457 526
   diet, effect of 10 526 568 605
   drugs, effect of cytotoxic 7
   enamel defects produced by experimental trauma 404
   enamel hypoplasia 438 439
   endocrines, hypophysectomy 528
      parathyroid ablation 438
   eruption, continuous 331
   growth rates, incisor 356
   inheritance 462
   lathyrism 333
   odontome 587
   overgrowth 360 361
   periodontal disease 525–528 568–569 Fig 24.3
   periodontium, age changes 527
   pseudo-odontome 333 605
   wedging of fibres, interdentally 568
   strains and mutants: black-hooded 462, incisor absent (ia) 333 605, N.I.H black 462, Osbourne-Mendel 462, Sprague-Dawley 462, toothless (H) 333, Wistar 462
Redunca, see Bovidae
reedbuck, *see* Bovidae
reindeer, *see* Cervidae
renal rickets 334 589 604–605
REPTILIA, *see also* non-mammalian dentitions
   developmental anomalies 158–159
   jaw deformities and malocclusion 158–159
   odontomes 506
   crocodiles
      injury, jaw 393
      malocclusion 158
      mineral deficiency, bendable jaws 158, translucent teeth 158
      polyphyodonty with age 505
      pig-head skull deformity 158
      regeneration of part of jaw 393
   lizards
      injury, fracture of jaw in fossil 393
      jaw disproportion with spinal curvature 159
      masticatory function of bare jaw-bone 523
      regeneration in 393
      replacement of teeth, continuous 523
      toothless in old age 505
resorption, *see* tooth resorption
reverse mutation 7
re-included teeth 332, in *Gorilla* 353–354 Fig 16.11
RHINOCEROTIDAE, rhinoceroses, *see also* Perissodactyla
   dental formulae 123 124
   injury, mandible 387
   normal dentition 123 311
   numerical variation 123 Fig 6.36
   position anomalies 312 352 Fig 13.79
   retained deciduous teeth 352 Fig 13.79
   tooth shape anomalies 123 Fig 6.36
   Diceros bicornis black rhinoceros 123 124, *Diceros simum* white rhinoceros 123 387, *Dicerorhinus sumatrensis* Sumatran rhinoceros 123, *Rhinoceros hemitoechus* fossil rhinoceros 312 352, *Rhinoceros sondaicus* Javan rhinoceros 123, *Rhinoceros unicornis* Indian rhinoceros 123 Fig 6.36
Rhinopithecus, see Cercopithecidae
Rhynchomeles, see Peramelidae
Rhynchonycteris, see Chiroptera
Rhyncogale, see Viverridae
rickets, *see* bone disorders
ringtail, *see* Phalangeridae
RODENTIA, mice, rats, gerbils, beavers, porcupines, *see also Rattus norvegicus*, laboratory rat, *Mesocricetus auratus* Syrian hamster and *Mus musculus* laboratory mouse
   actinomycosis 361 517–518
   captivity, effects of 461 483 568 569 Figs 21.28 21.29
   connation 9 133
   continuous growth, cheek teeth of 130
   dental calculus 362 527
   dental caries 461 483–484 526 527 595 Fig 25.26
   dental formulae 131
   dentigerous cysts 579
   dento-alveolar abscess 358 507 517–518 Figs 23.10 23.12
   diet, effects of 362 458 525 527
   enamel hypoplasia 439
   enamel fluorosis 362 450
   eruption, continuous 331
      sequence and chronology 345–346
      monophyodont (mice, rats and guinea pigs) 345
   experimental 404 439 461 525–528
   geographical variation 131 132 133 568
   growth, incisors 356 358 362 Fig 17.6
   molars 362
   incisor mechanism 360 Fig 17.5
   incisors, relation of growing end to molars 358 517 577 Figs 23.9 23.11
   inheritance 131 133
   injury, tooth 435–436 517 Figs 10.56 23.10
      tooth germs, experimental 404 439
   jaw mechanism 360–362 Fig 17.5
   lumpy jaw 361 517–518
   numerical variation 130–133
   odontomes 585 587 590 594–595 Figs 25.24 25.25 25.26
   overgrowth, cheek-tooth 362
      incisor 355–362 358 Fig 17.1
   periodontal disease 525–528 567–569
   periodontium, age changes in 527
   pigmentation of incisors 361
   position anomalies 316–338 Figs 14.3 14.6 14.7
      impaction 318 348 Fig 14.6
   retained deciduous teeth 133 352–353
   root anomalies 133 Tab 7.2
   slobbers 362
   teratoma 361
   tooth shape anomalies 131–133 Fig 7.1 Tab 7.1
   tooth wear, incisor 357–358 Fig 17.5
   Apodemus flavicollis yellow-necked mouse 133, *Apodemus sylvaticus* wood mouse 131 133, *Baiomys musculus* southern pygmy mouse 131, *Bandicota bengalensis* lesser bandicoot rat 361, Bathygeridae African mole rats 131, *Castor*

RODENTIA (cont.)
fiber beaver 131 133 351 353–358 360 517 Figs 17.1 17.5 23.11 23.12, *Cavia porcellus* domestic guinea pig 131 356 362 404 439 450 458 461 525 601, as experimental animal 461 525, *Chinchilla chinchilla* 361 362, *Clethrionomys gapperi* red-backed vole 317, *Clethrionomys glareolus* bank vole 133, *Coendu* tree porcupines 318 579 Figs 14.6 14.7, *Cricetus cricetus* common hamster 7 331 361, overgrowth, incisor, in jaw with actinomycosis 361, *Cricetulus migratorius* grey or Chinese hamster 527, periodontal disease, in hereditary diabetic strain 527, Ctenodactylidae, gundis 131, *Ctenomys* tuco-tuco 316, *Cynomys ludovicianus* prairie dog 595 Figs 25.25 25.26, *Dasyprocta* agouti 483 595 Fig 21.28, *Dicrostonyx groenlandicus* collared lemming 131, *Erethizion dorsatum* porcupine 131 595, as experimental animal 356, *Eutamias sibiricus* Asiatic chipmunk 133, *Gerbillus campestris* North African gerbil 133, Hydromyinae, island water rats 131, *Hystrix leucura* short-tailed porcupine 131, *Lemmus trimucronatus* brown lemming 568 *Marmota bobak* Bobak marmot 316 Fig 14.1, *Marmota caligata* hoary marmot 358 517 Figs 23.9 23.10, *Marmota marmota* alpine marmot 483 517 Figs 21.29 23.10, *Marmota monax* woodchuck 360 507 Fig 17.5, *Marmota* sp. marmot, odontome 594–595 Fig 25.24, *Meriones unguiculatus* Mongolian gerbil 461 527 568 569, as experimental animals 461 527 528, *Mesembriomys gouldi* tree rat 131, *Mesocricetus auratus* Syrian hamster, *see* separate entry, *Microtus* voles 131 133 517–518, *Mus musculus* house mouse, wild and commensal 131 132 319 568, *Mus musculus* laboratory mouse, *see under* separate heading, *Myocaster coypus* coypu and nutria 131 435–436 483–484 Fig 10.56, *Mysotromys albicaudatus* white-tailed rat, as experimental animals 461 527, *Ochtodontomys gliroides* mountain degu 317 Fig 14.3, *Ochotona rufescens* Afghan pika 134, Ochotonidae pikas 133, *Ondatra zibethicus* muskrat 360 361, *Oryzomys palustris* rice rat 9 133 461 527 569, as experimental animal 9 461 527, *Perognathus longimembris* little pocket mouse 568, *Peromyscus floridianus* deer mouse 131 132 568, *Petaurista* giant flying squirrel 316 Fig 14.2, *Proechimys* spiny rats 133 317 318 Figs 14.4 14.5, *Rattus exulans* Polynesian rat 131 360 361, *Rattus lutreolus* Australian swamp rat 131 Tab 7.1, *Rattus norvegicus* brown rat 133 567 Fig 17.9, *Rattus norvegicus* laboratory rat, *see under* separate heading, *Rattus rattus* black rat 133 358 Fig 17.6, *Saccostomus hildae* pouched rat 131, *Sciurus* squirrel 316, *Sigmodon hispidus* cotton rat, as experimental animal 461, *Spermophilus townsendi* Townsend's ground squirrel 358, *Tamasciurus hudsonicus* American red squirrel 358, *Tatera indica* Indian gerbil 133 352–353, *Zapus princeps* jumping mouse 131

Romer, A.S. 106
root anomalies, *see* tooth root anomalies
rorqual, *see* Mysteceti
RUMINANTIA, ruminant even-toed ungulates, *see also* Artiodactyla, Cervidae, Bovidae, Tragulidae and *Antilocapra*
  incisor mechanism 106
  mesial drift 288
  normal dentition 15 106–107 285–287 Fig 13.11
    arch disparity 286
    categories of obliquity 286
    echelon arrangement 286 Figs 13.35 13.44
    origins 106
  position anomalies 286–288
Rupicapra, *see* Bovidae

sabre tooth tiger, *see* Felidae
Saccopteryx, *see* Chiroptera
Saccostomus, *see* Rodentia
Saguinus, *see* Callitrichidae, and experimental animals
Saimiri, *see* Cebidae, and experimental animals
Sarcophilus, *see* Dasyuridae
Scalopus, *see* Insectivora
Scapanus, *see* Insectivora
Schoinobates, *see* Phalangeridae
Schultz, A.H. 20 23 52 177 392 463 465 469 507 536 540 541
Sciuromorpha 130
Sciurus, *see* Rodentia
Scotophilus, *see* Chrioptera
scurvy 362 525 526 528 539
sealions, *see* Otariidae
seals, eared, *see* Otariidae
seals, earless, *see* Phocidae
secondary cartilage 393
Selenarctos, *see* Ursidae
selenium
  deficiency 562
  in necrotizing ulcerative gingivitis 562
  in periodontal disease 560
  in soil compared with F 451
  toxicity 562
Selenka, E. 20 463
selenodont teeth in ruminants 110 111 114
Setifer, *see* Insectivora
Setonyx, *see* Macropodidae
shape anomalies, *see* tooth shape anomalies
shear-bite, in capuchin monkey 493, horse 487, howler monkey 493
sheep, domestic *see Ovis aries*
sheep, wild, *see* Bovidae
shrews, *see* Insectivora, and experimental animals
shrew-opossums, *see* Caenolestoidea
siamang, *see Symphalangus syndactylus*
sifaka, *see* Indriidae
Sigmodon, *see* Rodentia, and experimental animals
Simias, *see* Cercopithecidae
Sirenia, *see* Trichechidae
sitatunga, *see* Bovidae
size anomalies, *see* tooth size anomalies
skull deformities, *see* developmental anomalies
skunk, *see* Mustelidae
slobbers 362
sloth, *see* Edentata
Sminthopsis, *see* Dasyuridae
Sorex, *see* Insectivora
Spermophilus, *see* Rodentia
Spilogale, *see* Mustelidae
spingbok, *see* Bovidae
squalodont teeth, definition 101
squirrel, *see* Rodentia
steenbok, *see* Bovidae
Stenella, *see* Odontoceti
Steno, *see* Odontoceti
stoat, *see* Mustelidae
stomatitis, *see* necrotic stomatitis
Storer Bennett, W.C. ix
submerged teeth, *see* re-included teeth
SUIDAE, wild pigs, *see also Sus scrofa* domestic pigs
  alveolar bulb 155
  atrophic rhinitis 283
  dental caries 480
  dental formula 107
  enamel, absence of, on tusks 366
  eruption sequence 342 Tab 16.9
  mesial drift 107 284
  normal dentition 107 109 282 285 365 366 Fig 13.4
  numerical variation 107–109 Figs 6.1 6.3

SUIDAE (cont.)
  odontome 595
  osteometrics 282
  overgrowth of tusks 365–366 Fig 17.17
  position anomalies 282–285 Figs 13.1–13.4
  tooth wear 480
  tusks, in captivity 366
  Babyrousa babyrousa babirusa 109 284 366 Fig 13.4, *Hylochoerus meinertzhageni* giant forest hog 109 283 Fig 6.3, *Phacochoerus aethiopicus* warthog 107 284 365 Figs 13.2 13.3 17.7, *Potamochoerus porcus* bush pig 107 283 Figs 6.1 13.1, *Sus barbatus* bearded pig 109 282, *Sus scrofa* domestic pigs, *see* separate entry, *Sus scrofa* wild boar 108 364–365 Figs 17.15 17.16, *Sus verrucosus* Java and Celebes pigs 108
suni, *see* Bovidae
supernumerary teeth, *see* extra teeth
Suricata, *see* Viverridae
Sus scrofa domestic pigs, *see also* Suidae
  ancestry 283
  asymmetry of jaws 283
  atrophic rhinitis 283
  bone disorder, osteodystrophy 334
  cross-breeding 283
  Darwin on domestication 283
  dental calculus composition 551
  dental caries 480
  dental formula 107
  dentigerous cysts 578
  dento-alveolar abscess 513 550 Figs 23.7 24.32
  domestication 283
  eruption anomaly 334
    sequence and chronology 324 Tab 16.9
  experimental animals 480 550–551
  feral pigs 550–551 Fig 24.32
  long and short-muzzled breeds 283 Tab 13.1
  numerical variation 107–108 Fig 6.2
  osteometrics 283 Tab 13.1
  osteomyelitis from tusk clipping 513
  overgrowth of tusks 364–365
  periodontal disease 551
  position anomalies, impaction 282 348
  underjet 283
  skull deformity 283
  tooth wear 550
    crib-biting abrasion 494
  wedging, of plant material and foreign bodies, interdentally 551
  breeds: Chinese 283, large white Tab 13.1, middle white Tab 13.1, miniature 480 550–551, Vietnamese pot-bellied 107 480,

Welsh Tab 13.1
Sus, *see* Suidae and *Sus scrofa* domestic pigs
sutural bones, in gibbon 20
sutures 11
  premaxillary-maxillary inactivity at 168
  lack of growth at 200 221
  relationship to canine 11 23
  spheno-maxillary 184 187
Sylvicapra, *see* Bovidae
Symphalangus syndactylus siamang, *see also* Pongidae
  dental caries Tab 21.1
  eruption sequence 336
  normal dentition 20
  numerical variation 22 33 Fig 3.27 Tab 3.1
  disto-molars 33
  position anomalies 177 Fig 10.39
  tooth shape anomalies 33–34 Figs 3.26 3.27
Synceros, *see* Bovidae
synodonty, *see* connation

Tachyglossus, *see* Monotremata
Tadarida, *see* Chiroptera
Talpa, *see* Insectovora
tamarins, *see* Callitrichidae, and experimental animals
Tamasciurus, *see* Rodentia
TAPIRIDAE, tapirs, *see also* Perissodactyla
Tapirus tapirs
  dental formula 122
  lumpy jaw 516
  normal dentition 122–123
  position anomaly 311 Fig 13.78
  tooth shape anomaly 311
  tooth wear, incisor abrasion 495
TARSIDAE, tarsiers, *see also* Prosimii
  dental formula 61
  normal dentition 59
  numerical variation 61 Tab 3.8
  position anomalies 232–237 Tab 10.21
tarsiers, *see* Tarsidae
tartar, *see* dental calculus
Tasmanian devil, *see* Dasyuridae
Tasmanian wolf, *see* Dasyuridae
Tatera, *see* Rodentias
Taxidea, *see* Mustelidae
Tayassu, *see* Tayassuidae
TAYASSUIDAE, peccaries *see also* Artiodactyla
Tayassu peccary
  connation 109
  injury, mandible 388
  normal dentition 109 284 285
  numerical variation 109 Figs 6.4 6.5
  position anomalies 284–285 Figs 13.5 13.6
  tooth wear, incisor 495
tayra, *see* Mustelidae

tenrecs, *see* Insectivora
teratogenesis 10
teratomas 361 576 601 Fig 25.36
teratomatous odontomes 576 601
tetracycline compounds in teeth 438
Thalarctos, *see* Ursidae
thegosis, *see under* tooth wear
Theropithecus see Cercopithecidae
Thomas, O. 143 146–147
Thylacinus, *see* Dasyuridae
tiger, *see* Felidae
tiger-cat, *see* Dasyuridae
toad, leopard, *see Bufo pardalis*
Tomes, C.S. 13
Tomes, J. 8 426
TOOTH DEVELOPMENT, *see also* morphogenetic field concept
  between-cell influences 5
  initiation by field substance 3
  in tooth germs (cultured primordia) 4–6 8
  polarity reversal 5 7
  stem progenitor 3–4
TOOTH GERMS
  crowding 154
  culture of 4–6 8
  dichotomy, *see* connation
  injury, experimental 404 439
  pre-eruptive movements 154
  in cats 404 439, guinea pigs 404 439, mouse 4, rabbit 58, rats 404 439, sheep 404 439
tooth implantation 422
TOOTH RESORPTION 486–487 501–505
  burrowing 502 505, in cat 505, guenon monkey 502 Fig 22.17, horse 505, lemur 502 Fig 22.16, sheep 505
  root, in horse 564–565 Fig 24.45, in whales 501–502 Figs 22.14 22.15
  shedding of deciduous teeth and 501
  in elephants 504–505 Fig 22.18, whales 499 501–502 504, Figs 22.14 22.15
TOOTH ROOT ANOMALIES, *see also under* taxonomic headings
  in bears 78–82 Figs 4.23–4.39
  bone growth failure and 82 Fig 4.39
  distemper enamel hypoplasia in dogs and 445–447
  enamel pearls and root nodules 79 113 576 Fig 4.6
  epithelial root sheath and 102
  extra rootlets 19
  extra roots and extra cusps 22 44 396 430–431 Figs 3.43 3.45
  fusion 18 19 47
  in man 19
  odontome concept and 575 576
  terminology 18
  trauma-induced extra roots 396 430–431
  variability 18 19 67 Figs 4.6 4.7
  webbed 18 48 51 Fig 3.58

TOOTH SHAPE ANOMALIES, *see also* under taxonomic headings
  homoeosis 2
  odontomes and 1
  teeth of contralateral morphology 6
TOOTH SIZE ANOMALIES, *see also* under taxonomic headings
  absent teeth and, in gibbons 22
  diet and 10
  domestication and 10
  enamel thickness and 10
  field substance and 3
  geographic isolation and 10
  giant molars in rice rat 9
  in morphogenetic field concept 2 3 17 77
  normal range quantified in primates 18
  sex differences 10
  in tooth germs (cultured primordia) 5
  variation 1 9–10
tooth symbols 12
tooth types, *see* morphological definitions and nomenclature
TOOTH WEAR 486–501, *see also* under taxonomic headings
  abrasion and erosion 356–357 491 492 493 494–501
    burrowing soil organisms and 494 501
    definitions 486–487
    marine parasites and 494 498–499 501
    V-shaped 494
    wedge-shaped 494–495
    in cattle 494–495, man 494, monkeys 495–496, rodents and lagomorphs 356–357 Fig 17.5, seals, whales and manatees 494 496–501
  approximal 157 311 486 493 Fig 9.5, *see also* mesial drift and horizontal replacement
  attrition 356 362 454 472 486–494
    definition 486
    irregular and excessive 317 395 487–494 Figs 22.1–22.3 22.5 22.7, in sheep 491–493 552–559
  crib-biting abrasion, in bear 507, horses 494, mink 546, pigs 494, rhesus monkey 485, sheep 491–493
  disproportionate wear of M1, in macropods 320 494 Fig 22.6
  enamel fluorosis and 452–454 Figs 20.19 20.20
  erosion, *see* abrasion and erosion
  ice-sawing abrasion in Weddell's seal 497–498 512 Figs 22.11 22.12
  occlusal 486
    cupping in coati 471, in horses 477–488 Fig 22.4, in wart hog Fig 13.2
  periodontal disease and 492 531 540 541 545–546 550 560 570 Fig 21.15
  thegosis (tooth sharpening) 356–357
    in *Desmodus* vampire bat 357
    of rodent-type incisors 360 362
    tongue-to-tooth 357
Tragelaphus, *see* Bovidae
TRAGULIDAE, chevrotains, *see also* Ruminantia
Tragulus javanicus chevrotain
  canines 288
  dental formula 110
  normal dentition 110 288
  position anomalies 288 Fig 13.12
  tooth shape anomalies 110 Fig 6.9
tree porcupine, *see* Rodentia
Tremarctos, *see* Ursidae
tribosphenic crowns, definition 15 Fig 1.7
TRICHECHIDAE, manatees
  alveolar bulb 155 Fig 9.3
  cheek tooth replacement 155 345
  dental caries 480 501
  normal dentition 479–480
  numerical variation 345
  tooth wear 480
Trichechus, *see* Trichechidae
Trichosurus, *see* Phalangeridae
tuan, *see* Dasyuridae
tuco-tuco, *see* Rodentia
tumours, *see* odontogenic tumours
Tursiops, *see* Odontoceti

Ungulates, *see* Artiodactyla, Perissodactyla, Elephantidae, Hyracoidea
URSIDAE, bears, *see also* Carnivora
  bilaterality of anomalous development 77
  captivity, effects of 478 545 Fig 24.24
  connation 77 78 258 Fig 4.21
  dental calculus in captivity 545
  dental caries 477–478
  dental formula 77
  dento-alveolar abscess 507 Fig 23.1
  dilaceration 429 Fig 19.46
  enamel hypoplasia, traumatic 429 Fig 19.46
  enamel root nodules 79
  eruption, non-eruption 346 Fig 16.6
    precocious 341
    sequence and chronology 341
  injury, developing teeth 394–395 Fig 19.1
    mandible 379–380
    skull, gunshot 379–380 392 Fig 18.18
    zygomatic arch 392
  lumpy jaw 516
  mandibular joint, formation of new 392
  morphogenetic field 77
  normal dentition 77 Fig 4.22
  numerical variation 77 78 Figs 4.20 4.21
  odontocoele 430 Fig 19.46
  osteometrics 258 Tab 11.10
  periodontal disease 545–546 Fig 24.24
  position anomalies 258 Tab 11.1
    underjet 258
  root anomalies 77 79–82 Figs 4.23–4.39
  tooth wear 478 493 545–546
  Helarctos malayanus Malayan sun bear 81 429 Fig 19.46 Tab 11.10, *Melursus ursinus* sloth bear 78 258, *Selenarctos thibetanus* Asiatic black bear 77 78 80 81 82 258 341 Figs 4.21 4.28 4.32 4.33 4.38 Tab 11.10, *Thalarcto maritimus* polar bear 258 341 346 392 394–395 507 Figs 16.6 18.18 19.1 23.1 Tab 11.10, *Tremarctos ornatus* spectacled bear 78 545 Fig 24.24, *Ursus americanus* American black bear 77 79–80 341 380 477 546 Figs 4.24 4.27 4.29 Tab 11.10, *Ursus arctos* brown bear, grizzly bear 77 79 80 81 82 341 379–380 477–478 493 516 546 Figs 4.20 4.23 4.25 4.26 4.30 4.31 4.35–4.37 4.39 Tab 11.10
Ursus, *see* Ursidae

vanadium and caries resistance 456
VARIATION 1–2
  continuous 1–2
  discontinuous (discrete) 2
  domestication and 7
  environmental factors 1 9–11 131
  genetic mechanisms 1 2 3 9 11 131
  geographic, *see* separate entry
  homoeotic 2
  in man 11
  meristic 2
  normal 1
  normal vs abnormal 35
  numerical 1 6–7
  positional 1 11
  quasi-continuous 2 63 131
  tooth shape 1
  tooth size 1 9–10
  types of 1–2
vitamin A excess
  gingivitis in cats 544
  osteoporosis in dogs 550
  resorption in cats 505
  tooth fusion in mice 9
vitamin deficiency
  dental caries and, in dogs and rhesus monkeys 458
  enamel mineralization and, *see also* under enamel mineralization
    in dogs, rabbit, and guinea pigs 458–459
  enamel hypoplasia 438 442
  folic acid 528
  hyperparathryroidism and vitamin D 438

VARIATION (cont.)
  pantothenic acid 528
  periodontal disease and vitamins A and D in dogs 525
    and vitamin C in guinea pigs 525 528, in rhesus monkeys 539
  vitamin A deficient rats and pseudo-odontomes 605
  vitamin B_6 and enamel hypoplasia in rhesus monkey 442
  vitamin C and scurvy 362 525 526 528 539
  vitamin D complex 438 603
Viverra, see Viverridae
Viverricula, see Viverridae
VIVERRIDAE, viverrids, *see also* Carnivora
  captivity, effects of 247–248
  connation 68 70 586–587 Figs 4.11 4.12
  dental caries 477 Fig 21.20
  dental formulae 67 68 69 71
  diet 544–545
  enamel hypoplasia 444 Fig 20.13
  enamel root nodules Fig 4.6
  eruption sequence 339
  hamartoma/odontome 586–587 Fig 25.16
  injury, tooth 477
  normal dentition 67 69 241 242 545 Figs 11.10–11.12
  numerical variation 67–71 Figs 4.8 4.12 Tab 4.1
    disto-molars 68 69
  odontome 586–587 Fig 25.16
  osteometrics 243 Tab 11.6
  periodontal disease 544–545 Fig 24.23
  position anomalies 244–248 Figs 11.13–11.27 Tabs 11.1 11.5
    overjet 243 Fig 11.13
    underjet Fig 11.15
  root anomalies 67 Figs 4.6 4.7
  tooth shape anomalies 68 69 Figs 4.9–4.12
  tooth wear 477
  Arctictis binturong binturong 444,
  Arctogalidea trivirgata three-striped palm civet 444,
  Atilax paludinosus marsh mongoose 69 245 Figs 4.8 11.18, *Bdeogale nigripes* black-legged mongoose 246 247 Fig 11.21, *Crossarchus obscurus* long-nosed mongoose 69 70 586–587 Figs 4.10 4.11, *Cynictis penicillata* yellow mongoose 71, *Cynogale bennetti* otter-civet 444, *Genetta* genets 68 245 Fig 11.17, *Helogale parvula* 69, *Herpestes* mongoose 68–69 242 244 246 247 Figs 11.12–11.16 Tab 11.6, *Ichneumia albicauda* white-tailed mongoose 70 243 248 Figs 4.12 11.13 11.27, *Macrogalidea musschenbroekii* brown palm civet 444 Fig 20.13, *Mungos mungo* banded mongoose 69 587 Figs 4.9 25.16, *Paguma larvata* masked palm civet 247 544 Fig 11.24 24.23, *Paradoxurus* palm civets 68 242 245 339 Figs 11.11 11.20 11.23 11.26, *Rhyncogale melleri* Meller's mongoose 70 247 Fig 11.22, *Suricata* meerkats 68 544 545 Fig 24.24, *Viverra civetta* African civet 67 247–248 348 Figs 4.6 4.7 11.25, *Viverricula indica* small Indian civet 67–68 241 245 Figs 11.10 11.19
vole, *see* Rodentia

VOMBATIDAE, wombats, *see also* Marsupialia
  captivity, effects of 485
  dental caries 485
  dental formula 13 146
  numerical variation 146 493
  overgrowth of cheek teeth 493
  tooth wear 493
  Lasiorhinus hairy-nosed wombat 493, *Vombatus* common wombat 485 493
Vormela, see Mustelidae

Wallabia, see Macropodidae
wallaby, *see* Macropodidae
walruses, *see* Odobaenidae
wambenger, *see* Dasyuridae
wapiti, *see* Cervidae
warthog, *see* Suidae
waterbuck, *see* Bovidae
weasel, *see* Mustelidae
whales, toothed, *see* Odontoceti
whales, whalebone, *see* Mysticeti
wildebeest, *see* Bovidae
wolf-teeth in horses 121–122 352 Tab 6.2
wolverine, *see* Mustelidae
wolves, *see* Canidae
wombats, *see* Vombatidae
woodchuck, *see* Rodentia
woody tongue 563

yagouaroundi, *see* Felidae
yaws, in gorilla 392–393, *see also* bone disorders

Zalophus, see Otariidae
Zapus, see Rodentia
zebra, *see* Equidae
zinc deficiency and cara inchada 561